T0309296

Handbook of Complex Analysis

Handbook of Complex Analysis

Edited by
Steven G. Krantz

CRC Press
Taylor & Francis Group
Boca Raton London New York

CRC Press is an imprint of the
Taylor & Francis Group, an **informa** business

A CHAPMAN & HALL BOOK

First edition published 2022
by CRC Press
6000 Broken Sound Parkway NW, Suite 300, Boca Raton, FL 33487-2742

and by CRC Press
4 Park Square, Milton Park, Abingdon, Oxon, OX14 4RN

© 2022 Taylor & Francis Group, LLC

CRC Press is an imprint of Taylor & Francis Group, LLC

ISBN: 9781138064041 (hbk)
ISBN: 9781032202105 (pbk)
ISBN: 9781315160658 (ebk)

DOI: 10.1201/9781315160658

Typeset in font CMR10
by KnowledgeWorks Global Ltd.

Contents

Preface

Despite being nearly 500 years old, with work dating back to Cardano, Euler, Gauss, Cauchy, Riemann, and many others, the subject of complex analysis is still today a vital and active part of the mathematical sciences. In addition to all the exciting theoretical work being done today, there are important applications to physics, engineering, cosmology, and other aspects of technology. Many of the world's most distinguished and accomplished mathematicians conduct research in complex analysis. Several recent Fields Medalists study complex analysis.

Although a venerable subject, complex analysis continues to grow and prosper. New directions of development in the subject include dynamical systems, quasiconformal mappings, harmonic measure, automorphism groups, and the list can go on at some length. One of the sources of strength for the subject is its interaction with diverse parts of mathematics, including differential geometry, partial differential equations, functional analysis, algebra, combinatorics, and many other aspects of the subject.

This **Handbook of Complex Analysis** presents contributed chapters by several distinguished mathematicians, including a new generation of researchers. More than a compilation of recent results, this book offers a stepping stone for students to gain entry into the professional life of complex analysis. The essays presented here are all accessible to graduate students but will also be of considerable interest to the seasoned mathematician. Classes and seminars, of course, play a role in the maturation process that we are describing. But more is needed for the unilateral study. This handbook will play such a role.

As noted, this book will serve as a reference and a source of inspiration for mature mathematicians—both specialists in complex analysis and others who want to become acquainted with current modes of thought and investigation. And it will help the neophyte to become inured in the subject matter.

The chapters in this volume are authored by leading experts in the subject area, also gifted expositors. They are carefully crafted presentations of diverse aspects of the field, formulated for a broad and diverse audience. The editor intends this volume to be a touchstone for current ideas in the broadly construed subject area of complex analysis. It should enrich the literature and point to some new directions. The point here is not to present an epitaph for complex analysis but rather to provide an entree to a whole new life. We anticipate that the reader of this volume will be eager to explore other parts of complex analysis literature and to begin to play an active role in complex analysis research life.

Editor

Steven G. Krantz is a professor of mathematics at Washington University in St. Louis. He has previously taught at UCLA, Princeton University, and Pennsylvania State University. He has written more than 130 books and more than 250 scholarly papers and is the founding editor of the *Journal of Geometric Analysis*. An AMS Fellow, Dr. Krantz has been a recipient of the Chauvenet Prize, Beckenbach Book Award, and Kemper Prize. He received a PhD from Princeton University.

List of Contributors

David Barrett
University of Michigan
barrett@umich.edu

Steven R. Bell
Purdue University
bell@purdue.edu

Michael Bolt
Calvin Univeristy
mdb7@calvin.edu

Joseph A. Cima
University of North Carolina
cima@email.unc.edu

Buma Fridman
Wichita State University
buma.fridman@wichita.edu

Siqi Fu
Rutgers University
sfu@camden.rutgers.edu

Emily J. Gullerud
University of Minnesota
gulle069@umn.cdu

Dmitri Khavinson
University of South Florida
dkhavins@usf.edu

Christer Kiselman
Uppsala University
christer.kiselman@it.un.se

Steven G. Krantz
Washington University
sk@math.wustl.edu

Bingyuan Liu
University of California Riverside
bingyuan.liu@utrgv.edu

Daowei Ma
Wichita State University
dma@math.wichita.edu

T. H. Marshall
Massey University
T.Marshall1@massey.ac.nz

Gaven Martin
Massey University
g.j.martin@massey.ac.nz

Vicentiu Radulescu
University of Craiova
radulescu@inf.ucv.ro

Luis Reyna de la Torre
IUPUI
lereynad@iu.edu

James S. Walker
University of Wisconsin-Eau Claire
walkerjs@uwec.edu

1

Something about Poisson and Dirichlet

Steven R. Bell and Luis Reyna de la Torre

CONTENTS

1.1 Mathematical DNA

For reasons that have always been mysterious to the first author, he has often found himself thinking about the Dirichlet problem in the plane and has fought urges to look for explicit formulas for the Poisson kernels associated to various kinds of multiply connected domains, especially quadrature domains. He has obsessed about solutions to the Dirichlet problem with rational boundary data (see [4]), and he cannot stop thinking about the Khavinson-Shapiro conjecture about the same problem with polynomial data. His publication list is interspersed with papers where he has given in and found formulas for the Poisson kernel in terms of the Szegő kernel and in terms of Ahlfors maps. (See [2, 3] for an expository treatment of some of these results.) After thinking about the Dirichlet problem his entire adult life, he can solve it a different way every day of the week. He recently looked up his mathematical lineage at the Math Genealogy Project and found a possible explanation for his obsession. He is a direct mathematical descendent of both Poisson and Dirichlet. These problems are in his blood!

The authors worked together on a summer research project at Purdue University in 2018 to find a particularly elegant and simple way to approach these problems. We want to demonstrate here how Poisson and Dirichlet might solve

DOI: 10.1201/9781315160658-1

their famous problems today if they had lived another 200 years and developed a major lazy streak. We assume that our reader has seen a traditional approach to this subject in a course on complex analysis and so will appreciate the novelty and smooth sailing of the line of reasoning here, but just in case the reader hasn't, we have tried to present the material in a way that can be understood assuming only a background in basic analysis.

We would like to thank Harold Boas for reading an early draft of this work and making many valuable suggestions for improvement. Harold knows a lot about the family business because he, like the first author, was a student of Norberto Kerzman at MIT in the late 1970s.

1.2   Harmonic Functions

We define a harmonic function on a domain Ω in the complex plane to be a continuous complex valued function $u(z)$ on the domain that satisfies the *averaging property* on Ω, meaning that

$$u(a) = \frac{1}{2\pi} \int_0^{2\pi} u(a + re^{i\theta}) \, d\theta$$

whenever $D_r(a)$, the disc of radius r about a, is compactly contained in Ω. It is well known that this definition of harmonic function is equivalent to all the other standard definitions (see, for example, Rudin [7, Chap. 11]), and we will demonstrate this in very short order. Since we will explore other definitions of harmonic, we will emphasize that we are currently thinking of harmonic functions as being defined in terms of an averaging property by calling them harmonic-ave functions.

We first note that analytic polynomials are harmonic-ave. Indeed, if $P(z) = a_n z^n + a_{n-1} z^{n-1} + \cdots + a_1 z + a_0$, the averaging property is clear on discs centered at the origin because the constant function a_0 obviously satisfies the averaging property at the origin and so does z^n for $n \geq 1$ because

$$\int_0^{2\pi} (re^{i\theta})^n \, d\theta = r^n \int_0^{2\pi} \cos(n\theta) \, d\theta + ir^n \int_0^{2\pi} \sin(n\theta) \, d\theta = 0 = 0^n.$$

To see that the averaging property holds at a point $a \in \mathbb{C}$, write

$$P(z) = P((z - a) + a)$$

and expand to get a polynomial in $(z - a)$. Now the argument we used at the origin can be applied to discs centered at a.

It follows that the averaging property also holds for conjugates of polynomials in z, so polynomials in $\bar{z}$ are harmonic-ave. Note also that complex valued functions are harmonic-ave if and only if their real and imaginary parts are both harmonic-ave.

The *Dirichlet problem* on the unit disc is: given a continuous real valued function ϕ on the unit circle, find a real valued function u that is continuous on the closure of the unit disc with boundary values given by ϕ such that u is harmonic on $D_1(0)$. We will find a solution to this problem that is harmonic in the averaging sense in the next section. To do so, we will need to know the elementary fact that a continuous real valued function on the unit circle can be uniformly approximated on the unit circle by a real polynomial $p(x, y)$. We will now demonstrate this little fact, assuming the Weierstrass theorem about the density of real polynomials of one variable among continuous functions on closed subintervals of the real line.

Suppose we are given a continuous real valued function ϕ on the unit circle. We wish to find a real polynomial $p(x, y)$ that is uniformly close to ϕ on the unit circle. Note that we may assume that $\phi(\pm 1) = 0$ because we may subtract a polynomial function of the form $ax + b$ to make ϕ zero at ± 1. Define two continuous functions on $[-1, 1]$ via

$$h_{\text{top}}(x) = \phi(x + i\sqrt{1 - x^2})$$

and

$$h_{\text{bot}}(x) = \phi(x - i\sqrt{1 - x^2}).$$

We can uniformly approximate h_{top} and h_{bot} on $[-1, 1]$ by real polynomials $p_{\text{top}}(x)$ and $p_{\text{bot}}(x)$. Next, let $\chi_\epsilon(y)$ be a continuous function that is equal to one for $y > \epsilon$, equal to zero for $y < -\epsilon$ and follows the line connecting $(-\epsilon, 0)$ to $(\epsilon, 1)$ for $-\epsilon \leq y \leq \epsilon$. Let $p_\epsilon(y)$ be a polynomial in y that is uniformly close to χ_ϵ on $[-1, 1]$. Now, because ϕ vanishes at ± 1, the polynomial

$$p_{\text{top}}(x)p_\epsilon(y) + p_{\text{bot}}(x)p_\epsilon(-y)$$

approximates ϕ on the unit circle, and the approximation can be improved uniformly by shrinking ϵ and improving the approximations of the other functions involved.

We are now in position to solve the Dirichlet problem on the disc, but before we begin in earnest, this is a good place to emphasize that Green's theorem for the unit disc depends on nothing more than the fundamental theorem of calculus from freshman calculus. Indeed, let $P(x, y)$ be a C^1-smooth function. Let C_{top} denote the top half of the unit circle parameterized in the clockwise sense by $z(x) = x + iy_{\text{top}}(x)$, $-1 \leq x \leq 1$, where $y_{\text{top}}(x) = \sqrt{1 - x^2}$, and let C_{bot} denote the bottom half of the unit circle parameterized in the counterclockwise sense by $z(x) = x + iy_{\text{bot}}(x)$, $-1 \leq x \leq 1$, where $y_{\text{bot}}(x) = -\sqrt{1 - x^2}$. Note that the unit circle $C_1(0)$ parametrized in the counterclockwise sense is given by C_{bot} followed by $-C_{\text{top}}$. Drum roll. . .

$$\int_{C_1(0)} P \, dx = \left(\int_{C_{\text{bot}}} - \int_{C_{\text{top}}} \right) P \, dx = \int_{-1}^{1} [P(x, y_{\text{bot}}(x)) - P(x, y_{\text{top}}(x))] \, dx$$

$$= -\int_{-1}^{1} \left(\int_{y_{\text{bot}}(x)}^{y_{\text{top}}(x)} \frac{\partial P}{\partial y}(x, y) \, dy \right) dx = \iint_{D_1(0)} -\frac{\partial P}{\partial y} \, dx \wedge dy.$$

The other half of Green's formula follows by repeating the argument using the words left and right in place of top and bottom. This argument on the disc can be easily generalized to demonstrate Green's theorem on any region that can be cut up into regions that have a top boundary curve and a bottom curve and a left curve and a right curve. An annulus centered at the origin cut into four regions by the two coordinate axes is such a domain. We will need Green's theorem later in the paper when we study analytic functions from a philosophical point of view inspired by our observations about harmonic functions and the Dirichlet problem.

1.3   Solution of the Dirichlet Problem on the Unit Disc

The unit disc has the special feature that given polynomial data ϕ, it is straightforward to write down a polynomial solution to the Dirichlet problem with boundary values given by ϕ. Indeed, a polynomial $p(x, y)$ in the real variables x and y can be converted to a polynomial in z and $\bar{z}$ by replacing x by $(z + \bar{z})/2$ and y by $(z - \bar{z})/(2i)$ and expanding. It is now an easy matter to extend the individual terms in the sum to harmonic-ave functions on the disc by noting that

$$z^n \bar{z}^m$$

is equal to one on the unit circle if $n = m$, equal to z^{n-m} on the circle if $n > m$, and equal to $\bar{z}^{m-n}$ on the circle if $m > n$, each of which is harmonic-ave inside the unit circle.

Now, given a continuous real valued function ϕ on the unit circle, there is a sequence of real valued polynomials $p_n(x, y)$ that converges uniformly to ϕ on the unit circle. Let u_n be the polynomial harmonic-ave extension of p_n to the disc described in the paragraph above. Note that u_n can be expressed as a constant plus a polynomial in z that vanishes at the origin plus a polynomial in $\bar{z}$ that also vanishes at the origin. We now claim that the functions u_n converge uniformly on the closed disc to a solution of the Dirichlet problem. To see this, we must first show that real valued harmonic-ave functions u on the disc that extend continuously to the closure satisfy the *maximum principle* in the form

$$\max\{u(z) : |z| \leq 1\} = \max\{u(e^{i\theta}) : 0 \leq \theta \leq 2\pi\}.$$

Indeed, if the maximum value of such a function u occurs at a point z_0 inside the unit circle, we can express the value of u at z_0 as an average of u over a small circle centered at z_0. We can let the radius of that circle increase until the circle touches the unit circle at a single point. The averaging property holds on the limiting circle because of uniform continuity. Let M denote the maximum value $u(z_0)$. Now, in order for the average of a continuous function that is less than or equal to M over that circle to be equal to M, it must be that u is equal to M on the whole circle. Hence the value of u at the point where the inner circle touches the unit disc must also be M. This proves the maximum principle inequality for real harmonic-ave functions. The minimum principle follows by applying the maximum principle to $-u$.

Because the p_n converge to ϕ uniformly on the unit circle, the sequence $\{p_n\}$ is uniformly Cauchy on the unit circle, i.e., given $\epsilon > 0$, there is an N such that $|p_n - p_m| < \epsilon$ on the unit circle when n and m are greater than N. The maximum and minimum principle inequalities applied to the imaginary parts of u_n (which are zero on the unit circle) allow us to conclude that the functions u_n are *real* valued. Furthermore, the maximum and minimum principles applied to $u_n - u_m$ show that the uniformly Cauchy estimates for the sequence $\{p_n\}$ on the unit circle extend to hold for the sequence $\{u_n\}$ on the whole closed unit disc, showing that $\{u_n\}$ is uniformly Cauchy on the closed unit disc. Hence, since the u_n are continuous, they converge uniformly on the closed unit disc to a continuous function u that is equal to ϕ on the unit circle. Finally, it is clear that u is harmonic-ave on the inside of the unit circle because the averaging property is preserved under uniform limits. We have solved the Dirichlet problem in a purely existential manner without ever differentiating a function! We now turn to Poisson's problem of finding a *formula* for our solution.

1.4   Poisson's Formula

Notice that the set of harmonic-ave functions

$$\{1, z^n, \bar{z}^n : n = 1, 2, 3, \dots\}$$

is orthonormal under the inner product

$$\langle u, v \rangle = \frac{1}{2\pi} \int_0^{2\pi} u(e^{i\theta})\, \overline{v(e^{i\theta})}\, d\theta.$$

Define $K_N(z, w)$ via

$$K_N(z, w) := 1 + \sum_{n=1}^{N} z^n \bar{w}^n + \sum_{n=1}^{N} \bar{z}^n w^n,$$

and observe that if $u(z) = z^n$, then

$$u(z) = \frac{1}{2\pi} \int_0^{2\pi} K_N(z, e^{i\theta})\, u(e^{i\theta})\, d\theta \tag{1.1}$$

for $z \in D_1(0)$ if $N \geq n$ because of the orthonormality of the terms in the sum. The same is true if $u(z) \equiv 1$ or $u(z) = \bar{z}^n$. We conclude that if u is the solution of the Dirichlet problem for polynomial data p of degree n as constructed in the previous section, then formula (1.1) holds for $u(z)$ for $z \in D_1(0)$ when $N > n$.

Using the famous geometric series formula,

$$1 + \zeta + \cdots + \zeta^N = \frac{1}{1 - \zeta} - \frac{\zeta^{N+1}}{1 - \zeta},$$

we see that

$$K_N(z, w) = 1 + (z\,\bar{w}) \sum_{n=0}^{N-1} z^n \bar{w}^n + (\bar{z}\,w) \sum_{n=0}^{N-1} \bar{z}^n w^n$$

converges uniformly in $w = e^{i\theta}$ when $z \in D_1(0)$ to

$$K(z, w) := 1 + \frac{z\,\bar{w}}{1 - z\,\bar{w}} + \frac{\bar{z}\,w}{1 - \bar{z}\,w},$$

and the error $\mathcal{E}_N = |K_N - K|$ is controlled via

$$\mathcal{E}_N(z, w) \leq \frac{2|z|^{N+1}}{1 - |z|}$$

when $z \in D_1(0)$ and $|w| = 1$. Hence, it follows by taking uniform limits that formula (1.1) holds for $z \in D_1(0)$ with the N removed for the polynomials u_n that we constructed from polynomial boundary data p_n. We can now let the polynomials p_n tend uniformly to ϕ and use the fact proved above that the corresponding solutions u_n to the Dirichlet problem with boundary data p_n converge uniformly to a solution u of the Dirichlet problem to obtain Poisson's famous formula for the solution to the Dirichlet problem,

$$u(z) = \frac{1}{2\pi} \int_0^{2\pi} K(z, e^{i\theta}) \phi(e^{i\theta}) \, d\theta. \tag{1.2}$$

This formula reveals that the solution u can be written

$$u(z) = a_0 + h(z) + \overline{h(z)}$$

where a_0 is the (real valued) average of ϕ on the unit circle and h is an analytic function on $D_1(0)$ that vanishes at the origin given via

$$h(z) = \frac{1}{2\pi} \int_0^{2\pi} \frac{z e^{-i\theta}}{1 - z e^{-i\theta}} \phi(e^{i\theta}) \, d\theta = \frac{z}{2\pi i} \int_{C_1} \frac{\phi(w)}{w(w - z)} \, dw,$$

where C_1 denotes the unit circle parameterized in the standard sense using $w = e^{i\theta}$ and $dw = i e^{i\theta} \, d\theta$. It is now a rather easy exercise to take limits of complex difference quotients to see that complex derivatives in z can be taken under the integral sign in the definition of h. Hence, $h(z)$ is infinitely complex differentiable. It follows from the Cauchy-Riemann equations that u is a C^∞-smooth real valued function on $D_1(0)$ in x and y that satisfies the Laplace equation there and solves the Dirichlet problem. Furthermore, u is the real part of an infinitely complex differentiable function $H = a_0 + h/2$ on $D_1(0)$.

The maximum principle yields that u is the *unique* solution to the problem in the realm of harmonic functions understood in the sense of averaging. We will see in the next section that it is also the unique solution among harmonic functions defined in the traditional sense of satisfying the Laplace equation.

We remark here that rather simple algebra reveals the well-known formulas for the Poisson kernel,

$$K(z, w) = \operatorname{Re} \frac{1 + z\,\bar{w}}{1 - z\,\bar{w}} = \operatorname{Re} \frac{w + z}{w - z}$$

and

$$K(z, e^{i\theta}) = \frac{1 - |z|^2}{|z - e^{i\theta}|^2}.$$

It is a routine matter to extend the above line of reasoning to any disc (either by repeating the argument or making a complex linear change of variables $Az + B$). Since a complex valued function is harmonic-ave if and only if its real and imaginary parts are harmonic-ave, it follows from our work that a complex valued harmonic-ave function is given locally by $g + \overline{G}$ where g and G are infinitely complex differentiable functions.

At this point, it would be tempting to experiment with thinking of analytic functions as being harmonic-ave functions that do not involve the antianalytic $\overline{G}$ parts. The formula for h above reveals that the analytic g part is locally a uniform limit of analytic polynomials. Since $\{1, z^n : n = 1, 2, 3, \dots\}$ are orthonormal on the unit circle, we could let

$$k_N(z, w) = 1 + \sum_{n=1}^{N} z^n \bar{w}^n$$

and use the same line of reasoning that we used earlier in this section to conclude that

$$f(z) = \frac{1}{2\pi} \int_0^{2\pi} k_N(z, e^{i\theta}) f(e^{i\theta})\, d\theta \tag{1.3}$$

if $f(z)$ is equal to 1 or z^n with $n \leq N$. The geometric series estimate we used above shows that $k_N(z, w)$ converges uniformly in w on the unit circle for fixed z in $D_1(0)$ to

$$k(z, w) := \frac{1}{1 - z\,\bar{w}}.$$

Hence, we can let $N \to \infty$ in (1.3) to see that

$$f(z) = \frac{1}{2\pi} \int_0^{2\pi} k(z, e^{i\theta}) f(e^{i\theta})\, d\theta \tag{1.4}$$

when f is an analytic polynomial. Finally, if f is a uniform limit of analytic polynomials on an open set containing the closed unit disc, we may conclude that f satisfies identity (1.4), too. The identity can easily be seen to be the classical Cauchy integral formula on the unit disc, and from this point, the theory of analytic functions would gush forth. In particular, analytic functions would be seen to be infinitely complex differentiable and given locally by convergent power series. We will explore this idea and various other alternate ways of thinking about analytic functions after we verify that our definition of harmonic functions via an averaging property gives rise to the same set of functions as any of the more standard definitions.

1.5   Traditional Definitions of Harmonic Functions

Some complex analysis books define harmonic functions to be twice continuously differentiable functions that satisfy the Laplace equation. With this definition, one can use Green's identities on an annulus (which we pointed out in section 1 to be quite elementary) to show that such harmonic functions satisfy the averaging property. Hence, this class of functions could be seen to be the same as harmonic-ave functions. We won't pursue this idea here because we can easily prove something stronger with less effort.

It is most gratifying to define harmonic functions to be merely continuous functions u whose first partial derivatives exist and whose second partial derivatives $\partial^2 u/\partial x^2$ and $\partial^2 u/\partial y^2$ exist and satisfy the Laplace equation. Call such functions harmonic-pde. We will now adapt a classic argument to show that the class of harmonic-pde functions agrees with our class of continuous functions that satisfy the averaging property on an open set. Indeed, if u is a harmonic-pde function defined on an open set containing the closed unit disc, then u minus the harmonic-ave function U that we constructed solving the Dirichlet problem on the unit disc with the same boundary values as u on the unit circle, if not the zero function, would have either a positive maximum or a negative minimum. Suppose it has a positive maximum $M > 0$ at a point z_0 in $D_1(0)$. Choose ϵ with $0 < \epsilon < M$. Now

$$v(z) := u(z) - U(z) + \epsilon|z|^2$$

is equal to ϵ on the unit circle and attains a positive value at z_0 that is larger than ϵ. Hence, v attains a positive maximum at some point w_0 in $D_1(0)$. The Laplacian of v at w_0 is equal to 4ϵ, which is strictly positive. However, the one variable second derivative test from freshman calculus applied to v in the x-direction yields that $\partial^2 v/\partial x^2$ must be less than or equal to zero. (If it were positive, v could not have a local maximum at w_0.) Similarly, the second derivative test in the y-direction yields that $\partial^2 v/\partial y^2$ must be less than or equal to zero. We conclude that the Laplacian of v at w_0 must be less than or equal to zero, which is a contradiction because the Laplacian is strictly positive on $D_1(0)$. This shows that $u - U$ cannot have a positive value. If we replace $u - U$ by $U - u$, the same reasoning shows that $U - u$ cannot have a positive value. Hence $U - u \equiv 0$ on the unit disc, and we have shown that u, like U, is harmonic in the sense of being averaging. Finally, because the operations of translating and scaling preserve harmonic functions in both senses of the word, we can translate and scale any disc to the unit disc to be able to deduce the equivalence of the two definitions of harmonic on any open set. We now stop hyphenating harmonic and turn to hyphenating analytic.

1.6   Analytic Functions

When the 200+ year old Poisson and Dirichlet took a break from harmonic functions, they might have turned their attention to similar considerations applied to analytic functions, which satisfy both the averaging property on discs and Cauchy's theorem for circles.

In order to understand the rest of this section, the reader will need to know (or look up) the definition of the complex contour integral $\int_\gamma f \, dz$ along a curve γ, the basic estimate

$$\left| \int_\gamma f \, dz \right| \leq \sup\{|f(z)| : z \in \gamma\} \cdot \text{Length}(\gamma),$$

and the fundamental theorem of calculus $\int_\gamma f' \, dz = f(b) - f(a)$ for complex contour integrals, assuming that f is continuously complex differentiable and that γ starts at a and ends at b. We will also use the differential operators

$$\frac{\partial}{\partial z} = \frac{1}{2} \left(\frac{\partial}{\partial x} - i \frac{\partial}{\partial y} \right) \qquad \text{and} \qquad \frac{\partial}{\partial \bar{z}} = \frac{1}{2} \left(\frac{\partial}{\partial x} + i \frac{\partial}{\partial y} \right).$$

These two very important operators can be "discovered" by writing $dz = dx + i \, dy$ and $d\bar{z} = dx - i \, dy$ and manipulating

$$df = \frac{\partial f}{\partial x} \, dx + \frac{\partial f}{\partial y} \, dy$$

to appear in the form

$$df = \frac{\partial f}{\partial z} \, dz + \frac{\partial f}{\partial \bar{z}} \, d\bar{z}.$$

The condition $\frac{\partial f}{\partial \bar{z}} = 0$ is equivalent to writing the Cauchy-Riemann equations for the real and imaginary parts of f. If $h(z)$ is a complex differentiable function, then $\frac{\partial h}{\partial \bar{z}} = 0$ and $\frac{\partial h}{\partial z} = h'(z)$. Furthermore, $\frac{\partial \bar{h}}{\partial z} = 0$ and $\frac{\partial \bar{h}}{\partial \bar{z}} = \overline{h'(z)}$.

We will denote the boundary circle of the disc $D_r(a)$ parameterized in the counterclockwise sense by $C_r(a)$. Writing out the real and imaginary parts of the contour integral of a complex function f around $C_r(a)$ and applying Green's theorem for a disc to the real and imaginary parts yields, what we like to call, the *complex Green's theorem* for a disc,

$$\int_{C_r(z_0)} f \; dz = 2i \iint_{D_r(z_0)} \frac{\partial f}{\partial \bar{z}} \; dA,$$

where dA denotes the element of area $dx \wedge dy$. (Using the real Green's theorem to prove the complex Green's theorem is quick and easy, but a more civilized way to deduce the theorem would be to note that

$$d(f \; dz) = \frac{df}{dz} \; dz \wedge dz + \frac{df}{d\bar{z}} \; d\bar{z} \wedge dz = 0 + \frac{df}{d\bar{z}} \left(2i dx \wedge dy \right)$$

and to apply Stokes' theorem.)

Define a continuous complex valued function f to be analytic-circ on a domain Ω if it is harmonic on Ω (and so satisfies the averaging property) and, for each closed subdisc of Ω, the complex contour integral of f over the boundary circle is zero.

Complex polynomials are easily seen to be analytic-circ because the monomials are complex derivatives of monomials one degree higher and the fundamental theorem of calculus for complex contour integrals reveals that the integrals around closed curves are zero. It now follows that functions that are the uniform limit of complex polynomials (in z) on each closed subdisc of Ω are analytic. We will now show that functions that are analytic-circ must be the uniform limit of complex polynomials on each closed subdisc of Ω.

Suppose f is analytic-circ on Ω. Let $D_r(z_0)$ be a disc that is compactly contained in Ω and let $C_r(z_0)$ denote the boundary circle parameterized in the counterclockwise sense. The complex Green's theorem for a disc yields that

$$0 = \int_{C_r(z_0)} f \; dz = 2i \iint_{D_r(z_0)} \frac{\partial f}{\partial \bar{z}} \; dA,$$

where dA denotes the element of area $dx \wedge dy$. Since this integral is zero for every disc compactly contained in Ω, it follows that $\partial f / \partial \bar{z} \equiv 0$ on Ω, i.e., that the real and imaginary parts of f satisfy the Cauchy-Riemann equations. Since harmonic functions are infinitely differentiable, the textbook proof that f is complex differentiable can now be applied.

Since being analytic-circ is a local property and is invariant under changes of variables of the form $Az + B$, we may restrict our attention to a function f that is analytic-circ on a neighborhood of the closed unit disc. We can gain more insight into the implications of the definition by noting that such a harmonic function is given by $g + \overline{G}$ where g and G are infinitely complex differentiable. It follows from our complex Green's calculation above that $\partial f / \partial \bar{z} = \overline{G'} \equiv 0$ on Ω, and that f is therefore equal to the analytic part g plus a constant. From this point, it follows that f is given by a Cauchy integral formula, and we merge into the fast lane of the classical theory of analytic functions.

Of course, the traditional way to define analytic functions is as complex differentiable functions on open sets. Call such functions analytic-diff. We will now show that the traditional definition leads to the same class of functions that we have defined as being analytic-circ.

Suppose that f is an analytic-diff function defined on an open set containing the closed unit square $\mathcal{S} := [0,1] \times [0,1]$. We will show that the complex integral of f around the counterclockwise perimeter curve σ of the square must be zero. The well-known argument will be a beautiful bisection method tracing back to Goursat.

It follows from our assumption that f can be locally well approximated by a complex linear function in the following sense. Suppose a is a point in $\mathcal{S}$, and let $\epsilon > 0$ be given. Since $f'(a)$ exists, there is a $\delta > 0$ such that

$$\frac{f(z) - f(a)}{z - a} = f'(a) + E_a(z)$$

where $|E_a(z)| < \epsilon$ when $|z - a| < \delta$, $z \neq a$. Define $E_a(a)$ to be zero to make E_a continuous at a and so as to be able to assert that

$$f(z) = f(a) + f'(a)(z - a) + E_a(z)(z - a)$$

on the whole open set where f is defined and E_a is a continuous function in z on that set. Furthermore, $|E_a(z)| < \epsilon$ on $D_\delta(a)$. The complex integral of the polynomial $f(a) + f'(a)(z - a)$ around any square is zero because first-degree polynomials are derivatives of second-degree polynomials. If σ_h is the counterclockwise boundary curve of a small square $\mathcal{S}_h$ of side h contained in $D_\delta(a)$ that contains the point a, it follows that

$$\left| \int_{\sigma_h} f(z)\, dz \right| = \left| \int_{\sigma_h} E_a(z)(z - a)\, dz \right| \leq \epsilon(\sqrt{2}h)(4h).$$

Hence,

$$\left| \int_{\sigma_h} f\, dz \right| \leq (4\sqrt{2})\epsilon \operatorname{Area}(\mathcal{S}_h). \tag{1.5}$$

We will now follow a version of Goursat's famous argument to explain how this could be made *too small* if $\int_\sigma f\, dz$ were not zero.

Indeed, suppose that $I := \int_\sigma f\, dz$ is not equal to zero. Note that I is equal to the sum of the integrals around the four counterclockwise squares obtained by cutting the big square into four equal squares of side $1/2$ since the integrals along the common edges cancel. For these four integrals to add up to the non-zero value I, the modulus of at least one of them must be greater than or equal to $|I|/4$. Name such a square $\mathcal{S}_1$ and its counterclockwise boundary curve σ_1. Note that

$$\left| \int_{\sigma_1} f\, dz \right| \geq |I| \cdot \operatorname{Area}(\mathcal{S}_1).$$

We may now dice up $\mathcal{S}_1$ into four equal subsquares and repeat the argument to obtain a square $\mathcal{S}_2$ with boundary curve σ_2 such that the modulus of the integral of f around σ_2 is greater than or equal to $|I|$ times the area of $\mathcal{S}_2$. Continuing in this manner, we obtain a nested sequence of closed squares $\{\mathcal{S}_n\}_{n=1}^{\infty}$ with boundary curves σ_n, the diameters of which tend to zero as $n \to \infty$, such that

$$\left| \int_{\sigma_n} f \, dz \right| \geq |I| \cdot \text{Area}\,(\mathcal{S}_n). \tag{1.6}$$

There is a unique point a that belongs to all the squares. Now, given an ϵ less than $|I|/(4\sqrt{2})$, the squares that eventually fall in the disc $D_\delta(a)$ that we specified above satisfy both area inequalities (1.5) and (1.6), which are incompatible. This contradiction shows that I must be zero!

Since any square can be mapped to the unit square via mapping of the form $Az + B$, it follows from a simple change of variables that the integral of an analytic-diff function around any square must be zero. Furthermore, any rectangle can be approximated by a rectangle subdivided into a union of $n \times m$ squares. It follows that the integral of an analytic-diff function around any rectangle must be zero.

From this point, there are several standard arguments to prove the Cauchy integral formula on a disc for such functions (see Ahlfors [1, p. 109] or Stein [9, p. 37]). It then follows from the Cauchy integral formula that such a function would be locally the uniform limit of analytic polynomials, and so the function would be analytic-circ. However, we can simplify these standard arguments by using some of the power of our work on harmonic functions in the previous sections.

Suppose that f is analytic-diff on an open disc. Since f is complex differentiable, it is continuous. Define $F(z)$ at a point z in the disc by the integral of f along a horizontal "zig" from the center followed by a vertical "zag" connecting to the point z. The fundamental theorem of calculus reveals that

$$\frac{\partial}{\partial y} F(x + iy) = i f(x + iy).$$

Since the integral of f around rectangles is zero, we could also define F via an integral along a vertical zag followed by a horizontal zig. Using this definition, the fundamental theorem of calculus shows that

$$\frac{\partial}{\partial x} F(x + iy) = f(x + iy).$$

Hence, F is a continuously differentiable function whose real and imaginary parts satisfy the Cauchy-Riemann equations, and furthermore, is a complex differentiable function such that $F' = f$ on the disc. Repeat this construction to get a twice continuously complex differentiable function G such that $G'' = f$.

Now, since G is twice continuously complex differentiable, it is easy to use the Cauchy-Riemann equations to show that the real and imaginary parts of G are harmonic functions. Indeed, if $G(x + iy) = u(x, y) + iv(x, y)$, then the Cauchy-Riemann equations yield that

$$G' = u_x + iv_x = v_y - iu_y$$

and

$$G'' = u_{xx} + iv_{xx} = -u_{yy} - iv_{yy}$$

and we see that u and v satisfy the Laplace equation by equating the real and imaginary parts of G''. Our work in previous sections shows that these harmonic functions are C^∞-smooth, and it follows that the real and imaginary parts of f are C^∞-smooth and satisfy the Laplace equation and the Cauchy-Riemann equations. We may now use Green's theorem on a disc to prove the Cauchy theorem for f on discs. Hence, f is analytic-circ, and we have proved the equivalence of the definitions, revealing that f is locally the uniform limit of complex polynomials and is given by the Cauchy integral formula. The shortcuts we have revealed in the theory of analytic functions deliver us to page 114 of Ahlfors.

Before we conclude this section, we present one last alternate way to define analytic functions that might be of interest to experienced analysts. The result is known (see, for example, Springer [8] or Globevnik [5]), but we are in a position to prove it rather efficiently here. We now will define a continuous complex valued function to be analytic-ave on the unit disc if $f(z)(z - a)$ satisfies the averaging property on circles $C_r(a)$ contained in the disc, i.e., if

$$0 = \int_0^{2\pi} f(a + re^{i\theta})(re^{i\theta}) \, d\theta$$

whenever the closure of $D_r(a)$ is contained in the unit disc. Note that this condition is equivalent to the condition that

$$0 = \int_{C_r(a)} f \, dz$$

for each such circle. We will now prove that analytic-ave functions are analytic in the usual sense. This result can be viewed as a version of Morera's theorem saying that a continuous complex valued function that satisfies the Cauchy theorem on circles must be analytic.

Let $\chi(t)$ be an real valued non-negative function in $C^\infty[0, 1]$ that is equal to one for $t < \frac{1}{2}$ and equal to zero for $t > \frac{3}{4}$. Define

$$\phi(z) = c\chi(|z|^2)$$

where c is chosen so that $\int \phi \, dA = 1$. Let ϕ_ϵ denote the approximation to the identity given by

$$\phi_\epsilon(z) = \frac{1}{\epsilon^2}\phi(z/\epsilon).$$

The proof of our claim rests on a straightforward calculation that shows that

$$\frac{\partial}{\partial \bar{z}} \phi_\epsilon(z - w)$$

is $(z - w)$ times a function $\psi_\epsilon(z - w)$ that is radially symmetric about z in w.
In fact,

$$\psi_\epsilon(z) = \frac{c}{\epsilon^4} \chi'(|z|^2/\epsilon^2).$$

The calculation hinges on the chain rule plus the fact that

$$\frac{\partial}{\partial \bar{z}} |z^2| = \frac{\partial}{\partial \bar{z}} (z\,\bar{z}) = z.$$

Given a continuous function f on the unit disc such that $f(z)(z-a)$ satisfies the averaging property on circles $C_r(a)$ compactly contained in the disc, let $f_\epsilon = \phi_\epsilon * f$ for small $\epsilon > 0$. Note that f_ϵ is C^∞ smooth on $D_{1-\epsilon}(0)$ and that f_ϵ converges uniformly on compact subsets of the unit disc to f as $\epsilon \to 0$. One can differentiate under the integral in the convolution formula to see that

$$\frac{\partial f_\epsilon}{\partial \bar{z}} = \frac{\partial \phi_\epsilon}{\partial \bar{z}} * f = \iint_{w \in D_1(0)} \frac{\partial \phi_\epsilon}{\partial \bar{z}} (z - w)\, f(w)\, dA,$$

and the observation about the radially symmetric function and our hypothesis about f allows us to write the integral in polar coordinates about z to conclude that f_ϵ satisfies the Cauchy-Riemann equations, and so is analytic-diff on $D_{1-\epsilon}(0)$, and consequently, is analytic-circ there, too. It is easy to see that uniform limits of analytic-circ functions are analytic-circ. We conclude that f is analytic-circ and so analytic in any sense of the word.

1.7   The Dirichlet Problem in More General Domains

We solved the Dirichlet problem on the unit disc, given polynomial boundary data, by explicitly extending individual terms $z^n \bar{z}^m$ as harmonic polynomials. Another way to approach this problem is via linear algebra. Suppose a domain Ω is described via a real polynomial defining function $r(x, y)$ (meaning that $\Omega = \{x + iy : r(x, y) < 0\}$), where $r(x, y)$ is of degree two. For Ω to be a bounded domain, it is clear that the boundary of Ω, which is the zero set of r, must be a circle or an ellipse. Let Δ denote the Laplace operator. Now, the map $\mathcal{F}$ that takes a polynomial $p(x, y)$ to the polynomial $\Delta(rp)$ maps the finite-dimensional vector space $\mathcal{P}_N$ of polynomials of degree N or less to itself. (Multiplying by r increases the degree by two, and applying the second-order operator Δ brings it back down by two.) We claim that the map $\mathcal{F}$ is one-to-one on $\mathcal{P}_N$, and therefore *onto*. Indeed, if $\Delta(rp)$ is the zero polynomial, then

rp is a harmonic polynomial that vanishes on the boundary. The maximum principle implies that it must be the zero polynomial. Consequently, p must be the zero polynomial, and this proves that $\mathcal{F}$ is a one-to-one linear mapping of a finite-dimensional vector space into itself, and so also onto. Now, to solve the Dirichlet problem on Ω, given polynomial boundary data $q(x, y)$, we know there is polynomial p such that $\Delta(rp) = \Delta q$. The polynomial $q-rp$ is harmonic on Ω and equal to q on the boundary. It solves the Dirichlet problem. We could have solved the Dirichlet problem on the unit disc in the realm of polynomials without ever writing a formula down! Now we can solve the Dirichlet problem on an ellipse using the same procedure that we did on the disc. (However, the next obvious step, to try to write down a Poisson integral formula on the ellipse, gets more complicated because the monomials are not orthonormal in the boundary inner product of the ellipse.)

The Khavinson-Shapiro conjecture states that discs and ellipses are the only domains in the plane having the property that solutions to the Dirichlet problem with polynomial data must be polynomials. It is tantalizing that it seems so much harder to settle this question than the same problem with the word "polynomial" replaced by "rational." Only discs have the property that solutions to the Dirichlet problem with rational boundary data must be rational (see [4]).

For the remainder of this section, we will let our Poisson and Dirichlet urges run rampant and explain how the ideas in the previous sections might be used to solve the Dirichlet problem on more general domains. We will dispense with proofs and follow a line of bold declarations. The interested reader can find a more sober exposition of some of these ideas in Chapters 22 and 34 of [2] and in [3].

To solve the Dirichlet problem on a domain bounded by a Jordan curve, one can use Carathéodory's theorem (the theorem that states that the Riemann map associated to such a domain extends continuously to the boundary and maps the boundary one-to-one onto the unit circle) to be able to pull back solutions to the problem on the unit disc to the domain. But we wonder if there might be a more elementary way to do it.

Gustafsson's theorem [6] states that a bounded finitely connected domain with n continuous simple closed boundary curves can be mapped to an n-connected *quadrature domain* Ω with smooth real analytic boundary via a conformal mapping that is continuous up to the boundary and as close to the identity map in the uniform topology of the closure of the domain as desired. Such a "nearby" quadrature domain has the property that the average of an analytic function over the domain with respect to area measure is finite linear combination of values of the function and its derivatives at finitely many points in the domain. The resulting "quadrature identity" is the same for all analytic functions that are square-integrable with respect to area measure on the domain. Smooth real analytic curves have "Schwarz functions" $S(z)$ that are analytic on a neighborhood of the curve and satisfy $S(z) = \bar{z}$ on the curve. The Schwarz functions associated to the boundary curves of our quadrature domain Ω have the following stronger properties. There is a function $S(z)$ that is meromorphic on an open set containing the closure of Ω that has no poles on the boundary of Ω and satisfies the identity $\bar{z} = S(z)$ on the boundary. Quadrature domains can be thought of as a generalization of the unit disc (which is a *one point* quadrature domain), and Gustafsson's conformal mapping as a generalization of the Riemann map in the n-connected setting.

Given a continuous function ϕ on the boundary of our quadrature domain Ω, we can approximate it by a rational function of x and y via the Stone-Weierstrass theorem since the family of such rational functions without singularities on the boundary forms an algebra of continuous functions that separates points. Writing such a rational function as a rational function of z and $\bar{z}$ and replacing $\bar{z}$ by $S(z)$ produces a meromorphic function on a neighborhood of the closure of Ω that has no poles on the boundary and that agrees with the given rational function on the boundary. If we can solve the Dirichlet problem on Ω with boundary data

$$\frac{1}{(z-a)^n}$$

for fixed a in Ω and positive integers n, then, by subtracting such solutions from the data, we would have harmonic functions that vanish on the boundary and have general pole behavior at $z = a$. We could then use these functions to subtract off the poles of our meromorphic function and obtain a solution to the Dirichlet problem with the given rational boundary data. Then we could take uniform limits and solve the Dirichlet problem for continuous boundary data ϕ just like we did in the unit disc.

In case Ω is simply connected, it is possible to solve the Dirichlet problem with boundary data $(z-a)^{-n}$ using a Riemann mapping function. Let $f : \Omega \to D_1(0)$ be a Riemann map. The Green's function $G(z,w)$ for Ω is a constant times

$$\ln \left| \frac{f(z) - f(w)}{1 - \overline{f(w)}\, f(z)} \right|,$$

and derivatives

$$\frac{\partial^m}{\partial w^m} G(z,w)$$

are harmonic on $\Omega - \{w\}$, are continuous up to the boundary in z and vanish on the boundary in z, and the singularity at w is precisely of the form a constant times $(z-w)^{-m}$. One does not need to know that the Riemann map is continuous up to the boundary to see that these functions extend continuously up to the boundary in z and vanish there. This follows from the fact that conformal mappings are *proper* mappings: the inverse image of a compact subset of the unit disc is a compact subset of Ω.

Hence, in the simply connected case, we have a method to solve the Dirichlet problem rather analogous to the method we used in the case of the unit disc. There is something appealing about taking a close approximation to our original domain, followed by a close approximation to the boundary data, to be able to find an elementary formula for the solution to the Dirichlet problem.

Riemann maps associated to simply connected quadrature domains can be expressed as rational combinations of z and the Schwarz function, so solutions to the Dirichlet problem with rational boundary data can also be expressed as rational combinations of z and the Schwarz function!

Another way to construct the Poisson kernel is to express it in terms of a normal derivative of the Green's function, which, on a simply connected quadrature domain, is also expressible in terms of a Riemann map, and hence, also expressible in terms of z and the Schwarz function. It follows that the Poisson kernel of a simply connected quadrature domain is expressible in terms of z and the Scwharz function. Could we do similar things in the multiply-connected setting? Could we use Ahlfors maps in place of a Riemann map? Might the Poisson kernel there be expressible in terms of z and a Schwarz function and the harmonic measure functions? We wonder.

References

[1] Ahlfors, L., *Complex analysis*, 3rd Edition, McGraw Hill, New York, 1979.

[2] Bell, S., *The Cauchy transform, potential theory, and conformal mapping*, 2nd Edition, CRC Press, Boca Raton, 2016.

[3] Bell, S., *The Dirichlet and Neumann and Dirichlet-to-Neumann problems in quadrature, double quadrature, and non-quadrature domains*, Analysis and Mathematical Physics **5** (2015), 113–135.

[4] Bell, S., P. Ebenfelt, D. Khavinson, H. Shapiro, *On the classical Dirichlet problem in the plane with rational data*, J. Anal. Math. **100** (2006), 157–190.

[5] Globevnik, J., *Zero integrals on circles and characterizations of harmonic and analytic functions*, Trans. Amer. Math. Soc. **317** (1990), 313–330.

[6] Gustafsson, B., *Quadrature domains and the Schottky double*, Acta Applicandae Math. **1** (1983), 209–240.

[7] Rudin, W., *Real and Complex Analysis*, McGraw Hill, New York, 1987.

[8] Springer, G., *On Morera's theorem*, Amer. Math. Monthly **64** (1957), 323–331.

[9] Stein, E. M. and R. Shakarchi, *Complex analysis*, Princeton Lectures in Analysis, Princeton Press, Princeton, 2003.

2

The Cauchy-Leray Operator for Convex Domains

David Barrett and Michael Bolt

CONTENTS

2.1 Introduction

One of the fundamental constructions for a region in the complex plane is the Cauchy integral—it has many applications. In one complex variable, for instance, it can be used to show that a function satisfying the Cauchy-Riemann equations can be represented locally as the sum of a power series. In several variables, it can be used to show in iterated fashion that holomorphic functions on non-pseudoconvex regions always extend holomorphically to larger regions. This is the Hartogs extension phenomenon of 1906. There are alternatives, though, to generalizing the Cauchy integral to several variables that do not

DOI: 10.1201/9781315160658-2

involve iteration. Often they and their related operators are characterized as canonical (e.g., Szegő), universal (e.g., Bochner-Martinelli), or geometric (e.g., Cauchy-Fantappiè).

Such constructions are foundational to the method of integral representations which is important for proving regularity of solutions to partial differential equations in several complex variables. That method (including the applications and all kernel functions discussed here) is described well in the Krantz contribution to this volume [24] as well as [23, 26].

The present article should be seen as an additional perspective on the method where the kernels themselves are considered geometrically and computationally. In particular, we present one of the alternatives as a natural cousin to the Cauchy integral. The Cauchy-Leray integral, in particular, is constructed at once as a Cauchy-Fantappiè integral associated to domains that are weakly linearly convex (as defined in §3) and have twice differentiable boundary. It is hoped that this special relationship between the Cauchy and Cauchy-Leray integrals might be an avenue by which insights for one variable leads to new insights for several variables.

Our study begins in one complex variable and explores similarities between the Cauchy integral and Szegő projector. We include examples not only to show the similarities, but also to illustrate how hidden symmetries can lead to calculations that suggest a general behavior. Specifically we exploit the fact that the Cauchy integral has Möbius symmetry where the Szegő projector has holomorphic symmetry.

We next introduce the kind of convexity needed to construct the Cauchy-Leray integral. By default, all regions in one complex variable have this property and the Cauchy-Leray integral simplifies to the Cauchy integral. (Such simplification is common to integrals of Cauchy-Fantappiè type!) We then introduce the Fefferman measure and show that if the Cauchy-Leray kernel and Szegő kernel are defined with respect to Fefferman measure, then they have the same Möbius symmetry and holomorphic symmetry as in the case of one variable. We conclude by introducing a domain studied recently by Barrett and Edholm where the L^2-norm of the Cauchy-Leray operator can be found explicitly [2]. The details of this higher dimensional case are outside the scope of this article, but there is a resemblance to some of the examples in one variable and the unit ball.

We mention that many of the constructions discussed in this article also appear in the contribution to this volume by Krantz [24]. The emphasis here, though, is on making further connections between complex analysis in one and several variables and on enhancing the collection of examples.

The Cauchy-Leray integral has been studied in other special cases, too. Barrett and Lanzani studied the essential norm for the case of convex Reinhardt domains generally and for the case of L^p-balls specifically [4]. As well, Barrett clarifies the vector bundle context in which the Cauchy-Leray kernel arises naturally. This leads to an interpretation of the Cauchy-Leray integral as a way to quantify a pairing between dual hypersurfaces in complex projective space [3].

2.2 Szegő Kernel and Cauchy Kernel under Transformation

In general, we assume $\Omega \subset \mathbb{C}$ is a bounded region with C^∞ smooth boundary. Most of the discussion applies to more general regions, too, as will be apparent in examples.

2.2.1 Szegő kernel

To begin, the Lebesgue space $L^2(\partial\Omega)$ is defined with respect to arc length measure using the Hermitian inner product

$$(g, h)_{\partial\Omega} = \int_{\partial\Omega} g\,\overline{h}\,ds.$$

The Hardy space $H^2(\partial\Omega)$ is loosely the subspace of functions that extend holomorphically to the region Ω. Such extensions are unique by the Cauchy integral formula. To be precise, $H^2(\partial\Omega)$ is the closure in $L^2(\partial\Omega)$ of the space of functions $A^\infty(\Omega)$ that arise as boundary values of functions holomorphic in Ω and smooth in $\overline{\Omega}$. Since $H^2(\partial\Omega)$ is closed, one has an orthogonal projection $S : L^2(\partial\Omega) \to H^2(\partial\Omega)$ called the Szegő projector.

By the Riesz Representation Theorem, one has that the Szegő projector is an integral operator. In particular, an estimate based on the Cauchy integral formula shows for each $z \in \Omega$ that the point evaluation $g \in H^2(\partial\Omega) \to g(z)$ is a continuous linear functional. So there is a unique element $S_z = S(\,\cdot\,, z) \in H^2(\partial\Omega)$ such that $Sg(z) = (g, S_z)_{\partial\Omega}$ for $g \in L^2(\partial\Omega)$. Alternatively, the Szegő kernel can be constructed by taking an orthonormal basis $\{\phi_j\}$ for $H^2(\partial\Omega)$ and demonstrating convergence of the series $S(w, z) = \sum_j \phi_j(w)\overline{\phi_j(z)}$. With both approaches one has readily that $S(w, z) = \overline{S(z, w)}$.

The second approach is helpful for computing the Szegő kernel in the case of a highly symmetric region like the unit disc $\Delta = \{z : |z| < 1\}$. By Taylor's theorem, a holomorphic function on Δ is the sum of its Maclaurin series. So a basis for $H^2(\partial\Omega)$ is $\{\psi_j\}_{j\geq 0}$ where $\psi_j(z) = z^j$. In fact, this is an orthogonal basis and upon normalization one has

$$S(w, z) = \sum_{j\geq 0} \frac{w^j}{\sqrt{2\pi}}\frac{\overline{z}^j}{\sqrt{2\pi}} = \frac{1}{2\pi}\frac{1}{1 - w\overline{z}}. \tag{2.1}$$

A third approach to the Szegő kernel is via the Greens function. For the case of a simply connected region, one can define

$$S(w, z) = \sqrt{-\frac{1}{2\pi^2}\frac{\partial^2 G(w, z)}{\partial w\,\partial\overline{z}}}$$

and subsequently verify the reproducing and orthogonality properties. There is a messy generalization of this for multiply connected regions that involves the associated harmonic measures of the region. Such formulas highlight that the Szegő projector and kernel belong to the complex analytic "canon" of a given region although the derivations are indirect and typically given via the corresponding theory of the Bergman kernel, see [5, 14].

For our purposes, we consider the effect of a biholomorphism $f : \Omega_1 \to \Omega_2$ on the Szegő projector and kernel. Given that Ω_1 and Ω_2 have smooth boundaries, it is known that f extends smoothly to $\overline{\Omega}_1$, f' does not vanish on Ω_1, and f' is the square of a holomorphic function (see [5], p. 42). This leads to an isometry $\Lambda_f : L^2(\partial\Omega_2) \to L^2(\partial\Omega_1)$ given by $g \to (g \circ f)\sqrt{f'}$ that also preserves Hardy spaces $\Lambda_f : H^2(\partial\Omega_2) \to H^2(\partial\Omega_1)$. It, too, yields the transformation law

$$S_1(w, z) = f'(w)^{1/2} \, S_2(f(w), f(z)) \, \overline{f'(z)^{1/2}}. \tag{2.2}$$

Notice that the reason for the isometry is due to how arc length is affected by a holomorphic map. If ds_1, ds_2 are the differentials of arc length on $\partial\Omega_1$, $\partial\Omega_2$, and if $z(t)$ parameterizes $\partial\Omega_1$, then $(f \circ z)(t)$ parameterizes $\partial\Omega_2$ with $ds_1 = |z'(t)| \, dt$ and $ds_2 = |(f \circ z)'(t)| \, dt = |f'(z(t))| \, |z'(t)| \, dt - |f'| \, ds_1$.

The transformation law enables one to write the Szegő kernel for a simply connected region in terms of any biholomorphism to the unit disc. In particular, if Ω is simply connected and $f : \Omega \to \Delta$ is biholomorphic, then the Szegő kernel for Ω is

$$S(w, z) = \frac{1}{2\pi} \frac{f'(w)^{1/2}\overline{f'(z)^{1/2}}}{1 - f(w)\overline{f(z)}}. \tag{2.3}$$

2.2.2   Cauchy kernel

In contrast to the Szegő projector, there is the explicit Cauchy operator defined initially for $g \in L^1(\partial\Omega)$ according to

$$\mathcal{C}g(z) = \frac{1}{2\pi i} \int_{\partial\Omega} \frac{g(w) \, dw}{w - z}$$

for $z \in \Omega$. From this definition, one sees that $\mathcal{C}g$ is holomorphic in Ω. By restricting to functions $g \in C^\infty(\partial\Omega)$, one can show that the holomorphic function $\mathcal{C}g$ extends smoothly to boundary. That is, $\mathcal{C}$ restricts to an operator $C^\infty(\partial\Omega) \to C^\infty(\partial\Omega)$. In fact, $\mathcal{C}$ is bounded on $L^2(\partial\Omega)$ and therefore extends continuously to a bounded projector $\mathcal{C} : L^2(\partial\Omega) \to H^2(\partial\Omega)$. (The reason that $\mathcal{C}$ reproduces functions in the Hardy space is the Cauchy integral formula.) In this way, the Cauchy projector arises as an explicitly defined alternative to the Szegő projector.

For further comparison with the Szegő projector, observe that the Cauchy projector can be expressed via $\mathcal{C}g(z) = (g, C_z)_{\partial\Omega}$ where

$$C_z(w) = C(w, z) = \overline{\frac{1}{2\pi i} \frac{T(w)}{w - z}} \tag{2.4}$$

and $T(w)$ is the positively oriented unit tangent vector at $w \in \partial\Omega$.

It is apparent from these representations that the Cauchy projector and kernel will behave well with respect to affine transformations $f(z) = az + b$ for $a, b \in \mathbb{C}$. In fact, they behave well with respect to the larger class of Möbius transformations $f(z) = (az + b)/(cz + d)$ for $a, b, c, d \in \mathbb{C}$, $ad - bc = 1$. In particular, for a Möbius transformation $f : \Omega_1 \to \Omega_2$ one has the same transformation law for the Cauchy kernel as for the Szegő kernel,

$$C_1(w, z) = f'(w)^{1/2} \, C_2(f(w), f(z)) \, \overline{f'(z)^{1/2}}. \tag{2.5}$$

This is a direct calculation that uses the fact that the tangent vectors at $w \in \partial \Omega_1$, $f(w) \in \partial \Omega_2$ are related according to the chain rule by

$$T(f(w)) = \frac{f'(w)}{|f'(w)|} \, T(w) = \frac{\overline{cw + d}}{cw + d} \, T(w).$$

Naturally, the kernel transformation laws imply operator transformation laws; specifically, (2.2) implies $\mathcal{S}_1 \circ \Lambda_f = \Lambda_f \circ \mathcal{S}_2$ for a biholomorphism, and (2.5) implies $\mathcal{C}_1 \circ \Lambda_f = \Lambda_f \circ \mathcal{C}_2$ for a Möbius transformation.

2.2.3 Examples

From the Szegő kernel transformation law and the fact that the Szegő kernel for a disc is known, it is immediate that the Szegő projector can be expressed in terms of the biholomorphism to the unit disc. Given its Möbius invariance, the Cauchy operator, too, should be computable for regions with enough Möbius symmetry. We consider three such examples bounded respectively by two cir-

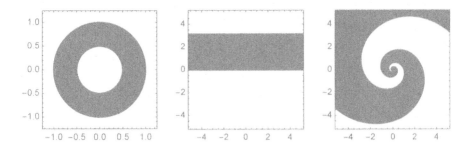

FIGURE 2.1
(i) Annulus, (ii) strip, and (iii) logarithmic sector.

cles, two lines, and two logarithmic spirals as in Figure 2.1. The behavior of Möbius transformations with respect to circles and lines is familiar, but they are well-behaved with respect to logarithmic spirals, too. In particular, logarithmic spirals have constant curvature and a natural parameter when studied invariantly with respect to Möbius transformation.

Example 1. The Szegő kernel for an annulus $\Omega = \{z \in \mathbb{C} : r < |z| < 1\}$ can be computed the same way as for a disc. In particular, holomorphic functions on an annulus can be written as the sum of a Laurent series, so a basis for $H^2(\partial\Omega)$ is $\{\psi_j\}_{j\in\mathbb{Z}}$ where $\psi_j = z^j$. This basis is orthogonal and upon normalization one finds

$$S(w, z) = \frac{1}{2\pi} \sum_{j\in\mathbb{Z}} \frac{w^j \bar{z}^j}{1 + r^{2j+1}}.$$

To show the utility of this representation, we compute the Szegő projector applied to a function $g \in L^2(\partial\Omega)$ supported on the outer circle. Let

$$g(w) = \begin{cases} e^{ij\theta} & \text{if } w = e^{i\theta} \\ 0 & \text{if } w = re^{i\theta}. \end{cases}$$

Then for $z \in \Omega$

$$\mathcal{S}g(z) = \int_0^{2\pi} e^{ij\theta} \left(\frac{1}{2\pi} \sum_{k\in\mathbb{Z}} \frac{e^{-ik\theta} z^k}{1 + r^{2k+1}} \right) d\theta = \frac{z^j}{1 + r^{2j+1}}.$$

Meanwhile, the Cauchy projector, too, is easily computed. In particular,

$$\overline{C(e^{i\theta}, z)} = \frac{1}{2\pi i} \frac{ie^{i\theta}}{e^{i\theta} - z} = \frac{1}{2\pi} \sum_{k\geq 0} z^k e^{-ik\theta}$$

so for $z \in \Omega$ one has $\mathcal{C}g(z) = \dfrac{1}{2\pi} \displaystyle\int_0^{2\pi} \sum_{k\geq 0} z^k e^{i(j-k)\theta}\, d\theta = z^j.$ $\qquad\square$

Example 2. For the strip $\Omega = \{z \in \mathbb{C} : 0 < \operatorname{Im} z < \pi\}$ one has a biholomorphism $f : \Omega \to \Delta$ given by

$$f(z) = \frac{e^z - i}{e^z + i}.$$

A computation using (2.3) shows therefore that

$$S(w, z) = \frac{1}{4\pi i} \operatorname{csch} \frac{\bar{z} - w}{2}.$$

As for the annulus, we show how this can be used to compute the Szegő projector applied to a function $g \in L^2(\partial\Omega)$ supported on the real axis. Take $\alpha > 0$ and let

$$g(w) = \begin{cases} (x^2 + \alpha^2)^{-1} & \text{if } w = x + i0 \text{ for } x \in \mathbb{R} \\ 0 & \text{if } w = x + i\pi \text{ for } x \in \mathbb{R}. \end{cases}$$

Then for $z \in \Omega$

$$\mathcal{S}g(z) = \int_{-\infty}^{+\infty} \frac{1}{x^2 + \alpha^2} \left(\frac{1}{4\pi i} \operatorname{csch} \frac{x - z}{2} \right) dx.$$

Standard methods of using residues to evaluate contour integrals are helpful for this integral. The poles for the integrand are at $x = \pm i\alpha$ and $x = z + 2\pi i k$ for $k \in \mathbb{Z}$. Then by integrating over squares with increasing side length $\operatorname{Im} z + (2k+1)\pi$ for $k = 0, 1, 2, \ldots$, and centered on the line $x = \operatorname{Re} z$ one finds after applying the residue theorem that

$$\mathcal{S}g(z) = \frac{1}{4i\alpha}\operatorname{csch}\frac{i\alpha - z}{2} + \sum_{k \geq 0} \frac{(-1)^k}{(z + 2\pi i k)^2 + \alpha^2}.$$

The Cauchy projector is done likewise but one can integrate over the boundary of upper half discs of increasing radii. The result for $z \in \Omega$ is

$$\mathcal{C}g(z) = \int_{-\infty}^{+\infty} \frac{1}{x^2 + \alpha^2}\left(\frac{1}{2\pi i}\frac{1}{x - z}\right)dx = \frac{1}{2i\alpha(i\alpha - z)} + \frac{1}{z^2 + \alpha^2}.$$

We mention that calculating the Szegő projector in this case is unremarkable. Indeed, in the presence of a biholomorphism $f : \Omega \to \Delta$, the Szegő projector for the strip can be represented as a conjugate to the Szegő projector for the disc. That is, $\mathcal{S}_\Omega = \Lambda_f \circ \mathcal{S}_\Delta \circ \Lambda_f^{-1}$ where $\Lambda_f : L^2(\partial\Delta) \to L^2(\partial\Omega)$ is the isometry from the earlier subsection. $\qquad\square$

Example 3. For the logarithmic sector $\Omega = \{z \in \mathbb{C} : \operatorname{Im}(z^{1+\pi i}) > 0\}$ one has a biholomorphism $f : \Omega \to \Delta$ expressed using the principal value of a complex power via

$$f(z) = \frac{z^{1+\pi i} - i}{z^{1+\pi i} + i}.$$

In this case,

$$S(w, z) = \frac{\sqrt{1 + \pi^2}}{2\pi i}\frac{w^{+\pi i/2}\bar{z}^{-\pi i/2}}{\bar{z}^{1-\pi i} - w^{1+\pi i}}.$$

As for the strip, the Szegő projector is understood immediately through a biholomorphism $f : \Omega \to \Delta$. Again, $\mathcal{S}_\Omega = \Lambda_f \circ \mathcal{S}_\Delta \circ \Lambda_f^{-1}$ where $\Lambda_f : L^2(\partial\Delta) \to L^2(\partial\Omega)$ is the isometry from the earlier subsection. The Cauchy projector, meanwhile, is best computed after change of parameter. Using a constant speed parameter, the boundary spirals have parameterizations $w_\pm$ given by

$$t > 0 \to w_\pm(t) \stackrel{\text{def}}{=} \pm t^{1-\pi i}$$

In fact, the parameter t is related to the arc length parameter via $t = s/\sqrt{1 + \pi^2}$. Consider a function $g \in L^2(\partial\Omega)$ supported on the boundary spiral w_+. Then for $z \in \Omega$

$$\mathcal{C}g(z) = \frac{1}{2\pi i}\int_0^\infty (g \circ w_+)(t)\frac{w_+'(t)\,dt}{w_+(t) - z}$$
$$= \frac{1 - \pi i}{2\pi i}\int_0^\infty (g \circ w_+)(t)\frac{t^{-\pi i}\,dt}{t^{1-\pi i} - z}.$$

After switching to an inversive arc length parameter this integral is computable like for the strip. Take $u = \log t$ so that

$$\mathcal{C}g(z) = \frac{1 - \pi \mathrm{i}}{2\pi \mathrm{i}} \int_{-\infty}^{+\infty} (g \circ w_+)(e^u) \frac{e^{u(1-\pi \mathrm{i})} \, du}{e^{u(1-\pi \mathrm{i})} - z}.$$

(Typically the inversive arc length parameter for these spirals is given as $u = \sqrt{\pi} \log s$; here any multiple will suffice.) To show the utility of this representation consider g defined for $\alpha > 0$ by

$$g(w) = \begin{cases} (u^2 + \alpha^2)^{-1} & \text{if } w = w_+(e^u) \text{ for } u \in \mathbb{R} \\ 0 & \text{if } w = w_-(e^u) \text{ for } u \in \mathbb{R}. \end{cases}$$

Again using a contour integral approach, one finds poles for the integrand at $u = \pm \mathrm{i}\alpha$ and $u = (\log z + 2\pi \mathrm{i}k)/(1 - \pi \mathrm{i})$ for $k \in \mathbb{Z}$. Restricting to the poles in the upper half plane and applying the residue theorem gives

$$\mathcal{C}g(z) = \frac{1 - \pi \mathrm{i}}{2\pi \mathrm{i}} \int_{-\infty}^{+\infty} \frac{1}{u^2 + \alpha^2} \frac{e^{u(1-\pi \mathrm{i})} \, du}{e^{u(1-\pi \mathrm{i})} - z}$$

$$= \frac{1 - \pi \mathrm{i}}{2\mathrm{i}\alpha} \frac{e^{\mathrm{i}\alpha(1-\pi \mathrm{i})}}{e^{\mathrm{i}\alpha(1-\pi \mathrm{i})} - z} + \sum_{k}' \frac{1}{\left(\frac{\log z + 2\pi \mathrm{i}k}{1 - \pi \mathrm{i}}\right)^2 + \alpha^2}$$

where the prime on the summation means to include only k for which $\mathrm{Im}\left((\log z + 2\pi \mathrm{i}k)/(1 - \pi \mathrm{i})\right) > 0$. For $z = e^{\mathrm{i}\alpha} t^{1-\pi \mathrm{i}}$ with $\alpha \in (0, \pi)$ this means $k > \lfloor \frac{\alpha}{2\pi} + \frac{1 - \log t}{2} \rfloor - \frac{\alpha}{2\pi}$. $\square$

2.2.4   Notes

- Our definition of the Hardy space is the one used in Bell's book [5]. The classical definition says that a function is in the Hardy space provided it is holomorphic and the supremum of its averages along curves approximating the boundary from the inside is finite. For the equivalence of these definitions, see [5, p. 17].

- As outlined in the Krantz contribution [24], the Szegő and Cauchy projectors are related via $\mathcal{S} = \mathcal{C} \left(\mathcal{I} + \mathcal{A}\right)^{-1}$ where the imaginary part of the Cauchy projector $\mathcal{A} \stackrel{\text{def}}{=} \mathcal{C} - \mathcal{C}^*$ is the Kerzman-Stein operator. This leads to the integral equation

$$S_a(z) - \int_{w \in \partial \Omega} \frac{1}{2\pi \mathrm{i}} \left(\frac{T(w)}{w - z} - \frac{\overline{T(z)}}{\overline{w} - \overline{z}}\right) S_a(w) \, ds_w = C_a(z)$$

giving the Szegő kernel as the unique solution to a Fredholm integral equation of the second kind. See [5, 22].

- From the last note it is apparent that $\mathcal{C} = \mathcal{S}$ only if $\mathcal{A} = 0$. Kerzman and Stein give an elementary geometric argument in terms of the kernel to show that this happens only for a disc or half-plane. See [21]. Using Taylor expansions of the kernel, one can show that this result persists when one uses measures besides arc length. See [9].

- Our examples show how the Szegő kernel can be expressed in terms of the Riemann map. Conversely, Kerzman and Trummer show how the Riemann map and Ahlfors maps can be recovered numerically from the Szegő kernel after solving the Kerzman-Stein integral equation. See [13, 22, 28].

- Using the identity $\|\mathcal{C}\| = \sqrt{1 + \|\mathcal{A}\|^2}$, one can find

  - $\|\mathcal{C}\| = \sqrt{1 + r}$ for Example 1,
  - $\|\mathcal{C}\| = \sqrt{2}$ for Example 2, and
  - $\|\mathcal{C}\| \geq \coth(1/2)$ for Example 3. (See [8].)

- Other example regions for which the Cauchy and Kerzman-Stein operators are at least partially computable include the wedge and ellipse. See [8, 10, 12, 16].

- The Szegő kernel transformation law (2.2) extends in a simple way for the case of a proper holomorphic map provided the target region is simply connected. If the target is multiply connected, the transformation law involves derivatives of the associated harmonic measure functions. See [15, 19, 20].

2.3 The Cauchy-Leray Operator

We now introduce a generalization of the Cauchy operator appropriate for convex domains in high dimensions. Take $n \geq 2$ and let $D \subset\subset \mathbb{C}^n$ be a domain with C^2 smooth boundary and defining function r. This means that $D = \{z \in \mathbb{C}^n : r(z) < 0\}$ for a continuously twice differentiable function $r : \mathbb{C}^n \to \mathbb{R}$ with $\nabla r|_{\partial D} \neq 0$.

2.3.1 Convexity and weak linear convexity

Our understanding of a convex domain is one supported by hyperplanes tangent at the boundary. In particular, the tangent space at any point of the boundary does not intersect the domain. In complex coordinates this is written Re $\langle \partial r(w), w - z \rangle > 0$ for $w = (w_1, \ldots, w_n) \in \partial D$, $z = (z_1, \ldots, z_n) \in D$. The brackets should be interpreted as

$$\langle \partial r(w), w - z \rangle = \sum_{j=1}^{n} \frac{\partial r}{\partial w_j}(w_j - z_j);$$

this is the natural pairing at w of the (1,0)-form ∂r with vector $w - z$. Our specific construction of the Cauchy-Leray operator allows that D has a weaker geometric property called weak linear convexity that can be expressed as $\langle \partial r(w), w - z \rangle \neq 0$ for all $w \in \partial D$, $z \in D$. This means that the maximal complex subspace of each tangent space at the boundary does not intersect the domain. The difference between these conditions can be seen in the following examples.

Example 4. Let $D = \{z \in \mathbb{C}^2 : 2\operatorname{Re} z_1 + |z_2|^2 < 0\}$. Then $w \in \partial D$, $z \in D$ mean that $w_1 + \overline{w}_1 + w_2\overline{w}_2 = 0$, $z_1 + \overline{z}_1 + z_2\overline{z}_2 < 0$. From this we estimate

$$
\begin{aligned}
2\operatorname{Re}\langle \partial r(w), w - z \rangle &= 2\operatorname{Re}\left(1(w_1 - z_1) + \overline{w}_2(w_2 - z_2)\right) \\
&= w_1 + \overline{w}_1 - z_1 - \overline{z}_1 + 2w_2\overline{w}_2 - \overline{w}_2 z_2 - w_2\overline{z}_2 \\
&> w_2\overline{w}_2 + z_2\overline{z}_2 - \overline{w}_2 z_2 - w_2\overline{z}_2 \\
&= |w_2 - z_2|^2 \\
&\geq 0.
\end{aligned}
$$

It follows that D is convex, and therefore, also weakly linearly convex. □

Example 5. Let $D = \{z \in \mathbb{C}^2 : -1 + 2\operatorname{Re} z_1^2 < 0\}$. Then $w \in \partial D$, $z \in D$ mean that $-1 + w_1^2 + \overline{w}_1^2 = 0$, $-1 + z_1^2 + \overline{z}_1^2 < 0$. From this we estimate

$$
\langle \partial r(w), w - z \rangle = 2w_1(w_1 - z_1) + 0(w_2 - z_2).
$$

So $\langle \partial r(w), w - z \rangle = 0$ means that $w_1 = 0$ or $w_1 - z_1 = 0$. The former case is impossible since $2\operatorname{Re} w_1^2 = 1$; the latter case is impossible since $2\operatorname{Re} w_1^2 = 1$ and $2\operatorname{Re} z_1^2 < 1$. It follows that $\langle \partial r(w), w - z \rangle \neq 0$ and D is weakly linearly convex.

To see that D is not convex, it is simplest to recognize that D is a product of a region bounded inside a hyperbola (in the z_1-plane) with a complex plane (in the z_2-plane) as in Figure 2.2. To check the claim algebraically take $w = (w_1, w_2) \in \partial D$ with $w_1 = u + iv$, $w_2 = 0$ so that $u^2 - v^2 = \frac{1}{2}$. Subsequently

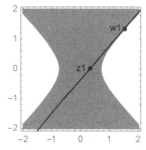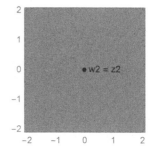

FIGURE 2.2
Product domain that is weakly linearly convex.

take $z = (z_1, z_2)$ with $z_1 = \frac{1}{2u} + i0$, $z_2 = 0$. Provided $u^2 > \frac{1}{2}$, then $z \in D$ since

$$-1 + 2\operatorname{Re} z_1^2 = -1 + 2\left(\frac{1}{2u}\right)^2 = \frac{-2u^2 + 1}{2u^2} < 0.$$

Meanwhile,

$$\operatorname{Re}\langle \partial r(w), w - z \rangle = \operatorname{Re}\left(2w_1(w_1 - z_1)\right) = \operatorname{Re}\left(2(u + iv)\left(\frac{v^2}{u} + iv\right)\right) = 0. \quad \square$$

2.3.2   The Cauchy-Leray operator

For a smooth, bounded, weakly linearly convex domain D, the Cauchy-Leray operator applied to a continuous function on the boundary is given by

$$\mathcal{L}g(z) = \frac{1}{(2\pi i)^n} \int_{\partial D} g(w) \frac{\partial r(w) \wedge (\overline{\partial}\partial r)^{n-1}}{\langle \partial r(w), w - z \rangle^n} \tag{2.6}$$

for $z \in D$. The hypothesis of weak linear convexity implies that the denominator is nonzero for $z \in D$. The additional hypotheses of boundedness and smoothness of D imply that both the $(2n - 1)$-form in the numerator is a continuous multiple of surface measure and the denominator is bounded away from zero (for fixed $z \in D$). These ensure that the integral is finite. Since also the integrand is holomorphic in z and the boundary is finite it follows that $\mathcal{L}g$ is holomorphic in D.

It is a harder fact that the Cauchy-Leray operator reproduces holomorphic functions in D. In particular, if g is a continuous function on ∂D that extends continuously to $\overline{D}$ and holomorphically in D, then $\mathcal{L}g = g$. The usual proof of this for convex domains involves writing the Cauchy-Leray kernel as a Cauchy-Fantappiè kernel using a generating form based on the pairing $\langle \partial r(w), w - z \rangle$. The argument is outlined in the Krantz contribution [24]; for more details, see also [23, 26].

It is important to note that the Cauchy-Leray operator reduces to the Cauchy operator in case $n = 1$. This can be seen easily by canceling the common non-zero factor $\partial r/\partial w$ in the numerator and denominator of (2.6). Notice that the maximal complex subspace of the line tangent at the boundary is the point of the boundary itself. So all smooth regions in $\mathbb{C}$ are weakly linearly convex.

An equivalent formulation of the Cauchy-Leray operator is due to Stanton and involves the Levi determinant

$$J(r) = -1 \cdot \det \begin{pmatrix} r & r_{\overline{k}} \\ r_j & r_{j\overline{k}} \end{pmatrix}.$$

Here the subscripts refer to partial derivatives; i.e., $r_j = \partial r/\partial w_j$, $r_{\overline{k}} = \partial r/\partial \overline{w}_k$. If $d\sigma^{\text{euc}}$ is the surface measure on ∂D (induced from $\mathbb{C}^n$), then

$$\mathcal{L}^S g(z) = \frac{(n-1)!}{\pi^n} \int_{\partial D} g(w) \frac{J(r)(w) \, d\sigma^{\text{euc}}(w)}{|\nabla r(w)| \, \langle \partial r(w), w - z \rangle^n} \tag{2.7}$$

for $z \in D$.

We point out that although the Cauchy-Leray operator is presented in terms of a defining function, the operator does not depend on the choice of defining function. For this, consider the second formulation and take defining functions r, $\tilde{r}$ related by $\tilde{r} = hr$ for a positive function h. From the product rule, we find that if $w \in \partial D$ (and so $r(w) = 0$) then $|\nabla \tilde{r}(w)| = h(w)|\nabla r(w)|$,

$$\langle \partial \tilde{r}(w), w - z \rangle = h(w)\langle \partial r(w), w - z \rangle,$$

and using elementary linear algebra,

$$
\det \begin{pmatrix} \tilde{r} & \tilde{r}_{\overline{k}} \\ \tilde{r}_j & \tilde{r}_{j\overline{k}} \end{pmatrix}
= \det \begin{pmatrix} hr & hr_{\overline{k}} \\ hr_j & hr_{j\overline{k}} + h_j r_{\overline{k}} + h_{\overline{k}} r_j \end{pmatrix}
$$

$$
= \det \begin{pmatrix} hr & hr_{\overline{k}} \\ hr_j & hr_{j\overline{k}} \end{pmatrix}
$$

$$
= h^{n+1} \det \begin{pmatrix} r & r_{\overline{k}} \\ r_j & r_{j\overline{k}} \end{pmatrix}.
$$

Combining these gives

$$
\frac{J(\tilde{r})(w)}{|\nabla \tilde{r}(w)| \, \langle \partial \tilde{r}(w), w - z \rangle^n} = \frac{J(r)(w)}{|\nabla r(w)| \, \langle \partial r(w), w - z \rangle^n}
$$

which verifies the claim.

Example 6. To illustrate the two formulations, consider the unit ball $D = \{z \in \mathbb{C}^2 : r(z) \stackrel{\text{def}}{=} |z_1|^2 + |z_2|^2 - 1 < 0\}$. It is straightforward to check that

$$
\mathcal{L}g(z) = \frac{1}{(2\pi i)^2} \int_{\partial D} \frac{g(w)\,(\overline{w}_1 dw_1 \wedge d\overline{w}_2 \wedge dw_2 + \overline{w}_2 dw_2 \wedge d\overline{w}_1 \wedge dw_1)}{(1 - \overline{w}_1 z_1 - \overline{w}_2 z_2)^2}.
$$

Meanwhile, $J(r)(w) = 1$ and $|\nabla r(w)| = 2$ when $w \in \partial D$. So then

$$
\mathcal{L}^S g(z) = \frac{1}{2\pi^2} \int_{\partial D} \frac{g(w)\,d\sigma^{\text{euc}}}{(1 - \overline{w}_1 z_1 - \overline{w}_2 z_2)^2}.
$$

To see that $\mathcal{L} = \mathcal{L}^S$ it is enough to check that

$$
d\sigma^{\text{euc}} = -\frac{1}{2}\left(\overline{w}_1 dw_1 \wedge d\overline{w}_2 \wedge dw_2 + \overline{w}_2 dw_2 \wedge d\overline{w}_1 \wedge dw_1\right).
$$

Since dr annihilates vectors tangent to ∂D, this follows from wedging both sides by $dr = \overline{w}_1 dw_1 + w_1 d\overline{w}_1 + \overline{w}_2 dw_2 + w_2 d\overline{w}_2$ and applying the general formula $dr \wedge d\sigma^{\text{euc}} = |\nabla r|\, dx_1 \wedge dy_1 \wedge dx_2 \wedge dy_2$ for an embedded boundary. □

2.3.3 Notes

- Our definition of weak linear convexity follows that of Andersson, Passare, and Sigurdsson [1, p. 16]. For the fact that any weakly linearly convex domain is pseudoconvex, see Proposition 2.1.8. This will be used in the next section to show that $J(r) \geq 0$ on ∂D.

- The Cauchy-Leray operator is known alternatively in the literature as the Leray operator (or transform), the Leray-Aïzenberg operator, or simply the Cauchy operator for convex domains.

- Using the Cauchy-Fantappiè method, Lanzani and Stein establish the reproducing property for the Cauchy-Leray operator in the case that D is strictly $\mathbb{C}$-linearly convex. This condition is stronger than weak linear convexity but weaker than convexity. The condition says $\langle \partial r(w), w - z \rangle \neq 0$ for $w \in \partial D$, $z \in \overline{D} \setminus \{w\}$. See [25, p. 276].

2.4 The Cauchy-Leray Transformation Formula

We begin this section by introducing the Fefferman surface measure that was constructed for the purpose of having a transformation law for the Szegő kernel under biholomorphic mappings in high dimensions. We then establish the same transformation law for the Cauchy-Leray kernel under Möbius transformations. We conclude with a summary of the properties of the Cauchy-Leray operator that distinguish it from other reproducing kernels in high dimensions.

2.4.1 Fefferman surface measure

The Fefferman surface measure $d\sigma^{\text{feff}}$ for a smooth pseudoconvex domain $D = \{z \in \mathbb{C}^n : r(z) < 0\}$ is defined as the measure on ∂D for which

$$d\sigma^{\text{feff}} \wedge dr = -i^n \, 2^{-n+1} J(r)^{1/(n+1)} dz_1 \wedge \overline{dz_1} \wedge \cdots \wedge dz_n \wedge \overline{dz_n}$$
$$= -2 \, J(r)^{1/(n+1)} dv$$

where as in the last section $J(r)$ is the Levi determinant

$$J(r) = -1 \cdot \det \begin{pmatrix} r & r_{\overline{k}} \\ r_j & r_{j\overline{k}} \end{pmatrix}$$

and $dv = dx_1 \wedge dy_1 \wedge \cdots \wedge dy_n$ is the volume measure in $\mathbb{C}^n$.

We explain briefly the context in which $J(r)$ is nonnegative on ∂D; this is needed for $J(r)^{1/(n+1)}$ to be well-defined. At any point $w \in \partial D$, one can make a unitary change of coordinates so that $r_j(w) = 0$, $1 \leq j < n$, and $r_n(w) \neq 0$. In these coordinates,

$$J(r)(w) = |r_n|^2 \, \det\left(r_{j\overline{k}}\right)_{j,k=1}^{n-1}$$

The latter determinant is the product of the eigenvalues of the Levi form restricted to the complex tangent space. For a pseudoconvex domain such eigenvalues are nonnegative and we conclude that $J(r) \geq 0$ on ∂D. From the notes of last section it follows that $J(r) \geq 0$, too, in the special case of weak linear convexity.

Any constant multiple of $d\sigma^{\text{feff}}$ also will be an invariant surface measure. Here, the constant was chosen so that the measure coincides with the Euclidean surface measure for the boundary of a ball as well as the arc length measure in one dimension.

With respect to Fefferman measure, we define function spaces $L^2(\partial D)$ using the inner product

$$(g, h)_{\partial D} = \int_{\partial D} g \, \overline{h} \, d\sigma^{\text{feff}}. \tag{2.8}$$

If we further assume D_1 is simply connected, then as in case $n = 1$ a biholomorphism $F = (f^1, \ldots, f^n) : D_1 \rightarrow D_2$ gives rise to an isometry $\Lambda_F : L^2(\partial D_2) \rightarrow L^2(\partial D_1)$ via $g \rightarrow (g \circ F) J_F^{n/(n+1)}$ where we denote the complex Jacobian by $J_F = \det(f_k^j)$. As before, subscripts refer to partial derivatives. Here, the condition on D_1 ensures that the fractional power makes sense. Notice, too, that the isometry preserves holomorphicity.

The reason for the fractional power in the isometry is due to the transformation of the Fefferman measure. On ∂D_1 and ∂D_2, respectively,

$$d\sigma^{\text{feff}} \wedge d(r \circ F) = -2J(r \circ F)^{1/(n+1)} \, dv$$
$$d\sigma^{\text{feff}} \wedge dr = -2J(r)^{1/(n+1)} \, dv.$$

It is standard for a biholomorphism that $F^*(dv) = |J_F|^2 \, dv$. The related identity $J(r \circ F) = J(r) \cdot |J_F|^2$ follows from taking the determinant of both sides of

$$\begin{pmatrix} r \circ F & (r \circ F)_{\overline{k}} \\ (r \circ F)_j & (r \circ F)_{j\overline{k}} \end{pmatrix} = \begin{pmatrix} 1 & 0 \\ 0 & f_j^l \end{pmatrix} \begin{pmatrix} r & r_{\overline{m}} \\ r_l & r_{l\overline{m}} \end{pmatrix} \begin{pmatrix} 1 & 0 \\ 0 & f_{\overline{k}}^{\overline{m}} \end{pmatrix}.$$

Since $F^*(dr) = d(r \circ F)$, we conclude $F^*(d\sigma^{\text{feff}}) = |J_F|^{2n/(n+1)} d\sigma^{\text{feff}}$.

It follows that if $\mathcal{S} : L^2(\partial D) \rightarrow H^2(\partial D)$ is the orthogonal projection (with respect to Fefferman measure) to the subspace of functions that extend holomorphically to D then one has a transformation law $\mathcal{S}_1 \circ \Lambda_F = \Lambda_F \circ \mathcal{S}_2$. As before, this Szegő projector is an integral operator whose kernel satisfies the transformation law

$$S_1(w, z) = J_F^{n/(n+1)}(w) \, S_2(F(w), F(z)) \, \overline{J_F^{n/(n+1)}(z)}. \tag{2.9}$$

In dimension one this reduces to the situation of the previous section (where the requirement that D_1 be simply connected was not necessary).

2.4.2 Cauchy-Leray transformation formula

We proceed to show that with respect to the Fefferman measure the Cauchy-Leray kernel is given by

$$C(w,z) = \overline{\frac{(n-1)!}{2\pi^n} \frac{J(r)^{n/(n+1)}(w)}{\langle \partial r(w), w-z \rangle^n}} \tag{2.10}$$

and this kernel satisfies the same transformation law as for the Szegő kernel provided $F : D_1 \to D_2$ is a Möbius transformation. In this situation there is no further assumption about D_1 since the fractional power of J_F makes sense. (See item (a) below.)

For the first claim, we start with the calculation

$$\frac{1}{(2\pi i)^n} \left(\partial r \wedge (\overline{\partial} \partial r)^{n-1} \right) \wedge dr = \frac{1}{(2\pi i)^n} \partial r \wedge \overline{\partial} r \wedge (\overline{\partial} \partial r)^{n-1}$$

$$= -\frac{(n-1)!}{\pi^n} J(r)\, dv$$

$$= \frac{(n-1)!}{2\pi^n} J(r)^{n/(n+1)}\, d\sigma^{\text{feff}} \wedge dr$$

where the third step uses the definition of Fefferman measure and the second step uses $\partial r \wedge \overline{\partial} r \wedge (\overline{\partial} \partial r)^{n-1} = -(2i)^n (n-1)! J(r)\, dv$. (The proof of this identity is essentially the same as what we used to explain why $J(r) \geq 0$.) Since dr acts transversely to the boundary, it follows that when restricting to ∂D and dividing by $\langle \partial r(w), w-z \rangle^n$,

$$\frac{(n-1)!}{2\pi^n} \frac{J(r)^{n/(n+1)}(w)}{\langle \partial r(w), w-z \rangle^n}\, d\sigma_w^{\text{feff}} = \frac{1}{(2\pi i)^n} \frac{\partial r(w) \wedge (\overline{\partial} \partial r)^{n-1}}{\langle \partial r(w), w-z \rangle^n}.$$

Referring back to (2.6) and (2.8), we have $\mathcal{L}g(z) = (g, C_z)_{\partial D}$ provided $C_z(w) \stackrel{\text{def}}{=} C(w,z)$ with $C(w,z)$ as expressed in (2.10).

For the second claim, we show that if $F = (f^1, \dots f^n) : D_1 \to D_2$ is a Möbius transformation with $f^j = g_j / g_{n+1}$ ($1 \leq j \leq n$), $g_j(w) = a_{j,1} w_1 + \cdots + a_{j,n} w_n + a_{j,n+1}$ ($1 \leq j \leq n+1$), and $\det(a_{j,k}) = 1$, then

$$C_1(w,z) = J_F^{n/(n+1)}(w)\, C_2(F(w), F(z))\, \overline{J_F^{n/(n+1)}(z)}. \tag{2.11}$$

Beyond what already is established, the proof requires the following:

(a) $J_F(z) = (1/g_{n+1}(z))^{n+1}$

(b) $\sum (r \circ F)_j(w)(w_j - z_j) = \sum r_j(F(w))(f^j(w) - f^j(z)) g_{n+1}(z)/g_{n+1}(w).$

For (a),

$$J_F(z) = \det\left(\frac{\partial f^j}{\partial z_k}\right) = \frac{1}{g_{n+1}^n}\det\left(a_{j,k} - a_{n+1,k}\frac{g_j}{g_{n+1}}\right)$$

$$= \frac{1}{g_{n+1}^n}\det\begin{pmatrix} a_{j,k} - a_{n+1,k}\,g_j/g_{n+1} & 0 \\ a_{n+1,k} & 1 \end{pmatrix}$$

$$= \frac{1}{g_{n+1}^n}\det\begin{pmatrix} a_{j,k} & g_j/g_{n+1} \\ a_{n+1,k} & 1 \end{pmatrix} = \frac{1}{g_{n+1}^{n+1}}\det\begin{pmatrix} a_{j,k} & g_j \\ a_{n+1,k} & g_{n+1} \end{pmatrix}$$

$$= \frac{1}{g_{n+1}^{n+1}}\det\begin{pmatrix} a_{j,k} & a_{j,n+1} \\ a_{n+1,k} & a_{n+1,n+1} \end{pmatrix} = \frac{1}{g_{n+1}^{n+1}}.$$

For (b),

$$\sum_{k=1\ldots n}(r\circ F)_k(w)(w_k - z_k)$$

$$= \sum_{j,k=1\ldots n} r_j(F(w))\,g_{n+1}(w)^{-1}\,[a_{j,k} - a_{n+1,k}\,g_j(w)/g_{n+1}(w)]\,(w_k - z_k)$$

$$= \sum_{j=1\ldots n} r_j(F(w))\,g_{n+1}(w)^{-1}\left[g_j(w) - g_j(z) - [g_{n+1}(w) - g_{n+1}(z)]\frac{g_j(w)}{g_{n+1}(w)}\right]$$

$$= \sum_{j=1\ldots n} r_j(F(w))\,g_{n+1}(w)^{-1}\,[g_{n+1}(z)\,g_j(w)/g_{n+1}(w) - g_j(z)]$$

$$= \sum_{j=1\ldots n} r_j(F(w))(f^j(w) - f^j(z))g_{n+1}(z)/g_{n+1}(w).$$

With these in hand, if $F : D_1 \to D_2$ is a Möbius transformation and r is a defining function for D_2, then

$$\overline{C_1(w,z)}\cdot 2\pi^n/(n-1)! = \frac{J(r\circ F)^{n/(n+1)}(w)}{\left(\sum(r\circ F)_j(w)(w_j - z_j)\right)^n}$$

$$= \frac{J(r)^{n/(n+1)}(F(w))\cdot |J_F(w)|^{2n/(n+1)}}{\left(\sum(r\circ F)_j(w)(w_j - z_j)\right)^n}$$

$$= \frac{J(r)^{n/(n+1)}(F(w))}{\left(\sum r_j(F(w))(f^j(w) - f^j(z))\right)^n}\cdot\frac{g_{n+1}(w)^n}{g_{n+1}(z)^n}\cdot |J_F(w)|^{2n/(n+1)}$$

$$= \overline{J_F^{n/(n+1)}(w)}\cdot\frac{J(r)^{n/(n+1)}(F(w))}{\left(\sum r_j(F(w))(f^j(w) - f^j(z))\right)^n}\cdot J_F^{n/(n+1)}(z)$$

$$= \overline{J_F^{n/(n+1)}(w)}\,\overline{C_2(F(w),F(z))}\,J_F^{n/(n+1)}(z)\cdot 2\pi^n/(n-1)!.$$

Taking conjugates establishes (2.11).   □

2.4.3 Comparison of kernels

Here we give a brief comparison of some of the more prominent boundary integral operators and kernels in complex analysis. These include the ones discussed in this article as well as the Bochner-Martinelli and Henkin-Ramirez integrals. For the latter operators, we refer to the Krantz contribution to this volume [24] as well as [23, 26]. Of the kernels under consideration, the Szegő kernel stands alone as a non-explicit kernel—its existence is known but it is known precisely only in special situations.

Beyond this, we have the Table 2.1 to summarize the key differences. *Class of domain* indicates geometric requirements for construction of the kernel.

	Cauchy $n = 1$	Szegő $n \geq 1$	Bochner-Martinelli $n > 1$	Cauchy-Leray $n > 1$	Henkin-Ramirez $n > 1$
Class of domain	General	General	General	Convex	Strictly pseudo-convex
Holomorphic	Yes	Yes	No	Yes	Yes
r-dependent	No	No	No	No	Yes
Invariance	Möbius	Biholo-morphism	Affine	Möbius	Affine

Table 2.1
Comparison of reproducing kernels

(The degree of smoothness typically is what is needed for the construction—twice differentiable for Cauchy-Leray and Henkin-Ramirez but only once differentiable for Cauchy and Bochner-Martinelli.) *Holomorphic* refers to the free variable; in particular, an operator with holomorphic kernel projects to a class of holomorphic functions even if integration is done over just a subset of the boundary. The *r-dependence* indicates whether the resulting operator depends on the defining function. Finally, *invariance* refers to symmetries or the existence of a transformation law for the kernel. As shown in this article, there is invariance for the Szegő and Cauchy-Leray kernels provided integration is done with respect to Fefferman measure.

2.4.4 Equivalence and example

We conclude with an explanation why $\mathcal{L} = \mathcal{L}^S$ and present an additional example (besides the sphere) that has enough symmetry to enable calculation of an L^2-norm.

The equivalence of (2.6) and (2.7) is a consequence of working out the wedge products and using standard formulas for surface area in terms of differential forms. To be precise, using the identity $\partial r \wedge \overline{\partial} r \wedge (\overline{\partial}\partial r)^{n-1} =$

$-(2i)^n (n-1)! \, J(r) \, dv$ from the earlier section, we have

$$\left(\partial r \wedge (\overline{\partial}\partial r)^{n-1} \right) \wedge dr = -(2\mathrm{i})^n (n-1)! \, J(r) \, \frac{dr}{|\nabla r|} \wedge d\sigma^{\mathrm{euc}}$$

$$= \frac{(2\mathrm{i})^n (n-1)! \, J(r)}{|\nabla r|} \, d\sigma^{\mathrm{euc}} \wedge dr.$$

So when restricting to ∂D and dividing by $(2\pi\mathrm{i})^n \, \langle \partial r(w), w - z \rangle^n$, we have

$$\frac{1}{(2\pi\mathrm{i})^n} \frac{\partial r(w) \wedge (\overline{\partial}\partial r)^{n-1}}{\langle \partial r(w), w - z \rangle^n} = \frac{(n-1)!}{\pi^n} \frac{J(r)(w) \, d\sigma^{\mathrm{euc}}(w)}{|\nabla r(w)| \, \langle \partial r(w), w - z \rangle^n}.$$

That $\mathcal{L} = \mathcal{L}^S$ then follows from (2.6) and (2.7). This is essentially the same proof that Stanton gave in [27].

Example 7. We conclude with the domain for which Barrett and Edholm compute the L^2-norm of the Cauchy-Leray transform as a singular integral operator on the boundary [2],

$$D = D_\beta = \{ z = (z_1, z_2) \in \mathbb{C}^2 : r(z) \overset{\text{def}}{=} |z_1|^2 + \beta \operatorname{Re}(z_1^2) - \operatorname{Im} z_2 < 0 \}$$

for $0 \le \beta < 1$. It is straightforward to check that D is strictly $\mathbb{C}$-linearly convex. (The definition of this property is given in the Notes of §3.) In particular, $\langle \partial r(w), w - z \rangle = (\overline{w}_1 + \beta w_1)(w_1 - z_1) - \frac{1}{2\mathrm{i}}(w_2 - z_2)$. So for $w \in \partial D$, $z \in \overline{D}$,

$$2 \operatorname{Re} \langle \partial r(w), w - z \rangle = \left(|w_1|^2 + \beta \operatorname{Re}(w_1^2) \right) - 2 \operatorname{Re}\left((\overline{w}_1 + \beta w_1) z_1 \right) + \operatorname{Im} z_2$$

$$\ge |w_1 - z_1|^2 + \beta \operatorname{Re}\left[(w_1 - z_1)^2 \right]$$

$$\ge 0.$$

The first step uses $w \in \partial D$ and the second step uses $z \in \overline{D}$. There is equality in the second and third steps only if $z \in \partial D$ and $w_1 = z_1$, respectively. To finish the claim, it is enough to see that $\langle \partial r(w), w - z \rangle \ne 0$ for $w, z \in \partial D$ with $w_1 = z_1$, $w_2 \ne z_2$. But then, $\langle \partial r(w), w - z \rangle = -\frac{1}{2\mathrm{i}}(w_2 - z_2)$ is clearly nonzero.

Using (2.6) we find that for a function g defined on ∂D,

$$\mathcal{L}g(z) = \frac{1}{(2\pi\mathrm{i})^2} \int_{\partial D} \frac{-\frac{1}{2\mathrm{i}} g(w) \, dw_2 \wedge d\overline{w}_1 \wedge dw_1}{\left[(\overline{w}_1 + \beta w_1)(w_1 - z_1) - \frac{1}{2\mathrm{i}}(w_2 - z_2) \right]^2}$$

for $z \in D$. Take $w_j = u_j + \mathrm{i}v_j$, $z_j = x_j + \mathrm{i}y_j$ with $u_j, v_j, x_j, y_j \in \mathbb{R}$ so that $v_2 = (1+\beta)u_1^2 + (1-\beta)v_1^2$ and $y_2 > (1+\beta)x_1^2 + (1-\beta)y_1^2$ since $w \in \partial D$, $z \in D$, respectively. In these coordinates the integral simplifies to

$$\mathcal{L}g(z) = \frac{1}{4\pi^2} \int_{\mathbb{R}^3} \frac{g(w) \, du_2 \wedge du_1 \wedge dv_1}{\left[(\overline{w}_1 + \beta w_1)(w_1 - z_1) - \frac{1}{2\mathrm{i}}(w_2 - z_2) \right]^2}.$$

The denominator is more complicated than our earlier examples, but still it is the square of a quadratic function. As shown in [2] (and indicated in the notes below), there are hidden symmetries that work to simplify the calculations. In fact, the first main result in [2] is the calculation

$$||\mathcal{L}||_{L^2(D_\beta, \sigma^{\text{feff}})} = \frac{1}{\sqrt[4]{1 - \beta^2}}.$$

Alternatively, one might attempt simpler calculations, for instance, by (i) considering a simplified function g (as in Examples 1–3), (ii) restricting to $\beta = 0$ (so D is the Siegel upper half-space), or (iii) evaluating only at $z = (0, iy_2)$ for $y_2 > 0$. □

2.4.5 Notes

- With respect to Fefferman measure, Hirachi proved the Szegő transformation formula for strictly pseudoconvex domains and for certain strictly pesudoconvex CR manifolds [18]. Fefferman introduced the measure in [17, p. 259].

- Like the case $n = 1$, the Cauchy-Leray and Szegő projectors are the same only when the domain is a Möbius image of a ball and only when one uses Fefferman measure. Similar results hold for the case of the Bochner-Martinelli kernel. See [6, 7, 9].

- A more detailed analysis of the Cauchy-Leray operator for the domain D_β in Example 7 leads to a representation that is reminiscent of that for the unit ball in Example 6. In particular, [2, §4] shows how our first representation of $\mathcal{L}$ for the ball is the same as that for D_β if one replaces the conjugate holomorphic integration variables (w_1, w_2) for the ball by "projective dual coordinates" for D_β. (The specifics of the duality construction are outside the scope of this article.) The domain D_β has been studied, too, using methods from differential geometry. See [11].

References

[1] Mats Andersson, Mikael Passare, and Ragnar Sigurdsson. *Complex convexity and analytic functionals*, volume 225 of *Progress in Mathematics*. Birkhäuser Verlag, Basel, 2004.

[2] David E. Barrett and Luke D. Edholm. The Leray transform: factorization, dual CR structures, and model hypersurfaces in $\mathbb{CP}^2$, November 2018.

[3] David E. Barrett. Holomorphic projection and duality for domains in complex projective space. *Trans. Amer. Math. Soc.*, 368(2):827–850, 2016.

[4] David E. Barrett and Loredana Lanzani. The spectrum of the Leray transform for convex Reinhardt domains in $\mathbb{C}^2$. *J. Funct. Anal.*, 257(9): 2780–2819, 2009.

[5] Steven R. Bell. *The Cauchy transform, potential theory, and conformal mapping*. Studies in Advanced Mathematics. CRC Press, Boca Raton, FL, 1992.

[6] Harold P. Boas. A geometric characterization of the ball and the Bochner-Martinelli kernel. *Math. Ann.*, 248(3):275–278, 1980.

[7] Harold P. Boas. Spheres and cylinders: a local geometric characterization. *Illinois J. Math.*, 28(1):120–124, 1984.

[8] Michael Bolt. Spectrum of the Kerzman-Stein operator for model domains. *Integral Equations Operator Theory*, 50(3):305–315, 2004.

[9] Michael Bolt. A geometric characterization: complex ellipsoids and the Bochner-Martinelli kernel. *Illinois J. Math.*, 49(3):811–826, 2005.

[10] Michael Bolt. Spectrum of the Kerzman-Stein operator for the ellipse. *Integral Equations Operator Theory*, 57(2):167–184, 2007.

[11] Michael Bolt. The Möbius geometry of hypersurfaces, II. *Michigan Math. J.*, 59(3):695–715, 2010.

[12] Michael Bolt. Spectrum of the Kerzman-Stein operator for a family of smooth regions in the plane. *J. Math. Anal. Appl.*, 413(1):242–249, 2014.

[13] Michael Bolt, Sarah Snoeyink, and Ethan Van Andel. Visual representation of the Riemann and Ahlfors maps via the Kerzman-Stein equation. *Involve*, 3(4):405–420, 2010.

[14] Young-Bok Chung. The Green function and the Szegő kernel function. *Honam Math. J.*, 36(3):659–668, 2014.

[15] Young-Bok Chung and Moonja Jeong. The transformation formula for the Szegő kernel. *Rocky Mountain J. Math.*, 29(2):463–471, 1999.

[16] Milutin R. Dostanić. The Kerzman-Stein operator for the ellipse. *Publ. Inst. Math. (Beograd) (N.S.)*, 87(101):1–7, 2010.

[17] Charles Fefferman. Parabolic invariant theory in complex analysis. *Adv. in Math.*, 31(2):131–262, 1979.

[18] Kengo Hirachi. Transformation law for the Szegő projectors on CR manifolds. *Osaka J. Math.*, 27(2):301–308, 1990.

[19] Moonja Jeong. *Approximation theorems on mapping properties of the classical kernel functions of complex analysis.* ProQuest LLC, Ann Arbor, MI, 1991. Thesis (Ph.D.)–Purdue University.

[20] Moonja Jeong. The Szegő kernel and the rational proper mappings between planar domains. *Complex Variables Theory Appl.*, 23(3-4):157–162, 1993.

[21] Norberto L. Kerzman and Elias M. Stein. The Cauchy kernel, the Szegö kernel, and the Riemann mapping function. *Math. Ann.*, 236(1):85–93, 1978.

[22] Norberto Kerzman and Manfred R. Trummer. Numerical conformal mapping via the Szegő kernel. *J. Comput. Appl. Math.*, 14(1-2):111–123, 1986. Special issue on numerical conformal mapping.

[23] Steven G. Krantz. *Function theory of several complex variables.* The Wadsworth & Brooks/Cole Mathematics Series. Wadsworth & Brooks/ Cole Advanced Books & Software, Pacific Grove, CA, second edition, 1992.

[24] Steven G. Krantz. Reproducing Kernels in Complex Analysis, 2021.

[25] Loredana Lanzani and Elias M. Stein. Cauchy-type integrals in several complex variables. *Bull. Math. Sci.*, 3(2):241–285, 2013.

[26] R. Michael Range. *Holomorphic functions and integral representations in several complex variables*, volume 108 of *Graduate Texts in Mathematics.* Springer-Verlag, New York, 1986.

[27] Nancy K. Stanton. Integral representations and the complex Monge-Ampère equation in strictly convex domains. *Proc. Amer. Math. Soc.*, 75(2):276–278, 1979.

[28] Manfred R. Trummer. An efficient implementation of a conformal mapping method based on the Szegő kernel. *SIAM J. Numer. Anal.*, 23(4):853–872, 1986.

3

Fractional Linear Maps and Some Applications. An "Augenblick".

Joseph A. Cima

CONTENTS

3.1 Introduction

It was a pleasure to get an invitation from Professor Krantz to add a chapter on some topics from function theory to this text that he is putting together. I have assumed that the reader is familiar with basic material about complex numbers, analytic functions and maybe a bit of linear algebra in the setting of $\mathbb{R}^n$ and $\mathbb{C}^n$. My goal is to introduce the reader to the basic material on fractional linear transformations in the setting of the complex plane. There are many extensions from this "scalar" setting to higher dimensional settings. I plan to segue from this complex analytic setting to extending some scalar functions (the fractional linear transformations) to operator-valued functions and eventually to prove one result about operators that are contractions. I have to leave the readers at a point where if they are interested they can get into some extremely useful and classical operator theory. I hope the reader will find the presentation readable and if not you might try one of the references that I have listed. Any errors in the presentation are my own, but I have tried

DOI: 10.1201/9781315160658-3

to carefully read over the material with the hopes of culling out any serious "faux pas".

3.2   Notations and Definitions

The letters $a, b, c,$ and d are fixed complex numbers, and the letters Z and W represent complex variables. The letter $\mathbb{C}$ stands for the complex plane and the notation $\mathbb{S} = \mathbb{C} \bigcup (\infty)$ represents the complex plane with the "point" at infinity added. This can be realized as the closed unit sphere, but we shall not need that information in our work.

The words "fractional linear maps" stand for the family of mappings $f(Z) = \frac{aZ+b}{cZ+d} = W$. These are also known as Mobius mappings. There are certain special cases (choices of $a, b, c,$ or d) that we wish to avoid and let us say a word about this matter. First, if c is zero and d is zero we can see there will be serious questions about the meaning of such an expression. To avoid this we assume (the determining number) $ad - bc \neq 0$.

For if this were not the case we observe that for different values Z_1 and Z_2 the expression

$$f(Z_1) - f(Z_2) = \frac{(Z_2 - Z_1)(bc - ad)}{(cZ_1 + d)(cZ_2 + d)} = 0$$

implying that f is a constant mapping. So for example the choice of $a = d = 1$ and $b = \frac{i}{2}$ and $c = 2i$ would yield

$$f(Z) = \frac{Z + \frac{i}{2}}{2iZ + 1} = \frac{i}{2}.$$

Further, in this case if $ad - bc = 0$ and assuming $c = 0$, we have the unpleasant situation that $a = 0$ or $d = 0$. We do not define a $f(Z) = \frac{aZ+b}{cZ+d} = W$ in these cases. We let the symbol $\mathbb{L}$ denote this family of mappings. Note also that if $\lambda \neq 0$ then $f(Z) = \frac{a\lambda Z+\lambda b}{c\lambda Z+\lambda d}$ as well, so the $a, b, c,$ and d constants are not unique.

It is easy to arrive at

$$f'(Z) = \frac{(ad - bc)}{(cZ + d)^2} \neq 0$$

which implies that f is locally one-to-one (a proof is given later), but of course more is true. We can find an inverse for $f(Z) = W$ which is $g(W) = \frac{b-dW}{cW-a}$. Note that a computation shows for f as above and $g(Z) = \frac{AZ+B}{CZ+D}$ the functional composition of g with f, $g \circ f(Z) = W(Z)$, yields another element of $\mathbb{L}$

$$W(Z) = \frac{(aA + cB)Z + (bA + dB)}{(aC + cD)Z + (bC + dD)}$$

and it will also follow that the determining number

$$(aA + cB)(bC + dD) - (bA + dB)(aC + cD) = (AD - BC)(ad - bc) \neq 0.$$

Hence, $\mathbb{L}$ is closed under the operation of composition. Further, each $f \in \mathbb{L}$ is one-to-one and has an inverse under composition in the set $\mathbb{L}$.

The reader may have noticed that our requirement on $ad - bc$ is reminiscent of the determinant of a two-by-two matrix. In fact, it is very useful to identify the element f in $f(Z) = \frac{aZ+b}{cZ+d} = W$ of $\mathbb{L}$ with the two by two matrix

$$\mathbf{T} \simeq \begin{pmatrix} a & b \\ c & d \end{pmatrix}.$$

Note first that if we consider the calculation with the composition of g and f as above, it is just the matrix multiplication of the two matrices associated to g and f,

$$\begin{pmatrix} A & B \\ C & D \end{pmatrix} \begin{pmatrix} a & b \\ c & d \end{pmatrix} = \begin{pmatrix} (aA + cB) & (bA + dB) \\ (aC + cD) & (bC + dD) \end{pmatrix}$$

Further, our requirement for non-degeneracy of the determining number is the requirement that the determinant of the associated matrix be non-zero (i.e., the matrices are invertible). Since matrix multiplication is associative the same holds for $\mathbb{L}$, $(h \circ g) \circ f = h \circ (g \circ f)$. Hence, we now can view $\mathbb{L}$ as a group with invertible elements. Note that since matrix multiplication is not commutative the group $\mathbb{L}$ is not commutative. Further, if one recalls the easy formula for the inverse of the two-by-two matrix

$$\begin{pmatrix} a & b \\ c & d \end{pmatrix}$$

then the inverse is just

$$\frac{1}{ad - bc} \begin{pmatrix} d & -b \\ -c & a \end{pmatrix}$$

and the associated form for the inverse of $f(Z) = W$ is

$$g(W) = \frac{1}{ad - bc} \frac{dW - b}{-cW + a}$$

and it has 1 for it determining number, and this is the same form for the inverse as we have calculated above with the scalar $\lambda = (bc - ad)^{-1}$. The elements of $\mathbb{L}$ play an important role in the theory of conformal mappings, one variable operator theory, and several other aspects of classical analysis and we shall discuss some of these uses later in this work.

It is possible to put an equivalence relation on $\mathbb{L}$ as follows.

Definition 1 *For each g and f in* $\mathbb{L}$, *we say that g is equivalent to f if there exists an element* $h \in \mathbb{L}$ *with*

$$g = h \circ f \circ m,$$

where m is the inverse of h.

It is easy to check that this is an equivalence relation and we shall make use of the fact later.

Let us view some paradigm examples of elements of the $\mathbb{L}$ with their defining numbers.

1. The identity map $f(Z) = Z$, with $ad - bc = 1$.
2. Magnifications, $m(Z) = aZ$, with $ad - bc = a$.
3. Linear mappings, $l(Z) = aZ + b$, with $ad - bc = a$.
4. Inversions, $v(Z) = \frac{1}{Z}$, with $ad - bc = -1$.

It is very useful to observe that a general element $f(Z) = \frac{aZ+b}{cZ+d} = W$ of $\mathbb{L}$ can be built up of these four maps. Namely, writing

$$\frac{aZ + b}{cZ + d} = \frac{a/c(cZ + d) - (\frac{ad}{c} - b)}{cZ + d} = \frac{a}{c} + \frac{b - \frac{ad}{c}}{cZ + d}$$

we can write $W_1 = cZ$, $W_2 = W_1 + d$, $W_3 = \frac{1}{W_2}, \ldots$ etc. to arrive at the decomposition from the elementary maps.

There is something very useful that comes out of the elementary mappings above about a general mapping from $\mathbb{L}$.

Namely, it is obvious that the identity mapping preserves geometric shapes, in particular lines and circles in the plane. Further, we see that translates also preserve geometric figures. The magnification map $m(Z) = aZ$ also maps a circle with center P and radius r to a circle with center aP and radius ar, and the reader can check that it maps affine lines to affine lines. Hence, the first three maps above preserve the families of circles and lines in the plane. More is true in that any map from $\mathbb{L}$ has this property. So, assuming that we identify lines as unbounded circles through infinity and circles in the plane as bounded circles, we have the following result.

Theorem 1 *Every element of* $\mathbb{L}$ *maps circles to circles.*

Proof: To prove this we have only to prove that the inversion map maps circles to circles. First assume we consider the (bounded) circle $C = (Z : Z = A + r \exp \imath\theta), \theta \in [0, 2\pi]$.

Applying the inversion formula to these values we find

$$|1/Z - A|^2 = \frac{1 + |A|^2|Z|^2 - AZ - \overline{AZ}}{|Z|^2} = r^2,$$

or in real coordinates with $Z = x + iy$ and $-2Re(AZ) = \alpha x + \beta y$

$$(|A|^2 - r^2)(x^2 + y^2) - \alpha x - \beta y + 1 = 0.$$

And this is the equation of a circle in the plane.

It may be useful for the reader to check that the circle

$$|Z|^2 + \bar{t}Z + t\bar{Z} + m = 0,$$

is mapped to the circle

$$|Z|^2 + \frac{t}{m}Z + t\bar{Z}/m + 1/m = 0, \quad m \neq 0,$$

and to

$$tZ + t\bar{Z} + 1 = 0,$$

if $m = 0$.

Note that if we consider an unbounded circle passing through infinity of the form $C = [Z = x + \imath m x]$ with m a real number, the inversion transformation maps C to the unbounded circle of the form

$$\frac{1}{x + \imath m x} = \frac{1 - \imath m}{x(1 + m^2)} = u + \imath v,$$

which is the circle in the u, v plane, $v = -mu$. Similarly, one can show that the image of any affine line in the plane is mapped to a circle under the inversion mapping and the discussion is left for the reader. More generally, if the equation of the circle is given as $\bar{t}Z + t\bar{Z} + m = 0$, then the inversion $\frac{1}{Z}$ takes the circle into the circle $|Z|^2 + \frac{t}{m}Z + \frac{t\bar{Z}}{m}$, $m \neq 0$, and if $m = 0$ it is mapped into the circle $tZ + t\bar{Z} = 0$.

3.3 Some Geometric Considerations

Assume Z_1, Z_2, and Z_3 are three distinct complex numbers. They determine a circle in our sense. Then the element $T \in \mathbb{L}$ of the form

$$f(Z) = \frac{(Z - Z_1)(Z_2 - Z_3)}{(Z - Z_3)(Z_2 - Z_1)} = W$$

has some interesting properties. Note that it maps Z_1 to zero, and Z_2 to the number 1, and Z_3 to infinity. Thus the circle containing Z_1, Z_2, and Z_3 is mapped to the circle containing $0, 1$, and the point at infinity. This

$f(Z) = \frac{aZ+b}{cZ+d} = W$ is known as the "cross ratio" and is denoted by the form (Z, Z_1, Z_2, Z_3). Now it follows from these properties if W_1, W_2, and W_3 are three distinct complex numbers the equation

$$f(Z) = (Z, Z_1, Z_2, Z_3) = (W, W_1, W_2, W_3) = g(W),$$

maps the circle determined by Z_1, Z_2, Z_3, one to one and onto the circle determined by the points W_1, W_2, W_3. This follows easily since setting $Z = Z_1$ gives the value number zero in the left terms and since W_1 is the unique point in the righthand term giving the value zero it must corresponding to Z_1, etc., for the two remaining terms.

Note that in the above we have assumed the chosen points were finite complex numbers. If one of these points is infinity we can still define the cross ratio as follows. For the case $Z_1 = \infty$, we set $(Z, \infty, Z_2, Z_3) = \frac{(Z_2 - Z_3)}{(Z - Z_3)}$, and if $Z_2 = \infty$ it becomes $(Z, Z_1, \infty, Z_3) = \frac{(Z - Z_1)}{(Z - Z_3)}$, similarly for the case where Z_3, equals ∞.

As an example, consider the $f(Z) = \frac{aZ+b}{cZ+d} = W$, $f(Z) = (Z, 0, 1, -1) = (W, \imath, 2, 4) = g(W)$. This takes the form

$$f(Z) = \frac{2Z}{Z+1} = \frac{-2(W - \imath)}{(W - 4)(2 - \imath)} = g(W).$$

For this example the one circle determined by $0, 1, -1$ is the circle $\text{Im}(Z) = 0$. The second is the circle with center at $(3, 9/2)$ and radius $r = \frac{\sqrt{85}}{2}$.

The solution is $W = \frac{(16 - 6\imath)Z + 2\imath}{(6 - 2\imath)Z + 2}$.

There are many more geometric properties enjoyed by the cross ratio and we mention one more and refer the interested reader to the classic text by Ahlfors [1].

Theorem 2 *The cross ratio* (Z_1, Z_2, Z_3, Z_4) *is real if and only if the four points* $[Z_j]$ *lie on a circle.*

There is an analytic proof given in Ahlfors but for those interested in some geometry consider the equation

$$\arg\left(\frac{(Z_1 - Z_2)(Z_3 - Z_4)}{(Z_1 - Z_4)(Z_3 - Z_2)}\right) = \arg\left(\frac{(Z_1 - Z_2)}{(Z_1 - Z_4)}\right) - \arg\left(\frac{(Z_3 - Z_4)}{(Z_3 - Z_2)}\right)$$

and draw some pictures.

One can check easily that elements $f(Z) = \frac{aZ+b}{cZ+d} = W$ do not preserve length between points in the plane. However, they have other important properties which we consider. Namely, suppose $\Gamma_1(t)$ and $\Gamma_2(t)$ are two differentiable curves in the complex plane, and $\Gamma_1(t_o) = \Gamma_2(t_o)$. The tangent lines to curves at the point $W_o = \Gamma(t_o)$ yield two vectors, say V_1 is a vector based at W_o and pointing in the direction of the tangent vector to $\Gamma_1(t_o)$, and similarly assume V_2 is a vector based at W_o and pointing in the tangent direction at $\Gamma_2(t_o)$. The angles are chosen consistently with the orientation of the two curves. The angle between the curves Γ_1 and Γ_2 for the value of t_o is the angle, say ϕ, through which V_1 must be rotated counterclockwise to lie in the V_2 direction. Now for an analytic function f defined in an open neighborhood of the point W_o we have two more curves given by $\Lambda_j(t) = f(\Gamma_j(t))$. The function f is said to be conformal at W_o if the angle between Λ_1 and Λ_2 at $f(W_o)$ is equal to the angle between Γ_1 and Γ_2 at the point in question (and this is true for all such curves).

Theorem 3 *An analytic function f is conformal at every point Z_o where $f'(Z_o)$ is not zero.*

Proof: By assumption $|f'(Z_o)| = A > 0$. We show that there is a disk, $C = [Z : |Z - Z_o| < r]$ on which f is a one-to-one function. Since f' is continuous, we may choose a disc $D = [Z \,|\, |Z - Z_o| < r]$ with $|f(Z) - f(Z_o)| < (A/2)$ for $Z \in C$, and so that $|f'(Z) - f'(Z_o)| \leq (A/2)$ for any $Z \in$ D.

Let Z_1 and Z_2 be distinct points of D and let $\gamma = \gamma(t)$ be the straight line segment in D joining them. Then

$$|f(Z_1) - f(Z_2)| = |\int_\gamma f'(Z)dZ| =$$

$$\left|\int_\gamma f'(Z_o)dZ - \int_\gamma [f'(Z_o) - f'(Z)]dZ\right| \geq$$

$$A\,|Z_2 - Z_1| - A/2\,|Z_1 - Z_1| = A/2\,|Z_2 - Z_1| \neq 0.$$

With this fact we show that every oriented, smooth curve through the point Z_o has its tangent turned through the same angle, under the mapping $f(Z)$. This implies that the angle between any two such curves passing through Z_o is preserved. So assume $\gamma(t)$ is such a curve, $\gamma(t_o) = Z_o$, and $\gamma'(t_o)$ is the tangent vector at the point in question. Now consider the image curve given by $f(\gamma(t)) = \Gamma(t)$. $\Gamma'(t_o) = f'(Z_o)\gamma'(t_o)$. Note that our assumption on $f'(Z_o)$ and the smoothness of γ implies that $\Gamma'(t_o) \neq 0$. Hence, the angles which tangent vectors $\gamma'(t_o)$ and $\Gamma'(t_o)$ make with the horizontal direction are related by the equation

$$\arg(\Gamma'(t_o)) = \arg f'(Z_o) + \arg \gamma'(t_o).$$

Thus every curve through Z_o is rotated through the same angle $\arg f'(Z_o)$ which is a constant (independent of the curve under consideration).

3.4   The Symmetry Principle

In the plane ($\mathbb{R}^2$) given a point $P = (x, y)$ (assume with out loss of generality that $y > 0$) there is a point naturally associated to P, namely the reflection of P in the x-axis. That point is $PQ = (x, -y)$. The x-axis is orthogonal (at right angles) to the segment joining P to PQ (and bisects it). Note also, that if C is any circle in the plane containing P and PQ, its center must lie on the real axis. Consequently, this circle must meets the x-axis at two points x_1, and x_2. The angles at which the tangent lines to the circle at points x_1, and x_2, where it meets the x-axis, are also ninety degree angles. That is, the circle meets the x-axis orthogonally. A picture will help at this point.

In a more general setting, we wish to define the idea of symmetry for any two (finite) points in the complex plane. So assume $\{Z_j, j = 1, 2\}$ are two distinct points in the plane. Let U be the line segment joining them, with midpoint of U being u. Assume the slope of U is m. Then the line G through u with slope $\frac{-1}{m}$ is orthogonal to U. We say the points Z_j are symmetric in the line $\mathbb{G}$. Again, since if we choose any circle C through Z_1, and Z_2, we see that U is a chord of this circle and that the chosen circle C will meet G at exactly two points, say P_1 and P_2. The angles formed by the tangent lines to C at P_1 and P_2 will be ninety degree angles. That is C meets G orthogonally.

Definition 2 *Two points Z_1 and Z_2 are said to be symmetric with respect to a line (circle) E if every circle through them intersects E orthogonally.*

Theorem 4 *Let E be a circle and let $f \in \mathbb{L}$. Then two points Z_1 and Z_2 are symmetric in E if and only if the images $f(Z_1)$ and $f(Z_2)$ are symmetric in the circle $f(E) = E'$.*

Proof: Consider the distinct points $W_1 = f(Z_1), W_2 = f(Z_2)$. Let V' be a circle containing W_1 and W_2 with $V = f^{-1}(V')$. Then V' meets E' at two points Q_1', and Q_2'. The inverse points under f, $f^{-1}(Q_j') = Q_j$ lie on V. Hence, V is a circle containing Z_1 and Z_2, and so the angle formed by the tangents to V at these points is ninety degrees. But f is conformal so the angles are preserved, implying that the tangent angle to the image circle V' is ninety degrees. This proves the sufficiency and we suggest the reader supply the necessity.

3.5   The Automorphism Group of the Unit Disc

One of the important subgroups of the group $\mathbb{L}$ is the set of one to one automorphisms of the unit disc $D = \{Z : |Z| < 1\}$. That is the holomorphic, one to one, onto mappings of D onto D. To develop this section it is useful to have a formula that relates two symmetric points in a given circle.

So assume E is a given circle (finite for this development) with center A and radius r. Let μ be a given finite point. The points $A - r, A + \imath r$, and $A + r$ are three points of E. Utilizing the cross ratio,

$$f(Z) = (Z, A - r, Z + \imath r, A + r) = \frac{\{Z - (A - r)\}\{\imath r - r\}}{\{Z - (A + r)\}\{\imath r + r\}} = \imath \frac{Z - (A - r)}{Z - (A + r)}.$$

We have a mapping from E onto the circle given by the real axis. By the symmetry principle if $(\mu)^*$ is the refection of μ in E then $T((\mu)^*)$ is symmetric to $T(\mu)$ in the real axis. In complex notation (recall if $Z = x + \imath y$, then $\overline{Z} = x - \imath y$) we have $T((\mu)^*) = \overline{T(\mu)}$.

From the immediate formula above this leads to the equation

$$\imath \frac{(\mu)^* - (A - r)}{(\mu)^* - (A + r)} = (-\imath) \frac{\overline{\mu} - (\overline{A} - r)}{\overline{\mu} - (\overline{A} + r)}.$$

If we solve this formula for $(\mu)^*$ we find that

$$(\mu)^* = A + \frac{r^2}{\overline{\mu} - \overline{A}}.$$

Note that

$$\frac{r^2}{\overline{\mu} - \overline{A}} = \frac{r^2(\mu - A)}{|A - \mu|^2} \equiv \mu_A$$

and so the argument of (μ_A) is the same as arg $((\mu)^* - A)$. It follows that μ and $(\mu)^*$ lie on the same ray originating at the point A and that

$$|(\mu)^* - A| = \frac{r^2}{|\mu - A|}.$$

To end this section, we prove an important and useful result about this special subgroup of $\mathbb{L}$. This subgroup will be denoted by $\mathbb{A}$.

Definition 3 *A one-to-one holomorphic mapping from the unit disc D onto itself is called an automorphism of the disc.*

The next theorem will show that all such mappings are in $\mathbb{L}$.

Theorem 5 *A holomorphic mapping $f \in \mathbb{L}$ is in $\mathbb{A}$ if and only there is a number, $c \in D$ and a number $t, |t| = 1$ with*

$$f(Z) = t \frac{Z - c}{1 - \overline{c}Z}.$$

Proof: First if f has the given form and $Z \in D$, then it is easy to check that

$$|Z - c|^2 = |Z|^2 + |c|^2 - 2Re(\bar{c}Z) \leq |1 - \bar{c}Z|^2 = 1 + |cZ|^2 - 2Re(\bar{c}Z),$$

so f maps D into D, and is one to one. Note that f has an inverse and from this f will map D onto D and the unit circle onto the unit circle. Hence, $f \in \mathbb{A}$.

Now assume f is in $\mathbb{L}$ and f is in $\mathbb{A}$. Then, by the ontoness assumption, there is a point a in D with $f(a) = 0$. By the formula for the symmetric point we see that $a^* = \frac{1}{\bar{a}}$. Now the symmetric point of 0 is ∞ and so $f(a^*) = \infty$ by the reflection principle above. Such an automorphism has the form (recall the cross ratio formula)

$$f(Z) = \frac{(Z - a)(Z_2 - \frac{1}{\bar{a}})}{(Z - \frac{1}{\bar{a}})(Z_2 - a)} \ , \quad \cdot$$

Now, using the fact that $|f(\exp(it)| = 1, t \in (0, 2\pi)$, we conclude that there is a number $\lambda, |\lambda| = 1$, so that

$$f(Z) = \lambda \frac{(Z - a)}{1 - \bar{a}Z}.$$

Thus f has the desired form.

3.6   Some Linear Algebra

In this section, we try to proceed to a bit of material, part of which is usually covered in an undergraduate course but some of which will be a bit advanced. In giving the background we hope that it will be partial review and that examples will fill in the missing pieces. We assume you are familiar with the term "vector spaces" and if not you can find the definition in many undergraduate texts. Think of $\mathbb{R}^n$, or $\mathbb{C}^n$ as paradigm examples. A vector space V is finite dimensional if there are a finite number of non-zero vectors from V, say $\{v_j\}_1^n$ for which given a vector X in V we may express it uniquely as the sum $X = \sum_1^n C_j v_j$, where the C_j are from the scalar field (in our case usually the complex numbers). If this is not the case we say that V is infinite dimensional. For example the polynomials of degree less than or equal to n considered on the interval $[0, 1]$ with pointwise addition and scalar multiplication are a finite dimensional vector space of dimension (n+1). A basis is given by the set $\{t^k : k = 0, 1, 2, \ldots, n\}$. However, if we consider polynomials on the same interval of all degrees it is a vector space but it is not finite dimensional. As another example from basic calculus the Riemann integrable functions on the interval $[0.1]$ is also a vector space and it is not finite dimensional. A norm on a vector space is a non-negative, real valued function $f(\circ)$ usually written as

$f(\circ) = \| \circ \|$, and it has essentially the same properties as the absolute value function on the real line. So for example considering $\mathbb{C}^n$ a norm is given for $Z = (z_1, ..., z_n)$ as $\|Z\|^2 = \sum_1^n |z_j|^2$, or in the case of the polynomials on $[0, 1]$ as $\|p(t)\| = \max\{|p(t)| \,|\, t \in [0, 1]\}$. So we have $\|x + y\| \leq \|x\| + \|y\|$ for vectors x and y in V, and with C a scalar $\|Cx\| = |C|\|x\|$. Such vector spaces are called normed vector spaces. If, whenever X_n is a sequence in a normed vector space V and if $\|X_k - X_n\| \to 0$ (sequences exhibiting such behavior are called Cauchy sequences) for k and n tending to infinity, we know that there is a unique vector $X^* \in V$ with $\|X_n - X^*\| \to 0$ as n tends to infinity then we say V is a complete normed vector space. Further, if the operations in V are continuous in the sense that for x and y in V we have the mapping from $V \times V$ into V by the addition operation $x + y = z$ and the scalar multiplication operation from $\mathbb{C} \times V$ by $(C, x) \to Cx$ then this a topological normed vector space.

The examples above are topological normed vector spaces. We shall refer to them as normed spaces. Note that the example given above of all polynomials on the interval $[0, 1]$ is not complete in that the absolute value function $|t - 1/2|$ is the limit of a Cauchy sequence from the polynomials. Try to get a picture of how this might happen. This is a non-trivial statement and requires proof, which we do not include. Now consider a few more examples. The first is called "little l^2" and is the vector space of vectors consisting of an infinite number of complex numbers $Z = (z_1, z_2, z_3, ..., z_n, ...)$ which satisfy the requirement $\sum_1^\infty |z_j|^2 < \infty$. For a specific example take the terms to satisfy $z_j = 1/j$. The norm is just $\|Z\|^2 = \sum_1^\infty |z_j|^2$, and with this norm it is a normed space. The set of polynomials of degree less than or equal to n is also such a space with the sup norm we defined above for the space of all polynomials. Finally, a normed space is a Hilbert space $\mathbb{H}$ if the norm on $\mathbb{H}$ is induced by an "inner product". That is a mapping from $V \times V$, usually written as $<U, W>$, into $\mathbb{C}$ which is linear in the first entry and anti-linear in the second and for which $\|u\|^2 = <U, U>$. That is, for $U, W, Z \in \mathbb{H}$, and C a scalar,

$$<U + W, Z> = <U, Z> + <W, Z>,$$

and $\overline{<U, Z>} = <Z, U>$, and $<CU, Z> = C <U, Z>$. Normed spaces and Hilbert spaces are very useful in theoretical aspects of analysis, but we only touch on one of the reasons why. Mappings between normed spaces is one of the important aspects of analysis. For example, the map on l^2 given by $Z = (z_1, z_2, z_3, ...) \to (z_1^2, \sin z_2/2, z_3, ..)$ is an interesting mapping however it may not be easy to work with. It is what is called a nonlinear mapping because the terms in it are not linear. The study of nonlinear analysis is an important and challenging field but we are not going into that area. We study another set of mappings of equal importance called "linear mappings".

Definition 4 *A mapping T on a normed space* $\mathbb{V}$ *into a normed space* $\mathbb{W}$ *is called linear if for vectors A and B in* $\mathbb{V}$ *and a scalar* $C \in \mathbb{C}$ *we have*

$$T(A + CB) = T(A) + CT(B).$$

Note that the vectors in the argument of T are in the space $\mathbb{V}$ whereas the vectors in the range of T are in $\mathbb{W}$. Some easy examples on say l^2 are given by the identity $Z = (z_1, z_2, z_3, ...) \to T_1(Z) = (z_1, z_2, z_3, ...)$, $Z \to T_2(Z) = (0, z_1, z_z, z_3, , ,)$ the shift operator, $Z \to (z_2, z_1, z_3, ...) = T_3(Z)$, or finally $Z \to (z_1, z_2/2^2, z_3/3^2, ...) = T_4(Z)$. Note that the identity mapping T_1 is one to one and maps all of l^2 onto itself. In contrast T_2 does not have any of the vectors of the form $(C, 0, 0, \ldots), C \neq 0$ in its range. The linear operator T_3 on the other hand is one-to-one and maps onto all of l^2, and finally given the vector $W = (1, 1/2, 1/3, \ldots) \in l^2$ you can check that no vector will map onto it under T_4. These different behaviors are the norm in that the map T_3 is special. It has what we call an inverse. That is, there is another linear operator (on l^2) ,say S, and we see that $S(W) = S(w_1, w_2, w_3, ..) = (w_2, w_1, w_3, ..)$ has the property that $S \circ T_3(Z) = T_3 \circ S(Z) = Z$ for all $Z \in l^2$. We write this inverse as $S = T^{-1}$. Note that, in the first few sections, I avoided using this notation and wrote out the word "inverse" or used a different symbol. The reason for this will come up later.

Definition 5 *An operator T from a normed space X into a normed space Y is invertible if it is one-to-one and onto.*

So if T is such an operator with $T(X) = Y$ then its inverse is $S(Y) = X$. It is easy to see that $S \circ T(X) = S(Y) = X = I(X)$, and that now $S = T^{-1}$ is also linear. This follows for example if when we choose $X_j, j = 1, 2$ in X and $C \in \mathbb{C}$ with $T(X_j) = Y_j$, then $S(Y_1 + CY_2) = S(T(X_1) + CT(X_2)) = S(T(X_1 + CX_2)) = X_1 + CX_2 = S(Y_1) + CS(Y_2)$.

Definition 6 *An operator T on a normed space X into a normed space Y is continuous at a point $X^* \in X$ if, whenever $X_n \in X$ with $||X_n - X^*|| \to 0$, then $||T(X_n) - T(X^*)|| \to 0$.*

Note that for such operators continuity at one point implies continuity at all points of X. This is easily checked by showing that for T as above it will be continuous at the origin. For if $W_n \to 0$ in X the mapping $W = X^* - X$ is continuous by definition of normed space and so $T(W_n) = T(X^* - X_n) = T(X^*) - T(X_n) \to 0$, where X_n corresponds with the W_n under the given map. Further there is another condition that is equivalent to continuity.

Definition 7 *A liner mapping T as above is said to bounded on X into Y if there is a number $M > 0$ for which*

$$||T(x)|| \leq M||x||,$$

for all $x \in X$.

The infimum of all such numbers M is called the norm of the operator T. Now the reader can show that the concept of continuity is equivalent to the concept of boundedness for such linear operators on normed spaces. All the operators we have introduced above are bounded linear operators. T_1, and T_3 are invertible but T_2, and T_4 are not invertible. The operator $T =$ differentiation on the polynomials on $[0, 1]$ is linear but not bounded since with $P_n(t) = \frac{t^n}{n}$, one has $||P_n|| = 1/n \to 0$, whereas $||T(P_n)|| = ||t^{n-1}|| = 1$.

There is an idea used in such spaces which is referred to as completeness.

Definition 8 *If every Cauchy sequence $\{x_n\}$ in a normed vector space X has a limit x in the space then X is a complete normed vector space, or for short a Banach space. That is, if $||x_n - x_k|| \to 0; n, k \to \infty$ then $||x_n - x|| \to 0, n \to \infty$.*

Having discussed (albeit briefly) the idea of a linear operator on normed spaces we wish to conclude with an idea that is useful in operators on Hilbert spaces.

3.7 Some Topology

The word topology has to do with a family of subsets of a given fixed set. In the hopefully familiar setting of $\mathbb{R}$, $\mathbb{R}^n$ or $\mathbb{C}$ the paradigm of open sets are the open intervals, or open balls. These are not all the open sets but any open set in the given space can be built up by unions of these basic objects. These open sets come from the norms on the space, given a fixed P in one of these spaces, the open ball with P as center is $\{Q \mid ||P - Q|| < \epsilon\}$ where ϵ is a positive number and Q is from the space. One of the things that is basic to a first course in analysis (dealing with $\mathbb{R}^n$ or $\mathbb{C}$ and the norm topology) is the idea of compactness. The idea of sequence has already been discussed and the ideas related to them (e.g. subsequences, limit infimum, etc.) are important for the concept of compactness in this setting. So a set in one of these finite dimensional normed spaces is compact if whenever K is a subset of such a space and all the elements of K are bounded by one fixed number then there is a sequence, say X_k in K, which converges to a point of K. Another important aspect of this is when K is the (norm) closed unit ball B in the space. That is the closed unit ball in a finite dimensional normed space is compact. That is there is sequence from B which converges (the points get close to each other and to the limit point).

In the infinite dimensional situation this fails for the examples above with the norm topology. For example consider l^2 and the sequence $e_k = (0, 0, .., 0, 1, 0, 0...)$ with all zeros except in the kth position where we have placed a one. These are in the closed unit ball but

$$||e_k - e_j||^2 = 2 \quad j \neq k.$$

Hence no subsequence of these terms can ever be close, when close means less than $\sqrt{2}$. The critical point here is that the closed unit ball in an infinite dimensional normed space can never be compact (again we are thinking of the topology induced by the norm).

This lack of compactness of the unit ball makes the behavior of the linear operators appearing in the next section somewhat difficult to handle in many situations. Although we will not get into many examples we wanted the reader to be aware that there are deep topological difficulties in the study of these operators in the infinite dimensional normed spaces.

3.8   Functions of Linear Operators

In this section I will do my best to sketch some properties of bounded linear operators acting on Hilbert spaces (incidentally just to avoid any bad situations I am always thinking of spaces which are separable). There is an old facetious saying concerning what one reads in print "Believe what you read for it is true!". I am going to apply this below in a few places (for short BWYRFIIT).

We see that it is possible to add linear operators and multiply by scalars and still remain within the realm of (bounded) linear operators. We also have the concept of multiplication played by composition. This operation is not commutative. For example on l^2 set $T_5(Z) = (z_2, z_3, z_4, ...)$ the backward shift. Then $T_5 \circ T_2(Z) = Z$ but $T_2 \circ T_5(Z)$ is not the identity. However one way to obtain more interesting examples of such operators is by taking nice functions f and considering the formal expression $f(T)$ when T is a bounded linear operator. We want this to be another bounded linear operator (on the same space as T). For a simple example if $f(z) = z^2 + (4 + 2i)$ and T is a bounded linear operator what is the operator $f(T)$? Of course we want the z^2 term to give us $T^2 = T \circ T$. What about the constant term? We think of powers of T (under composition) as above and $T^0 = I$ the identity operator. So in this case $f(T)$ is the operator

$$f(T) = T^2 + (4 + 2i)I = S.$$

Now what about another "nice" function like $f(z) = \exp(z)$. We will write $\beta(X)$ to represent all the bounded operators on X. Note we could consider bounded linear operators from X into another normed space Y but considering X and Y as the same will make our point (and $\beta(X, Y)$ as all the bounded operators from X into Y). The more general case is pretty much the same.

At this point recall our definition of the word "complete". We invoke our BWYRFIIT assumption and assert the fact that $\beta(X)$ is complete and we ask the reader to accept that. That is if T_n is a Cauchy sequence of operators (in the operator norm) then there is an operator T with T_n converging to T (again in the operator norm $||T_n - T_k|| \to 0, n, k \to \infty$ implies there is a unique operator T with $||T_n - T|| \to 0$). Some of these ideas will appear as we continue with this example so we will not develop them any further.

So let us consider the partial sums of f and then of $f(T)$.

$$\left|\left| \sum_{n=0}^{N} \frac{1}{n!} T^n - \sum_{n=0}^{M} \frac{1}{n!} T^n \right|\right| =$$

$$\left|\left| \sum_{n=N}^{M} \frac{1}{n!} T^n \right|\right| \leq \sum_{n=N}^{M} ||T||^n \frac{1}{n!}.$$

Now $||T|| = C$ is a fixed real (non-negative) number in this inequality and it is known that the series

$$\sum_{n=0}^{\infty} \frac{1}{n!} C^n$$

converges for any fixed C. So the partial sums above $\sum_{n=N}^{M} ||T||^n \frac{1}{n!}$ are the partial sums of a convergent series and so are a Cauchy sequence. This implies the terms $\sum_{n=0}^{N} \frac{1}{n!} T^n$ are Cauchy sequence in $\beta(X)$ and so by completeness of $B(X)$ they converge to an operator $S \in \beta(X)$. If we had looked at $\sum_{n=0}^{N} \frac{1}{n!} T^n(x)$ with $x \in X$ we would get an idea of how to prove the statement about the completness of $\beta(X)$.

At this juncture we need some even more technical material on Hilbert spaces and operators on them. First assume that $\mathbb{H}_j, j = 1, 2$ are Hilbert spaces. We wish to make a sum of these spaces (which we can do as vector spaces) into another Hilbert space. Namely, writing

$$\mathbb{H} = \mathbb{H}_1 \bigoplus \mathbb{H}_2 = \{(x, y) \ : \ x \in \mathbb{H}_1, y \in \mathbb{H}_2\}$$

this is a vector space with pointwise addition and scalar multiplication. That is

$$(x, y) + (u, v) = (x + u, y + v),$$

and for a number c we have $c(x, y) = (cx, cy)$. So that handles the vector space part of the definition; now for the inner product part we define

$$<(x, y), (u, v)>_{\mathbb{H}} = <x, u>_{\mathbb{H}_1} + <y, v>_{\mathbb{H}_2},$$

where the inner products are in the appropriate spaces (and we will drop the subscript notation when it is obvious which space we are working with). This yields for the norm

$$<(x, y), (x, y)> = ||x||^2 + ||y||^2,$$

again the norms are in the appropriate spaces. Further using the idea of sesquilinear forms (this is a use of BWYRFIIT) one can prove the following. For T a bounded operator on $\mathbb{H}_1$ into $\mathbb{H}_2$ there is another operator S from $\mathbb{H}_2$ into $\mathbb{H}_1$ for which the following holds

$$<Tx, y> = <x, Sy>.$$

This operator S is called the adjoint operator to T and is customarily written as T^*.

Definition 9 *An operator T on a Hilbert space $\mathbb{H}$ is called unitary if $TT^* = T^*T = I$ (the identity).*

As an example recall our operator $T_3(Z) = (z_2, z_1, z_3, ..., z_j, ..)$ on l^2. If we write a basis for l^2 as $E_j = (0, .., 0, 1, 0, 0, ..)$, where 1 is in the jth position and all other positions have a zero then it is easy to check that

$$T_1(E_1) = E_2, \; T_1(E_2) = E_1, \; T_1(E_j) = E_j, j > 2.$$

Further, using the definition we have

$$T_1^*(E_1) = E_2, T_1^*(E_2) = E_1, T_1^*(E_j) = E_j, j > 2,$$

so that one can check $T_1^* T_1 = T_1 T_1^* = I$, so T is a unitary operator (and its inverse is its adjoint T_1^*).

Definition 10 *An operator T on a Hilbert space H into a Hilbert space M is called a contraction if $||T(x)|| \leq ||x||$ for all $x \in X$. That is the norm $||T|| \leq 1$, and is a strict contraction if $||T|| < 1$.*

Note that if T is any operator in $\beta(H, M)$ then the operator $S = T/||T||$ is a contraction.

3.9   An Operator Equation

Recall the form of an $f \in \mathbb{L}$, $f(Z) = \frac{aZ+b}{cZ+d} = W$. Also recall that there are many different ways to express this rational function. Some of the most obvious are as follows

$$f(Z) = (aZ + b)(cZ + d)^{-1} = (cZ + d)^{-1}(aZ + b) = a' + \frac{b'Z}{cZ + d},$$

as well as many others. So how do we lift this expression to an operator theoretic setting? If we just assume A, B, C, D and T are operators on various spaces and try to plug these operators into one of these expressions we have some immediate problems. Namely, since operators do not commute the second equality gives pause for thought. Which should we choose? Next the denominator will be an operator and we do not have (as yet) a way of thinking about such expressions. One of the ways to get a useful recipe to overcome and deal with these problems is to choose the following operator-theoretic form

$$B - AT(I + CT)^{-1}D,$$

where the letters A, B, C, D and T represent bounded linear operators on various Hilbert spaces. The importance of this choice is related to preserving contraction operators (see Theorem 7). Note that if we return to the scalar form of this expression we have

$$\frac{b(1 + cZ) - adZ}{1 + cZ} = \frac{(bc - ad)Z + b}{1 + cZ}$$

Now recalling the definition of the determining number of the linear fractional map in this form (see page 1) we require $(bc - ad) - bc \neq 0$. This makes sense in that if $ad = 0$ then in the operator setting this would mean $AD = 0$, and this would imply our selected operator would be the constant operator B. So let us return to the Hilbert space $\mathbb{H} = \mathbb{H}_1 \oplus \mathbb{H}_2$ discussed above. If T is in $\beta\mathbb{H}$ then we observe the following

$$T((x, 0)) = (u, v) = (A(x), C(x)) \in \mathbb{H},$$

where we claim the letters A and C represent bounded linear operators. A is a bounded operator in $\beta(\mathbb{H}_1)$, and C is in $\beta(\mathbb{H}_1, \mathbb{H}_2)$. Let us check this claim. Choose $x_j, j = 1, 2$ in $\mathbb{H}_1$. Consider the expression

$$T((x_1, 0) + (x_2, 0)) = T((x_1 + x_2), 0)) = (A(x_1 + x_2), C(x_1 + x_2))$$

and using the linearity of T we deduce

$$T((x_1, 0)) + T((x_2, 0)) = ((A(x_1), C(x_1)) + (A(x_2, 0), C(x_2)).$$

This of course implies that A and C are linear on their respective spaces. Similarly one shows that $A(mx) = mA(x), C(mx) = mC(x)$ for scalars m. Now performing a similar argument with the expression

$$T((0, y)) = (u, v) = (B(y), D(y)),$$

it can be shown that $B \in \beta(\mathbb{H}_2, \mathbb{H}_1)$, and $C \in \beta(\mathbb{H}_2)$.

The discussion above shows that we may express an operator T on $\mathbb{H}$ as a matricial equation. In this expression we identify a point $(x, y) \in \mathbb{H}$ as a "vector"

$$\begin{pmatrix} x \\ y \end{pmatrix}$$

and we identify T with the "two by two" operator matrix

$$\mathbb{T} = \begin{pmatrix} A & B \\ C & D \end{pmatrix}.$$

With this identification we have as a result

$$T((x,y)) = (A(x) + B(y), C(x) + D(y)).$$

With this definition of T we define an operator expression

$$M(X;T) \equiv B - AX(I + CX)^{-1}D \equiv B - AXQ,$$

where $Q = (I + CX)^{-1}D$, and $X \in \beta(\mathbb{H}_2, \mathbb{H}_1)$, and $C \in \beta(\mathbb{H}_1, \mathbb{H}_2)$, and I is the identity on $\mathbb{H}_2$. Thus $(I + CX)^{-1}$ when it exists is a bounded operator on $\mathbb{H}_2$ into itself, and so the operator expression M maps $\mathbb{H}_2$ into $\mathbb{H}_1$. One more piece of notation. Let

$$M_T \equiv \{X \in \beta(\mathbb{H}_2, \mathbb{H}_1) \; : \; I + CX \text{ is invertible in } \mathbb{H}_2\}.$$

A specific example for which this is true is when $||C|| < 1$ and X is a contraction. Motivation for this comes from the scalar case of the formula

$$\frac{1}{(1-t)} = \sum_o^\infty t^n$$

which converges when $|t| < 1$. Hence, if we write

$$\frac{1}{I + CX} = \sum_o^\infty (-1)^n (CX)^n$$

this series will converge in the operator norm and this then provides the operator-theoretic meaning of the fraction on the left above.

Now we come to our last BWYRFIIT statement but before stating it we have need of another definition.

Definition 11 *An operator T on a Hilbert space H is said to be positive if* $\langle Tx, x \rangle \geq 0, \; all \; x \in H.$

So for example if T is a contraction then we see that

$$< (I - T^*T)x, x > = ||x||^2 - ||Tx||^2 \geq 0$$

and so the operator $(I - T^*T)$ is a positive operator. The next statement is a robust lemma requiring several steps of computation and you might be able to muddle through it or you can check reference [3], Chapter 7. The notation is that as above.

Theorem 6 *Assume T is a unitary operator on $\mathbb{H}$. Then, for each X in M_T we have*

$$(I - (M(X;T))^* M(X;T)) = Q^*(I - X^*X)Q.$$

With this result in hand we culminate this section with the following theorem

Theorem 7 *Assume $X \in M_T$. If X is a contraction, then $M(X;T)$ is a contraction.*

Proof: We have noted that if X is a contraction then the operator $(I - X^*X)$ is a positive operator. Now consider the expression

$$Q^*(I - X^*X)Q.$$

We claim that this is also a positive operator. To check this assume $x \in H_1$

$$< Q^*(I - X^*X)Qx, x >= ||Qx||^2 - ||X(Qx)||^2 \geq 0.$$

But by Theorem 6 this implies that $(I-(M(X:t)^*)M(X:T))$ is a contraction and thus $M(X:T)$ is a contraction.

Now where are we in terms of taking this path into operator theory? Well at this stage we are at the brink of lots of interesting and deep operator theory and we are stopping here. One can take "square roots" of positive operators (that is if A is a positive operator there is another operator B with $B^2 = A$, and using these operators it is possible to make operator valued isometries for the setting of Theorem 7. If you have found any of this material interesting and want to pursue more of it I strongly suggest the references Fabian et. al. [2]. There is a great deal of material dealing with positive operators, the Wold decomposition and it is possible of course to get into this type of material for Banach spaces.

References

1. L. Ahlfors, *Complex Analysis*, 3rd Ed. McGraw-Hill, New York, 1979.

2. M. Fabian, P. Habala, P. Hajek, V.M. Santalucia, J. Pelant, V. Zizler, *Functional Analysis and Infinite Dimensional Geometry*, Canadian Mathematical Society, Springer-Verlag, 2001.

3. E. Fricain, J. Mashreghi, *The Theory of $\mathbb{H}(b)$ Spaces.* Vol.1., Cambridge Univ. Press, New Mathematical Mongraphs,20, Oct.31, 21015.

4. R.E. Greene, S.G. Krantz, *Function Theory of One Complex Variable*, John Wiley and Sons, New York. 1997.

4

Biholomorphic Transformations

Buma Fridman and Daowei Ma

CONTENTS

4.1   Introduction

This is a short overview for the beginners (graduate students and advanced undergraduates) on some aspects of biholomorphic maps.

A mathematical theory identifies objects that are "similar in every detail" from the point of view of the corresponding theory. For instance, in algebra, we use the notion of isomorphism of algebraic objects (groups, rings, vector spaces, etc.), while in topology homeomorphic topological spaces are

DOI: 10.1201/9781315160658-4

considered similar in every detail. The idea is to *classify* the objects under consideration.

In complex analysis the corresponding notion is the *biholomorphism* of the main objects in the theory: complex manifolds. Two complex manifolds M_1, M_2 are biholomorphic if there is a bijective holomorphic map $F : M_1 \to M_2$. Like in those other theories one wants to find the biholomorphic classification of complex manifolds. This attempt is certainly an important motivation to study biholomorphic transformations. The classification problem appears to be a very difficult problem, and only partial results (though very important and highly interesting) are known.

The Riemann Mapping Theorem states that any two proper simply connected domains in the complex plane are conformally equivalent. This is an exception. The general case is: for any $n \geq 1$ in $\mathbb{C}^n$ any two "randomly" picked domains are not biholomorphic. We will mention three examples to support this viewpoint. We also are pointing out that in case $n = 1$ the classification problem is well researched and mostly understood. For $n \geq 2$ it is way more complicated, and by now the pursuit of it has produced many very interesting and deep results and also created useful tools for SCV. Our intention is to put together various related problems, reflecting the authors interests, and in most cases refer the reader to some published material where details can be found. There are many surveys on most topics we consider. Quick search on MathSciNet shows that there are over two thousand papers published on biholomorphic maps. By no means this article is a comprehensive review. It should be considered as a glance into some of the results in the theory of these transformations.

The approximate outline of this exposition follows.

We start with three examples of non-equivalence for some "similar" domains. Then we introduce invariant metrics. Later we mention some important biholomorphic invariants and the related question of the extension of a biholomorphic mapping to the boundary. After that we consider the automorphism group of a domain in $\mathbb{C}^n$, which is a biholomorphic invariant, and related results. In the end we introduce and analyze "approximate" biholomorphic relations.

So we start with some examples.

4.2   Three Examples

The Riemann Mapping Theorem guaranties the biholomorphism of two proper simply connected domains in $\mathbb{C}$. But once we look at non-simply connected domains the situation drastically changes. Here's a result for annuli.

I). (F.H. Schottky, 1877) [52] Two annuli in $\mathbb{C} : \Omega(r_i, R_i) = \{r_i <\mid z \mid< R_i\}; i = 1, 2$ are conformally equivalent if and only if $r_1/R_1 = r_2/R_2$.

Proof. (sketch) Denote the disc of radius r with center at the origin by $\Delta(r)$. Without any loss of generality we may assume that $r_1 = r_2 = 1$, $R_1 > R_2$, and that $f : \Omega(1, R_1) \to \Omega(1, R_2)$ is continuous to the boundary (we'll address the necessity of this in a later section), and maps the unit circle $\partial\Delta(1)$ of $\partial\Omega(1, R_1)$ into the unit circle of $\partial\Delta(1) \subset \partial\Omega(1, R_2)$. We will use the Schwarz reflection principle. The reflection over $\partial\Delta(1)$ extends f to an analytic map from $\Omega(R_1^{-1}, R_1)$ to $\Omega(R_2^{-1}, R_2)$. We now continue to reflect over the smaller circle $\partial\Delta(R_1^{-1})$ and get f extended to $\Omega(R_1^{-3}, R_1)$. Continue this process indefinitely we get f extended to a conformal map of two punctured discs $\Omega(0, R_1)$ to $\Omega(0, R_2)$. Since f is bounded it can be extended to $\{0\}$; $f : \overline{\Omega}(0, R_1) \to \overline{\Omega}(0, R_2)$. Since f is now holomorphic, near the origin it will satisfy $|f(z)| \leq C|z|$ for some constant C. This means that for large n we have $| R_2^{-n} |\leq C \mid R_1^{-n} \mid$. This cannot happen since by assumption $R_1 > R_2$. □

For $n > 1$ in $\mathbb{C}^n$, a theorem analogous to the Riemann Mapping Theorem does not hold even in the following case.

II). (Henri Poincaré, 1907) For $n \geq 2$ the unit ball $B^n = \{(z_1, ..., z_n) \mid \sum_{j=1}^{n} | z_j |^2 < 1\}$ and the polydisc $\Delta^n = \{(z_1, ..., z_n) \mid | z_i |< 1, i = 1, ..., n\}$ are not biholomorphically equivalent.

Later on several proofs of this theorem will be provided.

In $\mathbb{C}^n$, $n > 1$ small perturbations of the unit ball may create a continuum of non-biholomorphic domains.

III). (Burns-Shnider-Wells thm, 1978) We'll present one consequence of the main theorem from [7].

Consider the unit ball $B^n \subset \mathbb{C}^n$, $n \geq 2, \epsilon > 0$. Then in the $\epsilon-$neighborhood of B^n there is an uncountable number of mutually non-biholomorphic domains with real analytic boundaries, containing the ball and homeomorphic to the ball.

4.3   Some Invariant Metrics

By introducing an invariant metric into a domain in $\mathbb{C}^n$ one can then consider every biholomorphic map from one domain to another as an isometry with all the consequences coming out of that. There are several invariant measures: Kobayashi, Caratheodory, Bergman, and others. We refer the reader to a number of books written on this. It is quite an extensive topic, and we will only touch on this subject here. It will end with a proof of Poincare's example mentioned in the previous section.

Definition 4.3.1. Let D be a domain in $\mathbb{C}^n$. The Carathéodory and Kobayashi infinitesimal pseudo-metrics are functions from $D \times \mathbb{C}^n$ to $[0, \infty)$ defined by

$$C_D(z, v) = \sup\{|dg(z)(v)| : g \in \mathscr{O}(D, \Delta), g(z) = 0\},$$
$$K_D(z, v) = \inf\{|u| : u \in \mathbb{C}, f \in \mathscr{O}(\Delta, D), f(0) = z, df(0)(u) = v\}.$$

The Kobayashi *indicatrix* of D at z is

$$I_{D,z} := \{v \in \mathbb{C}^n : K_D(z,v) < 1\}.$$

Kobayashi Extremal maps exist when D is bounded.

It was proved by Royden ([48] pp. 125–137). that every Kobayashi hyperbolic complex manifold is infinitesimally Kobayashi non-degenerate. The converse is false ([37] Remark 3.5.11).

Proposition 4.3.2. $C_D \leq K_D$.

Proof. Let $z \in D$ and $v \in \mathbb{C}^n$. Let $\varepsilon > 0$ be given. There is an $f \in \mathcal{O}(D, \Delta)$ and $u \in \mathbb{C}^n$ such that $f(0) = z$, $df(0)(u) = v$, and $|u| < K_D(z,v) + \varepsilon$. There is a $g \in \mathcal{O}(D, \Delta)$ such that $g(z) = 0$, and $|dg(z)(v)| > C_D(z,v) - \varepsilon$. By Schwarz lemma, $|d(g \circ f)(0)(u)| \leq |u|$. It follows that

$$C_D(z,v) - \varepsilon < |dg(z)(v)| = |d(g \circ f)(0)(u)| \leq |u| < K_D(z,v) + \varepsilon.$$

Letting $\varepsilon \to 0$ yields that $C_D(z,v) \leq K_D(z,v)$. $\qquad\qquad\qquad\qquad$ $\square$

Theorem 4.3.3. *The Kobayashi infinitesimal pseudo-metric is decreasing under holomorphic maps in the sense that if $f \in \mathcal{O}(D, \Omega)$ then $K_\Omega(f(z), df(z)(v)) \leq K_D(z,v)$.*

Proof. Let $\varepsilon > 0$. Choose $h \in \mathcal{O}(\Delta, D)$ and $u \in \mathbb{C}$ so that $h(0) = z$, $dh(0)(u) = v$, and $K_D(z,v) + \varepsilon > |u|$. Since $f \circ h \in \mathcal{O}(\Delta, \Omega)$, and $d(f \circ h)(0)(u) = df(z)(v)$, we see that $|u| \geq K_\Omega(f(z), df(z)(v))$. Thus $K_D(z,v) + \varepsilon > K_\Omega(f(z), df(z)(v))$ for each $\varepsilon > 0$. $\qquad\qquad\qquad\qquad\qquad\qquad\qquad\qquad\qquad\qquad$ $\square$

Corollary 4.3.4. *The Kobayashi infinitesimal pseudo-metric is invariant under biholomorphic maps in the sense that if f is a biholomorphic mapping from D onto Ω then*

$$K_\Omega(f(z), df(z)(v)) = K_D(z,v).$$

Moreover $df(z)(I_{D,z}) = I_{\Omega, f(z)}$.

Theorem 4.3.5. *Let $n > 1$. Then Δ^n and B^n are not holomorphically equivalent.*

Proof. Suppose that g is a biholomorphic mapping from B^n onto Δ^n. Let $a = g(0)$. There is an $h \in \text{Aut}(\Delta^n)$ with $h(a) = 0$. Let $f = h \circ g$. Then $f(0) = 0$. By Corollary 4.3.4, $df(0)(B^n) = \Delta^n$. That is impossible because B^n has a smooth boundary and Δ^n does not. $\qquad\qquad\qquad\qquad\qquad$ $\square$

4.4 Boundary Extension of a Biholomorphic Map

A set of biholomorphic invariants has been introduced by S. S. Chern, J.K. Moser [12], and N. Tanaka [53]. These invariants can be used for C^∞ manifolds of co-dimension one in $\mathbb{C}^n$. To use them for the case of a biholomorphism $f : D_1 \to D_2$ of bounded domains in $\mathbb{C}^n$ with smooth boundaries one needs first to prove that f can be extended smoothly to the boundary. We will now discuss this question of extending a biholomorphic map between two bounded domains in $\mathbb{C}^n$ to a *diffeomorphism* or even a *homeomorphism* between the closures of these domains.

In the case of one complex variable the conformal map $f : D \to \Omega$ for bounded domains in $\mathbb{C}$ can be extended to a diffeomorphism $\bar{D} \to \bar{\Omega}$ if the boundaries $\partial D, \partial \Omega$ are smooth, and to a homeomorphism if these boundaries are piece-wise smooth simple closed curves. An examination of this problem in $\mathbb{C}$ one can find in [46].

In the case of $\mathbb{C}^n, n \geq 2$ the situation is much more complicated. A short counterexample to the extendability is given by Fridman [20]: two domains D, Ω are constructed in $\mathbb{C}^2$, both biholomorphic to the bidisk, both have piece-wise smooth boundaries, and there is a C^∞ diffeomorphism between them that extends smoothly to their closures. However, no biholomorphic mapping $F : D \to \Omega$, nor F^{-1} can be extended continuously to the boundary. If D and Ω are *strictly pseudoconvex* domains in $\mathbb{C}^n$ with C^∞ smooth boundaries the result of N. Vormoor (and independently G. Henkin) [35, 58] shows that any biholomorphic map between them extends to a *homeomorphism* of their closures.

The question of *smooth* extendability is a very difficult problem. A major breakthrough was made by C. Fefferman in 1974. He proved the following theorem [14].

Theorem 4.4.1. *Let f be a biholomorphic map from a domain $D \subset \mathbb{C}^n$ to another domain $\Omega \subset \mathbb{C}^n$. Suppose that D and Ω are bounded strongly pseudoconvex domains with C^∞-smooth boundaries. Then f extends C^∞ to the boundary.*

Fefferman's original proof consists of two parts. First he considers the behavior of the Bergman kernel near the boundary, which leads to precise estimates of the Bergman metric near the boundary (for definitions and main results regarding the Bergman kernel see the book by S. Krantz [41]). In the second part, he proves the theorem using those estimates and the above mentioned Vormoor's result. We would like to note that [14] is 65 pages long, and the introduction (8 pages), has a very clear sketch of both parts of the proof.

Following this major accomplishment, there have been a number of publications refining and presenting different proofs of the theorem. The book [41] has a clear and different proof of this theorem. We should also mention two more proofs that are widely cited: those by F. Forstneric [15] and S. Bell & E. Ligocka [4].

The proof of F. Forstneric is based on two classical results: the edge-of-the-wedge theorem and the Julia-Caratheodory theorem (for these classical results, see e.g. [49, 50]). In this sense it is considered elementary.

Bell and Ligocka's proof of Fefferman's theorem made use of the Bergman kernel. For a bounded domain $D \subset \mathbb{C}^n$, the Bergman kernel function $K(z, w)$ is defined by $K(z, w) = \sum_{j=1}^{\infty} \varphi_j(z)\overline{\varphi_j(w)}$, where $\{\varphi_j\}$ is an orthonormal basis for the Bergman space $B^2 := L^2(D) \cap H(D)$. The Bergman kernel does not depend on the orthonormal basis and it has the reproducing property that for each $f \in B^2(D)$, $f(z) = \int_D K(z, w)f(w) \, dw$.

Let W_0^s denote the closure of $C_0^{\infty}(D)$ in the Sobolev space $W^s(D)$. Let $H^s(D) := W^s(D) \cap H(D)$ be the subspace of $W^s(D)$ consisting of holomorphic functions. A domain $D \subset \mathbb{C}^n$ is said to satisfy *Condition R* if it is bounded with smooth boundary and if for every $s \geq 0$ there exist constants $M > 0$ and $C > 0$ such that the Bergman projection $P : L^2(D) \to B^2(D)$ satisfies $\|Pf\|_{H^s(D)} \leq C\|f\|_{W_0^{s+M}(D)}$.

Theorem 4.4.2. *(Bell and Ligocka [4]) If D_1 and D_2 satisfy Condition R, then any biholomorphic mapping between D_1 and D_2 extends smoothly to the boundary.*

It had been known that subelliptic estimates for the $\bar{\partial}$-Neumann operator imply Condition R and that the $\bar{\partial}$-Neumann operator satisfies the subelliptic estimates for bounded strictly pseudoconvex domains with smooth boundary. Thus Bell and Ligocka's theorem implies the Fefferman Theorem.

Bell and Ligocka firstly proved that Condition R implies the following two conditions:

Condition A. $K(\cdot, w) \in C^{\infty}(D)$ for each $w \in D$,

Condition B. For each $z_0 \in \partial D$, there are $n + 1$ points $a_0, \ldots, a_n$ in D such that $K(z_0, a_0) \neq 0$ and

$$\det \begin{pmatrix} K(z_0, a_j) \\ \frac{\partial K}{\partial z_i}(z_0, a_j) \end{pmatrix}_{\substack{i=1,\ldots,n \\ j=0,1,\ldots,n}} \neq 0.$$

Using Conditions A and B, they proved that for a given biholomorphic map h between domains D_1 and D_2 the Jacobian functions $Jh(z)$ and $(Jh(z))^{-1}$ are bounded on D_1. Assume that $(Jh(z))^{-1}$ is not bounded. Then there exists a sequence $z_n \to z_0 \in \partial D_1$ such that $Jh(z_n) \to 0$. The transformation rule for the Bergman kernel function

$$K_1(z_n, a) = K_2(h(z_n), h(a))Jh(z_n)\overline{Jh(a)}$$

and Condition A yield that $K_1(z_0, a) = 0$ for each $a \in D_1$, which contradicts Condition B.

Consider a point $z_0 \in \partial D_1$ and a sequence $\{z_n\}$ in D_1 with $z_n \to z_0$ such that $h(z_n)$ tends to a point $s_0 \in \partial D_2$. By Condition B, there are points $b_0, b_1, \ldots, b_n$ such that $K(s_0, b_0) \neq 0$ and

$$Q := \det \begin{pmatrix} K_2(s_0, b_j) \\ \frac{\partial K_2}{\partial s_i}(s_0, b_j) \end{pmatrix}_{\substack{i=1,\ldots,n \\ j=0,1,\ldots,n}} \neq 0.$$

Let $v_j(s) = K_2(s, b_j)/K_2(s, b_0)$, $j = 1, \ldots, n$. Then $\det(\frac{\partial v_j}{\partial s_i}(s_0)) = K_2(s_0, b_0)^{-n-1}Q \neq 0$. Set $a_j = h^{-1}(b_j)$ and $u_j(z) = K_1(z, a_j)/K_1(z, a_0)$, $j = 1, \ldots, n$. The relation

$$\left(\frac{\partial u_j}{\partial z_i}(z_n)\right) = \left(\frac{\partial v_j}{\partial s_i}(s_n)\overline{\frac{Jh(a_j)}{Jh(a_0)}}\right) \cdot \left(\frac{\partial h_j}{\partial z_i}(z_n)\right)$$

then tells us that $\frac{\partial h_j}{\partial z_i}(z_n)$ are bounded. This implies that $\frac{\partial h_j}{\partial z_i}$ are bounded on D_1. Therefore, h is continuous up to the boundary. In particular $h(z_0) = s_0$. Since

$$\det\left(\frac{\partial u_j}{\partial z_i}(z_0)\right) \neq 0, \text{ and } \det\left(\frac{\partial v_j}{\partial s_i}(s_0)\right) \neq 0,$$

the functions u_j and v_j are smooth local coordinates near z_0 and s_0 respectively. With respect to these coordinates the mapping h is expressed as a linear mapping given by $v_j = \overline{\frac{Jh(a_j)}{Jh(a_0)}}u_j$. It follows that h can be extended smoothly to the boundary in a neighborhood of z_0.

4.5 Symmetry: Automorphism Group of a Domain

4.5.1 Automophism groups

A biholomorphic map of a manifold M onto itself is called an automorphism of M. The set $\mathrm{Aut}(M)$ of all automorphisms of M forms a group. This group is clearly a biholomorphic invariant. There is a relatively recent survey of this group by S. Krantz [40], and we'll refer to this paper throughout the section.

A good example of using $\mathrm{Aut}(D)$ as invariant is Poincare's original proof of biholomorphic non-equivalence of the unit ball B^n and polydisk Δ^n (for $n > 1$) by comparison of their automorphisms groups; they happen to be Lie groups of different dimensions and therefore not isomorphic.

There has been a lot of effort devoted to the study of $\mathrm{Aut}(D)$. As with all known invarians, this one is interesting (giving an idea of how "symmetric" a domain is) but by no means defining (even for $n = 1$ two annuli have the same automorphism group but might not be conformally equivalent). Moreover, after many deep studies one may conclude that for a general "random" domain in $\mathbb{C}^n$ the $\mathrm{Aut}(D) = \{id\}$. However, there is a large set of domains with non-trivial group, and this is a justification for detailed study of $\mathrm{Aut}(D)$.

One of the first results was the paper by H. Cartan (1935) [10] which proved that $\mathrm{Aut}(D)$ is a real Lie group for any bounded domain D in $\mathbb{C}^n$. Bedford-Dadok and Saerens-Zame [3, 51], proved that every compact Lie group can be realized as an $\mathrm{Aut}(D)$ for some strictly pseudoconvex domain with a smooth boundary in a suitable $\mathbb{C}^n$.

4.5.2   Description of automorphism groups and characterization of domains, manifolds with a given group

Below for convenience we will use two notations for the holomorphic automorphism of D: Aut_D or $\mathrm{Aut}(D)$.

Let D be a complex manifold. Consider a fixed point $w \in D$. The automorphisms of D which fix w form the isotropy group at w,

$$\mathrm{Aut}_{D,w} := \{\varphi \in \mathrm{Aut}_D : \varphi(w) = w\}.$$

For $\varphi \in \mathrm{Aut}_D$ the left coset of $\mathrm{Aut}_{D,w}$ with respect to φ is defined to be

$$\varphi \, \mathrm{Aut}_{D,w} := \{\varphi g : g \in \mathrm{Aut}_{D,w}\}.$$

For $\varphi, \psi \in \mathrm{Aut}_D$, if $\psi^{-1}\varphi \notin \mathrm{Aut}_{D,w}$ then $\varphi \, \mathrm{Aut}_{D,w} \cap \psi \, \mathrm{Aut}_{D,w} = \emptyset$; if $\psi^{-1}\varphi \in \mathrm{Aut}_{D,w}$ then $\varphi \, \mathrm{Aut}_{D,w} = \psi \, \mathrm{Aut}_{D,w}$. The cosets space

$$\mathrm{Aut}_D \, / \, \mathrm{Aut}_{D,w} := \{\varphi \, \mathrm{Aut}_{D,w} : \varphi \in \mathrm{Aut}_D\}$$

is naturally a real analytic manifold. The map from the cosets space $\mathrm{Aut}_D \, / \, \mathrm{Aut}_{D,w}$ to the orbit $\mathrm{Aut}_D(w) := \{g(w) : q \in \mathrm{Aut}_D\}$ given by $\varphi \, \mathrm{Aut}_{D,w} \mapsto \varphi(w)$ is a bijective, real analytic map. Thus we have the following

Proposition 4.5.1. *The orbit* $\mathrm{Aut}_D(w)$ *is a real analytic submanifold of D. Its dimension is* $\dim \mathrm{Aut}_D(w) = \dim \mathrm{Aut}_D - \dim \mathrm{Aut}_{D,w}$.

Let D be a bounded domain in $\mathbb{C}^n$ or a Kobayashi hyperbolic manifold of complex dimension n. It follows from the normal family theory that $\mathrm{Aut}_{D,w}$ is compact. If $\varphi \in \mathrm{Aut}_{D,w}$ is an automorphism which fixes w, its tangent map $d\varphi_w$ is a member of $GL(n, \mathbb{C})$. When $d\varphi_w = \mathrm{id}$, by looking at iterations of series expansions of φ at w, we see that φ must be the identity map. This implies that the map $\mathrm{Aut}_{D,w} \to GL(n, \mathbb{C})$, $\varphi \mapsto d\varphi_w$, is injective. Thus $\mathrm{Aut}_{D,w}$ is isomorphic to a compact subgroup of $GL(n, \mathbb{C})$. Since each compact subgroup of $GL(n, \mathbb{C})$ is conjugate to a subgroup of the unitary group $U(n)$, it follows that $\mathrm{Aut}_{D,w}$ is isomorphic to a subgroup of $U(n)$. Thus $\dim \mathrm{Aut}_{D,w} \leq \dim U(n) = n^2$.

A few examples of the $\mathrm{Aut}(D)$:

1. The automorphism group of the projective space: $\mathrm{Aut}(\mathbb{P}^n)$ is isomorphic to $PGL(n + 1, \mathbb{C})$. The proof can be found in [1], p. 41.

2. The detailed construction of the automorphism group of the ball B^n and the polydisc Δ^n can be found in [1, p. 54]. As it is shown in that book the $\mathrm{Aut}(B^n)$ is a real Lie group of dimension $n^2 + 2n$, and $\dim(\mathrm{Aut}(\Delta^n)) = 3n$. So, these groups have different dimensions for $n > 1$, and therefore not isomorphic for $n \geq 2$.

Here's a way to describe the $\mathrm{Aut}(B^n)$.

Let $0 < \lambda < 1$ and let $f : B^n \to \mathbb{C}^n$ be defined by

$$f_1(z) = \frac{z_1 + \lambda}{1 + \lambda z_1},$$

$$f_j(z) = \frac{\sqrt{1 - \lambda^2}}{1 + \lambda z_1} z_j, \quad j = 2, \ldots, n. \tag{4.1}$$

Then $f \in \mathrm{Aut}(B^n)$ and $f(0) = (\lambda, 0, \ldots, 0)$. Fix a point $a \neq 0$ in B^n. There is a unitary transformation U such that $Ua = (|a|, 0, \ldots, 0)$. Let h be the transformation in (4.1) with λ replaced by $|a|$ and let $g = U^{-1} \circ h \circ U$. Then $g \in \mathrm{Aut}(B^n)$, $g(0) = a$, and $g(-a) = 0$. The transformation g can be expressed as

$$g(z) = \frac{P_a(z) + a}{1 + z \cdot \overline{a}} + \frac{\sqrt{1 - |a|^2}}{1 + z \cdot \overline{a}} Q_a(z), \tag{4.2}$$

where

$$P_a(z) = \frac{z \cdot a}{|a|^2} a, \quad Q_a(z) = z - P_a(z).$$

It follows that every point in B^n is mapped to the origin by some member of $\mathrm{Aut}(B^n)$, hence the $\mathrm{Aut}(B^n)$ action is transitive and B^n is a homogeneous domain. Thus $\dim \mathrm{Aut}_{B^n, 0} = 2n$. Since each element of the unitary group $U(n)$ is an automorphism of B^n that fixes the origin, we see that $U(n)$ is naturally a subgroup of $\mathrm{Aut}_{B^n, 0}$. Thus $\dim \mathrm{Aut}_{B^n, 0} = n^2$. Therefore $\dim \mathrm{Aut}(B^n) = n^2 + 2n$.

The group $U(n, 1)$ is the group of linear transformations under which the form $X_1 \overline{X}_1 + \cdots + X_n \overline{X}_n - X_0 \overline{X}_0$ is invariant. So

$$U(n, 1) = \{W \in GL(n + 1, \mathbb{C}) : W^T \begin{pmatrix} I_n & 0 \\ 0 & -1 \end{pmatrix} \overline{W} = \begin{pmatrix} I_n & 0 \\ 0 & -1 \end{pmatrix}.$$

In homogeneous coordinates, B^n is given by

$$X_1 \overline{X}_1 + \cdots + X_n \overline{X}_n - X_0 \overline{X}_0 < 0.$$

Each $W = \begin{pmatrix} A & b \\ c & d \end{pmatrix}$ in $U(n, 1)$ gives a map $g_W \in \mathrm{Aut}(B^n)$, where

$$g_W(z) = \frac{Az + b}{cz + d}. \tag{4.3}$$

The map g_W is the identity map if and only if $W \in S^1 := \{e^{it} I : t \in \mathbb{R}\}$. Thus $PU(n, 1) := U(n, 1)/S^1$ is considered a subgroup of $\mathrm{Aut}(B^n)$. Since $(n+1)^2 - 1 = \dim PU(n, 1) \leq \dim \mathrm{Aut}(B^n) = 2n + n^2$, we see that $\mathrm{Aut}(B^n) = PU(n, 1)$.

If $b = (t, 0, \ldots, 0)$, $t > 0$, $c = b^T$, then $A = \begin{pmatrix} \alpha\sqrt{1 + t^2} & 0 \\ 0 & V \end{pmatrix}$, $d = $

$\overline{\alpha}\sqrt{1+t^2}$, where $V \in U(n-1)$, $|\alpha| = 1$. Take $\alpha = 1$, $V = I_{n-1}$ to obtain
$A = \begin{pmatrix} \sqrt{1+t^2} & 0 \\ 0 & I_{n-1} \end{pmatrix}$ and $d = \sqrt{1+t^2}$. Then g_W is given by (4.1) with
$\lambda = t/\sqrt{1+t^2}$. The map in (4.3) is the transformation given by (4.2) with
$a = d^{-1}b$.

3. The automorphism group of the entire space $\mathbb{C}^n$, $n \geq 2$ is not finite
dimentional. Say, for $\mathbb{C}^2$ it contains shears ($w_1 = z_1; w_2 = z_2 + f(z_1)$) for any
entire function f.

4. More examples one can find in A. Isaev's book [36].

5. One more interesting result. It can be proved that if D is a bounded
domain in $\mathbb{C}^n$ and $\dim(\mathrm{Aut}(D)) = n^2 + 2n$, then D is biholomorphic to B^n.
A. Isaev gave explicit classification of all domains D in $\mathbb{C}^n$ for which $n^2 - 1 \leq \dim(\mathrm{Aut}(D)) < n^2 + 2n$. These results are fully presented in [36].

4.5.3 Greene-Krantz conjecture

In case of strictly pseudoconvex domains with smooth boundary only the
unit ball has a NON-compact group of holomorphic automorphisms. This was
proved by B. Wong [59]. The non-compactness of $\mathrm{Aut}(D)$ means that for at
least one point $z_0 \in D$ the orbit $\mathrm{Aut}(D)(z_0)$ has an accumulation point p on
the boundary. J. P. Rosay [47] extended Wong's theorem by proving that if p
is a strictly pseudoconvex point of ∂D then D is biholomorphic to the ball.
The proof of Wong-Rosey theorem can be done by the scaling method ([40],
p. 227). In the same survey one can find various versions of this theorem (even
for infinite dimensional cases).

Because of these theorems, the description of smoothly bounded domains
with non-compact automorphism group becomes interesting. All the known
examples of such domains reveal that the accumulation point on ∂D is of
"finite type in the sense of Kohn, D'Angelo, and Catlin". (A boundary point of
a domain in $\mathbb{C}^2$ is of finite type if the boundary has finite order of contact with
complex manifolds through the point; precise definition for $\mathbb{C}^n$ can be found
in [40]). In 1991, R. Greene and S. Krantz made the following conjecture [30]:

Conjecture. Let D be a smoothly bounded domain in $\mathbb{C}^n$. Suppose that
$x \in D$ has a boundary orbit accumulation point for the automorphism group
in the sense that there are automorphisms $\phi_j \in \mathrm{Aut}(D)$ and a point $p \in \partial D$
such that $\phi_j(x) \to p$ as $j \to \infty$. Then p is a point of finite type.

There have been many attempts to resolve this conjecture, only partial
results have been obtained by now; for a more detailed discussion on it see [40].

4.5.4 Narasimhan's question

In this section, we will again use the notation Aut_D instead of $\mathrm{Aut}(D)$.

A bounded domain $\Omega \subset \mathbb{C}^n$ is said to have *Property N* if there exists
a compact subset K of Ω with the property that for each $z \in \Omega$ there is
an $f \in \mathrm{Aut}_\Omega$ such that $f(z) \in K$. For $z \in \Omega$ and a subgroup Γ of Aut_Ω,

$\Gamma(z) := \{f(z) : f \in \Gamma\}$ is the Γ-orbit of z. If $S \subset \Omega$, then $\Gamma(S)$ denotes the union of the Γ-orbits of the points in S:

$$\Gamma(S) := \cup_{z \in S}\Gamma(z) = \{f(z) : z \in S, f \in \Gamma\}.$$

Thus, Property N is equivalently defined as follows: Ω is said to have property N if there is a $K \subset\subset \Omega$ such that $\mathrm{Aut}_\Omega(K) = \Omega$. The following is a question by R. Narasimhan:

Question 1. If Ω is a bounded domain in $\mathbb{C}^n$ with Property N, is Ω necessarily a homogeneous domain?

A *discrete* subgroup of Aut_Ω is a subgroup of Aut_Ω that is discrete in the compact open topology, the default topology of Aut_Ω. Recall that we say a subgroup Γ of Aut_Ω acts on Ω freely if each $f \in \Gamma$ is fixed-point free.

The following theorem is by S. Frankel [16].

Theorem 4.5.2. *Let Ω be a convex hyperbolic domain in $\mathbb{C}^n$ and suppose that there is a discrete subgroup $\Gamma \subset \mathrm{Aut}_\Omega$ such that Ω/Γ is compact. Then Ω is biholomorphic to a bounded symmetric domain.*

Frankel first proved a distortion theorem for convex holomorphic embeddings and used this to reduce a complex analysis problem to one in affine geometry. Then he applied rescaling to produce a continuous family of automorphisms. His particular technique of boundary localization was very different from what had gone before, and he called it the rescale blow-up.

Kazhdan conjectured that each irreducible bounded domain which admits both a compact quotient and a one-parameter group of holomorphic automorphisms must be biholomorphic to a bounded symmetric domain. Frankel [16] first confirmed the conjecture for bounded convex domains. Subsequent work by Nadel [44] and Frankel [17] proved it in general.

Theorem 4.5.3. *Let M be a compact complex manifold. Then the group Aut_M is a complex Lie transformation group and its Lie algebra consists of holomorphic vector fields on M.*

The above theorem is due to Bochner and Montgomery [5,6]. For bounded domains in $\mathbb{C}^n$, we have the following theorem of H. Cartan [9,11].

Theorem 4.5.4. *Let D be a bounded domain in $\mathbb{C}^n$. Then the group Aut_D is a Lie transformation group and the isotropy subgroup $\mathrm{Aut}_{D,z}$ at any point $z \in D$ is compact. If X is in the Lie algebra of Aut_D, then JX is not in the Lie algebra of Aut_D.*

Theorem 4.5.5. *(Kobayshi [39, p. 81]) Let M be a hyperbolic manifold. Then the group Aut_M is a Lie transformation group and the isotropy subgroup $\mathrm{Aut}_{M,x}$ of M at any point $x \in M$ is compact.*

The essential reason for the above three theorems is the following

Theorem 4.5.6. *(Bochner and Montgomery [6]) Let G be a locally compact group of differentiable transformations of a manifold M. Then G is a Lie transformation group.*

Theorem 4.5.5 is also based on the following early result of van Danzig and van der Waerden [13].

Theorem 4.5.7. *Let M be a connected, locally compact metric space and G_M the group of isometries of M. Then G_M is locally compact with respect to the compact-open topology.*

4.6 Determining Sets and Fixed Points

Let M be a complex manifold, $f : M \to M$ a holomorphic map. $z_0 \in M$ is a fixed point for f if $f(z_0) = z_0$.

The following is a result in the classical function theory [43, 45]: if $f : M \to M$ is a conformal self-mapping of a plane domain M which fixes three distinct points then $f(\zeta) = \zeta$.

This one-dimensional result is true even for endomorphisms of a bounded domain $D \subset\subset \mathbb{C}$. To prove this one needs to first use the well known theorem, stating that if an endomorphism of D fixes two distinct points, then it is an automorphism; and then use the above cited [43, 45] theorem.

We now introduce two notions to discuss how this result can be extended for higher dimensions.

For a complex manifold M let $H(M, M)$ be the set of holomorphic maps from M to M, i.e., the set of endomorphisms of M. The group of holomorphic automorphisms of M, $\mathrm{Aut}(M)$ is a subset of $H(M, M)$.

Definition 4.6.1. A set $K \subset M$ is called a *determining subset* of M with respect to $\mathrm{Aut}(M)$ ($H(M, M)$ resp.) if, whenever g is an automorphism (endomorphism resp.) such that $g(k) = k$ $\forall k \in K$, then g is the identity map of M.

So, any three points of a plane domain D form a determining set for $\mathrm{Aut}(D)$ as well as for $H(D)$.

The other notion is $Fix(f)$, it denotes the set of *fixed points* $\{x \in M \mid f(x) = x\}$ of f.

So, if M is a plane domain, and $Fix(f)$ is discrete, then the cardinality of this set $\#Fix(f) \leq 2$ for any f according to our remarks at the beginning of this section.

We now have two classes of problems to investigate. First: the description and properties of determining sets for various complex manifolds. There are many results in classical analysis and topology proving the non-emptiness of $Fix(f)$, or finding it for a given function. Our second class of problems to discuss is different from those: for a given M which sets can be $Fix(f)$ for some holomorphic $f \in H(M, M)$.

4.6.1 Determining sets

The notion of determining sets was first introduced in [25]. That paper was an attempt to find a higher dimensional analog of the above one-dimensional result. Determining sets (for automorphisms and endomorphisms) in case of bounded domains in $\mathbb{C}^n$ have been further investigated in the following papers [26–29, 38, 54, 55].

Let's first look now at some examples of discrete non-determining and determining sets (from [25]).

Example 4.6.2. Let $A = \{z \in \mathbb{C} : 1/2 < |z| < 2\}$. This is an annulus in the plane. The map $\tau(z) = 1/z$ has two fixed points (i.e., 1 and -1), yet τ is not the identity mapping, so these points are not a determining set. Since the $Aut(A)$ is well known one can check that any two points in general position in A do form a determining set.

Example 4.6.3. In $\mathbb{C}^2$ consider a shear of the form $\tau(z, w) = (z, w + \phi(z))$, where ϕ is any entire function on the plane. Then τ is a biholomorphic map of $\mathbb{C}^2$. If ϕ has infinitely many distinct zeros then τ will have infinitely many fixed points, even though τ is not the identity. So, the set of these zeros does not form a determining set for $Aut(\mathbb{C}^2)$.

By contrast, any biholomorphic (conformal) map of $\mathbb{C}$ that fixes two points must be the identity. So, any two distinct points in $\mathbb{C}$ form a determining set for the $Aut(\mathbb{C})$.

Example 4.6.4. It can be shown that a biholomorphic map of the unit ball B^n in $\mathbb{C}^n$ that fixes $n + 1$ points in general position (in the usual sense of topology) must be the identity. One may check this by using the description of the automorphism group of the ball given in the previous section. We leave the details to the interested reader. So, this set of $n+1$ points forms a determining set for the ball.

We also note that no set of n points in B^n forms a determining set. Indeed, let $p_1, \ldots, p_n$ be the n points. Since the ball is a homogeneous domain, we may consider $g \in \mathrm{Aut}(B^n)$ such that $g(p_1) = 0$. Now the set $(g(p_1), g(p_2) \ldots, g(p_n))$ lies in a linear space L of dimension $dim(L) \leq n - 1$. Therefore, there is a rotation $f \in \mathrm{Aut}(B^n)$ that keeps all the points of L fixed. So, the automorphism $h = g^{-1}fg \in \mathrm{Aut}(B^n)$ is not an identity and fixes all the n points $(p_1, \ldots, p_n)$.

Example 4.6.5. Consider the domain $U_m = \{(z_1, z_2) \in \mathbb{C}^2 : |z_1|^{2m} + |z_2|^{2m} < 1\}$, any integer $m \geq 2$. Then any automorphism of U_m that fixes two points in general position must be the identity. This result follows because the automorphism group of U_m is well-known to consist only of rotations in each variable separately.

Contrast this example with the result from the last example (for the unit ball in $\mathbb{C}^n$).

Example 4.6.6. Let U_m be one of the domains from the last example. Let V be any rigid domain in $\mathbb{C}^n$ (here rigid means that the domain has no automorphisms except the identity). Then, for an arbitrary chosen pair of points $z \in U_m, w \in U_m$, and an arbitrary $x \in V$, any automorphism of $U_m \times V$ which fixes both (z, x) and (w, x) will be the identity. For instance, the points $z = ((1/2, 0), x)$ and $w = ((0, 1/2), x)$ will do.

We will present now a few results when the determining set K is *discrete* and refer the interested reader to the above mentioned papers for more results and unsolved problems.

We will need two more notions. Let M be a complex manifold. Let $W_s(M)$ denote the set of s-tuples $(x_1, \ldots, x_s)$, where $x_j \in M$, such that $\{x_1, \ldots, x_s\}$ is a determining set with respect to $\mathrm{Aut}(M)$. Similarly, $\widehat{W}_s(M)$ denotes the set of s-tuples $(x_1, \ldots, x_s)$ such that $\{x_1, \ldots, x_s\}$ is a determining set with respect to $H(M, M)$. So $\widehat{W}_s(M) \subseteq W_s(M) \subseteq M^s$. We now introduce two values $s_0(M)$ and $\widehat{s}_0(M)$. In case $\mathrm{Aut}(M) - id$, $s_0(M) - 0$, otherwise $s_0(M)$ is the least integer s, such that $W_s(M) \neq \emptyset$. If $W_s(M) = \emptyset$ for all s then $s_0(M) = \infty$. Analogously symbol $\widehat{s}_0(M)$ denotes the least integer s such that $\widehat{W}_s(M) \neq \emptyset$, if no such integer exists (i.e. $\widehat{W}_s(M) = \emptyset$ for all s) then $\widehat{s}_0(M) = \infty$. In all cases $s_0(M) \leq \widehat{s}_0(M)$.

1. For hyperbolic manifolds of dimension n the following estimate holds $\widehat{s}_0(M) \leq n + 1$.

First we note that this estimate for bounded domains in $\mathbb{C}^n$ was proved by J.P. Vigue (see [55]). In that paper a more precise theorem is proved. Let a be a point in a bounded domain D. Then there is an open set $U \subset D^n$ such that $(a, ..., a) \in U$ and for all $(z_1, ..., z_n) \in U$, $(z_1, ..., z_n) \in \widehat{W}_s(M)$.

For a general hyperbolic manifold M one may consider a small Kabayashi ball $b \in M$ such that it is biholomorphic to a domain $D \subset \mathbb{C}^n$. Pick a point $a \in M$. Let $f : M \to M$ be a holomorphic map such that $f(a) = a$. Consider a small Kobayashi ball $b = b(a, \epsilon)$ whose closure is compact in M, and such that b is biholomorphic to a bounded domain $D \subset \mathbb{C}^n$; let $h : b \to D$ be such a biholomorphic map. Note that since the Kobayashi distance is non-increasing under holomorphic maps, we have $f : b \to b$, and therefore $g = hfh^{-1} : D \to D$. By using the above mentioned result, one can pick n points $z_1, ..., z_n \in D$, such that $Z = (h(a), z_1, ..., z_n) \in \widehat{W}_{n+1}(D)$. Consider the set of $n + 1$ points $h^{-1}(Z) = (a, h^{-1}(z_1), ..., h^{-1}(z_n)) \subset b$. If our function $f \in H(M, M)$(in addition to a) is also fixing all points $h^{-1}(z_j)$, i.e. $f|_{h^{-1}(Z)} = id$, then $g|_Z = id$ and therefore $g = id$. We conclude that $f|_b = id$, and consequently $f = id$. So, $h^{-1}(Z) \in \widehat{W}_{n+1}(M)$, and therefore $\widehat{s}_0(M) \leq n + 1$.

2. The above statement implies same inequality for *automorphisms* of a hyperbolic manifold M, $s_0(M) \leq n + 1$.

However, for automorphisms much more information can be provided. $s_0(M)$ depends on how large the group $\mathrm{Aut}(M)$ is. If M is a bounded domain in $\mathbb{C}^n$ $(dim(M) = n)$ then the general estimate $(s_0(M) \leq n + 1)$ can be refined to $s_0(M) \leq n$ for domains that are *not* biholomorphic to the unit ball $B^n \subset \mathbb{C}^n$, and the only hyperbolic manifolds for which $s_0(M) = n + 1$ are those biholomorphic to the ball. This gives a characterization of the ball in $\mathbb{C}^n$.

3. If a positive integer $s \geq s_0(M)$, then $W_s(M) \neq \emptyset$, so there are s points such that if an automorphism of M fixes these points it will fix any point of M. Now the question arises whether the choice of these s points is generic. The answer is positive for any hyperbolic manifold M: $W_s(M) \subseteq M^s$ is open and dense if not empty. Similar topological properties for the determining sets of *endomorphisms* of a general hyperbolic manifold do not hold.

4.6.2 Isolated fixed point sets and cardinality

In classical mechanics the following Euler's theorem is well known: *the general displacement of a rigid body with one point fixed is a rotation about some axis.* So, if one considers an orientation-preserving isometry of a domain in $\mathbb{R}^3$ fixing one point, the fixed point set of this isometry will necessarily contain at least a segment, so the fixed point set cannot be a discrete set. In the euclidean space $\mathbb{R}^n$, one can always find a domain which has a euclidean isometry with exactly one fixed point, however for any n, if an isometry of a domain in $\mathbb{R}^n$ has two fixed points it will force the existence of at least a segment to belong to the fixed point set, and so this set will be at least one dimensional.

Switching to complex analysis, we remark that any holomorphic automorphism of a bounded domain in $\mathbb{C}^n$ (or in general, hyperbolic manifold) is an isometry in an invariant metric, so an Euler type statement is certainly meaningful, that is if this automorphism has a discrete fixed point set one can inquire what its cardinality and structure might be. To describe this more precisely, let $f : M \to M$ be a holomorphic self-map of a complex manifold M. Suppose that $Fix(f)$ is discrete. We shall examine mostly two questions. First, how large this set can be for specific cases: M is a bounded domain in $\mathbb{C}^n$, a hyperbolic manifold, etc., while f is a holomorphic automorphism or endomorphism. Second, the structure of $Fix(f)$, namely which points of M could form such a set for some holomorphic self-map of M. Everywhere below we consider only holomorphic self-maps (automorphisms or endomorphisms) of various complex manifolds, and for the sake of compactness the word holomorphic may be omitted.

In examining the cardinality of a discrete fixed point set, let's first consider the situation in one dimension. For a bounded domain $D \subset \mathbb{C}$ the discrete fixed point set of a holomorphic map $f : D \to D$ can have no more than two points. This follows from the above mentioned observation: any set of three points in D must be a determining set for endomorphisms. An annulus gives an example of a domain that has an automorphism with exactly two fixed points.

In $\mathbb{C}^n$ the situation is not yet completely clear. Here are a few statements we know.

1. For a *convex* domain one has the following theorem: *the isolated fixed point set of any endomorphism consists of at most one point.* This statement follows from the main theorem in [56]: such a set has to be connected. The proof is based on establishing that in Bergman metric there is a unique geodesic connecting two fixed points, and it (the geodesic) will then also belong to the fixed point set.

2. For a *bounded strictly pseudoconvex domain D in $\mathbb{C}^n$ with real analytic boundary the number of points in a discrete fixed point set of an automorphism is finite. Moreover, there is a number $m = m(D)$ such that $\#(Fix(f)) \leq m$.*

Here's a proof of this statement. If D is biholomorphic to the ball or if $n = 1$, then the statement is clear. Assume that $n \geq 2$ and D is not biholomorphic to the ball. By a theorem in [57], there is a neighborhood U_1 of D such that each automorphism of D extends to be an injective holomorphic map on U_1. Consider a $g \in Aut(D)$. Choose domains U_2, U_3 with smooth boundaries so that $D \subset\subset U_3 \subset\subset U_2 \subset\subset U_1$. For every $h \in Aut(D)$ in some neighborhood of g, $h(\partial U_2)$ is so close to $G(\partial U_2)$ that $h(\partial U2) \cap g(U3) = \emptyset$. Since $h(U2)$ is a connected component of $\mathbb{C}^n \backslash h(\partial U_2)$ and since $h(U_2) \supset D$, we see that $h(U_2) \supset g(U_3)$ for every $h \in Aut(D)$ in some neighborhood of g. Since $Aut(D)$ is compact, there is a neighborhood Q of D such that $Q \subset g(U_1)$ for each $g \in Aut(D)$. Let U be the interior of the intersection of the sets $g(U_1), g \in Aut(D)$. Then $U \supset Q$ and $g(U) = U$ for each $g \in Aut(D)$. That is, each automorphism of D is also an automorphism of U. There is a finite cover of open sets $\{Vj : j = 1, ..., m\}$ of D such that each pair of points in a Vj is connected by a unique distance-minimizing geodesic with respect to the Bergman metric of U. Let $f \in Aut(D)$. If f fixes two points in a Vj, f must fix each point on the unique distance-minimizing geodesic connecting the two points. Consequently, each Vj contains at most one isolated fixed point of f. Therefore, the number of isolated fixed points of f is $\leq m$.

3. Must the cardinality of an isolated fixed point set of an automorphism or endomorphism be bounded by a number depending only on the dimension of the manifold under consideration? For endomorphisms of bounded domains in $\mathbb{C}^n$ the answer is negative. It is also negative for automorphisms of a general hyperbolic manifold and the entire $\mathbb{C}^n$. However, for an automorphism of a bounded domain in $\mathbb{C}^n$ the answer is not yet clear. Let's consider several examples demonstrating some of the results.

Example 4.6.7. *For any $k \in \mathbb{N}$, there exists a bounded domain $D \subset \mathbb{C}^n$, $n \geq 2$, and a holomorphic endomorphism $f : D \to D$, such that $\#(Fix(f)) = k$.*

Proof. Without any loss of generality we can present an example for $n = 2$. Let S be the open Riemann surface in $\mathbb{C}^2$: $S = \{(x, y) \in \mathbb{C}^2 | y^2 = (x - a_1)...(x - a_k)\}$, where $a_1, ..., a_k$ are k distinct points in $\mathbb{C}$. The restriction g of $(x, y) \to (x, -y)$ to S has exactly k fixed points. Following ([34], VIII, C8, p. 257) there exists a holomorphic retraction $\rho : V \to S$ of an open neighborhood V of S onto S. Now the mapping $f := g \circ \rho : V \to V$ has exactly

k fixed points. Of course V is not bounded, but we can consider a bounded open set $W \subset V, (a_s, 0) \in W$ for all $s = 1, ..., k$ and such that $g(W) = W$. This bounded domain will have the same property.

Example 4.6.8. *There exists a hyperbolic manifold with a holomorphic automorphism whose fixed point set is discrete and consists of an infinite number of points.*

Proof. Consider the submanifold X of D^2 defined by $y^2 = B(x)$, where D is the open unit disc and B is a Blaschke product with an infinite number of zeroes, the restriction to X of the map $(x, y) \to (x, -y)$ is an automorphism of X and has an infinite number of isolated fixed points.

Example 4.6.9. *For any $n \geq 2$ and any $k \in N$, there exists a polynomial automorphism f of $\mathbb{C}^n$, such that $\#(Fix(f)) = k$. Moreover, let $n \geq 2; p_1, p_2, ..., p_k$ are k distinct points in $\mathbb{C}^n$. Then there exists a polynomial automorphism $g \in Aut(\mathbb{C}^n)$ such that $Fix(g) = \{p_1, p_2, ..., p_k\}$.*

Proof. Let $a_1, ..., a_k$ be k distinct complex numbers. Consider the map $H : \mathbb{C}^n \to \mathbb{C}^n$ given by
$$w_1 = z_1 + z_2 + (z_1 - a_1)(z_1 - a_2)...(z_1 - a_k)$$
$$w_2 = z_2 + (z_1 - a_1)(z_1 - a_2)...(z_1 - a_k)$$
$$w_s = iz_s \text{ for all } s = 3, ..., n$$
This map is an automorphism, whose fixed point set is the set of the following k points: $(a_1, 0, ..., 0), (a_2, 0, ..., 0), ..., (a_k, 0, ..., 0)$.

Now $p_j = (a_j, b_j), a_j \in \mathbb{C}, b_j \in \mathbb{C}^{n-1}$. Without any loss of generality we assume that the a'_js are all distinct (in case they are not, one can first use an invertible linear transformation of $\mathbb{C}^n$ to achieve this). Consider the polynomial transformation $F : w_1 = z_1, w' = z' + f(z_1)$, where $f : \mathbb{C} \to \mathbb{C}^{n-1}$ is the Lagrange interpolation polynomial map satisfying $f(a_j) = b_j$, $w' = (w_2, ..., w_n)$. Then $F(a_j, 0) = p_j, j = 1, ..., k$, and $F \in Aut(\mathbb{C}^n)$. Now the automorphism $g = F \circ H \circ F^{-1}$ is such that $Fix(g) = \{p_1, p_2, ..., p_k\}$.

Some problems

1. Let D be a bounded domain in $\mathbb{C}^n, f \in Aut(D)$, and $Fix(f)$ is a discrete set. Can $\#(Fix(f)) = \infty$?

If one considers the domain $D \subset \mathbb{C}^n$ which is a direct product of n annuli, one can then find an $f \in Aut(D)$ with $\#(Fix(f)) = 2n$. So, the next natural question is

2. Let $n \geq 2, D$ be a bounded domain in $\mathbb{C}^n$, with a piecewise smooth boundary, $f \in Aut(D)$, and $Fix(f)$ is a set of isolated points. Can $\#(Fix(f)) \geq 2n + 1$? (As noted earlier, for $n = 1$ the answer is negative). A more restricted version of this question is

3. Is there a number m such that for any strongly pseudoconvex domain $D \subset\subset \mathbb{C}^n, \partial D \in C^\infty$, and $f \in Aut(D)$, if $Fix(f)$ is a set of isolated points, then $\#(Fix(f)) \leq m$, where $m = m(n)$ (i.e. m depends on the dimension only)?

4.6.3   Fixed point sets consisting of one or two points

In this section we'll discuss which subsets of a manifold D can be $Fix(f)$ for some holomorphic automorphism or endomorphism $f : D \to D$. As the title suggests we'll consider two cases.

First we consider the case when every *single* point of a domain is the $Fix(f)$ for a suitable holomorphic f.

Theorem 4.6.10. *[28, Theorem 2.1] If every point of a hyperbolic manifold D is a fixed point set for some holomorphic automorphism of D, then D is a homogeneous manifold.*

Proof. 1. First we note that the theorem will follow from a local statement: let $x \in D$, then there exists a neighborhood U_x of x such that for any $y \in U_x$ there is a $g \in Aut(D)$ such that $g(y) = x$. Indeed, if this is true consider two arbitrary points $a, b \in D$, connect them by a compact path L, cover L by a finite number of $U_x, x \in L$, and one can obtain an $f \in Aut(D)$, such that $f(a) = b$.

2. We now prove the local statement. Let $x \in D$. By [29], for each point $x \in D$ there is an invariant Hermition metric in some neighborhood of the orbit $G(x)$, where $G = Aut(D)$. Consider a small enough ball $b(x, \epsilon)$ in that metric with center x and radius ϵ, $\epsilon > 0$ will be determined by the construction later. Let $y \in b(x, \epsilon)$; consider the orbit $O(y) = \{z \in D : \exists g \in Aut(D), g(y) = z\}$. Consider now a point $p \in O(y)$, such that $d(x, p) = d(x, O(y))$, where $d(\cdot, \cdot)$ denotes the distance function induced by the local invariant metric. Clearly, $p \in b(x, \epsilon)$. If $p = x$, there is nothing to prove; otherwise consider a small ball b_1 of radius $< d(x, p)$ that lies inside $b(x, d(x, p))$, and such that $\partial b_1 \cap \partial b(x, d(x, p)) = p$. This construction is possible if ϵ is small enough, fixing such an $\epsilon = \epsilon(x)$, we denote $b(x, \epsilon) = U_x$.

We observe that $O(y) \cap b(x, d(x, p)) = \emptyset$. Let q denote the center of the ball b_1. By the assumption of the theorem there exists an $h \in Aut(D)$ whose fixed point set is q. Now $h(p) \neq p$, and $h(p) \in \partial b_1$, since $h(\partial b_1) = \partial b_1$. We now conclude that $h(p) \in O(y) \cap b(x, d(x, p))$, which contradicts the previous observation that this intersection is empty. Therefore $x = p \in O(y)$, and the theorem has been proved. $\square$

We now provide the following example.

Theorem 4.6.11. *There exists a domain D in $\mathbb{C}$ with infinite number of points each of which is the fixed point set for a holomorphic automorphism of D.*

Proof. Consider $D = \mathbb{C} \setminus \bigcup_{n \in \mathbb{Z}} \Delta(n, 1/3)$ where $\Delta(n, 1/3)$ is a disk with center at $n \in \mathbb{Z}$ and radius $1/3$. Consider $f_k : z \to (-z + (2k + 1))$. Then for any $k \in \mathbb{Z}, f_k \in Aut(D)$, and its fixed point set consists of one point $Fix(f) = \{k + 1/2\}$. $\square$

Let's now consider *pairs of points* as fixed point sets. Though such domains exist, no domain can have too many pairs of distinct points as a fixed point set for an automorphism.

Theorem 4.6.12. *Let* $D \subset\subset \mathbb{C}^n$. *The set* $N \subset D^2$ *of all pairs, each of which cannot be a fixed point set for a holomorphic automorphism of* D, *contains a full measure set in* D^2.

It follows from the following two Lemmas. The first is a classical statement (see [8] p. 80; also [54] thm 2.3) proving that for $z \in D$ and its isotropy subgroup $I_z = \{f(z) = z, f \in Aut(D)\}$ there is a local system of coordinates where each $f \in I_z$ is a linear map.

Lemma 4.6.13. *There exists a holomorphic map* $\phi : D \to \mathbb{C}^n$ *such that* $\phi(z) = 0, \phi'(z) = id$, *and for all* $f \in I_z$ *one has* $\phi \circ f = f'(z) \circ \phi$.

The theorem will now follow from the second lemma.

Lemma 4.6.14. *Let* $D \subset\subset \mathbb{C}^n, a \in D$. *Then there exists a complex analytic set* $Z \subset D, (dim Z < n)$, *such that if* $b \in D \backslash Z$ *then the two points* $\{a, b\}$ *are such that for any automorphism* f *fixing these two points, the fixed point set of* f *is at least one (complex) dimensional.*

Proof. Using the previous lemma we first find the function ϕ for the point a. Let $Z = \{z \in D | \phi(z) = 0\}$. If $b \in D \backslash Z$, then suppose $f \in Aut(D)$ and f fixes both points a and b. We have $f'(a) \cdot \phi(b) = \phi(f(b)) = \phi(b)$. Since by choice $\phi(b) \neq 0$, and ϕ is biholomorphic in the neighborhood U of a, for a number $\lambda, |\lambda| > 0$, small enough, there exists a point $c \in U \subset D, c \neq a, \phi(c) = \lambda\phi(b)$, and $f(c) \in U$. Now $\phi(f(c)) = f'(a) \cdot \phi(c) = f'(a) \cdot \lambda\phi(b) = \lambda\phi(b) = \phi(c)$.

Since ϕ is biholomorphic in U, we have $f(c) = c$. $\square$

4.7 Approximate Biholomorphisms

4.7.1 "Approximate" biholomorphism: exhaustion

As noted in the introduction two randomly picked domains in $\mathbb{C}^n$ are likely to be non-equivalent (=non-biholomorphic). Can they be "approximately" equivalent? Let's make this question precise. Consider a sequence of bounded domains $\{D_k\}_{k=1}^\infty \subset \mathbb{C}^n$, such that all $D_k \subset D$ and $\lim_{k \to \infty} D_k = D$ in some topology of domains in $\mathbb{C}^n$. So, D can be approximated by D_k for large k. Suppose now that there is a domain $G \subset\subset \mathbb{C}^n$ such that each D_k is a biholomorphic image of G, $f_k : G \to D_k$ a biholomorphic map onto D_k. In this case we will say that D can be *exhausted* by G. So, G is approximately equivalent to D.

So, if given a bounded domain G in $\mathbb{C}^n$ what can it exhaust? The list $\Lambda(G)$ of these domains is of course a biholomorphic invariant. The range varies widely depending on G, and we'll mention several examples. By definition $\Lambda(G)$ contains G. We will show that if G is homogeneous then $\Lambda(G)$ consists only of G itself:

Theorem 4.7.1. *If G, D are bounded domains in $\mathbb{C}^n$, G is homogeneous and D can be exhausted by G (in the Hausdorff topology), then D is biholomorphic to G.*

By the way, since the ball and the polydisc in $\mathbb{C}^n$ for $n > 1$ are non-biholomorphic it follows that neither can exhaust the other and there should be "largest" imbedding of each into the other. The precise estimates have been obtained by H. Alexander [2], which can be considered as another proof of the Poincare's Example.

If G is a smooth, bounded strictly pseudoconvex domain, then $\Lambda(G)$ has only the following elements: G and the unit ball B^n [21]. Comparing this example with the above theorem shows that the notion of exhaustion is not symmetric: it happens that $D \in \Lambda(G)$ but $\Lambda(D)$ does not contain G.

For half-the ball in $\mathbb{C}^n$, $n \geq 2$, $S = \{z : z = (z_1, ..., z_n) \in B^n, \, Re(z_n) > 0\}$, $\Lambda(S)$ contains the unit ball and the polydisk [19]. This, by the way shows that S is not biholomorphically equivalent to any smooth, bounded strictly pseudoconvex domain (though, of course, this can be proved by other means).

The above theorem shows that a homogeneous domain exhausts only itself. On the other side of the spectrum is the following domain.

Theorem 4.7.2. *(Fridman [18]). There exists a universal domain $U \subset\subset \mathbb{C}^n$ which can exhaust any other bounded domain in $\mathbb{C}^n$ (in the Hausdorff topology).*

So, this domain is "almost" equivalent to any other domain in $\mathbb{C}^n$. Any domain/set that has the approximation property described in this theorem we will call a *universal* domain/set.

As we noticed at the start of this exposition, there is no version of the Riemann mapping theorem in $\mathbb{C}^n$. The above statement can be considered the approximate Riemann mapping theorem for any $\mathbb{C}^n$.

A short explicit construction one can find in [18], it will not be repeated here. We'll make a few remarks concerning the universal domain.

There is a great flexibility of constructing this domain. That is, such a domain can have the automorphism group isomorphic to $\mathbb{Z}_k$, for any $k \in \mathbb{N}$. Also one can construct many such non-biholomorphic domains; this gives an example of two domains that can exhaust each other but are non-equivalent.

There is a great variety of universal domains. And most of them have one curious property we are about to describe. The exposition is short and elementary, but the statements are useful when dealing with general biholomorphic mappings and therefore we are including them here. It would be helpful for the interested reader to know the construction before reading about the following property of U. It is as follows. There is a designated point $p \in \partial U$ such that to approximate a domain $G \subset \mathbb{C}^n$ with a given precision ϵ, one has to find a small δ and a biholomorphic imbedding $T : U \to G$, such that the δ-neighborhood $W_\delta = B(p, \delta) \cap U$ will "blow up" to cover most of G while T will squeeze everything else outside that neighborhood in U almost to a point. This will accomplish the goal: $T(W_\delta) \subset G$ will be ϵ-close to G, while the rest of $T(U \setminus W_\delta)$ will be small enough to not create a larger approximation mistake. We use the notation $B(p, \delta)$ for a ball at center p and radius δ.

So, by an obvious association we can call p a "source" of all domains in $\mathbb{C}^n$, and the described property a "Big Bang property". One can also express this property by stating that for any $\delta > 0$ the set $B(p, \delta) \cap U$ is also a universal domain. Is the existence of such a "source" necessary for a universal domain? We prove the following

Theorem 4.7.3. *Let U be a universal domain. If the boundary ∂U does not contain any complex analytic variety of dimension one then ∂U contains a "source", that is such a point $p \in \partial U$ that $B(p, \delta) \cap U$ is a universal set for any $\delta > 0$.*

Corollary 4.7.4. *If $n = 1$ then any universal domain has a "source" on the boundary.*

The proof of the Theorem is based on a generalization for several complex variables of a one-dimensional Hurvitz theorem. For completeness, we include the proof.

Lemma 4.7.5. *Let G, D be bounded domains in $\mathbb{C}^n$. Suppose that there is a sequence of domains $\{V_k\}$, $V_k \subset G$, and maps $F_k : V_k \to D$ such that*
    1. $F_k(V_k) = D$ and F_k is a biholomorphic mapping.
    2. For any compact $K \subset G$
    a) there exists a number m such that $V_s \supset K$, $s \geq m$, and
    b) the sequence $\{F_s\}$ for $s \geq m$ tends uniformly on K to a map $F : K \to$ $\mathbb{C}^n$.
    If $F(G)$ contains a point $z_0 \in D$ then F is a biholomorphism between G and D.

Proof. 1 Evidently $F(G) \subset \overline{D}$. We want to show that $F(G) \subset D$. Let $w_0 \in G$ be such a point that $F(w_0) = z_0$ and $\epsilon > 0$ be so small that the balls in Kobayashi's metric $B_1 = B_D(z_0, \epsilon) \subset\subset D$ and $B_2 = B_G(w_0, 2\epsilon) \subset\subset G$. Let $z \in B_1$ for a large enough k, $F_k(w_0) \in B_1$. Therefore, $F_k^{-1}(z) \in B_2$. Let w be the limit point of $\{F_k^{-1}(z)\}$. Evidently, $F(w) = z$. We have proved that $F(G) \supset B_1$. F is a limit of regular holomorphic mappings. Since G is connected F can be either regular at every point in G or the Jacobian of F vanishes on G. In the latter case $F(G)$ could not contain any open set (by Sard's theorem). Since $F(G) \supset B_1$, F is regular on G. This implies that F is an open mapping, so $F(G) \subset D$.

2. We will show now that F is one-to-one. Let $w', w'' \in G$. For a large number k and the Kobayashi metric ρ we have

$$
\begin{aligned}
\rho_G(w', w'') &= \rho_G(F_k^{-1} \circ F_k(w'), F_k^{-1} \circ F_k(w'')) \\
&\leq \rho_D(F_k(w'), F_k(w'')) \\
&\leq \rho_D(F_k(w'), F(w')) + \rho_D(F(w'), F(w'')) + \rho_D(F(w''), F_k(w'')).
\end{aligned}
$$

When $k \to \infty$ we obtain $\rho_G(w', w'')) \leq \rho_D(F(w'), F(w''))$. Hence, if $F(w') = F(w'')$ then $w' = w''$.

3. To finish the proof we have to show now that $F(G) \supset D$. Without any loss of generality we may assume, passing to a subsequence if necessary, that $\{F_k^{-1}\}$ converges uniformly on compacta to $f : D \to \overline{G}$. Repeating the first step of the proof for this mapping we obtain $f(D) \subset G$. For the mapping $F \circ f : D \to D$ and any $z \in D$ we have $F \circ f(z) = \lim_{k \to \infty} [F_k \circ F_k^{-1}(z)] = z$.

Hence, $F(G) \supset D$ completing the proof of the Lemma. $\square$

Proof of the Theorem 4.7.3: In [18], the existence of two universal holomorphically non-equivalent domains is proved. Let G be one of them that is not biholomorphically equivalent to U. G can be represented as $G = \overset{\infty}{\underset{k=1}{\cup}} V_k$ where open sets $V_k \subset V_{k+1} \subset\subset G$ for all k. Since U is universal there exists a sequence of biholomorphic imbeddings $f_k : U \to G$ such that $f_k(U) \supset V_k$. Consider now $U_k = f_k(U)$ and $F_k = f_k^{-1}$. Since $\{F_k\}$ is a sequence of bounded holomorphic maps we may assume, taking a subsequence if necessary, that $\{F_k\}$ converges uniformly on any compact $K \subset G$. Let $F = \lim F_k$, $F : G \to \overline{U}$. Since U is not equivalent to G, $F : G \to \partial U$ according to the lemma. Since ∂U does not contain any analytic curves, $F(G) = p$ is a point on ∂U. Let $\delta > 0$. We are going to prove that $U' = B(p, \delta) \cap U$ is a universal set. Let M be any domain in $\mathbb{C}^n$, K a compact in M. Since G is universal, there exists a biholomorphic imbedding $g : G \to M$ such that $g(G) \supset K$. Denote $K_1 = g^{-1}(K)$—compact in G. $\{F_k\}$ converges uniformly on K_1 to p. Therefore, there exists a number N such that $F_N(K_1) \subset U'$. Consider now $f_N = F_N^{-1} : U' \to G$ and $h = g \circ f_N : U' \to M$. According to the construction $h(U') \supset K$. This completes the proof of the Theorem. $\square$

Using the same Lemma, we now prove the Theorem 4.7.1.

Proof. Pick two points: $z_0 \in G$, $w_0 \in D$. Since G is homogeneous then we may assume that for all k, $f_k(z_0) = w_0$. Now, one can see that the sequence $\{f_k\}$ converges uniformly on compacta and its limit is $f : G \to D$ a biholomorphism.

$\square$

4.7.2   Upper semicontinuity of automorphism groups

It is a general geometric observation that "small perturbations can destroy symmetry but not create symmetry". In case of domains in $\mathbb{C}^n$, one may interpret this statement more precisely. Let $\{D_k\}_{k=1}^{\infty}$ be bounded domains in

$\mathbb{C}^n$ and this sequence tends to a bounded domain $D \subset \mathbb{C}^n$ in some topology on domains in $\mathbb{C}^n$. Loosely put, is it possible that $\text{Aut}(D_k)$ is "larger" than $\text{Aut}(D)$ for all k large enough? If that was possible, perturbation (in the given topology) of D of any small size can create domains with more symmetry. This question is referred to as non-semi-continuity property for automorphism groups.

We will mention here several statements and examples pertaining to the question; for more detailed discussion on semi-continuity and open questions see [22–24, 40].

In the early eighties, R. Greene and S.G. Krantz ([31–33]) examined this question. They proved an upper semicontinuity result in the C^2 topology, and gave the first counterexample to the upper semicontinuity principle. In [23, 24, 42] the semicontinuity question has been examined further for various other topologies. We also note here that in Riemannian geometry results of this kind have been obtained by various authors.

Here's the statement in the C^∞ topology: the semi-continuity holds in this case; for a more precise statement see [40, p. 232].

Theorem 4.7.6. *(Green, Krantz). Let $U \subset \mathbb{C}^n$ be a smoothly bounded, strongly pseudoconvex domain. Let U' be another smoothly bounded domain whose boundary is sufficiently close to ∂U in the C^∞ topology. Then $\text{Aut}(U')$ is a subgroup of $\text{Aut}(U)$.*

In [32], the same authors provide the following counterexample of the failure of upper semicontinuity in the $C^{1-\epsilon}$ topology (for any $\epsilon \in (0, 1)$).

Example 4.7.7. There are pseudoconvex domains $\{D_j\}_{j=1}^\infty$ and D_0, each of which is C^∞ and strongly pseudoconvex except at one point, such that $\text{Aut}(D_j) \neq \{id\}$ for all $j \geq 1$, $\text{Aut}(D_0) = \{id\}$, and $D_j \to D_0$ in the $C^{1-\epsilon}$ topology, any $\epsilon \in (0, 1)$.

The other statements and examples below are in the topology induced by the Hausdorff metric; we denote the corresponding metric space by H^n, space of all bounded domains in $\mathbb{C}^n$ with the metric equal to the Hausdorff distance between boundaries of domains.

Example 4.7.8. There is a bounded C^∞ domain D in $\mathbb{C}$ that is not simply connected, and such that for any neighborhood U of D in the Hausdorff metric there is a C^∞ domain $\tilde{D}$ in U such that $\text{Aut}(\tilde{D})$ is not isomorphic to any subgroup of $\text{Aut}(D)$.

Construction. $B(z_0, r)$ denotes an open disk with center z_0 and radius r. Consider an $(N + 1)$-connected domain $D = \Delta \setminus \bigcup_{s=1}^{N} \overline{\Delta}_s$ where $\Delta = B(0, 1)$ is the open unit disk and all $\overline{\Delta}_s$ are smaller non-intersecting closed disks, whose boundaries lie entirely in Δ. For a given $1 > \epsilon > 0$, fix a positive $\epsilon_1 \leq \epsilon$ and such that the set $S = \{z \in \Delta | Re(z) > -1 + \epsilon_1\}$ contains all $\overline{\Delta}_s$. Suppose a natural number $j > 1$ is also given. We now choose a positive δ

such that $L(S) \subset B(1, 1/2^j)$ where L is a Möbius transformation $L(z) = \frac{z+a}{1+za}$, and $a = 1 - \delta$. We observe that $L(\Delta) = \Delta$. Consider now $M = \overset{j-1}{\underset{k=0}{\cap}}(L(D) \cdot \exp(\frac{2\pi k}{j}i)$ (each term is a rotation of $L(D)$ by angle $\frac{2\pi k}{j}$). One can verify that by construction $\mathbb{Z}_j$ acts on M. We define $\tilde{D}_j = L^{-1}(M)$. Then $\tilde{D}_j \subset D$, and since $Aut(\tilde{D}_j)$ is isomorphic to $Aut(M)$, $\mathbb{Z}_j$ is isomorphic to a subgroup of $Aut(\tilde{D}_j)$. Also the difference $D \backslash \tilde{D}_j \subset \Delta \backslash S$ and therefore the Hausdorff distance between D and $\tilde{D}_j$ is less than ϵ. If $N \geq 2$ the group $Aut(D)$is finite. Since j could be chosen to be arbitrarily large, the statement has been proved.

Remarks. The above construction works for any finitely connected domain: in any neighborhood of this domain and any integer j there is a domain whose automorphism group contains $\mathbb{Z}_j$. A similar construction can be done in $\mathbb{C}^n$ for any $n \geq 1$.

Theorem 4.7.9. *(Fridman and Poletsky [23]) Let M be any domain in $\mathbb{C}^n$. Then there exists an increasing sequence of bounded domains $M_k \subset M_{k+1} \subset\subset M$ such that $M = \cup M_k$ and $\mathrm{Aut}(M_k)$ contains a subgroup isomorphic to $\mathbb{Z}_k$.*

This statement can be proved by using the remark in the first sub-section: for each $n \in \mathbb{N}$ there exists a universal domain whose automorphism group has a subgroup isomorphic to $\mathbb{Z}_n$.

So, for any domain (even a rigid one, i.e. with $Aut(D) = \{id\}$) one can make a perturbation of less than a given size and obtain a domain with a large cyclic group. The natural question arises: which Lie groups will a similar statement hold for? It will hold for any finite group:

Theorem 4.7.10. *Let G be a group of order $m < \infty$. For any $n \geq m$ the set of bounded domains in $\mathbb{C}^n$ whose automorphism group contains a subgroup isomorphic to G is everywhere dense in H^n.*

A detailed proof of this theorem is given in [22].

So arbitrarily small perturbation of a domain in $\mathbb{C}^n$ may create a domain with a larger automorphism group. But in provided examples the groups are discrete, of dimension zero. The natural question arises: can small perturbation in H^n create domains with larger *dimensions* of automorphism groups? The following answer is "no".

Theorem 4.7.11. *(Fridman, Ma, Poletsky [24]). The function $\dim(Aut(D))$ is upper semicontinuous on H^n.*

References

[1] D. Akhiezer, *Lie group actions in complex analysis*, Aspects of Mathematics, Vol. E27, Vieweg and Sohn, Braunschweig, 1995.

[2] H. Alexander, Holomorphic mappings from the ball and polydisc, *Math. Ann.*, 209 (1974), 249–256.

[3] E. Bedford, J. Dadok, Bounded domains with prescribed group of automorphisms, *Comment. Math. Helv.*, 62 (1987), 561–572.

[4] S.R. Bell, E. Ligocka, A simplification and extension of Fefferman's theorem on biholomorphic mappings, *Invent. Math.*, 57 (1980), 283–289.

[5] S. Bochner, D. Montgomery, Groups of differentiable and real or complex analytic transformations, *Ann. of Math.*, 46 (1945), 685–694.

[6] S. Bochner, D. Montgomery, Locally compact groups of differentiable transformations, *Ann. of Math.*, 47 (1946), 639–653.

[7] D. Burns, Jr, S. Shnider, O. Wells, Deformations of strictly pseudoconvex domains, *Invent. Math.*, 46 (1978), 237–253.

[8] H. Cartan, Les fonctions de deux variables complexeses et le problème de la représentation analytique, *J. Math. pures et appl.*, 9e série, 11 (1931) 1–114.

[9] H. Cartan, Sur les fonctions de plusieurs variables complexes, L'itération des transformations intérieures d'un domaine borné, *Math. Z.*, 35 (1932), 760–773.

[10] H. Cartan, Sur l'itération des transformations conformes ou pseudoconformes, *Compositio Mathematica*, 1 (1935), 223–227.

[11] H. Cartan, *Les groupes de transformations analytiques*, Actualités Sci. Indust., No. 198, Hermann, Paris, 1935.

[12] S.S. Chern, J.K. Moser, Real hypersurfaces in complex manifolds, *Acta Math.*, 133 (1974), 219–271.

[13] D. van Dantzig, B.L. van der Waerden, Über metrisch homogene Räume, *Abh. Math. Sem. Univ. Hamburg*, 6 (1928), 374–376.

[14] C. Fefferman, The Bergman kernel and biholomorphic mappings of pseudoconvex domains, *Invent. Math.*, 26 (1974), 1–65.

[15] F. Forstneric, An elementary proof of Fefferman's theorem, *Exposition. Math.*, 10 (1992), 135–150.

[16] S. Frankel, Complex geometry of convex domains that cover varieties, *Acta Math.*, 163 (1989), 109–149.

[17] S. Frankel, Locally symmetric and rigid factors for complex manifolds via harmonic maps, *Ann. of Math.*, 141 (1995), 285–300.

[18] B. Fridman, A Universal Exhausting Domain, *Proc. AMS*, 98 (1986), 267–270.

[19] B. Fridman, An Approximate Riemann Mapping Theorem in $\mathbb{C}^n$, *Math. Ann*, 275 (1986), 49–55.

[20] B. Fridman, One example of the boundary behavior of biholomorphic transformations, *Proc. AMS*, 89 (1983), 226–228.

[21] B. Fridman, Biholomorphic invariants of a hyperbolic manifold and some applications, *Trans. Amer. Math. Soc.*, 276 (1983), no. 2, 685–698.

[22] B. Fridman, D. Ma, Perturbation of domains and automorphism groups, *J. Korean Math. Soc.*, 40 (2003), 487–501.

[23] B. Fridman, E. Poletsky, Upper semicontinuity of automorphism groups, *Math. Ann.*, 299 (1994), 615–628.

[24] B. Fridman, D. Ma, E. Poletsky, Upper semicontinuity of the dimensions of automorphism groups of domains in $\mathbb{C}^n$, *Amer. J. Math.*, 125 (2003), 289–299.

[25] B. Fridman et al, On fixed points and determining sets for holomorphic automorphisms, *Michigan Math. J.*, 50 (2002), 507–515.

[26] B. Fridman et al, On Determining Sets for Holomorphic Automorphisms, *Rocky Mountain Journal of Mathematics*, 36 (2006), 947–956.

[27] B. Fridman, D. Ma, Properties of Fixed Point Sets and a Characterization of the Ball in $\mathbb{C}^n$, *Proc. AMS*,135 (2007), 229–236.

[28] B. Fridman, D. Ma, J.-P. Vigué, Isolated fixed point sets for holomorphic maps, *J. Math. Pures. Appl.*, 86 (2006), 80–87.

[29] B. L. Fridman, D. Ma, J.-P. Vigué, Fixed Points and Determining Sets for Holomorphic Self-Maps of a Hyperbolic Manifold, *Michigan Math J.*, 55 (2007), 229–238.

[30] R. Greene and S. Krantz, Invariants of Bergman geometry and the automorphism groups of domains in $\mathbb{C}^n$, *Geometrical and Algebraic Aspects in Several Complex Variables (Cetraro, 1989)*, Sem. Conf. 8, EditEl, Rende, 1991, 107–136.

[31] R. Greene and S. Krantz, The Automorphism groups of strongly pseudo-convex domains, *Math. Ann.*, 261 (1982), 425–446.

[32] R. Greene and S. Krantz, Stability of the Caratheodory and Kobayashi metrics and applications to biholomorphic mappings, *Proceedings of Symposia in Pure Math., Providence: AMS*, 41 (1984), 77–94.

[33] R. Greene and S. Krantz, Normal Families and the Semicontinuity of Isometry and Automorphism Groups, *Math. Z.*, 190 (1985), 455–467.

[34] R. Gunning, H. Rossi, Analytic Functions of Several Complex Variables, Prentice-Hall, Inc., Englewood Cliffs, N.J. 1965.

[35] G. Henkin, An analytic polyhedron is not holomorphically equivalent to a strictly pseudoconvex domain,*Dokl. Akad. Nauk SSSR*, 210 (1973), 1026–1029.

[36] A. Isaev, *Lectures on the automorphism groups of Kobayashi-hyperbolic manifolds*, Lecture Notes in Mathematics, 1902, Springer, Berlin Heidelberg, 2007.

[37] M. Jarnicki, P. Pflug, *Invariant distances and metrics in complex analysis*, Walter de Gruyter & Co., Berlin, 1993.

[38] K. Kim, S. Krantz, Determining Sets and Fixed Points for Holomorphic Endomorphisms, *Contemporary Math.* 328 (2003), 239–246.

[39] S. Kobayshi, *Transformation groups in differential geometry*, Springer-Verlag, New York, 1972.

[40] S. Krantz, The automorphism groups of domains in complex space: a survey, *Quaestiones Mathematicae*, 36 (2013), 225–251.

[41] S. Krantz, *Geometric analysis of the Bergman kernel and metric*, Graduate Texts in Mathematics, 268, Springer, New York, 2013.

[42] D. Ma, Upper semicontinuity of isotropy and automorphism groups, *Math. Ann.*, 292 (1992), 533–545

[43] B. Maskit, The conformal group of a plane domain, *Amer. J. Math.*, 90 (1968), 718–722.

[44] A. Nadel, Semisimplicity of the group of biholomorphisms of the universal covering of a compact complex manifold with ample canonical bundle, *Ann. Math.*, 132 (1990), 193–211.

[45] E. Peschl and M. Lehtinen, A conformal self-map which fixes 3 points is the identity, *Ann. Acad. Sci. Fenn., Ser. A I Math.*, 4 (1979), 85–86.

[46] C. Pommerenke, *Boundary behaviour of conformal maps*, Springer-Verlag, 1992.

[47] J.-P. Rosay, Sur une caracterisation de la boule parmi les domaines de $\mathbb{C}^n$ par son groupe d'automorphismes, *Ann. Inst. Fourier (Grenoble)*, 29 (1979), 91–97.

[48] H.L. Royden, Remarks on the Kobayshi metric, Lecture Notes in Mathematics, 175, Springer, Berlin Heidelberg, 1971.

[49] W. Rudin, *Function theory in the unit ball of* $\mathbb{C}^n$, Springer-Verlag, 2008.

[50] W. Rudin, *Lectures on the edge-of-the-wedge theorem*, CBMS Regional Conference Ser. in Math., 6 in Amer. Math. Soc., Providence, RI, 1971.

[51] R. Saerens, W. Zame, The isometry groups of manifolds and the automorphism groups of domains, *Trans. Amer. Math. Soc.* 301 (1987), 1, 413–429.

[52] F.H. Schottky, Über konforme Abbildung von mehrfach zusammenhängenden Fläche, *J. für Math.*, 83, (1877).

[53] N. Tanaka, On the pseudo-conformal geometry of hypersurfaces of the space of n complex variables, *J. Math. Soc. Japan* 14 (1962), 397–429.

[54] J.-P. Vigué, Sur les ensembles d'unicité pour les automorphismes analytiques d'un domaine borné, *C. R. Acad. Sci. Paris, Ser. I*, 336 (2003), 589–592.

[55] J.-P. Vigué, Ensembles d'unicité pour les automorphismes et les endomorphismes analytiques d'un domaine borné, *Annales Institut Fourier*, 55 (2005), 147–159.

[56] J.-P. Vigué, Points fixes d'applications holomorphes dans un domaine born convexe de $\mathbb{C}^n$, *Trans. Amer. Math. Soc.*, 289 (1985), 345–353.

[57] A. Vitushkin, V. Ezhov, N. Kruzhilin, Continuation of holomorphic mappings along real-analytic hypersurfaces, *Trudy Mat. Inst. Steklov*, 167 (1985), 60–95.

[58] N. Vormoor, Topologische Fortsetzung biholomorpher Funktionen auf dem Rande bei beschr″ankten streng-pseudokonvexen Gebieten im $\mathbb{C}^n$ mit C^∞-Rand, *Math. Ann.*, 204 (1973), 239–261.

[59] B. Wong, Characterization of the unit ball in $\mathbb{C}^n$ by its automorphism group, *Invent. Math.* 41 (1977), 253–257.

5

Positivity in the $\bar{\partial}$-Neumann Problem

Siqi Fu

CONTENTS

5.1 Introduction

Since the foundational work of Kohn on subelliptic theory of the $\bar{\partial}$-Neumann Laplacian on smoothly bounded strongly pseudoconvex domains ([Ko63, Ko64]) and that of Hörmander on the L^2-estimates of the $\bar{\partial}$-operator on bounded pseudoconvex domains in $\mathbb{C}^n$ ([H65]), it has been known that, unlike the classical Dirichlet or Neumann Laplacian, existence and regularity of the $\bar{\partial}$-Neumann Laplacian closely depend on geometry of the underlying domains (see the surveys [BSt99, Ch99, DK99, FS01] and the monographs [CS99, St09]). It is natural to expect that spectral behavior of the $\bar{\partial}$-Neumann Laplacian is also more sensitive to geometric properties of the underlying domains than the classical Laplacians.

In spectral analysis of self-adjoint differential operators, two problems stand out: discreteness and positivity of their spectra. Both are widely studied problems in physical sciences with important ramifications. Discrete spectra are seen in many physical phenomena such as light emission and string vibration, while positivity is related to whether a physical system has positive ground state energy.

Positivity of the classical Dirichlet Laplacian is well understood. As a consequence to the classical Hardy inequality, the bottom of the spectrum of the Dirichlet Laplacian on a domain in $\mathbb{R}^n$ that satisfies the outer cone condition is positive if and only if the inradius of the domain is finite (cf. [D95]). Whereas positivity of the Dirichlet Laplacian is not sensitive to boundary geometry of the underlying domain, positivity of the $\bar{\partial}$-Neumann Laplacian, as we will see, is.

In this Chapter, we study the $\bar{\partial}$-Neumann problem from spectral theoretic perspective. Our emphasis is on the interplay between spectral behavior of the $\bar{\partial}$-Neumann Laplacian and geometry of the underlying domains. This is evidently motivated by Marc Kac's famous question "Can one hear the shape of a drum?" ([Ka66]). Here we are interested in determining geometry of a domain in $\mathbb{C}^n$ from positivity of the $\bar{\partial}$-Neumann Laplacian. (See [Fu05, Fu08] for related results.)

The plan of this paper is as follows: In Section 5.2, we review the classical setup of the $\bar{\partial}$-Neumann problem. In Section 5.3, we present a spectral theoretic setup for the $\bar{\partial}$-Neumann Laplacian and prove that these two setups are equivalent, and as a consequence, we establish the self-adjointness of the $\bar{\partial}$-Neumannh Laplacian. In Section 5.4, we review the concept of pseudoconvexity. In Section 5.5, we provide a treatise of Hörmander's L^2-estimates of the $\bar{\partial}$-operator through the lens of spectral theory. In Section 5.6, we prove the converse to Hörmander's theorem.

We have made an effort to present a treatment that is accessible and self-contained, modulus extensive usage of spectral theorems for self-adjoint operators. An excellent treatise on spectral theory of differential operators can be in Davies' book ([D95]).

5.2   The $\bar{\partial}$-Neumann Laplacian

In this section, we review the classical setup of the $\bar{\partial}$-Neumann problem. In the subsequent section, we will present a spectral theoretic setup for the $\bar{\partial}$-Neumann Laplacian and prove that these two setups are equivalent. As a consequence, we establish the self-adjointness of the $\bar{\partial}$-Neumannh Laplacian.

5.2.1   The Cauchy-Riemann operator

The $\bar{\partial}$-operator, also known as the Cauchy-Riemann operator, is arguably the most important differential operator in complex analysis. Let

$$\frac{\partial}{\partial z_j} = \frac{1}{2}\left(\frac{\partial}{\partial x_j} - i\frac{\partial}{\partial y_j}\right) \quad \text{and} \quad \frac{\partial}{\partial \bar{z}_j} = \frac{1}{2}\left(\frac{\partial}{\partial x_j} + i\frac{\partial}{\partial y_j}\right).$$

Let Ω be a domain in $\mathbb{C}^n$ and let $f \in C^\infty(\Omega)$. Define

$$\bar{\partial}f = \sum_{j=1}^n \frac{\partial f}{\partial \bar{z}_j}d\bar{z}_j \quad \text{and} \quad \partial f = \sum_{j=1}^n \frac{\partial f}{\partial z_j}dz_j.$$

Thus $\bar{\partial}f: C^\infty(\Omega) \to C^\infty_{(0,1)}(\Omega)$ and $\partial f: C^\infty(\Omega) \to C^\infty_{(1,0)}(\Omega)$. It is easy to check that the exterior differential operator $d = \partial + \bar{\partial}$. Let $\varphi = \sum_{j=1}^n \varphi_j d\bar{z}_j \in \mathcal{D}_{0,1}(\Omega)$, the space of $(0,1)$-forms with smooth coefficients compactly supported on Ω. Then

$$\langle \bar{\partial}f, \varphi \rangle = \sum_{j=1}^n \langle \frac{\partial f}{\partial \bar{z}_j}, \varphi_j \rangle = - \sum_{j=1}^n \langle f, \frac{\partial \varphi_j}{\partial z_j} \rangle = \langle f, \vartheta\varphi \rangle,$$

where

$$\vartheta\varphi = - \sum_{j=1}^n \frac{\partial \varphi_j}{\partial z_j}$$

is the formal adjoint of $\bar{\partial}$. Let $L^{loc}_{(0,1)}(\Omega)$ be the space of $(0,1)$-forms with locally integrable coefficients. Let $f \in L^{loc}(\Omega)$ and $g \in L^{loc}_{(0,1)}(\Omega)$. We say that

$$\bar{\partial}f = g$$

in the sense of distribution if

$$\langle f, \vartheta\varphi \rangle = \langle g, \varphi \rangle, \quad \forall \varphi \in \mathcal{D}_{(0,1)}(\Omega).$$

A locally integrable function f is *holomorphic* on Ω if $\bar{\partial}f = 0$ in the sense of distribution. Such a function is necessarily smooth. In fact, it has power series representation at any point in Ω.

Let

$$L^2_{(0,1)}(\Omega) = \{f = \sum_{j=1}^n f_j d\bar{z}_j \mid f_j \in L^2(\Omega)\}$$

be the space of $(0,1)$-forms with L^2-coefficients, equipped with the standard Hermitian inner product with the corresponding norm given by

$$\|f\|^2 = \sum_{j=1}^n \int_\Omega |f_j|^2 \, dV.$$

Note that here we use the convention that under the standard Hermitian metric on $\mathbb{C}^n$, $\langle dz_j, dz_k \rangle = \delta_{jk}$ where $\delta_{jk} = 1$ if $j = k$ and $\delta_{jk} = 0$ if $j \neq k$.

We now extend the $\bar{\partial}$-operator to be a densely defined closed operator

$$\bar{\partial}: L^2(\Omega) \to L^2_{(0,1)}(\Omega)$$

with

$$\text{Dom}\,(\bar{\partial}) = \{f \in L^2(\Omega) \mid \exists C > 0, |\langle f, \vartheta\varphi \rangle| \le C\|\varphi\|, \; \forall \varphi \in \mathcal{D}_{(0,1)}(\Omega)\}.$$

Namely, the domain of $\bar{\partial}$ is the space of all L^2-functions f such that $\bar{\partial}f \in L^2_{(0,1)}(\Omega)$ in the sense of distribution.

The $\bar{\partial}$-operator acts on (p,q)-forms, $0 \le p,q \le n$, is defined as follows. Let $u \in C^\infty_{(p,q)}(\Omega)$. Write

$$u = \frac{1}{p!q!} \sum_{I,J} u_{IJ} dz_I \wedge d\bar{z}_J = \sideset{}{'}\sum_{I,J} u_{I\bar{J}} dz_I \wedge d\bar{z}_J. \tag{5.1}$$

where $I = (i_1,\dots,i_p)$ is a p-tuple and $J = (j_1,\dots,j_q)$ a q-tuple of integers in $[1,\ n]$, $dz_I = dz_1 \wedge \dots \wedge z_{i_p}$, $d\bar{z}_J = d\bar{z}_{j_1} \wedge \dots \wedge d\bar{z}_{j_q}$, and the prime indicates that the sum is taken over strictly increasing tuples. Then $\bar{\partial}_q : C^\infty_{(p,q)}(\Omega) \to C^\infty_{(p,q+1)}(\Omega)$ is given by

$$\bar{\partial}_q u = \sideset{}{'}\sum_{I,J} \bar{\partial}u_{I\bar{J}} \wedge dz_I \wedge d\bar{z}_J.$$

(We may suppress the subscript q when it is clear from the context.) Let $\mathcal{D}_{(p,q)}(\Omega)$ be the space of (p,q)-forms with smooth compactly supported coefficients. For $\varphi \in \mathcal{D}_{(p,q)}(\Omega)$, the formal adjoint $\vartheta_q : \mathcal{D}_{(p,q+1)}(\Omega) \to \mathcal{D}_{(p,q)}(\Omega)$ of $\bar{\partial}_q$ is given by

$$\vartheta_q u = (-1)^{p+1} \sum_{k=1}^n \sideset{}{'}\sum_{I,K} \frac{\partial u_{I,kK}}{\partial z_k} dz_I \wedge d\bar{z}_K.$$

where K runs over all strictly increasing q-tuples.

Exercise 5.2.1 Show that

$$\langle \bar{\partial}_q f, \varphi \rangle = \langle f, \vartheta_q \varphi \rangle, \quad \forall f \in C^\infty_{(p,q)}(\Omega), \; \varphi \in \mathcal{D}_{(p,q+1)}(\Omega).$$

Let $L^{loc}_{(p,q)}(\Omega)$ be the space of (p,q)-forms with locally integrable coefficients. Let $u \in L^{loc}_{(p,q)}(\Omega)$ and $v \in L^{loc}_{(p,q+1)}(\Omega)$. We say that

$$\bar{\partial}_q u = v \tag{5.2}$$

in the sense of distribution if

$$\langle v, \varphi \rangle = \langle u, \vartheta_q \varphi \rangle, \quad \forall \varphi \in \mathcal{D}_{(p,q+1)}(\Omega). \tag{5.3}$$

Let $L^2_{(p,q)}(\Omega)$ be the space of (p,q)-forms u with L^2-coefficients, equipped with the standard Euclidean metric:

$$\|u\|^2 = \frac{1}{p!q!} \sum_{I,J} \|u_{IJ}\|^2 = \sideset{}{'}\sum_{I,J} \|u_{IJ}\|^2$$

for u given by (5.1).

The $\bar{\partial}$-*problem* is to solve the $\bar{\partial}$-equation (5.2) and to study existence and regularity of the solutions.

Definition 5.2.2 The operator $\bar{\partial}_q \colon L^2_{(p,q)}(\Omega) \to L^2_{(p,q+1)}(\Omega)$ is defined in the sense of distribution. More precisely,

$$\mathrm{Dom}\,(\bar{\partial}_q) = \big\{ u \in L^2_{(p,q)}(\Omega) \mid \exists C > 0, |\langle u, \vartheta\varphi \rangle| \leq C\|\varphi\|, \tag{5.4}$$
$$\forall \varphi \in \mathcal{D}_{(p,q+1)}(\Omega) \big\}$$

and hence by the Riesz representation theorem, there exists $v \in L^2_{(p,q)}(\Omega)$ such that $\bar{\partial}_q u = v$ in the sense of distribution as defined by (5.3).

Lemma 5.2.3 (1) *The operator $\bar{\partial}_q \colon L^2_{(p,q)}(\Omega) \to L^2_{(p,q+1)}(\Omega)$ is densely defined and closed.* (2) $\bar{\partial}_{q+1} \circ \bar{\partial}_q = 0$.

Proof: (1) Since $\mathcal{D}_{(p,q)}(\Omega) \subset \mathrm{Dom}\,(\bar{\partial}_q) \subset L^2_{(p,q)}(\Omega)$ and $\mathcal{D}_{(p,q)}(\Omega)$ is dense in $L^2_{(p,q)}(\Omega)$, so is $\mathrm{Dom}\,(\bar{\partial}_q)$. Hence, $\bar{\partial}_q$ is densely defined in $L^2_{(p,q)}(\Omega)$. Suppose that $u_j \in \mathrm{Dom}\,(\bar{\partial}_q)$ and $u_j \to u$ in $L^2_{(p,q)}(\Omega)$ and $\bar{\partial}_q u_j \to v$ in $L^2_{(p,q+1)}(\Omega)$. Then for any $\varphi \in \mathcal{D}_{(p,q+1)}(\Omega)$,

$$\langle u, \vartheta_q\varphi \rangle = \lim_{j \to \infty} \langle u_j, \vartheta_q\varphi \rangle = \lim_{j \to \infty} \langle \bar{\partial}_q u_j, \varphi \rangle = \langle v, \varphi \rangle.$$

Thus, $u \in \mathrm{Dom}\,(\bar{\partial}_q)$ and $\bar{\partial}_q u = v$. Therefore, $\bar{\partial}_q$ is closed.

(2) This follows from

$$d \circ d = (\bar{\partial}_{q+1} + \partial_{q+1}) \circ (\bar{\partial}_q + \partial_q)$$
$$= \bar{\partial}_{q+1} \circ \bar{\partial}_q + \bar{\partial}_{q+1} \circ \partial_q + \partial_{q+1} \circ \bar{\partial}_q + \partial_{q+1} \circ \partial_q = 0$$

and form degrees consideration. ∎

Since here we will work only in $\mathbb{C}^n$ with the flat Euclidean metric. There is no difference between the theory for (p, q)-forms and that for $(0, q)$-forms. For simplicity of notation, we will restrict ourselves to $(0, q)$-forms.

Definition 5.2.4 The operator $\bar{\partial}_q^* \colon L^2_{(0,q+1)}(\Omega) \to L^2_{(0,q)}(\Omega)$ is the adjoint of $\bar{\partial}_q$. Thus its domain is given by

$$\mathrm{Dom}\,(\bar{\partial}_q^*) = \big\{ u \in L^2_{(0,q+1)}(\Omega) \mid \exists C > 0, |\langle u, \bar{\partial}_q v \rangle| \leq C\|v\|, \tag{5.5}$$
$$\forall v \in \mathrm{Dom}\,(\bar{\partial}_q) \big\}.$$

As we will show shortly, $u \in \mathrm{Dom}\,(\bar{\partial}_q^*)$ imposes a boundary condition on u when Ω has C^1-smooth boundary $b\Omega$ and $u \in C^1_{(0,q)}(\overline{\Omega})$.

REMARK 5.2.5 The $\bar{\partial}_q$-operator defined by Definition 5.2.2 can be considered as the adjoint of the formal adjoint $\vartheta_q \colon L^2_{(p,q+1)}(\Omega) \to L^2_{(p,q)}(\Omega)$ whose domain

$\text{Dom}(\vartheta_q) = \mathcal{D}_{(p,q+1)}(\Omega)$. It is sometimes referred as the *maximal extension* of $\bar{\partial}_q \colon \mathcal{D}_{(p,q)}(\Omega) \to \mathcal{D}_{(p,q+1)}(\Omega)$. The $\bar{\partial}_q^*$-operator, the adjoint of $\bar{\partial}$, defined by Definition 5.2.4, is then the closure of ϑ_q. It is sometimes referred as the *minimal extension* of ϑ_q.

5.2.2   Integration by parts

Let Ω be a bounded domain in $\mathbb{R}^n$ with C^1-boundary. Let $r(x)$ be a *defining function* for Ω, i.e., $r \in C^1(\mathbb{R}^n)$, $\Omega = \{x \in \mathbb{R}^n \mid r(x) < 0\}$, and $dr \neq 0$ on $b\Omega$. The outward normal direction is then given by $\vec{n} = \nabla r / |\nabla r|$. By replacing r by $r/|\nabla r|$, we may assume that $|\nabla r| = 1$ on $b\Omega$. By the divergence theorem, we have for any $F = (F_1, \dots, F_n) \in C^1(\overline{\Omega})$,

$$\int_\Omega \nabla \cdot F \, dx = \int_{b\Omega} F \cdot \vec{n} \, dS. \tag{5.6}$$

Let $u, v \in C^1(\overline{\Omega})$. Applying (5.6) to F with $F_j = u\bar{v}$ and $F_k = 0$ for all $k \neq j$, we have

$$\int_\Omega \frac{\partial}{\partial x_j}(u\bar{v}) \, dx = \int_{b\Omega} u\bar{v} \frac{\partial r}{\partial x_j} \, dS.$$

Therefore,

$$\int_\Omega \frac{\partial u}{\partial x_j} \bar{v} \, dx = -\int_\Omega u \frac{\partial \bar{v}}{\partial x_j} \, dx + \int_{b\Omega} u \cdot \bar{v} \frac{\partial r}{\partial x_j} \, dS. \tag{5.7}$$

Now let $\Omega \subset\subset \mathbb{C}^n$ be a domain with C^1-boundary. Let $r = r(z)$ be a defining function of Ω such that $|dr| = 1$ on $b\Omega$.

The following formulas are complex versions of (5.7). Note that we will slightly abuse the notation and use $\langle \cdot, \cdot \rangle$ to denote both the pointwise inner product associated with the metric and the integral of the pointwise inner product over the domain.

Lemma 5.2.6 *Let $u, v \in C^1(\overline{\Omega})$. Then*

$$\left\langle \frac{\partial u}{\partial z_j}, v \right\rangle = -\left\langle u, \frac{\partial v}{\partial \bar{z}_j} \right\rangle + \int_{b\Omega} u \cdot \bar{v} \frac{\partial r}{\partial z_j} \, dS. \tag{5.8}$$

and

$$\left\langle \frac{\partial u}{\partial \bar{z}_j}, v \right\rangle = -\left\langle u, \frac{\partial v}{\partial z_j} \right\rangle + \int_{b\Omega} u \cdot \bar{v} \frac{\partial r}{\partial \bar{z}_j} \, dS. \tag{5.9}$$

To generalize the integration by parts formula to forms, it is convenient to study the contraction operator. For a k-form u and a vector ξ, the contraction $\xi \lrcorner u$ is a $(k-1)$-form defined by

$$\xi \lrcorner u(\eta_1, \dots, \eta_{k-1}) = u(\xi, \eta_1, \dots, \eta_{k-1}). \tag{5.10}$$

It is easy to check that

$$\xi \lrcorner (u \wedge v) = (\xi \lrcorner u) \wedge v + (-1)^{\deg u} u \wedge (\xi \lrcorner v).\qquad(5.11)$$

Let $I = (i_1, \ldots, i_p)$ and $J = (j_1, \ldots, j_q)$. Let K be a tuple of positive integers. We will use $J \setminus K$ to denote the tuple obtained by deleting entries in K from J. We have

$$\frac{\partial}{\partial z_j} \lrcorner dz_k = \delta_{jk}.$$

Moreover,

$$\frac{\partial}{\partial z_l} \lrcorner (dz_I \wedge d\bar z_J) = \begin{cases} 0, & l \notin I; \\ (-1)^{s-1} dz_{I\setminus\{l\}} \wedge d\bar z_J, & l = i_s, \end{cases}$$

and

$$\frac{\partial}{\partial \bar z_l} \lrcorner (dz_I \wedge d\bar z_J) = \begin{cases} 0, & l \notin J; \\ (-1)^{p+s-1} dz_I \wedge d\bar z_{J\setminus\{l\}}, & l = j_s. \end{cases}$$

For a vector ξ, we define its dual form by

$$\xi^* = \langle \cdot, \xi \rangle,$$

where $\langle \cdot, \cdot \rangle$ denotes the pointwise standard Hermitian inner product on $\mathbb{C}^n$. Thus if

$$\xi = \sum_{j=1}^n \left(a_j \frac{\partial}{\partial z_j} + b_j \frac{\partial}{\partial \bar z_j} \right),$$

then

$$\xi^* = \sum_{j=1}^n \left(\bar a_j dz_j + \bar b_j d\bar z_j \right).$$

Similarly, we can define the dual vector of a 1-form. Let ξ be a $(0,1)$-vector, it follows from straightforward computation that for any (p,q)-form u and $(p, q-1)$-form v,

$$\langle \xi \lrcorner u, v \rangle = \langle u, \xi^* \wedge v \rangle.$$

Given a (p,q)-form u and (p',q')-form f with $p' \geq p$, $q' \geq q$, the interior product $u \vee f$ is a $(p'-p, q'-q)$-form defined by

$$\langle u \vee f, g \rangle = \langle f, u \wedge g \rangle,$$

for any $(p'-p, q'-q)$-form g. Therefore, we have

$$\xi \lrcorner u = \xi^* \vee u.$$

Using these notations, we have:

$$\bar{\partial} = \sum_{j=1}^{n} d\bar{z}_j \wedge \frac{\partial}{\partial \bar{z}_j},$$

$$\vartheta = -\sum_{j=1}^{n} \frac{\partial}{\partial \bar{z}_j} \lrcorner \frac{\partial}{\partial z_j}.$$

Lemma 5.2.7 *Let $u \in C^1_{(0,q)}(\overline{\Omega})$. Then*

$$\langle u, \bar{\partial}\varphi \rangle = \langle \vartheta u, \varphi \rangle + \int_{b\Omega} \langle (\bar{\partial}r)^* \lrcorner u, \varphi \rangle \, dS \qquad (5.12)$$

Proof: This follows from simple computations:

$$\langle u, \bar{\partial}\varphi \rangle = \langle u, \sum_{j=1}^{n} d\bar{z}_j \wedge \frac{\partial}{\partial \bar{z}_j}\varphi \rangle = \sum_{j=1}^{n} \langle \frac{\partial}{\partial \bar{z}_j} \lrcorner u, \frac{\partial}{\partial \bar{z}_j}\varphi \rangle$$

$$= \langle \vartheta u, \varphi \rangle + \sum_{j=1}^{n} \int_{b\Omega} \langle \frac{\partial}{\partial \bar{z}_j} \lrcorner u, \varphi \rangle \frac{\partial r}{\partial z_j} \, dS$$

$$= \langle \vartheta u, \varphi \rangle + \int_{b\Omega} \langle (\bar{\partial}r)^* \lrcorner u, \varphi \rangle \, dS.$$

∎

As a consequence of Lemma 5.2.7, we have:

Lemma 5.2.8 *Let $\Omega \subset\subset \mathbb{C}^n$ be a bounded domain with a C^1-smooth defining function r. Write*

$$T = (\bar{\partial}r)^* = \sum_{j=1}^{n} \frac{\partial r}{\partial z_j} \frac{\partial}{\partial \bar{z}_j}.$$

Let $u \in C^1_{(0,q)}(\overline{\Omega})$. Then $u \in \mathrm{Dom}\,(\bar{\partial}^*)$ if and only if $T \lrcorner u = 0$ on $b\Omega$, which is

$$\sum_{k=1}^{n} u_{kK} \frac{\partial r}{\partial z_k} = 0 \qquad (5.13)$$

on $b\Omega$ for every $(q-1)$-tuple K. In this case, $\bar{\partial}^* u = \vartheta u$.

Definition 5.2.9 For $1 \le q \le n-1$, the $\bar{\partial}$-Neumann Laplacian $\square_q \colon L^2_{(0,q)}(\Omega) \to L^2_{(0,q)}(\Omega)$ is given by

$$\square_q = \bar{\partial}_{q-1}\bar{\partial}^*_{q-1} + \bar{\partial}^*_q\bar{\partial}_q$$

with

$$\mathrm{Dom}\,(\square_q) = \{u \in L^2_{(0,q)}(\Omega) \mid u \in \mathrm{Dom}\,(\bar{\partial}^*_{q-1}),\ \bar{\partial}^*_{q-1}u \in \mathrm{Dom}\,(\bar{\partial}_{q-1}),$$

$$u \in \mathrm{Dom}\,(\bar{\partial}_q),\ \bar{\partial}_q u \in \mathrm{Dom}\,(\bar{\partial}^*_q)\}.$$

On the bottom degree forms, the $\bar{\partial}$-Neumann Laplacian is given by $\Box_0 = \bar{\partial}_0^* \bar{\partial}_0$ with

$$\text{Dom}(\Box_0) = \{u \in L^2(\Omega) \mid u \in \text{Dom}(\bar{\partial}_0), \bar{\partial}_0 u \in \text{Dom}(\bar{\partial}_0^*)\}$$

and on the top degree forms $\Box_n = \bar{\partial}_{n-1} \bar{\partial}_{n-1}^*$ with

$$\text{Dom}(\Box_n) = \{u \in L^2_{(0,n)}(\Omega) \mid u \in \text{Dom}(\bar{\partial}_{n-1}^*), \bar{\partial}_{n-1}^* u \in \text{Dom}(\bar{\partial}_{n-1})\}.$$

The $\bar{\partial}$-*Neumann problem* is to study existence and regularity of solution to the equation

$$\Box u = f. \tag{5.14}$$

REMARK 5.2.10 Suppose $\Omega = \{r(z) < 0\}$ be a bounded domains with C^1-boundary. By Lemma 5.2.8, if $u \in C^2(\overline{\Omega})$, then $f \in \text{Dom}(\Box_0)$ if and only if

$$\sum_{j=1}^{n} \frac{\partial u}{\partial \bar{z}_j} \frac{\partial r}{\partial z_j} = 0$$

on $b\Omega$. This boundary condition resembles the Neumann boundary condition but they are not the same. For $u = f d\bar{z}_1 \wedge \ldots \wedge d\bar{z}_n$ with $f \in C^2(\overline{\Omega})$, $u \in \text{Dom}(\Box_n)$ if and only if $u = 0$ on $b\Omega$, which is exactly the Dirichlet boundary condition.

By Lemma 5.2.8, for a smooth form u in $\text{Dom}(\bar{\partial}^*)$, $\bar{\partial}^* u = \vartheta u$. Therefore, on a smooth form that is in the domain of $\Box$, we have

$$\Box = \bar{\partial}\bar{\partial}^* + \bar{\partial}^*\bar{\partial} = \bar{\partial}\vartheta + \vartheta\bar{\partial}$$

$$= -\sum_{j,l} d\bar{z}_l \wedge \frac{\partial}{\partial \bar{z}_l}(\frac{\partial}{\partial \bar{z}_j} \lrcorner \frac{\partial}{\partial z_j}) - \sum_{j,l} \frac{\partial}{\partial \bar{z}_j} \lrcorner \frac{\partial}{\partial z_j}(d\bar{z}_l \wedge \frac{\partial}{\partial \bar{z}_l})$$

$$= -\sum_{j} \frac{\partial^2}{\partial z_j \partial \bar{z}_j} = -\frac{1}{4}\Delta.$$

Therefore, as a differential operator, the $\bar{\partial}$-Neumann Laplacian in $\mathbb{C}^n$ is just a constant multiple of the usually Laplacian, acting coefficient-wise on forms. (Indeed, as we remarked earlier, on top degree forms, it is one-fourth of the Dirichlet Laplacian.) What distinguishes it from the classical Dirichlet or Neumann Laplacian is *the $\bar{\partial}$-Neumann boundary conditions*:

$$u \in \text{Dom}(\bar{\partial}^*) \quad \text{and} \quad \bar{\partial}u \in \text{Dom}(\bar{\partial}^*). \tag{5.15}$$

It does not follow readily from the above definition whether the $\bar{\partial}$-Neumann Laplacian is densely defined or self-adjoint. This will be answered in the next section after we introduce a spectral theoretic setup for the $\bar{\partial}$-Neumann Laplacian.

5.3   Spectral Theoretic Setup

We now present the spectral theoretic setup for the $\bar{\partial}$-Neumann Laplacian. We will show that the $\bar{\partial}$-Neumann Laplacian defined by its associated quadratic form is consistent with the one defined through Definition 5.2.9. As a consequence, we establish the self-adjoint property of the $\bar{\partial}$-Neumann Laplacian and show that the domain of its square root is the same as that of its associated quadratic form. We will use $\mathcal{R}(T)$ and $\mathcal{N}(T)$ to denote respectively the range and kernel of the operator T.

Lemma 5.3.1 *Let $T \colon \mathbb{H}_1 \to \mathbb{H}_2$ be a densely defined and closed operator and let T^* be its adjoint. Then the following are equivalent:*

  1. $\mathcal{R}(T)$ is closed.

  2. There exists a positive constant C such that $\|f\| \leq C\|Tf\|$ for all $f \in \mathrm{Dom}\,(T) \cap \mathcal{N}(T)^{\perp}$.

  3. $\mathcal{R}(T^*)$ is closed.

  4. There exists a positive constant C such that $\|f\| \leq C\|T^*f\|$ for all $f \in \mathrm{Dom}\,(T^*) \cap \mathcal{N}(T^*)^{\perp}$.

Proof: We first prove the implication $(1) \Rightarrow (2)$. In this case, $T \colon \mathrm{Dom}\,(T) \cap \mathcal{N}(T)^{\perp} \to \mathcal{R}(T)$ is a bijective closed map. Its inverse is also a closed map from the closed subspace $\mathcal{R}(T)$ into $\mathbb{H}_1$. Applying the closed graph theorem, we thus have (2). We now prove $(2) \Rightarrow (1)$. Suppose $Tf_j \to g$. Write $f_j = f_j^1 + f_j^2$ where $f_j^1 \in \mathcal{N}(T)$ and $f_j^2 \in \mathcal{N}(T)^{\perp}$. Then by (2), f_j^2 is a Cauchy sequence in $\mathbb{H}_1$. Assume $\lim_{j \to \infty} f_j^2 = f$. Then $Tf = g \in \mathcal{R}(T)$ since T is closed. The proof of $(3) \Leftrightarrow (4)$ follows the same lines as that of $(1) \Leftrightarrow (2)$.

   We proceed to prove $(2) \Rightarrow (4)$. For any $f \in \mathrm{Dom}\,(T) \cap \mathcal{N}(T)^{\perp}$ and $g \in \mathrm{Dom}\,(T^*) \cap \mathcal{N}(T^*)^{\perp}$,

$$|\langle g, Tf \rangle| = |\langle T^*g, f \rangle| \leq \|T^*g\| \|f\| \leq C\|T^*g\| \|Tf\|.$$

Since $g \in \mathcal{N}(T^*)^{\perp} = \overline{\mathcal{R}(T)}$, the above inequality implies that (4). The proof of the implication $(4) \Rightarrow (2)$ is similar.   ∎

Lemma 5.3.2 *Let $T \colon \mathbb{H}_1 \to \mathbb{H}_2$ be a densely defined closed operator. Let F be a closed subspace of $\mathbb{H}_2$ such that $F \supset \mathcal{R}(T)$. Then $\mathcal{R}(T) = F$ if and only if there exists an positive constant C such that*

$$\|f\| \leq C\|T^*f\|, \quad \forall f \in \mathrm{Dom}\,(T^*) \cap F.$$

In this case, for any $g \in \mathcal{N}(T)^{\perp}$, there exists $f \in \mathrm{Dom}\,(T^*)$ such that $\bar{\partial}^* f = g$ and $\|f\| \leq C\|g\|$.

Proof: We first prove the forward implication. It suffices to prove that

$$B = \{f \in \mathrm{Dom}\,(T^*) \cap F \mid \|T^*f\| \le 1\}$$

is a bounded set in $\mathbb{H}_2$. Let $f \in B$ and let $g \in F$. Assume that $g = Th$ for some $h \in \mathrm{Dom}\,(T)$. Then

$$|\langle f, g \rangle| = |\langle f, Th \rangle| = |\langle T^*f, h \rangle| \le \|h\| < \infty.$$

Thus by the uniform boundedness theorem, the set B is bounded.

We now prove the backward direction. Let $g \in F$. Consider the linear functional $T^*f \mapsto \langle f, g \rangle$ on $\mathcal{R}\,(T^*)$. To prove that this is well-defined, it suffices to show

$$|\langle f, g \rangle| \le C\|T^*f\|\|g\|, \quad \forall f \in \mathrm{Dom}\,(T^*).$$

This obviously holds when $f \in F^\perp \subset \mathcal{R}\,(T)^\perp = \mathcal{N}\,(T^*)$. When $f \in F$, this follows from the Schwarz inequality and the assumption. By the Hahn-Banach theorem, there exists an $h \in \mathbb{H}_1$ such that $\langle f, g \rangle = \langle T^*f, h \rangle$. Therefore, $g = Th$. Furthermore, since $F^\perp \subset \mathcal{N}\,(T^*)$, we have $\mathcal{R}\,(T^*) = \mathcal{R}\,(T^*_{F \cap \mathrm{Dom}\,(T^*)})$, which is closed by the assumption. The last statement in the lemma then follow from Lemma 5.3.1. ∎

Let Q be a non-negative, densely defined, and closed sesquilinear form on a complex Hilbert space $\mathbb{H}$ with domain $\mathrm{Dom}\,(Q)$. Then Q uniquely determines a non-negative and self-adjoint operator S such that $\mathrm{Dom}\,(S^{1/2}) = \mathrm{Dom}\,(Q)$ and

$$Q(u,v) = \langle S^{1/2}u,\ S^{1/2}v \rangle \tag{5.16}$$

for all $u, v \in \mathrm{Dom}\,(Q)$ (see, for example, Theorem 4.42 in [D95]).

For any subspace $L \subset \mathrm{Dom}\,(Q)$, let $\lambda(L) = \sup\{Q(u,u) \mid u \in L, \|u\| = 1\}$. For any positive integer j, let

$$\lambda_j(Q) = \inf\{\lambda(L) \mid L \subset \mathrm{Dom}\,(Q), \dim(L) = j\}. \tag{5.17}$$

Since S is self-adjoint and non-negative, its spectrum $\sigma(S)$ is a non-empty closed subset of $[0,\ \infty)$. The bottom of the spectrum is given by $\inf \sigma(S) = \lambda_1(Q)$. The essential spectrum $\sigma_e(S)$ is a closed subset of $\sigma(S)$ that consists of isolated eigenvalues of infinite multiplicity and accumulation points of the spectrum. Furthermore, $\sigma_e(T)$ is empty if and only if $\lambda_j(Q) \to \infty$ as $j \to \infty$. In this case, $\lambda_j(Q)$ is the jth eigenvalue of S, wherein the eigenvalues are arranged in increasing order and repeated according to multiplicity. The bottom of the essential spectrum $\inf \sigma_e(T)$ is the limit of $\lambda_j(Q)$ as $j \to \infty$ (see [D95, Chapter 4]). We set $\inf \sigma_e(S) = \infty$ if $\sigma_e(S) = \emptyset$.

Let $T_k \colon \mathbb{H}_k \to \mathbb{H}_{k+1}$, $k = 1, 2$, be densely defined and closed operators on Hilbert spaces. Assume that $\mathcal{R}\,(T_1) \subset \mathcal{N}\,(T_2)$, Let T_k^* be the adjoint of T_k. Then T_k^* is also densely defined and closed. Let

$$Q(u,v) = \langle T_1^*u, T_1^*v \rangle + \langle T_2u, T_2v \rangle$$

with its domain given by $\mathrm{Dom}\,(Q) = \mathrm{Dom}\,(T_1^*) \cap \mathrm{Dom}\,(T_2)$. The Proposition 5.3.3 shows that the $\bar{\partial}$-Neumann Laplacian as defined by Definition 5.2.9 is consistent with the one defined using the quadratic form as above. As the consequence, the $\bar{\partial}$-Neumann Laplacian defined by Definition 5.2.9 is self-adjoint (see Theorem 5.3.9 below).

Proposition 5.3.3 $Q(u, v)$ *is a densely defined, closed, non-negative sesquilinear form. The associated self-adjoint operator* $\square$ *defined through* (5.16) *is identical to the one given by*

$$\mathrm{Dom}\,(\square) = \{f \in \mathbb{H}_2 \mid f \in \mathrm{Dom}\,(Q), \atop T_2 f \in \mathrm{Dom}\,(T_2^*), T_1^* f \in \mathrm{Dom}\,(T_1)\} \tag{5.18}$$

and

$$\square = T_1 T_1^* + T_2^* T_2.$$

Proof: The closedness of Q follows easily from that of T_1 and T_2. The non-negativity follows from the definition. We now prove that $\mathrm{Dom}\,(Q)$ is dense in $\mathbb{H}_2$. Since $\mathcal{N}\,(T_2)^\perp = \overline{\mathcal{R}\,(T_2^*)} \subset \mathcal{N}\,(T_1^*)$ and

$$\mathrm{Dom}\,(T_2) = \mathcal{N}\,(T_2) \oplus \big(\mathrm{Dom}\,(T_2) \cap \mathcal{N}\,(T_2)^\perp\big),$$

we have

$$\begin{aligned} \mathrm{Dom}\,(Q) &= \mathrm{Dom}\,(T_1^*) \cap \mathrm{Dom}\,(T_2) \\ &= \big(\mathcal{N}\,(T_2) \cap \mathrm{Dom}\,(T_1^*)\big) \oplus \big(\mathrm{Dom}\,(T_2) \cap \mathcal{N}\,(T_2)^\perp\big). \end{aligned}$$

Since $\mathrm{Dom}\,(T_1^*)$ and $\mathrm{Dom}\,(T_2)$ are dense in $\mathbb{H}_2$, $\mathrm{Dom}\,(Q)$ is dense in $\mathcal{N}\,(T_2) \oplus \mathcal{N}\,(T_2)^\perp = \mathbb{H}_2$.

Recall that $f \in \mathrm{Dom}\,(\square)$ if and only if $f \in \mathrm{Dom}\,(Q)$ and there exists a $g \in \mathbb{H}_2$ such that

$$Q(u, f) = \langle u, g \rangle, \quad \text{for all } u \in \mathrm{Dom}\,(Q) \tag{5.19}$$

(see, for example, Lemma 4.4.1 in [D95]). Thus

$$\mathrm{Dom}\,(\square) \supset \{f \in H_2 \mid f \in \mathrm{Dom}\,(Q), T_2 f \in \mathrm{Dom}\,(T_2^*), T_1^* f \in \mathrm{Dom}\,(T_1)\}.$$

We now prove the opposite containment. Suppose $f \in \mathrm{Dom}\,(\square)$. For any $u \in \mathrm{Dom}\,(T_2)$, we write $u = u_1 + u_2 \in (\mathcal{N}\,(T_1^*) \cap \mathrm{Dom}\,(T_2)) \oplus \mathcal{N}\,(T_1^*)^\perp$. Note that $\mathcal{N}\,(T_1^*)^\perp \subset \mathcal{R}\,(T_2^*)^\perp = \mathcal{N}\,(T_2)$. It follows from (5.19) that

$$|\langle T_2 u, T_2 f \rangle| = |\langle T_2 u_1, T_2 f \rangle| = |Q(u_1, f)| = |\langle u_1, g \rangle| \le \|u\| \cdot \|g\|.$$

Hence $T_2 f \in \mathrm{Dom}\,(T_2^*)$. The proof of $T_1^* f \in \mathrm{Dom}\,(T_1)$ is similar. For any $w \in \mathrm{Dom}\,(T_1^*)$, we write $w = w_1 + w_2 \in (\mathcal{N}\,(T_2) \cap \mathrm{Dom}\,(T_1^*)) \oplus \mathcal{N}\,(T_2)^\perp$. Note that $\mathcal{N}\,(T_2)^\perp = \overline{\mathcal{R}\,(T_2^*)} \subset \mathcal{N}\,(T_1^*)$. Therefore, by (5.19),

$$|\langle T_1^* w, T_1^* f \rangle| = |\langle T_1^* w_1, T_1^* f \rangle| = |Q(w_1, f)| = |\langle w_1, g \rangle| \le \|w\| \cdot \|g\|.$$

Hence, $T_1^* f \in \mathrm{Dom}\,(T_1^{**}) = \mathrm{Dom}\,(T_1)$. It follows from the definition of $\square$ that for any $f \in \mathrm{Dom}\,(\square)$ and $u \in \mathrm{Dom}\,(Q)$,

$$\langle \square f, u \rangle = \langle \square^{1/2} f, \square^{1/2} u \rangle = Q(f, u)$$
$$= \langle T_1^* f, T_1^* u \rangle + \langle T_2 f, T_2 u \rangle = \langle (T_1 T_1^* + T_2^* T_2) f, u \rangle.$$

Hence, $\square = T_1 T_1^* + T_2^* T_2$. ∎

Definition 5.3.4 Let T be a self-adjoint operator on a Hilbert space. (1) T is *positive* if its spectrum $\sigma(\square) \subset [c, \infty)$ for some $c > 0$. (2) T is *essentially positive* if there is a constant $c > 0$ such that $\sigma_e(\square) \subset [c, \infty)$. (3) T is *gap positive* if there is a constant $c > 0$ such that $\sigma(\square) \cap (0, c) = \emptyset$.

Thus for a self-adjoint operator T, we have:

$$\text{positivity} \implies \text{essential positivity} \implies \text{gap positivity}.$$

The following proposition is due to Hörmander [H65, Theorems 1.1.2 and 1.1.4]. Let $\mathcal{N}(Q) = \mathcal{N}(T_1^*) \cap \mathcal{N}(T_2)$. Note that when it is non-trivial, $\mathcal{N}(Q)$ is the eigenspace of the zero eigenvalue of $\square$. When $\mathcal{R}(T_1)$ is closed, $\mathcal{N}(T_2) = \mathcal{R}(T_1) \oplus \mathcal{N}(Q)$.

Proposition 5.3.5 *The following statements are equivalent:*

1. $\square$ *is gap positive.*

2. $\square$ *has closed range* $\mathcal{R}(\square)$.

3. *Both* T_1 *and* T_2 *have closed ranges* $\mathcal{R}(T_1)$ *and* $\mathcal{R}(T_2)$.

4. *There exists a positive constant such that*

$$\|f\|_2^2 \leq C^2 Q(f, f), \quad \forall f \in \mathrm{Dom}\,(T_1^*) \cap \mathrm{Dom}\,(T_2), \ f \perp \mathcal{N}(Q). \tag{5.20}$$

Proof: Since $\mathcal{R}(T_1) \subset \mathcal{N}(T_2)$,

$$\mathbb{H}_2 = \mathcal{N}(T_2)^\perp \oplus \mathcal{N}(T_2) = \mathcal{N}(T_2)^\perp \oplus \overline{\mathcal{R}(T_1)} \oplus (\mathcal{N}(T_2) \ominus \mathcal{R}(T_1)^\perp)$$
$$= \overline{\mathcal{R}(T_2^*)} \oplus \overline{\mathcal{R}(T_1)} \oplus (\mathcal{N}(T_2) \cap \mathcal{N}(T_1^*)).$$

The proposition then follows from Lemma 5.3.1. ∎

The following proposition is well-known (compare [H65, Theorem 1.1.2 and Theorem 1.1.4], [C83, Proposition 3], and [Sh92, Proposition 2.3]).

Proposition 5.3.6 *The operator* $\square$ *is positive if and only if* $\mathcal{R}(T_1) = \mathcal{N}(T_2)$ *and* $\mathcal{R}(T_2)$ *is closed.*

Proof: Assume $\Box$ is positive. Then 0 is in the resolvent set of $\Box$ and hence $\Box$ has a bounded inverse $G\colon \mathbb{H}_2 \to \operatorname{Dom}(\Box)$. For any $u \in H_2$, write $u = T_1 T_1^* Gu + T_2^* T_2 Gu$. If $u \in \mathcal{N}(T_2)$,

$$0 = (T_2 u,\ T_2 Gu) = (T_2 T_2^* T_2 Gu,\ T_2 Gu) = (T_2^* T_2 Gu,\ T_2^* T_2 Gu).$$

Hence, $T_2^* T_2 Gu = 0$ and $u = T_1 T_1^* Gu$. Therefore, $\mathcal{R}(T_1) = \mathcal{N}(T_2)$. Similarly, $\mathcal{R}(T_2^*) = \mathcal{N}(T_1^*)$. Therefore, T_2^* and hence T_2 have closed range. To prove the opposite implication, for any $u \in \operatorname{Dom}(Q)$, we write $u = u_1 + u_2 \in \mathcal{N}(T_2) \oplus \mathcal{N}(T_2)^\perp$. Note that $u_1, u_2 \in \operatorname{Dom}(Q)$. Since $\mathcal{N}(T_2) = \mathcal{R}(T_1) = \mathcal{N}(T_1^*)^\perp$ and T_2 has closed range, it follows from Lemma 5.3.1 that that there exists a positive constant c such that $c\|u_1\|^2 \le \|T_1^* u_1\|^2$ and $c\|u_2\|^2 \le \|T_2 u_2\|^2$. Thus

$$c\|u\|^2 = c(\|u_1\|^2 + \|u_2\|^2) \le \|T_1^* u_1\|^2 + \|T_2 u_2\|^2 = Q(u,u).$$

Hence, $\inf \sigma(\Box) \ge c > 0$. ∎

For a subspace $L \subset \mathbb{H}_2$, denote by $P_{L^\perp}$ the orthogonal projection onto $L^\perp$ and $T_2|_{L^\perp}$ the restriction of T_2 to $L^\perp$. The next proposition generalizes Proposition 5.3.6 (see [Fu05]).

Proposition 5.3.7 *The following statements are equivalent:*

1. $\Box$ *is essentially positive.*

2. $\mathcal{R}(T_1)$ *and* $\mathcal{R}(T_2)$ *are closed and* $\mathcal{N}(Q)$ *is finite dimensional.*

3. $\mathcal{N}(Q)$ *is finite dimensional,* $\mathcal{N}(T_2) \cap \mathcal{N}(Q)^\perp = \mathcal{R}(T_1)$, *and* $\mathcal{R}(T_2|_{\mathcal{N}(Q)^\perp})$ *is closed.*

4. *There exists a finite dimensional subspace* $L \subset \operatorname{Dom}(T_1^*) \cap \mathcal{N}(T_2)$ *such that* $\mathcal{N}(T_2) \cap L^\perp = P_{L^\perp}(\mathcal{R}(T_1))$ *and* $\mathcal{R}(T_2|_{L^\perp})$ *is closed.*

Proof: We first prove (1) implies (2). Suppose $a = \inf \sigma_e(\Box) > 0$. If $\inf \sigma(\Box) > 0$, then $\mathcal{N}(Q)$ is trivial and (2) follows from Proposition 5.3.6. Suppose $\inf \sigma(\Box) = 0$. Then $\sigma(\Box) \cap [0, a)$ consists only of isolated points, all of which are eigenvalues of finite multiplicity of $\Box$ (see [D95, Theorem 4.5.2]). Hence $\mathcal{N}(Q)$, the eigenspace of the eigenvalue 0, is finite dimensional. Choose a sufficiently small $c > 0$ so that $\sigma(\Box) \cap [0, c) = \{0\}$. By the spectral theorem for self-adjoint operator, there exists a finite regular Borel measure μ on $\sigma(\Box) \times \mathbb{N}$ and a unitary transformation $U\colon \mathbb{H}_2 \to L^2(\sigma(\Box) \times \mathbb{N}, d\mu)$ such that $U\Box U^{-1} = M_x$, where $M_x \varphi(x, n) = x\varphi(x, n)$ is the multiplication operator by x on $L^2(\sigma(\Box) \times \mathbb{N}, d\mu)$ (see [D95, Theorem 2.5.1]). Let $P_{\mathcal{N}(Q)}$ be the orthogonal projection onto $\mathcal{N}(Q)$. For any $f \in \operatorname{Dom}(Q) \cap \mathcal{N}(Q)^\perp$,

$$U P_{\mathcal{N}(Q)} f = \chi_{[0,c)} U f = 0,$$

where $\chi_{[0,c)}$ is the characteristic function of $[0, c)$. Hence Uf is supported on $[c, \infty) \times \mathbb{N}$. Therefore,

$$Q(f,f) = \int_{\sigma(\Box) \times \mathbb{N}} x|Uf|^2\, d\mu \ge c\|Uf\|^2 = c\|f\|^2.$$

It then follows from Proposition 5.3.5 that both T_1 and T_2 have closed range.

To prove (2) implies (1), we use Proposition 5.3.5 in the opposite direction: There exists a positive constant c such that

$$c\|f\|^2 \le Q(f,f), \quad \text{for all } f \in \mathrm{Dom}\,(Q) \cap \mathcal{N}\,(Q)^\perp. \tag{5.21}$$

Proving by contradiction, we assume $\inf \sigma_e(\square) = 0$. Let ε be any positive number less than c. Since $L_{[0,\varepsilon)} = \mathcal{R}\,(\chi_{[0,\varepsilon)}(\square))$ is infinite dimensional, there exists a non-zero $g \in L_{[0,\varepsilon)}$ such that $g \perp \mathcal{N}\,(Q)$. However,

$$Q(g,g) = \int_{\sigma(\square) \times \mathbb{N}} x \chi_{[0,\varepsilon)} |Ug|^2 \, d\mu \le \varepsilon \|Ug\|^2 = \varepsilon \|g\|^2,$$

contradicting (5.21).

We do some preparations before proving the equivalence of (3) with (1) and (2). Let L be any finite dimensional subspace of $\mathrm{Dom}\,(T_1^*) \cap \mathcal{N}\,(T_2)$. Let $\mathbb{H}_2' = \mathbb{H}_2 \ominus L$. Let $T_2' = T_2|_{\mathbb{H}_2'}$ and let $T_1^{*\prime} = T_1^*|_{\mathbb{H}_2'}$. Then $T_2' \colon \mathbb{H}_2' \to \mathbb{H}_3$ and $T_1^{*\prime} \colon \mathbb{H}_2' \to \mathbb{H}_1$ are densely defined, closed operators. Let $T_1' \colon \mathbb{H}_1 \to \mathbb{H}_2'$ be the adjoint of $T_1^{*\prime}$. It follows from the definitions that $\mathrm{Dom}\,(T_1) \subset \mathrm{Dom}\,(T_1')$. The finite dimensionality of L implies the opposite containment. Thus, $\mathrm{Dom}\,(T_1) = \mathrm{Dom}\,(T_1')$. For any $f \in \mathrm{Dom}\,(T_1)$ and $g \in \mathrm{Dom}\,(T_1^{*\prime}) = \mathrm{Dom}\,(T_1^*) \cap L^\perp$,

$$\langle T_1' f, g \rangle = \langle f, T_1^{*\prime} g \rangle = \langle f, T_1^* g \rangle = \langle T_1 f, g \rangle.$$

Hence $T_1' = P_{L^\perp} \circ T_1$ and $\mathcal{R}\,(T_1') = P_{L^\perp}(\mathcal{R}\,(T_1)) \subset \mathcal{N}\,(T_2')$. Let

$$Q'(f,g) = \langle T_1'^* f, T_1'^* g \rangle + \langle T_2' f, T_2' g \rangle$$

be the associated sesquilinear form on $\mathbb{H}_2'$ with $\mathrm{Dom}\,(Q') = \mathrm{Dom}\,(Q) \cap L^\perp$.

We are now in position to prove that (2) implies (3). In this case, we take $L = \mathcal{N}\,(Q)$ in the above settings. By Proposition 5.3.5, there exists a positive constant c such that

$$Q(f,f) = Q'(f,f) \ge c\|f\|^2, \quad \text{for all } f \in \mathrm{Dom}\,(Q').$$

We then obtain (3) by applying Proposition 5.3.6 to T_1', T_2', and $Q'(f,g)$.

Finally, we prove (4) implies (1). By Proposition 5.3.6, we know that there exists a positive constant c such that

$$Q(f,f) \ge c\|f\|^2, \quad \text{for all } f \in \mathrm{Dom}\,(Q) \cap L^\perp.$$

The rest of the proof follows the same lines of the above proof of the implication (2) $\Rightarrow$ (1), with $\mathcal{N}\,(Q)$ there replaced by L. ∎

Let

$$H = \mathcal{N}\,(T_2)/\mathcal{R}\,(T_1)$$

be the quotient space of $\mathcal{N}\,(T_2)$ and $\mathcal{R}\,(T_1)$. This is the *cohomology* of the complex

$$\mathbb{H}_1 \xrightarrow{T_1} \mathbb{H}_2 \xrightarrow{T_2} \mathbb{H}_3.$$

It measures the extent to which one cannot solve the equation $T_1 u = f$ for $f \in \mathcal{N}(T_2)$. When $\mathcal{R}(T_1)$ is closed, H inherits the following metric structure from $\mathbb{H}_2$:

$$\|[f]\| = \inf\{\|f + g\|_2 \mid g \in \mathcal{R}(T_1)\},$$

with which H is complete. Furthermore, as we noted before,

$$\mathcal{N}(T_2) = \mathcal{R}(T_1) \oplus (\mathcal{N}(T_2) \cap \mathcal{R}(T_1)^\perp) = \mathcal{R}(T_1) \oplus \mathcal{N}(Q).$$

Thus H is isomorphic to $\mathcal{N}(Q)$ in this case. Together with Proposition 5.3.7, we obtain:

Proposition 5.3.8 □ *is essentially positive if and only if* $\mathcal{R}(T_1)$ *and* $\mathcal{R}(T_2)$ *are closed and* H *is finite dimensional.*

For $1 \leq q \leq n - 1$, let

$$Q_q(u, v) = \langle \bar{\partial}_q u, \bar{\partial}_q v \rangle + \langle \bar{\partial}_{q-1}^* u, \bar{\partial}_{q-1}^* v \rangle$$

with $\mathrm{Dom}(Q_q) = \mathrm{Dom}(\bar{\partial}_q) \cap \mathrm{Dom}(\bar{\partial}_{q-1}^*)$. The bottom and top degree cases are defined similarly with $Q_0(u, v) = \langle \bar{\partial}_0 u, \bar{\partial}_0 v \rangle$ and $Q_n(u, v) = \langle \bar{\partial}_n^* u, \bar{\partial}_n^* v \rangle$. By Proposition 5.3.3, Q_q is a densely defined, non-negative, closed sesquilinear form and the associated self-adjoint operator is the $\bar{\partial}$-Neumann Laplacian. More precisely, we have:

Theorem 5.3.9 *The* $\bar{\partial}$-*Neumann Laplacian* $\Box_q : L^2_{(0,q)}(\Omega) \to L^2_{(0,q)}(\Omega)$ *is the densely defined self-adjoint operator such that*

$$\mathrm{Dom}(\Box^{1/2}) = \mathrm{Dom}(\bar{\partial}_q) \cap \mathrm{Dom}(\bar{\partial}_{q-1}^*)$$

and

$$Q_q(u, v) = \langle \Box_q^{1/2} u, \Box_q^{1/2} v \rangle.$$

Furthermore, $u \in \mathrm{Dom}(\Box_q)$ *if and only if* $u \in \mathrm{Dom}(Q_q)$ *and there exists a* $g \in L^2_{(0,q)}(\Omega)$ *such that*

$$Q_q(u, v) = \langle g, v \rangle$$

for all $v \in \mathrm{Dom}(Q_q)$.

5.4   Pseudoconvexity

Pseudoconvexity is a central concept in several complex variables. In this section, we will review the rudiments of this concept. Interested readers can find a more extensive treatment in [Kr01]. One of the most striking differences between one complex variable and several complex variables is the Hartogs extension phenomenon, of which the following is the simplest example:

Example 5.4.1 *Suppose* $f(z_1, z_2)$ *is a holomorphic function on* $\Omega = \{(z_1, z_2) \in \mathbb{C}^2 \mid |z_1| < 2, |z_2| < 2\} \setminus \{(z_1, z_2) \in \mathbb{C}^2 \mid |z_1| < 1, |z_2| < 1\}$. *Then* f *has a holomorphic extension* $\tilde{f}$ *to* $\widehat{\Omega} = \{(z_1, z_2) \mid |z_1| < 2, |z_2| < 2\}$.

Proof: For $|z_1| < 1$, expanding $f(z_1, z_2)$ as a Laurent series in z_2, we have

$$f(z_1, z_2) = \sum_{j=-\infty}^{\infty} a_j(z_1) z_2^j,$$

where

$$a_j(z_1) = \frac{1}{2\pi i} \int_{|z_2|=3/2} \frac{f(z_1, \zeta)}{\zeta^{j+1}} \, d\zeta.$$

Since $f(z_1, z_2)$ is holomorphic on $\{|z_1| < 2\}$ when $|z_2| = 3/2$, it follows that $a_j(z_1)$ is holomorphic on $\{|z_2| < 2\}$. Moreover, since $f(z_1, z_2)$ is holomorphic on $\{|z_1| < 2\}$ when $1 < |z_2| < 2$, we have $a_j(z_1) = 0$ on $\{1 < |z_1| < 2\}$ when $j < 0$. Hence $a_j(z_1) = 0$ on $\{|z_1| < 2\}$. Therefor, f is holomorphic on $\widehat{\Omega}$. ∎

Convexity is not preserved under a biholomorphic map and whence not a natural notion in several complex variables. Pseudoconvexity is the analogous notion in several complex variable that is invariant under biholomorphic maps.

Recall that a domain Ω in $\mathbb{R}^n$ is *convex* if the line segment joining every pair of points in Ω also lies in Ω. For a C^2-smooth real-valued function ρ, the *real Hessian* of ρ is the symmetric matrix:

$$H_\rho^{\mathbb{R}}(p) = \left(\frac{\partial^2 \rho(p)}{\partial x_j \partial x_k} \right)_{1 \leq j, k \leq n}.$$

For economy of notations, we will identify the tangent space $T_p(\mathbb{R}^n)$ with $\mathbb{R}^n$.

Proposition 5.4.2 *Let* $\Omega = \{\rho < 0\}$ *be a bounded domain in* $\mathbb{R}^n$ *with* C^2-*boundary. Then* Ω *is convex if and only if*

$$H_\rho^{\mathbb{R}}(p)(\xi, \xi) = \sum_{j,k=1}^{n} \frac{\partial^2 \rho(p)}{\partial x_j \partial x_k} \xi_j \xi_k \geq 0 \qquad \forall \xi \in T_p(b\Omega) \qquad (5.22)$$

where $T_p(b\Omega) = \{\xi \in \mathbb{R}^n \mid \sum_{j=1}^{n} \frac{\partial \rho}{\partial x_j}(p)\xi_j = 0\}$ *is the real tangent space of* $b\Omega$ *at* p.

Proof: We first prove the necessity. Let $p \in b\Omega$. After a translation and a rotation, we may assume that p is the origin and the positive x_n-axis is the outward normal direction for $b\Omega$ at p. Thus

$$\rho(x) = x_n + \frac{1}{2} \sum_{j,k=1}^{n} \frac{\partial^2 \rho(p)}{\partial x_j \partial x_k} x_j x_k + o(|x|^2).$$

Proving by contradiction, we assume that

$$\left(\partial^2 \rho(p)/\partial x_j \partial x_k\right)_{j,k=1}^{n-1}$$

is not semi-positive definite. After a rotation, we may assume that the above matrix is diagonalized and

$$\rho(x) = x_n + \sum_{j=1}^{n-1} \lambda_j |x_j|^2 + \sum_{j=1}^{n-1} \frac{\partial^2 \rho(p)}{\partial x_j \partial x_n} x_j x_n + \frac{1}{2} \frac{\partial^2 \rho(p)}{\partial x_n^2} x_n^2 + o(|x|^2),$$

where one of the λ_j's, say λ_1, is negative. However, this leads to a contradiction as we will then have $(-\varepsilon, 0, \ldots, 0)$ and $(\varepsilon, 0, \ldots, 0)$ in Ω for sufficiently small $\varepsilon > 0$ but $(0, \ldots, 0)$ is not in Ω.

The proof of sufficiency is left as an exercise.   ∎

A domain Ω is said to be *strictly convex* at a boundary point p if the inequality (5.22) is strict when $\xi \neq 0$. A domain is strictly convex if it is strictly convex at any of its boundary points.

Exercise 5.4.3 Let $\Omega = \{\rho(x) < 0\} \subset\subset \mathbb{R}^n$ is a convex domain with C^2-boundary. Let $\tilde{\rho}_M = \rho e^{-M|x|^2}$. Show that $\Omega_\varepsilon = \{\tilde{\rho}_M < \varepsilon\}$ is strictly convex when M is sufficiently large and ε is sufficiently small. Use this to prove the sufficiency in Proposition 5.4.2.

Definition 5.4.4 Let $\Omega = \{\rho < 0\} \subset \mathbb{C}^n$ be a bounded domain with C^2-boundary. Then Ω is *Levi-pseudoconvex* at $p \in b\Omega$ if

$$H_\rho(p)(\xi, \xi) = \sum_{j,k=1}^{n} \frac{\partial^2 \rho(p)}{\partial z_j \partial \bar{z}_k} \xi_j \bar{\xi}_k \geq 0,$$

$$\forall \xi \in T_p^{1,0}(b\Omega) = \{\xi \in \mathbb{C}^n \mid \sum_{j=1}^{n} \frac{\partial \rho}{\partial z_j}(p)\xi_j = 0\}.$$

It is *strictly pseudoconvex* at p if the above inequality is strict when $\xi \neq 0$. The domain Ω is said to be Levi-pseudoconvex or strictly pseudoconvex if it is Levi-pseudoconvex or respectively strictly pseudoconvex at every boundary point. The Hermitian matrix

$$H_\rho(p) = \left(\frac{\partial^2 \rho(p)}{\partial z_j \partial \bar{z}_k}\right)$$

is called the *complex Hessian* of ρ.

REMARK 5.4.5 (1) The notions of Levi-pseudoconvexity and strict pseudo-convexity do not depend on the choice of local holomorphic coordinates. More precisely, if $b\Omega$ is Levi-pseudoconvex (respectively, strictly pseudoconvex) at

$p \in b\Omega$ and $w = \Psi(z)$ is a biholomorphic map from a neighborhood U of p into $\mathbb{C}^n$, then $\Psi(b\Omega \cap U)$ is Levi-pseudoconvex (strictly pseudoconvex) at $\Psi(p)$. This is a consequence of the following functorial property of vectors and forms: Let $\tilde{\xi} = \Psi_*(\xi)$, $\tilde{p} = \Psi(p)$, and $\tilde{\rho} = \rho \circ \Psi^{-1}$. Then

$$H_{\tilde{\rho}}(\tilde{p})(\tilde{\xi}, \tilde{\xi}) = \partial\bar{\partial}\tilde{\rho}(\tilde{\xi}, \bar{\tilde{\xi}})(\tilde{p}) = \partial\bar{\partial}\rho(\xi, \bar{\xi})(p) = H_\rho(p)(\xi, \xi).$$

(2) Every bounded convex domain Ω with C^2-smooth boundary is Levi-pseudoconvex. The converse is of course not true as a smoothly bounded planar domain is always Levi-pseudoconvex but not necessarily convex. The proof of the first statement is as follows. Let $\xi \in \mathbb{C}^n$. Write $\xi_j = t_j + is_j$. Then

$$\xi \in T(b\Omega) \iff \operatorname{Re}\left(\sum_{j=1}^{n} \xi_j \frac{\partial\rho}{\partial z_j}\right) = 0$$

and the convexity of Ω implies that the real Hessian

$$H_\rho^{\mathbb{R}}(p)(t, s) = \sum_{j,k=1}^{n} \frac{\partial^2\rho(p)}{\partial x_j \partial x_k} t_j t_k + \sum_{j,k=1}^{n} \frac{\partial^2\rho(p)}{\partial y_j \partial y_k} s_j s_k + 2\sum_{j,k=1}^{n} \frac{\partial^2\rho(p)}{\partial x_j \partial y_k} t_j s_k$$

$$= 2\operatorname{Re}\left(\sum_{j,k=1}^{n} \frac{\partial^2\rho(p)}{\partial z_j \partial z_k} \xi_j \xi_k\right) + 2\sum_{j,k=1}^{n} \frac{\partial^2\rho(p)}{\partial z_j \partial \bar{z}_k} \xi_j \bar{\xi}_k \geq 0$$

Now let $\xi \in T^{1,0}(b\Omega)$. Replacing ξ by $i\xi$ in the above inequality and then adding the result to the above inequality, we then obtain

$$\sum_{j,k=1}^{n} \frac{\partial^2\rho}{\partial z_j \partial \bar{z}_k} \xi_j \bar{\xi}_k \geq 0.$$

Proposition 5.4.6 *Let $\Omega \subset\subset \mathbb{C}^n$ be strictly pseudoconvex with C^2-boundary. Then*

  1. There exist a neighborhood U of $b\Omega$, a defining function of Ω, and a positive constant $c > 0$ such that

$$\sum_{j,k=1}^{n} \frac{\partial^2\rho(z)}{\partial z_j \partial \bar{z}_k} \xi_j \bar{\xi}_k \geq c|\xi|^2, \quad \forall z \in U, \; \forall \xi \in \mathbb{C}^n. \tag{5.23}$$

  2. For any $p \in b\Omega$, there exist a neighborhood U and a biholomorphic map from U into $\mathbb{C}^n$ such that $b\Omega \cap U$ is strictly convex.

Proof: (1) Let $\tilde{\rho} = \rho + M\rho^2$. It follows from direct computations that

$$\sum_{j,k=1}^{n} \frac{\partial^2\tilde{\rho}(z)}{\partial z_j \partial \bar{z}_k} \xi_j \bar{\xi}_k = (1 + 2M\rho)\sum_{j,k=1}^{n} \frac{\partial^2\rho(z)}{\partial z_j \partial \bar{z}_k} \xi_j \bar{\xi}_k + 2M|\langle\partial\rho, \xi\rangle|^2.$$

From the strict pseudoconvexity of Ω, we know that there exist a neighborhood U of $b\Omega$, a positive constant $c > 0$, and a sufficiently small $\varepsilon > 0$ such that

$$\sum_{j,k=1}^{n} \frac{\partial^2 \rho(z)}{\partial z_j \partial \bar{z}_k} \xi_j \bar{\xi}_k \geq c|\xi|^2, \ \forall z \in U, \xi \in \mathbb{C}^n \text{ with } |\langle \partial\rho, \xi \rangle|^2 \leq \varepsilon |\xi|^2.$$

By choosing M sufficiently large, we then obtain (5.23).

(2) Let ρ be a defining function of Ω that satisfies (5.23). After a translation and a unitary transformation, we may assume that p is the origin and the positive x_n-axis is the real outward direction. Thus,

$$\rho = \text{Re } z_n + \text{Re } \left(\sum_{j,k=1}^{n} \frac{\partial^2 \rho(p)}{\partial z_j \partial z_k} z_j z_k \right) + \sum_{j,k=1}^{n} \frac{\partial^2 \rho(p)}{\partial z_j \partial \bar{z}_k} z_j \bar{z}_k + o(|z|^2).$$

After a change of holomorphic coordinates in the form of $w = \Psi(z)$:

$$w_j = z_j, \ 1 \leq j \leq n-1; \quad w_n = z_n + \sum_{j,k=1}^{n} \frac{\partial^2 \rho(z)}{\partial z_j \partial z_k} z_j \bar{z}_k,$$

we have

$$\tilde{\rho}(w) = \text{Re } w_n + \sum_{j,k=1}^{n} \frac{\partial^2 \tilde{\rho}(w)}{\partial w_j \partial \bar{w}_k} w_j \bar{w}_k + o(|w|^2).$$

It follows from (5.23) that

$$\left(\frac{\partial^2 \tilde{\rho}(w)}{\partial w_j \partial \bar{w}_k} \right)$$

is positive definite. It then follows that after a unitary transformation in the w-variables, the quadratic term in the Taylor expansion of $\tilde{\rho}$ is

$$\sum_{j=1}^{n} \lambda_j |w_j|^2,$$

where the λ_j's are positive. This implies that $b\Omega$ is strictly convex at p. After possible shrinking of U, we then obtain that $b\Omega \cap U$ is strictly convex in the w-coordinates. ∎

Definition 5.4.7 Let $f: \Omega \to [-\infty, \infty)$. It is an *exhaustion function* of Ω if $\{z \in \Omega \mid f(z) < c\} \subset\subset \Omega$ for any $c \in \mathbb{R}$. It is *plurisubharmonic* on Ω if

1. f is upper-semicontinuous on Ω;

2. For any $z \in \Omega$ and $\xi \in \mathbb{C}^n$, $u(t) = f(z + t\xi)$ is subharmonic on $\{t \in \mathbb{C} \mid z + t\xi \in \Omega\}$.

We say f is *strictly plurisubharmonic* at $p \in \Omega$ if there exist a neighborhood U of p and a positive constant $c > 0$ such that $f(z) - c|z|^2$ is plurisubharmonic on U. It is strictly plurisubharmonic on Ω if it is strictly plurisubharmonic at every point in Ω.

Exercise 5.4.8 Let $f \in C^2(\Omega)$. Show that f is plurisubharmonic function if and only if its complex Hessian $H_f(z)$ is semi-positive definite at any $z \in \Omega$ and f is strictly plurisubharmonic if and only if $H_f(z)$ is positive definite for any $z \in \Omega$.

Definition 5.4.9 (1) A domain $\Omega \subset \mathbb{C}^n$ is called *pseudoconvex* if it can be exhausted from inside by smoothly bounded strictly pseudoconvex domains. More precisely, there exists a sequence of smoothly bounded strictly pseudoconvex domains Ω_j such that $\Omega_j \subset \Omega_{j+1}$ and $\cup_{j=1}^{\infty}\Omega_j = \Omega$. (2) A domain $\Omega \subset \mathbb{C}^n$ is a *domain of holomorphy* if there is no domain $\widetilde{\Omega}$ such that $\Omega \subsetneqq \widetilde{\Omega}$ and every holomorphic function on Ω extends holomorphically to $\widetilde{\Omega}$.

Evidently, every domain in $\mathbb{C}$ is pseudoconvex. Indeed, every bounded domain in $\mathbb{C}$ with C^2-boundary is strictly pseudoconvex according to Definition 5.4.4. In this case, $T^{1,0}(b\Omega)$ is trivial.

Theorem 5.4.10 *Let Ω be a domain in $\mathbb{C}^n$. Then the following statements are equivalent:*

> *1. Ω is pseudoconvex.*
>
> *2. $-\log \delta(z)$ is plurisubharmonic on Ω, where $\delta(z)$ is the Euclidean distance from z to the boundary $b\Omega$ of Ω.*
>
> *3. Ω has a smooth strictly plurisubharmonic exhaustion function.*

Furthermore, if Ω has C^2-smooth boundary, then the above statements are equivalent to Ω being Levi-pseudoconvex.

We skip the proof and refer the interested reader to [Kr01].

Theorem 5.4.11 *Let $\Omega \subset \mathbb{C}^n$. Then Ω is pseudoconvex if and only if Ω is a domain of holomorphy.*

The proof of sufficiency is not very difficult. The opposite direction is usually referred to as the *Levi problem*, and it was solved in the 1950's by Oka. We again refer the reader to [H91, Kr01] for detail.

5.5 Hörmander's L^2-Estimates

In this section, we present Hörmander's L^2-estimates for the $\bar{\partial}$-operator through the lens of spectral theory of the $\bar{\partial}$-Neumann operator. The following main theorem in this section is due to Hörmander ([H65]).

Theorem 5.5.1 *Let Ω be a bounded pseudoconvex domain in $\mathbb{C}^n$ with diameter D. Then*

$$\inf \sigma(\Box_q) \geq \frac{q}{D^2 e}, \qquad 1 \leq q \leq n-1. \tag{5.24}$$

Consequently, for any $f \in \mathcal{N}(\bar{\partial}_q)$, there exists $u \in \mathrm{Dom}(\bar{\partial}_{q-1})$ such that

$$\bar{\partial} u = f \quad and \quad \|u\|^2 \leq \frac{D^2 e}{q}\|f\|^2. \tag{5.25}$$

We divide the proof into several steps, presented in the following subsections.

5.5.1   The Morrey-Kohn-Hörmander formula

We first introduce some notations. For $(0,q)$-forms

$$u = {\sum_{|J|=q}}' u_J d\bar{z}_J \quad \text{and} \quad v = {\sum_{|J|=q}}' v_J d\bar{z}_J$$

on a domain Ω in $\mathbb{C}^n$ and a real-valued function $\varphi \in C^2(\overline{\Omega})$, we write:

$$H_\varphi(u,v) = {\sum_K}' \sum_{j,k=1}^n \frac{\partial^2 \varphi}{\partial z_j \partial \bar{z}_k} u_{jK} \bar{v}_{kK}; \quad H_\varphi(u) = H_\varphi(u,u).$$

We also write:

$$|\overline{\nabla} u|^2 = {\sum_{|J|=q}}' \sum_{k=1}^n \left|\frac{\partial u_J}{\partial \bar{z}_k}\right|^2.$$

Let $L^2_{(0,q)}(\Omega, e^{-\varphi})$ be the space of $(0,q)$-forms with weighted L^2-coefficients equipped with the norm

$$\|u\|_\varphi^2 = {\sum_{|J|=q}}' \int_\Omega |u_J|^2 e^{-\varphi}\, dV,$$

and the inner product $\langle u, v \rangle_\varphi = \langle u, v e^{-\varphi} \rangle$. Let $\bar{\partial}^*_{\varphi,q}$ be the adjoint of $\bar{\partial}_q \colon L^2_{(0,q)}(\Omega, e^{-\varphi}) \to L^2_{(0,q+1)}(\Omega, e^{-\varphi})$. Since

$$\langle \bar{\partial} u, v \rangle_\varphi = \langle \bar{\partial} u, v e^{-\varphi} \rangle = \langle u, \bar{\partial}^*(e^{-\varphi} v) \rangle = \langle u, e^{\varphi} \bar{\partial}^*(e^{-\varphi} v) \rangle_\varphi,$$

we obtain that $\mathrm{Dom}(\bar{\partial}^*) = \mathrm{Dom}(\bar{\partial}^*_\varphi)$ and

$$\bar{\partial}^*_\varphi = e^{\varphi} \bar{\partial}^* e^{-\varphi} = \bar{\partial}^* + (\bar{\partial}\varphi)^* \lrcorner. \tag{5.26}$$

The following integration by parts formula, due to Morrey-Kohn-Hörmander, plays an important role in the L^2-theory of the $\bar{\partial}$-problem.

Theorem 5.5.2 *Let* $\Omega = \{z \in \mathbb{C}^n \mid \rho(z) < 0\} \subset\subset \mathbb{C}^n$ *be a bounded domain with C^2-boundary and let ρ be a defining function of Ω such that $|\nabla \rho| = 1$ on $b\Omega$. Let $\varphi \in C^2(\overline{\Omega})$ be a real-valued functions. Then for any* $u \in C^2_{(0,q)}(\overline{\Omega}) \cap \mathrm{Dom}\,(\overline{\partial}^*)$,

$$\|\overline{\partial}u\|^2_\varphi + \|\overline{\partial}^*_\varphi u\|^2_\varphi = \int_{b\Omega} H_\rho(u) e^{-\varphi}\, dS + \int_\Omega \left(|\overline{\nabla} u|^2 + H_\varphi(u) \right) e^{-\varphi}\, dV. \quad (5.27)$$

Proof: We will (5.27) step by step. We first prove it for the simplest case when $\varphi = 0$ and $q = 1$. (This case is due to Morrey. The formula for a general q is due to Kohn. With the weight φ, it is due to Hörmander.) In this case, since

$$\overline{\partial}u = \sum_{1 \le j < k \le n} \left(\frac{\partial u_k}{\partial \bar{z}_j} - \frac{\partial u_j}{\partial \bar{z}_k} \right) d\bar{z}_j \wedge d\bar{z}_k,$$

we have

$$
\begin{aligned}
\|\overline{\partial}u\|^2 &= \sum_{j<k} \left\| \frac{\partial u_k}{\partial \bar{z}_j} - \frac{\partial u_j}{\partial \bar{z}_k} \right\|^2 \\
&= \sum_{j<k} \left(\left\| \frac{\partial u_k}{\partial \bar{z}_j} \right\|^2 + \left\| \frac{\partial u_j}{\partial \bar{z}_k} \right\|^2 - \langle \frac{\partial u_k}{\partial \bar{z}_j}, \frac{\partial u_j}{\partial \bar{z}_k} \rangle - \langle \frac{\partial u_j}{\partial \bar{z}_k}, \frac{\partial u_k}{\partial \bar{z}_j} \rangle \right) \\
&= \sum_{j,k=1}^n \left(\left\| \frac{\partial u_j}{\partial \bar{z}_k} \right\|^2 - \langle \frac{\partial u_j}{\partial \bar{z}_k}, \frac{\partial u_k}{\partial \bar{z}_j} \rangle \right)
\end{aligned}
\quad (5.28)
$$

and

$$\|\overline{\partial}^* u\|^2 = \left\| \sum_{j=1}^n \frac{\partial u_j}{\partial z_j} \right\|^2 = \sum_{j,k=1}^n \langle \frac{\partial u_j}{\partial z_j}, \frac{\partial u_k}{\partial z_k} \rangle. \quad (5.29)$$

Applying integration by parts formula (5.8), we have

$$
\begin{aligned}
\langle \frac{\partial u_j}{\partial \bar{z}_k}, \frac{\partial u_k}{\partial \bar{z}_j} \rangle &= -\langle \frac{\partial^2 u_j}{\partial z_j \partial \bar{z}_k}, u_k \rangle + \int_{b\Omega} \frac{\partial u_j}{\partial \bar{z}_k} \bar{u}_k \frac{\partial \rho}{\partial z_j}\, dS \\
&= \langle \frac{\partial u_j}{\partial z_j}, \frac{\partial u_k}{\partial z_k} \rangle + \int_{b\Omega} \left(\frac{\partial u_j}{\partial \bar{z}_k} \bar{u}_k \frac{\partial \rho}{\partial z_j} - \frac{\partial u_j}{\partial z_j} \bar{u}_k \frac{\partial \rho}{\partial \bar{z}_k} \right) dS
\end{aligned}
$$

Summing up from $j, k = 1$ to n and using the boundary condition

$$\sum_{k=1}^k u_k \frac{\partial \rho}{\partial z_k} = 0 \quad \text{on } b\Omega,$$

we then have

$$\sum_{j,k=1}^n \langle \frac{\partial u_j}{\partial \bar{z}_k}, \frac{\partial u_k}{\partial \bar{z}_j} \rangle = \sum_{j,k=1}^n \langle \frac{\partial u_j}{\partial z_j}, \frac{\partial u_k}{\partial z_k} \rangle + \int_{b\Omega} \sum_{j,k=1}^n \frac{\partial u_j}{\partial \bar{z}_k} \bar{u}_k \frac{\partial \rho}{\partial z_j}\, dS. \quad (5.30)$$

We now play the "Morrey trick" to convert the boundary term into a desirable form: Since $\sum_{j=1}^{n} u_j \partial\rho/\partial z_j = 0$ on $b\Omega$, we can write

$$\sum_{j=1}^{n} u_j \frac{\partial\rho}{\partial z_j} = f \cdot \rho \tag{5.31}$$

for some $f \in C^1$ near $b\Omega$. Differentiating both sides with respect to $\bar{z}_k$, multiplying the result by $\overline{u_k}$, and then summing up over $k = 1$ to n, we have

$$\sum_{k=1}^{n} \left(\sum_{j=1}^{n} \left(\frac{\partial u_j}{\partial \bar{z}_k} \frac{\partial \rho}{z_j} + u_j \frac{\partial^2 \rho}{\partial \bar{z}_k \partial z_j} \right) \right) = \sum_{k=1}^{n} \bar{u}_k \left(\rho \frac{\partial f}{\partial \bar{z}_k} + f \frac{\partial \rho}{\partial \bar{z}_k} \right).$$

Therefore, on $b\Omega$, we have

$$\sum_{j,k=1}^{n} u_j \bar{u}_k \frac{\partial^2 \rho}{\partial z_j \partial \bar{z}_k} = - \sum_{j,k=1}^{n} \bar{u}_k \frac{\partial u_j}{\partial \bar{z}_k} \frac{\partial \rho}{\partial z_j}. \tag{5.32}$$

Combining (5.28), (5.29), (5.30), and (5.32), we then obtain (5.27) when $\varphi = 0$ and $q = 1$.

We now prove (5.27) when $q = 1$ but with a general real-valued $\varphi \in C^2(\overline{\Omega})$. We write

$$\partial_j = \partial/\partial z_j, \quad \partial_j^{\varphi} = -e^{\varphi} \partial_j e^{-\varphi} = -\partial_j + \partial_j \varphi.$$

For the economy of notation, we also write

$$\rho_j = \partial_j \rho, \quad \varphi_{j\bar{k}} = \bar{\partial}_k \partial_j \varphi,$$

and so on. It is easy to check that

$$[\bar{\partial}_k, \, \partial_j^{\varphi}] = \varphi_{j\bar{k}}. \tag{5.33}$$

As in the proof of the previous case, we have

$$\|\bar{\partial}u\|_{\varphi}^2 + \|\bar{\partial}_{\varphi}^* u\|_{\varphi}^2 = \|\overline{\nabla}u\|_{\varphi}^2 + \sum_{j,k=1}^{n} \left(\langle \partial_j^{\varphi} u_j, \partial_k^{\varphi} u_k \rangle - \langle \bar{\partial}_k u_j, \bar{\partial}_j u_k \rangle_{\varphi} \right). \tag{5.34}$$

It follows from the integration by part formula (5.8) and commutative identity (5.33) that

$$\langle \bar{\partial}_k u_j, \bar{\partial}_j u_k \rangle_{\varphi} = \langle u_j, \partial_k^{\varphi} \bar{\partial}_j u_k \rangle_{\varphi} + \int_{b\Omega} u_j \partial_j \bar{u}_k \bar{\partial}_k \rho e^{-\varphi} \, dS$$

$$= \langle u_j, \bar{\partial}_j \partial_k^{\varphi} u_k \rangle_{\varphi} - \langle \partial_j \bar{\partial}_k \varphi u_j, u_k \rangle_{\varphi} + \int_{b\Omega} u_j \partial_j \bar{u}_k \bar{\partial}_k \rho e^{-\varphi} \, dS$$

$$= \langle \partial_j^{\varphi} u_j, \partial_k^{\varphi} u_k \rangle_{\varphi} - \langle \partial_j \bar{\partial}_k \varphi u_j, u_k \rangle_{\varphi}$$

$$+ \int_{b\Omega} u_j \partial_j \bar{u}_k \bar{\partial}_k \rho \, e^{-\varphi} \, dS - \int_{b\Omega} u_j \overline{\partial}_j^{\varphi} \bar{u}_k \partial_j \rho \, e^{-\varphi} \, dS$$

Summing up over $j, k = 1$ to n and using the boundary condition $\sum u_j \partial_j \rho = 0$, we then obtain that the last boundary term vanishes. Using the Morrey trick (5.31) to the first boundary term, we then obtain formula (5.27) when $q = 1$.

We now prove Theorem 5.5.2 in its full generality. We first fix some notations. Suppose A and B are tuples of integers between 1 and n of same length. We let ϵ^A_B be the sign of permutation from A to B if they contain the same set of integers and be zero if otherwise. Throughout the proof, we will use the capital letter J and L to denote strictly increasing q-tuples, K strictly increasing $(q-1)$-tuples, and M strictly increasing $(q+1)$-tuples. The lower case letters j, k, and l will denote integers running from 1 to n. If $J = (j_1, \ldots, j_q)$, then jJ denotes the $(q+1)$-tuple $(j, j_1, \ldots, j_q)$. Note that

$$\epsilon^{jJ}_M \epsilon^{lL}_M = \epsilon^{jJ}_{lL} \tag{5.35}$$

Let

$$u = {\sum_J}' u_J \, d\bar{z}_J.$$

Then

$$\bar{\partial} u = {\sum_J}' \sum_j \frac{\partial u_J}{\partial \bar{z}_j} d\bar{z}_j \wedge d\bar{z}_J = {\sum_M}' \left(\sum_j {\sum_J}' \frac{\partial u_J}{\partial \bar{z}_j} \epsilon^{jJ}_M \right) d\bar{z}_M.$$

It follows that

$$
\begin{aligned}
|\bar{\partial} u|^2 &= {\sum_M}' \left({\sum_{J,L}}' \sum_{j,l} \frac{\partial u_J}{\partial \bar{z}_j} \frac{\partial \bar{u}_L}{\partial z_l} \epsilon^{jJ}_M \epsilon^{lL}_M \right) = {\sum_{J,L}}' \sum_{j,l} \frac{\partial u_J}{\partial \bar{z}_j} \frac{\partial \bar{u}_L}{\partial z_l} \epsilon^{jJ}_{lL} \\
&= {\sum_{J,L}}' \sum_{j=l} \frac{\partial u_J}{\partial \bar{z}_j} \frac{\partial \bar{u}_L}{\partial z_l} \epsilon^{jJ}_{lL} + {\sum_{J,L}}' \sum_{j \neq l} \frac{\partial u_J}{\partial \bar{z}_j} \frac{\partial \bar{u}_L}{\partial z_l} \epsilon^{jJ}_{lL} \\
&= {\sum_J}' \sum_{j \notin J} \left| \frac{\partial u_J}{\partial \bar{z}_j} \right|^2 - {\sum_K}' \sum_{j \neq l} \frac{\partial u_{lK}}{\partial \bar{z}_j} \frac{\partial \bar{u}_{jK}}{\partial z_l} \\
&= {\sum_J}' \sum_j \left| \frac{\partial u_J}{\partial \bar{z}_j} \right|^2 - {\sum_K}' \sum_{j,l} \frac{\partial u_{lK}}{\partial \bar{z}_j} \frac{\partial \bar{u}_{jK}}{\partial z_l}. \tag{5.36}
\end{aligned}
$$

Since

$$\bar{\partial}^* u = - {\sum_K}' \sum_k \frac{\partial u_{kK}}{\partial z_k} d\bar{z}_K,$$

we have

$$|\bar{\partial}^* u|^2 = {\sum_K}' \sum_{k,l} \frac{\partial u_{kK}}{\partial z_k} \frac{\partial \bar{u}_{lK}}{\partial \bar{z}_l}. \tag{5.37}$$

Therefore, we have

$$\|\bar{\partial}u\|_\varphi^2 + \|\bar{\partial}_\varphi^* u\|_\varphi^2 = \||\bar{\nabla}u|\|_\varphi^2$$
$$+ \sideset{}{'}\sum_K \sum_{j,l} \left(-\langle \bar{\partial}_j u_{lK}, \bar{\partial}_l u_{jK} \rangle_\varphi + \langle \partial_k^\varphi u_{kK}, \partial_l^\varphi u_{lK} \rangle_\varphi \right). \quad (5.38)$$

The rest of the proof follows along the same lines as in the previous case, modulus index complications. Details are given below. It follows from the integration by part formula (5.8) that

$$\langle \bar{\partial}_j u_{lK}, \ \bar{\partial}_l u_{jK} \rangle_\varphi = \langle u_{lK}, \ \partial_j^\varphi \bar{\partial}_l u_{jK} \rangle_\varphi - \int_{b\Omega} u_{lK} \overline{\bar{\partial}_l u_{jK}} \rho_{\bar{j}} \, e^{-\varphi} dS. \quad (5.39)$$

Using commutative identity (5.33), we then obtain that

$$\langle u_{lK}, \partial_j^\varphi \bar{\partial}_l u_{jK} \rangle_\varphi = \langle u_{lK}, \bar{\partial}_l \partial_j^\varphi u_{jK} \rangle_\varphi - \langle u_{lK}, \varphi_{l\bar{j}} u_{jK} \rangle_\varphi$$
$$= \langle \partial_l^\varphi u_{lK}, \partial_j^\varphi u_{jK} \rangle_\varphi \quad (5.40)$$
$$- \int_{b\Omega} u_{lK} \overline{\partial_j^\varphi u_{jK}} \rho_l \, e^{-\varphi} dS - \langle u_{lK}, \varphi_{j\bar{l}} u_{jK} \rangle_\varphi.$$

Notice that since $u \in \mathrm{Dom}\,(\bar{\partial}^*)$,

$$\sum_j u_{jK} \rho_j = 0 \quad \text{on } b\Omega. \quad (5.41)$$

We now repeat the Morrey trick: From (5.41), we have

$$\sum_j \bar{u}_{jK} \rho_{\bar{j}} = f\rho$$

for some C^1 function f in a neighborhood of $b\Omega$. Applying ∂_l to both sides, multiplying them by u_{lK}, then summing up over l, we then have

$$\sum_{j,l} u_{lK} \partial_l \bar{u}_{jK} \rho_{\bar{j}} = -\sum_{j,l} u_{lK} \bar{u}_{jK} \rho_{l\bar{j}} \quad \text{on } b\Omega. \quad (5.42)$$

Back-substituting (5.40) into (5.39), then (5.38), and using boundary conditions (5.41) and (5.42), we then obtain the desirable (5.27). ∎

The following formula, sometimes referred as the *twisted* Morrey-Kohn-Hörmander formula, is a consequence of Theorem 5.5.2 (compare [BSt99] and [Ch13]).

Corollary 5.5.3 *Let $\Omega = \{z \in \mathbb{C}^n \mid \rho(z) < 0\} \subset\subset \mathbb{C}^n$ be a bounded domain with C^2-boundary and let ρ be a defining function of Ω such that $|d\rho| = 1$*

on $b\Omega$. Let $a, \varphi \in C^2(\overline{\Omega})$ be real-valued functions with $a \geq 0$. Then for any $u \in C^2_{(0,q)}(\overline{\Omega}) \cap \text{Dom}(\overline{\partial}^*)$,

$$\|\sqrt{a}\,\overline{\partial}u\|^2_\varphi + \|\sqrt{a}\,\overline{\partial}^*_\varphi u\|^2_\varphi = \int_{b\Omega} aH_\rho(u)e^{-\varphi}\,dS$$

$$+ \int_\Omega \left(a\left|\overline{\nabla}u\right|^2 + aH_\varphi(u) - H_a(u)\right)e^{-\varphi}\,dV + 2\,\text{Re}\,\langle(\overline{\partial}a)^*\lrcorner u, \overline{\partial}^*_\varphi u\rangle_\varphi. \quad (5.43)$$

Proof: Let $\varepsilon > 0$. Applying Theorem 5.5.2 with φ replaced by $\tilde{\varphi}_\varepsilon = \varphi - \log(a + \varepsilon)$ and using the identities

$$\overline{\partial}^*_{\tilde{\varphi}} = \overline{\partial}^* + (\overline{\partial}\tilde{\varphi})^*\lrcorner = \overline{\partial}^*_\varphi - (a + \varepsilon)^{-1}(\overline{\partial}a)^*\lrcorner$$

and

$$H_{\tilde{\varphi}}(u) = H_\varphi(u) - (a + \varepsilon)^{-1}H_a(u) + (a + \varepsilon)^{-2}|\overline{\partial}a)^*\lrcorner u|^2,$$

we obtain

$$\|\sqrt{a + \varepsilon}\,\overline{\partial}u\|^2_\varphi + \|\sqrt{a + \varepsilon}\,\overline{\partial}^*_\varphi u\|^2_\varphi = \int_{b\Omega} (a + \varepsilon)H_\rho(u)e^{-\varphi}\,dS$$

$$+ \int_\Omega \left((a + \varepsilon)\left|\overline{\nabla}u\right|^2 + (a + \varepsilon)H_\varphi(u) - H_a(u)\right)e^{-\varphi}\,dV + 2\,\text{Re}\,\langle(\overline{\partial}a)^*\lrcorner u, \overline{\partial}^*_\varphi u\rangle_\varphi.$$

Letting $\varepsilon \to 0$, we then obtain the desirable identity (5.43). ∎

The following estimate is due to Catlin [Ca87]. The proof presented here is from [BSt99]:

Theorem 5.5.4 *Let $\Omega = \{z \in \mathbb{C}^n \mid \rho(z) < 0\} \subset\subset \mathbb{C}^n$ be a bounded domain with C^2-boundary and let ρ be a defining function of Ω such that $|d\rho| = 1$ on $b\Omega$. Let $b \in C^2(\overline{\Omega})$ with $b \leq 0$. Then*

$$Q_q(u, u) \geq \int_\Omega e^b H_b(u)\,dV, \quad \forall u \in \text{Dom}(\overline{\partial}^*) \cap C^2_{(0,q)}(\overline{\Omega}). \quad (5.44)$$

Proof: Taking $a = 1 - e^b$ and $\varphi = 0$ in the twisted Morrey-Kohn-Hörmander formula (5.43), we have

$$\|\sqrt{a}\,\overline{\partial}u\|^2 + \|\sqrt{a}\,\overline{\partial}^* u\|^2 \geq \int_\Omega a\left|\overline{\nabla}u\right|^2\,dV - 2\,\text{Re}\,\langle(\overline{\partial}a)^*\lrcorner u, \overline{\partial}^* u\rangle$$

$$+ \sum_K' \sum_{j,k=1}^n \int_\Omega e^b \left(\frac{\partial^2 b}{\partial z_j \partial \bar{z}_k} + \frac{\partial b}{\partial z_j}\frac{\partial b}{\partial \bar{z}_k}\right) u_{jK}\bar{u}_{kK}\,dV. \quad (5.45)$$

Applying Schwarz inequality to the second term on the right hand side, we have

$$\|\sqrt{a}\,\overline{\partial}u\|^2 + \|\overline{\partial}^* u\|^2 \geq \int_\Omega a\left|\overline{\nabla}u\right|^2\,dV +$$

$$\sum_K' \sum_{j,k=1}^n \int_\Omega e^b \frac{\partial^2 b}{\partial z_j \partial \bar{z}_k} u_{jK}\bar{u}_{kK}\,dV. \quad (5.46)$$

The desirable inequality (5.44) then follows. ∎

Proposition 5.5.5 *Let Ω be a bounded pseudoconvex domain with C^2-boundary. Let D be the diameter of Ω. Then*

$$Q_q(u, u) \geq \frac{q}{eD^2} \|u\|^2, \quad \forall u \in \mathrm{Dom}\,(\bar{\partial}^*) \cap C^2_{(0,q)}(\overline{\Omega}). \tag{5.47}$$

Proof: Let p be a point in Ω. Let $b = -1 + |z - p|^2/D^2$. We then obtain (5.47) from (5.44). ∎

In the next subsection, we will transform (5.47) from an *apriori* estimate to a *bona fide* one: namely, we will establish (5.47) for all $u \in \mathrm{Dom}\,(Q_q)$. This is done by applying the Friederichs lemma (see Lemma 5.5.6 below).

5.5.2   The density lemmas

Density is a subtle issue in the $\bar{\partial}$-Neumann problem. This issue was overlooked in some early literature and was addressed by Hörmander in [H65]. (We refer the reader to [H03] for a fascinating history about the $\bar{\partial}$-Neumann problem, especially the density issue in the problem.) The presentation here is similar to that of [CS99].

We begin with the Friedrichs Lemma. Let $\chi \in C_c^\infty(\mathbb{R}^m)$ be a radial symmetric function such that $\chi > 0$ on $|x| < 1$, $\chi = 0$ on $|x| \geq 1$, and $\int_{\mathbb{R}^m} \chi \, dx = 1$. Let $\chi_\varepsilon(x) = (1/\varepsilon^m)\chi(x/\varepsilon)$ denote the standard Friedrichs mollifiers. Before we proceed to prove this important lemma, we first recall the following simple facts.

1. The product of a Lipschitz continuous function and the first order derivative of an L^2-function is a well-defined distribution:

$$aD_j(f)(\psi) = -\int_{\mathbb{R}^m} f D_j(a\psi) \, dx, \quad \psi \in \mathcal{D}(\mathbb{R}^m),$$

   where $D_j = \partial/\partial x_j$.

2. Let $\phi \in \mathcal{D}(U)$. Let $\phi_y(\cdot) = \phi(\cdot - y)$ and

$$U_\phi = \{y \in U \mid \mathrm{Supp}\, \phi_y \subset U\}.$$

   Let $T \in \mathcal{D}'(U)$ be a distribution. Then $y \mapsto T(\phi_y)$ is $C^\infty(U_\phi)$. This is due to the fact that $\phi_y(x) = \phi(x - y)$ depends smoothly on y.

3. Let $T \in \mathcal{D}'(\mathbb{R}^m)$ and let $g \in C_0^\infty(\mathbb{R}^m)$. Then for any test function $\phi \in \mathcal{D}(\mathbb{R}^m)$,

$$(T * g)(\phi) = T(g^- * \phi) = \int_{\mathbb{R}^m} t(x)\phi(x) \, dx$$

   where $g^-(x) = g(-x)$ and $t(x) = T(g_x^-) = T(g(x - \cdot))$. Note that $t(x) \in C^\infty(\mathbb{R}^m)$. Thus a convolution of a distribution and a compactly supported smooth function results in a smooth function.

Lemma 5.5.6 (Friederichs) *Let $f \in L^2(\mathbb{R}^m)$ be compactly supported in an open set $U \subset \mathbb{R}^m$.*

    1. *Let a be a Lipschitz function on U. Then*

$$aD_j(f * \chi_\varepsilon) - (aD_j(f)) * \chi_\varepsilon \to 0 \quad \text{in } L^2(\mathbb{R}^m). \qquad (5.48)$$

    2. *Let $L = \sum_{j=1}^m a_j D_j + a_0$ where a_j, $1 \le j \le m$, are Lipschitz on U and $a_0 \in C(U)$. If $Lf \in L^2(\mathbb{R}^m)$, then*

$$f * \chi_\varepsilon \to f, \quad L(f * \chi_\varepsilon) \to Lf \quad \text{in } L^2(\mathbb{R}^m). \qquad (5.49)$$

Proof: (1) Note that

$$a(x)D_j(f * \chi_\varepsilon)(x) = -a(x) \int_{\mathbb{R}^m} f(y) \frac{\partial \chi_\varepsilon}{\partial y_j}(x - y)\, dy$$

$$= \int_{\mathbb{R}^m} a(x)f(x - \varepsilon y) \frac{1}{\varepsilon} \frac{\partial \chi}{\partial y_j}(y)\, dy$$

and

$$(aD_j f) * \chi_\varepsilon = (aD_j f)(\chi_\varepsilon(x - \cdot))$$

$$= -\int_{\mathbb{R}^m} f(y) \frac{\partial}{\partial y_j}\big(a(y)\chi_\varepsilon(x - y)\big)\, dy$$

$$= -\int_{\mathbb{R}^m} f(y) \left(\frac{\partial a}{\partial y_j}(y)\chi_\varepsilon(x - y) + a(y)\frac{\partial \chi_\varepsilon}{\partial y_j}(x - y) \right) dy$$

$$= -\int_{\mathbb{R}^m} f(x - \varepsilon y) \left(\frac{\partial a}{\partial y_j}(x - \varepsilon y)\chi(y) - a(x - \varepsilon y)\frac{1}{\varepsilon}\frac{\partial \chi}{\partial y_j}(y) \right) dy.$$

Therefore,

$$aD_j(f * \chi_\varepsilon) - (aD_j f) * \chi_\varepsilon = -\int \frac{\partial a}{\partial y_j}(x - y)f(x - y)\chi_\varepsilon(y)\, dy$$

$$+ \int (a(x) - a(x - \varepsilon y))f(x - \varepsilon y)\frac{1}{\varepsilon}\frac{\partial \chi}{\partial y_j}(y)\, dy.$$

Since a is Lipschitz continuouis on U, we have that

$$|a(x) - a(x - y)| \le M|y| \quad \text{and} \quad |\partial a / \partial y_j| \le M$$

almost everywhere on the support of f. Therefore, by Young's inequality, we have

$$\|aD_j(f * \chi_\varepsilon) - (aD_j f) * \chi_\varepsilon\| \le M\|f\| \left(\int \chi_\varepsilon\, dy + \int |y \frac{\partial \chi_\varepsilon}{\partial y_j}|\, dy \right) \le C\|f\|.$$

Thus we have now proved that there exists a positive constant C such that

$$\|aD_j(f * \chi_\varepsilon) - (aD_j f) * \chi_\varepsilon\| \le C\|f\|, \quad \forall f \in L^2(\mathbb{R}^m). \qquad (5.50)$$

We now show how this implies (1). Let $f_k \in C_c^\infty(\mathbb{R}^m)$ be a sequence of functions such that $f_k \to f$ in $L^2(\mathbb{R}^m)$. Then

$$\|aD_j(f * \chi_\varepsilon) - (aD_j f) * \chi_\varepsilon\| \leq \|aD_j((f - f_k) * \chi_\varepsilon) - (aD_j(f - f_k)) * \chi_\varepsilon\|$$
$$+ \|aD_j(f_k * \chi_\varepsilon) - (aD_j f_k) * \chi_\varepsilon\|.$$

By (5.50), the first term on the right hand side tends to 0 as $k \to \infty$. The second term tends to 0 as $\varepsilon \to 0$ because $f_k \in C_c^\infty(\mathbb{R}^m)$.

(2) By the first part, we know that $\|L(f * \chi_\varepsilon) - (Lf) * \chi_\varepsilon\| \to 0$. Since $Lf \in L^2(\mathbb{R}^m)$, we have $\|(Lf) * \chi_\varepsilon - Lf\| \to 0$. We then obtain (5.49).   ∎

In proving the density of $C^k(\overline{\Omega}) \cap \text{Dom}\,(\bar{\partial}^*)$ in $\text{Dom}\,(Q_q)$, We will localize the problem near the boundary of Ω by using a partition of unity. The following formula is convenient for this purpose:

Lemma 5.5.7 *Let $\chi_j \in C^2$, $1 \leq j \leq m$, be partition of unity such that $\sum_{j=1}^m \chi_j^2 = 1$ on $\overline{\Omega}$. Then*

$$\sum_{j=1}^m Q_\varphi(\chi_j u, \chi_j u) = Q_\varphi(u, u) + \sum_{j=1}^m \langle |\bar{\partial}\chi_j|^2 u, u\rangle_\varphi$$

Proof: By straightforward computations, we have

$$\langle \bar{\partial}(\chi_j u), \bar{\partial}(\chi_j u)\rangle_\varphi = \langle \chi_j^2 \bar{\partial}u, \bar{\partial}u\rangle_\varphi + 2\,\text{Re}\,\langle \chi_j \bar{\partial}\chi_j \wedge u, \bar{\partial}u\rangle_\varphi + \langle \bar{\partial}\chi_j \wedge u, \bar{\partial}\chi_j \wedge u\rangle_\varphi$$

and

$$\langle \bar{\partial}_\varphi^*(\chi_j u), \bar{\partial}_\varphi^*(\chi_j u)\rangle_\varphi = \langle \chi_j^2 \bar{\partial}^* u, \bar{\partial}^* u\rangle_\varphi + \langle \chi_j^2 (\bar{\partial}\varphi)^* \lrcorner u, (\bar{\partial}\varphi)^* \lrcorner u\rangle_\varphi$$
$$+ \langle (\bar{\partial}\chi_j)^* \lrcorner u, (\bar{\partial}\chi_j)^* \lrcorner u\rangle_\varphi - 2\,\text{Re}\,\langle \chi_j(\bar{\partial}\chi_j)^* \lrcorner u, \bar{\partial}^* u\rangle_\varphi$$
$$+ 2\,\text{Re}\,\langle \chi_j^2 \bar{\partial}^* u, (\bar{\partial}\varphi)^* \lrcorner u\rangle_\varphi - 2\,\text{Re}\,\langle \chi_j(\bar{\partial}\chi_j)^* \lrcorner u, (\bar{\partial}\varphi)^* \lrcorner u\rangle_\varphi$$

where $(\bar{\partial}\varphi)^*$ denotes the $(0,1)$-vector that is dual to $\bar{\partial}\varphi$ with respect to the ambient metric h and $\lrcorner$ denotes the contraction operator. Summing up and using the fact that $\sum \chi_j \bar{\partial}\chi_j = 0$, we then have

$$\sum_{j=1}^m Q_\varphi(\chi_j u,\ \chi_j u)$$

$$= Q_\varphi(u, u) + \sum_{j=1}^m \left(\langle \bar{\partial}\chi_j \wedge u,\ \bar{\partial}\chi_j \wedge u\rangle_\varphi + \langle (\bar{\partial}\chi_j)^* \lrcorner u,\ (\bar{\partial}\chi_j)^* \lrcorner u\rangle_\varphi \right)$$

$$= Q_\varphi(u, u) + \sum_{j=1}^m \left(\langle (\bar{\partial}\chi_j)^* \lrcorner (\bar{\partial}\chi_j \wedge u),\ u\rangle_\varphi + \langle \bar{\partial}\chi_j \wedge ((\bar{\partial}\chi_j)^* \lrcorner u),\ u\rangle_\varphi \right)$$

$$= Q_\varphi(u,\ u) + \sum_{j=1}^m \langle |\bar{\partial}\chi_j|^2 u,\ u\rangle_\varphi$$

∎

We can now state and prove the following density theorem due to Hörmander ([H65]).

Theorem 5.5.8 *Let k be a positive integer. Let $\Omega = \{\rho < 0\}$ be a bounded domain in $\mathbb{C}^n$ with C^{k+1}-boundary. Then $\mathrm{Dom}\,(\overline{\partial}^*_{q-1}) \cap C^k(\overline{\Omega})$ is dense in $\mathrm{Dom}\,(Q_q)$ in the graph norm $\|u\|_Q = (\|u\|^2 + Q_q(u,u))^{1/2}$.*

We divide the proof into several steps by first proving the following lemmas.

Lemma 5.5.9 (1) *For any domain $\Omega \subset\subset \mathbb{C}^n$, $\mathcal{D}_{(0,q-1)}(\Omega)$ is dense in $\mathrm{Dom}\,(\overline{\partial}^*_{q-1})$ in the graph norm $\|f\|_{\overline{\partial}^*} = (\|f\|^2 + \|\overline{\partial}^* f\|^2)^{1/2}$. (2) For $f \in L^2_{(0,q-1)}(\Omega)$, $f \in \mathrm{Dom}\,(\overline{\partial}^*_{q-1})$ if and only if $\vartheta \tilde{f} \in L^2_{(0,q)}(\mathbb{C}^n)$ where $\tilde{f} = f$ on Ω and $\tilde{f} = 0$ on $\mathbb{C}^n \setminus \Omega$.*

Proof: The first part follows directly from the fact that $\overline{\partial}^*$, being the adjoint of maximally defined operator $\overline{\partial}$, is minimally defined (see the Remark after Definition 5.2.4). Thus $\mathcal{D}_{(0,q-1)}(\Omega)$ is dense in $\mathrm{Dom}\,(\overline{\partial}^*_{q-1})$ in the graph norm $\|f\|_{\overline{\partial}^*} = (\|f\|^2 + \|\overline{\partial}^* f\|^2)^{1/2}$. To prove the second part, we let $f \in \mathrm{Dom}\,(\overline{\partial}^*_{q-1})$. Set $\tilde{f} = f$ on Ω and $\tilde{f} = 0$ on $\mathbb{C}^n \setminus \Omega$. For any $\varphi \in \mathcal{D}_{(0,q-1)}(\mathbb{C}^n)$,

$$\langle \tilde{f}, \overline{\partial}\varphi \rangle_{\mathbb{C}^n} = \langle f, \overline{\partial}\varphi \rangle_\Omega = \langle \vartheta f, \varphi \rangle_\Omega.$$

It follows that $\vartheta f \in L^2_{(0,q)}(\mathbb{C}^n)$. Conversely, suppose $f \in L^2_{(0,q-1)}(\Omega)$ with $\vartheta \tilde{f} \in L^2_{(0,q-1)}(\mathbb{C}^n)$. Then for any $\varphi \in \mathcal{D}_{(0,q-1)}(\Omega)$,

$$\langle f, \overline{\partial}\varphi \rangle_\Omega = \langle \tilde{f}, \overline{\partial}\varphi \rangle_{\mathbb{C}^n} = \langle \vartheta\tilde{f}, \varphi \rangle_{\mathbb{C}^n} = \langle \vartheta f, \varphi \rangle_\Omega.$$

Hence $f \in \mathrm{Dom}\,(\overline{\partial}^*_{q-1})$. ∎

Lemma 5.5.10 *Let $\Omega \subset\subset \mathbb{C}^n$ be a bounded domain with C^1-boundary. Then $C^\infty_{(0,q)}(\overline{\Omega})$ is dense in $\mathrm{Dom}\,(Q_q)$ in the $\|\cdot\|_Q$-norm.*

Proof: Let $f \in \mathrm{Dom}\,(Q_q)$. Let $\{\Omega_j\}_{j=0}^m$ be a covering of $\overline{\Omega}$ with the following properties:

1. $\Omega_0 \subset\subset \Omega$.

2. $\Omega_j \cap b\Omega \neq \emptyset$, $1 \leq j \leq m$.

3. There exist a sufficiently small constant ε_0 and (outward) unit vectors $\vec{n}_j \in \mathbb{C}^n$ such that for any ε with $0 < \varepsilon < \varepsilon_0$,

$$\cup_{j=1}^m \{z - 2\varepsilon\vec{n}_j \mid z \in \Omega_j \cap \Omega\} \cup \Omega_0 \subset\subset \Omega^-_\varepsilon$$

and

$$\cup_{j=1}^m \{z + 2\varepsilon\vec{n}_j \mid z \in \Omega_j \cap \Omega\} \cup \Omega_0 \supset\supset \Omega^+_\varepsilon$$

where

$$\Omega_\varepsilon^- = \{z \in \Omega \mid \text{dist}\,(z, b\Omega) > \varepsilon\} \quad \text{and} \quad \Omega_\varepsilon^+ = \{z \in \mathbb{C}^n \mid \text{dist}\,(z, \Omega) < \varepsilon\}.$$

Let $\{\psi_j\}$ be a partition of unity subordinated this covering. We may assume that $\text{Supp}\,\psi_j \subset\subset \Omega_j$. Let

$$f_\varepsilon(z) = \psi_0(z)f(z) + \sum_{j=1}^m \psi_j(z)f(z - 2\varepsilon\vec{n}_j). \tag{5.51}$$

Then $\bar{\partial}f_\varepsilon, \vartheta f_\varepsilon \in L^2(\Omega_\varepsilon^+)$. Furthermore, $\bar{\partial}f_\varepsilon \to \bar{\partial}f$ and $\vartheta f_\varepsilon \to \vartheta f$ on $L^2(\Omega)$. Regularizing f_ε with the Friederichs mollifiers (here the convolutions of forms with the Friederichs' mollifier χ_ε are done component-wise), we have the desirable $f_\varepsilon^* = f_\varepsilon * \chi_{\delta_\varepsilon} \in C_{(0,q)}^\infty(\overline{\Omega})$ with $\delta_\varepsilon > 0$ sufficiently small, depending on ε, such that $f_\varepsilon^* \to f$ in the $\|\cdot\|_Q$-norm as $\varepsilon \to 0$ by Lemma 5.5.6. $\blacksquare$

Let $\Omega \subset\subset \mathbb{C}^n$ be a bounded domain with C^1-boundary. Let $\rho \in C^1$ be a normalized defining function of Ω. For $f \in L_{(0,q)}^2(\Omega)$, write

$$f^\nu = \bar{\partial}\rho \wedge f_N, \quad f^\tau = f - f^\nu,$$

where

$$f_N := (\bar{\partial}\rho)^* \lrcorner f = {\sum_{|K|=q-1}}' \left(\sum_{j=1}^n \frac{\partial\rho}{\partial z_j} f_{jK} \right) d\bar{z}_K.$$

Then $f^\nu, f^\tau \in L_{(0,q)}^2(\Omega)$. Note that for $f \in C_{(0,q)}^1(\overline{\Omega})$,

$$(\bar{\partial}\rho)^* \lrcorner f^\tau = 0.$$

Thus $f^\tau \in \text{Dom}\,(\bar{\partial}_{q-1}^*)$.

Lemma 5.5.11 *Let $\Omega \subset \mathbb{C}^n$ be a bounded domain with $C^{1,1}$-boundary. Let $f \in \text{Dom}\,(Q_q)$. Let $\widetilde{f^\nu}$ be extension of f^ν to 0 outside of Ω. Then* (1) $\bar{\partial}\widetilde{f^\nu} \in L_{(0,q+1)}^2(\mathbb{C}^n)$; (2) $f^\tau \in \text{Dom}\,(\bar{\partial}_{q-1}^*)$.

Proof: We first prove (1). Notice that since $\rho \in C^{1,1}$, we have $f^\nu \in \text{Dom}\,(\bar{\partial})$. From Lemma 5.5.10, we may assume $f \in C_{(0,q)}^\infty(\overline{\Omega})$. By Lemma 5.2.7, for any $\varphi \in \mathcal{D}_{(0,q+1)}(\mathbb{C}^n)$, we have

$$\langle \widetilde{f^\nu}, \vartheta\varphi \rangle_{\mathbb{C}^n} = \langle f^\nu, \vartheta\varphi \rangle_\Omega$$
$$= \langle \bar{\partial}f^\nu, \varphi \rangle_\Omega - \int_{b\Omega} \langle \bar{\partial}\rho \wedge f^\nu, \varphi \rangle \, dS$$
$$= \langle \bar{\partial}f^\nu, \varphi \rangle_\Omega.$$

It follows that $\overline{\partial}\widetilde{f^\nu} = \widetilde{\overline{\partial}f^\nu} \in L^2(\mathbb{C}^n)$, where $\widetilde{\overline{\partial}f^\nu}$ is the extension of $\overline{\partial}f^\nu$ to 0 outside of Ω. The proof of (2) is similar. $\blacksquare$

We are now in position to prove Theorem 5.5.8. Let Ω be a bounded domain in $\mathbb{C}^n$ with C^{k+1}-boundary. Let ρ be a normalized defining function of Ω. Let $\{\Omega_j\}_{j=0}^m$ be a covering of $\overline{\Omega}$ and let $\{\psi_j\}_{j=0}^m$ be a partition of unity subordinated to the covering as in the proof of Lemma 5.5.10. Let $f \in \text{Dom}\,(Q_q)$. For sufficiently small ε, set

$$f_\varepsilon^{\nu^-}(z) = \psi^0(z)\widetilde{f^\nu}(z) + \sum_{j=1}^m \psi_j(z)\widetilde{f^\nu}(z + 2\varepsilon\vec{n}_j).$$

Note that $\text{Supp}\, f_\varepsilon^{\nu^-} \subset\subset \Omega_\varepsilon^-$. Let

$$f_{\varepsilon,\nu} = f_\varepsilon^{\nu^-} * \chi_{\delta_\varepsilon} \in \mathcal{D}_{(0,q)}(\Omega),$$

when δ_ε is sufficiently small. By Lemma 5.5.9, $\vartheta\widetilde{f^\nu} \in L^2_{(0,q-1)}(\mathbb{C}^n)$ and by Lemma 5.5.11, $\overline{\partial}\widetilde{f^\nu} \in L^2_{(0,q+1)}(\mathbb{C}^n)$. It follows from Lemma 5.5.6 that

$$f_{\varepsilon,\nu} \to f^\nu \quad \text{in the } \|\cdot\|_Q\text{-norm.} \tag{5.52}$$

Let $f_j(z) = \widetilde{f}(z - 2\varepsilon\vec{n}_j)$ and let

$$f_\varepsilon^{\tau^+}(z) = \psi^0(z)f^\tau(z) + \sum_{j=1}^m \psi_j(z)\big(f_j(z)\big)^\tau.$$

Then $\overline{\partial}f_\varepsilon^{\tau+}(z) \in L^2_{(0,q+1)}(\Omega_\varepsilon^+)$ and $\vartheta f_\varepsilon^{\tau+} \in L^2_{(0,q-1)}(\Omega_\varepsilon^+)$. let

$$f_{\varepsilon,\tau}(z) = \psi^0(z)(f^\tau * \chi_{\delta_\varepsilon}(z)) + \sum_{j=1}^m \psi_j(z)\big(f_j * \chi_{\delta_\varepsilon}(z)\big)^\tau,$$

where δ_ε is sufficiently small. Then $f_{\varepsilon,\tau}(z) \in C^k_{(0,q)}(\overline{\Omega}) \cap \text{Dom}\,(\overline{\partial}^*)$. Furthermore, by Lemma 5.5.6,

$$f_{\varepsilon,\tau} \to f^\tau \quad \text{in the } \|\cdot\|_Q\text{-norm.} \tag{5.53}$$

Therefore, $f_\varepsilon = f_{\varepsilon,\nu} + f_{\varepsilon,\tau} \in C^k_{(0,q)}(\overline{\Omega}) \cap \text{Dom}\,(\overline{\partial}^*)$ and converges f in the $\|\cdot\|_Q$-norm as $\varepsilon \to 0$. This concludes the proof of Theorem 5.5.8.

We now prove Theorem 5.5.1. Let Ω_j be a sequence of smooth bounded pseudoconvex domains such that $\Omega_j \subset\subset \Omega_{j+1}$ and $\cup_{j=1}^\infty \Omega_j = \Omega$. By Proposition 5.5.5 and Theorem 5.5.8, we have

$$Q_q^{\Omega_j}(f,f) \geq \frac{q}{D^2 e}\|f\|^2, \qquad f \in \text{Dom}\,(Q_q^{\Omega_j}),$$

where Q^{Ω_j} is the quadratic form associated to the $\overline{\partial}$-Neumann Laplacian on

Ω_j. Let $f \in L^2_{(0,q)}(\Omega)$ and $\bar{\partial}f = 0$. By Proposition 5.3.6 (and its proof), there exists $u_j \in L^2_{(0,q-1)}(\Omega_j)$ such that

$$\bar{\partial}u_j = f \text{ on } \Omega_j, \text{ and } \|u_j\|^2_{\Omega_j} \leq \frac{D^2 e}{q}\|f\|^2_{\Omega_j}.$$

Passing to a subsequence, we may assume that u_j converges to some $u \in L^2_{(0,q-1)}(\Omega)$ weakly. It follows that

$$\bar{\partial}u = f \text{ on } \Omega \text{ and } \|u\|^2 \leq \frac{D^2 e}{q}\|f\|^2.$$

Since the above estimates hold for any $1 \leq q \leq n-1$, we obtain Theorem 5.5.1 by applying Proposition 5.3.6 in the reverse direction.

5.6 Hearing Pseudoconvexity

In this section, we establish the following characterization of pseudoconvexity via spectrum of the $\bar{\partial}$-Neumann Laplacian. Thus in Kac's language, one can "hear" pseudoconvexity via the $\bar{\partial}$-Neumann Laplacian. The presentation follows closely those in [Fu05, Fu08].

Theorem 5.6.1 *Let Ω be a bounded domain in $\mathbb{C}^n$ such that $\text{int}(\text{cl}(\Omega)) = \Omega$. Then the following statements are equivalent:*

> *1. Ω is pseudoconvex.*
>
> *2. $\inf \sigma(\Box_q) > 0$, for all $1 \leq q \leq n-1$.*
>
> *3. $\inf \sigma_e(\Box_q) > 0$, for all $1 \leq q \leq n-1$.*

We first introduce some terminologies. The Dolbeault and L^2-cohomology groups on Ω are defined respectively by

$$H^{0,q}(\Omega) = \frac{\{f \in C^\infty_{(0,q)}(\Omega) \mid \bar{\partial}_q f = 0\}}{\{\bar{\partial}_{q-1}g \mid g \in C^\infty_{(0,q-1)}(\Omega)\}}$$

and

$$H^{0,q}_{L^2}(\Omega) = \frac{\{f \in L^2_{(0,q)}(\Omega) \mid \bar{\partial}_q f = 0\}}{\{\bar{\partial}_{q-1}g \mid g \in L^2_{(0,q-1)}(\Omega)\}}.$$

These cohomology groups are in general not isomorphic. For example, when a complex variety is deleted from Ω, the L^2-cohomology group remains the same but the Dolbeault cohomology group could change from trivial to infinite dimensional. As noted in the paragraph preceding Proposition 5.3.8, when $\mathcal{R}(\bar{\partial}_{q-1})$ is closed in $L^2_{(0,q)}(\Omega)$, $H^{0,q}_{L^2}(\Omega) \cong \mathcal{N}(\Box_q)$.

The implication $(1) \Rightarrow (2)$ is Hörmander's L^2-estimates of the $\bar{\partial}$-operator [H65], and it holds without the assumption $\text{int}\,(\text{cl}\,(\Omega)) = \Omega$ (see Theorem 5.5.1 in the previous section). The implication $(2) \Rightarrow (3)$ is trivial.

It remains to prove the implication $(3) \Rightarrow (1)$. This is a consequences of the sheaf cohomology theory dated back to Oka and Cartan (cf. [Se53, L66, Siu67, Br83, O82]). An elementary proof of (3) implying (1), following [Fu05], is given below. The proof uses ideas from sheaf cohomology theory in [L66]. When adapting Laufer's method to the L^2-settings, one encounters a difficulty: While the restriction to the complex hyperplane of the smooth function resulting from the sheaf cohomology arguements for the Dolbeault cohomology groups is well-defined, the restriction of the corresponding L^2 function is not. This difficulty was overcome in [Fu05] by appropriately modifying the construction of auxiliary $(0, q)$-forms. In showing that (3) implies (1), we will actually prove the following statement:

Proposition 5.6.2 *Let Ω be a bounded domain in $\mathbb{C}^n$ such that $\text{int}\,(\text{cl}\,(\Omega)) = \Omega$. If $\dim H^{(0,q)}_{L^2}(\Omega) < \infty$ for all $1 \leq q \leq n-1$, then Ω is pseudoconvex.*

To illustrate the idea of the proof, we first show the case when $n = 2$. Proving by contradiction, we assume that Ω is not pseudoconvex. Then there exists a domain $\widetilde{\Omega} \supsetneq \Omega$ such that every holomorphic function on Ω extends to $\widetilde{\Omega}$. Since $\text{int}\,(\text{cl}\,(\Omega)) = \Omega$, $\widetilde{\Omega} \setminus \text{cl}\,(\Omega)$ is non-empty. After a translation and a unitary transformation, we may assume that the origin is in $\widetilde{\Omega} \setminus \text{cl}\,(\Omega)$ and there is a point z^0 in the intersection of z_2-plane with Ω that is in the same connected component of the intersection of the z_2-plane with $\widetilde{\Omega}$ as the origin. For any $\alpha \in \mathbb{N}$, let

$$u_\alpha = \frac{(\alpha+1)!\bar{z}_2^\alpha}{|z|^{2(\alpha+2)}}(-\bar{z}_1 d\bar{z}_2 + \bar{z}_2 d\bar{z}_1).$$

Since the origin is not in $\overline{\Omega}$, $u_\alpha \in C^\infty_{(0,1)}(\overline{\Omega})$. In particular, $u_\alpha \in L^2_{(0,1)}(\Omega)$. Furthermore, one checks easily that $\bar{\partial} u_\alpha = 0$. In fact, we have

$$\bar{\partial}\left(-\frac{\alpha!\bar{z}_2^{\alpha+1}}{|z|^{2(\alpha+1)}}\right) = z_1 u_\alpha.$$

Let M be an integer such that $M > \dim \widetilde{H}_{(0,1)}(\Omega)$. Then there exists a non-trivial linear combination of the u_α's

$$u = \sum_{\alpha=1}^M c_\alpha u_\alpha$$

such that $[u] = [0]$. Namely, there exists an $v \in L^2(\Omega)$ such that $\bar{\partial} v = u$. Let

$$F(z_1, z_2) = z_1 v(z_1, z_2) + \sum_{\alpha=1}^M c_\alpha \frac{\alpha!\bar{z}_2^{\alpha+1}}{|z|^{2(\alpha+1)}}. \tag{5.54}$$

Then $\bar{\partial}F = 0$. Notice that since Ω is bounded, $F \in L^2(\Omega)$. Thus F is holomorphic on Ω. However,

$$F(0, z_2) = \sum_{\alpha=1}^{M} \frac{c_\alpha \alpha!}{z_2^{\alpha+1}}. \tag{5.55}$$

This contradicts the fact the F extends to a holomorphic function on $\widetilde{\Omega}$ and the origin and a point $z^0 \in \Omega$ are in the same connected component of $\widetilde{\Omega} \cap \{z_1 = 0\}$.

The above proof follows the lines of arguments of Laufer [L66] for the Dolbeault cohomology groups. In this case, this proof works for all $n \geq 2$ after an inductive argument, even without the assumption $\mathrm{int}\,(\mathrm{cl}\,(\Omega)) = \Omega$ because here $u \in C^\infty_{(0,1)}(\Omega)$ and $v \in C^\infty(\Omega)$, and the restriction of v to $\Omega \cap \{z_1 = 0\}$ post no problem. While the above proof still works for the L^2-cohomology groups, there is a subtlety needed to be addressed: The restriction of an L^2-function to a hyperplane is not well-defined. In fact, when we obtain (5.55), we have actually utilized a version of the following lemma:

Lemma 5.6.3 *Let $v_1, \ldots, v_{n-1} \in L^2(\Omega)$ and let $m \in \mathbb{N}$. Assume that G is a continuous function on Ω such that*

$$G(z) = \sum_{j=1}^{n-1} z_j^m v_j(z). \tag{5.56}$$

If $m \geq n - 1$, then $G(0, \ldots, 0, z_n) = 0$ for all $(0, \ldots, 0, z_n) \in \Omega$.

Proof: Let $(0, \ldots, 0, z_n^0) \in \Omega$. Write $z' = (z_1, \ldots, z_{n-1})$. Then for a sufficiently small positive numbers a_1 and a_2, we have

$$D(a_1, a_2) := \{|z'| < a_1\} \times \{|z_n - z_n^0| < a_2\} \subset \Omega.$$

For any $\delta \in (0,\ 1)$, we have

$$\left(\int_{D(a_1,a_2)} |G(\delta z', z_n)|^2 \, dV \right)^{1/2} \leq a_1^m \delta^m \sum_{j=1}^{n-1} \left(\int_{D(a_1,a_2)} |v_j(\delta z', z_n)|^2 \, dV \right)^{1/2}$$

$$\leq a_1^m \delta^{m-n+1} \sum_{j=1}^{n-1} \left(\int_{D(a_1\delta,a_2)} |v_j(z', z_n)|^2 \, dV \right)^{1/2}$$

$$\leq a_1^m \delta^{m-n+1} \sum_{j=1}^{n-1} \left(\int_{\Omega} |v_j(z)|^2 \chi_{D(a_1\delta,a_2)}(z) \, dV \right)^{1/2}.$$

Since $m \geq n - 1$, letting $\delta \to 0$, we obtain from the Lebesgue dominated convergence theorem that

$$\int_{D(a_1,a_2)} |G(0', z_n)|^2 \, dV = 0.$$

Thus $G(0', z_n) = 0$ for $|z_n - z_n^0| < a_2$.   ∎

We now return to the proof of Proposition 5.6.2. The above lemma illustrates the difficulty when one tries to adopt the arguments of the Dolbeault cohomology groups for the L^2-cohomology groups to higher dimension $n > 2$. In order for the proof to work, we need to obtain a version of (5.54) but with the term z_1 in front of v replaced by z_1^m with $m \geq n - 1$. This was done in [Fu05]. We return to the case for $n = 2$ to illustrate the idea. For $\alpha, m \in \mathbb{N}$, let

$$u_{\alpha,m} = \frac{(\alpha + 1)! \bar{z}_2^{m\alpha} (\bar{z}_1 \bar{z}_2)^{m-1}}{(|z_1|^{2m} + |z_2|^{2m})^{\alpha+2}} (-\bar{z}_1 d\bar{z}_2 + \bar{z}_2 d\bar{z}_1).$$

Note that $u_{\alpha,1} = u_\alpha$. It is straightforward to check that

$$\bar{\partial} \left(-\frac{\alpha! \bar{z}_2^{m(\alpha+1)}}{(|z_1|^{2m} + |z_2|^{2m})^{\alpha+1}} \right) = m z_1^m u_{\alpha,m}.$$

Proceeding as before, we then obtain a non-trivial linear combination of $u_{\alpha,m}$ (with m fixed and α runs from 1 to a sufficiently large integer M) and a function $v \in L^2(\Omega)$ such $\bar{\partial} v = u$. Thus

$$F(z_1, z_2) = m z_1^m v(z_1, z_2) + \sum_{\alpha=1}^{M} c_\alpha \frac{\alpha! \bar{z}^{m(\alpha+1)}}{(|z_1|^{2m} + |z_2|^{2m})^{\alpha+1}}$$

is an L^2-holomorphic function on Ω in a desirable form. By choosing $m \geq n-1$, we can then make this proof works for all $n \geq 2$.

We now provide a proof of Proposition 5.6.2 for all n but assuming the L^2-cohomology groups $H_{L^2}^{(0,q)}(\Omega)$ are trivial. We again prove by contradiction and assume that Ω is not pseudoconvex. Then there exists a domain $\tilde{\Omega} \supsetneq \Omega$ such that every holomorphic function on Ω extends to $\tilde{\Omega}$. After a translation and a unitary transformation, we may assume that the origin is in $\tilde{\Omega} \setminus \text{cl}(\Omega)$ and there is a point z^0 in the intersection of z_n-plane with Ω that is in the same connected component of $\tilde{\Omega} \cap \{z_1 = \ldots = z_{n-1} = 0\}$ as the origin.

Let m be a positive integer (to be specified later). Let $k_q = n$. For any $\{k_1, \ldots, k_{q-1}\} \subset \{1, 2, \ldots, n - 1\}$, we define

$$u(k_1, \ldots, k_q) = \frac{(q - 1)! (\bar{z}_{k_1} \cdots \bar{z}_{k_q})^{m-1}}{r_m^q} \times$$

$$\sum_{j=1}^{q} (-1)^j \bar{z}_{k_j} d\bar{z}_{k_1} \wedge \ldots \wedge \widehat{d\bar{z}_{k_j}} \wedge \ldots \wedge d\bar{z}_{k_q}, \quad (5.57)$$

where $r_m = |z_1|^{2m} + \ldots + |z_n|^{2m}$. As usual, $\widehat{d\bar{z}_{k_j}}$ indicates the deletion of $d\bar{z}_{k_j}$ from the wedge product. Evidently, $u(k_1, \ldots, k_q) \in L^2_{(0,q-1)}(\Omega)$ is a smooth form on $\mathbb{C}^n \setminus \{0\}$. Moreover, $u(k_1, \ldots, k_q)$ is skew-symmetric with respect to the indices $(k_1, \ldots, k_{q-1})$. In particular, $u(k_1, \ldots, k_q) = 0$ when two k_j's are identical.

We now fix some notional conventions. Let $K = (k_1, \ldots, k_q)$ and J a collection of indices from $\{k_1, \ldots, k_q\}$. Write $d\bar{z}_K = d\bar{z}_{k_1} \wedge \ldots \wedge d\bar{z}_{k_q}$, $\bar{z}_K^{m-1} = (\bar{z}_{k_1} \cdots \bar{z}_{k_q})^{m-1}$, and $\widetilde{d\bar{z}_{k_j}} = d\bar{z}_{k_1} \wedge \ldots \wedge \widehat{d\bar{z}_{k_j}} \wedge \ldots \wedge d\bar{z}_{k_q}$. Denote by $(k_1, \ldots, k_q \setminus J)$ the tuple of remaining indices after deleting those in J from $(k_1, \ldots, k_q)$. For example,

$$(4, 6, 3, 1 \setminus (4, 1, 6 \setminus 4, 6)) = (4, 6, 3).$$

It follows from a straightforward calculation that

$$\bar{\partial}u(k_1, \ldots, k_q) = -\frac{q! m \bar{z}_K^{m-1}}{r_m^{q+1}}\left(r_m d\bar{z}_K + \left(\sum_{\ell=1}^n \bar{z}_\ell^{m-1} z_\ell^m d\bar{z}_\ell\right) \wedge \left(\sum_{j=1}^q (-1)^j \bar{z}_{k_j} \widetilde{d\bar{z}_{k_j}}\right)\right)$$

$$= -\frac{q! m \bar{z}_K^{m-1}}{r_m^{q+1}} \sum_{\substack{\ell=1 \\ \ell \neq k_1, \ldots, k_q}}^n z_\ell^m \bar{z}_\ell^{m-1}\left(\bar{z}_\ell d\bar{z}_K + d\bar{z}_\ell \wedge \sum_{j=1}^q (-1)^j \bar{z}_{k_j} \widetilde{d\bar{z}_{k_j}}\right)$$

$$= m \sum_{\ell=1}^{n-1} z_\ell^m u(\ell, k_1, \ldots, k_q). \tag{5.58}$$

It follows that $u(1, \ldots, n)$ is a $\bar{\partial}$-closed $(0, n-1)$-form.

By Proposition 5.3.6, we have $\mathcal{R}(\bar{\partial}_{q-1}) = \mathcal{N}(\bar{\partial}_q)$ for all $1 \leq q \leq n-1$. We now solve the $\bar{\partial}$-equations inductively, using $u(1, \ldots, n)$ as initial data. Let $v \in L^2_{(0,n-2)}(\Omega)$ be a solution to $\bar{\partial}v = u(1, \ldots, n)$. For any $k_1 \in \{1, \ldots, n-1\}$, define

$$w(k_1) = -m z_{k_1}^m v + (-1)^{1+k_1} u(1, \ldots, n \setminus k_1).$$

Then it follows from (5.58) that $\bar{\partial}w(k_1) = 0$. Let $v(k_1) \in L^2_{(0,n-3)}(\Omega)$ be a solution of $\bar{\partial}v(k_1) = w(k_1)$.

Suppose for any $(q-1)$-tuple $K' = (k_1, \ldots, k_{q-1})$ of integers from $\{1, \ldots, n-1\}$, $q \geq 2$, we have constructed $v(K') \in L^2_{(0,n-q-1)}(\Omega)$ such that it is skew-symmetric with respect to the indices and satisfies

$$\bar{\partial}v(K') = m \sum_{j=1}^{q-1} (-1)^j z_{k_j}^m v(K' \setminus k_j) + (-1)^{q-1+|K'|} u(1, \ldots, n \setminus K') \tag{5.59}$$

where $|K'| = k_1 + \ldots + k_{q-1}$ as usual. We now construct a $(0, n-q-2)$-forms $v(K)$ satisfying (5.59) for any q-tuple $K = (k_1, \ldots, k_q)$ of integers from $\{1, \ldots, n-1\}$ (with K' replaced by K). Let

$$w(K) = m \sum_{j=1}^q (-1)^j z_{k_j}^m v(K \setminus k_j) + (-1)^{q+|K|} u(1, \ldots, n \setminus K).$$

Then it follows from (5.58) that

$$\bar{\partial}w(K) = m \sum_{j=1}^q (-1)^j z_{k_j}^m \bar{\partial}v(K \setminus k_j) + (-1)^{q+|K|} \bar{\partial}u(1, \ldots, n \setminus K)$$

$$= m \sum_{j=1}^q (-1)^j z_{k_j}^m \left(m \sum_{1 \leq i < j} (-1)^i z_{k_i}^m v(K \setminus k_j, k_i) + m \sum_{j < i \leq q} (-1)^{i-1} z_{k_i}^m v(K \setminus k_j, k_i)\right)$$

$$-(-1)^{q+|K|-k_j}u(1,\ldots,n\setminus(K\setminus k_j)))+(-1)^{q+|K|}\overline{\partial}u(1,\ldots,n\setminus K)$$

$$=(-1)^{q+|K|}\Big(-m\sum_{j=1}^{q}(-1)^{j-k_j}z_{k_j}^m u(1,\ldots,n\setminus(K\setminus k_j))+\overline{\partial}u(1,\ldots,n\setminus K)\Big)$$

$$=(-1)^{q+|K|}\Big(-m\sum_{j=1}^{q}z_{k_j}^m u(k_j,(1,\ldots,n\setminus K))+\overline{\partial}u(1,\ldots,n\setminus K)\Big)=0$$

Therefore, by the hypothesis, there exists a $v(K)\in L^2_{(0,n-q-2)}(\Omega)$ such that $\overline{\partial}v(K)=w(K)$. Since $w(K)$ is skew-symmetric with respect to indices K, we may likewise choose $v(K)$ so that it is skew-symmetric with respect to its indexes. This then concludes the inductive step.

Now let

$$F=w(1,\ldots,n-1)=m\sum_{j=1}^{n-1}z_j^m v((1,\ldots,n-1)\setminus j)-(-1)^{n+\frac{n(n-1)}{2}}u(n),$$

where $u(n)=-\overline{z}_n^m/r_m$, as given by (5.57). Then $F(z)\in L^2(\Omega)$ and $\overline{\partial}F(z)=0$. By the hypothesis, $F(z)$ has a holomorphic extension to $\widetilde{\Omega}$. By Lemma 5.6.3, we have

$$F(0,z_n)=-(-1)^{n+\frac{n(n-1)}{2}}u(n)(0,z_n)=(-1)^{n+\frac{n(n-1)}{2}}z_n^{-m}.$$

for z_n near z_n^0. (Recall that $z^0\in\Omega$ is in the same connected component of $\{z'=0\}\cap\widetilde{\Omega}$ as the origin.) This contradicts the analyticity of F near the origin.

Finally, we prove Proposition 5.6.2 in its full generality. The proof essentially combine those of the previous two cases. For any integers $\alpha\geq 0$, $m\geq 1$, and $q\geq 1$, and for any $\{k_1,\ldots,k_{q-1}\}\subset\{1,2,\ldots,n-1\}$, let

$$u_{\alpha,m}(k_1,\ldots,k_q)=\frac{(\alpha+q-1)!\overline{z}_n^{m\alpha}(\overline{z}_{k_1}\cdots\overline{z}_{k_q})^{m-1}}{r_m^{\alpha+q}}\sum_{j=1}^{q}(-1)^j\overline{z}_{k_j}\widetilde{d\overline{z}_{k_j}}$$

Then as in the previous case, we have

$$\overline{\partial}u_{\alpha,m}(k_1,\ldots,k_q)=-\frac{(\alpha+q)!m\overline{z}_n^{m\alpha}\overline{z}_K^{m-1}}{r_m^{\alpha+q+1}}\Big(r_m d\overline{z}_K$$

$$+\Big(\sum_{\ell=1}^{n}\overline{z}_\ell^{m-1}z_\ell^m d\overline{z}_\ell\Big)\wedge\Big(\sum_{j=1}^{q}(-1)^j\overline{z}_{k_j}\widetilde{d\overline{z}_{k_j}}\Big)\Big)$$

$$=m\sum_{\ell=1}^{n-1}z_\ell^m u_{\alpha,m}(\ell,k_1,\ldots,k_q).$$

In particular, $u_{\alpha,m}(1,\ldots,n)$ is $\overline{\partial}$-closed. Our next goal is to solve the $\overline{\partial}$-equation in L^2-spaces inductively with the $(0,n-1)$-forms $u_{\alpha,m}(1,\ldots,n)$ as the initial data, and eventually produce an L^2-holomorphic function on $b\Omega$. This holomorphic function has a holomorphic extension to Ω. By the method of the construction, the extension has a singularity at the origin, which leads to a contradiction with the hypothesis. We now provide the details.

Let $S_q = \mathcal{N}(Q_q)$. By Proposition 5.3.7, $\mathcal{R}(\bar{\partial}_{q-1}) = \mathcal{N}(\bar{\partial}_q) \cap S_q^\perp$, $1 \leq q \leq n - 1$. Let M be a integer such that $M > \dim S_q$ for all $1 \leq q \leq n - 1$. Fix an integer $m \geq n - 1$. Let $\mathcal{F}_0$ be the linear span of $\{u_{\alpha,m}(1,\ldots,n);\ \alpha = 1,\ldots,M^{n-1}\}$. For any $u \in \mathcal{F}_0$ and for any $\{k_1,\ldots,k_{q-1}\} \subset \{1,\ldots,n-1\}$, we set

$$u(k_1,\ldots,k_{q-1},n) = \sum_{j=1}^{k} c_j u_{\alpha_j,m}(k_1,\ldots,k_{q-1},n)$$

if $u = \sum_{j=1}^{k} c_j u_{\alpha_j,m}(1,\ldots,n)$. We decompose $\mathcal{F}_0$ into a direct sum of M^{n-2} subspaces, each of which is M-dimensional. Since $\dim(S_{n-1}) < M$ and $u_{\alpha,m}(1,\ldots,n) \in \mathcal{N}(\bar{\partial}_{n-1})$, there exists a non-zero form u in each of the subspaces such that $\bar{\partial}v_u(\emptyset) = u$ for some $v_u(\emptyset) \in L^2_{(0,n-2)}(\Omega)$. Let $\mathcal{F}_1$ be the M^{n-2}-dimensional linear span of all such u's. We extend $u \mapsto v_u(\emptyset)$ linearly to all $u \in \mathcal{F}_1$.

For $0 \leq q \leq n - 1$, we use induction on q to construct an M^{n-q-2}-dimensional subspace $\mathcal{F}_{q+1}$ of $\mathcal{F}_q$ with the properties that for any $u \in \mathcal{F}_{q+1}$, there exists $v_u(k_1,\ldots,k_q) \in L^2_{(0,n-q-2)}(\Omega)$ for all $\{k_1,\ldots,k_q\} \subset \{1,\ldots,n-1\}$ such that $v_u(k_1,\ldots,k_q)$ depends linearly on u; $v_u(k_1,\ldots,k_q)$ is skew-symmetric with respect to indices $K = (k_1,\ldots,k_q)$; and

$$\bar{\partial}v_u(K) = m\sum_{j=1}^{q} (-1)^j z_{k_j}^m v_u(K \setminus k_j) + (-1)^{q+|K|} u(1,\ldots,n \setminus K),$$

where $|K| = k_1 + \ldots + k_q$.

We now show how to construct $\mathcal{F}_{q+1}$ and $v_u(k_1,\ldots,k_q)$ for $u \in \mathcal{F}_{q+1}$ and $\{k_1,\ldots,k_q\} \subset \{1,\ldots,n-1\}$ once $\mathcal{F}_q$ has been constructed. For any $u \in \mathcal{F}_q$ and any $\{k_1,\ldots,k_q\} \subset \{1,\ldots,n-1\}$, write $K = (k_1,\ldots,k_q)$, and let

$$w_u(K) = m\sum_{j=1}^{q} (-1)^j z_{k_j}^m v_u(K \setminus k_j) + (-1)^{q+|K|} u(1,\ldots,n \setminus K).$$

Then as in the previous case,

$$\bar{\partial}w_u(K) = (-1)^{q+|K|}\Big(-m\sum_{j=1}^{q} z_{k_j}^m u(k_j,(1,\ldots,n \setminus K)) + \bar{\partial}u(1,\ldots,n \setminus K)\Big) = 0.$$

We again decompose $\mathcal{F}_q$ into a direct sum of M^{n-q-2} linear subspaces, each of which is M-dimensional. Since $\dim(S_{n-q-2}) < M$ and $\bar{\partial}w_u(K) = 0$, there exists a non-zero form u in each of these subspaces such that $\bar{\partial}v_u(K) = w_u(K)$ for some $v_u(K) \in L^2_{(0,n-q-2)}(\Omega)$. Since $w_u(K)$ is skew-symmetric with respect to indices K, we may choose $v_u(K)$ to be skew-symmetric with respect to K as well. The subspace $\mathcal{F}_{q+1}$ of $\mathcal{F}_q$ is then the linear span of all such u's.

Note that $\dim(\mathcal{F}_{n-1}) = 1$. Let u be any non-zero form in $\mathcal{F}_{n-1}$ and let

$$F(z) = w_u(1, \ldots, n-1) = m \sum_{j=1}^{n-1} z_j^m v_u(1, \ldots, n-1 \setminus j) - (-1)^{n + \frac{n(n-1)}{2}} u(n).$$

Then $F \in L^2(\Omega)$ and $\bar{\partial} F = 0$. Therefore, F is holomorphic on Ω and hence has a holomorphic extension to $\widetilde{\Omega}$. Restricting to z_n-plane as in the previous case, we then arrive at a contradiction to the analyticity of F near the origin. We therefore conclude the proof of Proposition 5.6.2. ∎

Positivity of the $\bar{\partial}$-Neumann Laplacian can also be used to characterize pseudoconvexity on domains with holes in $\mathbb{C}^n$. We refer the reader to [FLS] for the following and other related results.

Theorem 5.6.4 *Let $\widetilde{\Omega}$ be a bounded domain with connected complement in $\mathbb{C}^n$ and let $D \subset\subset \widetilde{\Omega}$ be a relatively compact open subset with connected complement in $\widetilde{\Omega}$. Let $\Omega = \widetilde{\Omega} \setminus \overline{D}$. Suppose $\widetilde{\Omega}$ has Lipschitz boundary and D has C^2-boundary. Then both $\widetilde{\Omega}$ and D are pseudoconvex if and only if the $\bar{\partial}$-Neumann Laplacian $\Box_q$ on $(0, q)$-forms on Ω is positive for $1 \leq q \leq n-2$ and gap positive for $q = n-1$.*

REMARK 5.6.5 (1) It can be proved, using the close graph theorem and the fact that $\bar{\partial}$ is a closed operator, that if $H^{0,q}_{L^2}(\Omega)$ is finite dimensional then $\mathcal{R}(\bar{\partial}_{q-1})$ is closed. Thus the assumption in Proposition 5.6.2 is equivalent to statement (3) in Theorem 5.6.1.

(2) It follows from Theorem 3.1 in [H04] that for a domain Ω in $\mathbb{C}^n$ with C^3-boundary, if $\inf \sigma_e(\Box_q) > 0$ for some q between 1 and $n-1$, then Levi-form of $b\Omega$ cannot have exactly $n-q-1$ positive and q negative eigenvalues. Characterization of a smooth bounded domain in $\mathbb{C}^n$ with $\inf \sigma(\Box_q) > 0$ for a fixed q is still unsolved.

(3) For any domain in $\mathbb{C}^n$, if the Dolbeault cohomology group on $(0, q)$-form for a fix q is finite dimensional, then it is trivial (cf. [L66]). The L^2-analogue of this result–even on *bounded* domains–remains open. The following related conjecture is also open: If Ω is an *unbounded* pseudoconvex domain Ω in $\mathbb{C}^n$, then the Bergman space $A^2(\Omega)$–the space of square integrable holomorphic functions on Ω–is either trivial or infinite dimensional. This conjecture holds for domains in $\mathbb{C}$ but fails in $\mathbb{C}^2$ if the pseudoconvexity assumption is removed ([W84]; see [PZ17] and references therein for related results).

Acknowledgments: This work was supported in part by the National Science Foundation grant DMS-2055538.

References

[AG62]   Aldo Andreotti and Hans Grauert, *Théorème de finitude pour la co-homologie des espaces complexes*, Bull. Soc. Math. France **90** (1962), 193–259.

[BSt99] Harold P. Boas and Emil J. Straube, *Global regularity of the $\bar{\partial}$-Neumann problem: a survey of the L^2-Sobolev theory*, Several Complex Variables (M. Schneider and Y.-T. Siu, eds.), MSRI Publications, vol. 37, 79–112, 1999.

[B02]   Judith Brinkschulte, *Laufer's vanishing theorem for embedded CR manifolds*, Math. Z. **239**(2002), 863–866.

[Br83]   Thorsten Broecker, *Zur L^2-Kohomologie beschränkter Gebiete*, Bonner Mathematische Schriften, vol. 145, Universität Bonn, 1983.

[C83]   David Catlin, *Necessary conditions for subellipticity of the $\bar{\partial}$-Neumann problem*, Ann. Math. **117** (1983), 147–171.

[Ca87]   _____ , *Subelliptic estimates for the $\bar{\partial}$-Neumann problem on pseudoconvex domains*, Ann. of Math. (2) **126** (1987), no. 1, 131–191.

[Ch13]   Bo-Yong Chen, *The Morrey-Kohn-Hörmander formula implies the twisted Morrey-Kohn-Hörmander formula* (unpublished notes), arXiv:1310.3816, 2013.

[CS99]   So-Chin Chen and Mei-Chi Shaw, *Partial differential equations in several complex variables*, AMS/IP, 2000.

[Ch99]   Michael Christ, *Remarks on global irregularity in the $\bar{\partial}$-Neumann problem*, Several Complex Variables (M. Schneider and Y.-T. Siu, eds.), MSRI Publications, vol. 37, 161–198, 1999.

[DK99]   J. D'Angelo and J. J. Kohn, *Subelliptic estimates and finite type*, Several Complex Variables (M. Schneider and Y.-T. Siu, eds.), MSRI Publications, vol. 37, 199–232, 1999.

[D95]   E. B. Davies, *Spectral theory and differential operators*, Cambridge Studies in advanced mathematics, vol. 42, Cambridge University Press, 1995.

[FK72]   G. B. Folland and J. J. Kohn, *The Neumann problem for the Cauchy-Riemann complex*, Annals of Mathematics Studies, no. 75, Princeton University Press, 1972.

[FS87]   John Erik Fornæss and Berit Stensones, *Lectures on counterexamples in several complex variables*. Reprint of the 1987 original. AMS Chelsea Publishing, Providence, RI, 2007.

[Fu05]   Siqi Fu, *Hearing pseudoconvexity with the Kohn Laplacian*, Math. Ann. **331** (2005), 475–485.

[Fu08]   _____ , *Hearing the type of a domain in C2 with the d-bar-Neumann Laplacian*, Adv. in Math. **219** (2008), 568–603.

[FLS]   Siqi Fu, Christine Laurent-Thiébaut, and Mei-Chi Shaw, *Hearing pseudoconvexity in Lipschitz domains with holes via $\bar{\partial}$*, Math. Zeit. **287**(2017), 1157–1181.

[FS01]   Siqi Fu and Emil J. Straube, *Compactness in the $\bar{\partial}$-Neumann problem*, Complex Analysis and Geometry, Proceedings of Ohio State University Conference, vol. 9, 141-160, Walter De Gruyter, 2001.

[H65]   Lars Hörmander, *L^2 estimates and existence theorems for the $\bar{\partial}$ operator*, Acta Math. **113** (1965), 89–152.

[H91]   _____ , *An introduction to complex analysis in several variables*, third ed., Elsevier Science Publishing, 1991.

[H03]   _____ , *A history of existence theorems for the Cauchy-Riemann complex in L^2 spaces*, J. Geom. Anal. **13** (2003), no. 2, 329–357.

[H04]   _____ , *The null space of the $\bar{\partial}$-Neumann operator*, Ann. Inst. Fourier (Grenoble) **54** (2004), 1305–1369.

[Ka66]   Marc Kac, *Can one hear the shape of a drum?* Amer. Math. Monthly **73** (1966), 1–23.

[Ko63]   J. J. Kohn, *Harmonic integrals on strongly pseudo-convex manifolds, I*, Ann. Math. **78** (1963), 112–148.

[Ko64]   _____ , *Harmonic integrals on strongly pseudo-convex manifolds, II*, Ann. Math. **79** (1964), 450–472.

[KN65]   J. J. Kohn and L. Nirenberg, *Non-coercive boundary value problems*, Comm. Pure Appl. Math. **18** (1965), 443–492.

[Kr01]   Steven G. Krantz, *Function theory of several complex variables*, AMS Chelsea Publishing, Providence, RI, 2001.

[L66]   Henry B. Laufer, *On sheaf cohomology and envelopes of holomorphy*, Ann. Math. **84** (1966), 102–118.

[L75]   _____ , *On the finite dimensionality of the Dolbeault cohomology groups*, Proc. Amer. Math. Soc. **52**(1975), 293–296.

[O82]   T. Ohsawa, *Isomorphism theorems for cohomology groups of weakly 1-complete manifolds*, Publ. Res. Inst. Math. Sci. **18** (1982), no. 1, 191–232.

[O88]   Takeo Ohsawa, *Complete Kähler manifolds and function theory of several complex variables*, Sugaku Expositions **1** (1988), 75–93.

[PZ17]   P. Pflug and W. Zwonek, L_h^2-*functions in unbounded balanced domains*, J. Geom. Anal. **27** (2017), 2118–2130.

[Se53]   J.-P. Serre, *Quelques problèmes globaux relatifs aux variétés de Stein*, Colloque sur les Fonctions de Plusieurs Variables, 57-68, Brussels, 1953.

[Sh92]   Mei-Chi Shaw, *Local existence theorems with estimates for $\bar{\partial}_b$ on weakly pseudo-convex CR manifolds*, Math. Ann. **294** (1992), no. 4, 677–700.

[Sh09]   _____, *The closed range property for $\bar{\partial}$ on domains with pseudo-concave boundary*, preprint, 2009, in this volume.

[Siu67] Yum-Tong Siu, *Non-countable dimensions of cohomology groups of analytic sheaves and domains of holomorphy*, Math. Z. **102** (1967), 17–29.

[St09]   Emil Straube, *Lectures on the L^2-Sobolev theory of the $\bar{\partial}$-Neumann problem*, to appear in the series ESI Lectures in Mathematics and Physics, European Math. Soc.

[W84]   Jan Wiegerinck, *Domains with finite dimensional Bergman space*, Math. Z. **187** (1984), 559–562.

6

Symmetry and Art

Emily Gullerud and James S. Walker

CONTENTS

6.1 Introduction

The connection between symmetry and artistic design has a long history (see e.g., Weyl [30]). In this paper, we describe the application of domain coloring to the creation of symmetric designs and animations. First, we examine

DOI: 10.1201/9781315160658-6

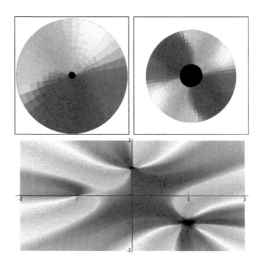

FIGURE 6.1

Top left: Color plot for $w = z$ over $[-2, 2] \times [-2, 2]$; a basic color map $\mathcal{C}$. Top right: Color plot for $w = z^2$. Bottom: SageMath color plot for $w = 3(z + 1)(z - i)^2(z - 1 + i)^3$. (First published in "Creating Symmetry: The Artful Mathematics of Wallpaper Patterns" by James S. Walker, Notices of the Amer. Math. Soc., Vol 62, No 11 (Dec. 2015), pp. 1350–1354, published by the American Mathematical Society. ©2015 American Mathematical Society.)

domain coloring in its original setting as a method for sketching graphs of complex-valued functions. Then we will describe the creation of designs in the Euclidean complex plane and the non-Euclidean hyperbolic upper half-plane.

At the very end of the twentieth century, Farris [11] played a principal role in developing the use of domain coloring to sketch graphs of functions, $f: \mathbb{C} \to \mathbb{C}$. The idea is to use a *color wheel*, a well-known tool in the visual arts. A basic example of a color wheel is shown at the top left of Figure 6.1. This color wheel is used to mark locations in the complex plane. For each value of w, the value of $\mathcal{C}(w)$ is a unique color (at least in principle). For example, on the top left of Figure 6.1, the values of w that are near i are colored greenish-yellow, while values near -1 have a light blue tint. As values approach zero they turn black, and beyond a certain radius they are all colored white. By composing a function $f(z)$ with $\mathcal{C}$, we get a function $\mathcal{C}(f(z))$ that gives a color portrait of $f(z)$. For instance, on the top right of Figure 6.1, a color map for $w = z^2$ is shown. Notice, how the colors near $w = 0$ cycle twice through the rainbow as we move once around $w = 0$. The connection to winding numbers is thereby made visually evident. Also, contour lines meet at right angles away from $w = 0$, just as they do for the function $w = z$. This illustrates conformality of $w = z^2$ away from the origin. There is a software now that produces color plots with great ease [23, 27]. At the bottom of Figure 6.1, we show a color plot of the function $w = 3(z+1)(z-i)^2(z-1+i)^3$. This plot was produced with the free SageMath system [23, 24]. Just these two commands were needed:

```
3*(z+1)*(z-i)^2*(z-1+i)^3
complex_plot(f, (-2, 2), (-2, 2))
```

The plot that SageMath produced clearly marks the location of the zeros at -1, i, and $1 - i$ and their multiplicities of 1, 2, and 3, respectively. All this information is encoded in the number of times the colors of the rainbow are cycled through in the neighborhood of each zero. There are several nice examples of color plots at the web site created by Crone [7], including plots of branching in Riemann surfaces. The reader may also wish to explore color plotting using the free software, SYMMETRYWORKS [27].

6.1.1 Pseudocode for creating color plots

We have found the most challenging part of producing color plots is the creation of a domain colored image using a specific color map. In Figure 6.2, we provide pseudocode for this procedure. It is usually a straightforward task to supplement this procedure with additional code for creating symmetric designs using various functions. For color maps, we will adopt the clever idea of Farris of using color photos of natural scenes, rather than a simple color wheel. The

Domain Coloring Procedure

```
CMp = loadimage(filename) %load the RGB image color map
R = number of rows of CMp and C = number of columns of CMp
ScaleFactor = 10 %a scaling factor for units in color map
mCent = R / 2 and nCent = C / 2 %coordinates of image center
xinc = (2*ScaleFactor) / R and yinc = (2*ScaleFactor) / C %x, y increments
%Create color-mapped display.
M = 1000 %increase M for higher resolution
L = 2  %increase/decrease L to zoom out/in
incr = 2*L/M  and  [x, y] = rectangular grid of points from -L to L spaced by incr
Initialize wImg %Matrix wImg, all zeros for RGB values, M rows, M columns.
z = x + iy  %create array of complex values
w = f(z) %compute values of the function f over grid
s = real(w) and t = imag(w) %real and imaginary parts of function values
for m = 1 to M
  for n = 1 to M
    %Calculate coordinates  p and q for color map
    p = mCent + round(s(m,n)/xinc)  and  q = nCent + round(t(m,n)/yinc)
    if p>0 and p<R then
      if q>0 and q<C then
      wImg(m,n) = CMp(p,q) %assign RGB values from color map
      end if
    end if
  end for
end for
Display(wImg) %Display RGB image wImg
```

FIGURE 6.2
Pseudocode for applying a color map to function $w = f(z)$.

FIGURE 6.3
Images for color maps, from public domain stock images found on the Internet. Left to right: *Waratah flower* with yellow border, *Reptile, Buoy.*

color images we used for the designs in this paper are shown in Figure 6.3. For those readers not wishing to reinvent the wheel, the software [27] can be used for artistic designs like the ones we discuss in the remainder of this paper.

6.2 Designs in the Euclidean Complex Plane

In the geometry of the Euclidean plane, there are three basic symmetry operations (congruences). They are (1) translation, (2) rotation, and (3) reflection. For a complex variable, $z = x + iy$, these symmetry operations are exemplified by the following mappings:

1. Translation: $T_c : z \longrightarrow z + c$, where $c = a + ib$ is a complex constant.

2. Reflection through the x-axis: $R_x : z \longrightarrow \overline{z}$, i.e., $z = x + iy \longrightarrow x - iy$.

3. Rotation by θ around the origin: $\rho_\theta : z \longrightarrow e^{i\theta} z$, where $e^{i\theta} = \cos\theta + i\sin\theta$.

Any Euclidean symmetry (congruence) can be expressed as a finite composition of these basic mappings. This set of all symmetries of the plane is a mathematical group, the **Euclidean group**, denoted by E_2.

For creating symmetric designs, we will apply domain coloring to a function $w = f(z)$ that has symmetry with respect to a group of transformations. If S is a transformation of the complex plane $\mathbb{C}$, then the function f is symmetric with respect to S if

$$f\big(S(z)\big) = f(z) \quad \text{for all } z \in \mathbb{C} \tag{6.1}$$

The set of symmetries of a function f is a mathematical group, S_f. Our method for creating designs is to apply domain coloring to functions that have some pre-assigned symmetry group. For that purpose, the Euclidean group is too large. The only functions symmetric under all mappings in E_2 are constant functions. To obtain non-trivial symmetric designs we restrict to discrete subgroups of E_2. We will examine symmetric designs created from symmetries with respect to these three discrete subgroups:

1. Symmetry with respect to a lattice of two-dimensional translations

$$T_{mu+nv} : z \longrightarrow z + mu + nv \quad \text{all } m, n \in \mathbb{Z},$$

where u and v are non-collinear complex numbers. Enforcing this symmetry creates a tessellation over the lattice in $\mathbb{C}$ generated by u and v.

2. Symmetry with respect to rotation about the origin by an nth root of unity:

$$\rho_{2\pi/n} : z \longrightarrow e^{i2\pi/n} z \quad \text{for fixed } n \in \mathbb{Z}, n \neq 0$$

Enforcing this symmetry creates a **rosette**, having n-fold rotational symmetry about the origin.

3. Symmetry with respect to both a lattice of two-dimensional translations and rotation by an nth root of unity about the origin. Symmetric designs of this type are only possible for rotations $\rho_{2\pi/n}$ with $n = 2, 3, 4$, or 6. We shall refer to a design with this symmetry as a **rotationally symmetric wallpaper pattern**.

All of these basic examples can be modified to add more symmetry to a design, including various reflective symmetries. We will discuss these modifications as we examine these three basic symmetric designs.

6.2.1 Translational symmetry

On the left of Figure 6.4, we show a design having symmetry with respect to a lattice of two-dimensional translations. The lattice is rectangular, and is generated by the non-collinear complex numbers $u = 2100$ and $v = 1700i$. The positive integers, 2100 and 1700, are the width and height in pixels of the image used for the color map. That color map image is the *Waratah flower* image shown in Figure 6.3. The function $f(z)$ that we used to create the design is, for $z = x + iy$,

$$f(x + iy) = [x \bmod 2100] + i[y \bmod 1700] \tag{6.2}$$

This formula assumes that the lower left corner of the *Waratah flower* image is located at the origin in $\mathbb{C}$, and uses mod-equivalence from elementary number theory to ensure that f is symmetric with respect to translations over the lattice generated by u and v. The image on the left of Figure 6.4 resembles a sheet of postage stamps, rather than a complex artistic design. Although if one began with an image of a person (say Marilyn Monroe), or a commodity (say a Campbell's soup can) then one could easily produce designs reminiscent of some of Andy Warhol's famous pictures.

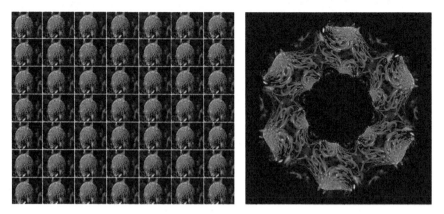

FIGURE 6.4
Left: Design having translational symmetry over a square lattice. Right: Rosette design having 6-fold rotational symmetry.

6.2.2 Rotational symmetry

6.2.2.1 Rosettes

On the right of Figure 6.4, we show an example of a design having symmetry with respect to rotation about the origin by a 6th root of unity. In other words, it has 6-fold rotational symmetry. The color map used to create the image is the *Waratah flower* image shown in Figure 6.3, *but without the yellow border*. The function we used to create the design is

$$f(z) = (1 + i) + i\,\frac{z^6}{4} + z^{-6} \tag{6.3}$$

Since $\left(e^{i2\pi/6}\right)^6 = 1$, it follows that $f(z)$ is symmetric with respect to the rotation $\rho_{2\pi/6}$, i.e., it has 6-fold rotational symmetry. We choose array values for $z = x + iy$ so that the rosette design is a square image of dimensions 1000×1000 pixels. The black colored regions in the design correspond to values of f lying outside the region in $\mathbb{C}$ corresponding to the color map. Later, we will show an interesting alternative way to color these values.

On the left of Figure 6.5, we show an example of a design having 5-fold rotational symmetry. It also has symmetry with respect to reflection through the x-axis. The color map used to create the image is the *Reptile* image shown in Figure 6.3. The function we used to create the design is

$$f(z) = (2 + 3i)\left(z^5 + \overline{z}^5\right) + i\left(z^6\,\overline{z}^1 + \overline{z}^6 z^1\right) + \frac{i}{2000}\left(z^4\,\overline{z}^{-6} + z^{-6}\,\overline{z}^{-4}\right) \tag{6.4}$$

The terms of $f(z)$ are grouped so that $f(\overline{z}) = f(z)$ clearly holds. Therefore, f is symmetric with respect to the x-axis reflection R_x. To verify that f is symmetric with respect to the rotation $\rho_{2\pi/5}$, we note that f is an example of a *finite* sum of the form:

$$f(z) = \sum_{\substack{m, n \\ m \equiv n \bmod 5}} a_{m,n} z^m \bar{z}^n \tag{6.5}$$

which is symmetric with respect to 5-fold rotation about the origin. The relation between Equations (6.4) and (6.5) is that we can insure symmetry with respect to reflection through the x-axis by requiring $a_{m,n} = a_{n,m}$. This condition on the coefficients, $\{a_{m,n}\}$, follows by considering $f(z) = f(\bar{z})$ in terms of the coefficients of the basis functions $z^m \bar{z}^n$ in Equation (6.5).

FIGURE 6.5
Left: Rosette having 5-fold rotational symmetry and reflection symmetries. Right: Rectangular view of rosette design with 6-fold rotational symmetry plus curved lattice symmetry.

On the right of Figure 6.5, we show another design which combines rotational symmetry with the mod operation used in Section 6.2.1. The function we used is the following composition:

$$z \xrightarrow{f} (1+i) + \frac{i}{4} z^6 + z^{-6} = u + iv \xrightarrow{g} [u \bmod 1000] + i[v \bmod 1000] \tag{6.6}$$

We are applying the function g to the square image of the rosette design on the right of Figure 6.4. This is, in effect, using an infinite version of the tessellation design in Figure 6.4 as color map. (This is the alternative way, mentioned above, for coloring pixels lying outside of a finite color map.) The composite function satisfies $g\Big(f\big(e^{i2\pi/6}z\big)\Big) = g\Big(f(z)\Big)$. Hence, the design on the right of Figure 6.5 has 6-fold rotational symmetry. This design also has several interesting features from both artistic and mathematical perspectives:

(1) For large magnitude z, we have $f(z) \sim iz^6/4$. Consequently, when composed with g, the lattice of images of the *Waratah* flower are (approximately) pre-images of $iz^6/4$ acting on the cells of a rectangular lattice of *Waratah* flower images (as shown on the left of Figure 6.4 but without the yellow grid lines). That explains the rotated, shrunken appearance of the flowers along a curved lattice. These flowers are most easily visible at the four corners of the square image. The curved grid of the lattice is shown clearly on the right of Figure 6.6, where we included the yellow borders in the *Waratah flower* image for our color map.

(2) The square framing of the design, along with the appearance of multiple small flowers approaching the four corners of the frame, draws the viewer's attention to the 4-fold rotational symmetry of the design's frame. This may lead to an ambiguity in the viewer's mind as to the validity of 6-fold rotational symmetry for the design itself. From an artistic perspective, we like this ambiguity. In fact, the 6-fold symmetry only holds for points in the image that remain within the image boundaries upon rotation by $2\pi/6$. On the left of Figure 6.6 we show a larger scale view of the design, within a circular frame. This latter design clearly displays 6-fold rotational symmetry.

(3) For small magnitude z, we have $f(z) \sim z^{-6}$. Consequently, the curved lattice contains an infinity of pre-images approaching the pole of multiplicity 6 at $z = 0$. The sizes of the pre-images decrease rapidly to 0 as $z \to 0$, hence they become incapable of realization with the digital images that we are working with here. This explains the random looking colored pixels lying within a region interior to the six flower-like parts of image. Those flower-like parts lie across the unit-circle, where $|z| \approx 1$. Some of the curved lattice within the unit-circle is visible in the image on the right of Figure 6.6.

6.2.2.2 Animations

We have also created animations involving our symmetric designs. We will just give one illustration of the ideas involved. The animation we describe can be found at this link:

https://www.youtube.com/watch?v=Pf1vJPywXWs

We produced this animation in the following way. First, we selected a unit-length complex number $p = e^{i\theta}$ for small positive θ. We then created a succession of designs using the mappings:

$$z \longrightarrow (1+i) + p^n \, (i/4)z^6 + \frac{1}{p^n} \frac{1}{z^6} = u + iv \xrightarrow{\;c\;} [u \bmod 1000] + i[v \bmod 1000]$$

$$(6.7)$$

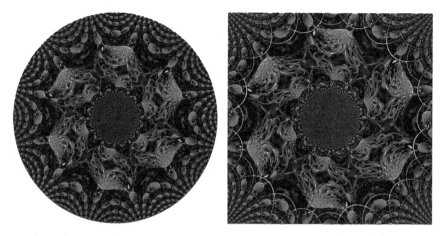

FIGURE 6.6
Left: Circular view of rosette design with 6-fold rotational symmetry plus curved lattice symmetry. Right: Curved lattice lines marked on rosette design with curved lattice symmetry.

for $n = 0, 1, 2, \ldots, 99$. This produced 100 separate designs that are displayed one after another to create the animation. The interesting thing about this animation is that it will appear to rotate in the positive angular direction on its outer part versus a negative angular rotation on its inner part. The reason for this is that p^n has angle $n\theta$ for $n = 0, 1, 2, \ldots, 99$, which is a sequence of increasing angles in the counter-clockwise direction. The outer part of the design corresponds to values of z that have lengths larger than 1, and for larger values of $|z|$ we have

$$(1 + i) + p^n\,(i/4)z^6 + \frac{1}{p^n}\,\frac{1}{z^6} \approx p^n\,(i/4)z^6$$

so the outer part of the design successively rotates through the angles $n\theta = \arg(p^n)$, $n = 0, 1, 2, \ldots, 99$. A similar argument shows that for z close to 0, the design simultaneously rotates through the angles $-n\theta$.

6.2.3 Rotationally symmetric wallpaper designs

6.2.3.1 Four-fold and two-fold wallpaper symmetry

In Figure 6.7, we show a wallpaper design that is symmetric with respect to 4-fold symmetry and various reflection symmetries. For this design, we used the *Reptile* image. To create a function f having square lattice symmetry and four-fold rotational symmetry, we follow the procedure described by Farris [9]. First, we use as a basis the set of complex exponentials $\{E_{m,n}(z) = e^{2\pi i(mx+ny)}\}$ for all $m, n \in \mathbb{Z}$ and $z = x + iy \in \mathbb{C}$. Any finite, or

convergent, sum $\sum a_{m,n} E_{m,n}(z)$ is guaranteed to have the required translational symmetry. To obtain rotational symmetry, we employ group averaging.
In this case, the group is the rotations given by powers of $i = e^{2\pi i/4}$. The
group average $W_{m,n}$ of $E_{m,n}$ is defined by $W_{m,n}(z) = \frac{1}{4} \sum_{k=0}^{3} E_{m,n}(i^k z)$. Any
finite, or convergent, linear combination $f(z) = \sum a_{m,n} W_{m,n}(z)$ is guaran-
teed to have both square lattice symmetry and four-fold rotational symmetry
about the origin.

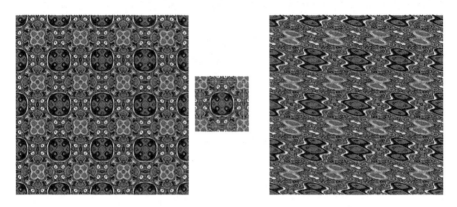

FIGURE 6.7
Left: 4-fold rotationally symmetric design, with additional symmetries. Unit cell
on its right marked in green. Right: 2-fold rotationally symmetric design with no
additional symmetries.

The specific function f that we employed is

$$f(z) = [W_{1,0}(z) + W_{0,-1}(z)] + 0.5 [W_{1,5}(z) + W_{-5,-1}(z)] \\ + 0.1i [W_{-2,4}(z) + W_{-4,2}(z)] - 0.05i [W_{-6,3}(z) + W_{-3,6}(z)] \tag{6.8}$$

In this equation, the functions $\{W_{m,n}(z)\}$ are grouped in order to enforce
reflectional symmetry through the x-axis. To see this, we first rewrite the
terms of $W_{m,n}$ to remove the powers of i. The term $E_{m,n}(iz)$ satisfies

$$E_{m,n}(iz) = E_{m,n}(-y + ix) \\ = e^{i2\pi(-my+nx)} \\ = E_{n,-m}(z)$$

Iterating this relation yields $E_{m,n}(i^2 z) = E_{-m,-n}(z)$ and $E_{m,n}(i^3 z) = E_{-n,m}(z)$. Therefore,

$$W_{m,n}(z) = \frac{1}{4} [E_{m,n}(z) + E_{n,-m}(z) + E_{-m,-n}(z) + E_{-n,m}(z)] \tag{6.9}$$

Now, if we apply reflection through the x-axis to $E_{m,n}(z)$, we obtain

$$E_{m,n}(\bar{z}) = E_{m,n}(x - iy)$$
$$= E_{m,-n}(z)$$

Iterating this relation, we obtain $E_{n,-m}(\bar{z}) = E_{n,m}(z)$, $E_{-m,-n}(\bar{z}) = E_{-m,n}(z)$, and $E_{-n,m}(\bar{z}) = E_{-n,-m}(z)$. Therefore,

$$W_{m,n}(\bar{z}) = \frac{1}{4}\left[E_{m,-n}(z) + E_{n,m}(z) + E_{-m,n}(z) + E_{-n,-m}(z)\right]$$
$$= W_{-n,-m}(z)$$

Consequently, terms of the form

$$a\left[W_{m,n}(z) + W_{-n,-m}(z)\right]$$

appearing in Equation (6.8), are symmetric with respect to reflection through the x-axis.

The symmetry with respect to R_x is apparent in the design in Figure 6.7, but there are other symmetries as well. For example, it has reflectional symmetry R_y about the y-axis. Since the y-axis is a rotation by $2\pi/4$ of the x-axis, the function f must be symmetric with respect to R_y. More precisely, R_y equals the conjugation operation $\rho_{2\pi/4}^{-1} R_x \rho_{2\pi/4}$ in the symmetry group S_f. Other reflection symmetries follow from group operations, such as reflection through the line $y = x$ and reflection through the line $y = -x$.

This wallpaper design is symmetric with respect to translation by the two independent vectors 1 and i in $\mathbb{C}$. The cells of the lattice generated by 1 and i are clearly evident. To the right of the design, we show a **unit cell**, the contents of which generate the entire design via translations by $m \cdot 1 + n \cdot i$, for $m, n \in \mathbb{Z}$. One interesting feature of the design is that the center of the unit cell is a point of 4-fold rotational symmetry for the design. This can be proved as follows. Assuming this unit cell has its lower left corner at the origin, its center is $\frac{1}{2} + \frac{1}{2}i$. The following mappings

$$\frac{1}{2} + \frac{1}{2}i \xrightarrow{\cdot i} \frac{-1}{2} + \frac{1}{2}i \xrightarrow{\tau_1} \frac{1}{2} + \frac{1}{2}i$$

imply that $\tau_1 \circ \rho_{2\pi/4}$ preserves $\frac{1}{2} + \frac{1}{2}i$ and rotates the unit cell by $2\pi/4$. Consequently, $\frac{1}{2} + \frac{1}{2}i$ is a center of 4-fold rotational symmetry. A similar argument shows that the midpoint of the top side of the unit cell is a center of 2-fold rotational symmetry. Hence, by four-fold rotation, the midpoints of each side of the unit cell are centers of 2-fold symmetry.

Two-fold rotationally symmetric wallpaper patterns can be generated in multiple ways. One, rather elementary, way is to just stretch a 4-fold pattern to create a rectangular lattice. The rectangular lattice for a 2-fold design, that is not 4-fold, is generated by the basis vectors 1 and ri, where $r > 0$ and $r \neq 1$. Such a design can be generated from a 4-fold design by simply stretching or shrinking the 4-fold design in the vertical direction. With the digital images we are creating, that can be done by even the most rudimentary image processing programs. A second method would be to use a group average approach, as we did with the 4-fold case. For instance, we can use group averages $W_{m,n}(z) = \frac{1}{2}\big(E_{m,n}(z) + E_{m,n}(-z)\big)$. A finite, or convergent, linear combination $f(z) = \sum a_{m,n} W_{m,n}(z)$ will then have both 2-fold rotational and square lattice symmetry. An example is shown on the right of Figure 6.7. The function $f(z)$ that we used for this design is

$$f(z) = W_{1,0}(z) + W_{0,-1}(z) + 0.5W_{1,5}(z) + 0.1iW_{-2,4}(z) - 0.05iW_{-6,3}(z) \quad (6.10)$$

Unlike the design on the left of Figure 6.7, this design has no additional symmetries. This corresponds to the absence of grouping of related terms in Equation (6.10), of the kind that we have in Equation (6.8).

6.2.3.2 Three-fold wallpaper symmetry

On the top left of Figure 6.8, we show a wallpaper design that is symmetric with respect to 3-fold symmetry. Next to the design is an image of a unit cell for the lattice of the design, a rhombus with sides constructed from the complex numbers 1 and $\omega = e^{i2\pi/3}$. The rhombic lattice for translational symmetry is defined by the vectors $m + n\omega$, for all $m, n \in \mathbb{Z}$ and $\omega = e^{i2\pi/3}$. Every complex number z can be expanded uniquely as $z = u + v\omega$ for unique $u, v \in \mathbb{R}$. The functions, $E_{m,n}(z)$ for $m, n \in \mathbb{Z}$, are defined by $E_{m,n}(z) = E_{m,n}(u + v\omega) = e^{i2\pi(mu+nv)}$ for all $u, v \in \mathbb{R}$. These functions are periodic over the cells of the rhombic lattice generated by 1 and ω. The group averages, $\{W_{m,n}\}$, for the group of 3-fold rotations about the origin, are defined by

$$W_{m,n}(z) = \frac{1}{3}\left[E_{m,n}(z) + E_{m,n}(\omega z) + E_{m,n}(\omega^2 z)\right] \quad (6.11)$$

This definition of $W_{m,n}(z)$ ensures that it has 3-fold rotational symmetry. To verify that it has translational symmetry, we need to further examine the terms $E_{m,n}(\omega z)$ and $E_{m,n}(\omega^2 z)$. Since $z = \omega$ satisfies the factored equation $0 = (z^3 - 1) = (z - 1)(z^2 + z + 1)$, it follows that $\omega^2 = -1 - \omega$. Therefore,

$$\begin{aligned}
E_{m,n}(\omega z) &= E_{m,n}(u\omega + v\omega^2) \\
&= E_{m,n}\big(-v + (u - v)\omega\big) \\
&= e^{i2\pi\big(-mv+n(u-v)\big)} \\
&= e^{i2\pi\big(nu+(-m-n)v\big)} \\
&= E_{n,-m-n}(z)
\end{aligned}$$

Thus, $E_{m,n}(\omega z) = E_{n,-m-n}(z)$. Iterating this relation, we obtain $E_{m,n}(\omega^2 z) = E_{-m-n,m}(z)$. These relations show that

$$W_{m,n}(z) = \frac{1}{3}\left[E_{m,n}(z) + E_{n,-m-n}(z) + E_{-m-n,m}(z)\right] \quad (6.12)$$

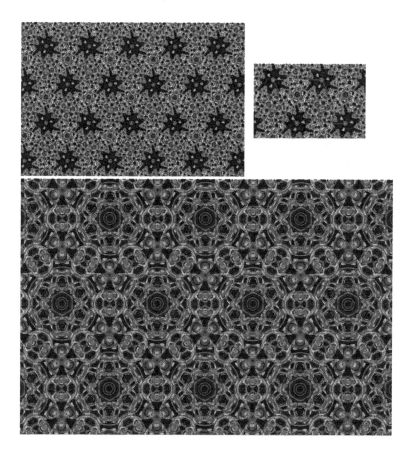

FIGURE 6.8
Top Left: Wallpaper design with 3-fold rotational symmetry. Top Right: a rhombic unit cell, marked in green. Bottom: Wallpaper design with 6-fold rotational symmetry and additional reflection symmetries.

Hence, $W_{m,n}$ is a linear combination of the basis functions $\{E_{m,n}\}$, so it also enjoys translational symmetry over the rhombic lattice. Consequently any function f defined by a finite, or convergent, sum $\sum_{m,n} a_{m,n} W_{m,n}(z)$ has both 3-fold rotational symmetry and rhombic lattice symmetry. Furthermore, because of the group operation $E_{m,n} \cdot E_{j,k} = E_{m+j,n+k}$ for the basis functions $\{E_{m,n}\}$, we can also include products of the form $W_{m,n} \cdot W_{j,k}$ in the terms for f. Specifically, the function we used to create the design on the top left of Figure 6.8 is

$$f(z) = 2W_{2,3}(z) \cdot W_{1,4}(z) + W_{1,0}(z)$$

It has 3-fold rotational symmetry because each of the factors, $W_{2,3}$ and $W_{1,4}$, and the term $W_{1,0}$ have that symmetry. Moreover, it has translational symmetry over the lattice because f is a linear combination of the basis functions $\{E_{m,n}\}$.

Remark 6.2.1. In this example, we were indeed fortunate that a rotation by $2\pi/3$ maps each basis function $E_{m,n}$, periodic over the lattice, into another such basis function. That fact allowed us to create a 3-fold rotationally symmetric wallpaper design. The **Crystallographic Restriction Theorem** below implies that this can not happen for most n-fold rotational symmetries.

6.2.3.3   Six-fold wallpaper symmetry

In Figure 6.8, we show a wallpaper design that is symmetric with respect to 6-fold symmetry and various reflection symmetries. Before we discuss the mathematics of creating this design, we will take a moment to comment on its artistic features. The design presents an appearance of interlocking circular regions. The larger regions have a central, three-armed cross that appears to be a center of three-fold rotational symmetry. Slightly smaller circular regions have dark blue, hexagonally shaped figures about their centers, which appear to have 6-fold rotational symmetry. If you relax your focus slightly, the whole design appears to float in the background as your attention rapidly shifts between these interlocking circles with different symmetries. These features slightly disguise the location of the rhombus shaped cells that form the lattice for the design.

We used the *Buoy* image for the color map of this design. To create a function f having rhombic lattice symmetry and six-fold rotational symmetry, we note that 6-fold rotational symmetry about the origin is equivalent to 3-fold rotational symmetry combined with 2-fold rotational symmetry. Two-fold rotational symmetry about the origin in $\mathbb{C}$ corresponds to the mapping $\rho_\pi : z \to -z$. Since $E_{m,n}(-z) = E_{-m,-n}(z)$, it follows that $W_{m,n}(-z) = W_{-m,-n}(z)$ for all $m, n \in \mathbb{Z}$. Consequently, if f is a finite, or convergent, sum of terms of the form

$$a\left[W_{m,n}(z) + W_{-m,-n}(z)\right]$$

then f is symmetric with respect to R_π and $R_{2\pi/3}$ and thus symmetric with respect to $R_{2\pi/6}$.

The function f that we used to create the design in Figure 6.8 is somewhat complicated. Suffice it to say that it had the following form:

$$\begin{aligned}
f(z) = {}& a\left[W_{2,3}(z) + W_{3,2}(z)\right] + a\left[W_{-2,-3}(z) + W_{-3,-2}(z)\right] \\
& + b\left[W_{1,5}(z) + W_{5,1}(z)\right] + b\left[W_{-1,-5}(z) + W_{-5,-1}(z)\right] \quad (6.13) \\
& + c\left[W_{3,4}(z) + W_{4,3}(z)\right] + c\left[W_{-3,-4}(z) + W_{-4,-3}(z)\right]
\end{aligned}$$

for certain complex constants, a, b, and c. The terms for $f(z)$ can be regrouped so that f is a linear combination of terms of the form $W_{m,n}(z) + W_{-m,-n}(z)$. Hence, f has 6-fold rotational symmetry. However, the terms for f were grouped in the pairs shown in Equation (6.13) in order to enforce reflection symmetry R_x about the x-axis. To see that this symmetry holds, we examine the effect of $R_x : z \to \bar{z}$ on $E_{m,n}$:

$$
\begin{aligned}
E_{m,n}(\bar{z}) &= E_{m,n}(u + v\bar{\omega}) \\
&= E_{m,n}(u + v(-1 - \omega)) \qquad \text{[since } \bar{\omega} = \omega^2\text{]} \\
&= e^{i2\pi\left(mu + (-m-n)v\right)} \\
&= E_{m,-m-n}(z)
\end{aligned}
$$

Thus, $E_{m,n}(\bar{z}) = E_{m,-m-n}(z)$. Iterating this relation, we obtain

$$
E_{n,-m-n}(\bar{z}) = E_{n,m}(z) \qquad \text{and} \qquad E_{-m-n,m}(\bar{z}) = E_{-m-n,n}(z)
$$

Combining these relations with Equation (6.12), we obtain $W_{m,n}(\bar{z}) = W_{n,m}(z)$. Consequently, terms of the form

$$
a\left[W_{m,n}(z) + W_{n,m}(z)\right]
$$

are symmetric with respect to reflection through the x-axis. Since f is a finite sum of such terms, it has this symmetry, too. Our design has other symmetries as well. For example, since $\rho_\pi \circ R_x : z \to -\bar{z}$, the function f is mirror symmetric through the y-axis.

At the beginning of our discussion of this 6-fold symmetric design, we mentioned circles having centers of 3-fold rotational symmetry within the rhombic cells of the design's lattice. The point $\frac{2}{3} + \frac{1}{3}\omega$ within the rhombic unit cell is a center of 3-fold symmetry. The following mappings

$$
\frac{2}{3} + \frac{1}{3}\omega \xrightarrow{\;\cdot\,\omega\;} \frac{-1}{3} + \frac{1}{3}\omega \xrightarrow{\;\tau_1\;} \frac{2}{3} + \frac{1}{3}\omega
$$

imply that $\tau_1 \circ \rho_{2\pi/3}$ preserves $\frac{2}{3} + \frac{1}{3}\omega$ and rotates the rhombic unit cell by $2\pi/3$. Consequently, $\frac{2}{3} + \frac{1}{3}\omega$ is a center of 3-fold rotational symmetry. Furthermore,

$$
\frac{1}{3} + \frac{2}{3}\omega \xrightarrow{\;\cdot\,\omega\;} \frac{-2}{3} - \frac{1}{3}\omega \xrightarrow{\;\tau_{1+\omega}\;} \frac{1}{3} + \frac{2}{3}\omega
$$

implies that $\tau_{1+\omega} \circ \rho_{2\pi/3}$ preserves $\frac{1}{3} + \frac{2}{3}\omega$, and so it is a second point of 3-fold symmetry within the rhombic unit cell.

6.2.3.4 The crystallographic restriction

We have shown that wallpaper designs can be generated with n-fold rotational symmetry when $n = 2, 3, 4,$ and 6. In fact, these are the only possible n-fold rotationally symmetric wallpaper designs.

Theorem 6.2.2 (The Crystallographic Restriction). *An n-fold rotationally symmetric wallpaper design is only possible if $n = 2, 3, 4,$ or 6.*

Proof. In the standard basis for $\mathbb{R}^2 \equiv \mathbb{C}$, the n-fold rotation $\rho_{2\pi/n}$ has matrix form

$$\begin{pmatrix} \cos(2\pi/n) & -\sin(2\pi/n) \\ \sin(2\pi/n) & \cos(2\pi/n) \end{pmatrix}$$

The trace of this matrix is $2\cos(2\pi/n)$. However, the trace is invariant under change of basis. For the basis that generates the lattice of cells for the wallpaper design, the trace of the matrix for $\rho_{2\pi/n}$ must be an integer. Hence, $2\cos(2\pi/n) = k$ for some $k \in \mathbb{Z}$. Therefore, we have

$$\cos(2\pi/n) = k/2, \quad k \in \mathbb{Z} \tag{6.14}$$

The only possible integers k for which Equation (6.14) can hold are $k = 0, \pm1, \pm2$, which yield $n = 2, 3, 4$, and 6. $\qquad\square$

There is an interesting analysis of this restriction when $n = 5$ in [10]. Some Matlab® programs for creating designs when $n = 5$ are in [28, p. 98]. These designs are related to *quasicrystals* [25, 28].

6.2.4 Summary

We have shown a number of symmetric designs generated by the application of complex analysis to the geometry of the Euclidean plane. The mathematics we have used is widely employed in crystallography [18, 25]. Many more designs, and a more thorough treatment including the relation to crystallography, can be found in the book by Farris [9]. In the next section, we describe symmetric designs that use properties of complex analysis in hyperbolic geometry.

6.3 Designs in the Hyperbolic Upper Half-Plane

We have also created designs using the symmetries in the geometry of the hyperbolic upper half-plane. Some of the designs we have created are shown in Figure 6.9. The symmetries in these designs are much different than those we discussed above for the Euclidean plane. The design at the top of Figure 6.9 is entitled *Blugold Fireworks*. It was exhibited as part of the 2018 MATHEMATICAL ART EXHIBITION held in San Diego [13]. The other two designs are more recent creations. One quite interesting feature of these designs, from a mathematical perspective, is that they display some of the principal geometric objects in the hyperbolic upper half-plane. We will describe what these principal objects are, and how the designs are constructed. But in order to do so, we first recount the basic mathematics underlying the geometry of the hyperbolic upper half-plane. References for additional details are [5, 9, 16, 20, 28].

FIGURE 6.9
Top: *Blugold Fireworks* design, using *Buoy* as color map. Middle and Bottom: Two designs using *Reptile*.

6.3.1 Geometry of the hyperbolic upper half-plane

The hyperbolic upper half-plane, $\mathbb{H}$, is the subset of $\mathbb{C}$ defined as follows:

$$\mathbb{H} = \{x + iy \mid x \in \mathbb{R}, y > 0\} \tag{6.15}$$

with differential metric

$$ds = \frac{\sqrt{dx^2 + dy^2}}{y} \tag{6.16}$$

With this metric, the **length** $\ell(\gamma)$ of a smooth curve $\gamma(t) = x(t) + iy(t)$, $a \leq t \leq b$, is defined as

$$\ell(\gamma) = \int_a^b \frac{\sqrt{x'(t)^2 + y'(t)^2}}{y(t)} \, dt = \int_a^b \frac{|z'(t)|}{\operatorname{Im} z(t)} \, dt$$

This metric is related to the metric $ds_{\mathbb{E}^+}$ for the Euclidean plane, restricted to $\mathbb{E}^+ = \{(x,y) \mid x \in \mathbb{R}, y > 0\}$, defined by

$$ds_{\mathbb{E}^+} = \sqrt{dx^2 + dy^2} \tag{6.17}$$

In other words, $ds = ds_{\mathbb{E}^+}/y$. This relation is crucial to verifying a number of important facts about the geometry of $\mathbb{H}$. We begin by discussing the isometries of $\mathbb{H}$.

6.3.1.1   Isometries of $\mathbb{H}$

The isometries of $\mathbb{H}$ are mappings $f \colon \mathbb{H} \to \mathbb{H}$ that preserve the differential metric ds. We will show in the next theorem that the set

$$S_{\mathbb{H}} = \left\{ f(z) = \frac{az+b}{cz+d} \ : \ a,b,c,d \in \mathbb{R} \text{ with } ad - bc = 1 \right\} \tag{6.18}$$

contains all the holomorphic isometries of $\mathbb{H}$.

There is also a notion of area in $\mathbb{H}$. The **area** of a region $\mathcal{U}$ will be denoted by $A(\mathcal{U})$. The area differential dA in $\mathbb{H}$ is given by

$$dA = \frac{1}{y^2} \, dx \, dy \tag{6.19}$$

and so we compute $A(\mathcal{U})$ by

$$A(\mathcal{U}) = \int_{\mathcal{U}} \frac{1}{y^2} \, dx \, dy \tag{6.20}$$

for a suitable region $\mathcal{U} \subset \mathbb{H}$. By a suitable region, we mean any region $\mathcal{U} \subset \mathbb{H}$ for which the integral in (6.20) is defined, say, as a Riemann integral. The area differential dA makes sense, by a dimensional argument, when viewed as $dA = dA_{\mathbb{E}^+}/y^2$ and noting that the length differential satisfies $ds = ds_{\mathbb{E}^+}/y$. It also follows from basic facts of Riemannian geometry:

$$ds^2 = \sum_{i,j=1}^{n} g_{i,j} dx^i dx^j \qquad \Longrightarrow \qquad d\Omega = \sum_{i,j=1}^{n} \sqrt{\det(g_{i,j})} \, dx^1 \wedge \cdots \wedge dx^n$$

(differential metric squared)   (differential volume element)

as shown in [2, p. 188, 241]. In this case, $(g_{i,j}) = \begin{pmatrix} 1/y^2 & 0 \\ 0 & 1/y^2 \end{pmatrix}$, and we obtain ds and dA as defined in (6.16) and (6.19). We will show that isometries in $S_{\mathbb{H}}$ also preserve area in $\mathbb{H}$.

Theorem 6.3.1. *The set $S_{\mathbb{H}}$ defined in (6.18) contains all the holomorphic isometries of $\mathbb{H}$. In addition to preserving the metric differential ds, these isometries also preserve the area differential dA. The isometries in $S_{\mathbb{H}}$ are also described by*

$$S_{\mathbb{H}} = \left\{ f(z) = \frac{az+b}{cz+d} \ : \ a,b,c,d \in \mathbb{R} \text{ with } ad - bc > 0 \right\} \tag{6.21}$$

Proof. For $f(z) = \dfrac{az + b}{cz + d}$ in $\mathsf{S}_{\mathbb{H}}$, we have

$$2i \operatorname{Im} f(z) = \frac{(ad - bc)(z - \bar{z})}{|cz + d|^2}$$

$$= \frac{1}{|cz + d|^2} \, 2i \operatorname{Im} z$$

and therefore $\operatorname{Im} f(z) = \operatorname{Im} z \, / \, |cz + d|^2$. Consequently, $f(z) \in \mathbb{H}$ if and only if $z \in \mathbb{H}$. (Note: $cz + d = 0$ is only possible when $z = -d/c \in \mathbb{R}$ and such z are not in $\mathbb{H}$.)

Now, for $f(z) = u + iv$ with $u \in \mathbb{R}$ and $v > 0$, we have $v = y/|cz + d|^2$. We also have

$$du^2 + dv^2 = |J|(dx^2 + dy^2)$$

$$= \left[\left(\frac{\partial u}{\partial x} \right)^2 + \left(\frac{\partial u}{\partial x} \right)^2 \right] (dx^2 + dy^2)$$

$$= |f'(z)|^2 \, (dx^2 + dy^2)$$

$$= \frac{1}{|cz + d|^4} \, (dx^2 + dy^2)$$

where we made use of the Cauchy-Riemann equations to simplify the Jacobian $|J|$ for the change of variables in the first line. Consequently, we obtain

$$\frac{du^2 + dv^2}{v^2} = \frac{|cz + d|^4}{y^2} \cdot \frac{1}{|cz + d|^4} \, (dx^2 + dy^2)$$

$$= \frac{dx^2 + dy^2}{y^2}$$

and that shows that $f(z)$ is an isometry of $\mathbb{H}$. Moreover, in our calculations we computed $f'(z)$ for $z \in \mathbb{H}$, so f is holomorphic on $\mathbb{H}$.

To prove preservation of dA, we calculate as above:

$$du \, dv = = |J| \, dx \, dy$$

$$= \frac{1}{|cz + d|^4} \, dx \, dy$$

and $du \, dv/v^2 = dx \, dy/y^2$ follows just as above. Thus, the isometry f also preserves the area differential dA.

Now, suppose that $f(z) = \dfrac{az + b}{cz + d}$ with $ad - bc > 0$. Let $r^2 = ad - bc$. Then,

$$f(z) = \frac{r^2}{r^2} \frac{a'z + b'}{c'z + d'}$$

$$= \frac{a'z + b'}{c'z + d'}$$

with $a'd' - b'c' = 1$. Therefore, $f(z)$ is in $\mathsf{S}_{\mathbb{H}}$, as defined in (6.18). Since the reverse inclusion obviously holds, it follows that $\mathsf{S}_{\mathbb{H}}$ is described by both (6.18) and (6.21).

Finally, by [4, Theorem 5], all holomorphic mappings from the disc $\mathbb{D} = \{z : |z| < 1\}$ to itself have the form $F(z) = t(z - c)/(1 - \bar{c}z)$, for some $t, c \in \mathbb{C}$

with $|t| = 1$ and $|c| < 1$. Conjugating each F with the conformal map $g \colon \mathbb{D} \to \mathbb{H}$ given by $g(z) = \dfrac{z+i}{iz+1}$, we obtain all the functions $f = g \circ F \circ g^{-1}$ that belong to $S_{\mathbb{H}}$. We omit the details for verifying this last statement, because we will not be using the fact that $S_{\mathbb{H}}$ consists of **all** the holomorphic isometries of $\mathbb{H}$. Complete details are in [26, Theorem 2.4, p. 222]. □

Remark 6.3.2. Theorem 6.3.1 deals with the holomorphic isometries of $\mathbb{H}$. There are other isometries. For example, the function $f(x + iy) = -x + iy$ is an isometry, since it clearly preserves ds. However, it is not holomorphic on $\mathbb{H}$ due to its failure to satisfy the Cauchy-Riemann equations.

Theorem 6.3.1 tells us that these sets of mappings are all isometries:

1. $\mathcal{M}_\rho \colon z \to \rho z$, for $\rho > 0$. In $\mathbb{E}^+$, this mapping would be a similarity transformation when $\rho \neq 1$, not an isometry. In $\mathbb{H}$, however, this mapping is an isometry for all $\rho > 0$.

2. $\tau_u \colon z \to z + u$, for $u \in \mathbb{R}$. Thus, all horizontally oriented translations are isometries of $\mathbb{H}$.

3. $\mathcal{I}_r \colon z \to \dfrac{-r}{z/r}$ for $r > 0$. This operation is called *inversion* through the circle of radius r, center 0 in $\mathbb{C}$. However, it is also inversion through the upper semicircle in $\mathbb{H}$ defined by $S_0^r = \{z \in \mathbb{H} : |z| = r\}$. In other words, $S_0^r = \{x + iy : x^2 + y^2 = r, y > 0\}$. In subsequent work, we shall also deal with upper semicircles of radius r and center $u \in \mathbb{R}$, which we denote by S_u^r. Note that $S_u^r = \tau_u\left(S_0^r\right)$ and $S_0^r = \mathcal{M}_r\left(S_0^1\right)$.

These special isometries generate all the isometries in $S_{\mathbb{H}}$ through composition. In fact, if $c \neq 0$, then by long division we obtain

$$\frac{az + b}{cz + d} = \frac{a}{c} + \frac{b - da/c}{cz + d}$$

$$= \frac{a}{c} + \frac{1}{c^2}\frac{-1}{z + d/c}$$

hence $f = \tau_{a/c} \circ \mathcal{M}_{1/c^2} \circ \mathcal{I}_1 \circ \tau_{d/c}$. While if $c = 0$, then

$$\frac{az + b}{d} = \frac{a}{d}z + \frac{b}{d}$$

Since $c = 0$, $ad - bc = 1$ reduces to $ad = 1$, and we have

$$f(z) = a^2(z + b/a)$$

Thus, $f = M_{a^2} \circ \tau_{b/a}$.

6.3.1.2 Geodesics

A **geodesic** in $\mathbb{H}$, connecting two points z_1 and z_2, is a (piecewise) smooth curve $\gamma(t) = x(t) + iy(t)$ for $a \le t \le b$ that satisfies $\gamma(a) = z_1$, $\gamma(b) = z_2$, and which has minimum length $\ell(\gamma)$. We shall now prove that, in $\mathbb{E}^+$, these geodesics lie along vertical rays or semicircles.

Theorem 6.3.3. *The geodesics in $\mathbb{H}$ lie along the following two types of curves in $\mathbb{E}^+$: (1) vertical rays $\mathcal{R}_u$ emanating from $\mathbb{R}$: $\mathcal{R}_u = \{u+iy : y > 0, \text{ fixed } u \in \mathbb{R}\}$, or (2) open semicircles S_u^r centered on $\mathbb{R}$: $S_u^r = \{x + iy : (x - u)^2 + y^2 = r^2, y > 0, \text{ fixed } u \in \mathbb{R}, r > 0\}$.*

Proof. First, we consider $z_1 = iy_1$ and $z_2 = iy_2$, choosing subscripts so that $y_2 > y_1 > 0$. For a smooth curve $\gamma(t) = x(t) + iy(t)$ satisfying $\gamma(a) = y_1$ and $\gamma(b) = y_2$, we have

$$\ell(\gamma) = \int_a^b \frac{\sqrt{x'(t)^2 + y'(t)^2}}{y(t)} \, dt$$
$$\ge \int_a^b \frac{y'(t)}{y(t)} \, dt$$
$$= \ln(y_2/y_1)$$

Moreover, this lower bound of $\ln(y_2/y_1)$ is realized for $\gamma(t) = [y_1 + (y_2 - y_1)t]i$ for $0 \le t \le 1$. Therefore, this function γ is a geodesic in $\mathbb{H}$, and clearly it lies on the ray $\mathcal{R}_0$. The minimum property also extends to the class of all continuous, piecewise smooth curves, by splitting integrals over $[a, b]$ into finite sums of integrals. Since the horizontal translation τ_u is an isometry, conjugation with τ_u implies that geodesics also lie along each vertical ray $\mathcal{R}_u$.

Second, we consider two points $z_1 \ne z_2$ on the open semicircle S_0^1, having $\arg(z_2) = \theta_2 > \theta_1 = \arg(z_1)$. The isometry $f(z) = (z + 1)/(-z + 1)$ maps S_0^1 to $\mathcal{R}_0$, with $iy_2 = f(z_2)$, $iy_1 = f(z_1)$, and $y_2 > y_1$. Given a geodesic $\gamma(t) = i[(y_1 + (y_2 - y_1)t]$ for $0 \le t \le 1$, connecting iy_1 and iy_2 on $\mathcal{R}_0$, we apply the isometry f^{-1} to obtain $f^{-1} \circ \gamma$ as a geodesic on S_0^1 connecting z_1 and z_2. Thus, S_0^1 contains geodesics in $\mathbb{H}$. Since the isometry $\mathcal{M}_r$ maps S_0^1 to S_0^r it follows that S_0^r contains geodesics in $\mathbb{H}$. Finally, since the isometry τ_u maps S_0^r to S_u^r, it follows that S_u^r contains geodesics in $\mathbb{H}$. $\square$

Remark 6.3.4. The **distance** $d(z_1, z_1)$ between two points $z_1, z_2 \in \mathbb{H}$ is defined to be the length of a geodesic that connects z_1 and z_2. For example, we found above that $d(x+iy_1, x+iy_2) = \ln(y_2/y_1)$ for $y_2 > y_1 > 0$. In general, for $x + iy_1, x + iy_2 \in \mathbb{H}$, we have $d(x + iy_1, x + iy_2) = |\ln(y_2/y_1)|$. It is important to note that $|\ln(y_2/y_1)| \to \infty$ if either $y_1 \to 0$ or $y_2 \to 0$. Consequently, the real line $\mathbb{R}$ is a *line at infinity* for all points in $\mathbb{H}$.

There is a distance formula for all $z, w \in \mathbb{H}$, given by

$$d(z, w) = \ln \frac{|z - \overline{w}| + |z - w|}{|z - \overline{w}| - |z - w|} \tag{6.22}$$

but we will not need it. Interested readers will find a proof of (6.22) in Katok [16, Theorem 1.2.6, p. 6].

For simplicity, in the rest of the paper, we shall refer to rays of type $\mathcal{R}_u$ and open semicircles of type S_u^r as *geodesics*. Strictly speaking, they contain geodesics, but there is little chance of confusion and our language is more straightforward if we simply call them geodesics as well. These geodesics in $\mathbb{H}$ can be interpreted as a model for the undefined term *lines* referred to in postulates of geometry. In fact, these geodesics do satisfy the first four of Euclid's postulates. However, they violate the notion of Euclidean parallelism. For example, in the image at the top of Figure 6.10, the two geodesics on the left intersect at a point. Yet, they fail to intersect the vertical geodesic on the right of the image. This situation violates the uniqueness of a parallel line, through a point not on a line, required in Euclidean geometry.

Returning to the artworks in Figure 6.9, it is interesting that parts of these designs correspond to geodesics. On the top of Figure 6.10 we have shown that geodesics of both types, S_u^r and $\mathcal{R}_u$, are evident within the design shown at the bottom of Figure 6.9. Parts of the other two designs also correspond to these types of geodesics. For instance, on the bottom right image in Figure 6.10 we have shown how a part of the middle design in Figure 6.9 corresponds to both types of geodesics in $\mathbb{H}$.

The images at the bottom of Figure 6.10 also contain geometric objects related to geodesics in $\mathbb{H}$. These objects are circles that are tangent to $\mathbb{R}$ at one point and have all other points lying in $\mathbb{H}$. To be specific, for $u \in \mathbb{R}$ and $r > 0$, a **horocycle** $H_r(u)$ is defined by

$$H_r(u) = \{x + iy \,:\, (x - u)^2 + (y - r)^2 = r^2, y > 0\}$$

so in $\mathbb{E}^+$ it is a circle with center $(u, r) \in \mathbb{R}^2$, and radius r, but omitting the point $(u, 0)$ on the x-axis. These horocycles are **not** geodesics. However, we will now discuss how the family of all horocycles are orthogonal curves in $\mathbb{H}$ for the family of all geodesics.

6.3.1.3   Angles and conformality in $\mathbb{H}$, horocycles and geodesics

An **angle** in $\mathbb{H}$ is defined to be an angle between tangent vectors of two curves meeting at a point. The following theorem shows that these angles are the same in both $\mathbb{H}$ and $\mathbb{E}^+$.

Theorem 6.3.5. *Let $\theta_{\mathbb{H}}$ and $\theta_{\mathbb{E}^+}$ be the angles between two curves at some intersection point in $\mathbb{H}$ and $\mathbb{E}^+$, respectively. Then, $\theta_{\mathbb{H}} = \theta_{\mathbb{E}^+}$.*

Proof. We can write the infinitesimal quadratic form $ds^2 = (dx^2 + dy^2)/y^2$ as

$$ds^2 = \frac{\langle [dx, dy], [dx, dy] \rangle_{\mathbb{E}^+}}{y^2}$$
$$= \|[dx, dy]\|_{\mathbb{H}}^2$$

where $\langle [dx, dy], [dx, dy] \rangle_{\mathbb{E}^+}$ stands for the standard inner product of the vector of differentials $[dx, dy]$ with itself, and $\|[dx, dy]\|_{\mathbb{H}}^2$ is our notation for ds^2 thought of

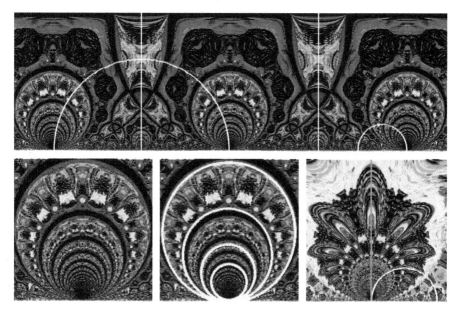

FIGURE 6.10

Top: Four geodesics, drawn in yellow over one of our hyperbolic designs. The two intersecting geodesics on the left are of types $\mathcal{R}_u$ and S_u^r. The two disjoint geodesics on the right are also of types $\mathcal{R}_u$ and S_u^r. Bottom Left: Portion of the same hyperbolic design that contains horocycles. Bottom Middle: Yellow circles indicating some of these horocycles. These horocycles are orthogonal to geodesics. Bottom Right: Yellow semicircle and green vertical line indicating geodesics on the middle design in Figure 6.9.

as a quadratic form of the vector $[dx, dy]$. The inner product $\langle [dx, dy], [d\tilde{x}, d\tilde{y}] \rangle_{\mathbb{H}}$ corresponding to this quadratic form is then

$$\langle [dx, dy], [d\tilde{x}, d\tilde{y}] \rangle_{\mathbb{H}} = \frac{dx\, d\tilde{x} + dy\, d\tilde{y}}{y^2}$$

Consequently, $\cos \theta_{\mathbb{H}}$ satisfies

$$\cos \theta_{\mathbb{H}} = \frac{\langle [dx, dy], [d\tilde{x}, d\tilde{y}] \rangle_{\mathbb{H}}}{\|[dx, dy]\|_{\mathbb{H}} \, \|[d\tilde{x}, d\tilde{y}]\|_{\mathbb{H}}}$$

$$= \frac{dx\, d\tilde{x} + dy\, d\tilde{y}}{\sqrt{dx^2 + dy^2} \, \sqrt{d\tilde{x}^2 + d\tilde{y}^2}}$$

$$= \cos \theta_{\mathbb{E}^+}$$

Thus, we must have $\theta_{\mathbb{H}} = \theta_{\mathbb{E}}^+$. $\qquad\qquad\square$

Since angles in $\mathbb{H}$ and $\mathbb{E}^+$ always correspond to angles between tangent vectors, we have proved that angles in the two geometries are always the same. The two geometries are said to be **conformal**.

Corollary 6.3.6. *The isometries in* $S_{\mathbb{H}}$ *preserve angles in* $\mathbb{H}$.

Proof. If $f(z) = (az + b) / (cz + d) \in S_{\mathbb{H}}$, then $f'(z) = 1/|cz + d|^2 \neq 0$. Therefore, f is a conformal mapping on $\mathbb{E}^+$ by [4, Theorem 3]. Hence, Theorem 6.3.5 above implies that f is a conformal mapping on $\mathbb{H}$.   □

This corollary can also be proved using the identity

$$2\langle \mathbf{v}, \mathbf{w} \rangle = \|\mathbf{v} + \mathbf{v}\|^2 - \|\mathbf{v}\|^2 - \|\mathbf{w}\|^2$$

relating inner products and quadratic forms, and the fact that an isometry preserves the quadratic form ds^2. However, our proof highlights the relation between the geometries of $\mathbb{E}^+$ and $\mathbb{H}$.

We now return to the concept of horocycles in $\mathbb{H}$, and how they are illustrated in the designs shown in Figure 6.9. The simplest type of horocycles are the sets of form, $\{t + iv : t \in \mathbb{R}\}$, parameterized by varying iv with $v > 0$. These sets are horizontal lines in $\mathbb{E}^+$, but in $\mathbb{H}$ they are **not** geodesics. Each geodesic ray $\mathcal{R}_u$, for $u \in \mathbb{R}$, lies orthogonal in $\mathbb{E}^+$ at each of its points to one of these horizontal lines, and therefore each geodesic ray $\mathcal{R}_u$ also lies orthogonal in $\mathbb{H}$ at each of its points to one of these horocycles $\{t + iv, t \in \mathbb{R}\}$.

The second type of horocycles are those that lie orthogonal to points of open semicircle geodesics. The isometry $\mathcal{I}_r(z) = -r^2/z$ maps the geodesic ray $\mathcal{R}_0$ to itself (with ri held fixed), and maps the geodesic ray $\mathcal{R}_{-r}$ to the open semicircular geodesic $S_{r/2}^{r/2}$. The horocycles for $\mathcal{R}_{-r}$, expressed as $\{t + ryi : t \in \mathbb{R}\}$ for each $y > 0$, are mapped by $\mathcal{I}_r$ to sets of the form $\{w \in \mathbb{H} : \left| w - \frac{r}{2y}i \right|^2 = \left(\frac{r}{2y}\right)^2\}$, which are circles in $\mathbb{E}^+$ except for the one point $0 + 0i \notin \mathbb{H}$. By Corollary 6.3.6, these horocycles are orthogonal to the open semicircular geodesic $S_{r/2}^{r/2}$ at all of its points. They are circles in $\mathbb{E}^+$ that are tangent to the point $(0,0)$, and all their points excepting $(0,0)$ lie in $\mathbb{H}$. Conjugating with horizontal translation τ_u for any fixed $u \in \mathbb{R}$, we find that the horocycles for all open semicircular geodesics in $\mathbb{H}$ are circles in $\mathbb{E}^+$ except for one point that is tangent to $\mathbb{R}$. Since these horocycles are all tangent to $\mathbb{R}$ at the same point u, with all radii $r > 0$, it follows that each family of horocycles is also orthogonal to the geodesic ray $\mathcal{R}_u$ at each of its points. On the bottom left and bottom middle of Figure 6.10, we illustrate a collection of such horocycles in one of our designs. The geodesics drawn on the design at the top of this figure are orthogonal at each of their points to such horocycles. On the bottom right of Figure 6.10, we show a part of the middle design in Figure 6.9 that exhibits both horocycles and geodesics.

Creating designs with hyperbolic symmetry

Our method for creating designs with hyperbolic symmetries is similar to our method for Euclidean symmetries. We symmetrize a given function f with domain $\mathbb{H}$. The symmetries will be a subgroup of $S_{\mathbb{H}}$. We cannot use

$\mathsf{S}_\mathbb{H}$ itself because the only functions on $\mathbb{H}$, symmetric with respect to all the transformations in $\mathsf{S}_\mathbb{H}$, are constant functions. Following Farris [9], we will use the subgroup Γ known as the *modular group*. The modular group Γ is defined as

$$\Gamma = \left\{ f(z) = \frac{jz+k}{mz+n} \; : \; j,k,m,n \in \mathbb{Z}, jn-mk=1 \right\} \quad (6.23)$$

Note that $jn-mk$ is the determinant of the matrix $\begin{pmatrix} j & k \\ m & n \end{pmatrix}$ of coefficients

of $f(z) = \dfrac{jz+k}{mz+n}$. The set Γ is a group because composition of two members

$f(z) = \dfrac{jz+k}{mz+n}$ and $g(z) = \dfrac{j'z+k'}{m'z+n'}$ satisfies

$$(f \circ g)(z) = \frac{(jj'+km')z + (jk'+kn')}{(mj'+nm')z + (mk'+nn')}$$

which corresponds to multiplication of the matrices of coefficients of f and g:

$$\begin{pmatrix} j & k \\ m & n \end{pmatrix} \begin{pmatrix} j' & k' \\ m' & n' \end{pmatrix} = \begin{pmatrix} jj'+km' & jk'+kn' \\ mj'+nm' & mk'+nn' \end{pmatrix}$$

and we know that determinants of matrices respect multiplication. So the determinants of each of the matrices in the equation above satisfy $1 \cdot 1 = 1$, hence $f \circ g$ is a member of Γ. Furthermore, $f^{-1}(z) = \dfrac{nz-k}{-mz+j}$, and therefore $f^{-1} \in \Gamma$.

Remark 6.3.7. The group $\mathsf{S}_\mathbb{H}$ is isomorphic to a subgroup of the matrix factor group:

$$\mathrm{PSL}(2,\mathbb{R}) = \mathrm{SL}(2,\mathbb{R}) \, / \, \{\mathrm{Id}, -\mathrm{Id}\}$$

where $\mathrm{SL}(2,\mathbb{R})$ is the **special linear group** consisting of all 2 by 2 matrices over $\mathbb{R}$ having determinant 1, and Id is the 2 by 2 identity matrix. The group $\mathrm{PSL}(2,\mathbb{R})$ is related to the projective geometry of all lines through the origin [5, p. 179]. It is called the **projective special linear group** over $\mathbb{R}$. The group Γ is isomorphic to $\mathrm{PSL}(2,\mathbb{Z}) = \mathrm{SL}(2,\mathbb{Z}) \, / \, \{\mathrm{Id}, -\mathrm{Id}\}$, where $\mathrm{SL}(2,\mathbb{Z})$ consists of matrices in $\mathrm{SL}(2,\mathbb{R})$ with integer coefficients.

The special isometries, $\mathcal{M}_r$, τ_u, $\mathcal{I}_r$, mostly are not members of Γ due to the requirement that their coefficients belong to $\mathbb{Z}$. In fact, the special isometries that belong to Γ are

1. Translations: $T^n \colon z \to z+n$, for $n \in \mathbb{Z}$. The unit-translation T^1 will be written as just T. These translations obey the group operation in Γ: $T^m \circ T^n = T^{m+n}$ for all $m,n \in \mathbb{Z}$.

2. Inversion: $\mathcal{I} \colon z \to -1/z$

Compositions with these isometries are sufficient to generate all the isometries in Γ.

Theorem 6.3.8. *The unit translation T and inversion $\mathcal{I}$ generate Γ. More precisely, if $f \in \Gamma$, then f can be written as some interlaced composition of $\mathcal{I}$ with various translations T^p:*

$$f = \left(\prod_{\ell=1}^{L} T^{p_\ell} \mathcal{I} \right) T^{p_0} \tag{6.24}$$

Proof. Let $f(z) = \frac{jz+k}{mz+n}$ be an arbitrary function in Γ. First, suppose $m = 0$. Then $f(z) = (j/n)z + (k/n)$ and we have $jn = 1$. Consequently, $(j, n) = (1, 1)$ or $(-1, -1)$. Hence, either $f(z) = z + k$ or $f(z) = z - k$, and so $f = T^k$ or $f = T^{-k}$. Now, if $m \neq 0$, we reduce to the first case as follows. We have, where $\ell \in \mathbb{Z}$,

$$\mathcal{I}f(z) = \frac{-mz - n}{jz + k}, \qquad T^\ell f(z) = \frac{(j + \ell m)z + (k + \ell n)}{mz + n}$$

If $|j| > |m|$, the division algorithm gives $j = mq + r$ with $0 \leq |r| < |m|$. Hence $T^{-q}f(z) = \frac{rz+(k+\ell n)}{mz+n}$. Then apply $\mathcal{I}$ to obtain $\mathcal{I}T^{-q}f(z) = \frac{-mz-n}{rz+(k+\ell n)}$ with $|r| < |m|$. Applying powers of T, followed by $\mathcal{I}$, eventually results in a remainder $r = 0$ as coefficient of z in the denominator. That is, we arrive at T^{p_0} for some $p_0 \in \mathbb{Z}$. Thus, we obtain

$$\left(\prod_{s=1}^{L} \mathcal{I}T^{-q_s} \right) f = T^{p_0}$$

Solving for f, we obtain the result in Equation (6.24). $\qquad\qquad\qquad\square$

The symmetrization f_S of a function f, defined on $\mathbb{H}$, can be done as follows

$$f_S(z) = \sum_{g \in \Gamma} f\big(g(z)\big) \tag{6.25}$$

Since Γ is a group, the function f_S is guaranteed to satisfy the symmetry condition $f(g(z)) = f(z)$ for all $g \in \Gamma$. In practice, of course, we can typically only create partial sums of the infinite series for f_S. Nevertheless, as shown above, our designs using such partial sums display many important features of the geometry of $\mathbb{H}$.

To create designs by domain coloring of $f_S(z)$, we need to express the series defining f_S in a more convenient form. To do that, we observe that the condition $jn - mk = 1$ can be rewritten as

$$nj + (-m)k = 1 \tag{6.26}$$

Equation (6.26) is a famous one from Number Theory. It is equivalent to the integers j and k being *relatively prime*, i.e., their greatest common divisor is 1, which we write as $\gcd(j, k) = 1$. The numbers n and $-m$ are called *Bézout coefficients* for j and k, and they ensure that Equation (6.26) holds. Because

$$\det \begin{pmatrix} j & k \\ m & n \end{pmatrix} = 1$$

will hold if j and k are perturbed by $j = \ell m$ and $k = \ell n$ for $\ell \in \mathbb{Z}$, these Bézout coefficients determine all transformations g having the form

$$g(z) = \ell + \frac{jz + k}{mz + n}$$

where $(n, -m)$ is a single pair of Bézout coefficients for (j, k). Therefore, we will assume that the initial function f has period 1 in the x-variable, and write f_S as

$$f_S(z) = \sum_{\gcd(j,k)=1} f\left(\frac{jz + k}{mz + n}\right) \tag{6.27}$$

To efficiently calculate the series in Equation (6.27), we use a recursive, tree-based method for computing relatively prime pairs of positive integers and associated pairs of Bézout coefficients. Randall [22] proves that pairs of relatively prime positive integers can be computed using $F(j, k) = (2j + k, j)$ in the following recursive formulas:

$$F(j, k) = (2j + k, j)$$
$$F(k, j) = (2k + j, k)$$
$$F(j, -k) = (2j - k, j)$$

starting from either $(2, 1)$ or $(3, 1)$ as initial pair. This recursive calculation generates two distinct trinary trees with roots $(2, 1)$ and $(3, 1)$, as illustrated in Figure 6.11. We have found that there is a similar recursive computation for finding associated Bézout coefficients that works for **both** of the nodes (j, k)

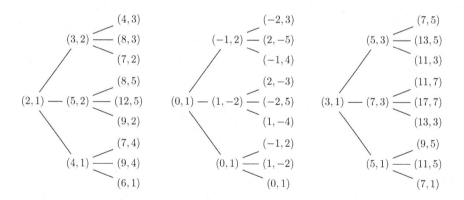

FIGURE 6.11
Left: Trinary tree of relatively prime positive integers generated by $(2, 1)$. Right: Trinary tree of relatively prime positive integers generated by $(3, 1)$. Middle: Trinary tree of Bézout coefficients generated by $(0, 1)$, corresponding to both of the other trees. All trees are shown to a depth of 2.

in these trinary trees. In [14], we show that using $G(u, v) = (v, u - 2v)$ and the recursive equations

$$G(u, v) = (v, u - 2v)$$
$$G(v, u) = (u, v - 2u)$$
$$G(u, -v) = (-v, u + 2v),$$

starting from $(u, v) = (0, 1)$, generates Bézout coefficients (u, v) satisfying $uj + vk = 1$ for each node (j, k) in both trees of relatively prime positive integers. See the tree in the middle of Figure 6.11. From these Bézout coefficients, and their associated pairs (j, k) of relatively prime positive integers, we obtain transformations $g(z) = \frac{jz+k}{mz+n} \in \Gamma$.

Based on this recursive, tree-based organization of elements of Γ we compute symmetrized designs by the following procedure:

Hyperbolic Symmetrized Design Procedure

1. Start with a function $f(z)$ having period 1 in x. It corresponds to $(j, k) = (1, 0)$ and $(m, n) = (0, 1)$. Then add $f(-1/z)$, corresponding to $(j, k) = (0, -1)$, and $(m, n) = (1, 0)$. For this step, we have $f(z) + f(-1/z)$.

2. Run through the trinary trees of relatively prime integers (j, k), starting at the roots $(2, 1)$ and $(3, 1)$. For each pair (j, k), and associated Bézout coefficients $(m, -n)$ at the corresponding node in the Bézout tree, add the terms $f\left(\frac{jz+k}{mz+n}\right)$ and $f\left(\frac{-jz+k}{mz-n}\right)$ to the terms already summed.

3. After adding a large number of terms—we typically used about 400 terms—create a symmetrized design using the **Domain Coloring Procedure** in Figure 6.2.

The designs in Figure 6.9 were all created using this method, starting with various functions f. For example, the design in the middle of Figure 6.9 was created using

$$f(x + iy) = 2i\, y \cos(2\pi x) + 2y \sin(2\pi y/3)$$

An animation illustrating the steps in the method above, in the construction of the *Blugold Fireworks* design, can be found at the link given in [15].

6.3.1.4 Rotational symmetry and tessellation of $\mathbb{H}$

The designs we have created relate to a number of other additional aspects of the geometry of $\mathbb{H}$. For example, on the left of Figure 6.12 we show a clip from the third design in Figure 6.9. Overlayed on this clip is a circle surrounding a point of 3-fold rotational symmetry in $\mathbb{H}$. It is important to note that this 3-fold rotational symmetry exists within $\mathbb{H}$ and not within $\mathbb{E}^+$. To see that we do have 3-fold symmetry in $\mathbb{H}$, in contrast to what we are used to seeing with Euclidean geometry, we need to discuss some further ideas from hyperbolic geometry. We first begin by describing the significance of the yellow circle on the left of Figure 6.12. Its center in $\mathbb{E}^+$ is marked by a blue dot. This circle is the locus of points that are a fixed distance ρ from the yellow dot in $\mathbb{H}$. To be precise, we have the following theorem.

Theorem 6.3.9. *For fixed $\rho > 0$, the locus of points $C_\rho(x + iy)$ that are distance ρ in $\mathbb{H}$ from a fixed point $x + iy$ is equal to the circle in $\mathbb{E}^+$ with center $x + iy \cosh \rho$ and radius $\sinh \rho$. (See Figure 6.12.)*

Proof. Begin by supposing that the fixed point is i. Using the distance formula $d(iy_1, iy_2) = |\ln(y_2/y_1)|$, the two points ie^ρ and $ie^{-\rho}$ are both distance ρ from i in $\mathbb{H}$. So, $ie^\rho, ie^{-\rho} \in C_\rho(i)$. The midpoint on the i-axis in $\mathbb{E}^+$ between these two points is $i \cosh \rho$, and it is Euclidean distance $\sinh \rho$ from both points. Now, map $C_\rho(i)$ in $\mathbb{H}$ to $C_\rho(0)$ in the unit disc $\mathbb{D}$, using the isometry $f(z) = (iz + 1) / (z + i)$ from $\mathbb{H}$ to $\mathbb{D}$ described below in Section 6.3.1.5. As discussed in that same section, the metric differential $ds_\mathbb{D}$ has rotational invariance about 0, and therefore $C_\rho(0)$ is a Euclidean circle about 0 (although its radius is **not** equal to ρ). Then map $C_\rho(0)$ back to $C_\rho(i)$ in $\mathbb{H}$ using the isometry $f^{-1}(z) = (z + i) / (iz + i)$ from $\mathbb{D}$ to $\mathbb{H}$. Because $f^{-1}(z)$ is a linear fractional transformation, it maps Euclidean circles to Euclidean circles, hence $C_\rho(i)$ is a circle in $\mathbb{E}^+$. Since we found its center and radius must be $i \cosh \rho$ and $\sinh \rho$, we have proved the result for fixed point i. Conjugating with $T_x M_y$, we get the result for fixed point $x + iy$. $\square$

 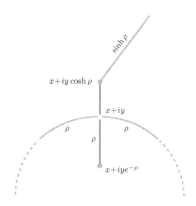

FIGURE 6.12
Left: Illustration of circular region in $\mathbb{H}$ with center at point of 3-fold rotational symmetry in $\mathbb{H}$. Right: Hyperbolic circle of radius ρ centered at the point $x + iy \in \mathbb{H}$, and its Euclidean center and radius in $\mathbb{E}^+$.

The yellow dot, located within a triangular region with curved edges on the left of Figure 6.12, is a center for a 3-fold rotation in $\mathbb{H}$. It is located at $1/2 + (\sqrt{3}/6)i$. Before we show that it is a center for a 3-fold rotation in $\mathbb{H}$, we show that the point $z_1 = 1/2 + (\sqrt{3}/2)i$ is also a center for a 3-fold rotation in $\mathbb{H}$. This point z_1 is marked by a green dot on the graph on the top of

Figure 6.13. It is related to a tessellation of $\mathbb{H}$ that we will discuss shortly. For now, observe that it lies at a vertex of a region labeled $T\mathcal{I}$ in the tessellation. We will show that z_1 is a fixed point for $T\mathcal{I}$, and that $T\mathcal{I}$ has order 3 in the group Γ. We have

$$T\mathcal{I}(z_1) = \frac{-1}{1/2 + (\sqrt{3}/2)i} + 1 = 1/2 + (\sqrt{3}/2)i$$

so z_1 is a fixed point for $T\mathcal{I}$. Moreover, we have

$$T\mathcal{I}(z) = \frac{z-1}{z}, \qquad (T\mathcal{I})^2(z) = \frac{-1}{z-1}, \qquad (T\mathcal{I})^3(z) = z$$

which shows that $T\mathcal{I}$ has order 3 in Γ.

We now turn to $z_2 = 1/2 + (\sqrt{3}/6)i$. We observe that $z_2 = \mathcal{I}T^{-2}z_1$. But, $\mathcal{I}T^{-2}$ is an isometry. Hence we can apply the following Lemma:

Lemma 6.3.10. *Suppose z is a fixed point for $g \in \mathsf{S}_{\mathbb{H}}$, and that g has finite order k. If $f \in \mathsf{S}_{\mathbb{H}}$, then $f \circ g \circ f^{-1}$ has fixed point $f(z)$ and order k.*

Proof. We find that $(f \circ g \circ f^{-1}) \circ f(z) = (f \circ g)(z) = f(z)$ so $f(z)$ is a fixed point. Moreover, $(f \circ g \circ f^{-1})^j = f \circ g^j \circ f^{-1}$ for any integer $j \geq 0$. When $j = k$, we have $g^k = \text{Id}$, so $(f \circ g \circ f^{-1})^k = f \circ f^{-1} = \text{Id}$. Also, when $j < k$, if $f \circ g^j \circ f^{-1} = \text{Id}$, then we would have $g^j = f^{-1} \circ f = \text{Id}$ and that would contradict k being the order of g. Consequently, $f \circ g \circ f^{-1}$ has order k. □

Applying the Lemma, we see that $1/2 + (\sqrt{3}/6)i = \mathcal{I}T^{-2}z_1$ is a fixed point for $h = \mathcal{I}T^{-2} \circ T\mathcal{I} \circ (\mathcal{I}T^{-2})^{-1}$ and h has order 3.

Returning to the left of Figure 6.12, the significance of the yellow dot and the circle enclosing it can now be explained in terms of rotation in $\mathbb{H}$. Since the yellow dot corresponds to the fixed point z_2 for the isometry h of order 3, it is analogous to a 3-fold rotation in $\mathbb{E}^+$. In fact, for points sufficiently close to z_2, the metric differential $ds = ds_{\mathbb{E}^+}/y$ is approximately equal to a multiple of $ds_{\mathbb{E}^+}$. Consequently, the isometry h is acting like a 3-fold rotation in $\mathbb{E}^+$ in the limit of approaching z_2. The enlargement of parts of the design as one rotates towards the vertical corresponds to what we see in $\mathbb{E}^+$. By Theorem 6.3.1, we know that area is preserved by isometries. Hence, in $\mathbb{H}$, the upper arm of the figure along the vertical direction, has exactly the same area as each of the two lower arms of the figure (extending out from the edges of the curved triangle).

There are also many other centers of 3-fold rotation in $\mathbb{H}$ that are illustrated in this design. If we let $z_k = \mathcal{I}T^k z_1$ for $k = -3, -4, -5, \ldots$, then Lemma 6.3.10 implies that we have centers of 3-fold rotations at each z_k. Since each point $T^k z_1 = z_1 + k$ lies along a horizontal horocycle in $\mathbb{E}^+$, applying the inversion $\mathcal{I}$ maps them to a circular horocycle in $\mathbb{E}^+$. We can see some of these points z_k in the image on the bottom right of Figure 6.13. They are lying above the blue curve, extending downwards towards the bottom left corner, which is a slightly lower horocycle belonging to the same family of horocycles tangent to 0 in $\mathbb{R}$. Finally, let $z_k = T\mathcal{I}T^k z_1$ for $k = 2, 3, 4, \ldots$. This produces another collection of centers of 3-fold rotations that are on a second horocycle that moves away to the right of z_1 towards 1 in $\mathbb{R}$. Some of these centers are visible in the figure as well. For all of these centers, we observe repetitions of the curved triangle containing z_1 but at smaller scale. They are only smaller scale in $\mathbb{E}^+$. In $\mathbb{H}$, isometries preserve area, so these curved triangles all have the same area in $\mathbb{H}$ as the one containing z_1, i.e., they are of the same scale in $\mathbb{H}$.

In addition to 3-fold centers, there are numerous centers of 2-fold rotational symmetries in $\mathbb{H}$. The point $z = i$ is a fixed point for $\mathcal{I}$ which has order 2. It is shown as a red dot in the tessellation at the top of Figure 6.13. Applying isometries to i we obtain sequences of 2-fold centers lying along horocycles. These centers of symmetry are plotted as red dots on the image shown on the right of Figure 6.13.

We have discussed the relation between our design and the tessellation shown at the top of Figure 6.13. We shall now discuss this tessellation in more detail. Equation (6.24) shows how any $f \in \Gamma$ can be written in terms of powers of T interlaced with $\mathcal{I}$. Starting from a *fundamental domain*, indicated by the shaded region $\mathcal{F}$ in the figure, and applying compositions of $\mathcal{I}$ with powers of T generates this tessellation of $\mathbb{H}$. In fact, the construction of the tessellation reproduces all the possible combinations of powers of T interlaced with $\mathcal{I}$ in Equation (6.24). We now make all of these ideas precise with the following theorem.

Theorem 6.3.11. *The modular group Γ generates a tessellation of $\mathbb{H}$ via*

$$\mathbb{H} = \bigcup_{f \in \Gamma} f(\mathcal{F}) \tag{6.28}$$

where $\mathcal{F} = \{z \in \mathbb{H} : |z| \geq 1, |\operatorname{Re} z| \leq 1/2\}$, and every pair of regions $f(\mathcal{F})$ and $g(\mathcal{F})$ for $f \neq g$ have disjoint interiors.

Proof. First, we show that $\mathbb{H} = \bigcup_{f \in \Gamma} f(\mathcal{F})$. Let z be an arbitrary element in $\mathbb{H}$. We will show that there is a $g(z) = \frac{jz+k}{mz+n} \in \Gamma$ for which $w = g(z) \in \mathcal{F}$. Then we will have $f(w) = z$ for $f = g^{-1}$, which will establish the decomposition of $\mathcal{F}$ in Equation (6.28). Our main tool is

$$\operatorname{Im} g(z) = \operatorname{Im} z \, / \, |mz + n|^2 \tag{6.29}$$

which was shown for all isometries at the beginning of the proof of Theorem 6.3.1. Since there are only finitely many $m, n \in \mathbb{Z}$ for which $|mz + n| \leq 1$, it follows from (6.29) that there are only finitely many $g \in \Gamma$ for which $\operatorname{Im} g(z) \geq \operatorname{Im} z$. Therefore, we can choose a $g \in \Gamma$ for which $\operatorname{Im} g(z)$ is maximal. If $|\operatorname{Re} g(z)| > 1/2$, then we may compose g with some power of T so that $|\operatorname{Re} g(z)| \leq 1/2$. Therefore, without loss of generality, we assume that $|\operatorname{Re} g(z)| \leq 1/2$. Then we must have $|g(z)| \geq 1$, because if $|g(z)| < 1$ we would have $0 < \operatorname{Im} g(z) < 1$, hence $\operatorname{Im}\left[\mathcal{I}g(z)\right] = \operatorname{Im}\left[\frac{-1}{g(z)}\right] > \operatorname{Im} g(z)$ and that contradicts the maximality of $\operatorname{Im} g(z)$. Thus, $|\operatorname{Re} g(z)| \leq 1/2$ and $|g(z)| \geq 1$, and so $w = g(z) \in \mathcal{F}$.

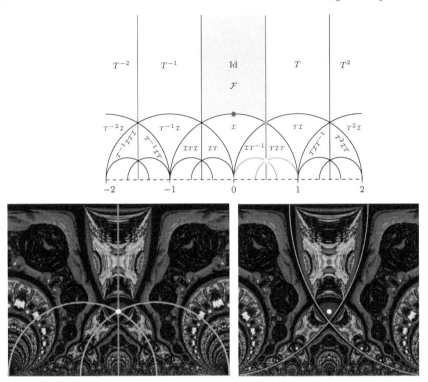

FIGURE 6.13
Top: Tessellation of $\mathbb{H}$ generated by T and $\mathcal{I}$. Regions are labeled by transformations that produce them from shaded region $\mathcal{F}$. Bottom Left: Location of center of 3-fold symmetry (yellow dot). This center lies at the intersection of three geodesics, shown in blue and orange in this image and in the tessellation above it. Bottom Right: Centers of 3-fold (yellow dots) and 2-fold (red dots) symmetries located along horocycles.

Second, we prove that $f(\mathcal{F})$ and $g(\mathcal{F})$ have disjoint interiors when $f \neq g$. But this is equivalent to $\mathcal{F}$ and $g(\mathcal{F})$ having disjoint interiors for every $g(z) = \frac{jz+k}{mz+n} \in \Gamma$ not equal to Id. Suppose first that $m = 0$. Then $g(z) = z \pm k$, for $k \neq 0$, hence $\mathcal{F}$ and $g(\mathcal{F})$ have disjoint interiors. Now, suppose $m \neq 0$. Let z be in the interior of $\mathcal{F}$. Hence $|\operatorname{Re} z| < 1/2$ and $|z| > 1$. We then have

$$|mz + n|^2 = m^2|z|^2 + 2(\operatorname{Re} z)mn + n^2$$
$$> m^2 + n^2 - |mn| = (|m| - |n|)^2 + |mn|$$

Since $m \neq 0$, the strict lower bound $(|m| - |n|)^2 + |mn|$ is a positive integer. Consequently, $|mz + n|^2 > 1$ and so $\operatorname{Im} g(z) < \operatorname{Im} z$. Since $g^{-1}(w) = \frac{nw-k}{-mw+j}$, if $w = g(z)$ were in the interior of $\mathcal{F}$, then the same argument yields $\operatorname{Im} g^{-1}(w) < \operatorname{Im} w$. Hence $\operatorname{Im} z < \operatorname{Im} g(z)$, and this contradiction shows that the interiors of $\mathcal{F}$ and $g(\mathcal{F})$ are disjoint.   □

A remarkable feature of the animation in [15] is how new features are added near the bottom of the screen that clearly correspond to the bottom portions of the tessellation shown in Figure 6.13.

6.3.1.5 Mappings to the disk $\mathbb{D}$

We have emphasized the hyperbolic upper half-plane as a model for non-Euclidean geometry. Another model, which is equivalent to $\mathbb{H}$, uses the unit disc $\mathbb{D} = \{z \in \mathbb{C} : |z| < 1\}$ as underlying set of points. Needham [20, pp. 317–318] shows that the map $f(z) = \dfrac{iz+1}{z+i}$ is a conformal map from $\mathbb{H}$ to $\mathbb{D}$, and it induces a metric differential $ds_\mathbb{D}$ on $\mathbb{D}$ given by

$$ds_\mathbb{D}^2 = \frac{2}{1 - |z|^2} \, ds_\mathbb{E}^2 \tag{6.30}$$

where $ds_\mathbb{E}^2 = dx^2 + dy^2$ is the Euclidean metric differential for $\mathbb{D}$ as a subset of the Euclidean plane $\mathbb{E}$. With this metric for $\mathbb{D}$, the map $f : \mathbb{H} \to \mathbb{D}$ is an isometry. Consequently, it maps geodesics in $\mathbb{H}$ to geodesics in $\mathbb{D}$. The geodesics in $\mathbb{D}$ lie on either arcs of circles that intersect the unit circle at right angles, or diameters in $\mathbb{D}$ (which also intersect the unit circle at right angles). An excellent treatment of this disc model for non-Euclidean geometry can be found in Krantz [17]. More details about the geometry, including tessellations of $\mathbb{D}$, are in Climenhaga and Kotek [5].

There is an abundance of artistic designs in $\mathbb{D}$ that have already been created. The designs by Escher, using curved polygonal tessellations of $\mathbb{D}$ are surely the most famous [8]. The mathematics of creating curved polygonal tessellations of $\mathbb{D}$ was worked out by Coxeter [6]. An entertaining app for creating your own designs, using Coxeter's tessellations of $\mathbb{D}$, can be found at the link in [3]. In Figure 6.14 we show two designs we created with this app. Design 1 used one of our *Waratah* flower rosettes as source image. It clearly retains the original 6-fold rotational symmetry of the original rosette. More importantly, from an artistic standpoint, it contains an ambiguity of *figure-ground* relations. If you view Design 1 from a far distance, the design features a star with six arms emanating from the central region of the design. However, when you move up close, this star fades into the background and six green/red flower-like regions closer to the circular boundary of the disc are more prominent.

Design 2 used one of our 6-fold symmetric wallpaper designs as source image. This latter design is particularly interesting in that it includes curved polygons (shown in white) that form a tessellation of $\mathbb{D}$. The app [3] created this design by loading a clip of the source image into a central curved hexagon, shown in white at the center of Design 2. During the design process, it successively displays the iteration of isometries of this central, fundamental, hexagonal region to fill out the rest of the curved hexagons that are tessellating the disc. As with the tessellation of $\mathbb{H}$ we discussed above, there are centers of 2-fold and 3-fold symmetries in the completed design. It is easy to spot centers for 2-fold and 3-fold *hyperbolic* rotational symmetry near the top and bottom of the image. The 3-fold centers are located at intersections of the tessellating curves (see the two zooms on the right of Figure 6.14), just as they occur at such intersections for the tessellation of $\mathbb{H}$. Notice that the original 6-fold symmetry of the source image is lost in Design 2, due to the clipping of only a part of the source image within the central curved hexagon. From an artistic standpoint, we like this *symmetry breakage* in Design 2.

It is also an important fact that both $1/(1-|z|^2)$ and $ds_{\mathbb{E}}^2 = dx^2 + dy^2$ are invariant under rotations centered at 0. Since $|e^{i\theta}z|^2 = 1$, we have the invariance of $1/(1-|z|^2)$ by rotation by θ, and the invariance of $ds_{\mathbb{E}}^2$ under rotation holds because rotation by θ can be expressed as an orthogonal matrix. Since $ds_{\mathbb{D}}^2$ is the product of $1/(1-|z|^2)$ and $ds_{\mathbb{E}}^2$, it is invariant under rotation about the origin. This rotational invariance is exhibited near the centers of both Design 1 and Design 2 in Figure 6.14, where these designs retain the rotational symmetry about the origin enjoyed by the rosette and wallpaper designs used as their source images. Similar 3-fold and 2-fold hyperbolic rotational symmetries, and 4-fold Euclidean rotational symmetry about the origin, are even more clearly evident in Escher's *Circle Limit III* woodcut [8] (if one ignores the different colors of the fishes).

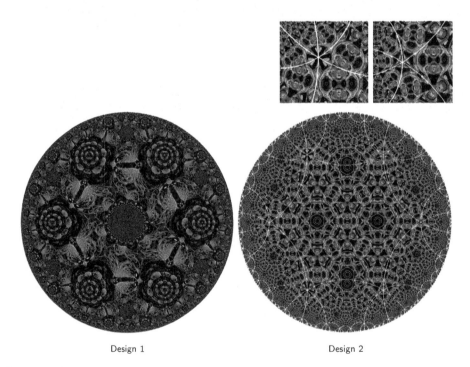

Design 1   Design 2

FIGURE 6.14
Two designs created with [3]. Design 2 includes tessellating curves, shown in white. Above Design 2 are two zooms showing 3-fold symmetries in the hyperbolic disc $\mathbb{D}$.

FIGURE 6.15
Design created using $\log z + \log(-z)$.

We have also produced images using conformal maps from $\mathbb{H}$ to $\mathbb{D}$. For example, Stein [26, p. 211] shows that $\log z$ defines a conformal map from the half-disc $\{z = x + iy : |z| < 1, y > 0\} \subset \mathbb{D}$ to the half-strip $\{w = u + iv : u < 0, 0 < v < \pi\} \subset \mathbb{H}$. With this idea in mind, we created a design in $\mathbb{D}$ using a multiple of $f(z) = \log z + \log(-z)$ and *Blugold Fireworks* as color map. It is shown in Figure 6.15. This design clearly shows 6-fold rotational symmetry about the origin in $\mathbb{D}$ and reflection symmetries through six diameter geodesics. We intend to continue exploring mappings from $\mathbb{H}$ to $\mathbb{D}$, including ones that more fully exploit the isometric equivalence of $\mathbb{H}$ and $\mathbb{D}$.

Conclusion

We have shown a number of symmetric designs generated by the application of complex analysis to the non-Euclidean geometry of the hyperbolic upper-half plane and the hyperbolic disc. The history of the mathematics of hyperbolic geometry is a fascinating one. References for this history include Needham [20] and Greenberg [12]. Penrose [21, Chap. 2] has some fascinating insights. Mathematics related to hyperbolic geometry continues right up to the present day, see e.g., Adams [1].

References

[1] C. Adams, What is a Hyberbolic 3-Manifold. *Notices of the A.M.S.*, **65**(5), May 2018, 544–546. Available at https://www.ams.org/journals/notices/201805/rnoti-p544.pdf

[2] W. Boothby, **An Introduction to Differentiable Manifolds and Riemannian Geometry, Second Edition**. Academic Press, 1986.

[3] Christersson, M., Make Hyperbolic Tilings of Images, Internet app, 2015. Available at http://www.malinc.se/m/ImageTiling.php

[4] J.A. Cima, Fractional Linear Maps and Some Applications. An "Augenblick." Chapter 3 of this Handbook.

[5] V. Climenhaga and A. Kotek, **From Groups to Geometry and Back**. Amer. Math. Soc., Providence, RI, 2017.

[6] H.S.M. Coxeter, The Non-Euclidean Symmetry of Escher's Picture 'Circle Limit III', *Leonardo*, **12**(1) (Winter, 1979), pp. 19–25. Available at http://jstor.org

[7] L. Crone web page on complex variable color plots: http://fs2.american.edu/lcrone/www/ComplexPlot.html

[8] M.C. Escher, *Circle Limit III*, image accessible at https://mcescher.com/gallery/recognition-success/#iLightbox[gallery_image_1]/31

[9] F. Farris, **Creating Symmetry: The Artful Mathematics of Wallpaper Patterns**. Princeton University Press, 2015.

[10] F. Farris, Forbidden Symmetries, *Notices of the Amer. Math. Soc.*, **59**(10), Nov. 2012, 1386–1390.

[11] F. Farris web page on complex variable color plots: http://www.maa.org/ visualizing-complex-valued-functions-in-the-plane

[12] M.J. Greenberg, **Euclidean and Non-Euclidean Geometries: Development and History, Third Edition**. W.H. Freeman and Co., 1993.

[13] E. Gullerud and J.S. Walker, *Blugold Fireworks*. Exhibited at the MATHEMATICAL ART EXHIBITION, Joint Mathematical Meetings, San Diego, CA, January 10–13, 2018.

[14] E. Gullerud and J.S. Walker, Generating Bézout trees for Pythagorean pairs. Preprint available at http://arxiv.org/abs/1803.04875

[15] E. Gullerud and J.S. Walker, Animation illustrating the construction of *Blugold Fireworks*. Available at https://www.youtube.com/ embed/umpgXcBr4zY. Best viewed in HD-quality.

[16] S. Katok, **Fuchsian Groups**, University of Chicago Press, Chicago, IL, 1992.

[17] S. Krantz, **Complex Analysis: The Geometric Viewpoint, Second Edition**. The Mathematical Association of America, 2004.

[18] M. Ladd and R. Palmer, **Structure Determination by X-ray Crystallography, Fifth Edition**. Cambridge University Press, 1995.

[19] MATLAB. A language and environment for technical computing. Product of MathWorks.

[20] T. Needham, **Visual Complex Analysis**. Oxford University Press, 1997.

[21] R. Penrose, **The Road to Reality: A Complete Guide to the Laws of the Universe**. Oxford University Press, 2004.

[22] T. Randall and R. Saunders, The family tree of the Pythagorean triplets revisited. *Math. Gaz.* **78**(482), 190–193, 1994.

[23] SageMath web page: http://www.sagemath.org/

[24] SageMath documentation on `complex_plot` procedure: https://doc.sagemath.org/html/en/reference/plotting/sage/plot/complex_plot.html

[25] M. Senechal, **Quasicrystals and Geometry**. Cambridge University Press, 1995.

[26] E.M. Stein and R. Shakarachi, **Complex Analysis**. Princeton University Press, 2015.

[27] SYMMETRYWORKS, free open-source software for creating color plots. C++ code available at https://github.com/imrimt/symmetryworks-research-bowdoin

[28] A. Terras, **Harmonic Analysis on Symmetric Spaces—Euclidean Space, the Sphere, and the Poincaré Upper Half-Plane, Second Edition**. Springer, 2013.

[29] J. S. Walker, Review of F. Farris, Creating Symmetry: The Artful Mathematics of Wallpaper Patterns, *Notices of the American Mathematical Society*, **62**(11), Dec. 2015, 1350–1354. Available at http://www.ams.org/notices/201511/rnoti-p1350.pdf

[30] H. Weyl, **Symmetry**. Princeton University Press, 1952.

7

A Glimpse into Invariant Distances in
Complex Analysis

Marek Jarnicki and Peter Pflug

CONTENTS

DOI: 10.1201/9781315160658-7

One of the most exciting results in the classical complex analysis is the Riemann mapping theorem which says that, except the whole complex plane, every simply connected plane domain is biholomorphically equivalent to the unit disc. Thus, the topological property "simply connected" is already sufficient to describe, up to biholomorphisms, a large class of plane domains. On the other hand, the Euclidean ball and the bidisc in $\mathbb{C}^2$ are topologically equivalent simply connected domains but they are not biholomorphic. This observation, which (under some additional assumptions) was made by H. Poincaré as early as at the end of 19th century, shows that even inside the class of bounded simply connected domains there is no single model (up to biholomorphisms) as it is in the plane case. To make the situation at that time more clear let us quote Carathéodory who wrote: *"Die Übertragung der Riemannschen Theorie auf Funktionen von mehreren Veränderlichen ist bisher aber nicht gelungen: Es liegen nur ganz vereinzelte und ziemlich dürftige Versuche vor, aus denen man hauptsächlich das eine ersehen kann, daß nämlich die Verhältnisse hier ganz anders liegen, so daß man nicht einmal recht weiß, wo man den Spaten anzusetzen hat, um die vermutete Goldader zu finden."*

If two domains $G, D \subset \mathbb{C}^n$ are biholomorphically equivalent, then they have the same bounded holomorphic functions, the same plurisubharmonic functions with a certain singularity, and the same amount of analytic discs (up to a biholomorphic map). Based on this observation it is important to associate with domains in $\mathbb{C}^n$ tractable objects that are invariant under biholomorphic mappings. Provided that these objects are sufficiently concrete, one can hope to be able to decide, at least in principle, whether two given domains are biholomorphically distinct. We say that a system of functions $(d_D)_{D \subset \mathbb{C}^n \text{ a domain}, \, n \in \mathbb{N}}$, $d_D : D \times D \longrightarrow \mathbb{R}_+$, is a *Schwarz-Pick system* (or a *holomorphically contractible system*) *of functions* if it satisfies the following two conditions:

a) $d_{\mathbb{D}}$ equals the Poincaré distance of the unit disc;

b) if $D \subset \mathbb{C}^n$, $G \subset \mathbb{C}^m$ are domains and $F : D \longrightarrow G$ a holomorphic mapping, then $d_G(F(z), F(w)) \leq d_D(z, w)$, $z, w \in D$, i.e. F operates as an contraction.

An object of this kind was introduced by C. Carathéodory in the thirties of the last century using the set of bounded holomorphic functions. In fact he defined a family of pseudodistances on domains via a "generalized" Schwarz Lemma. Thus, in particular, biholomorphic mappings operate as isometries. For such objects the name "invariant pseudodistances" has become popular. With the help of this pseudodistance he was able to give a simple proof of the fact that the ball and the bidisc in $\mathbb{C}^2$ are not holomorphically equivalent (without the additional assumption which was needed in Poincaré's argument)(see Section 7.1.3). Moreover, N. Kritikos exploiting the Carathéodory distance has shown that every biholomorphic mapping f from the domain $D := \{z \in \mathbb{C}^2 : |z_1| + |z_2| < 1\}$ onto itself has the origin as a fixed point, i.e., $f(0) = 0$, (see Section 7.1.9).

It turns out that this family is the smallest possible Schwarz-Pick system of functions but also that it is suffers from some unpleasant properties which makes its discussion more difficult. For example: (1) the topology induced by the Carathéodory distance on a domain does not necessarily coincide with the standard topology of that domain, if the domain is unbounded; (2) the Carathéodory distance is, in general, not an inner distance which causes (at least at the moment) two different notions of "completeness", even for a bounded domain (see Section 7.1.7).

Discussing the family of plurisubharmonic function has led to the Green function which is not a pseudodistance but keeps the property that holomorphic mappings act as contractions with respect to the Green functions. Although the Green function plays an important role in modern pluripotential theory, we will only mention this family during the article but omit a detailed discussion (see Chapter 7.4).

Besides using families of functions to associate (via an extremal problem) tractable objects with domains in $\mathbb{C}^n$, one can consider sets of analytic discs as new biholomorphic invariants. This idea is due to S. Kobayashi. While the family of Carathéodory pseudodistances is the smallest Schwarz-Pick system, the family of Kobayashi pseudodistances is the greatest one. Moreover, the difficulties for the Carathéodory distance mentioned above does not appear for the Kobayashi distance.

For certain types of domains (convex domains or strongly linearly convex domains) it turned out that the Carathéodory and the Kobayashi pseudodistance coincide, a deep result which is due L. Lempert. Whether this equality would imply the convexity of the discussed domain was a long standing problem which only recently had a negative answer. At the end of Chapter 7.3 we will shortly present and discuss this example.

Moreover, all these objects admit infinitesimal versions associating to any "tangent vector" a specific length contractible under holomorphic mappings.

The main goal of this article is to introduce these concepts of *"holomorphically contractible families of functions"* for domains in $\mathbb{C}^n$ and to emphasize how to apply them.

Both authors thank S. Krantz for his kind invitation to write this survey.

7.1 The Carathéodory Pseudodistance

7.1.1 The hyperbolic geometry of the unit disc $\mathbb{D}$

The unit planar disc $\mathbb{D} := \{\lambda \in \mathbb{C} : |\lambda| < 1\}$ is the classical model domain for many problems of Complex Analysis. Recall that any simply connected domain whose boundary consists of at least two different points is biholomorphic to $\mathbb{D}$ (Riemann theorem, cf. e.g. [Con78]). For $\lambda', \lambda'' \in \mathbb{D}$ let

$$\boldsymbol{p}(\lambda', \lambda'') := \frac{1}{2} \log \frac{1 + \boldsymbol{m}(\lambda', \lambda'')}{1 - \boldsymbol{m}(\lambda', \lambda'')} = \tanh^{-1}(\boldsymbol{m}(\lambda', \lambda'')) \geq \boldsymbol{m}(\lambda', \lambda''), \text{where}$$

$$\boldsymbol{m}(\lambda', \lambda'') := \left| \frac{\lambda' - \lambda''}{1 - \lambda'\bar{\lambda}''} \right|, \quad \lambda', \lambda'' \in \mathbb{D}.$$

Remark 7.1.1.

(a) One can check that $p : \mathbb{D} \times \mathbb{D} \longrightarrow \mathbb{R}_+$ is a distance. It is called the *Poincaré (hyperbolic) distance*. Moreover, $m : \mathbb{D} \times \mathbb{D} \longrightarrow [0, 1)$ is also a distance (the *Möbius distance*).

(b) Recall that the group $\operatorname{Aut}(\mathbb{D})$ of holomorphic automorphisms of $\mathbb{D}$ consists of all mappings of the form $\mathbb{D} \ni \lambda \longmapsto e^{i\theta} h_a(\lambda) \in \mathbb{D}$, where $\theta \in \mathbb{R}$, $a \in \mathbb{D}$, and $h_a(\lambda) := \frac{\lambda - a}{1 - \bar{a}\lambda}$. Observe that $h_a(a) = 0$ and $h_a^{-1} = h_{-a}$. In particular, the group $\operatorname{Aut}(\mathbb{D})$ acts *transitively* on $\mathbb{D}$.

(c) The topology generated by p (resp. m) on $\mathbb{D}$ coincides with the standard Euclidean topology of $\mathbb{D}$ and the metric space $(\mathbb{D}, p)$ (resp. $(\mathbb{D}, m)$) is complete.

(d) $\displaystyle\lim_{\lambda' \neq \lambda'',\, \lambda', \lambda'' \to \lambda} \frac{p(\lambda', \lambda'')}{|\lambda' - \lambda''|} = \lim_{\lambda' \neq \lambda'',\, \lambda', \lambda'' \to \lambda} \frac{m(\lambda', \lambda'')}{|\lambda' - \lambda''|} = \frac{1}{1 - |\lambda|^2} =: \gamma(\lambda)$, $\lambda \in \mathbb{D}$.

Lemma 7.1.2 (Schwarz–Pick lemma, cf. [Rud74]). *Let* $f \in \mathcal{O}(\mathbb{D}, \mathbb{D})$ ([1]). *Then:*

(a) $p(f(\lambda'), f(\lambda'')) \leq p(\lambda', \lambda'')$, λ', $\lambda'' \in \mathbb{D}$, *i.e. f is a contraction of the metric space $(\mathbb{D}, p)$.*

(b) *The following statements are equivalent:*
  (i) $f \in \operatorname{Aut}(\mathbb{D})$;
  (ii) $p(f(\lambda'), f(\lambda'')) = p(\lambda', \lambda'')$, λ', $\lambda'' \in \mathbb{D}$, *i.e. f is an isometry of the metric space $(\mathbb{D}, p)$;*
  (iii) $p(f(\lambda_0'), f(\lambda_0'')) = p(\lambda_0', \lambda_0'')$ *for some* λ_0', $\lambda_0'' \in \mathbb{D}$ *with* $\lambda_0' \neq \lambda_0''$.

Remark 7.1.3.

(a) Note that for any $a, b \in \mathbb{D}$, $a \neq b$ there exists a unique $h = h_{a,b} \in \operatorname{Aut}(\mathbb{D})$ such that $h(a) = 0$ and $h(b) \in (0, 1)$.

(b) For $a, b \in \mathbb{D}$, $a \neq b$, let $\alpha_{a,b}(t) := h^{-1}(th(b))$, $t \in [0, 1]$, where $h := h_{a,b}$. Note that $\alpha_{a,b} : [0, 1] \longrightarrow \mathbb{D}$ is a curve, $\alpha_{a,b}(0) = a$, and $\alpha_{a,b}(1) = b$. Observe that $I_{a,b} := \alpha_{a,b}([0, 1])$ lies on the unique circle $C_{a,b}$ that passes through a and b and is orthogonal to $\mathbb{T} := \partial \mathbb{D}$ (if $0 \in I_{a,b}$, then $C_{a,b}$ is a line through 0).

(c) One can prove that the *p-segment* $\{c \in \mathbb{D} : p(a, c) + p(c, b) = p(a, b)\}$ coincides with $I_{a,b}$.
  Indeed, since p is invariant under $\operatorname{Aut}(\mathbb{D})$, we may assume that $a = 0$, $b \in (0, 1)$. We have to prove that $\{c \in \mathbb{D} : p(0, c) + p(c, b) = p(0, b)\} = [0, b]$. In other words, $\frac{1+|c|}{1-|c|} \frac{1+m(c,b)}{1-m(c,b)} = \frac{1+b}{1-b}$ if and only if $c \in [0, b]$. Direct calculations show that the left hand side is equivalent to $|c| \leq b$ and $|c| = \operatorname{Re} c$.

(d) The sets $C_{a,b} \cap \mathbb{D}$ play in the geometry of $(\mathbb{D}, p)$ the role of straight lines. Thus the fifth Euclidean axiom is not fulfilled in $(\mathbb{D}, p)$. Consequently,

---

([1]) $\mathcal{O}(X, Y)$ denotes the space of all holomorphic mappings $f : X \longrightarrow Y$; $\mathcal{O}(X) := \mathcal{O}(X, \mathbb{C})$.

$(\mathbb{D}, \boldsymbol{p})$ is a model of a non-Euclidean geometry (the *Poincaré disc model* or *Klein–Beltrami model*).

To see more geometric properties of the space $(\mathbb{D}, \boldsymbol{p})$ we apply the following procedure (cf. [Rin61]), which is known from metric analysis. Let $d \in \{\boldsymbol{p}, \boldsymbol{m}\}$. We associate to any curve $\alpha : [0,1] \longrightarrow \mathbb{D}$ (*we always assume that curves are continuous*) its d-length $L_d(\alpha) \in [0, +\infty]$,

$$L_d(\alpha) := \sup \Big\{ \sum_{j=1}^{N} d(\alpha(t_{j-1}), \alpha(t_j)) : N \in \mathbb{N}, \ 0 = t_0 < t_1 < \cdots < t_N = 1 \Big\}.$$

Note that $L_d(\alpha) \geq d(\alpha(0), \alpha(1))$. We say that α is d-*rectifiable* if $L_d(\alpha) < +\infty$. We define $d^i : \mathbb{D} \times \mathbb{D} \longrightarrow [0, +\infty]$,

$$d^i(\lambda', \lambda'') := \inf\{L_d(\alpha) : \alpha : [0,1] \longrightarrow \mathbb{D} \text{ is a curve}, \ \lambda' = \alpha(0), \ \lambda'' = \alpha(1)\},$$
$$\lambda', \lambda'' \in \mathbb{D}.$$

Clearly, $d^i \geq d$. We say that d is *inner* if $d^i = d$.

Remark 7.1.4. In view of Remark 7.1.1(d) we get:

(a) $L_{\boldsymbol{p}} = L_{\boldsymbol{m}}$ and therefore $\boldsymbol{p}^i = \boldsymbol{m}^i$.
  Indeed, by Remark 7.1.1(d), for any compact $K \subset \mathbb{D}$ and $\varepsilon > 0$ there exists a $\delta > 0$ such that $0 \leq \boldsymbol{p}(\lambda', \lambda'') - \boldsymbol{m}(\lambda', \lambda'') \leq \varepsilon |\lambda' - \lambda''|$ for all $\lambda', \lambda'' \in K$ with $|\lambda' - \lambda''| \leq \delta$. In particular, $L_{\boldsymbol{p}}(\alpha) \leq L_{\boldsymbol{m}}(\alpha) + \varepsilon L_{\| \ \|}(\alpha)$ for any curve $\alpha : [0,1] \longrightarrow K$. Hence $\boldsymbol{p}^i \leq \boldsymbol{m}^i$.
(b) A curve $\alpha : [0,1] \longrightarrow \mathbb{D}$ is $\boldsymbol{p}$-rectifiable if and only if α is $\boldsymbol{m}$ rectifiable if and only if α is rectifiable in the Euclidean sense. In particular, d^i is a distance.

Moreover, we associate to any piecewise $\mathcal{C}^1$-curve $\alpha : [0,1] \longrightarrow \mathbb{D}$ its γ-length $L_\gamma(\alpha) := \int_0^1 \gamma(\alpha(t))|\alpha'(t)|dt$.

Theorem 7.1.5. $\boldsymbol{p}(a,b) = \boldsymbol{p}^i(a,b) = L_{\boldsymbol{p}}(\alpha_{a,b})$, $a, b \in \mathbb{D}$, $a \neq b$. In particular, $\boldsymbol{p}$ is inner, but $\boldsymbol{m}$ is not inner. Moreover, $L_{\boldsymbol{p}}(\alpha) = L_\gamma(\alpha)$ for every piecewise $\mathcal{C}^1$-curve $\alpha : [0,1] \longrightarrow \mathbb{D}$.

The following proposition shows that the non-Euclidean geometry of $(\mathbb{D}, \boldsymbol{p})$ fits perfectly with the holomorphic structure of $\mathbb{D}$.

Proposition 7.1.6. *For* any *mapping* $f : \mathbb{D} \longrightarrow \mathbb{D}$ *the following conditions are equivalent:*

(i) f *is a* $\boldsymbol{p}$-*isometry;*
(ii) *either* $f \in \mathrm{Aut}(\mathbb{D})$ *or* $\overline{f} \in \mathrm{Aut}(\mathbb{D})$.

7.1.2   The Carathéodory pseudodistance. The general Schwarz–Pick lemma.

We are going to extend the functions m and p to arbitrary domains $G \subset \mathbb{C}^n$. There are several ways to proceed. Here, we use the method based on bounded holomorphic functions (this way was the first also from the historical point of view; cf. [Car26], [Car27], [Car28]). For any domain $G \subset \mathbb{C}^n$, $n \geq 1$, put

$$m_G(z', z'') := \sup\{m(f(z'), f(z'')) : f \in \mathcal{O}(G, \mathbb{D})\}, \quad z', z'' \in G,$$
$$c_G(z', z'') := \sup\{p(f(z'), f(z'')) : f \in \mathcal{O}(G, \mathbb{D})\}, \quad z', z'' \in G.$$

A standard Montel argument shows that c_G is finite. It is clear that $c_G = \tanh^{-1}(m_G) \geq m_G$ and (by the Schwarz–Pick lemma) $m_{\mathbb{D}} = m$ and $c_{\mathbb{D}} = p$. Observe that $m_{\mathbb{C}^n} \equiv 0$ (by the Liouville theorem). Obviously, we can always pass from m_G to c_G or conversely. In the sequel we will use both m_G and c_G.

We are not going to discuss the infinitesimal form γ_G of m_G and c_G (like γ in § 7.1.1) in details. Nevertheless, in the sequel some proofs will require methods based on γ_G. Therefore, we define

$$\gamma_G(z; X) := \sup\{\gamma(f(z))|f'(z)X| : f \in \mathcal{O}(G, \mathbb{D})\}, \quad z \in G, \ X \in \mathbb{C}^n.$$

Since m, p, and γ are invariant under $\operatorname{Aut}(\mathbb{D})$, we get

$$m_G(z', z'') = \sup\{|f(z'')| : f \in \mathcal{O}(G, \mathbb{D}), f(z') = 0\}, \quad z', z'' \in G,$$
$$c_G(z', z'') = \sup\{p(0, f(z'')) : f \in \mathcal{O}(G, \mathbb{D}), f(z') = 0\}, \quad z', z'' \in G,$$
$$\gamma_G(z; X) = \sup\{|f'(z)X| : f \in \mathcal{O}(G, \mathbb{D}), \ f(z) = 0\}, \quad z \in G, \ X \in \mathbb{C}^n.$$

Applying Montel's theorem, we find that for any z', $z'' \in G$ there exists an $f \in \mathcal{O}(G, \mathbb{D})$ such that $f(z') = 0$, $|f(z'')| = m_G(z', z'')$. Any such a function f will be called an *extremal function for* $m_G(z', z'')$. Since m and p are distances, the functions m_G and c_G are pseudodistances; m_G is called the *Möbius pseudodistance for G*; c_G is the *Carathéodory pseudodistance for G*. As a direct consequence of the definitions, we get

Theorem 7.1.7 (General Schwarz–Pick lemma). *For arbitrary domains $G \subset \mathbb{C}^n$, $D \subset \mathbb{C}^m$ and for any holomorphic mapping $F : G \longrightarrow D$ we have*

$$m_D(F(z'), F(z'')) \leq m_G(z', z''), \quad c_D(F(z'), F(z'')) \leq c_G(z', z''), \quad z', z'' \in G.$$

In particular, if F is biholomorphic, then the equalities hold.

In other words, the systems $(m_G)_G$, $(c_G)_G$ are *holomorphically contractible*.

Remark 7.1.8.

(a) Observe that from the point of view of the general Schwarz–Pick lemma the Carathéodory pseudodistance is minimal in the following sense:

If $(d_G)_G$ is any system of functions $d_G : G \times G \longrightarrow \mathbb{R}$, where G runs on all domains in all $\mathbb{C}^n$'s, such that $d_{\mathbb{D}}(F(z'), F(z'')) \leq d_G(z', z'')$, $z', z'' \in G$, $F \in \mathcal{O}(G, \mathbb{D})$, and $d_{\mathbb{D}} = \boldsymbol{m}$ (resp. $d_{\mathbb{D}} = \boldsymbol{p}$), then $\boldsymbol{m}_G \leq d_G$ (resp. $\boldsymbol{c}_G \leq d_G$).

(b) Note that in Theorem 7.1.7 we do not claim that $\boldsymbol{m}_D(F(z_0'), F(z_0'')) = \boldsymbol{m}_G(z_0', z_0'')$ for some $z_0', z_0'' \in G, z_0' \neq z_0''$, implies that F is biholomorphic (cf. Lemma 7.1.2(b). This is not true even for $D = G \subsetneq \mathbb{C}^n$ and even under more restrictive assumptions on z_0', z_0''—take for instance $D = G = \mathbb{D}^2$; then using Theorem 7.1.10(b) we easily conclude that $|z_1| = \boldsymbol{m}_{\mathbb{D}^2}((0,0), (z_1, z_2)) = \boldsymbol{m}_{\mathbb{D}^2}(F(0,0), F(z_1, z_2))$ provided that $|z_1| \geq |z_2|$, where $F(z_1, z_2) := (z_1, 0)$.

(c) Using the Montel argument one can easily prove the following continuity property:

If $(G_k)_{k=1}^{\infty}$ is a sequence of subdomains of G such that $G_k \subset G_{k+1}$, $k \in \mathbb{N}$, and $G = \bigcup_{k=1}^{\infty} G_k$, then $\boldsymbol{m}_{G_k} \searrow \boldsymbol{m}_G$, $\boldsymbol{c}_{G_k} \searrow \boldsymbol{c}_G$, and $\gamma_{G_k} \searrow \gamma_G$.

(d) Similar to Remark 7.1.1(d), one can check that γ_G is a "strong" derivative of $\boldsymbol{c}_G$ in the following sense: for any $a \in G$ and $X \in \mathbb{C}^n$ with $\|X\| = 1$ $(^2)$ we have $\gamma_G(a; X) = \lim_{\substack{z',z'' \to a,\ z' \neq z'' \\ \frac{z'-z''}{\|z'-z''\|} \to X}} \frac{\boldsymbol{c}_G(z', z'')}{\|z' - z''\|}$. Note that for $n = 1$ the above formula is equivalent to the following one: $\gamma_G(a; 1) = \lim_{z' \neq z'',\ z', z'' \to a} \frac{\boldsymbol{c}_G(z', z'')}{|z' - z''|}$.

7.1.3 Carathéodory pseudodistance in balanced domains

One of the natural questions of complex analysis of several variables is to find a sufficiently rich category of domains $G \subset \mathbb{C}^n$ for which one may perform some effective calculations/estimates etc. The most natural one is the category of balanced domains. We say that a domain $G \subset \mathbb{C}^n$ is *balanced* if for any $z \in G$ and $\lambda \in \overline{\mathbb{D}}$ the point λz belongs to G.

If G is balanced, then denote by $\mathfrak{h}$ the *Minkowski function* of G, $\mathfrak{h}(z) := \inf\{t > 0 : \frac{1}{t} z \in G\}$. Note that

- $\mathfrak{h}(\lambda z) = |\lambda| \mathfrak{h}(z)$;
- $\mathfrak{h}$ is upper semicontinuous;
- $G = G_{\mathfrak{h}} = \{z \in \mathbb{C}^n : \mathfrak{h}(z) < 1\}$.

It is clear that in the case where $n = 1$ the only balanced domains are discs centered at 0, e.g. $\mathbb{D} = \mathbb{D}_{\mathfrak{h}}$ with $\mathfrak{h}(\lambda) = |\lambda|$.

Remark 7.1.9. Let $G \subset \mathbb{C}^2$ be an arbitrary convex balanced domain. Then ∂G can be described in the following non-standard way which will be used in Section 7.1.9.

$(^2)$ Here and in the sequel $\|X\| := (|X_1|^2 + \cdots + |X_n|)^{1/2}$ denotes the Euclidean norm in $\mathbb{C}^n$.

We associate to each point $a = (a_1, a_2) \in \partial G$ either the point $(|a_1|, a_2 \frac{|a_1|}{a_1}) \in \mathbb{R}_+ \times \mathbb{C}$ provided that $a_1 \neq 0$, or the circle $\{0\} \times \mathbb{T}a_2$ provided that $a_1 = 0$. Observe that ∂G, and then also G, is completely determined by its image ∂G^* under the above mapping. The set ∂G^* is called the *parametrization* of ∂G.

Using mainly the Hahn-Banach theorem one obtains the following result which helps to write down formulas for the Carathéodory pseudodistance for some domains.

Theorem 7.1.10. *Let $G = G_{\mathfrak{h}} \subset \mathbb{C}^n$ be a balanced domain.*

(a) *For every $f \in \mathcal{O}(G, \mathbb{D})$ with $f(0) = 0$, we have $|f| \leq \mathfrak{h}$ in G and $|f'(0)z| \leq \mathfrak{h}(z)$, $z \in \mathbb{C}^n$. In particular, $\boldsymbol{m}_G(0, \cdot) \leq \mathfrak{h}$ in G.*
(b) *The following conditions are equivalent:*

    (i) $\boldsymbol{m}_G(0, \cdot) = \mathfrak{h}$ *in G;*
    (ii) $\boldsymbol{\gamma}_G(a; \cdot) = \mathfrak{h}$ *in $\mathbb{C}^n$;*
    (iii) $\mathfrak{h}$ *is a seminorm, i.e. $\mathfrak{h}(z + w) \leq \mathfrak{h}(z) + \mathfrak{h}(w)$;*
    (iv) *G is convex.*

Note that for $G = \mathbb{D}$ part (a) is nothing else than the classical Schwarz lemma.

Remark 7.1.11. For a balanced domain $G \subset \mathbb{C}^n$ and a point $a \in G$ denote by h_a an automorphism of G with $h_a(a) = 0$ (if it exists). Recall that:

(a) if $G = \mathbb{D}^n$, then we may take $h_a(z_1, \ldots, z_n) := (\boldsymbol{h}_{a_1}(z_1), \ldots, \boldsymbol{h}_{a_n}(z_n))$;
(b) if $G = \mathbb{B}_n$ and $a \neq 0$, then we may take

$$h_a(z) := \frac{1}{\|a\|^2} \frac{\sqrt{1 - \|a\|^2}(\|a\|^2 z - \langle z, a \rangle a) - \|a\|^2 a + \langle z, a \rangle a}{1 - \langle z, a \rangle},$$

where $\langle \cdot, \cdot \rangle$ stands for the standard complex scalar product in $\mathbb{C}^n$.

Corollary 7.1.12. *If $G = G_{\mathfrak{h}}$ is a balanced convex domain in $\mathbb{C}^n$, $a \in G$, and if $h_a \in \mathrm{Aut}(G)$ is such that $h_a(a) = 0$, then $\boldsymbol{m}_G(a, z) = \mathfrak{h}(h_a(z))$, $z \in G$. In particular, $\boldsymbol{m}_{\mathbb{D}^n}(a, z) = \max\{\boldsymbol{m}(a_j, z_j) : j = 1, \ldots, n\}$ and $\boldsymbol{m}_{\mathbb{B}_n}(a, z) = \left(1 - \frac{(1 - \|a\|^2)(1 - \|z\|^2)}{|1 - \langle z, a \rangle|^2}\right)^{\frac{1}{2}}$.*

Remark 7.1.13.

(a) Note that $\boldsymbol{m}_{\mathbb{D}^n}$ is calculated via the maximum of the Möbius distances of the corresponding factors. In fact, it turns out that this kind of product property is true for arbitrary product-domains, i.e. the so called *product property* $\boldsymbol{m}_{D \times G}((a, b), (z, w)) = \max\{\boldsymbol{m}_D(a, z), \boldsymbol{m}_G(b, w)\}$ holds for arbitrary domains $D \subset \mathbb{C}^n$, $G \subset \mathbb{C}^m$ (cf. [JP89]).

(b) In the context of Corollary 7.1.12 one should point out that there are only a few classes of domains G for which c_G may be effectively calculated. Besides the unit polydisc and ball, the formulas are known for example for complex ellipsoids, the symmetrized bidisc, and the Neile parabola (cf. [JP13], §§ 2.11, 7.1, 16.6).

Corollary 7.1.14.

(a) *If* $\mathbb{B}(a, 3r) \subset G$, *then* $m_G(z', z'') \leq m_{\mathbb{B}(z', 2r)}(z', z'') = \frac{\|z' - z''\|}{2r}$, $z', z'' \in \mathbb{B}(a, r)$. *In particular,* c_G *is continuous.*
(b) *If* G *is a bounded domain,* $R := \operatorname{diam}(G)$ *(in the Euclidean sense), then* $m_G(z', z'') \geq m_{\mathbb{B}(z', R)}(z', z'') = \frac{\|z' - z''\|}{R}$, $z', z'' \in G$. *Consequently, if* G *is biholomorphic to a bounded domain, then the topology induced by* c_G *is equivalent to the Euclidean topology of* G, *i.e.* $\operatorname{top} c_G = \operatorname{top} G$.

Note that $\mathbb{B}_n$ and $\mathbb{D}^n$ ($n \geq 2$) are topologically equivalent. Nevertheless, they are not biholomorphically equivalent as the next corollary shows. So the situation in higher dimensions differs strictly from the one dimensional case where the unit disc can be treated as a biholomorphically equivalent model for almost all simply connected domains.

Corollary 7.1.15 (Poincaré theorem, cf. [Car26], [Rei21]). *For* $n \geq 2$ *there is no biholomorphic mapping of* $\mathbb{D}^n$ *onto* $\mathbb{B}_n$.

Proof. Suppose that $F = (F_1, \ldots, F_n) : \mathbb{D}^n \longrightarrow \mathbb{B}_n$ is biholomorphic. Since $\operatorname{Aut}(\mathbb{B}_n)$ acts transitively on $\mathbb{B}_n$, we may assume that $F(0) = 0$. Then, by Theorem 7.1.10(b) and Corollary 7.1.12, we have $\max\{|z_1|^2, \ldots, |z_n|^2\} = (m_{\mathbb{D}^n}(0, z))^2 = (m_{\mathbb{B}_n}(0, F(z))) = |F_1(z)|^2 + \cdots + |F_n(z)|^2$, $z \in \mathbb{D}^n$. The right hand side is $\mathcal{C}^\infty$, the left hand side is even not differentiable; contradiction. $\square$

Remark 7.1.16. Poincaré has started the first steps into proving the corollary under the additional assumption that the biholomorphic mapping is regular up to the boundary. The above result has been proved by Reinhardt (see [Rei21]) without assuming this boundary condition. According to Reinhardt, Poincaré studied the question of the biholomorphically equivalence of three dimensional surfaces in $\mathbb{C}^2$ via establishing their groups of biholomorphism. Along this discussion he observed that there are domains in $\mathbb{C}^2$ bounded by such three dimensional surfaces which cannot mapped biholomorphically onto the other one if the mapping is also assumed to be holomorphic up to the boundary. The proof above which as Carathéodory wrote does not use any calculation was, in fact, given by Carathéodory (see [Car26]).

The above Poincaré theorem may be generalized in the following way. For $\alpha = (\alpha_1, \ldots, \alpha_k) \in \mathbb{N}^k$ let $\mathbb{B}_\alpha := \mathbb{B}_{\alpha_1} \times \cdots \times \mathbb{B}_{\alpha_k}$.

Theorem 7.1.17 (cf. [JP08], Theorem 2.1.17). *Let* $\alpha = (\alpha_1, \ldots, \alpha_k) \in \mathbb{N}^k$, $\beta = (\beta_1, \ldots, \beta_\ell) \in \mathbb{N}^\ell$. *Then the following conditions are equivalent:*

(i) *there exists a biholomorphism* $\mathbb{B}_\alpha \longrightarrow \mathbb{B}_\beta$;
(ii) $\ell = k$ *and* $\alpha = \beta$ *up to a permutation.*

Moreover, every biholomorphic mapping $F : \mathbb{B}_\alpha \longrightarrow \mathbb{B}_\beta$ *is, up to a permutation of* $\mathbb{B}_{\beta_1}, \ldots, \mathbb{B}_{\beta_k}$, *of the form* $F(z) = (F_1(z_1), \ldots, F_k(z_k))$, $z = (z_1, \ldots, z_k) \in \mathbb{B}_\alpha$, *where* $F_\mu \in \mathrm{Aut}(\mathbb{B}_{\alpha_s})$, $s = 1, \ldots, k$.

In the case where $k = n \geq 2$, $\alpha_1 = \cdots = \alpha_n = 1$, $\ell = 1$, $\beta_1 = n$ the result reduces to the Poincaré theorem (Corollary 7.1.15).

Remark 7.1.18. The above theorem may be generalized to the case of Carathéodory isometries $F : \mathbb{B}_\alpha \longrightarrow \mathbb{B}_\beta$ (cf. Proposition 7.1.6). For more complicated domains the situation becomes unknown.

7.1.4   Carathéodory hyperbolicity

In general, the pseudodistance c_G need not be a distance, e.g. $c_{\mathbb{C}^n} \equiv 0$. Note that $c_G \equiv 0$ if and only if $\mathcal{H}^\infty(G) \simeq \mathbb{C}$, where "$\mathcal{H}^\infty(G) \simeq \mathbb{C}$" means that all bounded holomorphic functions on G are constant, i.e. G is a *Liouville domain*. On the other hand, c_G is a distance if and only if the space $\mathcal{H}^\infty(G)$ of all bounded holomorphic functions on G separates points in G. If c_G is a distance, then we say that G is *c-hyperbolic*. By Corollary 7.1.14(b), if G is biholomorphic to a bounded domain, then G is *c*-hyperbolic.

If $G \subset \mathbb{C}^1$, then G is *c*-hyperbolic if and only if $\mathcal{H}^\infty(G) \not\simeq \mathbb{C}$. In other words, if a domain $G \subset \mathbb{C}^1$ is not a Liouville domain, then it is *c*-hyperbolic.

If $n \geq 2$, then there are domains such that $c_G \not\equiv 0$ but G is not *c*-hyperbolic. For example, take the convex balanced domain $G := \{z \in \mathbb{C}^2 : |z_1 + z_2| < 1\}$.

7.1.5   The Carathéodory topology

Recall (Corollary 7.1.14(b)) that if G is biholomorphic to a bounded domain, then $\mathrm{top}\, c_G = \mathrm{top}\, G$. In $\mathbb{C}^1$ the situation is extremely simple, namely we have

Proposition 7.1.19. *If* $G \subset \mathbb{C}^1$ *is* *c-hyperbolic, then* $\mathrm{top}\, c_G = \mathrm{top}\, G$.

Proof. Let $G \ni a_s \longrightarrow a \in G$ in the Euclidean sense. Observe that $|f - f(a)| \leq \|f - f(a)\|_G \cdot m_G(a, \cdot)$, $f \in \mathcal{H}^\infty(G)$. Consequently, $f(a_s) \longrightarrow f(a)$ for any $f \in \mathcal{H}^\infty(G)$. Since $\mathcal{H}^\infty(G) \not\simeq \mathbb{C}$, there exists an $f_0 \in \mathcal{H}^\infty(G)$, $f_0 \not\equiv 0$, with $f_0(a) = 0$. Write $f_0(z) = (z - a_0)^k g(z)$, $z \in G$, where $g(a_0) \neq 0$. Clearly, $g \in \mathcal{H}^\infty(G)$. Since $f_0(a_s) \longrightarrow 0$ and $g(a_s) \longrightarrow g(a) \neq 0$, we have $a_s \longrightarrow a$ in $\mathrm{top}\, G$.   □

Unfortunately, for $n \geq 3$ there exist *c*-hyperbolic domains with $\mathrm{top}\, c_G \neq \mathrm{top}\, G$.

Theorem 7.1.20 (cf. [JPV91]). *For any* $n \geq 3$ *there exists a* *c-hyperbolic domain* $G \subset \mathbb{C}^n$ *such that* $\mathrm{top}\, c_G \subsetneqq \mathrm{top}\, G$.

?│For $n = 2$ we do not know whether such a domain exists │?│(^{3})

7.1.6 The inner Carathéodory pseudodistance

Let $d \in \{c_G, m_G\}$. Similarly as in § 7.1.1, for a curve $\alpha : [0,1] \longrightarrow G$ we may define its *d-length* $L_d(\alpha)$. If $\alpha : [0,1] \longrightarrow G$ is a piecewise $\mathcal{C}^1$-curve, then we can also define its γ_G-*length* by the formula $L_{\gamma_G}(\alpha) := \int_0^1 \gamma_G(\alpha(t); \alpha'(t))dt$. Moreover, we define

$$d^i(z', z'') = \inf\{L_d(\alpha) : \alpha : [0,1] \longrightarrow G, \ \alpha \text{ is a curve}$$

rectifiable in the Euclidean sense with $\alpha(0) = z'$, $\alpha(1) = z''\}$, $\quad z', z'' \in G$.

Observe that by Remark 7.1.4(b) in the case $G = \mathbb{D}$ the above formula gives $\boldsymbol{p}^i$ (resp. $\boldsymbol{m}^i$).

Theorem 7.1.21.

(a) $L_{\boldsymbol{m}_G} = L_{\boldsymbol{c}_G}$ and, moreover, if $\alpha : [0,1] \longrightarrow G$ is a curve rectifiable in the Euclidean sense, then $L_{\boldsymbol{c}_G}(\alpha) < +\infty$.

(b) If α is piecewise $\mathcal{C}^1$, then $L_{\boldsymbol{c}_G}(\alpha) = L_{\gamma_G}(\alpha)$.

(c) If $\alpha : [0,1] \longrightarrow G$ is a curve rectifiable in the Euclidean sense, then for any $\varepsilon > 0$ there exists a piecewise $\mathcal{C}^1$-curve $\beta : [0,1] \longrightarrow G$ such that $\beta(0) = \alpha(0)$, $\beta(1) = \alpha(1)$, and $|L_{\boldsymbol{c}_G}(\alpha) - L_{\boldsymbol{c}_G}(\beta)| \leq \varepsilon$.

Corollary 7.1.22.

(a) $c_G^i = m_G^i$.

(b) c_G^i is a pseudodistance and for any $F \in \mathcal{O}(G, D)$ we have $c_D^i(F(z'), F(z'')) \leq c_G^i(z', z'')$, $z', z'' \in G$, with equality for biholomorphic mappings.

(c) $c_G^i(z', z'') = \inf\{L_{\gamma_G}(\alpha) : \alpha : [0,1] \longrightarrow G, \ \alpha \text{ is a piecewise } \mathcal{C}^1\text{-curve}$ joining z' and $z''\}$, $z', z'' \in G$.

We say that $\boldsymbol{c}_G^i$ is the *inner Carathéodory pseudodistance for* G. Notice that the definition of c_G^i is a little bit different than the one in § 7.1.1. Here we take only those curves that are rectifiable in the Euclidean sense. │?│ We do not know whether in the definition of c_G^i the Euclidean rectifiability of α may be omitted. │?│ It is known that this is possible in the case where G is γ-*hyperbolic*, i.e. $\gamma_G(a; X) > 0$ for all $a \in G$ and $X \in \mathbb{C}^n \setminus \{0\}$ (e.g. G is biholomorphic to a bounded domain); cf. [Bar95] for a general discussion.

In view of the equality $\boldsymbol{p} = \boldsymbol{p}^i$ one could conjecture that $\boldsymbol{c}_G = \boldsymbol{c}_G^i$. Unfortunately, this not true as the following theorem shows.

Theorem 7.1.23 (cf. [Vig83]). *Let* $a, b \in G$, $a \neq b$. *Suppose that there exists an* $f \in \mathcal{O}(G, \mathbb{D})$ *which is extremal for* $m_G(a, b)$ ($f(a) = 0$) *and such that* $|f'(a)X| < \gamma_G(a; X)$ *for all* $X \in \mathbb{C}^n \setminus \{0\}$. *Then* $c_G(a, b) < c_G^i(a, b)$.

(^{3}) Open problems are marked "│?│...│?│".

Example 7.1.24 (cf. [JP90]). Let $P := \{z \in \mathbb{C} : 1/R < |z| < R\}$ $(R > 1)$. If $g \in \mathcal{O}(P, \mathbb{D})$ is an extremal function for $m_P(1, -1)$, then the function $f(z) = \frac{1}{2}(g(z) + g(1/z))$, $z \in P$, is also extremal for $m_P(1, -1)$ and $f'(1) = 0$. Hence by Lemma 7.1.23 we get $c_P(1, -1) < c_P^i(1, -1)$.

Notice that even the following better result is true (cf. [JP91b]).

For $a \in (1/R, R)$ we have: $c_P^i(a, z) = c_P(a, z)$ if and only if $z \in (1/R, R)$.

7.1.7 Completeness—general discussion

Recall that in the theory of Riemannian manifolds the concept of completeness for the distance induced by length has a successful history to clarify via this metric notion geometric properties of the manifold itself and vice versa (e.g. the result of Hopf and Rinow later in this section). So the study of completeness in the context of invariant distances may give some hope to get a better understanding of complex analysis problems via metric properties.

It is well-known that the theory of holomorphic functions of several complex variables essentially differs from the theory of one variable. One of the main differences lies in the fact that each domain $G \subset \mathbb{C}$ is a domain of existence of a holomorphic function, i.e. there exists an $f \in \mathcal{O}(G)$ such that there are no $a \in G$ and $0 < r < R$ with $\mathbb{D}(a, r) \subset G$, $\mathbb{D}(a, R) \not\subset G$, such that the function $f|_{\mathbb{D}(a,r)}$ extends holomorphically to $\mathbb{D}(a, R)$ ($\mathbb{D}(a, \tau)$ stands for the disc centered at a and of radius τ). If $G = \operatorname{int} \overline{G}$, then the above function f may be even chosen in the space $\mathcal{H}^\infty(G)$ (cf. e.g. [JP00], §§ 1.7, 1.8). This is not longer true for domains $G \subset \mathbb{C}^n$ with $n > 1$. For example, if $G_0 \subset \mathbb{C}^n$ $(n > 1)$ is a domain and $K \subset G_0$ a compact subset such that $G := G_0 \setminus K$ is a domain, then any function holomorphic in G extends holomorphically to G_0 (cf. [JP00], Theorem 2.6.6). For instance, one can take $G_0 := \mathbb{B}(R)$, $K := \overline{\mathbb{B}}(r)$ with $0 < r < R$ (Hartogs' Kugelsatz). These types of phenomena have led to the following three important notions.

Let $G \subset \mathbb{C}^n$ be a domain and let $\varnothing \neq \mathcal{F} \subset \mathcal{O}(G)$. We say that G is:

- an *$\mathcal{F}$-domain of holomorphy* if there are no $a \in G$ and $0 < r < R$ with $\mathbb{B}(a, r) \subset G$, $\mathbb{B}(a, R) \not\subset G$, such that for every function $f \in \mathcal{F}$ there exists an $\widetilde{f} \in \mathcal{O}(\mathbb{B}(a, R))$ such that $\widetilde{f} = f$ on $\mathbb{B}(a, r)$ (cf. [JP00], § 1.7); if $\mathcal{F} = \mathcal{O}(G)$, then we say that G is a *domain of holomorphy*; if $\mathcal{F} = \mathcal{H}^\infty(G)$, then we say that G is an *$\mathcal{H}^\infty$-domain of holomorphy*;
- *$\mathcal{F}$-holomorphically convex* if for every compact $K \subset G$ the set $\widehat{K}_{\mathcal{F}} := \{z \in G : \forall_{f \in \mathcal{F}} : |f(z)| \leq \|f\|_K\}$ is also compact (cf. [JP00], § 1.10); if $\mathcal{F} = \mathcal{O}(G)$, then we say that G is *holomorphically convex*; if $\mathcal{F} = \mathcal{H}^\infty(G)$, then we say that G is *$\mathcal{H}^\infty$-convex*;
- *pseudoconvex* if the function $G \ni z \longmapsto -\log \operatorname{dist}(z, \partial G)$ is plurisubharmonic.

A geometric background of pseudoconvexity may be based on the following two facts (cf. [Hör94]):

— If $\Omega \subsetneq \mathbb{R}^N$ is a convex domain, then the function $\Omega \ni x \longmapsto -\log \text{dist}(x, \partial\Omega)$ is a convex function, where the $\text{dist}(x, \partial\Omega)$ is taken in the Euclidean sense.

— A function $u \in \mathcal{C}^2(\Omega, \mathbb{R})$ is convex iff $\sum_{j,k=1}^{N} \frac{\partial^2 u}{\partial x_j \partial x_k}(x) X_j X_k \geq 0$, $x \in \Omega$, $X = (X_1, \ldots, X_N) \in \mathbb{R}^N$.

Summarizing, the property "pseudoconvex" can be seen as a kind of complex analogue of standard convexity.

One can easily check that each domain $G \subset \mathbb{C}$ is a domain of holomorphy, holomorphically convex, and pseudoconvex.

It is well known that in the case where $\mathcal{F} = \mathcal{O}(G)$ the above three notions are equivalent (cf. [JP00], Theorem 2.5.7).

Observe (see Theorem 7.1.25) that in the case where $\mathcal{F} = \mathcal{H}^\infty(G)$ the first two notions are strictly connected with completeness of the space $(G, \mathbf{c}_G)$.

The implication (pseudoconvexity $\Longrightarrow$ domain of holomorphy) is called the *Levi problem* and it was formulated by E. E. Levi in 1910 (cf. [Lev10]). The first positive answer has been given 30 years later in 1942 by K. Oka (cf. [Oka42]—the case $n = 2$) and in 1953 ([Oka53])—the general case). In the general case the Levi problem has been independently solved by H.-J. Bremermann (cf. [Bre54]) and F. Norguet (cf. [Nor54]).

Proposition 7.1.25. *Let $G \subset \mathbb{C}^n$ be a $\mathbf{c}$-hyperbolic domain.*

(a) *Assume that any Cauchy sequence in the sense of $\mathbf{c}_G$ converges to a point in G with respect to the Euclidean topology. Then G is an $\mathcal{H}^\infty$-domain of holomorphy.*

(b) *Assume that all balls (with finite radius) in the sense of $\mathbf{c}_G$ are relatively compact in G in the Euclidean sense. Then G is $\mathcal{H}^\infty$-convex and is an $\mathcal{H}^\infty$-domain of holomorphy.*

In particular, in both cases G is pseudoconvex.

Observe that the assumption in (b) is stronger than in (a).

In the general situation let G be an arbitrary domain in $\mathbb{C}^n$ equipped with a continuous distance d_G, e.g. $d_G = \mathbf{c}_G$ when G is $\mathbf{c}$-hyperbolic. We point out that the d_G-topology may be different from top G. We distinguish four different notions of completeness.

(C1) G is *weakly d_G-complete* if the metric space (G, d_G) is complete.

(C2) G is *d_G-complete* if any d_G-Cauchy sequence in the sense of d_G converges to a point in G with respect to top G.

(C3) G is *d_G-finitely compact* if all d_G-balls (with finite radius) are relatively compact in G in the sense of top G.

(C4) G is *weakly d_G-finitely compact* if all d_G-balls (with finite radius) are relatively compact in G in the sense of d_G.

Remark 7.1.26.

(a) (C3) $\Longrightarrow$ top d_G = top G.
(b) (C3) $\Longrightarrow$ (C2) $\Longrightarrow$ (C1).
(c) In the case of the Carathéodory distance it is not known whether (C1) $\Longrightarrow$ (C3). A counterexample is known in the category of complex spaces (cf. [JPV93]).
(d) The only well understood case is the planar case which was studied by M. A. Selby (cf. [Sel74]) and N. Sibony (cf. [Sib75]).
   Let $G \subset \mathbb{C}$ be *c*-hyperbolic. Then G is *c*-complete if and only if G is *c*-finitely compact.

As it is known from differential geometry, a theorem of H. Hopf asserts the equivalence of Cauchy-completeness and finite compactness. This result was generalized by W. Rinow (cf. [Rin61]) and S. Cohn-Vossen (cf. [CV35]) to the situation we are interested in.

Let d_G be a continuous distance on G. Analogously as for the case $d_G = c_G$, we say that d_G is *inner* if $d_G = d_G^i$, where

$$d_G^i(z', z'') = \inf\{L_{d_G}(\alpha) : \alpha : [0,1] \longrightarrow G, \ \alpha \text{ is a curve}$$
rectifiable in the Euclidean sense with $\alpha(0) = z', \ \alpha(1) = z''\}, z', z'' \in G$.

Remark 7.1.27. Let d_G be a continuous inner distance on G. Then

$$d_G(z', z'') = \inf\{L_{d_G}(\alpha) : \alpha : [0,1] \longrightarrow G, \ \alpha \text{ is continuous}$$
as a mapping into metric space $(G, d_G), \ \alpha(0) = z', \ \alpha(1) = z''\}, \ z', z'' \in G$.

Consequently, d_G is also inner the sense of Rinow (cf. [Rin61]).

Theorem 7.1.28. *Let d_G denote a continuous inner distance on G. Then* top G = top d_G *and* (C1) $\Longleftrightarrow$ (C2) $\Longleftrightarrow$ (C3) $\Longleftrightarrow$ (C4).

Recall that unfortunately c_G is in general not inner.

7.1.8 Carathéodory completeness

Proposition 7.1.25(b) can be made more precise during the next result.

Theorem 7.1.29 (cf. [Pfl84]). *For a *c*-hyperbolic domain $G \subset \mathbb{C}^n$ the following statements are equivalent:*

(i) *G is *c*-finitely compact;*
(ii) *for any $z_0 \in G$ and for any sequence $(z_\nu)_{\nu \in \mathbb{N}} \subset G$ without accumulation points (w.r.t top G) in G there exists an $f \in \mathcal{O}(G, \mathbb{D})$ with $f(z_0) = 0$ and $\sup_{\nu \in \mathbb{N}} |f(z_\nu)| = 1$.*

We say that a domain $G \subset \mathbb{C}^n$ is $\mathbb{C}$-*convex* if for any complex line $L = a + b\mathbb{C}$ $(a, b \in \mathbb{C}^n,\ b \neq 0)$ with $L \cap G \neq \varnothing$ the intersection $L \cap G$ is connected and simply connected (as a planar domain). Any convex domain is $\mathbb{C}$-convex. It is known (cf. [Hör94], [APS04]) that if G is $\mathcal{C}^2$-smooth, then G is $\mathbb{C}$-convex iff $(\mathcal{L}r)(z; X) := \sum_{j,k=1}^n \frac{\partial^2 u}{\partial z_j \partial \overline{z}_k}(z) X_j \overline{X}_k \geq |\sum_{j,k=1}^n \frac{\partial^2 r}{\partial z_j \partial z_k}(z) X_j X_k|$, $z \in \partial G$, $X \in T_z^{\mathbb{C}}(\partial G) := \{X \in \mathbb{C}^n : \sum_{j=1}^n \frac{\partial r}{\partial z_j}(z) X_j = 0\}$, where r is a *defining function* for ∂G (i.e. r is a $\mathcal{C}^2$-function on an open neighborhood U of ∂G satisfying $U \cap G = \{z \in U : r(z) < 0\}$, $U \setminus \overline{G} = \{z \in U : r(z) > 0\}$, and $\operatorname{grad} r(z) \neq 0$ for every $z \in \partial G$).

Theorem 7.1.30 (cf. [KZ13]). *For any bounded $\mathbb{C}$-convex domain $G \subset \mathbb{C}^n$ and any boundary point $a \in \partial G$ there exists an $f \in \mathcal{O}(G, \mathbb{D})$ with $\lim_{G \ni z \to a} f(z) = 1$. In particular, any bounded $\mathbb{C}$-convex domain $G \subset \mathbb{C}^n$ is c-finitely compact.*

It is known that if G is $\mathcal{C}^2$-smooth, then G is pseudoconvex iff $(\mathcal{L}r)(z; X) \geq 0$, $z \in \partial G$, $X \in T_z^{\mathbb{C}}(\partial G)$, where r is a defining function (cf. [JP00], § 2.2).

A bounded $\mathcal{C}^2$-smooth domain $G \subset \mathbb{C}^n$ is called *strongly pseudoconvex* (cf. [JP00], Definition 2.2.4) if $(\mathcal{L}r)(z; X) > 0$, $z \in \partial G$, $X \in T_z^{\mathbb{C}}(\partial G) \setminus \{0\}$, where r is a defining function.

A boundary point a of a bounded domain $G \subset \mathbb{C}^n$ is called a *peak-point with respect to a family* $\mathcal{F} \subset \mathcal{C}(\overline{G})$ if there exists an $f \in \mathcal{F}$ such that $|f(a)| = 1$ and $|f(z)| < 1$ for $z \in \overline{G} \setminus \{a\}$. It is known that if G is strongly pseudoconvex, then any boundary point is a peak-point with respect to $\mathcal{O}(\overline{G})$. If G is a bounded pseudoconvex domain in $\mathbb{C}^2$ with real analytic boundary, then any boundary point is a peak point with respect to $\mathcal{C}(\overline{G}) \cap \mathcal{O}(G)$ (cf. [BF78]).

Remark 7.1.31. Using peak-functions we conclude that:

(a) Any bounded strongly pseudoconvex domain in $\mathbb{C}^n$ and any bounded pseudoconvex domain in $\mathbb{C}^2$ with real analytic boundary is c-finitely compact (cf. [BF78], [FM94], [FS81]).

(b) Any bounded convex domain is c-finitely compact.

Remark 7.1.32. (a) In the context of Proposition 7.1.25(b) $\boxed{?}$ it is not known whether the property "c-complete" implies that G is $\mathcal{H}^\infty$-convex $\boxed{?}$

(b) Observe that although the notions "$\mathcal{H}^\infty$-domain of holomorphy" and "$\mathcal{H}^\infty$-convex" coincide for bounded plane domains (cf. [AS75]) they are not comparable in higher dimensions (cf. [Sib75]).

(c) There exists a bounded pseudoconvex balanced domain G in $\mathbb{C}^3$ with a continuous Minkowski function which is not c-complete although it is an $\mathcal{H}^\infty$-convex $\mathcal{H}^\infty$-domain of holomorphy (cf. [JP91a]). Other examples of domains sharing these properties were given by P.R. Ahern and R. Schneider (cf. [AS75]) and N. Sibony (cf. [Sib75]).

(d) It is well known that any bounded pseudoconvex domain with smooth $\mathcal{C}^\infty$-boundary is $\mathcal{H}^\infty$-convex and an $\mathcal{H}^\infty$-domain of holomorphy. $\boxed{?}$ Despite much efforts no example has been constructed of a domain of this type not being $\boldsymbol{c}$-finitely compact $\boxed{?}$

We say that a domain $G \subset \mathbb{C}^n$ is *Reinhardt (n-circled)* if $(e^{i\theta_1} z_1, \dots, e^{i\theta_n} z_n) \in G$ for any $(z_1, \dots, z_n) \in G$ and $\theta_1, \dots, \theta_n \in \mathbb{R}$.

In the class of Reinhardt domains the $\boldsymbol{c}$-completeness is completely understood and even expressed by a geometric condition.

Theorem 7.1.33 ([Pfl84]), [Fu94]), [Zwo00a]). *Let $G \subset \mathbb{C}^n$ be a pseudoconvex Reinhardt domain. Then the following conditions are equivalent:*

   (i) *G is $\boldsymbol{c}$-finitely compact;*
   (ii) *G is $\boldsymbol{c}$-complete;*
   (iii) *G is bounded and fulfills the following so called* Fu-condition: *if $\overline{G} \cap V_j \neq \varnothing$, then $G \cap V_j \neq \varnothing$, where $V_j = \{z \in \mathbb{C}^n : z_j = 0\}$.*

7.1.9   An application

Shortly after Carathéodory had introduced his pseudodistance, Kritikos (see [Kri27b]) was able to prove the following result using this at that time new object. Looking at his proof one finds already a lot of ideas like complex geodesics and indicatrices which were also later used and made more explicit. Therefore we believe that the interested reader may enjoy to see some details of the proof.

Theorem 7.1.34. *Let $\mathfrak{D} := \{z \in \mathbb{C}^2 : |z_1| + |z_2| < 1\}$ and let $F \in \text{Aut}(\mathfrak{D})$, then $F(0) = 0$.*

Remark 7.1.35. (1) Observe that this result was received before (also by Kritikos) but under some additional assumption on the boundary behavior of F (see [Kri27a]). With this new tool at hand he was able to argue now only with the inner geometry of the domain $\mathfrak{D}$.

(2) In the meantime there exists the same result for every domain D of the following form: $D = \{z \in \mathbb{C}^n : \sum_{j=1}^{n} |z_j|^{2p_j} < 1\}$, where $p_j \in (0, \infty) \setminus \{1\}$, $j = 1, \dots, n$ (see [JP13], Corollary 16.4.6). Moreover, using Lie theory the general case of bounded Reinhardt domains containing the origin is due to T. Sunada (see [Sun78]).

(3) There are other examples of balanced domains which are even not Reinhardt such that any automorphism of it has the origin as a fix point, e.g. the *minimal ball* $\{z = (z_1, \dots, z_n) \in \mathbb{C}^n : \|z\|^2 + |z_1^2 + \cdots + z_n^2| < 1\}$ (see [Kim91], [Zwo96]).

(4) Even more obscure there exists a bounded convex domain in $\mathbb{C}^2$ with C^∞-smooth boundary such that $\text{Aut}(D) = \{\text{id}_D\}$ (see [Eic61]); this example solves a question posed by H. Behnke and P. Thullen in [BT70], chapter VII, §9.1.

The proof of Theorem 7.1.34 needs several steps. The ones for which the Carathéodory distance is used will be explained in detail while the other steps mainly consisting of some calculations will be skipped.

Lemma 7.1.36. *Let* $G \subset \mathbb{C}^n$ *be a domain and let* $\varphi \in \mathcal{O}(\mathbb{D}, G)$ *be a complex geodesic, i.e.* $\boldsymbol{c}_G(\varphi(\lambda'), \varphi(\lambda'')) = \boldsymbol{c}_{\mathbb{D}}(\lambda', \lambda'')$, $\lambda', \lambda'' \in \mathbb{D}$, *then* $\boldsymbol{\gamma}_G(\varphi(\lambda); \varphi'(\lambda)(1 - |\lambda|^2)) = 1$, $\lambda \in \mathbb{D}$.

Proof. We first assume that $\lambda_0 = 0$. We know that $\varphi(\lambda) \neq \varphi(0)$ for all $\lambda \in \mathbb{D}_*$. Take any sequence $(\lambda_s)_{s=1}^\infty \subset \mathbb{D}_*$ such that $\lambda_s \longrightarrow 0$. We know that $\frac{\boldsymbol{m}(\varphi(\lambda_s), \varphi(0))}{|\lambda_s|} \longrightarrow 1$. Passing to a subsequence we may assume that $\lambda_s/|\lambda_s| \longrightarrow e^{i\theta}$ and $\frac{\varphi(\lambda_s) - \varphi(0)}{\|\varphi(\lambda_s) - \varphi(0)\|} \longrightarrow X$. Then

$$1 = \lim_{s \to +\infty} \frac{\boldsymbol{m}(\varphi(\lambda_s), \varphi(0))}{|\lambda_s|} = \lim_{s \to +\infty} \frac{\boldsymbol{m}(\varphi(\lambda_s), \varphi(0))}{\|\varphi(\lambda_s) - \varphi(0)\|} \cdot \left\| \frac{\varphi(\lambda_s) - \varphi(0)}{\lambda_s} \right\|$$
$$= \boldsymbol{\gamma}_G(\varphi(0); X) \|\varphi'(0)\|.$$

In particular, $\varphi(0) \neq 0$. Now we have

$$\frac{\varphi(\lambda_s) - \varphi(0)}{\|\varphi(\lambda_s) - \varphi(0)\|} = \frac{\lambda_s}{|\lambda_s|} \frac{\frac{\varphi(\lambda_s) - \varphi(0)}{\lambda_s}}{\left\| \frac{\varphi(\lambda_s) - \varphi(0)}{\lambda_s} \right\|} \longrightarrow e^{i\theta} \frac{\varphi'(0)}{\|\varphi'(0)\|} = X.$$

Finally, $1 = \boldsymbol{\gamma}_G(\varphi(0); X) \|\varphi'(0)\| = \boldsymbol{\gamma}_G(\varphi(0); e^{i\theta} \varphi'(0)) = \boldsymbol{\gamma}_G(\varphi(0); \varphi'(0))$.

Now let $\lambda_0 \in \mathbb{D}$ be an arbitrary point. Then put $\psi(\lambda) := \varphi(\frac{\lambda + \lambda_0}{1 + \lambda \bar{\lambda}_0})$, $\lambda \in \mathbb{D}$. Thus $\psi(0) = \varphi(\lambda_0)$. Therefore,

$$1 = \boldsymbol{\gamma}_G(\psi(0); \psi'(0)) = \boldsymbol{\gamma}_G(\varphi(\lambda_0); \varphi'(\lambda_0))(1 - |\lambda_0|^2),$$

which gives the claim in the lemma. $\qquad\square$

Let G, φ be as in the lemma and let $a \in G$. We define the *c-indicatrix at a* as $I(G; a) := \{X \in \mathbb{C}^n : \boldsymbol{\gamma}_G(a; X) < 1\}$. Note that the indicatrix is a convex balanced domain in $\mathbb{C}^n$ and if $\lambda_0 \in \mathbb{D}$, then $\varphi'(\lambda_0)(1 - |\lambda_0|^2) \in \partial I(G; \varphi(\lambda_0))$.

Recall (cf. Theorem 7.1.10(b)) that if $D \subset \mathbb{C}^n$ is a convex balanced domain, then for every boundary point $a \in \partial D$ there is the following complex geodesic $\varphi_a(\lambda) := \lambda a, \lambda \in \mathbb{D}$. In particular, we have $I(D; 0) = D$.

Lemma 7.1.37. *Let* $G \subset \mathbb{C}^n$ *be a domain,* $a \in G$, *and* $F \in \text{Aut}(G)$, *then* $F'(a)(I(G; a)) = I(G; F(a))$; *in particular,* $F'(a)$ *is a linear isomorphism from* $I(G; a)$ *onto* $I(G; f(a))$.

Proof. Note that $F'(a)$ defines a linear isomorphism from $\mathbb{C}^n$ onto $\mathbb{C}^n$. Fix an $X \in I(G; a)$. Then $1 > \boldsymbol{\gamma}_G(a; X) = \boldsymbol{\gamma}_G(F(a); F'(a)X)$. Thus $F'(a)X \in I(G; F(a))$. Arguing now with F^{-1} instead of F, we see the inverse conclusion. $\qquad\square$

To get a part of the indicatrix of $\mathfrak{D}$ at a point $a \in \mathfrak{D}$ we will find some complex geodesics through a. From now on let $I(a) := I(\mathfrak{D}; a)$.

Lemma 7.1.38. *If $r \neq s$ are real numbers of the closed unit interval $([-1, 1]$, then the map $\psi_{r,s} : \mathbb{D} \longrightarrow \overline{\mathbb{C}}$,*

$$\psi_{r,s}(\lambda) := \frac{(r-s)\lambda^2 + 2(r+s)\lambda + r - s}{4\lambda}, \quad \lambda \in \mathbb{D},$$

is a biholomorphic mapping from $\mathbb{D}$ onto $\overline{\mathbb{C}} \setminus [r, s]$ sending 0 to ∞. Moreover, $\psi_{r,s}$ maps $\mathbb{T}$ onto the closed interval $[r, s]$.

Proof. Recall that the Joukowski transform $\mathbb{D} \ni \lambda \xmapsto{h} \frac{\lambda^2 + 1}{2\lambda}$ is a conformal map from $\mathbb{D}$ onto $\overline{\mathbb{C}} \setminus [-1, 1]$ sending $\mathbb{T}$ onto $[-1, 1]$. Thus the map $\lambda \longmapsto r + (h(\lambda) - 1)\frac{r-s}{2}$ does the job and this is exactly the map given by the formula in the above lemma. $\square$

Let $\Phi : \mathbb{C}^2 \to \mathbb{C}^2$ be given by $\Phi(z) := (z_1 + z_2, z_1 - z_2)$, $z = (z_1, z_2) \in \mathbb{C}^2$. Then ψ is a biholomorphic map from $\mathbb{C}^2$ onto $\mathbb{C}^2$. Set $\widetilde{\mathfrak{D}} := \Phi(\mathfrak{D})$. Observe that $\widetilde{\mathfrak{D}} \subset \mathbb{D}^2$.

Our next aim is to find some complex geodesics in $\widetilde{\mathfrak{D}}$. Put $\widetilde{\varphi}_{r,s}(\lambda) := (\lambda, \lambda\psi_{r,s}(\lambda))$, $\lambda \in \mathbb{D}$. Note that $\widetilde{\varphi}_{r,s}$ is a holomorphic mapping. Then we have the following result.

Lemma 7.1.39. *If $r \neq s$ are real points inside of $[-1, 1]$, then $\widetilde{\varphi}_{r,s}(\mathbb{D}) \subset \widetilde{\mathfrak{D}} \subset \mathbb{D}^2$ and $\widetilde{\varphi}_{r,s}$ is a complex geodesic in $\widetilde{\mathfrak{D}}$.*

Proof. (a) Note that $u(z) := |z_1| + |z_2|$, $z \in \mathbb{C}^2$, is plurisubharmonic. Then $h := u(\Phi^{-1} \circ \widetilde{\varphi}_{r,s})$ is a subharmonic function on $\mathbb{D}$ which is not identically constant. Applying that $h(e^{it}) = 1$ for all real t and the maximum principle leads to the fact that $h < 1$ on $\mathbb{D}$. Hence, $\widetilde{\varphi}_{r,s}(\mathbb{D}) \subset \widetilde{\mathfrak{D}}$.

(b) Fix now two points $\lambda', \lambda'' \in \mathbb{D}$. Then

$$c_{\mathbb{D}}(\lambda', \lambda'') \geq c_{\widetilde{D}}(\widetilde{\varphi}_{r,s}(\lambda'), \widetilde{\varphi}_{r,s}(\lambda'')) \geq c_{\mathbb{D}^2}(\widetilde{\varphi}_{r,s}(\lambda'), \widetilde{\varphi}_{r,s}(\lambda''))$$
$$= \max\{c_{\mathbb{D}}(\lambda', \lambda''), c_{\mathbb{D}}(\lambda'\psi_{r,s}(\lambda'), \lambda''\psi_{r,s}(\lambda''))\} \geq c_{\mathbb{D}}(\lambda', \lambda'').$$

The last inequality follows using the Schwarz–Pick lemma. $\square$

Applying that the Carathéodory distance is invariant under biholomorphic mappings we finally have:

Corollary 7.1.40. *If $r \neq s$ are two real numbers in $[-1, 1]$, then*

$$\varphi_{r,s}(\lambda) := \left(\frac{\lambda + \lambda\psi_{r,s}(\lambda)}{2}, \frac{\lambda - \lambda\psi_{r,s}(\lambda)}{2} \right), \quad \lambda \in \mathbb{D},$$

defines a complex geodesic in $\mathfrak{D}$. Or more explicit,

$$\varphi_{r,s}(\lambda) = \left(\frac{r-s}{8}\lambda^2 + \frac{1}{2}\left(1 + \frac{r+s}{2}\right) + \frac{r-s}{8}, \frac{s-r}{8}\lambda^2 + \frac{1}{2}\left(1 - \frac{r+s}{2}\right) + \frac{s-r}{8} \right)$$

With a similar argument as above we obtain a few more complex geodesics for $\mathfrak{D}$.

Lemma 7.1.41. *Let $a \in \mathfrak{D}$. Then the mappings*

$$\mathbb{D} \ni \lambda \longmapsto (a_1, (1 - |a_1|)\lambda) \in \mathfrak{D}, \quad \mathbb{D} \ni \lambda \longmapsto ((1 - |a_2|)\lambda, a_2) \in \mathfrak{D}$$

are complex geodesics in $\mathfrak{D}$.

Proof. Because of the symmetry of $\mathfrak{D}$ it suffices to study the first map; call it φ.

Put $\Psi : \mathfrak{D} \longrightarrow \mathbb{C}^2$, $\Psi(z) := (z_1, \frac{z_2}{1-z_1})$. Note that ψ is an injective holomorphic mapping with $\Psi(\mathfrak{D}) =: \widetilde{\mathfrak{D}} \subset \mathbb{D}^2$. Then $\widetilde{\varphi}(\lambda) := (a_1, \frac{(1-|a_1|)\lambda}{1-a_1})$ maps $\mathbb{D}$ to $\widetilde{\mathfrak{D}}$ and also to $\mathbb{D}^2$. Therefore, if $\lambda', \lambda'' \in \mathbb{D}$, then

$$\boldsymbol{c}_{\mathbb{D}}(\lambda', \lambda'') \geq \boldsymbol{c}_{\widetilde{D}}(\widetilde{\varphi}(\lambda'), \widetilde{\varphi}(\lambda'')) \geq \boldsymbol{c}_{\mathbb{D}^2}(\widetilde{\varphi}(\lambda'), \widetilde{\varphi}(\lambda''))$$
$$= \max\{\boldsymbol{c}_{\mathbb{D}}(a_1, a_1), \boldsymbol{c}_{\mathbb{D}}(\lambda', \lambda'')\} = \boldsymbol{c}_{\mathbb{D}}(\lambda', \lambda'').$$

Using Ψ^{-1} finishes the proof. $\qquad\qquad\square$

Now fix a point $a \neq 0$ in $\mathfrak{D}$. We are going to describe certain complex geodesics through a. Because of the symmetry of $\mathfrak{D}$ we assume that the a_j's are non negative real values with $0 < a_1 + a_2 < 1$.

Lemma 7.1.42. *Let $a \in \mathfrak{D}$ as before. Whenever $s \in [-1, 1]$, $s \neq \frac{a_1 - a_2}{a_1 + a_2}$, then*

$$r(s) := \frac{4(a_1 - a_2) + s(1 - a_1 - a_2)^2}{(1 + a_1 + a_2)^2} \in [-1, 1], \quad r(s) \neq s, \quad \varphi_{r(s),s}(a_1 + a_2) = a,$$

i.e. there is a one parameter family $\varphi_{r(s),s}$ of complex geodesics in $\mathfrak{D}$ through the point a.

Proof. A simple calculation gives immediately the result. $\qquad\qquad\square$

Hence we obtain the following boundary points of the indicatrix $I(a)$.

Corollary 7.1.43. *Let a be as above and $\lambda_0 := a_1 + a_2$. Then all the following points*

(a) $\varphi'_{r(s),s}(\lambda_0)(1 - |\lambda_0|^2) \in \partial I(a)$, s as above,

(b) $\left(0, (1 - a_1)(1 - \frac{a_2^2}{(1-a_1)^2})\right)$,

(c) $\left((1 - a_2)(1 - \frac{a_1^2}{(1-a_2)^2}), 0\right)$

belong to $\partial I(a)$.

Moreover, a simple calculation gives the following explicit form of the points in (a):

$$\frac{1-a_1-a_2}{2}\left(1+3a_1-a_2+s(1-a_1-a_2), 1-a_1+3a_2-s(1-a_1-a_2)\right), \ s \text{ as above.}$$

Then the following set M, described in (a), (b), and (c) below, belong to the parametrization of $\partial I(a)$, where $0 < a_1 + a_2 < 1$, $a_j \in \mathbb{R}_+$:

(a) $\frac{1-a_1-a_2}{2}\left(1 + 3a_1 - a_2 + s(1 - a_1 - a_2), 1 - a_1 + 3a_2 - s(1 - a_1 - a_2)\right)$, s as above;

(b) $\left(0, \pm(1 - a_1)(1 - \frac{a_2^2}{(1-a_1)^2})\right)$;

(c) $\left((1 - a_2)(1 - \frac{a_1^2}{(1-a_2)^2}), 0\right)$.

Analyzing the set in (a) gives that it consists of an interval on the line ℓ given through the

$$\alpha = -\gamma + 1 - (a_1 + a_2)^2 \tag{7.1}$$

with the following endpoints

$$\left((1 + a_1 - a_2)(1 - a_1 - a_2), 2a_2(1 - a_1 - a_2)\right),$$

$$\left(2a_1(1 - a_1 - a_2), (1 - a_1 + a_2)(1 - a_1 - a_2)\right),$$

which lie in the open first quadrant except the case when $a_1 = 0$ (resp. $a_2 = 0$).

Moreover, the point in (c) lies on the γ-axis between the projections of the first endpoint and the point, where ℓ cuts the γ-axes. On the other hand, the point given in (b) with positive second coordinate lies on the positive α-axes between the projection of the second endpoint given above and the point, where ℓ cuts the α-axes. The remaining point in (b) is just opposite. Therefore, it needs at least three lines each of which cuts the set M to cover M, and no pair of these lines are symmetric with respect to 0.

The remaining step consist of the following claim which immediately implies Theorem 7.1.34.

Lemma 7.1.44. *There is no linear isomorphism* $A : I(0) \longrightarrow I(a)$, *if* $a \in \mathfrak{D}$, $a \neq 0$.

Proof. Assume the contrary. Then there exists a linear isomorphism Ψ of $\mathbb{C}^2$ onto itself given by the inverse of A, that sends $\partial I(a)$ onto $\partial I(0)$. Let us assume that

$$\Psi(z) = (pz_1 + qz_2, uz_1 + vz_2), \ z \in \mathbb{C}^2,$$

where $p = p_1 + ip_2, q = q_1 + iq_2, u = u_1 + iu_2, v = v_1 + iv_2 \in \mathbb{C}$. If $(\gamma, \alpha + i\beta) \in \partial I(a)^*$, then $\Psi(\gamma, \alpha + i\beta) \in \partial I(0)$. More concrete,

$$\Psi(\gamma, \alpha + i\beta) = (A + iB, C + iD), \quad \text{where}$$
$$A := p_1\gamma + q_1\alpha - q_2\beta, \quad B := p_2\gamma + q_1\beta + q_2\alpha,$$
$$C := r_1\gamma + v_1\alpha - v_2\beta, \quad D := u_2\gamma + v_2\alpha + v_1\beta.$$

Or, if $(A, B) \neq (0,0)$, then the point

$$\left(\sqrt{A^2 + B^2}, (C + iD)\left(\frac{A}{\sqrt{A^2 + B^2}} - i\frac{B}{\sqrt{A^2 + B^2}}\right)\right)$$

is the corresponding point in $\partial I(0)^*$. If $A = B = 0$, then the corresponding point is given by $(0, C + iD)\mathbb{T}$.

Using that this new point belongs to $\partial I(0)$ it follows that $C^2 + D^2 = (1 - \sqrt{A^2 + B^2})^2$, resp. $C^2 + D^2 = 1$.

Applying now this equation to those points in $\partial I(a)$ with $\beta = 0$ leads to the following equality

$$4((p_1\gamma + q_1\alpha)^2 + (p_2\gamma + q_2\alpha)^2)$$
$$= \left(1 + (p_1\gamma + q_1\alpha)^2 + (p_2\gamma + q_2\alpha)^2) - (u_1\gamma + v_1\alpha)^2 - (u_2\gamma + v_2\alpha)^2\right)^2. \quad (7.2)$$

Recall that a non trivial interval on some line, given by the equation $\alpha = -\gamma + c$, belongs to part $\beta = 0$ of $\partial I(a)^*$. Therefore the former equation after substituting α by $c - \gamma$ is true for all $\gamma \in \mathbb{R}$, i.e. its left side is the square of a polynomial. Hence one obtains that $0 = \det\begin{pmatrix} p_1 - q_1 & q_2 - q_2 \\ q_1 & q_2 \end{pmatrix} = \det\begin{pmatrix} p_1 & q_2 \\ q_1 & q_2 \end{pmatrix}$. Therefore, $(p_1\gamma + q_1\alpha)^2 + (p_2\gamma + q_2\alpha)^2 = (P\gamma + Q\alpha)^2$ for all α and γ. Then equation (7.2) is of the form

$$(u_1\gamma + v_1\alpha)^2 + (u_2\gamma + v_2\alpha)^2 = \left(1 \pm (P\gamma + Q\alpha)\right)^2.$$

A similar argument as before leads to

$$(u_1\gamma + v_1\alpha)^2 + (u_2\gamma + v_2\alpha)^2 = (R\gamma + Q\alpha)^2, \quad \gamma, \alpha \in \mathbb{R}.$$

Finally, (7.2) reads as follows: $(R\gamma + Q\alpha)^2 = (1 \pm (P\gamma + Q\alpha))^2$. So one obtains that $\partial I(a)^* \cap (\mathbb{R}_+ \times (\mathbb{R} + i0))$ consists of at most four lines pairwise symmetric to the origin. But this was already impossible for that part of $\partial I(a)^* \cap (\mathbb{R}_+ \times (\mathbb{R} + i0))$ which was discovered in (7.1); a contradiction. $\square$

As a final consequence there is the following complete description of $\mathrm{Aut}(\mathfrak{D})$.

Theorem 7.1.45. *If* $\mathfrak{D} := \{z \in \mathbb{C}^2 : |z_1| + |z_2| < 1\}$, *then* $\mathrm{Aut}(\mathfrak{D}) = \{\varphi_{\alpha,\beta}, \psi_{\gamma,\delta} : \alpha, \beta, \gamma, \delta \in \mathbb{R}\}$, *where*

$$\varphi_{\alpha,\beta}(z) := (e^{i\alpha}z_1, e^{i\beta}z_2), \quad \psi_{\gamma,\delta}(z) := (e^{i\gamma}z_2, e^{i\delta}z_1), \quad z = (z_1, z_2) \in \mathfrak{D}.$$

Note that the behavior of $\mathrm{Aut}(\mathfrak{D})$ is extremely opposite to the one of the bidisc or the ball.

To present the proof of Theorem 7.1.45 the following lemma is needed which, in fact, is a special case of a theorem of H. Cartan (cf. [Car30]).

Lemma 7.1.46. *If $F = (f_1, f_2) : \mathfrak{D} \longrightarrow \mathfrak{D}$ be holomorphic with $F(0,0) = (0,0)$, $\frac{\partial f_1}{\partial z_1}(0,0) = 1$, $\frac{\partial f_1}{\partial z_2}(0,0) = 0$, $\frac{\partial f_2}{\partial z_1}(0,0) = 0$, $\frac{\partial f_2}{\partial z_2}(0,0) = 1$, then $F = \mathrm{id}_{\mathfrak{D}}$.*

Proof of Theorem 7.1.45. Let $F \in \mathrm{Aut}(\mathfrak{D})$. Then the mapping $\widehat{F}_\zeta$ defined as $\widehat{F}_\zeta(z) := F^{-1}(\frac{1}{\zeta} F(\zeta z))$, $\zeta \in \mathbb{T}$ and $z \in \mathfrak{D}$, is an automorphism of $\mathfrak{D}$ which satisfies the assumptions of the former lemma. Hence, $\widehat{F}_\zeta = \mathrm{id}_{\mathfrak{D}}$. Or in other words one has that $F(\zeta z) = \zeta F(z)$ for all $\zeta \in \mathbb{T}$ and $z \in \mathfrak{D}$. Using the expansion of F into homogeneous polynomials leads to the fact that F is a linear isomorphism of $\mathfrak{D}$. Taking the form of $\mathfrak{D}$ into account the claim in the theorem is now an easy consequence. □

Finally, let us quote another result due to H. Cartan (see [Car36]) whose proof is also based on the Carathéodory pseudometric. To quote H. Cartan: "Pour la demonstration nous utiliserons la métrique de Carathéodory".

Theorem 7.1.47. *Let $G_j \subset \mathbb{C}^{n_j}$, $j = 1, 2$, be bounded domains. Then for every Φ in the connected component of $\mathrm{Aut}(G)$ that contains the identity, there exist $\Phi_j \in \mathrm{Aut}(G_j)$, $j = 1, 2$, such that $\Phi(z, w) = (\Phi_1(z), \Phi_2(w))$, $z \in G_1$, $w \in G_2$.*

Looking back one can already here say that certain problems in several complex variables could be successfully handled using the language of metric geometry.

7.2 The Kobayashi Pseudodistance and the Kobayashi–Royden Pseudometric

In the previous chapter, we discussed the Carathéodory pseudodistance for domains $G \subset \mathbb{C}^n$ using an extremal problem for $\mathcal{O}(G, \mathbb{D})$. About thirty years later, S. Kobayashi introduced another pseudodistance based on the family $\mathcal{O}(\mathbb{D}, G)$, the set of so-called *analytic discs in G*. We already learnt that the Carathéodory pseudodistances form the smallest family of pseudodistances being a Schwarz-Pick system. It will turn out that the new family is the largest possible one.

7.2.1 The Lempert function

First we introduce a family $(\ell_G)_G$ of functions $\ell_G : G \times G \longrightarrow \mathbb{R}_+$ from which the new pseudodistance will be derived as the largest pseudodistance below ℓ_G.

Before we are able to present the formal definition, we have to make the following observation.

Remark 7.2.1. Let G be a domain in $\mathbb{C}^n$ and $z', z'' \in G$. Then there exists a curve $\alpha : [0,1] \longrightarrow G$ connecting the points z', z''. Using the Weierstrass approximation theorem we find a polynomial map $P : [0,1] \longrightarrow G$ with $P(0) = z'$ and $P(1) = z''$. Then it is easy to choose a bounded simply connected domain $D \subset \mathbb{C}$, $[0,1] \subset D$, such that $P(D) \subset G$. By the Riemann mapping theorem we conclude that z', z'' lie on an analytic disc $\varphi : \mathbb{D} \longrightarrow G$ with $\varphi(0) = z'$ and $\varphi(\sigma) = z''$ for some $0 \leq \sigma < 1$. Summarizing we see that *any two points in a domain $G \subset \mathbb{C}^n$ lie on an analytic disc in G.*

Definition 7.2.2. Let G be a domain in $\mathbb{C}^n$ and $z', z'' \in G$. Put

$$\ell_G(z', z'') := \inf\{p(\lambda', \lambda'') : \lambda', \lambda'' \in \mathbb{D} : \exists_{\varphi \in \mathcal{O}(\mathbb{D}, G)} : \varphi(\lambda') = z', \varphi(\lambda'') = z''\}$$
$$= \inf\{p(0, \lambda'') : \lambda'' \in \mathbb{D} : \exists_{\varphi \in \mathcal{O}(\overline{\mathbb{D}}, G)} : \varphi(0) = z', \varphi(\lambda'') = z''\},$$
$$\ell_G^* := \tanh \ell_G.$$

ℓ_G is called to be the *Lempert function for G.*

Intuitively, the larger a disc in G containing the points z', z'' the smaller the value of $\ell_G(z', z'')$. For example, $\ell_{\mathbb{C}^n}(z', z'') = 0$ for all pairs of points $z', z'' \in \mathbb{C}^n$ ($\varphi_R(\lambda) := z' + R\lambda(z'' - z')$ is an analytic disc in $\mathbb{C}^n$ with $\varphi_R(0) = z'$ and $\varphi_R(1/R) = z''$).

Moreover observe that:

(a) $\ell_G : G \times G \longrightarrow \mathbb{R}_+$ is a symmetric function;

(b) $(\ell_G)_G$ is a contractible family of functions with respect to holomorphic mappings, i.e. if $F \in \mathcal{O}(G, D)$, then $\ell_D(F(z'), F(z'')) \leq \ell_G(z', z'')$, $z', z'' \in G$;

(c) in particular, we have $\ell_D(F(z'), F(z'')) = \ell_G(z', z'')$ whenever $F : G \longrightarrow D$ is a biholomorphic map;

(d) $\ell_{\mathbb{D}} = p$ (use the Schwarz–Pick lemma);

(e) $c_G \leq \ell_G$;

(f) if there is a family $(d_G)_G$ of functions $d_G : G \times G \longrightarrow \mathbb{R}_+$ contractible under holomorphic mappings with $d_{\mathbb{D}} = p$, then $d_G \leq \ell_G$. In other words, the system $(\ell_G)_G$ is *the largest Schwarz-Pick system of functions.*

The following example shows that, in general, the Lempert function does not fulfill the triangle inequality; thus it is not a pseudodistance.

Example 7.2.3 (cf. [PS89]). Let $G := \{z \in \mathbb{C}^2 : |z_1 z_2| < 1, |z_2| < 10\}$ and let $a := (1, 0)$, $b := (0, 1)$ be two points in G. Then using the above properties one easily sees that the following equality is true:

$$\ell_G(a, 0) + \ell_G(b, 0) = \ell_{\mathbb{C}}(0, 1) + \ell_{\mathbb{D}(10)}(0, 1) = \ell_{10\mathbb{D}}(0, 1) = \ell_{\mathbb{D}}(0, 1/10) = p(0, 1/10).$$

On the other hand there is the following lower estimate of $\ell_G(a, b)$: let $\varphi = (\varphi_1, \varphi_2) \in \mathcal{O}(\mathbb{D}, G)$ be an analytic disc with $\varphi(0) = a$ and $\varphi(s) = b$, where $s \in (0, 1)$. Since $\varphi_1(s) = 0$ and $\varphi_2(0) = 0$, the functions φ_j may be written as

$\varphi_1(\lambda) = (s - \lambda)\widetilde{\varphi}_1(\lambda)$ and $\varphi_2(\lambda) = \lambda\widetilde{\varphi}_2(\lambda)$, $\lambda \in \mathbb{D}$, where $\widetilde{\varphi}_j \in \mathcal{O}(\mathbb{D}, \mathbb{C})$. Note that $\widetilde{\varphi}_1(0) = 1/s$ and $\widetilde{\varphi}_2(s) = 1/s$.

Take an arbitrary $r \in (s, 1)$. Using that $|\varphi_1\varphi_2| < 1$ on G, the maximum principle leads to $|\widetilde{\varphi}_1(\lambda)\widetilde{\varphi}_2(\lambda)| \leq \frac{1}{(r-s)r}$ as long as $|\lambda| \leq r$. Or when r tends to 1, then $|\widetilde{\varphi}_1\widetilde{\varphi}_2| \leq \frac{1}{1-s}$.

Now observe that the function $\widetilde{\varphi}_2 - \widetilde{\varphi}_2(0) \in \mathcal{O}(\mathbb{D}, 20\mathbb{D})$ and vanishes at zero. Hence Schwarz's lemma implies that $|\widetilde{\varphi}_2(\lambda) - \widetilde{\varphi}_2(0)| \leq 20|\lambda|$ for all $\lambda \in \mathbb{D}$. Or $|\widetilde{\varphi}_2(0)| \geq |\widetilde{\varphi}_2(s)| - 20s = 1/s - 20s$. Moreover, $\frac{|\widetilde{\varphi}_2(0)|}{s} = |\widetilde{\varphi}_1(0)\widetilde{\varphi}_2(0)| \leq \frac{1}{1-s}$. Combining the last two inequalities leads to $1 - 20s^2 \leq \frac{s^2}{1-s}$ or after some simple calculations to $\boldsymbol{\ell}_G(a, b) \geq \boldsymbol{p}(0, 1/5) > \boldsymbol{p}(0, 1/10) = \boldsymbol{\ell}_G(a, 0) + \boldsymbol{\ell}_G(0, b)$, which contradicts the triangle inequality.

Remark 7.2.4. Already here we draw the attention of the reader to the fact that convex domains will play a non-expected important role during the discussion of invariant functions. For example, for a convex domain $G \subset \mathbb{C}^n$ the triangle inequality for the Lempert function $\boldsymbol{\ell}_G$ is true.

Indeed, fix a convex domain $G \subsetneq \mathbb{C}^n$ and three pairwise different points $z, w, s \in G$. Then, if $\varepsilon > 0$ is given, there exist analytic discs $\varphi, \psi \in \mathcal{O}(\overline{\mathbb{D}}, G)$ and numbers $\tau, \sigma \in (0, 1)$, $\tau < \sigma$, such that $\varphi(0) = z$, $\varphi(\tau) = w = \psi(\tau)$, $\psi(\sigma) = s$, and $\boldsymbol{\ell}_G(z, w) + \varepsilon > \boldsymbol{p}(0, \tau)$, $\boldsymbol{\ell}_G(w, s) + \varepsilon > \boldsymbol{p}(\tau, \sigma)$. Define $\chi(\lambda) := \frac{(\lambda-\sigma)(\lambda-1/\sigma)}{(\lambda-\tau)(\lambda-1/\tau)}$, $\lambda \in \mathbb{C} \setminus \{\tau, 1/\tau\}$. Obviously, χ is holomorphic on $U \setminus \{\tau\}$ for an open neighborhood U of $\overline{\mathbb{D}}$, having a simple pole at τ. Moreover, observe that $\chi(\mathbb{T}) \subset [0, 1)$. Finally, define the "convex" combination $h := (\varphi - \psi)\chi + \psi$. Since $\varphi(\tau) = \psi(\tau)$, it turns out that h is holomorphic on $\overline{\mathbb{D}}$ satisfying $h(\mathbb{T}) \subset G$ and $h(0) = z$. Applying some maximum principle, it follows that $h \in \mathcal{O}(\mathbb{D}, G)$ with $h(0) = z$, $h(\sigma) = s$. Therefore,

$$\boldsymbol{\ell}_G(z, s) \leq \boldsymbol{p}(0, \sigma) = \boldsymbol{p}(0, \tau) + \boldsymbol{p}(\tau, \sigma) \leq \boldsymbol{\ell}_G(z, w) + \boldsymbol{\ell}_G(w, s) + 2\varepsilon,$$

which proves, since ε was arbitrarily chosen, the triangle inequality.

Later on (cf. Chapter 7.3) we will learn that, due to a very deep result of L. Lempert [Lem81], [Lem82], [Lem84] and H. L. Royden & P. M. Wong [RW83] even more is true, namely $\boldsymbol{\ell}_G = \boldsymbol{c}_G$ for such domains.

Moreover, the Lempert function is in general not continuous.

Proposition 7.2.5. *Let* $G := \{z \in \mathbb{C}^n : \mathfrak{h}(z) < 1\}$ *denote a balanced pseudoconvex domain with Minkowski function* $\mathfrak{h}$, *i.e.* $\mathfrak{h} : \mathbb{C}^n \longrightarrow [0, +\infty)$ *is a plurisubharmonic function with* $\mathfrak{h}(\lambda z) = |\lambda|\mathfrak{h}(z)$ ($\lambda \in \mathbb{C}$, $z \in \mathbb{C}^n$). *Then* $\boldsymbol{\ell}_G(0, z) = \boldsymbol{p}(0, \mathfrak{h}(z))$, $z \in G$. *In particular,* $\boldsymbol{\ell}_{\mathbb{B}_n}^*(0, z) = \|z\|$, $z \in \mathbb{B}_n$, *and* $\boldsymbol{\ell}_{\mathbb{D}^n}^*(0, z) = \max\{|z_j| : j = 1, \ldots, n\}$, $z \in \mathbb{D}^n$.

The proof is based on the maximum principle for subharmonic functions.

Since there are many bounded balanced pseudoconvex domains whose Minkowski functions are not continuous, Proposition 7.2.5 shows that the Lempert function is in general not continuous, even not as a function of one variable.

But $\boldsymbol{\ell}_G$ is always upper semicontinuous.

Proposition 7.2.6. *The Lempert function ℓ_G is upper semicontinuous.*

Recall that ℓ_G may be defined via analytic discs $\varphi \in \mathcal{O}(\overline{\mathbb{D}}, G)$, i.e. discs with a compact image $\varphi(\overline{\mathbb{D}})$ in G. Hence a small deformation of such a φ easily leads to the former result.

Although those defects described above one has to emphasize that, in general, it is easier to handle ℓ_G than c_G. To conclude this paragraph some other properties of the Lempert function will be summarized. In order to be able to calculate the Lempert function the following results turn out to be helpful.

Proposition 7.2.7. *Suppose that two domains $G \subset \mathbb{C}^n$ and $D \subset \mathbb{C}^m$ are given. If $z', z'' \in G$ and $w', w'' \in D$, then the following formula*

$$\ell_{G \times D}((z', w'), (z'', w'')) = \max\{\ell_G(z', z''), \ell_D(w', w'')\}$$

is true.

It is said that the family of Lempert functions satisfies the *product property*. It should be emphasized, that the proof of Proposition 7.2.7 is much simpler than the one for the product property of the Carathéodory pseudodistances.

Finally, we mention how the Lempert family behaves under an increasing union of domains.

Proposition 7.2.8. *Let $G, G_j \subset G$, $j \in \mathbb{N}$, be domains with the property that $G_j \nearrow G$. Then $\ell_{G_j} \searrow \ell_G$.*

7.2.2 Tautness

Although the Lempert function need not to be continuous, there is a sufficiently rich family of pseudoconvex domains whose Lempert functions are continuous.

Definition 7.2.9. Let $\Omega \subset \mathbb{C}^n$ be a domain. Then Ω is called *taut* if the space $\mathcal{O}(\mathbb{D}, \Omega)$ is normal, i.e. whenever we start with a sequence $(f_j)_{j=1}^\infty \subset \mathcal{O}(\mathbb{D}, \Omega)$, then there exists a subsequence $(f_{j_\nu})_{\nu=1}^\infty$ with $f_{j_\nu} \longrightarrow f \in \mathcal{O}(\mathbb{D}, \Omega)$ locally uniformly in $\mathbb{D}$ or there exists a subsequence $(f_{j_\nu})_{\nu=1}^\infty$ which diverges uniformly on compact sets, i.e. for any two compact sets $K \subset \mathbb{D}$, $L \subset \Omega$ there is an index ν_0 such that $f_{j_\nu}(K) \cap L = \varnothing$ if $\nu \geq \nu_0$.

Before we continue our investigations, we make a small digression on tautness. We only collect the most important results on this subject omitting their proofs.

Remark 7.2.10.

(a) Any taut domain is pseudoconvex.

(b) Any bounded *hyperconvex* domain $G \subset \mathbb{C}^n$, i.e. any bounded domain G for which there exists a negative plurisubharmonic exhaustion function $\varphi : G \longrightarrow (-\infty, 0)$, is a taut domain. In particular, any bounded convex domain is a taut one. Note that a hyperconvex domain is pseudoconvex.

(c) Any bounded pseudoconvex domain with $\mathcal{C}^1$-boundary is taut. We point out that these domains are also hyperconvex (cf. [KR81]) so that (b) yields tautness, too.

(d) In the case of a balanced pseudoconvex domain $G := \{z \in \mathbb{C}^n : \mathfrak{h}(z) < 1\}$ with Minkowski function $\mathfrak{h}$, there is even a complete characterization of tautness (cf. [Bar83]), namely:
G is taut if and only if $\mathfrak{h}$ is continuous with $\mathfrak{h} \geq C\| \ \|$ for a suitable $C > 0$.

(e) Let $\Pi : D \longrightarrow G$ be a holomorphic covering, i.e. for any point $a \in G$ there exists an open neighborhood $U = U(a) \subset G$ such that $\Pi^{-1}(U)$ is the union of pairwise disjoint open sets $V_\alpha \subset D$ such that $\Pi|_{V_\alpha}$ is a biholomorphic mapping from V_α onto U. Then D is taut if and only if G is taut (cf. [Bar71]).

(f) The above property and the uniformization theorem in classic complex analysis immediately imply the following result.
A domain $G \subset \mathbb{C}$ is taut if and only if $\#(\mathbb{C} \setminus G) \geq 2$ if and only if there exists a holomorphic covering $\Pi : \mathbb{D} \longrightarrow G$.

Now, we come back to our discussion on the Lempert function. Namely, we want to study this function on taut domains. Our first result is the following one.

Proposition 7.2.11. *Let z', z'' be two points of a taut domain $G \subset \mathbb{C}^n$. Then there exist a holomorphic map $\varphi \in \mathcal{O}(\mathbb{D}, G)$ and a number $\sigma \in [0, 1)$ with $\varphi(0) = z'$, $\varphi(\sigma) = z''$, and $\boldsymbol{\ell}_G(z', z'') = \boldsymbol{p}(0, \sigma)$. Such a φ is called to be an extremal disc in G passing through the points z', z''.*

Proof. By definition we find a sequence $(\varphi_j)_{j=1}^\infty \subset \mathcal{O}(\mathbb{D}, G)$ satisfying

$$\varphi_j(0) = z', \ \varphi_j(\sigma_j) = z'' \ (0 \leq \sigma_j < 1), \quad \text{and} \quad \boldsymbol{p}(0, \sigma_j) \searrow \boldsymbol{\ell}_G(z', z'').$$

Since G is a taut domain and $\varphi_j(0) = z'$, we can choose subsequences $(\varphi_{j_\nu})_{\nu=1}^\infty \subset (\varphi_j)_{j=1}^\infty$ and $(\sigma_{j_\nu})_{\nu=1}^\infty \subset (\sigma_j)_{j=1}^\infty$ with $\varphi_{j_\nu} \longrightarrow \varphi \in \mathcal{O}(\mathbb{D}, G)$ locally uniformly in $\mathbb{D}$ and $\sigma_{j_\nu} \longrightarrow \sigma \in [0, 1)$. From this we conclude that $\varphi(\sigma) = z''$, $\varphi(0) = z'$, and $\boldsymbol{p}(0, \sigma) = \boldsymbol{\ell}_G(z', z'')$. Hence, we have proved that there always exist extremal discs through two given points. $\qquad\square$

Remark 7.2.12. (a) Note that the claim of Proposition 7.2.11 is not longer true if G is not taut. For example, take $G_0 := \mathbb{B}_2 \setminus \{(1/2, 0)\}$ and $z' := (0, 0)$, $z'' := (1/4, 0)$. Using the analytic discs $\mathbb{D} \ni \lambda \longmapsto \left(R\lambda, s(R)\lambda(\lambda - \frac{1}{4R})\right)$, $R < 1$, $s(R) \ll 1$, one can easily deduce that $\boldsymbol{\ell}_{G_0}(z', z'') \leq \boldsymbol{p}(0, 1/4)$.

Now suppose that there exists $\varphi = (\varphi_1, \varphi_2) \in \mathcal{O}(\mathbb{D}, G_0)$, $\varphi(0) = z'$, $\varphi(\sigma) = z''$ such that $\boldsymbol{\ell}_G(z', z'') = \boldsymbol{p}(0, \sigma)$. Thus $\sigma \leq 1/4$. On the other hand, the Schwarz lemma implies that $1/4 \leq \sigma$. Hence $\varphi_1(1/4) = 1/4$, i.e. $\varphi_1(\lambda) \equiv \lambda$.

Since $\varphi \in \mathcal{O}(\mathbb{D}, \mathbb{B}_2)$, it turns out that $\varphi_2 \equiv 0$. In particular, $\varphi(1/2) = (1/2, 0)$ which contradicts the definition of G.

(b) Let $D \subset \mathbb{C}^n$ be an arbitrary domain. Observe that if $\varphi \in \mathcal{O}(\mathbb{D}, D)$ is an extremal disc through two different points $z', z'' \in D$ with $\varphi(0) = z'$ and $\varphi(\sigma) = z''$, i.e. $0 < \sigma = \ell_D^*(z', z'')$, then $\varphi(\mathbb{D})$ cannot be relatively compact in D. Otherwise one could find a number $\theta \in (0, 1)$ and a new analytic disc $\widetilde{\varphi}(\lambda) := \varphi(\lambda) + \frac{\lambda}{\sigma\theta}(\varphi(\sigma) - \varphi(\theta\sigma))$, $\lambda \in \mathbb{D}$, in D. Then $\widetilde{\varphi}(0) = z'$ and $\widetilde{\varphi}(\theta\sigma) = z''$; a contradiction.

Moreover, we have the following result on the continuity on ℓ_G.

Proposition 7.2.13. *If G is a taut domain in $\mathbb{C}^n$, then the Lempert function ℓ_G is continuous on $G \times G$.*

7.2.3   The Kobayashi pseudodistance

To overcome the difficulty connected with the triangle inequality we modify the function ℓ_G in such a way that the new function becomes a pseudodistance, the largest one below of ℓ_G.

Definition 7.2.14. Let $G \subset \mathbb{C}^n$ be a domain and $z', z'' \in G$. Put

$$k_G(z', z'') :$$

$$= \inf \left\{ \sum_{j=1}^{N} \ell_G(z_{j-1}, z_j) : N \in \mathbb{N}, z_0 = z', z_1, \dots, z_{N-1} \in G, z_N = z'' \right\},$$

$k_G^* := \tanh k_G$. The function k_G is called the *Kobayashi pseudodistance for* G.

k_G was introduced in 1967 by S. Kobayashi (see [Kob67]).

Remark 7.2.15. Notice that the following properties hold for the system $(k_G)_G$:

(a) k_G is a pseudodistance on G;
(b) even more, k_G is the largest minorant of ℓ_G that satisfies the triangle inequality;
(c) if $F \in \mathcal{O}(G, D)$, then $k_D(F(z'), F(z'')) \le k_G(z', z'')$, i.e. the system $(k_G)_G$ is contractible with respect to holomorphic mappings;
(d) $k_{\mathbb{D}} = \ell_{\mathbb{D}} = p$. Even more, we have:
(e) if $(d_G)_G$ is a system of pseudodistances $d_G : G \times G \longrightarrow \mathbb{R}_+$ with the properties stated in (c) and (d), then $d_G \le k_G$;
(f) in particular, $c_G \le k_G$.

To be able to continue the discussion on the Lempert function and the Kobayashi pseudodistance we need at least a few examples for which these objects can be calculated.

Example 7.2.16. Let $\mathfrak{h}$ be a seminorm on $\mathbb{C}^n$. Denote by $G := \{z \in \mathbb{C}^n : \mathfrak{h}(z) < 1\}$ the associated open unit $\mathfrak{h}$-ball. Then $\boldsymbol{k}_G(0, z) = \boldsymbol{\ell}_G(0, z) = \boldsymbol{p}(0, \mathfrak{h}(z))$, $z \in G$. In fact, by Corollary 7.1.12 and Proposition 7.2.5 we have $\boldsymbol{p}(0, \mathfrak{h}(z)) = \boldsymbol{c}_G(0, z) \le \boldsymbol{k}_G(0, z) \le \boldsymbol{\ell}(0, z) \le \boldsymbol{p}(0, \mathfrak{h}(z))$.

In particular, we mention the following special cases.

Example 7.2.17. (a) $\boldsymbol{k}_{\mathbb{D}^n}(0, z) = \boldsymbol{\ell}_{\mathbb{D}^n}(0, z) = \max\{\boldsymbol{p}(0, |z_j|) : 1 \le j \le n\}$,
(b) $\boldsymbol{k}_{\mathbb{B}_n}(0, z) = \boldsymbol{\ell}_{\mathbb{B}_n}(0, z) = \boldsymbol{p}(0, \|z\|)$.

As a consequence of this example we obtain (cf. Corollary 7.1.14)

Proposition 7.2.18. *The function* $\boldsymbol{k}_G : G \times G \longrightarrow [0, +\infty)$ *is continuous.*

7.2.4   General properties of k

We already know that the $\|\ \|$-topology of a domain $G \subset \mathbb{C}^n$ is stronger than the $\boldsymbol{k}_G$-topology on G. We remember that in the case of the Carathéodory distance the $\|\ \|$-topology can be different from the $\boldsymbol{c}_G$-topology. To discuss the analogous question for the Kobayashi distance we need the following observation; see [Kob73].

Proposition 7.2.19. *The Kobayashi pseudodistance is* inner, *i.e. if* z', $z'' \in G$, *then*

$$\boldsymbol{k}_G(z', z'') = \inf\{L_{\boldsymbol{k}_G}(\alpha) : \alpha : [0, 1] \longrightarrow G$$
$$\text{is continuous and } \|\ \|\text{-rectifiable with } \alpha(0) = z', \ \alpha(1) = z''\}, \text{ where}$$

$$L_{\boldsymbol{k}_G}(\alpha) := \sup\left\{\sum_{j=1}^{N} \boldsymbol{k}_G(\alpha(t_{j-1}), \alpha(t_j)) : N \in \mathbb{N},\ 0 = t_0 < t_1 < \cdots < t_N = 1\right\}$$

denotes the $\boldsymbol{k}_G$-*length of* α.

Recall that, in general, the formula analogous to that of Proposition 7.2.19 fails to hold for the Carathéodory pseudodistance.

It is obvious that a necessary condition for the $\|\ \|$-topology and the $\boldsymbol{k}_G$-topology to be equal is that $\boldsymbol{k}_G$ is a distance. We say that a domain $G \subset \mathbb{C}^n$ is $\boldsymbol{k}$-*hyperbolic* if its Kobayashi pseudodistance is a distance.

Remark 7.2.20. By the well-known inequality $\boldsymbol{c}_G \le \boldsymbol{k}_G$ it is clear that any $\boldsymbol{c}$-hyperbolic domain is also $\boldsymbol{k}$-hyperbolic. Therefore, any bounded domain G is $\boldsymbol{k}$-hyperbolic.

So far we have shown that the Kobayashi pseudodistance is continuous and inner. These conditions suffice to prove the following comparison property of the topologies (cf. [Bar72]). Recall the situation for the Carathéodory pseudodistance which was more complicated.

Proposition 7.2.21. *If G is a k-hyperbolic domain in $\mathbb{C}^n$, then its $\| \|$-topology is equal to the k_G-topology.*

Proof. Since k_G is continuous, every k_G-ball is a $\| \|$-open set. Suppose now that there is a norm-ball $\mathbb{B}(a,r) \subset\subset G$ which is not k_G-open. Then there exists a point $b \in \mathbb{B}(a,r)$ containing no k_G-ball. Therefore one may find points $z_j \in G \setminus \mathbb{B}(a,r)$, $j \in \mathbb{N}$, with $s_j := k_G(b, z_j) \longrightarrow 0$.

Using the k-hyperbolicity of G and the continuity it follows that $k_G(b,z) \geq s > 0$, $z \in \partial\mathbb{B}(a,r)$, for some s. Now one applies the innerness. Thus for every j there is a curve γ_j in G connecting b with z_j such that $L_{k_G}(\gamma_j) \leq 2s_j$. Take a point $a_j \in \partial\mathbb{B}(a,r)$ lying on the j-th curve. Then $2s_j \geq k_G(b,a_j) + k_G(a_j, z_j) \geq s$; a contradiction.

It should be mentioned that for a bounded domain G this result follows directly from Exercise 7.2.17. $\qquad\square$

Applying that the Kobayashi pseudodistance k_G is the largest pseudodistance below of ℓ_G, Propositions 7.2.8 and 7.2.7 immediately lead to the two following properties.

Remark 7.2.22.

(a) Let $G = \bigcup_{\nu=1}^{\infty} G_\nu$, where $(G_\nu)_{\nu=1}^{\infty}$ is an increasing sequence of domains in $\mathbb{C}^n$. Then for z', $z'' \in G$ we have $\ell_G(z', z'') = \lim_{\nu\to\infty} \ell_{G_\nu}(z', z'')$.
(b) The Kobayashi pseudometric satisfies the product property.

In calculations of the Kobayashi pseudodistance holomorphic coverings often play an important role because of the following result of S. Kobayashi (cf. [Kob05]) whose proof is based on the lifting property for holomorphic coverings.

Theorem 7.2.23. *Let $\Pi : \widetilde{G} \longrightarrow G$ be a holomorphic covering. Then for x, $y \in G$ and $\widetilde{x} \in \widetilde{G}$, $\Pi(\widetilde{x}) = x$, the Lempert function and the Kobayashi pseudodistance for G satisfy the following formulas:*

$$\ell_G(x,y) = \inf\{\ell_{\widetilde{G}}(\widetilde{x}, \widetilde{y}) : \widetilde{y} \in \widetilde{G}, \ \Pi(\widetilde{y}) = y\},$$
$$k_G(x,y) = \inf\{k_{\widetilde{G}}(\widetilde{x}, \widetilde{y}) : \widetilde{y} \in \widetilde{G}, \ \Pi(\widetilde{y}) = y\}.$$

Remark 7.2.24.

(a) From the previous theorem we conclude that if all $k_{\widetilde{G}}$-balls with finite radii are relatively compact subsets of $\widetilde{G}$, then for x, $y \in G$, and $\widetilde{x} \in \widetilde{G}$, $\Pi(\widetilde{x}) = x$, there exists a point $\widetilde{y} \in \widetilde{G}$ with $\Pi(\widetilde{y}) = y$ and $k_G(x,y) = k_{\widetilde{G}}(\widetilde{x}, \widetilde{y})$ (the same statement is also true for the Lempert functions provided that $\widetilde{G}$ is taut).
(b) Notice that in general Theorem 7.2.23 is not true in the following strong sense, namely, that there exists a point $\widetilde{y} \in \widetilde{G}$ with $k_{\widetilde{G}}(\widetilde{x}, \widetilde{y}) = k_G(x,y)$ (this was a question posed by Kobayashi; the counterexample is due to W. Zwonek).

(c) Since any domain $G \subset \mathbb{C}$ has $\mathbb{D}$ or $\mathbb{C}$ as its universal covering, Theorem 7.2.23 also implies that $\boldsymbol{\ell}_G$ satisfies the triangle inequality; hence $\boldsymbol{k}_G = \boldsymbol{\ell}_G$.

(d) We also mention that for $G := \mathbb{C} \setminus \{0,1\}$ we have $\boldsymbol{c}_G \equiv 0$ whereas $\boldsymbol{k}_G(z', z'') > 0$ if $z' \neq z''$; i.e. G is $\boldsymbol{k}$-hyperbolic. The latter fact follows because of Theorem 7.2.23 and the well-known result that $\mathbb{D}$ is the universal covering of G.

(e) Moreover, the covering result may be used to obtain a formula of the Kobayashi distance for the punctured disc $\mathbb{D}_*$, namely: if $a \in (0,1)$ and $z \in \mathbb{D}_*$ with $z = |z|^{i\theta}$, $\theta \in (-\pi, \pi]$, then $\boldsymbol{k}_{\mathbb{D}_*}(a,z) = \tanh^{-1}\left(\frac{\theta^2 + (\log|z| - \log a)^2}{\theta^2 + (\log|z| + \log a)^2}\right)$.

So far we know only few examples of domains G for which $\boldsymbol{c}_G \neq \boldsymbol{k}_G$. For plane domains we have the following complete characterization of such domains.

Proposition 7.2.25.

(a) *Let G be a $\boldsymbol{c}$-hyperbolic domain in $\mathbb{C}$ and let us suppose that there is at least one pair of different points z', $z'' \in G$ with $\boldsymbol{k}_G(z', z'') = \boldsymbol{c}_G(z', z'')$. Then G is biholomorphically equivalent to $\mathbb{D}$ and so $\boldsymbol{k}_G \equiv \boldsymbol{c}_G$.*

(b) *If a plane domain G is not $\boldsymbol{c}$-hyperbolic, then $\boldsymbol{c}_G \equiv 0$ and either $\boldsymbol{k}_G \equiv 0$ or G is $\boldsymbol{k}$-hyperbolic.*

Proof. In the case when G is $\boldsymbol{c}$-hyperbolic there is the holomorphic covering $\Pi : \mathbb{D} \longrightarrow G$. According to Remark 7.2.24(a) there exist points λ', λ'' in the unit disc such that $\Pi(\lambda') = z'$, $\Pi(\lambda'') = z''$, and $\boldsymbol{k}_G(z', z'') = \boldsymbol{k}_{\mathbb{D}}(\lambda', \lambda'') = \boldsymbol{p}(\lambda', \lambda'')$. On the other hand, $\boldsymbol{c}_G(z', z'')$ can be written as $\boldsymbol{c}_G(z', z'') = \boldsymbol{p}(f(z'), f(z''))$ for a suitable $f \in \mathcal{O}(G, \mathbb{D})$. For the function $f \circ \Pi \in \mathcal{O}(\mathbb{D}, \mathbb{D})$ this implies

$$\boldsymbol{p}(f \circ \Pi(\lambda'), f \circ \Pi(\lambda'')) = \boldsymbol{c}_G(z', z'') = \boldsymbol{k}_G(z', z'') = \boldsymbol{p}(\lambda', \lambda'').$$

Now, the Schwarz–Pick lemma tells us that $f \circ \Pi$ is a biholomorphic map, and therefore Π is biholomorphic.

We turn to the proof of claim (b). Since G is not $\boldsymbol{c}$-hyperbolic, we have $\boldsymbol{c}_G \equiv 0$. In the case where the universal covering of G is given by $\mathbb{C}$, it is clear that $\boldsymbol{k}_G \equiv 0$. So we may assume that $\Pi : \mathbb{D} \longrightarrow G$ is the universal covering. Hence by Theorem 7.2.23 and Remark 7.2.24(a), we conclude that whenever $z', z'' \in G$, $z' \neq z''$, then there are points $\lambda', \lambda'' \in \mathbb{D}$, $\Pi(\lambda') = z'$, $\Pi(\lambda'') = z''$ with $\boldsymbol{k}_G(z', z'') = \boldsymbol{p}(\lambda', \lambda'') > 0$. $\square$

Corollary 7.2.26. *Let $P := \{\lambda : 1/R < |\lambda| < R\}$ $(R > 1)$. Then for z', $z'' \in P$, $z' \neq z''$, we have $\boldsymbol{c}_P(z', z'') < \boldsymbol{k}_P(z', z'')$.*

7.2.5 The Kobayashi–Royden pseudometric

In Chapter 7.1 we have already learned that for the Carathéodory pseudodistance there is an infinitesimal version, the Carathéodory–Reiffen pseudometric, which measures the lengths of tangent vectors. A similar notion with respect to the Kobayashi pseudodistance was introduced by Royden in 1997 (see [Roy71]). Because of its strong relation to the Kobayashi pseudometric it will be introduced here but only shortly investigated.

Let G be a domain in $\mathbb{C}^n$. The function $\varkappa_G : G \times \mathbb{C}^n \longrightarrow \mathbb{R}_+$ defined by $\varkappa_G(z; X) := \inf\{\gamma(\lambda)|\alpha| : \exists_{\varphi \in \mathcal{O}(\mathbb{D}, G)} \exists_{\lambda \in \mathbb{D}} : \varphi(\lambda) = z, \ \alpha\varphi'(\lambda) = X\}$ is called the *Kobayashi–Royden pseudometric*.

Remark 7.2.27. Observe that:

(1) $\varkappa_G(z; X) = \inf\{\alpha > 0 : \exists_{\varphi \in \mathcal{O}(\mathbb{D}, G)} : \varphi(0) = z, \ \alpha\varphi'(0) = X\}$
$\qquad\qquad = \inf\{\alpha > 0 : \exists_{\varphi \in \mathcal{O}(\overline{\mathbb{D}}, G)} : \varphi(0) = z, \ \alpha\varphi'(0) = X\}$;

(2) $\varkappa_G(z; \lambda X) = |\lambda| \varkappa_G(z; X), \ \lambda \in \mathbb{C}, \ X \in \mathbb{C}^n, \ z \in G \subset \mathbb{C}^n$;

(3) $\varkappa_D(F(z); F'(z)X) \leq \varkappa_G(z; X), \ F \in \mathcal{O}(G, D), \ z \in G \subset \mathbb{C}^n, \ X \in \mathbb{C}^n$.
Hence $\varkappa_G(z; \cdot)$ assigns a length to any tangent vector at z, and moreover (3) shows that the system $(\varkappa_G)_G$ is contractible with respect to holomorphic mappings. In particular, if $F : G \longrightarrow D$ is a biholomorphic map, then

(4) $\varkappa_D(F(z); F'(z)X) = \varkappa_G(z; X), \ z \in G, \ X \in \mathbb{C}^n$;

(5) $\varkappa_{\mathbb{D}}(\lambda; X) = \gamma(\lambda)|X| = \gamma_{\mathbb{D}}(z; X), \ \lambda \in \mathbb{D}, \ X \in \mathbb{C}$;

(6) for any domain $G \subset \mathbb{C}^n$ we have $\gamma_G \leq \varkappa_G$ (use the Schwarz–Pick lemma);

(7) for any domain $G \subset \mathbb{C}^n$ we have $\gamma_G \leq \varkappa_G$;

(8) $\varkappa_G$ is upper semicontinuous on $G \times \mathbb{C}^n$; note that, in general, $\varkappa_G$ is not continuous (see Proposition 7.2.29);

(9) $\varkappa_{G_1 \times G_2}(a; X) = \max\{\varkappa_{G_1}(a_1; X_1), \varkappa_{G_2}(a_2; X_2)\}$, where $G_j \subset \mathbb{C}^{n_j}$, $j = 1, 2$, are domains and $a = (a_1, a_2) \in G_1 \times G_2$ and $X = (X_1, X_2) \in \mathbb{C}^{n_1} \times \mathbb{C}^{n_2}$, i.e. $\varkappa$ satisfies the product property.

Moreover, it turns out that whenever there is a system $(\delta_G)_G$ of functions $\delta_G : G \times \mathbb{C}^n \longrightarrow \mathbb{R}_+$ $(G \subset \mathbb{C}^n)$ with the properties (3) and (5), then $\delta_G \leq \varkappa_G$ for any G. We recall that we already know that also $\gamma_G \leq \delta_G$ is true.

Remark 7.2.28. Recall that γ_G could be thought as a certain derivative of the Carathéodory pseudodistance (cf. Remark 7.1.8(d)). A similar result is true with respect to the Lempert function.

(a) Assume that $G \subset \mathbb{C}^n$ is a taut domain. Then (see [Pan94], [NP08])

$$\varkappa_G(a; X) = \lim_{\substack{\mathbb{C}\setminus\{0\} \ni \lambda \to 0 \\ z \to a \\ X' \to X}} \frac{1}{|\lambda|} \ell_G(z, z + \lambda X'), \quad a \in G, \ X \in \mathbb{C}^n.$$

(b) Without the assumption that G is taut this formula remains, in general, not true.

To see at least a few concrete examples, we calculate the Kobayashi–Royden pseudometric for balanced pseudoconvex domains.

Proposition 7.2.29. *Let $G = G_{\mathfrak{h}}$ be a balanced pseudoconvex domain in $\mathbb{C}^n$ given by its Minkowski function $\mathfrak{h}$. Then $\varkappa_G(0; X) = \mathfrak{h}(X)$, $X \in \mathbb{C}^n$.*

Proof. First of all observe that if $\mathfrak{h}(X) \neq 0$, then $\mathbb{D} \ni \lambda \overset{\varphi}{\longmapsto} \lambda X/\mathfrak{h}(X)$ is an analytic disc in G with $\varphi(0) = 0$, $\varphi'(0) = X/\mathfrak{h}(X)$; hence we obtain $\varkappa_G(0; X) \leq \mathfrak{h}(X)$.

The fact that the same inequality is also true if $\mathfrak{h}(X) = 0$ is left to the reader (cf. Example 7.2.16). On the other hand, let $\varphi \in \mathcal{O}(\mathbb{D}, G)$ with $\varphi(0) = 0$, $\alpha\varphi'(0) = X$ ($\alpha > 0$). As in Proposition 7.2.5, we observe that $\mathfrak{h}(\varphi(\lambda)) = \mathfrak{h}(\lambda\widetilde{\varphi}(\lambda)) < 1$, and therefore $\mathfrak{h} \circ \widetilde{\varphi} \leq 1$.

Thus we end up with $\mathfrak{h}(X) = \mathfrak{h}(\alpha\varphi'(0)) = \alpha\mathfrak{h} \circ \widetilde{\varphi}(0) \leq \alpha$, which guarantees the missing inequality. $\qquad\square$

We emphasize that the proof above is based on the information that G is pseudoconvex, i.e. that the Minkowski function is plurisubharmonic; in general it is false (cf. [FS87]).

Applying Proposition 7.2.29 leads to the following formulae.

Example 7.2.30.

(a) For $z \in \mathbb{B}_n$ and $X \in \mathbb{C}^n$, the following formula is true.

$$\varkappa_{\mathbb{B}_n}(z; X) = \left(\frac{\|X\|^2}{1 - \|z\|^2} + \frac{|\langle z, X \rangle|^2}{(1 - \|z\|^2)^2} \right)^{1/2}.$$

For the proof use the fact that $\varkappa_{\mathbb{B}_n}$ is invariant under $\mathrm{Aut}(\mathbb{B}_n)$.

In particular, this formula immediately yields to the following observation: if G is a bounded domain, then there exists a constant $C > 0$ such that for any $z \in G$, $X \in \mathbb{C}^n$ the following inequality $\varkappa_G(z; X) \geq C\|X\|$ is true.

(b) A similar argument leads to the corresponding formula for the unit polycylinder.

$$\varkappa_{\mathbb{D}^n}(z; X) = \max\left\{ \frac{|X_1|}{1 - |z_1|^2}, \dots, \frac{|X_n|}{1 - |z_n|^2} \right\}.$$

(c) Moreover it turns out that, in general, $\varkappa_G(z; \cdot)$ is not a norm on $\mathbb{C}^n$. For example put $G := \{ z \in \mathbb{D}^2 : |z_1 z_2| < 1/2 \}$; G is a bounded balanced pseudoconvex domain with Minkowski function $\mathfrak{h}(z) = \max\{|z_1|, |z_2|, \sqrt{2|z_1||z_2|}\}$. Therefore we know that $\varkappa_G(0; X) = \mathfrak{h}(X)$. In particular, $\varkappa_G(0; (\frac{3}{2}, \frac{3}{2})) = \frac{3}{\sqrt{2}} > 2 = \varkappa_G(0; (1, \frac{1}{2})) + \varkappa_G(0; (\frac{1}{2}, 1))$.

Moreover, Proposition 7.2.29 can be also used to describe biholomorphic mappings between balanced pseudoconvex domains.

Corollary 7.2.31. *Let $G_j = G_{\mathfrak{h}_j} \subsetneq \mathbb{C}^n$ be pseudoconvex balanced domains with Minkowski functions $\mathfrak{h}_j$, $j = 1, 2$. Then the following conditions are equivalent:*

(i) *there exists a biholomorphic mapping $F : G_1 \longrightarrow G_2$ with $F(0) = 0$;*
(ii) *there exists a $\mathbb{C}$-linear isomorphism $L : \mathbb{C}^n \longrightarrow \mathbb{C}^n$ such that $\mathfrak{h}_2 \circ L = \mathfrak{h}_1$, i.e. G_1 and G_2 are linearly equivalent.*

Remark 7.2.32.

(a) Observe that $\Psi := F^{-1} \circ F'(0)$ is a biholomorphic mapping of G_1 sending the origin to the origin. Moreover, $\Psi'(0) = \mathrm{id}$. In case where G_1 is bounded, a result similar to Lemma 7.1.46 finally yields that $\psi = \mathrm{id}_{G_1}$ or $F = F'(0)$, i.e. F is a linear isomorphism. Note that the assumption "G_1 is bounded" is essential–take for instance $G_1 = G_2 = \mathbb{C}$ and $\psi(z) := z(1 - z)$.
(b) Moreover, a deep result by Kaup–Upmeier (see [KU76]) says that whenever there exists a biholomorphic mapping $F : G_1 \longrightarrow G_2$, then there is another one sending $0 \in G_1$ to $0 \in G_2$. In particular, if G_1 is biholomorphically equivalent to G_2, then G_1 is linearly equivalent to G_2.

Similarly as in the case of the Lempert function (cf. § 7.2.3), we also obtain better results for the Kobayashi–Royden pseudometric on taut domains. Since the argument here is more or less the same, we only formulate the result.

Proposition 7.2.33. *Let G be a taut domain in $\mathbb{C}^n$. Then*

(a) *for any $z \in G$ and for any $X \in \mathbb{C}^n$ there exists an extremal analytic disc $\varphi \in \mathcal{O}(\mathbb{D}, G)$, i.e. $\varphi(0) = z$ and $\varkappa_G(z; X)\varphi'(0) = X$;*
(b) *the Kobayashi–Royden pseudometric is continuous on $G \times \mathbb{C}^n$.*

Recall that the γ_G–length has led to the inner Carathéodory pseudometric which, in general, is different to c_G. On the other hand the Kobayashi pseudodistance can be obtained via integration of the Kobayashi-Royden pseudometric.

Theorem 7.2.34. *Let $G \subset \mathbb{C}^n$ be an arbitrary domain. Then*

$$k_G(z', z'') = \inf\{L_{\varkappa_G}(\alpha) : \alpha \text{ a piecewise } C^1\text{-curve in } G \text{ connecting } z' \text{ and } z''\},$$

where $L_{\varkappa_G}(\alpha) := \int_0^1 \varkappa_G(\alpha(t); \alpha'(t))dt$.

As a direct consequence of Remark 7.2.27(9) and the former theorem one obtains the product property for the Kobayashi pseudometric.

Corollary 7.2.35. *Let $G_j \subset \mathbb{C}^{n_j}$, $j = 1, 2$, be given domains. Then*

$$k_{G_1 \times G_2}(a, b) = \max\{k_{G_1}(a_1, b_1), k_{G_2}(a_2, b_2)\},$$

where $a = (a_1, a_2), b = (b_1, b_2) \in G_1 \times G_2$.

7.2.6 k-hyperbolicity

Let M and N be complex manifolds with M a relatively compact open subset of N, and $f : \mathbb{D}_* \longrightarrow M$ a holomorphic mapping. When does f extend to a holomorphic mapping $\tilde{f} : \mathbb{D} \longrightarrow N$? The classical case of this question is when $N = \mathbb{P}^1$ is the Riemann sphere and $M = \mathbb{P}^1 \setminus \{0, 1, \infty\}$. Then extending f is equivalent to the big Picard theorem. Recall that M here is k-hyperbolic. On the other hand put $M := \{(z, w) \in \mathbb{D}_* \times \mathbb{C} : |w| < \sqrt{2}e^{1/|z|} < 1\} \subset \mathbb{P}^2$. Then (see [Kie70]) M is k-hyperbolic. (Using similar techniques, it can be shown that M is complete hyperbolic). Define $f : \mathbb{D}_* \longrightarrow M$ putting $f(z) := (z, e^{1/z})$. It is easy to see that f cannot be extended to a holomorphic map $\tilde{f} : \mathbb{D} \longrightarrow \mathbb{P}^2$. Hence, the big Picard theorem cannot be simply generalized to higher dimensions. According to this extension problem there is some need to discuss k-hyperbolicity.

The following generalization of the big Picard theorem is given by M. H. Kwack (cf. [Kwa69]).

Theorem 7.2.36. *Let $f : \mathbb{D}_* \longrightarrow G$ be a holomorphic map, where G is a k-hyperbolic domain in $\mathbb{C}^n$. Assume that for a sequence $(\lambda_k)_{k=1}^{\infty} \subset \mathbb{D}_*$ with $\lambda_k \longrightarrow 0$ the sequence $(f(\lambda_k))_{k=1}^{\infty}$ converges to a point $z_0 \in G$. Then f extends to a holomorphic map $\tilde{f} : \mathbb{D} \longrightarrow G$.*

Remark 7.2.37. Take $G := \mathbb{C} \setminus \overline{\mathbb{D}}$ and $f : \mathbb{D}_* \longrightarrow G$, $f(z) := 1/z$. Then G is hyperbolic, but f does not extend to the whole unit disc. Therefore, it is clear that an additional property of f is needed to get the extension result. On the other hand there is the following example by D. D. Thai and P. J. Thomas (see [TT98]): there exists a subharmonic function u on $\mathbb{D}$, such that the Hartogs domain $H_u := \{z \in \mathbb{D} \times \mathbb{C} : |z_2| < e^{-u(z_1)}\}$ is not Kobayashi hyperbolic, nevertheless any holomorphic $f \in \mathcal{O}(\mathbb{D}_*, H_u)$ extends to a holomorphic map $\tilde{f} : \mathbb{D} \longrightarrow H_u$.

Proof of Theorem 7.2.36. Without loss of generality, we may assume that the sequence $r_k := |\lambda_k|$ is strictly decreasing and that $z_0 = 0 \in G$. We are going to prove that the function $\tilde{f}$ given by $\tilde{f}(\lambda) := f(\lambda)$, $\lambda \in \mathbb{D}_*$, $\tilde{f}(0) := 0$ is continuous on $\mathbb{D}$. Fix an $\varepsilon \in (0, \text{dist}(0, \partial G))$. Since G is k-hyperbolic, we have $B_{\boldsymbol{k}_G}(0, \delta_\varepsilon) \subset \mathbb{B}(\varepsilon)$ with δ_ε being a suitable positive number. By assumption there is a $k_0 \in \mathbb{N}$ such that for $k \geq k_0$ we have $f(\lambda_k) \in B_{\boldsymbol{k}_G}(0, \delta_\varepsilon/2)$. Applying Remark 7.2.24(g) for $\lambda \in \mathbb{D}_*$, $|\lambda| = r_k$, we obtain

$$\boldsymbol{k}_G(0, f(\lambda)) \leq \boldsymbol{k}_G(0, f(\lambda_k)) + \boldsymbol{k}_G(f(\lambda_k), f(\lambda))$$

$$\leq \boldsymbol{k}_G(0, f(\lambda_k)) + \boldsymbol{k}_{\mathbb{D}_*}(\lambda_k, \lambda) < \frac{\delta_\varepsilon}{2} + \frac{\pi}{-\log r_k} < \delta_\varepsilon \ \text{ if } k \geq k_1 \geq k_0.$$

Therefore, $f(\partial \mathbb{D}(r_k)) \subset \mathbb{B}(\varepsilon)$ if $k \geq k_1$.

It remains to show that if $r_{k+1} < |\lambda| < r_k$, $k \gg 1$, then $f(\lambda) \in \mathbb{B}(\varepsilon)$. But this is an easy consequence of the fact that the Euclidean norm is a plurisubharmonic function and the maximum principle for the subharmonic function $\lambda \longmapsto \|f(\lambda)\|$. □

Hence, it is important to have tools to decide whether a given domain is **k**-hyperbolic.

In the class of convex domains containing the origin the following result due to T. J. Barth (cf. [Bar80]) is true.

Theorem 7.2.38. *Let $G \subset \mathbb{C}^n$ be a convex domain, $0 \in G$. Moreover, assume that no complex line through 0 stays inside G. Then G is biholomorphically equivalent to a bounded domain. In particular, G is **k**-hyperbolic.*

Remark 7.2.39.

(a) It seems to be still an open problem whether G is even biholomorphic equivalent to a bounded convex domain.
(b) In [BS09] one may find even a longer list of equivalent properties for a convex domain to be **k**-hyperbolic. Similar results for $\mathbb{C}$-convex domains are given in [NS07].

Moreover, the following criteria for hyperbolicity are extremely useful.

Theorem 7.2.40. *Let $\Pi : \widetilde{G} \longrightarrow G$ be a holomorphic covering. Then we have: $\widetilde{G}$ is **k**-hyperbolic if and only if G is **k**-hyperbolic.*

A similar result is due to A. Eastwood (see [Eas75]).

Theorem 7.2.41. *Let $G_j \subset \mathbb{C}^{n_j}$, $j = 1, 2$, be domains and let $F \in \mathcal{O}(G_1, G_2)$. Assume that G_2 is **k**-hyperbolic and that there exists an open covering $(U_\alpha)_{\alpha \in A}$ of G_2 such that each connected component of $F^{-1}(U_\alpha)$ is a **k**-hyperbolic domain, $\alpha \in A$. Then G_1 is **k**-hyperbolic.*

Since it is simpler to deal with $\varkappa_G$ than with k_G, it is important to express the property "hyperbolic" in terms of the associated metric. Let G be any domain in $\mathbb{C}^n$. Then G is $\varkappa$-*hyperbolic* if for any $z_0 \in G$ there exist a neighborhood $U = U(z_0) \subset G$ and a positive real number C such that $\varkappa_G(z; X) \geq C\|X\|$, $z \in U$, $X \in \mathbb{C}^n$.

There is the following characterization of **k**-hyperbolicity.

Theorem 7.2.42. *For a domain G in $\mathbb{C}^n$ the following properties are equivalent:*

(i) *G is **k**-hyperbolic;*
(ii) *$\operatorname{top} G = \operatorname{top} k_G$;*
(iii) *for any domain $G' \subset \mathbb{C}^m$, any $w' \in G'$, any $z' \in G$, and any neighborhood $U = U(z') \subset G$ there exist neighborhoods $V = V(w') \subset G'$ and $\widetilde{U} = \widetilde{U}(z') \subset U$ such that if $f \in \mathcal{O}(G', G)$ with $f(w') \in \widetilde{U}$, then $f(V) \subset U$;*
(iv) *condition (iii) is true for $G' = \mathbb{D}$ and $w' = 0 \in \mathbb{D}$;*

(v) G is $\varkappa$-hyperbolic;

(vi) for any $z' \in G$ there exists a Kobayashi-ball around z' with finite radius r, which is a bounded subset of $\mathbb{C}^n$;

(vii) any point $z' \in G$ has a neighborhood $U = U(z') \subset G$ such that, for z, $w \in U$, $\boldsymbol{k}_G(z,w) \geq M\|z - w\|$, where M is a suitable positive constant.

7.2.7   Examples

In the class of Reinhardt domains a complete characterization of hyperbolicity exists. To be able to formulate this theorem several notions are needed:

- $\boldsymbol{V}_j := \{z \in \mathbb{C}^n : z_j = 0\}, \quad j = 1, \ldots, n;$
- $\boldsymbol{D}_{\alpha,C} := \{z \in \mathbb{C}^n, z_j \neq 0 \text{ if } \alpha_j < 0 : |z_1|^{\alpha_1} \cdots |z_n|^{\alpha_n} < e^C\}$, where $\alpha \in \mathbb{R}^n \setminus \{0\}$ and $C \in \mathbb{R}$;
- $\mathbb{M}(n \times n; S) :=$ the set of all $n \times n$-matrices with entries in $S \subset \mathbb{C}$;
- for $A = (A_k^j)_{j,k=1,\ldots,n} \in \mathbb{M}(n \times n; \mathbb{Z})$, denote by A^j its j-th row;
- for $A \in \mathbb{M}(n \times n; \mathbb{Z})$ put $\Phi_A : (\mathbb{C} \setminus \{0\})^n \longrightarrow (\mathbb{C} \setminus \{0\})^n$, $\Phi(z) := (z^{A^1}, \ldots, z^{A^n})$.

Theorem 7.2.43 (cf. [Zwo99]). *Let G be a pseudoconvex Reinhardt domain in $\mathbb{C}^n$. Then the following properties are equivalent:*

(i) G is $\boldsymbol{c}$-hyperbolic;

(ii) G is $\boldsymbol{k}$-hyperbolic;

(iii) $\log G := \{x \in \mathbb{R}^n : (e^{x_1}, \ldots, e^{x_n}) \in G\}$ contains no affine lines, and either $\boldsymbol{V}_j \cap G = \varnothing$ or $\boldsymbol{V}_j \cap G$ (treated as a domain in $\mathbb{C}^{n-1}$) is $\boldsymbol{c}$-hyperbolic, $j = 1, \ldots, n;$

(iv) there exist $A \in \mathbb{M}(n \times n; \mathbb{Z})$, $|\det A| = 1$, and a vector $C \in \mathbb{R}^n$ such that

- $G \subset \boldsymbol{D}_{A,C} := \boldsymbol{D}_{A^1,C_1} \cap \cdots \cap \boldsymbol{D}_{A^n,C_n},$
- either $\boldsymbol{V}_j \cap G = \varnothing$ or $\boldsymbol{V}_j \cap G$ is $\boldsymbol{c}$-hyperbolic as a domain in $\mathbb{C}^{n-1}$, $j = 1, \ldots, n;$

(v) G is algebraically equivalent to a bounded domain (i.e. there exists a matrix $A \in \mathbb{M}(n \times n; \mathbb{Z})$ such that Φ_A is defined on G and gives a biholomorphic mapping from G to the bounded domain $\Phi_A(G)$);

(vi) G is $\boldsymbol{k}$-complete, (i.e. G is $\boldsymbol{k}$-hyperbolic and every $\boldsymbol{k}_G$-Cauchy sequence in G converges (in the standard topology) to a point in G).

7.2.8   Kobayashi completeness

Since $\boldsymbol{c} \leq \boldsymbol{k}$, every $\boldsymbol{c}$-complete domain is $\boldsymbol{k}$-complete. So section 7.1.8 provides a lot of examples of $\boldsymbol{k}$-complete domains. On the other hand, the following necessary condition shows that there are many domains which are not $\boldsymbol{k}$-complete.

Proposition 7.2.44. *Any **k**-complete domain is taut.*

Proof. Let $(\varphi_j)_{j=1}^{\infty} \subset \mathcal{O}(\mathbb{D}, G)$. Assume that $(\varphi_j)_{j=1}^{\infty}$ is not uniformly divergent. This implies that there are compact sets $K \subset \mathbb{D}$ and $L \subset G$ such that, without loss of generality, $\varphi_j(\lambda_j) \in L$ with $\lambda_j \in K$. Fix $z^* \in L$ and let $0 < r < 1$ with $K \subset \mathbb{D}(r)$. Then for $\lambda \in \mathbb{D}(r)$ we obtain

$$\boldsymbol{k}_G(\varphi_j(\lambda), z^*) \leq \boldsymbol{k}_G(\varphi_j(\lambda), \varphi_j(\lambda_j)) + \boldsymbol{k}_G(\varphi_j(\lambda_j), z^*)$$
$$\leq \boldsymbol{p}(\lambda, \lambda_j) + \sup\{\boldsymbol{k}_G(z, z^*) : z \in L\} =: C.$$

Hence, $\bigcup_{j \in \mathbb{N}} \varphi_j(\mathbb{D}(r)) \subset B_{\boldsymbol{k}_G}(z^*, C+1) \subset\subset G$. Therefore, Montel's theorem guarantees the existence of a subsequence $(\varphi_{j_\nu})_{\nu=1}^{\infty} \subset (\varphi_j)_{j=1}^{\infty}$ which converges locally uniformly to a map in $\mathcal{O}(\mathbb{D}, G)$. $\square$

Corollary 7.2.45. *Any **k**-complete domain is a domain of holomorphy.*

Remark 7.2.46. For a while there was the question whether tautness can imply $\boldsymbol{k}$-completeness. The first negative example was found by J.-P. Rosay (cf. [Ros82]). Later in this section we will present another example.

There is a simple example of a domain which is not $\boldsymbol{c}$-complete, but which is $\boldsymbol{k}$-complete, namely the punctured disc $\mathbb{D}_*$. This observation is a direct consequence of the next result due to S Kobayashi (cf. [Kob67], [Kob05]).

Theorem 7.2.47. *If $\Pi : \widetilde{G} \longrightarrow G$ denotes a holomorphic covering between domains in $\mathbb{C}^n$, then the following statements are equivalent:*

(i) $\widetilde{G}$ *is $\boldsymbol{k}$-complete;*
(ii) G *is $\boldsymbol{k}$-complete.*

Proof. (i) $\Longrightarrow$ (ii): According to Theorem 7.2.40, G is $\boldsymbol{k}$-hyperbolic. Fix a ball $B_{\boldsymbol{k}_G}(z_0, r)$ in G. By Theorem 7.2.23 it is clear that $B_{\boldsymbol{k}_G}(z_0, r) \subset \Pi(B_{\boldsymbol{k}_{\widetilde{G}}}(\widetilde{z}_0, r))$, where $\widetilde{z}_0$ is a point in $\widetilde{G}$ with $\Pi(\widetilde{z}_0) = z_0$. Recall that $\boldsymbol{k}_{\widetilde{G}}$ is inner and so this implies that $\boldsymbol{k}$-completeness is equivalent to $\boldsymbol{k}$-finite compactness. Hence, $B_{\boldsymbol{k}_{\widetilde{G}}}(\widetilde{z}_0, r) \subset\subset \widetilde{G}$, and therefore $B_{\boldsymbol{k}_G}(z_0, r) \subset\subset G$.

(ii) $\Longrightarrow$ (i): As above, $\widetilde{G}$ is $\boldsymbol{k}$-hyperbolic. Fix a $\boldsymbol{k}_{\widetilde{G}}$-Cauchy sequence $(\widetilde{z}_\nu)_{\nu=1}^{\infty} \subset \widetilde{G}$. Then obviously $(\Pi(\widetilde{z}_\nu))_{\nu=1}^{\infty}$ is a $\boldsymbol{k}_G$-Cauchy sequence. By assumption, this sequence converges to a point $z_0 \in G$. Using again Theorem 7.2.23 it is easy to construct a subsequence $(z_{\nu_\mu})_{\mu=1}^{\infty}$ of $(z_\nu)_{\nu=1}^{\infty}$ and points $\widetilde{z}_{0,\mu} \in \widetilde{G}$ with $\Pi(\widetilde{z}_{0,\mu}) = z_0$ and $\boldsymbol{k}_{\widetilde{G}}(\widetilde{z}_{\nu_\mu}, \widetilde{z}_{0,\mu}) < 1/\mu$. Thus $\boldsymbol{k}_{\widetilde{G}}(\widetilde{z}_{0,\mu}, \widetilde{z}_{0,\lambda}) \xrightarrow[\lambda,\mu\to\infty]{} 0$. On the other hand, there exist a neighborhood $B_{\boldsymbol{k}_G}(z_0, r)$ of z_0 and neighborhoods U_μ of $\widetilde{z}_{0,\mu}$ such that $\Pi|_{U_\mu} : U_\mu \longrightarrow B_{\boldsymbol{k}_G}(z_0, r)$ is biholomorphic, $\mu \in \mathbb{N}$. Put $V_\mu := \Pi^{-1}(B_{\boldsymbol{k}_G}(z_0, r/2)) \cap U_\mu$; then $\boldsymbol{k}_{\widetilde{G}}(\widetilde{z}_{0,\mu}, \partial V_\mu) \geq r/2$. This observation together with $\boldsymbol{k}_{\widetilde{G}} = \boldsymbol{k}_{\widetilde{G}}^i$ shows that for a sufficiently large μ_0 we obtain $\widetilde{z}_{0,\mu_0} = \widetilde{z}_{0,\mu}$, $\mu \geq \mu_0$. Put $\widetilde{z}_0 := \widetilde{z}_{0,\mu_0}$. Then a standard argument leads to $\lim_{\nu\to\infty} \boldsymbol{k}_{\widetilde{G}}(\widetilde{z}_\nu, \widetilde{z}_0) = 0$. $\square$

Example. $\mathbb{C} \setminus \{0, 1\}$ is a k-complete domain but it is even not c-hyperbolic.

We will see that the property of k-completeness is a local one (cf. [Eas75]) in contrast to the property of being c-finitely compact.

Theorem 7.2.48. *Let G be a bounded domain in $\mathbb{C}^n$. Suppose that any boundary point $z_0 \in \partial G$ permits a bounded neighborhood $U = U(z_0)$ such that any connected component of $G \cap U$ is k-complete. Then G itself is k-complete.*

Remark 7.2.49. The above theorem also provides a simple argument to show that any strongly pseudoconvex domain is k-complete using only the existence of local peak functions.

By Theorem 7.2.43 we know that any bounded pseudoconvex Reinhardt domain containing 0 is k-complete. Moreover, tautness, i.e. the continuity of its Minkowski function, is necessary for a balanced domain to be k-complete. Nevertheless, the following result shows that tautness, even in the case of a balanced domain, does not imply k-completeness.

Theorem 7.2.50 (cf. [JP91a]). *There exists a bounded taut balanced pseudoconvex domain $G = G_{\mathfrak{h}} = \{z \in \mathbb{C}^n : \mathfrak{h}(z) < 1\}$ ($n \geq 3$) with continuous Minkowski function $\mathfrak{h}$ which is not k-complete.*

Remark 7.2.51.

(a) ? It would be very interesting to know whether such an example could be also constructed in $\mathbb{C}^2$? We emphasize that the method used to prove Theorem 7.2.50 does not work in the two-dimensional case.
(b) ? So far it is totally unclear how to characterize the k-completeness (or the c-completeness) of a bounded pseudoconvex balanced domain via the properties of its Minkowski function ?

Up to now ? it is an open problem whether every bounded pseudoconvex domain with $\mathcal{C}^\infty$-smooth boundary is k-complete ? The strongest result in the negative direction is the following unpublished one due to N. Sibony; cf. [Sib81a].

Theorem 7.2.52. *There exists a pseudoconvex non k-complete domain $G \subset\subset \mathbb{B}_2$ given as a connected component of $\{z \in \mathbb{B}(3) : u(z) < 1\}$, where $u \in \mathcal{PSH}(\mathbb{B}(3)) \cap \mathcal{C}(\mathbb{B}(3)) \cap \mathcal{C}^\infty(\mathbb{B}(3) \setminus \{0\})$, $\operatorname{grad} u(z) \neq 0$ if $z \neq 0$, and $u(0) = 1$.*

To conclude this section one should emphasize that k-completeness is much easier to handle than c-completeness; nevertheless, there are still a lot of unsolved questions.

7.3 Lempert's Theorems and the Symmetrized Bidisc

In this last chapter, we discuss various results telling that for certain domains $G \subset \mathbb{C}^n$ the extremal invariant pseudodistances coincide.

7.3.1 Lempert's theorems

Theorem 7.3.1 (1st theorem of Lempert, [Lem81], [Lem82]). *If $G \subset \mathbb{C}^n$ is a domain which is the union of an increasing sequence of domains G_j each biholomorphically equivalent to a convex domain, then $\mathbf{c}_G = \boldsymbol{\ell}_G$. In particular, the Carathéodory pseudodistance and the Kobayashi pseudodistance coincide on such a G.*

To give an idea of the proof one first observes that it suffices to prove Theorem 7.3.1 in the case when G is bounded, convex, and $0 \in G$ (cf. Remark 7.1.8(c) and Proposition 7.2.8).

So from now on G is assumed to be bounded convex and $0 \in G$. Fix then two different points a and b in G. Then Proposition 7.2.11 allows to find an extremal disc $\varphi \in \mathcal{O}(\mathbb{D}, G)$ with $\varphi(0) = a$, $\varphi(\sigma) = b$, and $\boldsymbol{p}(0, \sigma) = \boldsymbol{\ell}_G(a, b)$. So it remains to find a holomorphic function $f \in \mathcal{O}(G, \mathbb{D})$ such that $f \circ \varphi = \mathrm{id}_{\mathbb{D}}$. Indeed, $\boldsymbol{p}(\lambda', \lambda'') \leq \mathbf{c}_G(\varphi(\lambda'), \varphi(\lambda'')) \leq \boldsymbol{\ell}_G(\varphi(\lambda'), \varphi(\lambda'')) \leq \boldsymbol{p}(\lambda', \lambda'')$, $\lambda', \lambda'' \in \mathbb{D}$; in particular, $\mathbf{c}_G(a, b) = \boldsymbol{\ell}_G(a, b)$.

To see how convexity enters the proof a few notions have to be repeated. Let $z \bullet w := \langle z, \overline{w} \rangle = z_1 w_1 + \cdots + z_n w_n$. Since G is convex, there is the Minkowski function $q_G(z) := \inf\{t \in (0, \infty) : z/t \in G\}$ and the dual Minkowski subnorm $\widehat{q}_G(z) = \sup\{\mathrm{Re}(z \bullet w) : w \in \partial G\}$, $z \in \mathbb{C}^n$. Note that a point z sits in G if and only if $q_G(z) < 1$.

Let $F := C(\mathbb{T}, \mathbb{C}^n)$. Then F together with the norm $\|f\|_F := \sup\{|f(\lambda)| : \lambda \in \mathbb{T}\}$ is a Banach space. If $\psi \in F$, put $Q(\psi) := \sup\{q_G(\psi(\lambda)) : \lambda \in \mathbb{T}\}$. Then Q is a Minkowski subnorm on F. Moreover, let $A := \{\psi \in F : \psi \in C(\overline{\mathbb{D}}, \mathbb{C}^n) \cap \mathcal{O}(\mathbb{D}, \mathbb{C}^n)\}$.

Now fix an arbitrary analytic disc $\varphi_0 \in \mathcal{O}(\mathbb{D}, \mathbb{C}^n)$ with $\varphi_0(\lambda') = z'$, $\varphi_0(\lambda'') = z''$ and define $V_0 := \{\psi \in A : \psi(\lambda') = \psi(\lambda'') = 0\}$. According to Remark 7.2.12(b) it follows for any $\psi \in V_0$ that $\sup\{q_G(\varphi_0(\lambda) + \psi(\lambda)) : \lambda \in \mathbb{D}\} \geq 1$. Consequently, $Q(\varphi_0 + \psi) \geq 1$.

Looking at the functional $\mathbb{R}\varphi_0 + V_0 \ni t\psi_0 + \psi \overset{l}{\longmapsto} t$ it is easily seen that $l \leq Q$ and so, by the Hahn-Banach theorem, it can be extended to a $\mathbb{R}$-linear functional L on F, $L \leq Q$. Finally, setting $\mu(\psi) := L(i\psi) - iL(i\psi)$ one ends up with a continuous linear functional $\mu \in F'$ with $\mu|_{V_0} = 0$; μ may be thought as a Borel measure on $\mathbb{T}$.

Using the F. and M. Riesz theorem one concludes that μ is given as $\mu = \frac{\lambda \widetilde{h}^* \tau}{(\lambda - \lambda')(\lambda - \lambda'')}$, $\lambda \in \mathbb{T}$, where $\widetilde{h}$ is a mapping from $\mathbb{D}$ to $\mathbb{C}^n$, each of its components belongs to the Hardy space on $\mathbb{D}$ ($\widetilde{h} \in H^1(\mathbb{D}, \mathbb{C}^n)$), the star means the boundary value componentwise, and τ stands for the normalized Lebesgue measure on $\mathbb{T}$.

Put $u := \frac{\lambda \tilde{h}}{(\lambda - \lambda')(\lambda - \lambda'')}$ on $\mathbb{D} \setminus \{\lambda', \lambda''\}$. Then it follows that $\hat{q}_G(\varphi^*(\lambda)) = 0$ and $\mathrm{Re}\,(\varphi^*(\lambda) \bullet u^*(\lambda)) = \hat{q}_G(u^*(\lambda))$ almost everywhere on $\mathbb{T}$. Modifying $\tilde{h}$ leads finally to a mapping $h \in H^1(\mathbb{D}, \mathbb{C}^n)$, $h \neq 0$, such that $\mathrm{Re}\,(\varphi^*(\lambda) \bullet \frac{h^*(\lambda)}{\lambda}) = \hat{q}_G(\frac{h^*(\lambda)}{\lambda})$ for almost all $\lambda \in \mathbb{T}$. So far the geometric property has been heavily exploited.

The final step starts with the condition that $\mathrm{Re}\,((z - \varphi^*(\lambda)) \bullet \frac{h^*(\lambda)}{\lambda}) < 0$ for $z \in G$ and almost all $\lambda \in \mathbb{T}$ which is a consequence of the result just before. Then pure complex analysis reasoning leads to the existence of the left inverse for φ which was above postulated to be found.

The proof above is based on the unpublished paper [RW83]; details may be found in [JP13].

Remark 7.3.2. To generalize the above situation, an analytic disc $\varphi \in \mathcal{O}(\mathbb{D}, G)$, G a domain in $\mathbb{C}^n$, is said to be *weak m-extremal for points* $\lambda_1, \ldots, \lambda_m \in \mathbb{D}$ (pairwise different and $m \geq 2$) if there is no analytic disc $h \in \mathcal{O}(\overline{\mathbb{D}}, G)$ with $\varphi(\lambda_k) = h(\lambda_k)$, $k = 1, \ldots, m$. And a mapping $f \in \mathcal{O}(\mathbb{D}, G)$ is called an *m-extremal map* if there exists a holomorphic mapping $F \in \mathcal{O}(G, \mathbb{D})$ with $F \circ f = B$, where B is a Blaschke product of order $m - 1$. So Lempert's result as described before may be read as: any weak 2-extremal disc is a 2-geodesic. There was some hope that the Lempert result may be generalized in this context claiming that if G is convex, then any weak m-extremal analytic disc $\varphi \in \mathcal{O}(\mathbb{D}, G)$ is also m-geodesic for all $m \geq 2$. But as it turned out such a proposition is already false for $G = \mathbb{B}_2$ and $m \geq 4$ (see [KZ16]). Moreover, there are convex domains for which this kind of result fails to hold for $m = 3$ (cf. [War15]).

To be able to formulate the second result the following notion is needed. A domain $G \subset \mathbb{C}^n$ with a C^2-defining function r is called to be *strongly linearly convex* if $(\mathcal{L}r)(a; X) > |\sum_{j,k=1}^n \frac{\partial^2 r}{\partial z_j \partial z_k}(a) X_j X_k|$, $a \in \partial G$, $X \in T_a^{\mathbb{C}}(\partial G) \setminus \{0\}$.

Theorem 7.3.3 (2nd Lempert theorem, [Lem84]). *Let $G \subset \mathbb{C}^n$ be a domain which can be written as the union of an increasing sequence of bounded strongly linearly convex domains G_j. Then again $\mathbf{c}_G = \boldsymbol{\ell}_G$.*

A readable and complete proof can be found in [KW12].

Later, D. Jacquet (see [Jac06]) was able to extend the above results to bounded $\mathbb{C}$-convex domains with a smooth C^2 boundary.

Theorem 7.3.4. *Let $G \subset \mathbb{C}^n$ be a bounded $\mathbb{C}$-convex domain with a smooth C^2-boundary. Then G can be exhausted by a sequence of bounded strongly linearly convex domains; in particular, $\mathbf{c}_G = \boldsymbol{\ell}_G$.*

A few words to the proof of Jacquet's result:

- A function $u \in C^2(D; R)$, D a domain in $\mathbb{C}^n$, is called to be $\mathbb{C}$-convex (resp. strongly $\mathbb{C}$-convex), if

$$\mathcal{L}u(a; X) \overset{\geq}{\underset{(\text{resp. }>)}{}} \left| \sum_{j,k=1}^{n} \frac{\partial^2 u}{\partial z_j \partial z_k}(a) X_j X_k - \sum_{j=1}^{n} \frac{\partial u}{\partial z_j}(a) X_j \right|,$$

$$a \in D, \ X \in \mathbb{C}^n \setminus \{0\}.$$

- Let $D \subset \mathbb{C}^n$ be a domain with C^2-boundary. Then:
  D is $\mathbb{C}$-convex if and only if $-\log \text{dist}^2(\cdot, \partial D)$ is a $\mathbb{C}$-convex function on $D \cap U$ for some open neighborhood $U = U(\partial D)$.
- If u is a $\mathbb{C}$-convex function, then $u_\varepsilon := -\log(e^u - \varepsilon(1 + \|z\|^2)) \searrow u$ as $\varepsilon \searrow 0$, and the u_ε are strongly $\mathbb{C}$-convex functions. In particular, this leads to an exhaustion by strongly linearly convex domains.

At the moment it is unclear whether the former theorem remains true without assumptions on the smoothness of the boundary.

7.3.2 The symmetrized bidisc

For a long time it was absolutely unclear whether Theorem 7.3.3 is eventually a consequence of Theorem 7.3.1. The surprising answer to that question was found by discussions which do not belong directly to the field of several complex variables; it stems from discussions of the so called μ-synthesis problem. That is an interpolation problem for analytic matrix valued functions which may be thought as a generalization of the classical problems of Nevanlinna–Pick. Here μ denotes a positive cost function that generalizes for example the operator norm for matrices. Then the μ synthesis problem is to find an analytic matrix valued function on $\mathbb{D}$ such that it satisfies a finite number of interpolation conditions together with the condition that $\mu(f) \leq 1$ on $\mathbb{D}$. The precise definition of the cost function μ will be omitted; for our purpose μ is taken as the spectral radius $r(A)$ of a 2×2 matrix A.

In this section as a special case of the μ-synthesis problem only the spectral Nevanlinna–Pick problem will be discussed. To be more precise: given pairwise different points $\lambda_1, \ldots, \lambda_N \in \mathbb{D}$ and $k \times k$ matrices $A_1, \ldots, A_N$, $r(A_j) < 1$, then the problem is to construct an analytic $k \times k$ matrix valued function F on $\mathbb{D}$ such that $F(\lambda_j) = A_j$, $j = 1, \ldots, N$, and $r(F(\lambda)) \leq 1$ for all $\lambda \in \mathbb{D}$, where $r(A) := \max\{|\lambda| : \lambda \text{ is an eigenvalue of } A\}$ denotes the spectral radius of A.

When $k = 1$ this is just the classical Nevanlinna–Pick problem, and it is well known that a suitable F exists if and only if a certain $n \times n$ matrix formed from the λ_j's and A_j's is positive definite (this is Pick's Theorem).

Put $\Omega_k := \{A \in \mathbb{M}(k \times k; \mathbb{C}) : r(A) < 1\}$ and call this set in $\mathbb{C}^{k^2}$ *the unit spectral ball*. So the above problem consists in finding certain holomorphic mappings into Ω_k with prescribed values at given points.

The unit spectral ball shares a lot of interesting properties. Only a few will be mentioned here:

- Ω_k is a pseudoconvex domain but it is neither convex nor bounded (in fact it contains even complex lines);
- Ω_k is balanced, but not taut (i.e. the Montel theorem does not hold for $\mathcal{O}(\mathbb{D}, \Omega_k)$).

Recall that if $A \in \Omega_k$, then $\det(x\mathbb{I}_k - A) = x^n + \sum_{j=1}^{k}(-1)^j \psi_{k,j}(A)x^{k-j}$, where the $\psi_{k,j}(A)$ is given via the elementary symmetric polynomials $\sigma_{k,j}$ and the eigenvalues of A. Hence, there are the following maps ψ_k : $\Omega_k \longrightarrow \mathbb{C}^k$, $A \longmapsto (\psi_{k,1}(A), \dots, \psi_{k,k}(A))$ and $\sigma_k : \mathbb{D}^k \longrightarrow \mathbb{C}^k$, $\lambda \longmapsto (\sigma_{k,1}(\lambda), \dots, \sigma_{k,k}(\lambda))$. Then it turns out that $\psi_k(\Omega_k) = \sigma_k(\mathbb{D}^k) =: \mathbb{G}_k$ is a domain in $\mathbb{C}^k$ which is called the *k–dimensional symmetrized polydisc*.

If the spectral Nevanlinna–Pick problem with data $\lambda_1, \dots, \lambda_N \in \mathbb{D}$ and $A_1, \dots, A_N \in \Omega_k$ can be solved, then using ψ_k the interpolation problem with data $\lambda_1, \dots, \lambda_N \in \mathbb{D}$ and $\psi_k(A_1), \dots, \psi_k(A_N)) \in \mathbb{G}_k$ allows a solution in $\mathbb{G}_k$. So the original problem may be studied in two steps:

(a) solve the corresponding interpolation problem in $\mathbb{G}_k$ (note that the number of parameters now involved is $N(1 + k)$, while in the spectral Nevanlinna–Pick problem it is $N(1 + k^2)$;

(b) try to lift the solution in $\mathbb{G}_k$ to a holomorphic mapping F from $\mathbb{D}$ to Ω_k with $F(\lambda_j) = A_j$.

From now on it is assumed that $k = 2$, $N = 2$. Then $\mathbb{G}_2$ is the so called *symmetrized bidisc* which shares the following interesting geometric properties most of them are found in [AY04]:

- $\psi_2(A) = (\operatorname{trace}(A), \det(A))$;
- $\mathbb{G}_2 = \{(s, p) \in \mathbb{C}^2 : |s - \bar{s}p| + |p|^2 < 1\}$; in particular, $\mathbb{G}_2$ is bounded, pseudoconvex, and taut;
- $\mathbb{G}_2$ is not convex and has no smooth boundary;
- $\mathbb{G}_2$ cannot be exhausted by domains G_j biholomorphic to convex domains (see [Cos04b], [Cos04a], [Edi03]); i.e. Theorem 7.3.1 cannot be applied;
- $\mathbb{G}_2$ is $\mathbb{C}$-convex (see [NPZ08]);
- $\mathbb{G}_2$ is not homogeneous, the orbit of 0 under $\operatorname{Aut}(\mathbb{G}_2)$ is given by $\{(2\lambda, \lambda^2) : \lambda \in \mathbb{D}\}$.

In order to study the spectral Nevanlinna–Pick problem one has to find conditions to solve the corresponding interpolation problem in the symmetrized bidisc. So let $\lambda_1, \lambda_2 \in \mathbb{D}$ and points $z', z'' \in \mathbb{G}_2$ be given. Without loss of generality one may assume that $z' \neq z''$ and $\lambda_1 = 0$ and $\lambda_2 = \sigma \in (0, 1)$. Obviously, a necessary condition for the existence of a solution is that $\ell_{\mathbb{G}_2}^*(z', z'') \leq \sigma$. On the other hand, if $\ell_{\mathbb{G}_2}^*(z', z'') < \sigma$, then there exists an analytic disc $\varphi \in \mathcal{O}(\mathbb{D}, \mathbb{G}_2)$ with $\varphi(0) = z'$ and $\varphi(\tau) = z''$ with a certain $\tau \in [\ell_{\mathbb{G}_2}^*(z', z''), \sigma]$. Then the disc $\widetilde{\varphi}(\lambda) := \varphi(\lambda\tau/\sigma)$ does the job of interpolation. So it remains to determine the number $\ell_{\mathbb{G}_2}^*(z', z'')$ or at least a lower bound. Note that a first lower bound is obviously given by the Möbius function $\boldsymbol{m}_{\mathbb{G}_2}(z', z'')$.

In fact, J. Agler -N. Young [AY01] and C. Costara [Cos04b], [Cos04a] were able to prove the following result.

Theorem 7.3.5 (Theorem of Agler-Young, Costara).

$$\boldsymbol{m}_{\mathbb{G}_2}(z', z'') = \max\{\boldsymbol{m}(\varPhi_\lambda(z'), \varPhi_\lambda(z'')) : \lambda \in \mathbb{T}\} = \boldsymbol{\ell}_{\mathbb{G}_2}(z', z''),$$

where $\varPhi_\lambda : \mathbb{G}_2 \longrightarrow \mathbb{D}$, $(s, p) \longmapsto \frac{2\lambda p - s}{2 - \lambda s}$. *In particular,* $\boldsymbol{\ell}_{\mathbb{G}_2}$ *is given by the following concrete formula*

$$\max_{\zeta \in \mathbb{T}} \left| \frac{(s_1 p_2 - s_2 p_1)\zeta^2 + 2(p_1 - p_2)\zeta + s_2 - s_1}{(\bar{s}_2 p_1 - s_1)\zeta^2 + 2(1 - p_1 \bar{p}_2)\zeta + s_1 \bar{p}_2 - \bar{s}_2} \right|.$$

An different proof was given by P. Pflug and W. Zwonek (see [PZ12]).

Theorem 7.3.6. *The symmetrized bidisc can be exhausted by bounded strongly linearly convex domains. In particular, by Theorem 7.3.3,* $\boldsymbol{c}_{\mathbb{G}_2} = \boldsymbol{\ell}_{\mathbb{G}_2}$.

As a byproduct this result together with former properties of $\mathbb{G}_2$ leads to the following consequence answering an old question by L. Aizenberg.

Corollary 7.3.7. *There exist a strongly linearly convex domain in* $\mathbb{C}^2$ *which cannot be exhausted by a sequence of domains biholomorphically equivalent to convex domains.*

The first idea to extend the above result to higher dimensions was without success. In fact, in [NPZ07] and [NPTZ08] the authors were able to show that the equality of $\boldsymbol{c}_{\mathbb{G}_n}$ and $\boldsymbol{\ell}_{\mathbb{G}_n}$ fails for $n \geq 3$. Nevertheless, there is still a great hope that this equality remains true for all $\mathbb{C}$-convex domains.

Remark 7.3.8. It should be mentioned that the symmetrized polydisc appears as a special case of a large family of other domains whose construction will be shortly presented (see [Zap15]). Fix natural numbers $n \geq 2$, $s \leq n$, and $r_1, \ldots, r_s$ such that $\sum_{j=1}^{s} r_j = n$. On the product space $A := \prod_{j=1}^{s} \{0, 1, \ldots, r_j\} \setminus \{(0, \ldots, 0)\}$ define an ordering saying for $\alpha, \beta \in A$ that $\alpha < \beta$ if there exists an $j_0 \in \{1, \ldots, s\}$ with $\alpha_{j_0} < \beta_{j_0}$ and $\alpha_j = \beta_j$ for $j_0 < j \leq s$. Then $A = \{\alpha^1, \ldots, \alpha^N\}$ for $N := \prod_{j=1}^{s}(r_j + 1) + 1$ where $\alpha^j < \alpha^{j+1}$ for all possible j. Then the following set

$$\mathbb{E} = \mathbb{E}_{n,s,r_1,\ldots,r_s} := \left\{ z \in \mathbb{C}^N : 1 + \sum_{j=1}^{N} (-1)^{|\alpha^j|} z_j \lambda^{\alpha^j} \neq 0 \text{ for all } \lambda \in \overline{\mathbb{D}}^s \right\}$$

is a bounded pseudoconvex domain satisfying the following properties:

- $\mathbb{E}$ cannot be exhausted by domains biholomorphically equivalent to convex ones;

- if one of the $r_j \geq 3$, then $c_{\mathbb{E}} \neq \ell_{\mathbb{E}}$ and $\mathbb{E}$ is not $\mathbb{C}$-convex;
- if $r_2 = \cdots r_s = 1$, then $\mathbb{E}$ is linearly convex and, consequently, pseudoconvex.

Moreover, $\mathbb{E}_{2,1,1,1} = \mathbb{G}_2$ and for the so-called tetrablock $\mathbb{E}_{2,2,1,1}$ (see [AWY07]) the Carathéodory pseudodistance and Lempert function coincide and both domains are $\mathbb{C}$-convex (see [EZ09], [EKZ13], and [Zwo13]).

So the discussion of this general class of domains gives hope that $\mathbb{C}$-convexity may lead to the equality of the Carathéodory pseudodistance and the Lempert function.

What remains is to answer problem (b) of the possibility to lift solutions of the Nevanlinna–Pick problem to solutions of the spectral Nevanlinna-Pick problem (see [NPT11]).

Theorem 7.3.9. *Let* $A_1 = \alpha_1 \mathbb{I}_2, \ldots, A_k = \alpha_k \mathbb{I}_2 \in \Omega_2, A_{k+1}, \ldots, A_\ell \in \Omega_2$ *non scalar matrices (i.e. matrices which have not the form of the previous* A_j*), and* $\varphi \in \mathcal{O}(\mathbb{D}, \mathbb{G}_2)$ *such that* $\varphi(\lambda_j) = \psi(A_j)$, $j = 1, \ldots, \ell$. *Then there exists a* $\widetilde{\varphi} \in \mathcal{O}(\mathbb{D}, \Omega_2)$ *satisfying* $\psi \circ \widetilde{\varphi} = \varphi$ *and* $\widetilde{\varphi}(\lambda_j) = A_j$ *for* $j = 1, \ldots, \ell$, *if and only if* $\varphi_2'(\lambda_j) = \alpha_j \varphi_1'(\lambda_j)$, $j = 1, \ldots, k$.

Hence, the spectral Nevanlinna–Pick problem is solved in dimension 2 and, as a byproduct, new information with respect to the equality of c and ℓ is obtained. Nevertheless, as it was seen a lot of unsolved problems remain for further research. Let us finish this article with the one of the most important conjecture in this area: *Let* $G \subset \mathbb{C}^n$ *be* $\mathbb{C}$-convex, *then* $c_G = \ell_G$. All experiences so far make it highly probable that the answer will be "yes".

7.4   Epilogue

7.4.1   The Green function

First let us recall the notion of the classical Green function (cf. [Ran95]). Let $G \subset \mathbb{C}$ be a domain and let $a \in G$. We say that a function $\mathfrak{g}_G(a, \cdot) : G \longrightarrow [-\infty, +\infty]$ is the *classical Green function of* G *with pole at* a if:

- $\mathfrak{g}_G(a, \cdot)$ is harmonic on $G \setminus \{a\}$;
- $\mathfrak{g}_G(a, a) = +\infty$;
- the function $G \setminus \{a\} \ni z \longmapsto \mathfrak{g}_G(a, z) + \log|z - a|$ extends to a harmonic function on G;
- there exists a polar set $F \subset \partial G$ $\left(^4\right)$, such that:
  — if $\zeta \in (\partial G) \setminus F$, then $\lim_{G \ni z \to \zeta} \mathfrak{g}_G(a, z) = 0$,
  — if $\zeta \in F$ or $\zeta = \infty \in \partial G$, then $\mathfrak{g}_G(a, \cdot)$ is bounded near ζ.

$\left(^4\right)$ A set $A \subset \mathbb{C}$ is said to be *polar*, if for every $a \in A$ there exist $r > 0$ and $u \in \mathcal{SH}(\mathbb{D}(a, r))$, $u \not\equiv -\infty$, such that $u = -\infty$ on $A \cap \mathbb{D}(a, r)$.

Observe that $\mathfrak{g}_{\mathbb{D}}(a, z) = -\log|\frac{z-a}{1-\bar{a}z}|$, $a, z \in \mathbb{D}$.

Theorem 7.4.1. (a) *If $G \subset \mathbb{C}$ is a domain such that ∂G is not polar, then for every $a \in G$ the Green function $\mathfrak{g}_G(a, \cdot)$ exists and is unique. Moreover, $\mathfrak{g}_G(a, z) > 0$ for $z \neq a$ and the function $\mathfrak{g}_G : G \times G \longrightarrow (0, +\infty]$ is symmetric and continuous.*

(b) *If $\mathfrak{g}_G(a, \cdot)$ exists, then, by the maximum principle for subharmonic functions, we get*

$$-\mathfrak{g}_G(a, z) = \sup\{v(z) : v \in \mathcal{SH}(G, [-\infty, 0)),$$
$$\sup_{w \in G \setminus \{a\}} (v(w) - \log|w - a|) < +\infty\}, \quad z \in G.$$

For an arbitrary domain $G \subset \mathbb{C}^n$, Theorem 7.4.1(b) suggests the following definition of the *pluricomplex Green function with pole at a*:

$$\boldsymbol{g}_G(a, z) := \sup\left\{u(z) : u : G \longrightarrow [0, 1), \log u \in \mathcal{PSH}(G),\right.$$
$$\left.\sup_{w \in G \setminus \{a\}} \frac{u(w)}{\|w - a\|} < +\infty\right\}, \quad z \in G.$$

With this notation Theorem 7.4.1(b) reads as follows: *for $G \subset \mathbb{C}$, if $\mathfrak{g}_G(a, \cdot)$ exists, then* $\mathfrak{g}_G(a, \cdot) = -\log \boldsymbol{g}_G(a, \cdot)$. Observe that $\boldsymbol{g}_{\mathbb{D}} = \boldsymbol{m}$ and for any holomorphic mapping $F : G \longmapsto D$ we have $\boldsymbol{g}_D(F(z), F(w)) \leq \boldsymbol{g}_G(z, w)$, $z, w \in G$. Thus the system $(\boldsymbol{g}_G)_G$ is holomorphically contractible and therefore $\boldsymbol{m}_G \leq \boldsymbol{g}_G \leq \boldsymbol{\ell}_G^*$.

Similar to $\boldsymbol{c}_G$ and $\boldsymbol{\ell}_G$, the pluricomplex Green function has its infinitesimal form $\boldsymbol{A}_G(a; X) := \limsup_{\lambda \to 0} \frac{1}{|\lambda|} \boldsymbol{g}_G(a, a + \lambda X)$, $a \in G$, $X \in \mathbb{C}^n$, which is called the *Azukawa pseudometric* (cf. [Azu86]). One can easily check that $\boldsymbol{A}_G(a; \lambda X) = |\lambda| \boldsymbol{A}_G(a; X)$, $\boldsymbol{A}_{\mathbb{D}}(a; 1) = \boldsymbol{\gamma}$, and $\boldsymbol{A}_D(F(z); F'(z)X) \leq \boldsymbol{A}_G(z; X)$ for any $F \in \mathcal{O}(G, D)$. In particular, $\boldsymbol{\gamma}_G \leq \boldsymbol{A}_G \leq \boldsymbol{\varkappa}_G$.

The following deep result corresponds to the harmonicity of $\mathfrak{g}_G(a, \cdot)$ on $G \setminus \{a\}$. It establishes an important link between the theory of the pluricomplex Green function and pluricomplex analysis.

Theorem 7.4.2 (cf. [Kli85]). *For each $a \in G$ the function $\log \boldsymbol{g}_G(a, \cdot)$ is a maximal plurisubharmonic function on $G \setminus \{a\}$. In particular, if $\log \boldsymbol{g}_G(a, \cdot) \in L_{\mathrm{loc}}^\infty(G \setminus \{a\})$ (e.g. if the set G is bounded), then $(dd^c \log \boldsymbol{g}_G(a, \cdot))^n = 0$ in $G \setminus \{a\}$, where $(dd^c \cdot)^n$ denotes the Monge–Ampère operator $(^5)$.*

We collect below various basic properties of the Green and Azukawa functions.

Remark 7.4.3 (cf. [JP13], Proposition 7.2.10). (a) $\log \boldsymbol{g}_G(a; \cdot) \in \mathcal{PSH}(G)$ and $\sup_{w \in G \setminus \{a\}} \frac{\boldsymbol{g}_G(a, w)}{\|w - a\|} < +\infty$.

$(^5)$ If u is $\mathcal{C}^2$, then $(dd^c u)^n = \det[\frac{\partial^2 u}{\partial z_j \partial \bar{z}_k}]_{j, k = 1, \dots, n}$.

(b) $\log A_G(a; \cdot) \in \mathcal{PSH}(G)$.
(c) If $G_s \nearrow G$, then $\boldsymbol{g}_{G_s} \searrow \boldsymbol{g}_G$ and $\boldsymbol{A}_{G_s} \searrow \boldsymbol{A}_G$.
(d) If $G = G_{\mathfrak{h}} \subset \mathbb{C}^n$ is a balanced domain, then the following conditions are equivalent:
    (i) $\boldsymbol{g}_G(0, \cdot) = \mathfrak{h}$ on G;
    (ii) $\boldsymbol{A}_G(0; \cdot) = \mathfrak{h}$ on $\mathbb{C}^n$;
    (iii) G is pseudoconvex.
(e) If G is bounded, then for any $z_0 \in G$ the function $G \ni z \longmapsto \boldsymbol{g}_G(z, z_0)$ is continuous. For unbounded domains G the function $G \ni z \longmapsto \boldsymbol{g}_G(z, z_0)$ need not be continuous; cf. Example 7.4.4.
(f) For any G the function $\boldsymbol{g}_G$ is upper semicontinuous on $G \times G$. Note that, by (d), the function $G \ni z \longmapsto \boldsymbol{g}_G(a, z)$ need not be continuous (even for bounded domains of holomorphy).
(g) For any G the function $\boldsymbol{A}_G$ is upper semicontinuous on $G \times \mathbb{C}^n$. Note that, by (d), the function $\mathbb{C}^n \ni X \longmapsto \boldsymbol{A}_G(a; X)$ need not be continuous (even for bounded domains of holomorphy).

Example 7.4.4 (cf. [JP13], Proposition 7.1.3). Consider the Reinhardt domain $G := \{(z_1, z_2) \in \mathbb{C}^2 : |z_1 z_2| < 1\}$. Then

$$\boldsymbol{g}_G(a, z) = \begin{cases} \boldsymbol{m}(a_1 a_2, z_1 z_2), & \text{if } a \neq 0 \\ |z_1 z_2|^{1/2}, & \text{if } a = 0 \end{cases}, \quad z \in G,$$

$$\boldsymbol{A}_G(a; X) = \begin{cases} \gamma(a_1 a_2)|a_2 X_1 + a_1 X_2|, & \text{if } a \neq 0 \\ |X_1 X_2|^{1/2}, & \text{if } a = 0 \end{cases}, \quad X = (X_1, X_2) \in \mathbb{C}^2.$$

If $b \in G_0 := \{(z_1, z_2) \in G : z_1 z_2 \neq 0\}$, then $\boldsymbol{g}_G(0, b) = |b_1 b_2|^{1/2} > |b_1 b_2| = \boldsymbol{g}_G(b, 0)$. Thus $\boldsymbol{g}_G$ is not symmetric. If D is a subdomain of G (with $0, b \in D$), then $\boldsymbol{g}_D(0, b) > \boldsymbol{g}_D(b, 0)$ provided that D is sufficiently close to G. Thus *there exist regular domains* $D \subset \mathbb{C}^2$ (*e.g. bounded Reinhardt domains with real analytic boundary*) *such that* $\boldsymbol{g}_D$ *is not symmetric*. Observe that the invariant functions discussed before are all symmetric ones.

If $G_0 \ni a_s \longrightarrow 0$, then

$$\boldsymbol{g}_G(a_s, b) = \boldsymbol{m}(a_{s,1} a_{s,2}, b_1 b_2) \longrightarrow |b_1 b_2| < |b_1 b_2|^{1/2} = \boldsymbol{g}(0, b).$$

Consequently, $\boldsymbol{g}_G$ *is not continuous in the first variable*.
    Let $X = (X_1, X_2) \in \mathbb{C}^2$, $X_1 X_2 \neq 0$. Then

$$\boldsymbol{A}(a_s; X) = \gamma(a_{s,1} a_{s,2})|a_{s,2} X_1 + a_{s,1} X_2| \longrightarrow 0 < |X_1 X_2|^{1/2} = \boldsymbol{A}_G(0; X).$$

Hence, $\boldsymbol{A}_G$ *is not continuous in the first variable*.

    In the case where G is hyperconvex the pluricomplex Green function and the Azukawa metric are much more regular than in the general case (cf. [Dem87], [Kli85], [Kli91], [Zwo00b]).

Theorem 7.4.5 (cf. [JP13], Proposition 4.2.10, Theorem 5.1.2). *Assume that G is bounded hyperconvex. Then:*

(a) $\lim_{z \to \partial G} \boldsymbol{g}_G(a, z) = 1$, $a \in G$.

(b) $\boldsymbol{g}_G$ *is continuous on* $G \times \overline{G}$, *where* $\boldsymbol{g}_G|_{G \times \partial G} := 1$.

(c) $\boldsymbol{A}_G$ *is continuous.*

(d) $\boldsymbol{A}_G(a; X) = \lim_{\lambda \to 0} \dfrac{1}{|\lambda|} \boldsymbol{g}_G(a, a + \lambda X)$, $\quad a \in G$, $X \in \mathbb{C}^n$.

Under some stronger assumptions $\boldsymbol{g}_G$ and $\boldsymbol{A}$ are even in some sense Lipschitz (cf. [NPT09]).

From a general point of view the invariant objects studied so far were defined via certain extremal problems related to:

(a) holomorphic mappings $f : G \longrightarrow \mathbb{D}$, e.g. $\boldsymbol{m}_G(a, z)$, $\boldsymbol{\gamma}_G(a; X)$;

(b) analytic discs $\varphi : \mathbb{D} \longrightarrow G$, e.g. $\boldsymbol{\ell}_G(a, z)$, $\boldsymbol{\varkappa}_G(a; X)$;

(c) logarithmically plurisubharmonic functions $u : G \longrightarrow [0, 1)$, e.g. $\boldsymbol{g}_G(a, z)$, $\boldsymbol{A}_G(a; X)$.

At the end of the eighties E.A. Poletsky invented and partially developed a general *holomorphic discs method*, which in the meantime became one of the important tools of modern complex analysis. This method permits to reduce in some sense problems of type (c) to (b). It has various important applications, due to A. Edigarian (cf. [Edi02]) and E.A. Poletsky (cf. [Pol91], [Pol93], [EP97]). In particular, one may express the pluricomplex Green function in the language of holomorphic discs by the following *Poletsky formula*.

Theorem 7.4.6 (cf. [PS89], [Pol91], [Pol93], [Edi97a]).

$$\boldsymbol{g}_G(a, z) = \inf_{\substack{\varphi \in \mathcal{O}(\overline{\mathbb{D}}, G) \\ \varphi(0) = z \\ a \in \varphi(\mathbb{D})}} \prod_{\lambda \in \varphi^{-1}(a)} |\lambda|^{\operatorname{ord}_\lambda(\varphi - a)} =: \widetilde{\boldsymbol{g}}_G(a, z), \quad a, z \in G,$$

where $\operatorname{ord}_c f$ stands for the order of zero of f at c.

The idea of the proof (cf. [Edi97a]). The maximum principle for subharmonic functions implies that $\boldsymbol{g}_G \leq \widetilde{\boldsymbol{g}}_G$. The system $(\widetilde{\boldsymbol{g}}_G)_G$ is holomorphically contractible. In particular, $\widetilde{\boldsymbol{g}}_G \leq \boldsymbol{\ell}_G^*$. Fix an $a \in G$ and let $u := \log \boldsymbol{\ell}_G^*(a, \cdot)$. Observe that $\log \boldsymbol{g}_G(a, \cdot) = \sup\{v \in \mathcal{PSH}(G) : v \leq u\}$. The main proof consists of two steps.

(*) If $\varphi \in \mathcal{O}(\overline{\mathbb{D}}, G)$ is such that $\varphi(0) = z$ and $a \notin \varphi(\mathbb{T})$, then

$$\log \widetilde{\boldsymbol{g}}_G(a, z) \leq \frac{1}{2\pi} \int_0^{2\pi} u(\varphi(e^{it}))dt.$$

(**) The function

$$\widetilde{u}(w) := \inf\left\{ \frac{1}{2\pi} \int_0^{2\pi} u(\varphi(e^{it}))dt : \varphi \in \mathcal{O}(\overline{\mathbb{D}}, G), \ \varphi(0) = w \right\}, \quad w \in G,$$

is plurisubharmonic and $\widetilde{u} = \sup\{v \in \mathcal{PSH}(G) : v \leq u\}$.

Property (**) implies that $\widetilde{u} = \log \boldsymbol{g}_G(a, \cdot)$. On the other hand, by (*), $\log \widetilde{\boldsymbol{g}}_G(a, z) \leq \widetilde{u}$, which finishes the proof. $\square$

Exploiting the Poletsky formula, A. Edigarian proved the *product property* for the Green function (cf. [Edi97b], [Edi01]; some special cases had been solved before in [JP91c], [JP91d], and [JP95]).

For any domains $G_j \subset \mathbb{C}^{n_j}$, $j = 1, 2$, *we have* $\boldsymbol{g}_{G_1 \times G_2}((a_1, a_2), (z_1, z_2)) = \max\{\boldsymbol{g}_{G_1}(a_1, z_1), \boldsymbol{g}_{G_2}(a_2, z_2)\}$, $a_j, z_j \in G_j$, $j = 1, 2$. *Consequently,* $\boldsymbol{A}$ *has the product property* $\boldsymbol{A}_{G_1 \times G_2}((a_1, a_2); (X_1, X_2)) = \max\{\boldsymbol{A}_{G_1}(a_1; X_1), \boldsymbol{A}_{G_2}(a_2; X_2)\}$, $a_j \in G_j$, $X_j \in \mathbb{C}^{n_j}$, $j = 1, 2$.

7.4.2 Sibony pseudometric

Besides the Green function and the Azukawa pseudometric there are other invariant objects defined via extremal problems for plurisubharmonic functions. The most important one is the *Sibony pseudometric*, introduced by N. Sibony in [Sib81b] in the context of $\varkappa$-hyperbolicity. It is defined as

$$\boldsymbol{S}_G(a; X) := \sup\{\left((\mathcal{L}u)(a; X)\right)^{\frac{1}{2}} : u \in \boldsymbol{\mathcal{S}}_G(a)\}, \quad a \in G, \ X \in \mathbb{C}^n,$$

where

$$\boldsymbol{\mathcal{S}}_G(a) := \{u : G \longrightarrow [0, 1), \ \log u \in \mathcal{PSH}(G), \ u(a) = 0, \ u \text{ is } \mathcal{C}^2 \text{ near } a\}.$$

Observe that the family $(\boldsymbol{S}_G)_G$ is holomorphically contractible and $\boldsymbol{\gamma}_G \leq \boldsymbol{S}_G \leq \boldsymbol{A}_G$. Moreover, $\boldsymbol{S}_G(a; \cdot)$ is a seminorm for any $a \in G$, and therefore, if $G = G_{\mathfrak{h}}$ is a non-convex pseudoconvex balanced domain (e.g. $\{(z_1, z_2) \in \mathbb{D}^2 : |z_1 z_2| < 1/2\}$), then $\boldsymbol{S}(0; \cdot) \not\equiv \boldsymbol{A}_G(0; \cdot) = \mathfrak{h}$ (see [Kli89]). On the other hand if $G := \{z \in \mathbb{C} : 1/2 < |z| < 2\}$, then $\boldsymbol{\gamma}_G(a; 1) < \boldsymbol{S}_G(a; 1) = \boldsymbol{A}_G(a; 1)$, $a \in G$ (cf. [JP13]). In contrast to the previous invariant pseudometrics, the Sibony pseudometric is in general not upper semicontinuous (cf. [JP13]). Recent research was focused on boundary behavior of $\boldsymbol{S}_G$, see e.g. [For09], [FL09]. In particular, if $G := \{z \in \mathbb{C}^2 : 1/2 < \|z\| < 1\}$, then $\frac{c_1}{t^{1/2}} \leq \boldsymbol{S}_G(P_t; X) \leq \frac{c_2}{t^{1/2}}$ for sufficiently small $t > 0$, where $P_t := (1/2 + t, 0)$, $X := (1, 0)$. Consequently, using [Kra92], we observe that $\boldsymbol{\gamma}_G(P_t; X) < \boldsymbol{S}_G(P_t; X) < \varkappa_G(P_t; X)$, $0 < t \ll 1$. Notice that the Sibony pseudometric is the "infinitesimal version" of the following holomorphically contractible family of functions

$$\boldsymbol{s}_G(a, z) := \sup\{\sqrt{u(z)} : u \in \boldsymbol{\mathcal{S}}_G(a)\}, \quad a, z \in G.$$

Unfortunately, we have to confess that up to now almost nothing is known on properties of $\boldsymbol{s}_G$.

7.4.3 Bergman metric

In the recent years the theory of the Bergman kernel and the Bergman metric became a very important tool for many questions in complex analysis (e.g. extension of biholomorphic mappings to the boundary). A detailed discussion of this topic is skipped in this survey by several reasons, mainly by lack of space. Nevertheless, for the convenience of the reader let us recall some basic definitions.

For a domain $G \subset \mathbb{C}^n$ let $L_h^2(G)$ be the Hilbert space of all holomorphic square integrable functions with the scalar product $\langle f, g \rangle := \int_G f \bar{g} d\mathcal{L}^{2n}$. Then there exists a unique *Bergman kernel* $\boldsymbol{K}_G : G \times G \longrightarrow \mathbb{C}$ such that

- $\boldsymbol{K}_G(\cdot, y) \in L_h^2(G)$, $y \in G$,
- $f(y) = \langle f, \boldsymbol{K}_G(\cdot, y) \rangle$, $f \in L_h^2(G)$, $y \in G$.

Examples of $\boldsymbol{K}_G$ for different G can be found in the article by S. G. Krantz in this book. Recall that the Bergman kernel $\boldsymbol{K}_G$ of a domain $G \subset \mathbb{C}^n$ is a $\mathcal{C}^\infty$-function on $G \times G$. From now on we assume that $\boldsymbol{K}_G(z, z) > 0$, $z \in G$ (e.g. G bounded), which guarantees that the function $z \longmapsto \log \boldsymbol{K}_G(z, z)$ is plurisubharmonic and $\mathcal{C}^2$. Then $\boldsymbol{K}_G$ leads to the following positive semidefinite Hermitian form

$$B_G(z; X) := \sum_{\nu, \mu = 1}^n \frac{\partial^2}{\partial z_\nu \partial \bar{z}_\mu} \log \boldsymbol{K}_G(z, z) X_\nu \overline{X}_\mu$$

and the *Bergman pseudometric* $\beta_G(z; X) := \sqrt{B_G(z; X)}$, $z \in G$, $X \in \mathbb{C}^n$. Finally, set

$$b_G(z', z'') := (\smallint \beta_G)(z', z'') = \inf \left\{ \int_0^1 \beta_G(\alpha(t); \alpha'(t)) dt : \right.$$

$$\left. \alpha : [0, 1] \longrightarrow G \text{ is a piecewise } \mathcal{C}^1\text{-curve with } \alpha(0) = z', \ \alpha(1) = z'' \right\},$$

$$z', z'' \in G;$$

b_G is the *Bergman pseudodistance* on G. We point out that the Bergman pseudodistance is invariant only under biholomorphic mappings, so from general point of view it is not holomorphically contractible. Nevertheless, we have $\gamma_G \leq \beta_G$ for bounded domains G ([Hah76]). On the other hand there are bounded very regular domains $G \subset \mathbb{C}^n$ such that $\beta_G \leq \varkappa_G$ fails to hold ([DF80]).

It has to be emphasized that the pluricomplex Green function turned out to be a very important tool in getting results of the Bergman theory. The reader who is interested in more details in the Bergman theory (for example, related to exhaustiveness and completeness) is invited to continue his study with diving into this new field, see e.g. [JP13], [Kra13], [Bło15], [Che15], [KZ15].

References

[APS04] M. Andersson, M. Passare, and R. Sigurdsson. *Complex convexity and analytic functions.* Birkhäuser, 2004.

[AS75]   P.R. Ahern and R. Schneider. Isometries of H^∞. *Duke Math. J.*, 42:321–326, 1975.

[AWY07]   A.A. Abouhajar, M.C. White, and N.J. Young. A Schwarz lemma for a domain related to μ-synthesis. *J. Geom. Anal.*, 17:717–749, 2007.

[AY01]   J. Agler and N.J. Young. A Schwarz lemma for the symmetrized bidisc. *Bull. London Math. Soc.*, 33:175–186, 2001.

[AY04]   J. Agler and N.J. Young. The hyperbolic geometry of the symmetrized bidisc. *J. Geom. Anal.*, 14:375–403, 2004.

[Azu86]   K. Azukawa. Two intrinsic pseudo-metrics with pseudoconvex indicatrices and starlike domains. *J. Math. Soc. Japan*, 38:627–647, 1986.

[Bar71]   T.J. Barth. Normality domains for families of holomorphic maps. *Math. Ann.*, 190:293–297, 1971.

[Bar72]   T.J. Barth. The Kobayashi distance induces the standard topology. *Proc. Amer. Math. Soc.*, 35:439–441, 1972.

[Bar80]   T.J. Barth. Convex domains and Kobayashi hyperbolicity. *Proc. Amer. Math. Soc.*, 79:556–558, 1980.

[Bar83]   T.J. Barth. The Kobayashi indicatrix at the center of a circular domain. *Proc. Amer. Math. Soc.*, 88:527–530, 1983.

[Bar95]   T.J. Barth. Topologies defined by some invariant pseudodistances. *Banach Center Publ.*, 31:69–76, 1995.

[BF78]   E. Bedford and J.E. Fornæss. A construction of peak functions on weakly pseudoconvex domains. *Ann. of Math.*, 107:555–568, 1978.

[Bło15]   Z. Błocki. Bergman kernel and pluripotential theory. In P.M.N. Feehan et al., editor, *Analysis, complex geometry, and mathematical physics: in honor of Duong H. Phong*, volume 644 of *Contemporary Mathematics*, pages 1–10. American Mathematical Society, 2015.

[Bre54]   H.-J. Bremermann. Über die äquivalenz der pseudokonvexen Gebiete und der Holomorphiegebiete in Raume von n komplexen Veränderlichen. *Math. Ann.*, 128:63–91, 1954.

[BS09]   F. Bracci and A. Saracco. Hyperbolicity in unbounded convex domains. *Forum Math.*, 21:815–825, 2009.

[BT70]   H. Behnke and P. Thullen. *Theorie der Funktionen mehrerer komplexer Veränderlichen*. Ergebnisse der Mathematik und ihrer Grenzgebiete 51, 2. erweit. Aufl., Springer Verlag, 1970.

[Car26] C. Carathéodory. Über das Schwarzsche Lemma bei analytischen Funktionen von zwei komplexen Veränderlichen. *Math. Ann.*, 97:76–98, 1926.

[Car27] C. Carathéodory. Über eine spezielle Metrik, die in der Theorie der analytischen Funktionen auftritt. *Atti Pontificia Acad. Sc., Nuovi Lincei*, 80:135–141, 1927.

[Car28] C. Carathéodory. Über die Geometrie der analytischen Abbildungen, die durch analytische Funktionen von zwei veränderlichen vermittelt werden. *Abh. Math. Sem. Hamburg*, 6:96–145, 1928.

[Car30] H. Cartan. Sur les fonctions de deux variables complexes. les transformations d'un domaine borné d en un domaine intérieur à d. *Bull. Soc. Math. France*, 200:199–219, 1930.

[Car36] H. Cartan. Sur les fonctions de n variables complexes: les transformations du produit topologique de deux domaines borneés. *Bull. Soc. Math. France*, 64:37–48, 1936.

[Che15] Bo-Yong Chen. A survey on Bergman completeness. In F. Bracci et al., editor, *Complex analysis and geometry*, volume 144 of *Springer Proceedings in Mathematics & Statistics*, pages 99–117, 2015. Proceedings of the 10th symposium, Gyeongju, Korea, August 7–11, 2014.

[Con78] J.B. Conway. *Functions of One Complex Variable*, volume 11 of *GTM*. Springer Verlag, 1978.

[Cos04a] C. Costara. *Le problème de Nevanlinna–Pick spectral*. PhD thesis, Laval Univ., 2004.

[Cos04b] C. Costara. The symmetrized bidisc as a counterexample to the converse of Lempert's theorem. *Bull. London Math. Soc.*, 36:656–662, 2004.

[CV35] S. Cohn-Vossen. Existenz kürzester Wege. *Comptes Rendus l'Acad. Sci. URSS, Vol. III (VIII)*, 8 (68):239–242, 1935.

[Dem87] J.-P. Demailly. Mesures de Monge–Ampère et mesures plurisousharmoniques. *Math. Z.*, 194:519–564, 1987.

[DF80] K. Diederich and J.E. Fornæss. Comparison of the Bergman and the Kobayashi metric. *Math. Ann.*, 254:257–262, 1980.

[Eas75] A. Eastwood. À propos des variétés hyperboliques completes. *C. R. Acad. Sci. Paris*, 280:1071–1075, 1975.

[Edi97a] A. Edigarian. On definitions of the pluricomplex Green function. *Ann. Polon. Math.*, 67:233–246, 1997.

[Edi97b]   A. Edigarian. On the product property of the pluricomplex Green function. *Proc. Amer. Math. Soc.*, 125:2855–2858, 1997.

[Edi01]   A. Edigarian. On the product property of the pluricomplex Green function, II. *Bull. Pol. Acad. Sci. Math.*, 49:389–394, 2001.

[Edi02]   A. Edigarian. Analytic discs method in complex analysis. *Dissertationes Math.*, 402:1–56, 2002.

[Edi03]   A. Edigarian. A note on Rosay's paper. *Ann. Polon. Math.*, 80:125–132, 2003.

[Eic61]   J. Eickel. Glatte starre Holomorphiegebiete vom topologischen Typ der Hyperkugel. *Schriftenr. Math. Inst. Univ. Münster*, 20, 1961.

[EKZ13]   A. Edigarian, Ł. Kosiński, and W. Zwonek. The Lempert theorem and the tetrablock. *J. Geom. Anal.*, 23:1818–1831, 2013.

[EP97]   A. Edigarian and E.A. Poletsky. Product property of the relative extremal function. *Bull. Sci. Acad. Polon.*, 45:331–335, 1997.

[EZ09]   A. Edigarian and W. Zwonek. Schwarz lemma for the tetrablock. *Bull. London Math. Soc.*, 41:506–514, 2009.

[FL09]   J.E. Fornæss and L. Lee. Kobayashi, Carathéodory and Sibony metric. *Complex Var. Elliptic Equ.*, 54:293–301, 2009.

[FM94]   J.E. Fornæss and J. D. McNeal. A construction of peak functions on some finite type domains. *Amer. J. Math.*, 116:737–755, 1994.

[For09]   J.E. Fornæss. Comparison of the Kobayashi-Royden and Sibony metrics on ring domains. *Sci. China, Ser. A*, 52:2610–2616, 2009.

[FS81]   J.E. Fornæss and N. Sibony. Increasing sequences of complex manifolds. *Math. Ann.*, 255:351–360, 1981.

[FS87]   J.E. Fornæss and B. Stensønes. *Lectures on Counterexamples in Several Complex Variables*. Mathematical Notes 33, Princeton University Press, 1987.

[Fu94]   S. Fu. On completeness of invariant metrics of Reinhardt domains. *Arch. Math. (Basel)*, 63:166–172, 1994.

[Hah76]   K.T. Hahn. On completeness of the Bergman metric and its subordinate metric. *Proc. Nat. Acad. Sci. USA*, 73:4294, 1976.

[Hör94]   L. Hörmander. *Notions of Convexity*. Progress in Mathematics 127, Birkhäuser, 1994.

[Jac06]   D. Jacquet. $\mathbb{C}$-convex domains with $\mathcal{C}^2$ boundary. *Complex Var. Elliptic Equ.*, 51:303–312, 2006.

[JP89]   M. Jarnicki and P. Pflug. The Carathéodory pseudodistance has the product property. *Math. Ann.*, 285:161–164, 1989.

[JP90]   M. Jarnicki and P. Pflug. The simplest example for the non-innerness of the Carathéodory distance. *Results in Math.*, 18:57–59, 1990.

[JP91a]   M. Jarnicki and P. Pflug. A counterexample for the Kobayashi completeness of balanced domains. *Proc. Amer. Math. Soc.*, 112:973–978, 1991.

[JP91b]   M. Jarnicki and P. Pflug. The inner Carathéodory distance for the annulus. *Math. Ann.*, 289:335–339, 1991.

[JP91c]   M. Jarnicki and P. Pflug. Invariant pseudodistances and pseudo-metrics — completeness and product property. *Ann. Polon. Math.*, 55:169–189, 1991.

[JP91d]   M. Jarnicki and P. Pflug. Some remarks on the product property. *Proc. Symp. Pure Math.*, 52 (Part 2):263–272, 1991.

[JP95]   M. Jarnicki and P. Pflug. Remarks on the pluricomplex Green function. *Indiana Univ. Math. J.*, 44:535–543, 1995.

[JP00]   M. Jarnicki and P. Pflug. *Extension of Holomorphic Functions*. de Gruyter Expositions in Mathematics 34, Walter de Gruyter, 2000.

[JP08]   M. Jarnicki and P. Pflug. *First Steps in Several Complex Variables: Reinhardt Domains*. European Mathematical Society Publishing House, 2008.

[JP13]   M. Jarnicki and P. Pflug. *Invariant Distances and Metrics in Complex Analysis, 2nd Edition*. de Gruyter Expositions in Mathematics 9, Walter de Gruyter, 2013.

[JPV91]   M. Jarnicki, P. Pflug, and J.-P. Vigué. The Carathéodory distance does not define the topology — the case of domains. *C. R. Acad. Sci. Paris*, 312:77–79, 1991.

[JPV93]   M. Jarnicki, P. Pflug, and J.-P. Vigué. An example of a Carathéodory complete but not finitely compact analytic space. *Proc. Amer. Math.*, 118:537–539, 1993.

[Kie70]   P. Kiernan. Some results concerning hyperbolic manifolds. *Proc. Am. Math. Soc.*, 25:588–592, 1970.

[Kim91]   K.-T. Kim. Automorphism group of certain domains in $\mathbb{C}^n$ with a singular boundary. *Pacific J. Math.*, 151:57–64, 1991.

[Kli85]   M. Klimek. Extremal plurisubharmonic functions and invariant pseudodistances. *Bull. Soc. Math. France*, 113:231–240, 1985.

[Kli89]   M. Klimek. Infinitesimal pseudometrics and the Schwarz Lemma. *Proc. Amer. Math. Soc.*, 105:134–140, 1989.

[Kli91]   M. Klimek. *Pluripotential Theory*. Oxford University Press, 1991.

[Kob67]   S. Kobayashi. Invariant distances on complex manifolds and holomorphic mappings. *J. Math. Soc. Japan*, 19:460–480, 1967.

[Kob73]   S. Kobayashi. Some remarks and questions concerning the intrinsic distance. *Tôhoku Math. J.*, 25:481–486, 1973.

[Kob05]   S. Kobayashi. *Hyperbolic Manifolds and Holomorphic Mappings. An introduction*. World Scientific Publishing Co. Pte. Ltd., Hackensack, NJ, second edition, 2005.

[KR81]   N. Kerzman and J. P. Rosay. Fonctions plurisousharmoniques d'exhaustion bornées et domaines taut. *Math. Ann.*, 257:171–184, 1981.

[Kra92]   S.G. Krantz. The boundary behavior of the Kobayashi metric. *Rocky Mountain J. Math.*, 22:227–233, 1992.

[Kra13]   S.G. Krantz. *Geometric analysis of the Bergman kernel and metric*, volume 268 of *Graduate Texts in Mathematics*. Springer, 2013.

[Kri27a] N. Kritikos. Über analytische Abbildungen des Gebietes $|x| + |y| <$ 1 auf sich. *Bulletin Soc. Math. Grèce*, 8:42–45, 1927.

[Kri27b] N. Kritikos. Über analytische Abbildungen einer Klasse von vierdimensionalen Gebieten. *Math. Ann.*, 99:321–341, 1927.

[KU76]   W. Kaup and H. Upmeier. Banach spaces with biholomorphically equivalent unit balls are isomorphic. *Proc. Amer. Math. Soc.*, 58:129–133, 1976.

[KW12]   Ł. Kosiński and T. Warszawski. Lempert theorem for strictly linearly convex domains with real analytic boundaries. *Ann. Polon. Math.*, 107:167–216, 2012.

[Kwa69]   M.H. Kwack. Generalization of the big Picard theorem. *Ann. of Math.*, 19:9–22, 1969.

[KZ13]   Ł. Kosiński and W. Zwonek. Proper holomorphic mappings vs. peak points and Silov boundary. *Ann. Polon. Math.*, 107:97–108, 2013.

[KZ15]   Ł. Kosiński and W. Zwonek. Bergman kernel in Complex Analysis. In D. Alpay, editor, *Operator theory*, pages 73–86. Springer, Basel, 2015.

[KZ16]   Ł. Kosiński and W. Zwonek. Extremal holomorphic maps in special classes of domains. *Ann. Sc. Norm. Super. Pisa, Cl. Sci.*, 16:159–182, 2016.

[Lem81]   L. Lempert. La métrique de Kobayashi et la représentation des domaines sur la boule. *Bull. Soc. Math. France*, 109:427–474, 1981.

[Lem82]   L. Lempert. Holomorphic retracts and intrinsic metrics in convex domains. *Analysis Mathematica*, 8:257–261, 1982.

[Lem84]   L. Lempert. Intrinsic distances and holomorphic retracts. *Complex Analysis and Applications '81, Sophia*, pages 341–364, 1984.

[Lev10]   E.E. Levi. Studii sui punti singolari essenziali delle funzioni analitiche di due o più variabili complesse. *Ann. Mat. Pura Appl.*, 17:61–87, 1910.

[Nor54]   F. Norguet. Sur les domaines d'holomorphie des fonctions uni- formes de plusieurs variables complexes (passage du local au global). *Bull. Soc. Math. France*, 82:139–159, 1954.

[NP08]   N. Nikolov and P. Pflug. On the derivatives of the Lempert func- tions. *Ann. Mat. Pura Appl.*, 187:547–553, 2008.

[NPT09]   N. Nikolov, P. Pflug, and P.J. Thomas. Lipschitzness of the Lem- pert and the Green function. *Proc. Amer. Math. Soc.*, 137:2027–2036, 2009.

[NPT11]   N. Nikolov, P. Pflug, and P.J. Thomas. Spectral Nevanlinna-Pick and Carathéodory-Fejér problems for $n \leq 3$. *Indiana Univ. Math. J.*, 60:883–894, 2011.

[NPTZ08]   N. Nikolov, P. Pflug, P.J. Thomas, and W. Zwonek. Estimates of the Carathéodory metric on the symmetrized polydisc. *J. Math. Anal. Appl.*, 341:140–148, 2008.

[NPZ07]   N. Nikolov, P. Pflug, and W. Zwonek. The Lempert function of the symmetrized polydisc in higher dimesions is not a distance. *Proc. Amer. Math. Soc.*, 135:2921–2928, 2007.

[NPZ08]   N. Nikolov, P. Pflug, and W. Zwonek. An example of a bounded ℂ- convex domain which is not biholomorphically to a convex domain. *Math. Scan.*, 102:149–155, 2008.

[NS07]   N. Nikolov and A. Saracco. Hyperbolicity of ℂ-convex domains. *C. R. Acad. Bulg. Sci.*, 60:935–938, 2007.

[Oka42]   K. Oka. Domaines pseudoconvexes. *Tôhoku Math. J.*, 49:15–52, 1942.

[Oka53]   K. Oka. Domaines finis sans point critique intérieur. *Japanese J. Math.*, 23:97–155, 1953.

[Pan94]   M.-Y. Pang. On infinitesimal behavior of the Kobayashi distance. *Pacific J. Math.*, 162:121–141, 1994.

[Pfl84]   P. Pflug. About the Carathéodory completeness of all Reinhardt domains. In G.I. Zapata, editor, *Functional Analysis, Holomorphy and Approximation Theory II*, volume 86 of *Math. Studies*, pages 331–337. North Holland, 1984.

[Pol91]   E.A. Poletsky. Plurisubharmonic functions as solutions of variational problems. *Proc. Sympos. Pure Math.*, 52:163–171, 1991.

[Pol93]   E.A. Poletsky. Holomorphic currents. *Indiana Univ. Math. J.*, 42:85–144, 1993.

[PS89]   E.A. Poletskiĭ and B.V. Shabat. Invariant metrics. In G.M. Khenkin, editor, *Several Complex Variables III*, pages 63–112. Springer Verlag, 1989.

[PZ12]   P. Pflug and W. Zwonek. Exhausting domains of the symmetrized bidisc. *Ark. Mat.*, 50:397–402, 2012.

[Ran95]   T. Ransford. *Potential Theory in the Complex Plane*. London Math. Soc. Students Texts 28, Cambridge University Press, 1995.

[Rei21]   K. Reinhardt. Über Abbildungen durch analytische Funktionen zweier Veränderlicher. *Math. Ann.*, 83:211–255, 1921.

[Rin61]   W. Rinow. *Die innere Geometrie der metrischen Räume.* Grundlehren der mathematischen Wissenschaften 105, Springer Verlag, 1961.

[Ros82]   J.-P. Rosay. Un example d'ouvert borne de $\mathbb{C}^3$ "taut" mais non hyperbolique complet. *Pacific J. Math.*, 98:153–156, 1982.

[Roy71]   H.L. Royden. Remarks on the Kobayashi metric. In *Several complex variables, II*, volume 189 of *Lecture Notes in Math.*, pages 125–137. Springer Verlag, 1971.

[Rud74]   W. Rudin. *Real and Complex Analysis*. McGraw-Hill Book Company, 1974.

[RW83]   H.L. Royden and P.-M. Wong. Carathéodory and Kobayashi metric on convex domains. Preprint, 1983.

[Sel74]   M.A. Selby. On completeness with respect to the Carathéodory metric. *Canad. Math. Bull.*, 17:261–263, 1974.

[Sib75]   N. Sibony. Prolongement de fonctions holomorphes bornées et metrique de Carathéodory. *Invent. Math.*, 29:205–230, 1975.

[Sib81a] N. Sibony. Personal communication, 1981.

[Sib81b] N. Sibony. A class of hyperbolic manifolds. In J. E. Fornæss, editor, *Recent developments in several complex variables*, volume 100 of *Ann. Math. Studies*, pages 347–372, 1981.

[Sun78]   T. Sunada. Holomorphic equivalence problem for bounded Reinhardt domains. *Math. Ann.*, 235:111–128, 1978.

[TT98]   D.D. Thai and P.J. Thomas. $\mathbb{D}^*$-extension property without hyperbolicity. *Indiana Univ. Math. J.*, 47:1125–1130, 1998.

[Vig83]   J.-P. Vigué. La distance de Carathéodory n'est pas intérieure. *Resultate d. Math.*, 6:100–104, 1983.

[War15]   T. Warszawski. (Weak) m-extremals and m-geodesics. *Complex Var. Elliptic Equ.*, 60:1077–1105, 2015.

[Zap15]   P. Zapalowski. Geometric properties of domains related to μ-synthesis. *J. Math. Anal. Appl.*, 430:126–143, 2015.

[Zwo96]   W. Zwonek. Automorphism group of some special domain in $\mathbb{C}^n$. *Univ. Iagel. Acta Math.*, 33:185–189, 1996.

[Zwo99]   W. Zwonek. On hyperbolicity of pseudoconvex Reinhardt domains. *Arch. Math. (Basel)*, 72:304–314, 1999.

[Zwo00a] W. Zwonek. On Carathéodory completeness of pseudoconvex Reinhardt domains. *Proc. Amer. Math. Soc.*, 128:857–864, 2000.

[Zwo00b] W. Zwonek. Regularity properties of the Azukawa metric. *J. Math. Soc. Japan*, 52:899–914, 2000.

[Zwo13]   Z. Zwonek. Geometric properties of the tetrablock. *Arch. Math.*, 100:159–165, 2013.

8

Variations on the (Eternal) Theme of Analytic Continuation

Dmitry Khavinson

CONTENTS

8.1   Introduction

Let's illustrate the above quote with a very simple example.

(i) Consider a perfect "bell-shaped graph" $f(x) = \frac{1}{1+x^2}$. If we take its Taylor series around the origin

$$f(x) = \sum_0^\infty (-1)^n x^{2n},$$

we note immediately that it diverges for all real $x : |x| \geq 1$. Why? Of course, the answer is clear, if we replace x by a complex variable z, the function $f(z) := \frac{1}{1+z^2}$ has two polar singularities at $z = \pm i$ on the boundary of the circle of convergence. Thus, since the Taylor series naturally converge in disks in the complex plane, the presence of complex singularities interferes with the behavior of the series in the real domain.

DOI: 10.1201/9781315160658-8

(ii) In the opposite direction, if we consider the Taylor series

$$f(x) = \sum_0^\infty \cos\left(\sqrt{n}\right) x^n$$

that converges only for $|x| < 1$, the function $f(x)$ extends as a smooth, in fact a real-analytic function, to all real $x < 1$. In fact, if we consider $f(z) := \sum_0^\infty \cos\left(\sqrt{n}\right) z^n$ as an analytic function of the complex variable z, it extends as an analytic function to the whole complex plane $\mathbb{C} \setminus \{1\}$ except one point $z = 1$, where $f(z)$ has an essential singularity. How do we know that?

There is an enormous amount of literature dedicated to studying the properties of analytic functions at large encoded in their local expansions in Taylor series—cf., the classical monographs by P. Dienes and L. Bieberbach ([3], [4, Ch. X]). One of the first results in this direction is the following beautiful theorem of L. Kronecker ([4, Ch. X]).

Theorem 8.1.1 (L. Kronecker, 1881 [12]). *The Taylor series $\sum_0^\infty a_n z^n$ represents a rational function $f(z) = P(z)/Q(z)$, P, Q are polynomials and $\max\left(\deg P, \deg Q\right) = N$ if and only if all the determinants*

$$\det \begin{pmatrix} a_0 \cdots a_n \\ a_1 \cdots a_{n+1} \\ \vdots \\ a_n \cdots a_{2n} \end{pmatrix} = 0 \text{ for all } n \geq N.$$

Our example (i) illustrates this theorem with $N = 2$. Since rational functions are obviously globally defined on the whole Riemann sphere, Kronecker's theorem is a very good example of a mandatory analytic continuation (single-valued as well) of a locally defined Taylor series.

The example (ii) is an illustration of a compilation of results of L. Leau—1899, S. Wigert—1900 and G. Faber—1903, cf. [4, p. 337 ff].

Theorem 8.1.2 (G. Faber, L. Leau, S. Wigert). *The Taylor series $\sum_0^\infty a_n z^n$ with the radius of convergence 1 extends to $\mathbb{C} \setminus \{1\}$ if and only if there exists a (unique) entire function $g(z)$ of order zero (minimal type) such that $a_n = g(n)$, $n = 1, 2, \ldots$. If $g(z)$ is a polynomial of degree m then 1 is a pole of order $m + 1$.*

(Recall that an entire function g is called of minimal type if for every $\varepsilon > 0$ there exists a constant C_ε such that $|f(z)| \leq C_\varepsilon e^{\varepsilon|z|}$.)

For example, the geometric series $\sum_0^\infty z^n = \frac{1}{1-z}$ illustrates the latter part of the theorem with $m = 0$.

To illustrate the ideas behind this and similar results, let's sketch a slightly more modern result of this type due to T. Qian ([14]) and D. Khavinson [8] (the latter with a different, much shorter proof).

Theorem 8.1.3 ([8, 14]). *Let* $f(z) = \sum\limits_{1}^{\infty} b_n z^n$ *be an analytic function in* $\mathbb{D} := \{|z| < 1\}$, *and* $b_n = g(n)$, *where* g *is of minimal type in the sector* $S_\varphi := \{z : |\arg z| < \varphi, 0 < \varphi \leq \frac{\pi}{2}\}$. *Then,* $f(z)$ *extends to the "heart-shaped" domain* $\Omega_\varphi := \{z = re^{i\theta}, 2\pi - \cot\varphi \cdot \log r > \theta > \cot\varphi \cdot \log r\}$.

Note that when $\varphi = \frac{\pi}{2}$, $\Omega_\varphi = \mathbb{C} \smallsetminus [1, +\infty)$.

Sketch of the proof [8] following ideas of Le Roy and Lindelöf [4, p. 340 ff]. By the residue theorem

$$\sum_{1}^{N} g(n)z^n = \int_{\gamma} \frac{g(w)z^w \, dw}{e^{2\pi i w} - 1}, \tag{8.1}$$

where γ is any contour in S_φ enclosing the integers $1, \ldots, N$ and no others. (Can choose, e.g., for γ the boundary of the sector $\{w : |\arg(w - \alpha)| \leq \varphi, |w - \alpha| \leq R, 0 < \alpha < \frac{1}{2}, R = N + \frac{1}{2}\}$.) $\gamma = \gamma_\varphi \cup \gamma_R$, where γ_R is a circular part of the boundary, while γ_φ comprises two sides of the angle with vertex at α.

An elementary argument yields that for all $w \in \mathbb{C} \smallsetminus \bigcup\limits_{n=-\infty}^{\infty} \{z : |z - n| <$ $\eta, \eta > 0\text{-small}\}$, $\left|e^{2\pi i w} - 1\right| \geq c > 0$, $c = c(\eta)$ is a constant ([4, p. 341]). Hence, the part of the integral (8.1) restricted to γ_R tends to zero when $R \to \infty$ (i.e., $N \to \infty$) for $z = -r$, $0 < r < 1$, i.e., z in $\mathbb{D}$ and on the negative radius. (This is seen from an elementary estimate $\left|e^{\pi i w} - e^{-\pi i w}\right| \geq c e^{\pi R |\sin\theta|} > c$, for $w = \alpha + Re^{i\theta}$, assuming at first $\varphi < \frac{\pi}{\alpha}$.) Thus, for $z \in (-1, 0]$ in $\mathbb{D}$,

$$f(z) = \int_{\Gamma_\varphi} \frac{g(w)z^w}{e^{2\pi i w} - 1} \, dw, \tag{8.2}$$

where $\Gamma_\varphi := \lim\limits_{R \to \infty} \gamma_\varphi = \{w \cdot \arg(w - \alpha) = \pm\varphi\}$. Thus, in order for the integrand in (8.2) to decay exponentially on Γ_φ, we obtain using the assumptions on g, that it suffices to have

$$\cot\varphi \cdot \log|z| < \arg z, \text{ for } \arg(w - \alpha) = \varphi$$

and

$$2\pi - \cot\varphi \cdot \log|z| > \arg z \text{ for } \arg(w - \alpha) = -\varphi.$$

This proves the statement for $\varphi < \frac{\pi}{2}$, since (8.2), an analytic function in z, converges for all $z \in \Omega_\varphi$ and coincides with f on $(-1, 0)$. Finally, $\Omega_{\frac{\pi}{2}} = \bigcup\limits_{\varphi < \frac{\pi}{2}} \Omega_\varphi$. $\square$

Remark. The assumptions on g can be relaxed somewhat further (to g being of exponential type less than π—cf. [4, pp. 341–342]). It is not known whether Ω_φ is the largest domain one can extend $f(z)$ to. Thus, the "only if" part is still missing, unlike for classical results of Kronecker and Leau (Thms. 8.1.1 and 8.1.2).

Of course, it is impossible in an article like this one to survey all beautiful topics investigated in the classical avenue of analytic continuation: monodromy, continuation of algebraic functions, over-convergence and gap series, universal Taylor series, and many others. We hope that an interested reader will be tempted to continue research on her own: [4, Ch. X, XI], [3] are good books to start. The more recent vast literature on universal Taylor series can be found on MathSciNet.

From these classical themes of continuation of Taylor series, let's make a leap to the problem of analytic continuation of solutions of most basic equations of mathematical physics.

8.2   Continuation of Solutions of Linear PDE

8.2.1   ODE vs. PDE

It is well known that the solution of the initial value problem for the linear ODE

$$w^{(n)}(z) + a_{n-1}(z)w^{(n-1)}(z) + \cdots + a_n(z)w'(z) + a_0(z)w(z) = f(z), \tag{8.3}$$
$$w(0) = w_0, \ldots, w^{(n-1)}(0) = w_{n-1},$$

with the coefficients a_j's and f analytic in a domain Ω containing the origin, extends as analytic function throughout Ω. (It might end up being multi-valued, if Ω is not simply connected, i.e., has holes, but nevertheless can be analytically continued everywhere in Ω.)

Yet, if we consider a very simple initial value problem for PDE:

$$\frac{\partial w}{\partial y} = x^2 \frac{\partial w}{\partial x}, \quad w(x,0) = x; \tag{8.4}$$

the easily found solution $w := \frac{x}{1-xy}$ blows up arbitrarily close to the initial line $\{y = 0\}$ on the hyperbola $y = 1/x$. How can we explain this?

Moreover, if we consider a more general initial value problem for (8.4) with arbitrary "data" $w(x,0) = f(x)$, where f is a polynomial, or an entire function, one easily checks that the solution is

$$w(x,y) := f\left(\frac{x}{1-xy}\right).$$

This is striking since it yields that the variety $\Gamma := \{xy = 1\}$ is the only possible carrier of singularities for all solutions to (8.4), independently of the data, as long as the data itself has no singularities.

Let's postpone the heuristic explanation of this fact till later and discuss another natural problem of analytic continuation coming from mathematical physics.

8.2.2 G. Herglotz' memoir of 1914

The following question was first tackled by G. Herglotz in 1914 (also, cf. [15, 18]). Imagine a solid in $\mathbb{R}^3$, or a "plate" (a domain) in $\mathbb{R}^2$, Ω, bounded by, say, a nice algebraic surface (or, a curve).

Let

$$u_\Omega(x) = \int_\Omega k_n(x, y)\, dy, \tag{8.5}$$

where

$$k_n(x, y) = \frac{1}{2\pi} \log \frac{1}{|x - y|}, \quad n = 2$$

$$= \frac{-1}{4\pi} \frac{1}{|x - y|}, \quad n = 3$$

be the "gravitational" or "electrostatic" potential of Ω. Obviously, u is harmonic outside Ω, and the natural question tackled by Herglotz was: how far can one harmonically continue $u(x)$ inside Ω before running into a singular point? (Herglotz did answer the question in $\mathbb{R}^2$ and made some headway in $\mathbb{R}^3$, but because his prize-winning memoir appeared on the brink of World War I it fell into oblivion while most of the results were rediscovered by other authors—cf. the references in [10, 16]. For example, if Ω is a disk or a ball, then, of course, (e.g., $n = 3$) by the mean value theorem, we have:

$$u_\Omega(x) = -\frac{1}{4\pi} \int_{\{|y|<1\}} \frac{dy}{|x - y|} = \frac{\text{const}}{|x|}$$

(dy, of course, stands for the Lebesgue measure in $\mathbb{R}^n$.)

So, $u_\Omega(x)$ extends as a harmonic function everywhere in $\mathbb{R}^3$ except for the center of the ball. This goes back to I. Newton and is well-known. What is perhaps less well-known is that conclusion stays true for

$$u_{\Omega,p}(x) = -\frac{1}{4\pi} \int_\Omega \frac{p(y)\, dy}{|x - y|}, \quad \Omega = \{y : |y| < 1\}$$

with an ARBITRARY polynomial, or even an entire density $p(y)$. In the latter case, all the symmetry associated with the ball goes out the window, and the conclusion that $u_{\Omega,p}$ extends harmonically to all of $\mathbb{R}^3 \smallsetminus \{0\}$, matches Leau's theorem from Section 8.1 in mystery and beauty.

Herglotz' problem is often restated in more "physical" terms: consider the exterior gravitational potential of an analytic mass density $p_0(y)$ in the region Ω. Find a smaller object E inside Ω and a different mass-density p_1 on E that is gravi-equivalent to p_0, i.e., such that the potential u_{E,p_1} and u_{Ω,p_0} coincide outside of Ω.

Example 8.2.1. For $\Omega = \{x : |x| < 1\} \subset \mathbb{R}^3$ (or, more generally, $\mathbb{R}^n$) $p_0 =$ polynomial of degree $\leq N$, $E = \{0\}$ and p_1 is the distribution of order $\leq N$ at the origin.

Example 8.2.2 (cf., [9, 10, 16])**.** An oblate spheroid

$$\Omega := \left\{ x \in \mathbb{R}^3 : \frac{x_1^2}{a} + \frac{x_2^2}{a^2} + \frac{x_3^2}{b^2} \leq 1, a > b > 0 \right\}$$

(planet Earth, e.g.). Then, for say, uniform density $p_0 \equiv 1$ (or any other polynomial, or, entire density) $u_\Omega(x)$ extends into $\Omega \smallsetminus E$, where

$$E := \left\{ x_3 = 0, x_1^2 + x_2^2 \leq a^2 - b^2 \right\}$$

is the caustic disc. The relevant density p_1 on E (relevant to $p_0 = 1$) is algebraic and equals const $\left(a^2 - b^2 - x_1^2 - x_2^2 \right)^{1/2}$—cf. [10, Ch. 15].

Example 8.2.3. A prolate spheroid $\Omega := \left\{ \frac{x_1^2}{a^2} + \frac{x_2^2}{b^2} + \frac{x_3^2}{b^2} \leq 1, a > b > 0 \right\}$ gives a completely different picture, an exciting mystery on its own. The potential u_Ω extends to $\Omega \smallsetminus E$, but E in this case is a 1-dimensional segment $\left\{ x_2 = x_3 = 0, |x_1| \leq \sqrt{a^2 - b^2} \right\}$, while the density $p_1 = \left(a^2 - b^2 - x_1^2 \right)$ is a polynomial. Moreover, below we shall touch upon the rather deep problem regarding the dramatic differences in singularities of u_Ω in the latter two examples: bounded, a square-root type singularity in the former, and unbounded—in the latter.

8.2.3 A further discussion of the Herglotz question

As one readily obtains directly from (8.5) via Green's theorem,

$$\Delta u_\Omega = \chi_\Omega, \tag{8.6}$$

where $\chi_\Omega(x) = \begin{cases} 1, & x \in \Omega \\ 0, & x \in \mathbb{R} \smallsetminus \overline{\Omega} \end{cases}$ and stands for the characteristic function of Ω. Denote by M, the so-called modified Schwarz potential (of $\partial\Omega$) the solution of the following initial value problem

$$\Delta M = 1 \text{ near } \Gamma := \partial\Omega;$$
$$M = \nabla M = 0 \text{ on } \Gamma. \tag{8.7}$$

(The solution exists and is unique by the Cauchy–Kovalevskaya theorem—cf. [10], e.g.) Then, the function

$$
u := \begin{cases} u_\Omega, & \text{outside } \Omega \\ u_\Omega - M & \text{inside } \Omega \end{cases}
\tag{8.8}
$$

gives the desired continuation. Indeed, $u_\Omega - M$ is harmonic in Ω near Γ and coincides with u_Ω on Γ together with its first derivatives. The statement then follows by a straightforward application of Green's formula—cf. [10, Thm. 6.1].

For an arbitrary polynomial or entire mass density p, we only need to modify (8.7) and define M_p as a solution of the initial value problem

$$
\begin{cases} \Delta M_p = p \text{ near } \Gamma; \\ M_p = \nabla M_p = 0 \text{ on } \Gamma. \end{cases}
\tag{8.9}
$$

If instead of M we consider the Schwarz potential u_Γ of Γ defined by

$$
\begin{cases} \Delta u_\Gamma = 0 & \text{near } \Gamma; \\ u_\Gamma = \frac{1}{2n} |x|^2 \text{ on } \Gamma; & \text{grad } u_\Gamma = \frac{1}{n} \vec{x} \text{ on } \Gamma \end{cases}
\tag{8.10}
$$

($n = 2$ or 3, as in our examples), then obviously, $M_\Gamma = \frac{1}{2n} |x|^2 - u_\Gamma$. Similarly, $M_p = Q - u_{\Gamma,p}$, where Q is a polynomial, an entire function such that $\Delta Q = p$, and u_Γ accordingly defined as a solution of the initial value problem similar to (8.10):

$$
\begin{cases} \Delta u_{\Gamma,p} = 0 \text{ near } \Gamma; \\ u_{\Gamma,p} = Q, \nabla u_{\Gamma,p} = \nabla Q \text{ on } \Gamma. \end{cases}
\tag{8.11}
$$

Thus, if we could show that the singularities of any initial value problem for the Laplace operator posed on Γ are only dictated by Γ itself, not by initial data ($\frac{1}{2n} |x|^2$, or Q), we would have achieved the high ground needed for understanding the Herglotz' problem. A deep and beautiful theory of Leray explains the origins for the appearance of singularities of initial value problems near initial surfaces—cf. [10, Ch. 13, 19–20] and Leray's original papers referenced there.

Indeed, generically, it asserts that the singularities appear and take off (locally, sic!), from the initial surfaces at the same places and along the same routs independently of data. Moreover, in dimension 2, and also, in higher dimensions, but only for quadratic surfaces, it has been proved that the local theory of Leray, first verified only near initial surfaces, holds globally—cf. [10]. Here, we will simply illustrate the Leray principle by a couple of straightforward examples. We only sketch the main steps, more details can be found in [10].

Example 8.2.4. Let $\Omega = \left\{ \frac{x^2}{a^2} + \frac{y^2}{b^2} - 1 < 0, a > b \right\}$ be an ellipse, $\Gamma := \partial\Omega$.

One can calculate u_Γ, and then further

$$\frac{1}{2} \nabla u_\Gamma = \frac{\partial u_\Gamma}{\partial \bar{z}} = \frac{a^2 + b^2}{a^2 - b^2} \bar{z} - \frac{2ab}{a^2 - b^2} \sqrt{\bar{z}^2 - c^2}, \quad c^2 - a^2 - b^2 \quad (8.12)$$

($z = x + iy$, as usual). So, the singularities of u_Γ are at the foci of Ω. The solution of the initial value problem (8.10) is fine by the Cauchy–Kovalevskaya theorem near complexified quadratic curve $\widehat{\Gamma}$ in $\mathbb{C}^2$,

$$\widehat{\Gamma} := \left\{ x, y \in \mathbb{C} : \frac{x^2}{a^2} + \frac{y^2}{b^2} - 1 = 0 \right\}$$

except for 4 points $\left\{ \left(\pm\frac{a^2}{c}, \pm\frac{b^2}{c} \right) \right\}$ on $\widehat{\Gamma}$ where C–K theorem breaks down. From those "bad" points, as Leray's theory asserts, the singularities travel along 4 complex (characteristic) lines $\{(x, y) \in \mathbb{C}^2 : x \pm iy = \text{const}\}$ tangent to $\widehat{\Gamma}$ at the above characteristic points. These lines, the "carriers" of singularities, reach the "real" space $\mathbb{R}^2$ at the foci $(\pm c, 0)$ of the ellipse. Since the carries of singularities depend on Γ only, logarithmic potentials of Ω with arbitrary polynomial, or entire densities exhibit the same behavior and might become singular only at the foci as well.

Moreover, when $b \uparrow a$, $c \downarrow 0$, an ellipse becomes a circle, the "bad" points on $\widehat{\Gamma}$ all move to infinity, and the singularities of (8.12) change from algebraic ($\sqrt{}$-type) to polar-$\frac{c}{z}$ at the origin, the limiting position of the collapsing foci.

A similar but technically much more demanding analysis provides the justification for Examples 8.2.1–8.2.3 in $\mathbb{R}^3$, and in general in $\mathbb{R}^n$—cf. [10] and the works of G. Johnsson referenced therein. However, we emphasize that for algebraic surfaces of degree ≥ 3 in $\mathbb{R}^n$, $n \geq 3$, the analysis of Herglotz' problem, i.e., the global version of Leray's principle, is still waiting to be discovered.

We shall finish this section with another transparent example illustrating Leray's theory.

Example 8.2.5. Consider the initial value problem

$$\frac{\partial^2 u}{\partial x \partial y} = 0, \text{ near } \Gamma := \{y = x^3\};$$

$$\frac{\partial u}{\partial x} = y, \quad \frac{\partial u}{\partial y} = x \text{ on } \Gamma. \tag{8.13}$$

One readily finds the solution $u(x, y) = \frac{x^4}{4} + \frac{3y^{4/3}}{4}$ that is "ramified" around $\{y = 0\}$. The latter is, in fact, Leray's characteristic tangent to Γ at the (unique, w.r.t. $\frac{\partial^2}{\partial x \partial y}$ operator) "bad" characteristic point $(0, 0)$.

In higher dimensions the situation is more complicated. In a nutshell, there are more "complex characteristic lines" tangent to the initial surface that we view as continuation of Γ into $\mathbb{C}^n$. These lines carry out singularities off the initial surface. The analytic functions having singularities on a piece of an analytic hypersurface however must be singular on the whole hypersurface by a celebrated theorem of Hartogs (in $\mathbb{C}^n$, $n > 1$, of course). In other words, the singularities propagate from "bad" points on the complexified (embedded into $\mathbb{C}^n$) surface Γ and then exhibit themselves in $\mathbb{R}^n$ at points where the Leray characteristic tangent, the carrier of singularities, hit $\mathbb{R}^n$. This is transparent and proved rigorously (cf. [10, Ch. 13], e.g.), by G. Johnsson for quadratic surfaces in $\mathbb{R}^n$—cf. [10, Ch. 19-20] and references to Johnsson's original papers contained therein. This also explains the difference in the nature of singularities in Examples (8.4)–(8.5). In the case of the oblate spheroid, each point on the circular caustic $\{x_1^2 + x_2^2 \le a^2 - b^2, x_3 = 0\}$ is a "meeting point" of true characteristic lines coming from $\mathbb{C}^3$ and tangent at characteristic points on the complexified surface $\left\{(x_1, x_2, x_3) \in \mathbb{C}^3 : \frac{x_1^2}{a^2} + \frac{x_2^2}{a^2} + \frac{x_3^2}{b^2} - 1 = 0, a > b\right\}$. For the prolate spheroid, each point of the caustic segment $\{|x_1| \le \sqrt{a^2 - b^2}, x_2 = x_3 = 0\}$ is a meeting point of infinitely many characteristics, thus causing unbounded singularities. So, intuitively, the idea that "more carriers of singularities meeting at a point in $\mathbb{R}^n$" should result in a "heavier" singular behavior is tempting and reasonable. However, essentially, nothing has been rigorously proved along these lines. A worthy and challenging avenue for further research.

8.3   Analytic Continuation and Problems of Uniqueness

Consider the spherical shell $\Omega := \{x \in \mathbb{R}^3 : r < |x| < R\}$ ($|x|^2 = x_1^2 + x_2^2 + x_3^2$, as usual). Let u be a harmonic function in Ω that vanishes on the segment $(-R, -r)$ of the x_1-axis. Then, let us pose the question (cf. [10, Ch. 9]).

Question 8.3.1. Must u also vanish on the segment $r < x_1 < R$, $x_2 = x_3 = 0$?

The same question in 2 dimensions can be easily settled by elementary complex analysis. To fix the ideas, let $r = 1$, $R = 2$, $n = 2$. Consider the harmonic function $v(z) := u(z) + u(\bar{z})$ in the annulus $\Omega := \{1 < |z| < 2\}$. By the Schwarz effection principle, $v \equiv 0$ in a small disk centered on $(-2, -1)$, say, $\left\{z : \left|z + \frac{3}{2}\right| < \frac{1}{4}\right\}$. Hence, $v \equiv 0$ in Ω and, accordingly, $2u(x) = v(x) = 0$ on $(1, 2)$. Thus, in this situation, the answer is "yes". Although, the answer is also "yes" in $\mathbb{R}^n$, $n \ge 3$, the above argument of course, fails. Moreover, the above argument doesn't work either if instead of a line through the center of the annulus, or a spherical shell, we consider an arbitrary line still cutting Ω in two disjoint segments. It might come as a surprise that the answer remains "yes" if $\frac{R}{r} > 3$ (a thick annulus, or shell), but becomes "no, not necessarily" if $\frac{R}{r} \le 3$ (a thin annulus) and the constant 3 is sharp (cf. [10, Ch. 9], [11]).

Even more intriguing ([10, Ch. 9]), the same question posed for a torus in $\mathbb{R}^3$, has always a negative answer in general.

What is the high ground for this question? The answer is analytic continuation. Indeed, a harmonic in a domain $\Omega \subset \mathbb{R}^n$, $n \geq 2$, function u automatically extends as a holomorphic function of n variables to a domain $\widehat{\Omega}$ in $\mathbb{C}^n$. $\widehat{\Omega}$ can be viewed in a rather simple way as follows. For all $x^o \in \mathbb{R}^n \setminus \Omega$, consider the isotropic cone $\Gamma_{x^o} := \left\{ z \in \mathbb{C}^n : \sum_1^n \left(z_j - x_j^0 \right)^2 = 0 \right\}$.

Then, $\widehat{\Omega} := \mathbb{C}^n \setminus \bigcup_{x^o \in \mathbb{R}^n \setminus \Omega} \Gamma_{x^o}$—cf. [1, Ch. 1].

The beautiful fact established in the theory of linear analytic PDE is that solutions of ALL PDEs ($\Delta^n +$ lower terms) $u = f$ with, say polynomial or entire coefficients in Ω automatically extend to $\widehat{\Omega}$. For example, if $\Omega = \{|x| < 1\}$ is the unit ball in $\mathbb{R}^n$, $n \geq 2$, $\widehat{\Omega} =$

$$\left\{ z \in \mathbb{C}^n : \left(\|z\|^4 - \left| \sum_1^n z_j^2 \right|^2 \right)^{1/2} + \|z\|^2 < 1 \right\}, \text{ the celebrated Lie ball (cf.}$$

[1, 2, 10]). (For $n = 2$, $\widehat{\Omega} = \{(X, Y) \in \mathbb{C}^2 : |X \pm iY|\}$, the bidisk (cf. [1, 2, 10]). Thus, a sufficient condition that would yield an affirmative answer to our question, is whether the intersection $\widehat{\Omega} \cap \{Y = c\}$ the harmonicity hull of the shell Ω and the complexified line $\{y = c\}$ (in dimension 2, e.g.) is connected or not. In the original question for the annulus, for example, this intersection becomes $\{(X, c) : r < |X \pm ic| < R\}$ and is disconnected if $\frac{R-r}{2} < c < r$, and connected if $0 < c \leq \frac{R-r}{2}$. If $\frac{R}{r} \geq 3$, e.g., $\frac{R-r}{2} \geq r > c$, so the intersection of $\widehat{\Omega}$ and the complex line $\{Y = c\}$ is connected. The fact that the constant 3 is sharp is seen (T. Ransford) by taking Ω to be an annulus separating $\{0, -i\}$ from i, and $u(z) := \mathrm{Re}\, \sqrt{z(z-i)(z+i)}$, where we can take any branch of the square root. Then, $u(x) = 0$, $x < 0$ and $u(x) > 0$, $x > 0$. $\frac{R}{r} < 3$ but can be made arbitrary close to 3—see [11], [10, Ch. 9] for more details.

As is remarked in [11], this simple consideration allows to answer questions similar to Q. 8.3.1 not only for solutions of linear PDE with the power of the Laplacian in the principle part but also for functions represented by arbitrary Riesz potentials, the latter, in general, need not satisfy any linear PDE.

8.4   Analytic Continuation of Series of Zonal Harmonics and Series of Orthogonal Polynomials

By analogy with analytic continuation of Taylor series, let's consider the problem of finding singularities of other series expansions. To fix the ideas, let

$$u := \sum_{n=0}^{\infty} a_n r^n P_n(\cos \theta) \tag{8.14}$$

be an axially symmetric harmonic function in the unit ball in $\mathbb{R}^3$. $a_n \in \mathbb{R}$, $\lim\limits_{n \to \infty} |a_n|^{1/n} = 1$, $P_n(x) = \frac{1}{2^n n!} \frac{d^n}{dx^n}\left[(x^2 - 1)^n\right]$ are Legendre polynomials (orthogonal on $[-1,1]$, $\|P_n\|_2^2 = \frac{2}{2n+1}$). $r = \left(x_1^2 + x_2^2 + x_3^2\right)^{1/2}$ the distance to the origin, θ is the usual azimuth angle in spherical coordinates. One can easily verify that the expansion (8.14) diverges for $r > 1$, so u must have singularities on the unit sphere $S^2 := \{r = 1\}$. The question is where? The following remarkable theorem was proved by G. Szegő in 1954 [17].

Theorem 8.4.1. $u(r, \theta)$ *extends harmonically across the circle* $(1, \theta_0 : 0 \le \varphi \le 2\pi)$ *on the sphere* S^2 *if and only if the Taylor series* (!) $f(\xi) := \sum\limits_{k=0}^{\infty} a_n \xi^n$ *extends across* $\xi_0 := e^{i\theta_o}$.

This is a truly amazing result, since at first glance, the expansions in zonal harmonics and the Taylor series built with the same coefficients should have nothing in common. Moreover, inspired by Szegő's theorem, Nehari proved the following beautiful follow-up [13].

Theorem 8.4.2. *Let* $\{a_n\} \in \mathbb{C}$, *satisfy* $\varlimsup\limits_{n \to \infty} |a_n|^{1/n} = \frac{1}{R}$, $R > 1$ *and let* $f(t) = \sum\limits_{n=0}^{\infty} a_n P_n(t)$ *be (as is easily checked) an analytic function inside the ellipse* $\mathcal{D}_R$ *with foci at* ± 1 *and sum of whose semiaxes equals* R. *In other words,* $\mathcal{D}_R := \left\{(x,y) : \frac{x^2}{a^2} + \frac{y^2}{b^2} < 1, a + b = R, a - b = \frac{1}{R}\right\}$. $P(t)$, *as in Szegő's theorem, denote Legendre polynomials. Then,* $f(t)$ *is analytically continuable across* $t_o \in \partial \mathcal{D}_R$ *if and only if the analytic function* $g(s) := \sum\limits_{n=0}^{\infty} a_n s^n$, $|s| < R$, *where* s *and* t *are related by the conformal map* $s = \varphi(t) = t + \sqrt{t^2 - 1}$, $\varphi : \widehat{\mathbb{C}} \smallsetminus [-1, 1] \to \{s \in \mathbb{C} : |s| > 1\}$, $\varphi(\infty) = \infty$, *is analytically continuable across the corresponding point* $s_0 = \varphi(t_0)$, $(t_0 = \frac{1}{2}\left(s_0 + s_0^{-1}\right))$, $s_0 \in \{s : |s| = R\}$.

Once again, the reader should observe how Nehari's result, unexpectedly, connects the singularities of the expansion in orthogonal polynomials with seemingly disjoint Taylor series.

Since there is a wide variety of results allowing one to identify singularities of Taylor series on the circle of convergence, the above two results provide a powerful tool for identifying singularities of harmonic functions and orthogonal polynomial expansions.

Both theorems can be significantly extended by replacing Legendre polynomials with arbitrary Jacobi orthogonal polynomials (R. P. Gilbert (1969), P. Ebenfelt–D. Khavinson–H. S. Shapiro (1996))—cf. [10, Ch. 10], [5], [6] and references therein.

Recall that for $\alpha, \beta > -1$, the Jacobi polynomials are orthogonal polynomials on $[-1, 1]$ with respect to the weight $(1 - x)^\alpha (1 + x)^\beta$, normalized by $P_n^{\alpha,\beta}(1) = \binom{n + \alpha}{n}$. So, $\alpha = \beta = 0$ corresponds to Legendre polynomials, $\alpha = \beta = -\frac{1}{2}$—Tschebysheff polynomials (that, in turn, under a suitable change of variables, correspond to monomials z^n, $n \geq 0$ in the unit disk—cf. [5, 6, 10]; $\alpha = \beta = \frac{k-3}{2}$, $k \geq 2$ being an integer corresponds to ultraspherical polynomials appearing in expansions of axially symmetric harmonic functions in $\mathbb{R}^k$ (Gegenbauer polynomials).

Now, the original proofs of Thms. 8.4.1 and 8.4.2 boil down to writing down the given expansions in terms of a certain integral and ingenious manipulation of the latter. The "high ground" approach advocated and developed in [5, 6] consists of noticing that ALL relevant expansions in general Jacobi polynomials can be interpreted as solutions of a Cauchy problem for a linear PDE in two variables with the same initial data. Of course, the partial differential operator corresponding to every particular expansion is different in each case. But all of them share the same principal part, the part of the differential operator that involves the senior derivatives. A deep result in the theory of linear PDE, based on the 1970 extension by M. Zerner (cf. [10, Ch. 4] of the classical Cauchy–Kovalevskaya theorem yields that the singularities of the solutions of the Cauchy problems locally depend exclusively on the principal part of the differential operator. Hence, all the expansions in Thms. 8.4.1 and 8.4.2 share the same singularities thus unveiling the mystery behind Szegő's and Nehari's results.

8.5 An Epilogue

This article's only intent is to initiate for the curious reader a few possible modern directions in the classical theme of analytic continuation. There is absolutely no way to cover all possible topics, thus our choices were limited to several topics the author felt most comfortable with.

There are so many themes that were left out entirely, e.g., beautiful results of Eisenstein regarding algebraic properties of Taylor series depending on properties of coefficients—[3, 4]. Essentially, no deep and subtle results, starting from Painlevé classical researchers, dealing with classification of singularities and analytic continuation of solutions to nonlinear ODE in the complex domain—cf., e.g., [7]. In classical potential theory, we left out beautiful modern generalizations due to A. Givental of the Newton's "no gravity in the ellipsoidal cavity" theorem, Ivory's theorem, MacLauren's mean value theorem for ellipsoids, viewed from the modern viewpoint of analytic continuation of Cauchy's problem for the Laplace equation—cf. [9, 10] and references therein. The theme of analytic continuation of solutions to the Dirichlet problem in domains with algebraic boundaries is far from developed and has an attractive array of important open problems, even in two dimensions—cf. [10, Ch. 18]. Even more basic open problems await an interested reader if one extends the search for singularities of solutions to the classical Dirichlet problem for the Laplacian to that for Helmholtz' equation. In other words, expanding the program to study possible singularities of the eigenfunctions of the Laplacian in domains with algebraic boundaries. The reigning open conjecture that ellipsoids are the ONLY domains for which *all* eigenfunctions are entire (and of exponential type) remains virtually untouched.

A fairly recent solution by P. Ebenfelt and D. Khavinson—cf. [10, Chs. 11, 12] of the problem of reflection of harmonic functions across analytic hypersurfaces in higher dimensions (or, why doesn't Schwarz reflection principle work in, say, $\mathbb{R}^3$?) opens up a new venue for investigations: the "antenna problem". In short, it is the question of the possibility of reflection from a point to a compact set vs. point-to-point reflection.

Once again, more important for applications is the reflection question for solutions of the Helmholtz equation, i.e., the eigenfunctions for the Laplacian. That playing field is widely open as well—cf. [10] and references therein.

Finally, we have only mentioned in passing the powerful methods of analytic continuations of solutions of linear analytic PDE combined with the modern techniques of several complex variables. The results culminate in Leray's theory of propagation of irregularities through $\mathbb{C}^n$. The underlying techniques based on the so-called method of "globalizing families" is both clear and quite powerful—cf. [10, Chs. 4–10, 19, 20] and references therein.

However, at present, the theory is more or less complete (we mean the **global** theory of propagation of singularities) mostly in two variables and also in $n \geq 3$-variables, but there exclusively for singularities initiated on quadratic surfaces [10, Ch. 19–20]. Once again, the importance of the remaining open problems is difficult to overestimate.

In conclusion, by this short survey, we wanted to demonstrate that the classical theme of analytic continuation of functions of one variable that has intrigued researchers for at least 200 years since the concept of an analytic function had come into focus, is alive, doing well and is quite rich with plenty of attractive and beautiful problems, conjectures, and attractive routes for further study. Thus, we hope that this small survey and the appended references will prompt the reader to invest time and effort in further research on these truly "eternal" topics.

Acknowledgments

The author gratefully acknowledges support from the Simons Foundation.

References

[1] Nachman Aronszajn, Thomas M. Creese, and Leonard J. Lipkin. *Polyharmonic functions*. Oxford Mathematical Monographs. The Clarendon

Press, Oxford University Press, New York, 1983. Notes taken by Eberhard Gerlach, Oxford Science Publications.

[2] Vazgain Avanissian. *Cellule d'harmonicité et prolongement analytique complexe*. Travaux en Cours. [Works in Progress]. Hermann, Paris, 1985.

[3] Ludwig Bieberbach. *Analytische Fortsetzung*. Ergebnisse der Mathematik und ihrer Grenzgebiete (N.F.), Heft 3. Springer-Verlag, Berlin-Göttingen-Heidelberg, 1955.

[4] P. Dienes. *The Taylor series: an introduction to the theory of functions of a complex variable*. Dover Publications, Inc., New York, 1957.

[5] P. Ebenfelt, D. Khavinson, and H. S. Shapiro. Extending solutions of holomorphic partial differential equations across real hypersurfaces. *J. London Math. Soc. (2)*, 57(2):411–432, 1998.

[6] Peter Ebenfelt, Dmitry Khavinson, and Harold S. Shapiro. Analytic continuation of Jacobi polynomial expansions. *Indag. Math. (N.S.)*, 8(1):19–31, 1997.

[7] E. L. Ince. *Ordinary Differential Equations*. Dover Publications, New York, 1944.

[8] D. Khavinson. A remark on a paper by T. Qian: "A holomorphic extension result" [Complex Variables Theory Appl. **32** (1997), no. 1, 59–77; MR1448480 (98f:42009)]. *Complex Variables Theory Appl.*, 32(4):341–343, 1997.

[9] Dmitry Khavinson and Erik Lundberg. A tale of ellipsoids in potential theory. *Notices Amer. Math. Soc.*, 61(2):148–156, 2014.

[10] Dmitry Khavinson and Erik Lundberg. *Linear holomorphic partial differential equations and classical potential theory*, volume 232 of *Mathematical Surveys and Monographs*. American Mathematical Society, Providence, RI, 2018.

[11] Dmitry Khavinson and Harold S. Shapiro. On a uniqueness property of harmonic functions. *Comput. Methods Funct. Theory*, 8(1-2):143–150, 2008.

[12] L. Kronecker. *Zur Theorie der Elimination einer Variabeln aus zwei algebraischen Gleichnungen*. Monatsber. König. Preuss. Akad. Wiss., Berlin, 1881.

[13] Zeev Nehari. On the singularities of Legendre expansions. *J. Rational Mech. Anal.*, 5:987–992, 1956.

[14] Tao Qian. A holomorphic extension result. *Complex Variables Theory Appl.*, 32(1):59–77, 1997.

[15] Erhard Schmidt. Bemerkung zur Potentialtheorie. *Math. Ann.*, 68(1):107–118, 1909.

[16] Harold S. Shapiro. *The Schwarz function and its generalization to higher dimensions*, volume 9 of *University of Arkansas Lecture Notes in the Mathematical Sciences*. John Wiley & Sons, Inc., New York, 1992. A Wiley-Interscience Publication.

[17] G. Szegö. On the singularities of zonal harmonic expansions. *J. Rational Mech. Anal.*, 3:561–564, 1954.

[18] R. Wavre. Sur l'identification des potentiels. *Vierteljschr. Naturforsch. Ges. Zürich*, 85(Beibl, Beiblatt (Festschrift Rudolf Fueter)):87–94, 1940.

9

Complex Convexity

Christer Oscar Kiselman

CONTENTS

DOI: 10.1201/9781315160658-9

9.1 Introduction to This Chapter

This chapter is devoted to complex convexity. Which are the most significant results presented here? This is the question that I answer in this first section of the chapter. But first I shall explain why real convexity is of interest, why complex convexity is important, and why mathematical morphology is a useful tool in the study of convexity.

What makes the approach in the present chapter different from other presentations of the subject? Also, this question will receive an answer.

9.1.1 Why is real convexity of interest?

A subset A of $\mathbf{R}^n$ is defined to be convex if for any pair $\{a, b\}$ of points in A the whole interval $[a, b]$ is also contained in A. A function $f \colon \mathbf{R}^n \to [-\infty, +\infty]$ is said to be convex if its finite epigraph

$$\mathbf{epi}^{\text{finite}}(f) = \{(x, t) \in \mathbf{R}^n \times \mathbf{R};\ f(x) \leqslant t\}, \tag{9.1}$$

is a convex set.

Convex functions possess a property of great importance in optimization theory: A local minimum of a convex function $\mathbf{R}^n \to \mathbf{R}$ is automatically a global minimum. In other words, if $f\colon \mathbf{R}^n \to \mathbf{R}$ is convex and $f(x) \geqslant c$ for all points x in a neighborhood of a, however small, then $f(x) \geqslant c$ for all x in $\mathbf{R}^n$.

Real convexity appears naturally also in complex analysis: the indicator function of the Fourier transform (defined in $\mathbf{C}^n$) of a function or distribution in $\mathbf{R}^n$ of compact support is convex.

We collect in Section 9.2 definitions and basic properties of convex sets and functions in vector spaces over the field of real numbers or over the field of complex numbers.

9.1.2   Why is complex convexity important?

In one complex variable, complex convexity of sets is not of great importance. Given any open set Ω of the complex plane $\mathbf{C}$ and a point p not belonging to Ω, we define a rational function $z \mapsto 1/(z - p)$ which cannot be extended as a holomorphic function across p.

But in two variables, the set of singularities of a rational function, indeed of any meromorphic function, cannot be just a singleton set. There are easy examples of two sets $\omega \subset \Omega$ such that any holomorphic function in ω can be extended to a holomorphic function in Ω.

This phenomenon gives rise to the concept of domains of holomorphy, which are domains such that there are holomorphic functions that cannot be continued to a larger domain, in a sense to be made precise. Related to these is the definition of a pseudoconvex domain. That a domain of holomorphy is pseudoconvex was proved by Eugenio Elia Levi (1883–1917); the converse, at the time an unsolved problem, came to be known as the Levi problem. It was solved by Kiyoshi Oka (1901–1978) in two variables, and later in any finite dimension by Oka, François Norguet (1929–2010) and Hans-Joachim Bremermann (1926–1996)—for a survey, see (Slatyer 2016).

So these phenomena point to the fact that there are great differences between the geometry of $\mathbf{C}$ and the geometry of $\mathbf{C}^2$. We can draw two-dimensional figures on a paper, and we can visualize objects in three dimensions. Nowadays there are even nice programs that create figures on the screen that can be rotated to exhibit all properties of an object in three-space. But two complex variables is a challenge because they correspond to four real coordinates.

Can you see in four dimensions? Yes, it is indeed possible to train one's inner eyes to see in four dimensions. A nontrivial but most rewarding sport. We can actually arrive at true stereoscopic vision ...   However, if you are not yet a master of four-dimensional landscapes, you will appreciate the Hartogs sets, named for Friedrich Moritz Hartogs (1874–1943), where we can be content with three real variables $(\operatorname{Re} z_1, \operatorname{Im} z_1, |z_2|)$ instead of the four $(\operatorname{Re} z_1, \operatorname{Im} z_1, \operatorname{Re} z_2, \operatorname{Im} z_2)$. An example is Figure 9.1 on page 281. To view

Reinhardt domains, named for Karl Reinhardt (1895–1941), we need only $(|z_1|, |z_2|) \in \mathbf{R}^2$.

9.1.3 Why is mathematical morphology a useful tool in the study of convexity?

Mathematical morphology can be superficially described as applied lattice theory. As such it is about the operations $(x, y) \mapsto x \wedge y = \min(x, y)$ and $(x, y) \mapsto x \vee y = \max(x, y)$ in an ordered set. These operations replace addition and multiplication in a ring, and an important example is the Boolean ring of all subsets of a given set, with $\wedge$ as intersection and $\vee$ as union.

In convexity theory, we see that the intersection of two convex sets is convex, while the union is, in general, not. But we still have a lattice, in that the convex hull of the union of two convex sets is the smallest convex set containing the two, and therefore is the supremum of the two. This can then be done analogously for convex functions, and, more generally for plurisubharmonic functions. It turns out that complete lattices are important; they are the ordered sets which allow infima and suprema also of infinite families.

Mathematical morphology provides us with important concepts in the theory of ordered sets that are helpful in understanding several related phenomena in mathematics. See Section 9.3 for more details.

9.1.4 Which are the most significant results reported in the present chapter?

An important observation is the non-local character of lineal convexity for general sets. As always, properties such that the local and global variants are different create difficulties—which may be challenging.

Because of the non-local character just mentioned, it is of importance to know that for bounded sets with a smooth boundary, the property of being locally lineally convex actually implies the global property. This is proved in Section 9.6.

9.1.5 What makes the approach in the present chapter different from other presentations of the subject?

Complex convexity is quite a well-studied field, and the ways to approach it are not many. However, we shall view convexity from the inside as well as from the outside, and this gives perhaps interesting perspectives. A set is concave if and only if its complement is convex, and the two notions should be studied together.

As mentioned, a subset A of $\mathbf{R}^n$ is defined to be convex if for any pair $\{a, b\}$ of points in A, the whole segment $[a, b]$ is also contained in A. This is what we can call convexity from the inside, i.e., looking at subsets of the given set. But we can also look at the set from the outside: If p does not belong to A and A is open or closed, then there is a half-space that contains A but not p. This is the Hahn–Banach theorem, of utmost important in convexity theory, both in finite dimension and infinite dimension. Explicitly, we say that a set in a vector space over **R** or **C** is *lineally concave* if it is a union of hyperplanes, and *lineally convex* if its complement is lineally concave. In one dimension, hyperplanes are just points, so every set is both lineally concave and lineally convex. In higher dimensions a convex set need not be lineally convex, but if it is open or closed, this is true. All this is true both in the real and the complex settings. For more details, see Section 9.4.

Acknowledgments

The topics presented in this chapter have evolved over a number of years—as is witnessed by the list of references. In particular, this is true of my own papers, and I would like to thank the persons who have been of great help during the writing of these.

I am grateful to Lê Hai Khôi for helpful discussions concerning the article resulting in Section 9.4, and equally grateful to Ragnar Sigurðsson for comments to an early version of Section 9.5.

Concerning Section 9.5, I am grateful to Yuriĭ Borisovič Zelinskij (Юрій Борисович Зелінський in Ukrainian, Юрий Борисович Зелинский in Russian) for several letters on this topic; to Peter Pflug for drawing my attention to a joint paper of his and Włodzimierz Zwonek, a paper which refers to the paper by Jim Agler and Nicholas John Young quoted in Subsections 9.2.13 and 9.5.2; to Jan Boman for pointing out an error in an earlier version; to Mats Andersson for helpful remarks on local weak lineal convexity; and to Nikolai Nikolov for drawing my attention to his thesis (2012).

Part of the work presented in Section 9.5, in particular Proposition 9.5.18, was done at the Institute for Mathematical Sciences (IMS) of the National University of Singapore (NUS) in November and December 2013. The author is most grateful for the invitation to IMS.

Section 9.7 is based on my lecture at Będlewo Palace on 2014 July 04. I am most grateful for the invitation to participate in this very successful conference on Constructive Approximation of Functions in honor of Wiesław Antoni Pleśniak.

With reference to Section 9.7, it has been a great experience to follow the work done on this topic by Mats Andersson, Mikael Passare (1959–2011), and Ragnar Sigurðsson, which led to their book, published in 2004. I am grateful to two anonymous referees for their corrections and advice concerning the article on which Section 9.7 is based.

As to Section 9.8, I am grateful to Lê Hai Khôi for many discussions concerning the topics presented in that section; thanks to him, the exposition could be considerably improved. My thanks go also to Stefan Halvarsson and Ragnar Sigurdsson for valuable help with the manuscript.

A special thank you to Erik Melin for drawing Figure 9.1 on page 281 and to Hania Uscka-Wehlou for drawing Figures 9.2 and 9.3 on page 306 and page 312, respectively.

9.2   Introduction to Convexity

9.2.1   Introduction to this section

The theory of convexity of sets and functions in vector spaces is a highly developed and very rich theory. We collect in this section definitions and basic properties of convex sets and functions in vector spaces over the field of real numbers or over the field of complex numbers. However, we present but a bare minimum of what is needed for the rest of this chapter. For fuller accounts, see the classical book by R. Tyrrell Rockafeller (1970, 1997), the books by Jean-Baptiste Hiriart-Urruty and Claude Maréchal (1993, 2002), and also my book manuscript (MS 2021).

9.2.2   Sets, mappings, and order relations

9.2.2.1   Notation for numbers, norms, and derivatives

We write $\mathbf{N} = \{0, 1, 2, \dots\}$ for the set of natural numbers, following Bourbaki (1963:67),[1] and $\mathbf{Z}$, $\mathbf{R}$, $\mathbf{C}$ for the ring of integers, the fields of real and complex numbers.

We shall use the l^p-norm $\|z\|_p = (\sum_j |z_j|^p)^{1/p}$, $1 \leqslant p < +\infty$, and the l^∞-norm $\|z\|_\infty = \sup_j |z_j|$ for $z \in \mathbf{C}^n$. When any norm can serve, we write only $\|z\|$.

The **bilinear inner product** of two vectors in $\mathbf{R}^m$ or $\mathbf{C}^n$ shall be denoted by a dot:

$$x{\cdot}y = x_1 y_1 + \cdots + x_m y_m; \quad z{\cdot}w = z_1 w_1 + \cdots + z_n w_n, \qquad x, y \in \mathbf{R}^m, z, w \in \mathbf{C}^n.$$

The Euclidean norm will be written like this:

$$\|x\|_2 = \sqrt{x \cdot x}; \qquad \|z\|_2 = \sqrt{z \cdot \bar{z}}, \qquad x \in \mathbf{R}^m, \quad z \in \mathbf{C}^n.$$

We shall denote by $B_<(c, r)$ and $B_\leqslant(c, r)$ the **open ball** and the **closed ball**, respectively, with center at $c \in \mathbf{C}^n$ and radius $r \in \mathbf{R}$ for any norm, thus

$$B_<(c, r) = \{z \in \mathbf{C}^n; \ \|z - c\| < r\} \text{ and } B_\leqslant(c, r) = \{z \in \mathbf{C}^n; \ \|z - c\| \leqslant r\}.$$

If $n = 1$, we shall write instead $D_<(c, r)$ and $D_\leqslant(c, r)$ for the disks.

The closure, interior and boundary of a subset A of a topological space will be denoted by $\overline{A}$, A° and ∂A, respectively. Thus $\overline{B_<(c, r)} = B_\leqslant(c, r)$ if r is positive, and $B_\leqslant(c, r)^\circ = B_<(c, r)$ for all real r.

For derivatives of functions we shall use the notation

$$f_{x_j} = \frac{\partial f}{\partial x_j}, \quad f_{y_j} = \frac{\partial f}{\partial y_j}, \quad f_{z_j} = \frac{\partial f}{\partial z_j} = \tfrac{1}{2}(f_{x_j} - \mathrm{i} f_{y_j}),$$

---

[1] Alfred Tarski (1956:121) calls 0 a natural number.

$$f_{\bar{z}_j} = \frac{\partial f}{\partial \bar{z}_j} = \tfrac{1}{2}(f_{x_j} + if_{y_j}), \quad f_{z_j \bar{z}_k} = \frac{\partial^2 f}{\partial z_j \partial \bar{z}_k}, \qquad j, k = 1, \ldots, n.$$

Differentials are written as

$$df = d'f + d''f, \text{ where } d'f = \sum f_{z_j} dz_j \text{ and } d''f = \sum f_{\bar{z}_k} d\bar{z}_k. \tag{9.2}$$

9.2.2.2   Counting with infinities

We shall also use a notation for the **extended real line**

$$\mathbf{R}_! = \mathbf{R} \cup \{-\infty, +\infty\} = [-\infty, +\infty],$$

thus creating the two-point compactification of $\mathbf{R}$ by adding two infinities, $+\infty$ and $-\infty = -(+\infty)$. We also add these infinities to the integers,

$$\mathbf{Z}_! = \mathbf{Z} \cup \{-\infty, +\infty\} = [-\infty, +\infty]_{\mathbf{Z}}.$$

Similarly we write $Y_! = Y \cup \{-\infty, +\infty\}$ for any subset Y of $\mathbf{R}$.

How shall we define a sum like $(+\infty) + (-\infty)$? Can addition $\mathbf{R} \times \mathbf{R} \ni (x, y) \mapsto x + y \in \mathbf{R}$ be extended to an operation $\mathbf{R}_! \times \mathbf{R}_! \ni (x, y) \mapsto x + y \in \mathbf{R}_!$ in a reasonable way? A convenient solution, pioneered by Jean-Jacques Moreau (1923–2014) in his paper (1970), is to define two extensions, **upper addition** and **lower addition**. The first is an upper semicontinuous mapping from $\mathbf{R}_! \times \mathbf{R}_!$ into $\mathbf{R}_!$; the second a lower semicontinuous mapping. They are denoted by $\dotplus$ and $+\!\!\cdot$ and are defined by the requirements of being commutative and to satisfy

$$\begin{aligned} x \dotplus (+\infty) &= +\infty \text{ for all } x \in \mathbf{R}_!; \\ x \dotplus (-\infty) &= -\infty \text{ for all } x \in [-\infty, +\infty[; \text{ and} \\ x + \!\!\cdot\, y &= -\big((-x) \dotplus (-y)\big) \text{ for all } x, y \in \mathbf{R}_!. \end{aligned} \tag{9.3}$$

When there are several terms we may use the summation symbol with a dot:

$$\sum_{j=1}^{m} {}^{\bullet} t_j = t_1 \dotplus \cdots \dotplus t_m, \qquad t_j \in \mathbf{R}_!. \tag{9.4}$$

A convenient rule is the following.

Lemma 9.2.1 *For any element $c \in \mathbf{R}_!$ and any function $f \colon X \to \mathbf{R}_!$ defined on an arbitrary set X we have*

$$\inf_{x \in X} \big(c \dotplus f(x) \big) = c \dotplus \inf_{x \in X} f(x). \tag{9.5}$$

Note that there are no exceptions to this formula.

Proof We just need to check all possibilities where our intuition is less reliable than usual, i.e., when $c = \pm\infty$ or X is empty.

We also note the equivalence

$$a \dot{+} b \geqslant c \Leftrightarrow a \geqslant c \dot{+} (-b), \qquad a, b, c \in \mathbf{R}_!. \tag{9.6}$$

9.2.2.3 Sets

The ***empty set***, the unique set with no elements, will be denoted by $\emptyset$. If A and B are sets such that every element of A belongs also to B, then A is said to be a ***subset*** of B and B a ***superset*** of A. This is written $A \subset B$ and $B \supset A$.

The family of all subsets of a set W is called the ***power set*** of W, and will denoted by $\mathscr{P}(W)$. Thus $A \in \mathscr{P}(W)$ if and only if $A \subset W$. We denote by $\mathscr{P}_{\text{finite}}(W)$ the family of all finite subsets of W.

We shall use the usual symbols for the ***intersection*** and ***union*** of a family $(A_j)_{j \in J}$ of sets:

$$\bigcap_{j \in J} A_j \text{ and } \bigcup_{j \in J} A_j, \qquad A_j \in \mathscr{P}(W). \tag{9.7}$$

When the index set J has only two elements, we write $A_1 \cap A_2$ and $A_1 \cup A_2$.

The set of all elements in A that are not elements of B is called the ***set-theoretical difference of A and B***, written $A \setminus B$. When A is equal to the whole set W we can write $\complement B$ for $W \setminus B$, the ***complement*** of a subset B of W.

The ***Cartesian product*** of two sets X and Y is the set of all pairs (x, y) with $x \in X$ and $y \in Y$. The Cartesian product of n sets X_j, $j = 1, \ldots, n$, denoted by

$$X_1 \times \cdots \times X_n = \prod_{j=1}^{n} X_j,$$

is the set of all n-tuples $(x_1, \ldots, x_n)$ with $x_j \in X_j$.

To any set $A \subset X$ we associate its ***characteristic function*** χ_A defined to take the value 1 in A and the value 0 in $X \setminus A$. We also define its ***indicator function*** indf_A, which takes the value 0 in A and the value $+\infty$ in its complement.

9.2.2.4 Graphs, epigraphs, and hypographs

Definition 9.2.2 *To any mapping $f \colon X \to Y$ we define its **graph**:*

$$\mathbf{graph}(f) = \{(x, y); \ f(x) = y\} \subset X \times Y,$$

a subset of the Cartesian product $X \times Y$.   □

Definition 9.2.3 *To any function $f\colon X \to \mathbf{R}_!$, we associate its **epigraph***

$$\mathbf{epi}(f) = \{(x,t) \in X \times \mathbf{R}_!;\ t \geqslant f(x)\}; \tag{9.8}$$

*and its **strict epigraph***

$$\mathbf{epi}_{\mathrm{strict}}(f) = \{(x,t) \in X \times \mathbf{R}_!;\ t > f(x)\}. \tag{9.9}$$

*We define its **finite epigraph** as*

$$\mathbf{epi}^{\mathrm{finite}}(f) = \{(x,t) \in X \times \mathbf{R};\ t \geqslant f(x)\}; \tag{9.10}$$

*and its **strict finite epigraph** as*

$$\mathbf{epi}^{\mathrm{finite}}_{\mathrm{strict}}(f) = \{(x,t) \in X \times \mathbf{R};\ t > f(x)\}. \tag{9.11}$$

The first two being subsets of the Cartesian product $X \times \mathbf{R}_!$ the following two of the product $X \times \mathbf{R}$. □

It is easy to pass from the finite epigraph to the strict finite epigraph as well as in the other direction:

$$\mathbf{epi}^{\mathrm{finite}}_{\mathrm{strict}}(f) = \bigcup_{c>0} \mathbf{epi}^{\mathrm{finite}}(f+c); \qquad \mathbf{epi}^{\mathrm{finite}}(f) = \bigcap_{c>0} \mathbf{epi}^{\mathrm{finite}}_{\mathrm{strict}}(f-c). \tag{9.12}$$

Similarly for $\mathbf{epi}_{\mathrm{strict}}$ and $\mathbf{epi}$.

The finite epigraph of a function $f\colon \mathbf{R}^n \to \mathbf{R}_!$ is contained in the closure (taken for the usual topology in $\mathbf{R}^n \times \mathbf{R}$) of $\mathbf{epi}^{\mathrm{finite}}_{\mathrm{strict}}(f)$, maybe strictly.

Definition 9.2.4 *Analogously we define the **hypograph** of a function:*

$$\mathbf{hypo}(f) = \{(x,t) \in X \times \mathbf{R}_!;\ t \leqslant f(x)\}, \tag{9.13}$$

*as well as the **strict hypograph**, $\mathbf{hypo}_{\mathrm{strict}}(f)$, the **finite hypograph**, $\mathbf{hypo}^{\mathrm{finite}}(f)$, and the **strict finite hypograph**, $\mathbf{hypo}^{\mathrm{finite}}_{\mathrm{strict}}(f)$.* □

It is often convenient to express properties of mappings in terms of their epigraphs or hypographs.

9.2.2.5 Inverse and direct images

To any mapping $f\colon X \to Y$ we define two mappings on a higher level,

$$f^*\colon \mathscr{P}(Y) \to \mathscr{P}(X) \text{ and } f_*\colon \mathscr{P}(X) \to \mathscr{P}(Y).$$

The first is defined by

$$f^*(B) = \{x \in X;\ f(x) \in B\}, \qquad B \in \mathscr{P}(Y). \tag{9.14}$$

Here $f^*(B)$ is called the **inverse image** of B. The second is defined by

$$f_*(A) = \{f(x);\ x \in A\}, \qquad A \in \mathscr{P}(X). \tag{9.15}$$

The set $f_*(A)$ is called the *(direct) image* of A. We write $\mathbf{im}(f)$, the *image of* f, for $f_*(X)$.

The mappings f^* and f_* are the simplest examples of the pullbacks and pushforwards used in differential geometry, and similarly in homology theory and distribution theory.

Definition 9.2.5 *To any function* $f\colon X \to \mathbf{R}_!$, *we associate its* ***effective domain***, *denoted by* $\mathbf{dom}(f)$ *and defined as the set where the function takes values less than* $+\infty$:

$$\mathbf{dom}(f) = \{x \in X;\ f(x) < +\infty\} = f^*([-\infty, +\infty[),$$

the inverse image of $\mathbf{R} \cup \{-\infty\}$. $\qquad\qquad\qquad\qquad\qquad\qquad \square$

9.2.3   Defining convex sets

It is most convenient to define convex functions with the help of convex sets. This also has the advantage that we can treat functions with infinite values without difficulty. That is why we start now with convex sets.

Definition 9.2.6 *A* ***rectilinear segment*** *in a vector space* E *is the set*

$$[a, b] = \{(1 - t)a + tb;\ t \in \mathbf{R},\ 0 \leqslant t \leqslant 1\},$$

where a *and* b *are its* ***endpoints***. *A subset* A *of a vector space is said to be* ***convex*** *if* $\{a, b\} \subset A$ *implies* $[a, b] \subset A$. $\qquad\qquad\qquad \square$

Every segment, every straight line, and all affine subspaces are convex sets.

It is worth noticing that to check convexity of a set, only its intersections with one-dimensional subspaces need to be considered. In other words, a set $A \subset E$ is convex if and only if the inverse image $f^*(A)$ is an interval (bounded or unbounded) in $\mathbf{R}$ for every mapping $f\colon \mathbf{R} \to E$ of the form $f(t) = ta + b$, $t \in \mathbf{R}$, $a, b \in E$.

A convex set A in $\mathbf{R}^n$ need not have interior points, but it does so if we consider it as a subset of the smallest affine space that contains it. We define the ***relative interior*** of A, denoted by $\mathbf{relint}(A)$, as the interior taken with respect to the topology in this affine subspace that is induced by the usual topology on $\mathbf{R}^n$. In this way every convex set, even if only a singleton set, has a nonempty relative interior. The set $\overline{A} \smallsetminus \mathbf{relint}(A)$ is the boundary of A taken in the smallest affine space containing A and is more interesting than the boundary of A if the space mentioned is not all of $\mathbf{R}^n$.

9.2.3.1   The convex hull

Definition 9.2.7 *Given a subset* A *of a vector space* E, *we define the* ***convex hull*** *of* A *as the smallest convex set containing* A. *It will be denoted by* $\mathbf{cvxh}(A)$. $\qquad\qquad\qquad\qquad\qquad\qquad\qquad\qquad\qquad\qquad\qquad \square$

The convex hull is well defined since any intersection of convex sets is convex.

9.2.4   The Hahn–Banach theorem

Among the affine subspaces we will pay attention to the **hyperplanes**, those of codimension 1, which means that they are defined by a single equation $\xi(x) = c$, where ξ is a nonzero linear form on the vector space. Other important sets are the **half-spaces**, which are defined by an inequality $\xi(x) \geqslant c$ or $\xi(x) > c$.

A **topological vector space** is a vector space E equipped with a topology such that both **addition**

$$E \times E \ni (x, y) \mapsto x + y \in E$$

and **multiplication by scalars**

$$\mathbf{C} \times E \ni (t, x) \mapsto tx \in E$$

are continuous.

After having restricted some variables, we see that all translations $x \mapsto x + a$ are continuous as well as all mappings $t \mapsto ta$ for a fixed a. The latter property implies that the inverse image of an open set in E under the mapping $t \mapsto ta$ is open for the usual topology in $\mathbf{C}$. Similarly of the field of scalars is $\mathbf{R}$.

Theorem 9.2.8 (The Hahn–Banach theorem) *Every open convex set in a topological vector space E is the intersection of a family of open half-spaces.*

Every closed convex set in a topological vector space is the intersection of a family of closed half-spaces—and also the intersection of a family of open half-spaces.

There are topological vector spaces with dual equal to zero. The separation can then seem like a paradox. It is resolved by the fact that in these vector spaces, the only open convex sets are the empty set and the whole space.

Here we shall accept this important theorem and refer to its proof in any of the many books on functional analysis.

Werner Fenchel (1952) characterized the sets that are intersections of a family of open half-spaces and called them **evenly convex**. As the Hahn–Banach theorem states, all open convex sets, and all closed convex sets are evenly convex. So are all strictly convex sets and a set like the closed triangle in $\mathbf{R}^2$ with one vertex removed defined by $0 < x_1 + x_2 \leqslant 1$, $x_1 \geqslant 0$, $x_2 \geqslant 0$. An open triangle with one boundary point added is not evenly convex. See Section 9.7, page 325 for the definition of refined half-spaces, which serve to represent general convex sets.

9.2.5   Supporting hyperplanes

Definition 9.2.9 *Given any set A in a topological vector space E, a **supporting hyperplane** is a hyperplane Y such that A is contained in one of the closed half-spaces defined by Y and such that the closure of A meets Y.* $\square$

Theorem 9.2.10 *Let A be a convex subset of a topological vector space E with $\emptyset \neq A \neq E$ and let a be any boundary point of A. Then there exists a supporting hyperplane of A passing through a.*

Proof In view of the Hahn–Banach theorem, there is a closed hyperplane Y that passes through a and such that A is in one of the closed half-spaces defined by Y.

A set A in a topological vector space is said to be ***bounded*** if for any neighborhood U of the origin there exists a number λ_0 such that λU contains A for all λ with $|\lambda| \geqslant \lambda_0$.

Corollary 9.2.11 *Let A be a bounded nonempty subset of a topological vector space and let a be any boundary point of $\mathbf{cvxh}(A)$. Then there exists a supporting hyperplane of A passing through a.*

Proof We apply the theorem to $\mathbf{cvxh}(A)$. A supporting hyperplane of $\mathbf{cvxh}(A)$ must also be a supporting hyperplane of A.

9.2.6 Defining convex functions

Definition 9.2.12 *A function $f\colon E \to \mathbf{R}_!$ defined in a vector space E is said to be **convex** if its finite epigraph is convex as a subset of $E \times \mathbf{R}$. We shall write $CVX(E, \mathbf{R}_!)$ for the set of these functions.* □

The functions that take one of the values $-\infty$, $+\infty$ identically are convex, since their finite epigraphs are, respectively, the whole space $E \times \mathbf{R}$ and the empty set. An affine function is convex, since its finite epigraph is a closed half-space.

It is easy to see that a function $f\colon E \to \mathbf{R}_!$ is convex if and only if it satisfies ***Jensen's inequality***

$$f((1-t)x + ty) \leqslant (1-t)f(x) \dotplus tf(y), \qquad x, y \in E, \quad 0 \leqslant t \leqslant 1. \quad (9.16)$$

Here we define $0 \cdot (+\infty) = 0 \cdot (-\infty) = 0$ (or else consider only $0 < t < 1$). There are more general inequalites to be described now.

9.2.7 Strict and strong convexity

Definition 9.2.13 *We shall say that a set A in a vector space is **strictly convex** if every supporting hyperplane cuts its closure in only one point. This means that the boundary of A does not contain any intervals of nonzero length. A function is said to be **strictly convex** if its finite epigraph is strictly convex.* □

Definition 9.2.14 *A function* $f\colon \mathbf{R}^n \to \mathbf{R}_!$ *is said to be **strongly convex** if for every point* $a \in E$ *there is a positive number* s *such that* $x \mapsto f(x) - s\|x\|_2^2$ *is convex in a neighborhood of* a *(we use a Euclidean norm here).*

A subset A *of finite-dimensional vector space* E *is said to be **strongly convex** if for every point* $a \in E$ *there is strongly convex function* f *so that* A *agrees with the set* $\{x \in \mathbf{R}^n;\ f(x_1,\ldots,x_{n-1}) < x_n\}$ *near* a *for some choice of coordinates in* E. $\square$

The function $\mathbf{R} \ni x \mapsto \sqrt{1+x^2}$ is strongly convex, but we see that the number s cannot be chosen independently of a.

The function $\mathbf{R} \ni x \mapsto x^4$ is strictly convex but not strongly convex.

9.2.8 The convex envelope

Definition 9.2.15 *Given a function* $f\colon A \to \mathbf{R}_!$, *where* A *is a subset of a vector space* E, *the largest convex function* $F\colon E \to \mathbf{R}_!$ *such that* $F\big|_A \leqslant f$ *is called the **convex envelope** of* f *and will be denoted by* $\mathbf{cvxe}(f)$. $\square$

Remark 9.2.16 *In general we have*

$$(\mathbf{cvxe}(f))(x) = \inf_{t \in \mathbf{R}} \big[t;\ (x,t) \in \mathbf{cvxh}(\mathbf{epi}^{\mathrm{finite}}(f))\big]. \tag{9.17}$$

The convex envelope of $f\colon E \to \mathbf{R}_!$ *evaluated at a point* x *is equal to the infimum of all expressions*

$$\sum_{j=1}^N \lambda_j f(x + a^{(j)}),$$

where the λ_j *and the* $a^{(j)}$ *satisfy* $\lambda_j \geqslant 0$, $\sum \lambda_j = 1$ *and* $\sum \lambda_j a^{(j)} = 0$. *We note that the points* $(x + a^{(j)}, f(x + a^{(j)}))$ *belong to* $\mathbf{epi}^{\mathrm{finite}}(f)$, *implying that the point*

$$\big(\sum \lambda_j (x + a^{(j)}), \sum \lambda_j f(x + a^{(j)})\big)$$

belongs to $\mathbf{cvxh}(\mathbf{epi}^{\mathrm{finite}}(f))$ *and that therefore* $(x, (\mathbf{cvxe}(f))(x) + t)$ *belongs to* $\mathbf{cvxh}(\mathbf{epi}^{\mathrm{finite}}(f))$ *for every positive number* t. $\square$

Remark 9.2.17 *We have inclusions*

$$\mathbf{epi}^{\mathrm{finite}}_{\mathrm{strict}}(\mathbf{cvxe}(f)) = \mathbf{cvxh}(\mathbf{epi}^{\mathrm{finite}}_{\mathrm{strict}}(f))$$
$$\subset \mathbf{cvxh}(\mathbf{epi}^{\mathrm{finite}}(f)) \subset \mathbf{epi}^{\mathrm{finite}}(\mathbf{cvxe}(f)). \tag{9.18}$$

The two inclusion relations here can be strict. $\square$

Definition 9.2.18 *Let* A *be any subset of a vector space* E *and let* $f\colon A \to \mathbf{R}_!$ *be any function defined on* A. *We shall say that* f *is **convex extensible** if it is the restriction to* A *of a convex function defined on all of* E, *i.e., if there is a convex function* $F\colon E \to \mathbf{R}_!$ *such that* $F\big|_A = f$. $\square$

There may exist more than one convex extension of a given function f. For example $F(x) = |x| - 1$ and $F^+(x) = \max(|x| - 1, 0)$, $x \in \mathbf{R}$, have the same restriction to $A = \mathbf{Z} \setminus \{0\}$. If there exists a convex extension, then $\mathbf{cvxe}(f)$ is the largest one.

9.2.9 Normed spaces

To any vector space E over the field of real numbers, we associate its *algebraic dual* $E^\star$, the vector space of all linear forms on E, i.e., the functions $\xi \colon E \to \mathbf{R}$ satisfying $\xi(x + ty) = \xi(x) + t\xi(y)$ for all $x, y \in E$ and all $t \in \mathbf{R}$.

We say that a function $E \ni x \mapsto \|x\| \in \mathbf{R}$ is a *norm* if $\|x\| \geqslant 0$ with equality if and only if $x = 0$; $\|x + y\| \leqslant \|x\| + \|y\|$ for all x and y; and finally $\|tx\| = t\|x\|$ for all $x \in E$ and all positive numbers t. This means that the function $E^2 \ni (x, y) \mapsto \|x - y\|$ is a metric with the extra property of being positively homogeneous. The subadditivity and the homogeneity together imply that $x \mapsto \|x\|$ is convex.

The space $\mathbf{R}^n$ of all n-tuples can be normed by the l^p *norm* $\|\cdot\|_p$, $1 \leqslant p \leqslant +\infty$, which is defined for $1 \leqslant p < +\infty$ by

$$\|x\|_p = \left(\sum |x_j|^p\right)^{1/p}, \qquad x = (x_1, \ldots, x_n) \in \mathbf{R}^n.$$

When $p = +\infty$ this has to be interpreted as a limit. More explicitly one defines

$$\|x\|_\infty = \max_j |x_j|, \qquad x \in \mathbf{R}^n.$$

In addition to the algebraic dual $E^\star$, which is defined for any vector space, we consider for any normed vector space E also its *dual space*, denoted by E' and consisting of all continuous linear forms on E. These are the linear mappings $\xi \colon E \to \mathbf{R}$ such that $|\xi(x)| \leqslant C\|x\|$ for some constant C. On the dual we define the norm *dual* to $\|\cdot\|$ by

$$\|\xi\|' = \sup_{\|x\| \leqslant 1} |\xi(x)|, \qquad \xi \in E'. \tag{9.19}$$

It follows that $|\xi(x)| \leqslant \|\xi\|' \cdot \|x\|$ for all $x \in E$ and all $\xi \in E'$.

When $E = \mathbf{R}^n$, we may identify both $E^\star$ and E' with $\mathbf{R}^n$, and the evaluation of ξ at the point x, i.e., the number $\xi(x)$, is then the *inner product*, defined by $\xi \cdot x = \xi_1 x_1 + \cdots + \xi_n x_n$. The Euclidean norm $\|\cdot\|_2$, defined by $\|x\|_2^2 = x \cdot x$, is dual to itself:

$$\|\xi\|_2' = \sup_{\|x\|_2 \leqslant 1} \xi(x) = \|\xi\|_2 = \sqrt{\sum \xi_j^2}.$$

It is not difficult to prove that the norm dual to $\|\cdot\|_1$ is $\|\cdot\|_\infty$ and vice versa. More generally, one can prove that the norm dual to $\|\cdot\|_p$ is $\|\cdot\|_q$, where $q = p/(p-1)$, $1 < p < +\infty$, with a natural interpretation also when $p = 1, +\infty$. This statement follows from Hölder's inequality and its converse.

A vector space provided with a Euclidean norm is called a *Euclidean space*.

9.2.10   Duality in convex analysis

By the term *duality* we aim at properties and results that involve a vector space and its dual. The most important examples are the support function and the Fenchel transformation, to be defined now.

9.2.10.1   The support function

Definition 9.2.19 *Given any subset A of a vector space E we define its **support function** H_A by*

$$H_A(\xi) = \sup_{x \in A} \xi(x), \qquad \xi \in E^\star.$$

Here $E^\star$ is the algebraic dual of E. ☐

Example 9.2.20 *If A is a ball, $A = B_{\leqslant}(c, r)$ with $r \geqslant 0$, or $B_<(c, r)$ with $r > 0$, then $H_A(\xi) = \xi(c) + r\|\xi\|'$, $\xi \in E^\star$, where $\|\cdot\|$ is an arbitrary norm in E and $\|\cdot\|'$ its dual norm, defined on $E^\star$ in (9.19) above.* ☐

9.2.10.2   The Fenchel transformation

Definition 9.2.21 *To any function $\varphi \colon E \to \mathbf{R}_!$ we define its **Fenchel transform** $\widetilde{\varphi}$ by*

$$\widetilde{\varphi}(\xi) = \sup_{x \in E} \big(\xi(x) - \varphi(x)\big),$$

defined for $\xi \in E^\star$, the algebraic dual of E. ☐

It follows that

$$\xi(x) \leqslant \varphi(x) \dotplus \widetilde{\varphi}(\xi), \quad x \in E, \quad \xi \in E^\star, \tag{9.20}$$

called ***Fenchel's inequality***.

It is evident that $\widetilde{\varphi}$ is the smallest function g such that $\xi(x) \leqslant \varphi(x) \dotplus g(\xi)$ holds.

For any family of functions $(\varphi_j)_{j \in J}$, $\varphi_j \in \mathscr{F}(E, \mathbf{R}_!)$, we clearly have

$$\sup_{j \in J} \widetilde{\varphi} = \Big(\inf_{j \in J} \varphi \Big)^{\sim}. \tag{9.21}$$

We see that the support function of a set is the Fenchel transform of its indicator function: $H_A = \widetilde{\mathbf{indf}}_A$.

Example 9.2.22 *If the graph of a function φ is a paraboloid, $\varphi(x) = a + \beta \cdot x + \frac{1}{2}c\|x\|_2^2$, $x \in \mathbf{R}^n$, where $a \in \mathbf{R}$, $\beta \in \mathbf{R}^n$ and $c > 0$, then the same is true of the graph of its transform: $\widetilde{\varphi}(\xi) = -a + \frac{1}{2}c^{-1}\|\xi - \beta\|_2^2$.* ☐

We now ask what happens if we apply the transformation again. We let Ξ be any nonempty subset of $E^\star$, and define the Fenchel transform of any function f defined on Ξ by

$$\widetilde{f}(x) = \sup_{\xi \in \Xi} \left(\xi(x) - f(\xi) \right), \qquad x \in E.$$

Here we can take $\Xi = \{0\}$ as well as $\Xi = E^\star$. In a normed space it is customary to take $\Xi = E'$, the dual of E.

The equivalence

$$\widetilde{\varphi} \leqslant f \text{ if and only if } \widetilde{f} \leqslant \varphi, \qquad \varphi \in \mathscr{F}(E, \mathbf{R}_!), \ f \in \mathscr{F}(\Xi, \mathbf{R}_!), \qquad (9.22)$$

follows easily from the definition.

We may form the second transform $\widetilde{\widetilde{\varphi}}$ of a function defined on E. The main result in the theory of the Fenchel transformation is the following.

Theorem 9.2.23 (Fenchel's theorem) *Let φ be a function defined on a vector space E and let Ξ be any nonempty subset of its algebraic dual $E^\star$. Then we always have $\widetilde{\widetilde{\varphi}} \leqslant \varphi$. Equality holds if and only if*

(A) *φ is convex;*

(B) *φ is lower semicontinuous for the weakest topology for which all linear forms in Ξ are continuous; and*

(C) *φ does not take the value $-\infty$ unless it is identically equal to $-\infty$.*

Property (B) here means that if $\varphi(a) > s$, then there are linear forms $\xi_1, \ldots, \xi_m$ in Ξ and a number $\theta > 0$ such that $\varphi(x) > s$ when $|\xi_j(x - a)| \leqslant \theta$, $j = 1, \ldots, m$. We shall denote this topology by $\sigma(E, \Xi)$. (We get the chaotic topology when $\Xi = \{0\}$, implying that the only lower semicontinuous functions are the constants.)

In $\mathbf{R}^n$ we usually choose $\Xi = \mathbf{R}^n$; the semicontinuity is then semicontinuity with respect to the usual topology of $\mathbf{R}^n$.

Before we go on, let us consider other expressions of this semicontinuity. The topology $\sigma(E, \Xi)$ on E gives rise to a topology in $E \times \mathbf{R}$, viz. the product topology, for which a basis for the neighborhoods of a point $(a, s) \in E \times \mathbf{R}$ are given by the sets

$$\{(x, t); \ |\xi_j(x - a)| < \theta, \ j = 1, \ldots, N, \ |t - s| < \theta\}, \qquad \xi_j \in \Xi, \quad \theta > 0.$$

Let us call this topology $\tau(E \times \mathbf{R}, \Xi)$.

So a function φ that satisfies the three conditions can be represented as the supremum of a family of affine functions $\varphi(x) = \sup_\xi \left(\xi(x) - \widetilde{\varphi}(\xi) \right)$. This can be most helpful in proving that certain functions are convex.

We accept Fenchel's Theorem without proof here.

It is now clear that Fenchel's inequality can be improved to

$$\xi(x) \leqslant \widetilde{\widetilde{\varphi}}(x) \dotplus \widetilde{\varphi}(\xi).$$

For Werner Fenchel's pioneering work on duality in convexity theory, see (1949, 1952, 1953, 1983).

9.2.11 Introduction to complex convexity

Every vector space over the field of complex numbers is at the same time a real vector space, obtained by simply restricting the multiplication by complex scalars to real scalars. So everything what we have said here about convexity applies also to complex vector spaces. But the presence of complex numbers gives birth to new phenomena. Let us list here several variants of complex convexity, in increasing order of strength. Most of them will be considered in detail in later sections of this chapter, some of them in other chapters.

1. An open set Ω in $\mathbf{C}^n$ is *pseudoconvex* if there is a continuous plurisubharmonic function defined in Ω that tends to $+\infty$ at the boundary of Ω. See, e.g., (Hörmander 1990: Theorem 2.6.7 and Definition 2.6.8.)

2. An open set Ω in $\mathbf{C}^n$ is a *domain of holomorphy* if there exists a holomorphic function defined in Ω that cannot be continued, in a precise sense, over the boundary of Ω. See, e.g., (Hörmander 1990: Definition 2.5.1).

3. A connected open set Ω in $\mathbf{C}^n$ is called *hyperconvex* if there is a continuos negative plurisubharmonic function u defined in Ω such that the sublevel set $\{z \in \Omega;\ u(z) \leqslant c\}$ is compact for every negative number c. See, e.g., (Kerzman & Rosay 1981).

There are several different notions of convexity related to lineal convexity. In increasing order of strength we have:

1. *Local weak lineal convexity in the sense of Yužakov & Krivokolesko*; see Definition 9.5.8 on page 303;

2. *Local weak lineal convexity*; see Definition 9.5.7 on page 302;

3. *Weak lineal convexity*, originally introduced as *Planarkonvexität* by Behnke & Peschl (1935:158, 162); see Definition 9.5.3 on page 301;

4. *Lineal convexity*, introduced as *convexité linéelle* by André Martineau (1966: 73; 1977:228); see Definition 9.5.1 on page 300;

5. **C**-*convexity*, originally introduced as *convexité linéelle forte* by Martineau (1967:400; 1968; 1977:265, 325).

   Hörmander (1994: Definition 4.6.6) defines an open susbset of $\mathbf{C}^n$ to be **C** *convex* if $\Omega \cap L$ is a connected and simply connected subset of L for every affine complex line L.

Andersson, Passare & Sigurdsson (2004: Definition 2.2.1) first defines a subset E of $\mathbf{P}$ to be **C-*convex*** if $E \neq \mathbf{P}$ and both E and its complement $\mathbf{P} \smallsetminus E$ are connected. A subset E of $\mathbf{P}^n$ is called **C-*convex*** if all its intersections with complex lines are **C**-convex.

6. Then we have the usual convex sets already discussed; see Definition 9.2.6) on page 256.

7. The strict convex sets; see Definition 9.2.13 on page 258.

8. The strong convex sets; see Definition 9.2.14 on page 259.

9.2.12   Notes on the history of the concepts discussed in this chapter

I learned about lineal convexity from André Martineau during the academic year 1967–1968 when I was in Nice with him. His premature death on 1972 May 04 was a great loss to world mathematics. He introduced also the notion of strong lineal convexity (1968), which, however, was not geometrically defined. Later Znamenskij (1979) found a geometric characterization; the property is now called **C**-convexity. Nowadays the most important sources for **C**-convexity are the book by Hörmander (1994) and the survey by Andersson, Passare & and Sigurdsson (2004). My earlier contributions to the field are to be found in (1978, 1996, 1997, 2016, 2019).

9.2.13   A note on terminology

Heinrich Behnke and Ernst Peschl (1935) introduced the notion which is now known as weak lineal convexity. They called it *Planarkonvexität*.

André Martineau used the terms *convexité linéelle* and *linéellement convexe*—see Martineau (1966:73) and (1968:427), reprinted in (Œuvres de André Martineau 1977:228) and (1977:323), respectively. In French there are two adjectives, *linéaire*, corresponding to the English *linear*; and *linéel*, which I rendered as *lineal*. (There is also an adjective *linéal*.) Martineau obviously wanted a distinctive term in order to signal the special meaning of his convexity, not to be misunderstood as ordinary convexity. Diederich & Fornæss (2003) and Diederich & Fischer (2006) write "lineally convex."

In Russian, the adjective *линейный* is most often used for both French terms *linéel* and *linéaire*, and this is the term used by Aĭzenberg, Krivokolesko, Yužakov, and others who write in Russian. In the translations into English of these Russian texts, there appears most often *linear convexity* and *linearly convex*.

Also Znamenskiĭ (1979:83; 1990:1037) used *линейный*, as did Znamenskiĭ & Znamenskaya (1996:359).

Later Znamenskiĭ (2001) used *линейчатый* (usually translated as 'ruled'; a common term is *линейчатая поверхность* 'ruled surface'). He thus established the distinction between *lineal, linéel* and *linear, linéaire* in Russian (Yuriĭ Zelinskij, personal communication 2013 March 26).

Hörmander (1994:290, Definition 4.6.1), Andersson, Passare & Sigurdsson (2004:16, Definition 2.1.2), and Jacquet (2008:8, Definition 2.1.2) used *linear* and *linearly* and thus did not keep the distinction introduced by Martineau. In my opinion, these authors unnecessarily copied the usage in the translations from Russian and did not pay attention to the pioneering work of Martineau. It should also be noted that the English *lineal* is actually the older of the two words, being attested since the fourteenth century, while *linear* is attested from 1706 (Webster 1983).

Another term is *hypoconvex*. The first appearance in this context[2] that I have found is in Helton & Marshall (1990:182), where it is used for sets with a boundary of class C^2 and has the meaning of 'strongly lineally convex' (satisfying the strict Behnke–Peschl differential condition as I called it in my paper (1998:3); later it was weakened to a synonym of *lineally convex* by Whittelsey (2000:678), and used in this sense by Agler & Young (2004:379). The term helpfully reminds us that it signifies a property weaker than convexity.

---

[2]Norberto Salinas (1976:144, 1979:327) used the term *hypoconvex* in a different sense.

9.3 Introduction to Mathematical Morphology

9.3.1 Introduction to this section

Lattice theory is a mature mathematical theory thanks to the pioneering work by Garrett Birkhoff (1911–1996), Øystein Ore (1899–1968), and others in the first half of the twentieth century. A standard reference is still Birkhoff's book (1995), first published in 1940.

Mathematical morphology is a branch of science that was created in the 1960s by Georges Matheron (1930–2000) and Jean Serra. It thrives in complete lattices.

Mathematical morphology can be described as lattice theory applied to several branches of science, in particular, to image analysis and image processing, where many concepts and procedures can be successfully described with concepts from mathematical morphology. In this section we give the most basic definitions and the simplest properties—those that will be useful in the coming parts of the chapter. For more complete treatments, see the books by Matheron and Serra mentioned in the list of references as well as my papers (2007, 2010) and my book manuscript (MS 2021).

With the arrival of tropical geometry, lattice theory can (now) be viewed as a tropicalization of other mathematical theories. Developments originate in several branches of mathematics, for instance algebra (Blyth & Janowitz 1972, Blyth 2005), logic (Stoltenberg-Hansen et al. 1994, Gierz et al. 2003), general topology and functional analysis (Gierz et al. 2003:xxx–xxxii), convexity theory (Singer 1997), and, for mathematical morphology with applications in image processing, books by Matheron (1975), Serra (1982), Serra, Ed. (1988), and Heijmans (1994); articles by Heijmans & Ronse (1990), Ronse (1990), Ronse & Heijmans (1991, 1998), Heijmans (1995), Serra (2006), and Ronse & Serra (2008, 2010). Other areas where concepts from lattice theory are used include semantics (abstract interpretation) of programming, the theory of fuzzy sets, fuzzy logic, and formal concept analysis (Ganter & Wille 1999). For general lattice theory a standard reference is Grätzer (1998).

This variety of sources for fundamental concepts has led to varying terminology and hence to difficulties in tracing history.

The two concepts of lattice and complete lattice must be carefully distinguished. This becomes obvious when we see that a complete lattice L can contain another complete lattice M with M as a sublattice of L ... but M is not a sub-complete-lattice of L (see Example 9.3.15).

A useful tool in the sequel will be generalized inverses and generalized quotients. They come in two versions, lower and upper.

To define an inverse of a general mapping seems to be a hopeless task. However, if the mapping is between preordered sets, there is some hope of constructing mappings that can serve in certain contexts just like inverses do.

There is an analogy between lattice theory and the theory of vector spaces. The theory of topological vector spaces was developed to a large extent because of the theory of distributions, which in turn was motivated by applications in partial differential equations. Developments in image processing motivated a renewed interest in lattice theory, in particular in complete lattices. Lattice theory was applied to switching circuits, and it was then enough, because of general finiteness conditions, to form models using lattices, but in image processing it is more convenient to assume completeness; for a motivation, see (Ronse 1990).

While vector spaces are useful in modelling linear problems, lattices seem to be more adapted to nonlinear problems. Auditory phenomena are often additive: all the instruments of an orchestra can be heard, while with visual phenomena this is not so: one object can block another from our view. This indicates that linear models may suffice for the first kind of phenomena (Fourier analysis and synthesis are successful for sound waves), while the visual ones are more in agreement with nonlinear operators, like maximum and minimum.

There are also analogies between topological spaces and preordered sets, in particular lattices. The continuous linear mappings in the first case correspond to increasing mappings in the second.

A comparison of the equations $a + x = b$ and $a \vee x = b$ shows that the second is more complicated than the first: The first has the unique solution $x = b - a$ for $a, b \in \mathbf{R}$, while the second has no solution if $a, b \in \mathbf{R}$ with $a > b$; a unique solution $x = b$ if $a < b$ and infinitely many solutions $x \leqslant a$ if $a = b$. In our schools we tend to prefer problems with a unique solution but in real life problems are more like $a \vee x = b$.

9.3.2 Preorders and orders

For the morphological operations on the computer screen, we need to consider families of sets. The family of all subsets of a given set is ordered by the inclusion relation, which is an example of an order relation. It is therefore convenient to introduce concepts that will be useful in the general theory of order relations. In this subsection we shall do so.

Definition 9.3.1 *A **preorder** in a given set X is a relation (a subset of $X \times X = X^2$) which is reflexive and transitive. A **preordered set** is a set together with a preorder.*

*An **order** is a preorder which is antisymmetric. An **ordered set** is a set together with an order.*

*Two elements x and y are said to be **comparable** if either $x \leqslant y$ or $y \leqslant x$.*

*An order is said to be **total** if any two elements are comparable.* ☐

The definition of preorder means, if we denote the relation by $\leqslant$, that

$$x \leqslant x, \qquad x \in X; \text{ and that} \tag{9.23}$$

$$x \leqslant y \text{ and } y \leqslant z \text{ implies } x \leqslant z, \qquad x, y, z \in X. \qquad (9.24)$$

The definition of order means that the relation shall in addition satisfy

$$x \leqslant y \ \& \ y \leqslant x \ \Rightarrow \ x = y, \qquad x, y \in X. \qquad (9.25)$$

We shall write $x < y$ if $x \leqslant y$ and $x \neq y$. We shall also write $x \geqslant y$ and $x > y$ for $y \leqslant x$ and $y < x$, respectively.

As already mentioned, a basic example of an ordered set is the power set of a set W with the order relation given by inclusion, thus $A \leqslant B$ being defined as $A \subset B$ for $A, B \in \mathscr{P}(W)$.

Suppose that we have two preorders defined in a set X; denote them by $\leqslant$ and $\preccurlyeq$. The preorder $\leqslant$ is said to be **finer** than the preorder $\preccurlyeq$, and $\preccurlyeq$ is said to be **coarser** than $\leqslant$, if $x \leqslant y$ implies $x \preccurlyeq y$ for all x, y.

There is a finest preorder in a set, viz. when we define $x \leqslant y$ to mean that $x = y$. This preorder is an order; let us call it the **discrete order**. There is also a coursest preorder in any set X, when we declare that $x \leqslant y$ for all $x, y \in X$. Let us call this the **chaotic preorder**. The set of all preorders on any set is thus an ordered set with a largest and a smallest element.

In a preordered set L, given $a, b \in L$, the **interval** $[a, b]$ is the set

$$\{x \in L; \ a \leqslant x \leqslant b\},$$

in particular to be used when $L = \mathbf{R}_!$. We shall write $[a, b]_{\mathbf{Z}}$ for an interval of integers, thus

$$[a, b]_{\mathbf{Z}} = [a, b] \cap \mathbf{Z} = \{x \in \mathbf{Z}; \ a \leqslant x \leqslant b\}, \qquad a, b \in \mathbf{Z}. \qquad (9.26)$$

Definition 9.3.2 *An **equivalence relation** is a preorder which is symmetric, i.e., such that $x \preccurlyeq y$ if and only if $y \preccurlyeq x$.* □

This means that $x \leqslant y$ implies $y \leqslant x$ for all $x, y \in X$.

9.3.3 Mappings between preordered sets

In preordered spaces the increasing mappings are of importance:

Definition 9.3.3 *If $f \colon X \to Y$ is a mapping from a preordered set X to a preordered set Y, then we say that f is **increasing** if*

for all $x, x' \in X$, the relation $x \leqslant_X x'$ implies $f(x) \leqslant_Y f(x')$.

*We shall write $\mathrm{incr}(X, Y)$ for the set of all increasing mappings $X \to Y$. We shall say that f is **decreasing** if*

the relation $x \leqslant_X x'$ implies $f(x) \geqslant_Y f(x')$

for all elements $x, x' \in X$. □

The increasing mappings play the same role in the context of ordered sets as the linear mappings in the theory of vector spaces and as the continuous mappings in the theory of topological spaces.

A preorder $\leqslant$ is finer than another preorder $\preccurlyeq$ if and only if the identity mapping $(X, \leqslant) \to (X, \preccurlyeq)$ is increasing.

If $f, g \colon X \to Y$ are increasing, then so are $f \wedge g$ and $f \vee g$. If also $h \colon Y \to Z$ is increasing, then so is $h \circ f \colon X \to Z$. In particular, when $X = Y$, we have the three operations $(f, g) \mapsto f \wedge g, f \vee g, g \circ f$, which all preserve the property of being increasing.

A comparison with topology is in order here. If $f \colon X \to Y$ is a mapping of a topological space X into a topological space Y with topologies (families of open sets) τ_X and τ_Y, we can define a new topology τ_f in X as the family of all sets

$$\{x \in X; \ f(x) \in V\}, \qquad V \in \tau_Y.$$

Then f is continuous if and only if τ_X is finer than τ_f.

For mappings $f \colon X \to X$ with target set equal to the domain we can form the iterations $f \circ f$, $f \circ f \circ f$ and so on, and among these mappings those that satisfy $f \circ f = f$ are of interest:

Definition 9.3.4 *We shall say that a mapping $f \colon X \to X$ is **idempotent** if $f \circ f = f$, i.e., if $f(f(x)) = f(x)$ for all $x \in X$. A mapping which is both increasing and idempotent will be called an **ethmomorphism**.* ☐

Definition 9.3.5 *We shall say that a mapping $f \colon X \to X$ is **extensive** if it is larger than the identity, i.e., $f(x) \geqslant x$ for all elements $x \in X$. We shall say that it is **antiextensive** if it is smaller than the identity, i.e., $f(x) \leqslant x$ for all $x \in X$.* ☐

We define the **invariance set of a function** $f \colon X \to X$ as the set of all $x \in X$ such that $f(x) = x$. We denote it by $\mathbf{invar}(f)$. For extensive mappings f the invariance set is decreasing in f, while it is increasing in f for antiextensive mappings.

9.3.4 Cleistomorphisms and anoiktomorphisms

Definition 9.3.6 *A **cleistomorphism** in an ordered set X is an ethmomorphism (see Definition 9.3.4) $\kappa \colon X \to X$ which is extensive (see Definition 9.3.5); in other words, which satisfies the following three conditions.*

$$x \leqslant x' \text{ implies } \kappa(x) \leqslant \kappa(x'); \qquad (9.27)$$

$$\kappa(\kappa(x)) = \kappa(x); \qquad (9.28)$$

$$x \leqslant \kappa(x), \qquad x \in X. \qquad (9.29)$$

for all elements $x, x' \in X$. ☐

The element $\kappa(x)$ is said to be the **closure** of x. Elements x such that $\kappa(x) = x$ are called **invariant** or **closed** (for this operator). An element is closed if and only if it is the closure of some element (and then it is the closure of itself).

In many applications the set X is the power set $\mathscr{P}(W)$ of some set W. Then the cleistomorphism is given as an intersection:

$$\overline{A} = \bigcap_Y \big(Y; Y \text{ is closed and } Y \supset A\big).$$

When W is a topological space, a basic example is the topological closure operator which associates to a set in a topological space its topological closure, i.e., the smallest closed set containing the given set, denoted by $A \mapsto \overline{A}$. In fact a cleistomorphism in $\mathscr{P}(W)$ defines a topology in W if and only if it satisfies, in addition to (9.27), (9.28), (9.29) above, two extra conditions, viz.

$$\overline{\varnothing} = \varnothing \text{ and } \overline{A \cup B} = \overline{A} \cup \overline{B} \text{ for all } A, B \subset W, \tag{9.30}$$

where $\varnothing$ denotes the empty set.

Another cleistomorphism of great importance is the operator which associates to a set A in $\mathbf{R}^n$ its convex hull, the smallest convex set containing the given set, denoted by $\mathbf{cvxh}(A)$. The composition $A \mapsto \overline{\mathbf{cvxh}(A)}$ is a cleistomorphism, whereas the composition in the other order, $A \mapsto \mathbf{cvxh}\big(\overline{A}\big)$ is not idempotent if $n \geqslant 2$. We see that the composition of two cleistomorphisms is sometimes, but not always, a cleistomorphism.

Dual to the concept of cleistomorphism is the concept of anoiktomorphism.

Definition 9.3.7 *An ethmomorphism $\alpha\colon X \to X$ is said to be an* **anoiktomorphism** *if it is antiextensive; in other words, if it satisfies the following three conditions.*

$$x \leqslant x' \text{ implies } \alpha(x) \leqslant \alpha(x');$$

$$\alpha \circ \alpha = \alpha;$$

$$\alpha(x) \leqslant x,$$

for all elements $x, x' \in X$. $\qquad\square$

The composition of a cleistomorphism and an anoiktomorphism is always idempotent:

Proposition 9.3.8 *Let $\alpha, \kappa\colon L \to L$ be an anoiktomorphism and a cleistomorphism. Then $\eta = \alpha \circ \kappa$ and $\theta = \kappa \circ \alpha$ are ethmomorphisms.*

Proof That η and θ are increasing is obvious. Since κ is extensive, we get

$$\eta \circ \eta = \alpha \circ \kappa \circ \alpha \circ \kappa \geqslant \alpha \circ \alpha \circ \kappa = \alpha \circ \kappa = \eta.$$

Since α is antiextensive, we get

$$\eta \circ \eta = \alpha \circ \kappa \circ \alpha \circ \kappa \leqslant \alpha \circ \kappa \circ \kappa = \alpha \circ \kappa = \eta,$$

so η is idempotent. The proof for θ is similar.

Example 9.3.9 *A typical example is when we take α and κ as the operations of taking the interior and the topological closure in a topological space, respectively; $\alpha(A) = A^\circ$, $\kappa(A) = \overline{A}$. Then a fixed point of the composition $\alpha \circ \kappa$ is called a **regular open set** and a fixed point of $\kappa \circ \alpha$ is called a **regular closed set**. These operations are neither extensive nor antiextensive in general.* □

Proposition 9.3.10 *The infimum of a family of cleistomorphisms in a complete lattice L is a cleistomorphism. The supremum of any family of anoiktomorphisms in L is an anoiktomorphism.*

Proof Let κ_j, $j \in J$, be cleistomorphisms and define $\kappa = \bigwedge_{j \in J} \kappa_j$, meaning that the value of κ at $x \in L$ equals $\bigwedge_{j \in J} \kappa_j(x)$. Clearly κ is increasing and larger than the identity. It follows that $\kappa \circ \kappa \geqslant \kappa$. To prove the opposite inequality, we note that $\kappa \circ \kappa \leqslant \kappa_j \circ \kappa_j = \kappa_j$. Taking the infimum over all j we get what we want.

The result for anoiktomorphisms follows by duality.

9.3.5 Lattices and complete lattices

Let L be an ordered set and A a subset of L. An element $b \in L$ is said to be the **infimum** of all elements $a \in A$ if b is the largest minorant of all $a \in A$. This means that $b \leqslant a$ for all elements $a \in A$, and that if $b' \leqslant a$ for all $a \in A$, then $b' \leqslant b$. The infimum, if it exists, is necessarily unique. The infimum of the empty set exists if and only if L possesses a largest element, and if so, the infimum is this largest element.

We shall write

$$b = \inf_{a \in A} a = \inf(a;\ a \in A) = \bigwedge_{a \in A} a$$

for the infimum of all elements in A; if A has only n elements we write $b = a_1 \wedge \cdots \wedge a_n$. If the infimum belongs to A, we call it a **minimum**. As an example, the set of all positive real numbers has 0 as its infimum, but 0 is not a positive number and therefore not a minimum.

Similarly we define the **supremum**

$$c = \sup_{a \in A} a = \sup(a;\ a \in A) = \bigvee_{a \in A} a$$

as the smallest majorant of all elements in A. If the supremum belongs to A, we call it a **maximum**. The supremum of all elements in the empty set, $\sup_{x \in \varnothing} x$, exists if and only if L has a smallest element.

Definition 9.3.11 *Let L be a nonempty set. If any subset consisting of two elements in L has an infimum, we shall call L an **inf-semilattice**; similarly, if any two-set of L has a supremum, we shall call L a **sup-semilattice**. If L is both an inf-semilattice and a sup-semilattice we shall call L a **lattice**.* □

Definition 9.3.12 *If any nonempty subset (finite or infinite) of a nonempty set L has an infimum, L will be said to be a* **complete inf-semilattice;** *analogously we define* **complete sup-semilattice** *and* **complete lattice.** □

We may denote the smallest element in a complete lattice by **0** and the largest by **1**. We have

$$\sup_{x \in L} x = \inf_{x \in \varnothing} x = \mathbf{1}; \quad \inf_{x \in L} x = \sup_{x \in \varnothing} x = \mathbf{0};$$

the infimum of the empty set exists and is **1**, and the supremum of the empty set is **0**.

A ***sublattice*** is defined just like a subgroup with respect to the operations $\wedge$ and $\vee$: that M is a sublattice of L means that for all $x, y \in M$, $x \wedge y$ and $x \vee y$, when calculated in L, are elements of M. A sublattice is therefore something more than a subset with the induced order; see the following examples.

Example 9.3.13 *The space of real-valued continuous functions on a topological space is a lattice with the usual order: $f \leqslant g$ if and only if $f(x) \leqslant g(x)$ for all x. The space $C^1(\mathbf{R}^n, \mathbf{R})$ of real-valued continuously differentiable functions on $\mathbf{R}^n$ is not a sublattice of $C(\mathbf{R}^n, \mathbf{R})$ if $n \geqslant 1$. It is not even a lattice on its own. (The functions $\mathbf{R} \ni x \mapsto \sqrt{t^2 + x^2}$, $t > 0$, converge to $x \mapsto |x|$ as $t \to 0$, but there is no infimum in $C^1(\mathbf{R}, \mathbf{R})$.)* □

Example 9.3.14 *The family $\mathscr{P}(W)$ of all subsets of a set W is a complete lattice, with $\bigwedge A_j = \bigcap A_j$ and $\bigvee A_j = \bigcup A_j$. The compact sets in $\mathbf{R}^n$ form a sublattice $\mathscr{K}(\mathbf{R}^n)$ of $\mathscr{P}(\mathbf{R}^n)$. This lattice is a complete inf-semilattice but not a complete sup-semilattice. The family $\mathscr{K}_{cvx}(\mathbf{R}^n)$, $n \geqslant 1$, of all convex compact sets is a lattice but not a sublattice of $\mathscr{K}(\mathbf{R}^n)$: the supremum of two convex compact sets is not always the same in the two lattices.* □

Example 9.3.15 *The family of all closed sets in $\mathbf{R}^n$, denoted by $\mathscr{C}(\mathbf{R}^n)$, is a sublattice of $\mathscr{P}(\mathbf{R}^n)$: the union and intersection of two closed sets are closed. But, although $\mathscr{C}(\mathbf{R}^n)$ is a complete lattice, it is not a sub-complete-lattice of the complete lattice $\mathscr{P}(\mathbf{R}^n)$ when $n \geqslant 1$. The union of a family of closed sets is not always closed, but there is a supremum, viz. the closure of the union. Thus, finite suprema agree with those in $\mathscr{P}(\mathbf{R}^n)$ while infinite suprema do not.* □

Example 9.3.16 *The set $\mathscr{F}(\mathbf{R}^n, \mathbf{R}_!)$ of all functions defined on $\mathbf{R}^n$ and with values in the extended real line $\mathbf{R}_!$ is a lattice under the usual order for real numbers, extended in an obvious way to the two infinities. The subset of all convex functions is ordered in the same way and is also a lattice under this order. However, the convex functions $CVX(\mathbf{R}^n, \mathbf{R}_!)$ do not form a sublattice of $\mathscr{F}(\mathbf{R}^n, \mathbf{R}_!)$ if $n \geqslant 1$. The supremum of two convex functions is equal to the pointwise supremum of them:*

$$(f \vee g)(x) = f(x) \vee g(x) = \max(f(x), g(x)),$$

but the infima are different in the two lattices: the infimum in the lattice of convex functions is

$$f \wedge_{cvx} g = \sup \left(h \in CVX(\mathbf{R}^n, \mathbf{R}_!); \ h \leqslant f, g \right) \leqslant f \wedge g = \min(f, g),$$

where the supremum of all $h \leqslant f, g$ is calculated in $\mathscr{F}(\mathbf{R}^n, \mathbf{R}_!)$ and has a sense because that lattice is complete. That the two infima may be different is shown by easy examples like $f(x) = e^x$, $g(x) = e^{-x}$, $x \in \mathbf{R}$. Here $\min(f, g)(x) = e^{-|x|}$, while $f \wedge_{cvx} g = 0$. □

9.3.5.1 Dilations and erosions in complete lattices

Mappings of the form $\mathscr{P}(\mathbf{R}^n) \ni A \mapsto A + B \in \mathscr{P}(\mathbf{R}^n)$ with a fixed set B are called dilations. It can be proved that a mapping $\mathscr{P}(\mathbf{R}^n) \to \mathscr{P}(\mathbf{R}^n)$ which commutes with translations and the formation of infinite unions is necessarily of this form. In lattice theory it is therefore natural to take the latter property as a definition:

Definition 9.3.17 *We say that a mapping $\delta \colon L \to M$, where L and M are complete lattices, is a **dilation** if it commutes with the formation of suprema, i.e.,*

$$\delta \left(\bigvee_{x \in A} x \right) = \bigvee_{x \in A} \delta(x)$$

for all subsets A of L. □

In particular we get $\delta(\mathbf{0}_L) = \mathbf{0}_M$ (take A empty), while

$$\delta(\mathbf{1}_L) = \bigvee_{x \in L} \delta(x) \leqslant \mathbf{1}_M. \tag{9.31}$$

Definition 9.3.18 *Similarly we shall say that ε is an **erosion** if it commutes with the formation of infinite infima,*

$$\varepsilon \left(\bigwedge_{x \in A} x \right) = \bigwedge_{x \in A} \varepsilon(x)$$

for all subsets A of L. □

We note that $\varepsilon(\mathbf{1}_L) = \mathbf{1}_M$ (take A empty), while $\varepsilon(\mathbf{0}_L) = \bigwedge_{x \in L} \varepsilon(x) \geqslant \mathbf{0}_M$.

Dilations and erosions are always increasing. Indeed, we have $\delta(x \vee y) = \delta(x) \vee \delta(y)$. If $x \leqslant_L y$, this equation simplifies to $\delta(y) = \delta(x) \vee \delta(y) \geqslant_M \delta(x)$, which shows that δ is increasing, $\delta \in \mathbf{incr}(L, M)$. A similar argument shows that erosions are increasing.

In $\mathbf{R}^n$ or $\mathbf{C}^n$ we define $\delta_B, \varepsilon_B \colon \mathscr{P}(G) \to \mathscr{P}(G)$ by $\delta_B(A) = A + B$ and $\varepsilon_B(A) = \{x \in G; \ x + B \subset A\}$. It is easily seen that $\delta_B(A) \subset C$ if and only if $A \subset \varepsilon_B(C)$. In a lattice this may be written as $\delta(x) \leqslant y$ iff $x \leqslant \varepsilon(y)$, equivalently as $\mathbf{epi}\,\delta = (\mathbf{hypo}\,\varepsilon)^{\vee}$, where the symbol $^{\vee}$ means that we swap the components: for a subset A of a Cartesian product $X \times Y$ we define

$$A^{\vee} = \{(y,x); \ (x,y) \in A\} \subset Y \times X. \tag{9.32}$$

May we use this as a model to define erosions from dilations and conversely in the more general lattice situation? Indeed this is the case, and we shall do so in the next subsection.

Example 9.3.19 *The mapping* $f^*\colon \mathscr{P}(Y) \to \mathscr{P}(X)$, *defined by* (9.14), *is both a dilation and an erosion, while* $f_*\colon \mathscr{P}(X) \to \mathscr{P}(Y)$, *defined in* (9.15), *is a dilation but in general not an erosion.* $\square$

9.3.6   Inverses and quotients of mappings

9.3.6.1   Introduction

In this subsection we shall study inverses and quotients of mappings between ordered sets which are analogous to inverses $1/y$ and quotients x/y of positive numbers. The theory of lower and upper inverses defined in Subsubsection 9.3.6.2 generalizes the theory of Galois connections as well as residuation theory and the theory of adjunctions. An interesting question is to what extent a generalized inverse can serve as a left inverse, as a right inverse, and how an inverse of an inverse relates to the identity mapping. These inverses and quotients can be used to create a convenient formalism for a unified treatment of dilations $\delta\colon L \to M$ and erosions $\varepsilon\colon M \to L$ as well as of cleistomorphisms $\kappa = \varepsilon \circ \delta\colon L \to L$ and anoiktomorphisms $\alpha = \delta \circ \varepsilon\colon M \to M$.

In general a mapping $g\colon X \to Y$ between sets does not have an inverse. If g is injective, we may define a left inverse $u\colon Y \to X$, thus with $u \circ g = \mathbf{id}_X$, where $\mathbf{id}_X$ denotes the identity mapping in X, defining $u(y)$ in an arbitrary way when y is not in the image of g. If g is surjective, we may define a right inverse $v\colon Y \to X$, thus with $g \circ v = \mathbf{id}_Y$. We then need to define $v(y)$ as an element of the preimage $\{x; \ g(x) = y\}$. In the general situation this has to be done using the axiom of choice. In a complete lattice, however, it could be interesting to define $v(y)$ as the supremum or infimum of all points x such that $g(x) = y$, even though this supremum or infimum need not belong to the set of points that are mapped to y. At any rate, the preimage of y is contained in the interval defined by the infimum and the supremum. However, for various reasons it is convenient to take instead the infimum of all x such that $g(x) \geqslant y$ or the supremum of all x such that $g(x) \leqslant y$. This yields better monotonicity properties. (The case $g(x) = y$ is allowed, since we can take the preorder in Y be the discrete order.)

If, given a mapping $g\colon L \to M$ from an ordered set L into an ordered set M, we can find mappings $u, v\colon M \to L$ such that $\mathbf{hypo}\,u = (\mathbf{epi}\,g)^{\vee}$ or $\mathbf{epi}\,v = (\mathbf{hypo}\,g)^{\vee}$, we would be content to have some kinds of inverses to g. However, usually the best we can do is to study mappings satisfying either

hypo $u \supset (\mathbf{epi}\, g)^{\vee}$ or **epi** $v \supset (\mathbf{hypo}\, g)^{\vee}$. This will be the approach in what follows, where we shall define not one but two inverses, viz. the lower (to be denoted by $g_{[-1]}$) and the upper (written as $g^{[-1]}$).

9.3.6.2 Defining inverses of mappings

Definition 9.3.20 *Let L be a complete lattice, M a preordered set, and $g\colon L \to M$ any mapping. We then define the **lower inverse** $g_{[-1]}\colon M \to L$ and the **upper inverse** $g^{[-1]}\colon M \to L$ as the mappings*

$$g_{[-1]}(y) = \bigvee_{x \in L} (x;\ g(x) \leqslant_M y)) = \bigvee_{x \in L} (x;\ (x,y) \in \mathbf{epi}\, g); \tag{9.33}$$

$$g^{[-1]}(y) = \bigwedge_{x \in L} (x;\ g(x) \geqslant_M y)) = \bigwedge_{x \in L} (x;\ (x,y) \in \mathbf{hypo}\, g), \tag{9.34}$$

where $y \in M$. ☐

As a first observation, let us note that these inverses are always increasing. If there exists a largest element $\mathbf{1}_M$, then $g_{[-1]}(\mathbf{1}_M) = \mathbf{1}_L$. Similarly, if M possesses a smallest element $\mathbf{0}_M$, then $g^{[-1]}(\mathbf{0}_M) = \mathbf{0}_L$. If M has the chaotic preorder, then both inverses are constant, $g_{[-1]} = \mathbf{1}_L$ and $g^{[-1]} = \mathbf{0}_L$ identically. Here the lower inverse is larger than the upper inverse.

Example 9.3.21 *For any mapping $f\colon X \to Y$ we defined in (9.14) and (9.15) the mappings $f^*\colon \mathscr{P}(Y) \to \mathscr{P}(X)$ and $f_*\colon \mathscr{P}(X) \to \mathscr{P}(Y)$. We note that, for all $A \in \mathscr{P}(X)$ and all $B \subset \mathscr{P}(Y)$ we have $f_*(A) \subset B$ if and only if $A \subset f^*(B)$. It follows that $(f^*)^{[-1]} = f_*$ and $(f_*)_{[-1]} = f^*$. From this we see that $((f^*)^{[-1]})_{[-1]} = f^*$ and that $((f_*)_{[-1]})^{[-1]} = f_*$.* ☐

9.3.6.3 Properties of inverses

We note that we always have

$$\mathbf{epi}\, g \subset (\mathbf{hypo}\, g_{[-1]})^{\vee}, \tag{9.35}$$

in other words, if $y \geqslant g(x)$, then $x \leqslant g_{[-1]}(y)$; and

$$\mathbf{hypo}\, g \subset (\mathbf{epi}\, g^{[-1]})^{\vee}, \tag{9.36}$$

in other words, if $y \leqslant g(x)$, then $x \geqslant g^{[-1]}(y)$. Here $R^{\vee}$ for a subset R of $X \times Y$ is defined by (9.32). In general, these inclusions are strict.

Example 9.3.22 *If $f\colon \mathscr{P}(G) \to \mathscr{P}(G)$ is the dilation $f(A) = \delta_U(A) = A + U$, $A \in \mathscr{P}(G)$, where G is an abelian group and U a fixed subset of G, called the **structuring element**, then*

$$(\delta_U)_{[-1]}(C) = \bigcup_{A \in \mathscr{P}(G)} (A;\ \delta_U(A) \subset C) = \varepsilon_U(C), \qquad C \in \mathscr{P}(G),$$

the erosion associated to δ_U. A most fundamental example.

The compositions $\alpha_U = \delta_U \circ \varepsilon_U$ and $\kappa_U = \varepsilon_U \circ \delta_U$ are the anoiktomorphism and the cleistomorphism associated to δ_U, respectively.

When $X = Y = \mathbf{R}^n$, the upper inverse of δ_U is not interesting, since

$$(\delta_U)^{[-1]}(C) = \varnothing$$

for all C if U has interior points, and equal to C if $U = \{0\}$. □

An ideal inverse u would satisfy $u \circ g = \mathrm{id}_L$, $g \circ u = \mathrm{id}_M$, and the inverse of u would be g. It is therefore natural to compare $g_{[-1]} \circ g$ and $g^{[-1]} \circ g$ with id_L; $g \circ g_{[-1]}$ and $g \circ g^{[-1]}$ with id_M; and inverses of inverses of g with g. We shall not do so here but refer the reader to the book manuscript (MS 2021).

9.3.6.4 Quotients of mappings

We shall now generalize the definitions of upper and lower inverses.

Definition 9.3.23 *Let a set X, a complete lattice M, and a preordered set Y, as well as two mappings $f \colon X \to M$ and $g \colon X \to Y$ be given. We define two mappings $f /_\star g, f /^\star g \colon Y \to M$ by*

$$(f /_\star g)(y) = \bigvee_{x \in X} (f(x); \; g(x) \leqslant_Y y), \qquad y \in Y;$$

$$(f /^\star g)(y) = \bigwedge_{x \in X} (f(x); \; g(x) \geqslant_Y y), \qquad y \in Y.$$

*We shall call them the **lower quotient** and the **upper quotient** of f and g.* □

$$
\begin{array}{ccc}
 & X & \\
{}^{g}\swarrow & & \searrow^{f} \\
Y & \longrightarrow & M \\
 & {\scriptstyle f /_\star g,\, f /^\star g} &
\end{array}
$$

We shall often assume that X, M and Y are all complete lattices, but this is not necessary for the definitions to make sense.

The mappings $f /_\star g, f /^\star g \in \mathscr{F}(Y, M)$ are always increasing.

The quotients $f /_\star g$ and $f /^\star g$ increase when f increases, and they decrease when g increases—just as with the division of positive numbers:

If $f_1 \leqslant_M f_2$ and $g_1 \geqslant_Y g_2$, then $f_1 /_\star g_1 \leqslant_M f_2 /_\star g_2$ and $f_1 /^\star g_1 \leqslant_M f_2 /^\star g_2$.

If $g(x) \leqslant_Y y$, then $f(x) \leqslant_M (f /_\star g)(y)$; if $g(x) \geqslant_Y y$, then $f(x) \geqslant_M (f /^\star g)(y)$. In particular,

if $g(x) = y$, then $(f /^\star g)(y) \leqslant_M f(x) \leqslant_M (f /_\star g)(y)$.

We note some special cases.

(1). If we specialize the definitions to the situation when $X = M$ and $f = \mathrm{id}_X$, then $f /_\star g = \mathrm{id}_X /_\star g = g_{[-1]}$ and $f /^\star g = \mathrm{id}_X /^\star g = g^{[-1]}$; cf. Definition 9.3.20. So inverses are quotients.

(2). A second special case is this: Taking $Y = M$ and $g = f$ in the definition we see that, for all mappings $f \colon X \to M$ we have

$$f /_\star f \leqslant \mathrm{id}_M \leqslant f /^\star f; \tag{9.37}$$

$$(f /_\star f) \circ f = f = (f /^\star f) \circ f. \tag{9.38}$$

9.3.7 Set-theoretical representation of dilations, erosions, cleistomorphisms, and anoiktomorphisms

We present here an easy result for translation-invariant operators on the family of subsets of an abelian group, putting several operations under a common roof.

Proposition 9.3.24 *Let S be a subset of an abelian group G. Then the dilation, erosion, cleistomorphism and anoiktomorphism with structuring element S can all be written in the form*

$$\varphi_{S,T,U}(A) = \bigcup_{x \in G} (x + S; \ x + T \subset A + U), \qquad A \in \mathscr{P}(G), \tag{9.39}$$

for special choices of the structuring elements S, T, U, viz.

$$\delta_S = \varphi_{S,\{0\},\{0\}}, \ \varepsilon_S = \varphi_{\{0\},S,\{0\}}, \ \alpha_S = \varphi_{S,S,\{0\}}, \ \kappa_S = \varphi_{\{0\},S,S}. \tag{9.40}$$

We can also write the mappings as

$$\varphi = (f /_\star g) \circ h,$$

where

$$f(B) = B \ \text{or} \ B + S, \ g(B) = B + S, \ h(A) = A \ \text{or} \ A + S,$$

for $A, B \in \mathscr{P}(G)$.

Proof The dilation $\delta = \delta_S$, the erosion $\varepsilon = \delta_{[-1]}$, the cleistomorphism $\kappa = \varepsilon \circ \delta$, and the anoiktomorphism $\alpha = \delta \circ \varepsilon$ can be written

$$\delta(A) = \bigcup_{B \in \mathscr{P}(G)} (B + S; \ B + S \subset A + S),$$

$$\varepsilon(A) = \bigcup_{B \in \mathscr{P}(G)} (B; \ B + S \subset A),$$

$$\kappa(A) = \bigcup_{B \in \mathscr{P}(G)} (B; \ B + S \subset A + S),$$

$$\alpha(A) = \bigcup_{B \in \mathscr{P}(G)} (B + S; \ B + S \subset A).$$

We now let

$f(B) = B + S$, $h(A) = A + S$ in the first case;

$f(B) = B$, $h(A) = A$ in the second case;

$f(B) = B$, $h(A) = A + S$ in the third case; and

$f(B) = B + S$, $h(A) = A$ in the fourth case;

while $g(B) = B + S$ in all four cases.

We can think of the points as atoms and the sets $x + S$ as molecules. Then $\delta(A)$ and $\alpha(A)$ consists of molecules, the latter of those that are contained in A; whereas $\varepsilon(A)$ and $\kappa(A)$ consists of centers of molecules (which makes sense if the structuring set S is symmetric).

9.4 Lineally Convex Hartogs Domains

Abstract of this section
We study lineally convex domains of a special type, viz. Hartogs domains, and prove that such sets can be characterized by local conditions if they are smoothly bounded.

9.4.1 Introduction to the present section

Lineal convexity is a kind of complex convexity intermediate between usual convexity and pseudoconvexity. More precisely, if A is a convex set which is either open or closed, then A is lineally convex (this is true also in the real category), and if Ω is a lineally convex open set in $\mathbf{C}^n$, the space of n complex variables, then Ω is pseudoconvex. Now pseudoconvexity is a local property in the sense that if any boundary point of an open set Ω has an open neighborhood ω such that $\Omega \cap \omega$ is pseudoconvex, then Ω is pseudoconvex; the analogous result holds for convexity. But it is well known that the property of lineal convexity is not a local property in this sense—for easy examples see Subsection 9.4.3. The purpose of this section is to investigate to what extent this is true for sets that are of a special form: the Hartogs domains.

Let us now give the main definition.

Definition 9.4.1 *A set A in $\mathbf{C}^n$ is said to be **lineally concave** if it is a union of hyperplanes. It is called **lineally convex** if its complement is lineally concave.* □

A lineally convex set whose boundary is sufficiently smooth satisfies a differential condition. Let ρ be a defining function for Ω (see Definition 9.4.18), and let H and L denote, respectively, the Hessian and the Levi form at a boundary point a of Ω. Then the differential condition says that

$$|H(s)| \leqslant L(s) \text{ for all vectors } s \in T_{\mathbf{C}}(a), \qquad (9.41)$$

where $T_{\mathbf{C}}(a)$ is the complex tangent space at the point a. See Subsection 9.4.5 for details. Every lineally convex domain of class C^2 satisfies the differential condition—for the converse, see Section 9.6. Here we shall prove that this is so in the special case of Hartogs domains, which we now proceed to define.

Definition 9.4.2 *A **Hartogs set** in $\mathbf{C}^n \times \mathbf{C}$ is a set which contains, along with a point $(z, t) \in \mathbf{C}^n \times \mathbf{C}$, also every point (z, t') with $|t'| = |t|$. It is said to be a **complete Hartogs set** if it contains, with (z, t), also (z, t') for all t' with $|t'| \leqslant |t|$.* □

Here we shall study open and complete Hartogs sets; they are always defined by a strict inequality $|t| < R(z)$, thus

$$\Omega = \{(z,t) \in \mathbf{C}^n \times \mathbf{C}; \ |t| < R(z)\}, \tag{9.42}$$

where R is a function on $\mathbf{C}^n$ with values in $\mathbf{R}_!$.

Given R, we define a set ω in $\mathbf{C}^n$ by $\omega = \{z \in \mathbf{C}^n; \ 0 < R(z) \leqslant +\infty\}$. We shall say that Ω is a Hartogs domain ***over*** ω, or that ω is the ***base*** of Ω, if (9.42) holds with $R(z)$ positive if and only if $z \in \omega$.

Most of our results will be concerned with the case $n = 1$, thus

$$\Omega = \{(z,t) \in \mathbf{C} \times \mathbf{C}; \ |t| < R(z)\} = \{(z,t) \in \omega \times \mathbf{C}; \ |t| < R(z)\}. \tag{9.43}$$

The main result here is the following theorem.

Theorem 9.4.3 *Let Ω be a bounded complete Hartogs domain in $\mathbf{C}^2$ with boundary of class C^2. If Ω satisfies the differential condition (9.41) at all boundary points, then Ω is lineally convex.*

Thus for complete Hartogs domains, the property of being lineally convex is a local property. Next we consider sets with R of class C^2 in ω but which do not necessarily have a smooth boundary at points (z,t) with $z \in \partial\omega$.

In this case we prove:

Theorem 9.4.4 *Let ω be a bounded open set in the complex plane $\mathbf{C}$. If the closure of ω is not a disk, then lineal convexity over ω is not a local condition: we can find a Hartogs domain Ω over ω and two open sets ω_0 and ω_1 such that the Hartogs domains Ω_j over ω_j are lineally convex, $j = 0, 1$, but their union $\Omega = \Omega_0 \cup \Omega_1$ is not. If on the other hand ω is a disk, and Ω is a Hartogs domain satisfying the differential condition (9.41) at all boundary points over ω, then Ω is lineally convex.*

Corollary 9.4.5 *Let ω be an open set in $\mathbf{C}$ which is equal to the interior of its closure, and let Ω be a Hartogs domain over ω. Then the differential condition (9.41) imposed on all boundary points over ω is equivalent to lineal convexity if and only if ω is a disk.*

9.4.2   Weak lineal convexity

There are several other notions related to lineal convexity:

Definition 9.4.6 *An open connected set is called **weakly lineally convex** if through any boundary point there passes a complex hyperplane which does not intersect the set. An open set is said to be **locally weakly lineally convex**[3] if through every boundary point $a \in \partial\Omega$ there is a complex hyperplane Y passing through a such that a does not belong to the closure of $Y \cap \Omega$.* □

---

[3] There are actually two versions of this concept: see Definitions 9.5.7 and 9.5.8.

It is not difficult to prove that local weak lineal convexity implies pseudoconvexity.

For complete Hartogs sets it is very easy to see that weak lineal convexity implies lineal convexity:

Lemma 9.4.7 *A complete Hartogs domain which is weakly lineally convex and has a lineally convex base is lineally convex.*

Proof Let $(z^0, t^0) \in \mathbf{C}^n \times \mathbf{C}$ be an arbitrary point in the complement of Ω, a Hartogs domain defined by (9.42). If $R(z^0) > 0$, then the point $(z^0, R(z^0)t^0/|t^0|)$ belongs to $\partial\Omega$, and if Ω is weakly lineally convex, there is a hyperplane passing through that point which does not cut Ω. Then the parallel plane through (z^0, t^0) does not cut Ω either. If $R(z^0) \leqslant 0$, then z^0 does not belong to the base, and a hyperplane with equation $\zeta \cdot z = \zeta \cdot z^0$ will do, since the base is lineally convex. This proves the lemma.

9.4.3 The non-local character of lineal convexity

The domain

$$V = \{(z,t) \in \mathbf{C}^2; \; |t| < |z|\} \tag{9.44}$$

is easily seen to be lineally convex. Indeed, if $(z_0, t_0) \notin V$ with $t_0 \neq 0$, then the complex line $\{(z,t); \; z_0 t = t_0 z\}$ passes through (z_0, t_0) and does not cut V; if on the other hand $t_0 = 0$, we can for instance take the line $\{0\} \times \mathbf{C}$. A simple example of a domain that is locally lineally convex but not lineally convex can be built up from this set as follows.

Example 9.4.8 *(Kiselman 1996, Example 2.1.)*

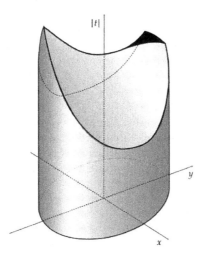

FIGURE 9.1
An open connected Hartogs set in $\mathbf{C}^2$ which is locally weakly lineally convex but not weakly lineally convex. Coordinates $(z,t) \in \mathbf{C}^2$; $(x, y, |t|) \in \mathbf{R}^3$.

Define first

$$\Omega_+ = \{(z,t); \ |z| < 1 \ and \ |t| < |z - 2|\};$$

$$\Omega_- = \{(z,t); \ |z| < 1 \ and \ |t| < |z + 2|\},$$

and then

$$\Omega_0 = \Omega_+ \cap \Omega_-; \qquad \Omega_0^r = \{(z,t) \in \Omega_0; \ |t| < r\},$$

where r is a constant with $2 < r < \sqrt{5}$. All these sets are lineally convex. The two points $(\pm i, \sqrt{5})$ belong to the boundary of Ω_0; in the three-dimensional space of the variables $(\operatorname{Re} z, \operatorname{Im} z, |t|)$, the set representing Ω_0 has two peaks, which have been truncated in Ω_0^r.

We now define Ω^r by glueing together Ω_0 and Ω_0^r: Define Ω^r as the subset of Ω_0 such that $(z,t) \in \Omega_0^r$ if $\operatorname{Im} z > 0$; we truncate only one of the peaks of Ω_0.

The point $(i - \varepsilon, r)$ for a small positive ε belongs to the boundary of Ω^r and the tangent plane at that point has the equation $t = r$ and so must cut Ω^r at the point $(-i + \varepsilon, r)$. Therefore Ω^r is not lineally convex, but it agrees with the lineally convex sets Ω_0 and Ω_0^r when $\operatorname{Im} z < \delta$ and $\operatorname{Im} z > -\delta$, respectively, for a small positive δ. The set has Lipschitz boundary; in particular it is equal to the interior of its closure. $\qquad\qquad\qquad\qquad\qquad\qquad\qquad\qquad\qquad$ □

Proposition 9.4.9 *Let ω_0 and ω_1 be two bounded open subsets in the complex plane such that none is contained in the closure of the other. Then there exists a Hartogs domain over $\omega = \omega_0 \cup \omega_1$ that is not lineally convex, but is such that the subsets Ω_j over ω_j are both lineally convex, $j = 0, 1$.*

Proof Take two points $a \in \omega_1 \setminus \overline{\omega}_0$ and $b \in \omega_0 \setminus \overline{\omega}_1$, which exist by hypothesis. It is no restriction to assume that $a = i$, $b = -i$. Then take $c > 0$ so large that ω is contained in the disk of radius $c - 1$ and with center at the origin. We then define as in Example 9.4.8,

$$\Omega_0 = \left\{ (z,t) \in \mathbf{C}^2; \ |t| < |z \pm c| \ and \ |t| < \left| z \pm i\big(1 + \sqrt{c^2 + 1}\big) \right| \right\},$$

and

$$\Omega_1 = \{(z,t) \in \Omega_0; \ |t| < r\},$$

where r is a number slightly smaller than $\sqrt{c^2 + 1}$ but so close to that number that the peak that we have truncated in Ω_1 near $i \in \mathbf{C}$ lies outside ω_0, and the peak near $-i$ lies outside ω_1. This is possible since we have assumed that $i \notin \overline{\omega}_0$, $-i \notin \overline{\omega}_1$, and Ω_0 and Ω_1 differ only above small neighborhoods of i and $-i$. These neighborhoods shrink to $\{i, -i\}$ as r increases to $\sqrt{c^2 + 1}$. We now define Ω to agree with Ω_j over ω_j, $j = 0, 1$. The conclusion is as in Example 9.4.8.

9.4.4   Smooth vs. Lipschitz boundaries

The lineally convex set Ω_0 constructed in Example 9.4.8 has the remarkable property that it cannot be approximated by lineally convex sets with smooth boundary. Its boundary, which is Lipschitz, cannot in any reasonable way be rounded off if we want to preserve lineal convexity. This is why we shall continue this investigation to see whether smoothly bounded sets admit a passage from the local to the global.

Before doing so, however, we shall illustrate the difference between domains which can be approximated by smoothly bounded lineally convex domains and those that have only Lipschitz boundary.

Let Ω be a complete Hartogs domain with R a function of class C^1, ω being the open set where $R > 0$. Often it will be convenient to use not R but $h = R^2$ to define the set, thus

$$\Omega = \{(z,t) \in \omega \times \mathbf{C}; \ |t| < R(z)\} = \{(z,t) \in \omega \times \mathbf{C}; \ |t|^2 < h(z)\}. \quad (9.45)$$

The complex tangent plane at a boundary point (z_0, t_0) with $z_0 \in \omega$ has the equation

$$t - t_0 = \alpha(z - z_0), \text{ where } \alpha = \frac{h_z(z_0)}{\overline{t_0}} = \frac{2t_0 R_z(z_0)}{R(z_0)}. \quad (9.46)$$

The tangent plane intersects the plane $t = 0$ in the point

$$b(z_0) = z_0 - \frac{h(z_0)}{h_z(z_0)} = z_0 - \frac{R(z_0)}{2R_z(z_0)}. \quad (9.47)$$

If $R_z(z_0) = 0$, the tangent plane has the equation $t = t_0$, and in this case we define $b(z_0) = \infty$, the infinite point on the Riemann sphere $S^2 = \mathbf{C} \cup \{\infty\}$.

Proposition 9.4.10 *Let $R \in C^1(\mathbf{C})$ and define Ω by (9.43). If Ω is bounded and lineally convex, then $b(z)$, defined by (9.47), does not belong to ω, so that b is a continuous mapping from ω into $S^2 \setminus \omega$. Its range contains $S^2 \setminus \overline{\omega}$.*

Proof Clearly b is continuous as a mapping into $\mathbf{C}$ except where $R_z = 0$. Near such points, however, $1/b$ is continuous. The point $(b(z_0), 0)$ cannot belong to Ω since Ω is lineally convex; thus $b(z_0) \notin \omega$. From every point $(z, 0)$ outside the closure of Ω we can draw a tangent to Ω: this shows that the range of b contains $\mathbf{C} \setminus \overline{\omega}$; clearly it also contains ∞.

Corollary 9.4.11 *If Ω is as in Proposition 9.4.10, then Ω is connected. The same is true if Ω is the union of an increasing family of bounded lineally convex sets Ω_j defined by functions $R_j \in C^1(\mathbf{C})$.*

Proof Let ω_1 be a component of ω and let Ω_1 be the set over ω_1. Then the image of the boundary of Ω_1 under b contains $S^2 \setminus \overline{\omega_1}$. Since $b(z_0) \notin \omega$ there can be no other component: we must have $\omega_1 = \omega$. The statement about $\bigcup \Omega_j$ is now immediate.

Corollary 9.4.11 should be compared with the following easy result for Lipschitz boundaries.

Proposition 9.4.12 *Let ω be any open set in $\mathbf{C}$. Then there exists a Lipschitz continuous function $R \in C(\mathbf{C})$ such that ω is the set where R is positive and the set Ω defined by R is lineally convex.*

Proof We define $R(z) = \inf_{a \notin \omega} |z - a|$. The set Ω is lineally convex since it is an intersection of sets of the type V defined in (9.44).

The set $M_{\sup R}$ where the function R assumes its maximum can be rather arbitrary as shown be the next proposition.

Proposition 9.4.13 *Given any closed set M in the complex plane such that its complement is a union of open disks of radius r there exists a Lipschitz continuous function R such that $M_{\sup R} = M$ and the domain Ω defined by (9.45) with this R is lineally convex.*

Proof Define $R(z) = \min\left(r, \inf_{a \in A} |z - a|\right)$, where A is the set of all centers of disks of radius r in the complement of M.

But when R is of class C^1, the set $M_{\sup R}$ is convex:

Theorem 9.4.14 *Let $R \colon \mathbf{C}^n \to \mathbf{R}$ be a function of class C^1 or more generally a continuous function which is the limit of an increasing sequence of functions R_j of class C^1 in the sets $\{z;\ R_j(z) > 0\}$. We assume that R is positive only in a bounded subset of the complex plane. The functions R_j define open sets Ω_j, which we assume to be lineally convex. Then the set $M_{\sup R} = \{z;\ R(z) = \sup_w R(w)\}$ is convex.* $\qquad \square$

This could be proved here, but it is more easily done with the methods of Section 9.7: see Theorems 9.7.29 and 9.7.30.

If a set does not have a boundary of class C^1, we cannot give a meaning to the notion of tangent plane. However, if the set is the union of an increasing family of sets with smooth boundaries, it is possible to use instead their tangent planes and then pass to the limit. Such limits of tangent planes can serve as well, as explained in the following lemma.

Lemma 9.4.15 *Let Ω be the union of an increasing family of open lineally convex sets Ω_j with boundaries of class C^1. Let (j_k) be a sequence tending to $+\infty$, and let Y_k be the complex tangent plane of $\partial \Omega_{j_k}$ at some point in the boundary of Ω_{j_k}, $k \in \mathbf{N}$. Assume that the Y_k converge to a hyperplane Y in the topology of hyperplanes. Then Y does not intersect Ω.*

Proof Suppose there is a point $z \in Y \cap \Omega$. Then also $z \in Y \cap \Omega_{j_k}$ for all large k. Since Ω_{j_k} is open, there is a ball $B_<(z, \varepsilon) \subset \Omega_{j_k}$ for large k, say for $k \geqslant k_0$. But then Y_k intersects $B_<(z, \varepsilon)$ for all large k, say for $k \geqslant k_1$. Thus $Y_k \cap \Omega_{j_k}$ is non-empty for all $k \geqslant \max(k_0, k_1)$, contradicting the lineal convexity of Ω_{j_k}.

To recognize such limits of tangent planes we shall use the concept in the following definition.

Definition 9.4.16 *Let X be any subset of $\mathbf{C}^n$ and a a point in the boundary ∂X. We shall say that a complex hyperplane Y is an **admissible tangent plane to** ∂X **at** a if there exists an open set A with boundary of class C^1 such that A and X are disjoint, a belongs to the boundary of A, and Y is the complex tangent plane to A at a.* $\qquad\square$

Proposition 9.4.17 *Let $\Omega \subset \mathbf{C}^n$ be the union of an increasing family of lineally convex open sets Ω_j with boundaries of class C^1. Then any admissible tangent plane Y to $\partial\Omega$ is the limit of a sequence of tangent planes Y_j to $\partial\Omega_j$. Therefore, in view of Lemma 9.4.15, Y cannot intersect Ω.*

Proof Let a and A be as in Definition 9.4.16. By a coordinate change we may suppose that $a = 0$, that the real tangent plane to ∂A at the origin has the equation $y_n = 0$, and that A is defined by an inequality $y_n > f(z_1, ..., z_{n-1}, x_n)$ near the origin for some function f of class C^1, which consequently vanishes at the origin together with its gradient. Write $z' = (z_1, ..., z_{n-1}) \in \mathbf{C}^{n-1}$. We then know that all points in Ω satisfy $y_n < f(z', x_n)$. Define $g(z', x_n) = f(z', x_n) + \|z'\|_2^2 + x_n^2$, and let A_c be the set of all points such that $y_n > g(z', x_n) - c$. We let $c = c_j$ be the largest real number such that A_c and Ω_j are disjoint. Now $0 \in \partial\Omega$ and $\Omega_j \nearrow \Omega$; therefore we can be sure that c_j tends to zero as $j \to \infty$. There is a point z^j which is common to the boundaries of A_{c_j} and Ω_j. Since A and Ω_j are disjoint, we have $\|(z^j)'\|_2^2 + (x_n^j)^2 \leqslant c_j$. The real tangent plane to ∂A_{c_j} at z^j is identical to the real tangent plane to $\partial\Omega_j$ at that point. We can control its slope, for the gradient of g is

$$\mathbf{grad}\, g = \mathbf{grad}\, f + \mathbf{grad}(\|z'\|_2^2 + x_n^2),$$

which is continuous and vanishes at the origin. Since $((z^j)', x_n^j)$ tends to the origin, this shows that the real tangent plane to ∂A_{c_j} at z^j must be close to the real hyperplane $y_n = 0$ if j is large, and then the complex tangent plane to ∂A_{c_j} at z^j is close to the complex hyperplane $z_n = 0$. The last statement now follows from Lemma 9.4.15.

9.4.5 Differential conditions

Definition 9.4.18 *Let Ω be an open set in $\mathbf{C}^n$ with boundary of class C^1. Then a function $\rho \in C^1(\mathbf{C}^n)$ is called a **defining function** for Ω, if $\mathrm{d}\rho \neq 0$ wherever $\rho = 0$ and if $\Omega = \{z \in \mathbf{C}^n;\ \rho(z) < 0\}$. (Here the differential d is defined in (9.2).)* $\qquad\square$

Definition 9.4.19 *The **complex tangent space** at a point a on the boundary of Ω is defined by*

$$\sum_{j=1}^{n} \frac{\partial\rho}{\partial z_j}(a)s_j = 0.$$

We shall denote it by $T_{\mathbf{C}}(a)$. The **real tangent space** is defined by

$$\mathrm{Re} \sum_{j=1}^{n} \frac{\partial \rho}{\partial z_j}(a)s_j = 0$$

and will be denoted by $T_{\mathbf{R}}(a)$. The **complex tangent plane** is then $a + T_{\mathbf{C}}(a)$; it is contained in the **real tangent plane** $a + T_{\mathbf{R}}(a)$. □

To be able to characterize sets by infinitesimal conditions, we shall describe boundaries and their curvature using defining functions and the Hesse and Levi forms. We now give the needed definitions.

Definition 9.4.20 *The* **complex Hessian** *(or* **complex Hesse form**) *of a function f of class C^2 is defined to be*

$$H_f^{\mathbf{C}}(z;t) = \sum \frac{\partial^2 f}{\partial z_j \partial z_k}(z)t_j t_k, \qquad z \in \mathbf{C}^n, \quad t \in \mathbf{C}^n, \qquad (9.48)$$

a quadratic form in the t_j. □

Definition 9.4.21 *The* **Levi form** *of f is the Hermitian form*

$$L_f(z;t) = \sum \frac{\partial^2 f}{\partial z_j \partial \bar{z}_k}(z)t_j \bar{t}_k, \qquad z \in \mathbf{C}^n, \quad t \in \mathbf{C}^n. \qquad (9.49)$$

We say that Ω satisfies the **Levi condition at** $a \in \partial\Omega$ *if*

$$L_\rho(a;t) \geqslant 0 \ \text{when } t \in T_{\mathbf{C}}(a), \qquad (9.50)$$

where ρ is a defining function for Ω; and that Ω satisfies the **strong Levi condition at** a *if strict inequality holds in* (9.50) *for $t \neq 0$.* □

Definition 9.4.22 *The* **real Hessian** *of a function f of real variables $x_1, \ldots, x_m$ is*

$$H_f^{\mathbf{R}}(x;s) = \sum \frac{\partial^2 f}{\partial x_j \partial x_k}(x)s_j s_k, \qquad x \in \mathbf{R}^m, \quad s \in \mathbf{R}^m, \qquad (9.51)$$

a quadratic form.

When a function of n complex variables is given, its real Hessian in the $2n$ real variables $(\mathrm{Re}\, z_1, \mathrm{Im}\, z_1, \ldots, \mathrm{Re}\, z_n, \mathrm{Im}\, z_n)$ can be expressed using its complex Hessian and its Levi form as

$$H_f^{\mathbf{R}}(z;s) = 2(\mathrm{Re}\, H_f^{\mathbf{C}}(z;t) + L_f(z;t)),$$

for $z \in \mathbf{C}^n$, $s \in \mathbf{R}^{2n}$, $t \in \mathbf{C}^n$, $t_j = s_{2j-1} + \mathrm{i}s_{2j}$.

Thus the characterization of convexity mentioned in the introduction is that $\mathrm{Re}\, H_\rho^{\mathbf{C}}(a;t) + L_\rho(a;t)$ be nonnegative for all $a \in \partial\Omega$ and all $t \in T_{\mathbf{R}}(a)$. For a lineally convex set the same inequality holds for all $t \in T_{\mathbf{C}}(a)$. It is then equivalent to $L(a;t) \geqslant |H(a;t)|$ for $a \in \partial\Omega$ and $t \in T_{\mathbf{C}}(a)$.

Definition 9.4.23 *We shall say that a set Ω with boundary of class C^2 satisfies* **the Behnke–Peschl differential condition** *at a boundary point a of Ω if*

$$|H_\rho^{\mathbf{C}}(a; s)| \leqslant L_\rho(a; s) \text{ for all vectors } s \in T_{\mathbf{C}}(a), \qquad (9.52)$$

where ρ is a defining function for Ω. We shall say that Ω satisfies **the strict Behnke–Peschl differential condition** *at a if there exists a positive number ε such that we have*

$$|H_\rho^{\mathbf{C}}(a; s)| \leqslant L_\rho(a; s) - \varepsilon \|s\|_2^2 \qquad (9.53)$$

for all $s \in T_{\mathbf{C}}(a)$. □

It is easy to prove that these conditions are invariant under complex affine mappings. They also do not depend on the choice of defining function. They were introduced for $n = 2$ by Behnke and Peschl (1935:169).

These conditions should be compared with the differential condition for convexity: $|H_\rho(s)| \leqslant L_\rho(s)$ for all vectors s in the real tangent space $T_{\mathbf{R}}(a)$. This is a local condition, and it is well known that it is equivalent to convexity of Ω. The proof of this fact most conveniently goes via approximation of the set by sets satisfying the corresponding strong condition, i.e., $|H_\rho(s)| \leqslant L_\rho(s) - \varepsilon \|s\|_2^2$ for a positive ε and for all $s \in T_{\mathbf{R}}(a)$.

The following two lemmas are well known; cf. (Zinov'ev 1971) and (Hörmander 1994: Corollary 4.6.5). We include them for ease of reference.

Lemma 9.4.24 *Let Ω be an open subset of $\mathbf{C}^n$ with boundary of class C^2. If Ω is locally weakly lineally convex, then Ω satisfies the Behnke–Peschl differential condition at every boundary point.*

Proof Let a be an arbitrary boundary point of a locally weakly lineally convex open set Ω. Then there exists a complex hyperplane through a that does not cut Ω close to a. This hyperplane cannot be anything but $T_{\mathbf{C}}(a)$ since the boundary is of class C^1. Therefore if we take an arbitrary vector $s \in T_{\mathbf{C}}(a)$ and consider the function $\varphi(t) = \rho(a + ts)$ of a real variable t, its second derivative must be non-negative at the origin. If we express the condition $\varphi''(0) \geqslant 0$ in terms of H and L we get $\operatorname{Re} H(s) + L(s) \geqslant 0$, which, since H is quadratic and L sesquilinear, is equivalent to $|H| \leqslant L$.

Lemma 9.4.25 *Let Ω be an open subset of $\mathbf{C}^n$ with boundary of class C^2. If Ω satisfies the strict Behnke–Peschl differential condition at every boundary point, then Ω is locally weakly lineally convex.*

Proof With φ as in the proof of the previous lemma we must have $\varphi''(0) > 0$ if Ω satisfies the strict Behnke–Peschl differential condition. This implies that $T_{\mathbf{C}}(a)$ cannot cut Ω close to a.

It is known that if Ω is a connected open set with boundary of class C^1 which is locally weakly lineally convex, then Ω is weakly lineally convex; see, e.g., (Hörmander 1994: Proposition 4.6.4). We shall come back to this result in Subsection 9.4.7.

9.4.6   Differential conditions for Hartogs domains

In this subsection we shall see what the differential conditions look like in the case of a complete Hartogs domain in $\mathbf{C}^2$. Let Ω be a complete Hartogs domain in $\mathbf{C}^2$ defined by (9.45). If h is of class C^1, we can choose as its defining function

$$\rho(z,t) = t\bar{t} - h(z), \qquad (z,t) \in \mathbf{C} \times \mathbf{C}.$$

It must satisfy $d'\rho \neq 0$ when $\rho = 0$, which means that $d'\rho = \bar{t}\,dt - h_z\,dz \neq 0$ when $|t|^2 = h(z)$. Since the first term of $d'\rho$ is $\bar{t}\,dt$, which is non-zero everywhere except in the plane $t = 0$, the only condition is that $h_z \neq 0$ when $h = 0$, i.e., that h itself shall be a defining function in $\mathbf{C}$. It defines a subset ω of the complex plane over which Ω is situated.

Lemma 9.4.26 *Let h be a defining function of an open set ω in $\mathbf{C}$ of class C^k, $k \geqslant 1$. Then the complete Hartogs domain in $\mathbf{C}^2$ defined by (9.45) has boundary of class C^k. When $k \geqslant 2$, it satisfies the Behnke–Peschl differential condition at every boundary point if and only if h satisfies the condition*

$$\frac{|h_z|^2}{h} \geqslant h_{z\bar{z}} + |h_{zz}| \ \text{wherever } h > 0. \tag{9.54}$$

Furthermore Ω satisfies the strict Behnke–Peschl differential condition if and only if there is strict inequality in (9.54).

Proof Let us look at the Hessian and Levi forms of $\rho(z,t) = |t|^2 - h(z)$. They are, respectively,

$$H(s) = -h_{zz}s_1^2 \ \text{and} \ L(s) = -h_{z\bar{z}}|s_1|^2 + |s_2|^2, \qquad s = (s_1, s_2) \in \mathbf{C}^2.$$

The differential condition $|H| \leqslant L$ takes the form

$$|h_{zz}||s_1|^2 \leqslant -h_{z\bar{z}}|s_1|^2 + |s_2|^2 \ \text{for all } s \in T_{\mathbf{C}}(a).$$

The tangent plane is defined by $-h_z s_1 + \bar{t}s_2 = 0$. When $t \neq 0$ we use this equation to eliminate s_2: the condition takes the form (9.54). Near $t = 0$ we eliminate instead s_1 and get

$$(h_{z\bar{z}} + |h_{zz}|)\frac{h}{|h_z|^2} \leqslant 1.$$

This inequality is satisfied, even strictly, at all boundary points sufficiently close to $t = 0$, provided $h_z \neq 0$ near $h = 0$. Therefore, if h is a defining function for ω, then ρ is a defining function for Ω and condition (9.54) implies the Behnke–Peschl differential condition at all boundary points of Ω, including those where $t = 0$. Conversely, if ρ is a defining function for Ω, then h is a defining function for ω, and the Behnke–Peschl differential condition for ρ implies the condition (9.54) for h.

Remark 9.4.27 *We can also express the Behnke–Peschl differential condition* (9.54) *in terms of the radius* $R = \sqrt{h}$. *It becomes*

$$|R_z|^2 \geqslant |R_z^2 + RR_{zz}| + RR_{z\bar{z}}, \tag{9.55}$$

which is less convenient to work with than (9.54). *If* h *is concave, then* $h_{z\bar{z}} + |h_{zz}| \leqslant 0$, *so that* (9.54) *holds. More generally, if* R *is concave, then* $R_{z\bar{z}} + |R_{zz}| \leqslant 0$, *which implies that* (9.55)) *holds. It is also possible to express the Behnke–Peschl differential condition in terms of the function* $f = -\log R$. *It then takes the form*

$$|f_{zz} - 2f_z^2| \leqslant f_{z\bar{z}}. \tag{9.56}$$

We note that $(z,t) \mapsto |t|^2 - h(z)$ *is convex if and only if* h *is concave, and that* $(z,t) \mapsto \log\|z\|_2 + f(z)$ *is plurisubharmonic if and only if* f *is.* □

9.4.7 Approximating bounded lineally convex Hartogs domains by smoothly bounded ones

Theorem 9.4.28 *Let*

$$\Omega = \{(z,t) \in \mathbf{C}^2;\ |t|^2 < h(z)\} \tag{9.57}$$

be a bounded complete Hartogs domain in $\mathbf{C}^2$ *with boundary of class* C^2. *Suppose* Ω *satisfies the Behnke–Peschl differential condition at all boundary points. Then* Ω *can be approximated from the inside by Hartogs domains*

$$\Omega_\varepsilon = \{(z,t);\ |t| < R_\varepsilon(z)\}$$

which satisfy the strict Behnke–Peschl differential condition all boundary points (z,t) *except those where* $h_z(z) = 0$. *In fact, we can take* $h_\varepsilon = h - \varepsilon$ *with* ε *positive and small enough.*

Proof This is an instance where it is more convenient to use h rather than R. The Behnke–Peschl differential condition (9.54) contains the value of h only at one place, and $h_\varepsilon = h - \varepsilon$ has the same derivatives as h, so we can write

$$\frac{|h_z|^2}{h-\varepsilon} > \frac{|h_z|^2}{h} \geqslant h_{z\bar{z}} + |h_{zz}|$$

except when $h_z = 0$. Thus the boundary of Ω_ε satisfies the strict Behnke–Peschl differential condition except at the points where $h_z = 0$. So far the argument is valid for all positive ε. We need to check that h_ε is a defining function; otherwise we cannot apply Lemma 9.4.26. But the gradient of h_ε is the same as that of h, which is non-zero when $h = 0$, hence also when $h_\varepsilon = 0$, provided ε is small enough. Thus h_ε is indeed a defining function for all small ε, proving the theorem.

We shall now see that the approximating sets Ω_ε that we constructed in Theorem 9.4.28 are, in fact, lineally convex. Let us agree to say that a complex plane with the equation $z = constant$ is **vertical** and a plane with the equation $t = constant$ is **horizontal**.

Proposition 9.4.29 *Let Ω be a bounded complete Hartogs domain in $\mathbf{C}^2$ with boundary of class C^2 satisfying the strict Behnke–Peschl differential condition except possibly at the points where the tangent plane is horizontal. Then Ω is lineally convex.*

We shall need the following three lemmas.

Lemma 9.4.30 *Let Ω be as in Proposition 9.4.29 and let L be a complex line in $\mathbf{C}^2$ which is not horizontal. Then $L \cap \Omega$ consists of a finite number of open sets bounded by C^2 curves obtained as transversal intersections of L and $\partial\Omega$ (and $L \cap \partial\Omega$ consists of these curves plus a finite number of isolated points).*

Proof Take an arbitrary boundary point a and let L be a complex line through a which is not horizontal. If L is the tangent plane, $L = a + T_\mathbf{C}(a)$, then the proof of Lemma 9.4.25 shows that L intersects $\overline{\Omega}$ near a only in the point a. If, on the other hand, L is not the tangent plane, then $L \cap (a + T_\mathbf{C}(a)) \neq L$, so $\partial\Omega$ cuts L transversally, and $\partial\Omega \cap L$ is a C^2 curve in L near a. Thus $L \cap \partial\Omega$ consists of a number of C^2 curves plus isolated points—by compactness there can only be finitely many curves and points.

Lemma 9.4.31 *Let Ω and L satisfy the hypotheses of the previous lemma. Then $\Omega \cap L$ is connected, and $\Omega \cap (a + T_\mathbf{C}(a))$ is empty for all $a \in \partial\Omega$.*

Proof We shall follow the proof of Proposition 4.6.4 in (Hörmander 1994)— we only have to be careful to avoid horizontal planes. Let (z_j, t_j), $j = 0, 1$, be two points in $L \cap \Omega$. We have to prove that they belong to the same component of $L \cap \Omega$. Suppose first that both t_0 and t_1 are non-zero. Since Ω is connected, there is a curve γ which goes from $\gamma(0) = (z_0, t_0)$ to $\gamma(1) = (z_1, t_1)$. We can actually do this in such a way that the complex line L_s that contains $\gamma(0)$ and $\gamma(s)$, $0 < s \leqslant 1$, is never horizontal. Indeed, we first go from (z_0, t_0) to $(z_0, 0)$ along a curve in the plane $z = z_0$ avoiding (z_0, t_1); then along a curve in the plane $t = 0$ from $(z_0, 0)$ to $(z_1, 0)$; and then finally from $(z_1, 0)$ to (z_1, t_1) along a curve in the plane $z = z_1$ avoiding (z_1, t_0). (We know that $t_0 \neq t_1$.) Thus none of the lines L_s is horizontal, and we can apply Lemma 9.4.30 to them. Consider the set C of all $s \in \,]0, 1]$ such that $\gamma(0)$ and $\gamma(s)$ belong to the same component of $L_s \cap \Omega$. Then certainly C contains all sufficiently small numbers, for $\gamma(0)$ and $\gamma(s)$ are then in the line $z = z_0$, whose intersection with Ω is a disk. The set C is open as a subset of $]0, 1]$ in view of Lemma 9.4.30, but so is its complement with respect to $]0, 1]$. Since it is non-empty, it must contain 1, i.e., (z_0, t_0) and (z_1, t_1) belong to the same component of $L \cap \Omega$. If one of t_0, t_1 is zero, we choose a point with non-zero second coordinate in the neighborhood and argue as above.

Consider now a tangent plane $L = a + T_{\mathbf{C}}(a)$ and planes $L_\varepsilon = a_\varepsilon + T_{\mathbf{C}}(a)$ parallel to it, where we write $a_\varepsilon = (z_0, (1 - \varepsilon)t_0)$ if $a = (z_0, t_0)$. We already know from Lemma 9.4.25 that L cannot intersect Ω close to a. However, it cannot cut Ω at all, for if it did, then a parallel plane L_ε for some small positive ε would intersect Ω in a component close to a and another nonempty set at some distance from a, thus in a disconnected set. This proves Lemma 9.4.31.

Lemma 9.4.32 *Let Ω be as in Proposition 9.4.29 and let $a \in \partial\Omega$ be such that the tangent plane is horizontal. Then $\Omega \cap (a + T_{\mathbf{C}}(a))$ is empty; in other words R has a global maximum at a. Consequently any horizontal plane L intersects Ω in finitely many open sets bounded by C^2 curves obtained as transversal intersections of L by $\partial\Omega$.*

Proof Let (z_0, t_0) be a boundary point such that the tangent plane is horizontal, i.e., $R_z(z_0) = 0$. Suppose the tangent plane cuts Ω in some point (z_1, t_1). We must then have $t_1 = t_0$. Since Ω and its base ω are connected, we can find a curve γ in ω connecting z_0 to z_1, say $\gamma(s) = z_s$, $s \in [0, 1]$. Consider now the tangent planes at the points $(z_s, R(z_s))$; we denote them by $L_s = (z_s, R(z_s)) + T_{\mathbf{C}}(z_s, R(z_s))$. It is no restriction to assume $t_0 > 0$, so that $R(z_0) = t_0$. We know that L_0 is horizontal, but certainly not all the L_s can be horizontal, since $R(z_1) > |t_1| = |t_0| = R(z_0)$. Let s_0 be the infimum of all s such that L_s is not horizontal; we must have $0 \leqslant s_0 < 1$. The planes L_s with $0 \leqslant s \leqslant s_0$ are identical and all intersect Ω in the point (z_1, t_1). It is now clear that there exists a tangent plane L_s with s just a little bit larger than s_0 which is not horizontal and still cuts Ω. This contradicts Lemma 9.4.31.

Proof of Proposition 9.4.29. We know from Lemma 9.4.31 that a tangent plane which is not horizontal does not intersect Ω; we obtain the same conclusion from Lemma 9.4.32 for a horizontal tangent plane. Thus Ω is weakly lineally convex. Lemma 9.4.7 shows that this implies lineal convexity.

We can now finally state the main result of this section:

Theorem 9.4.33 *Let Ω be a bounded complete Hartogs domain in $\mathbf{C}^2$ with boundary of class C^2. If Ω satisfies the Behnke–Peschl differential condition (9.52) at all boundary points, then Ω is lineally convex.*

Proof Using Theorem 9.4.28 we construct open sets Ω_ε, which tend to Ω. Also, if $R(z_0) > 0$, the tangent plane of $\partial\Omega_\varepsilon$ at $\left(z_0, \sqrt{R(z_0)^2 - \varepsilon}\right)$ tends to that of $\partial\Omega$ at $(z_0, R(z_0))$. The sets Ω_ε are lineally convex by Proposition 9.4.29. Then also their limit Ω is lineally convex. Indeed, if a tangent plane to $\partial\Omega$ intersected Ω, then it would cut also Ω_ε for all sufficiently small ε, and then also for ε small enough the corresponding tangent plane to $\partial\Omega_\varepsilon$ would cut Ω_ε. This is a contradiction.

9.4.8 The non-local character of lineal convexity, revisited

Having settled the question of lineal convexity of smoothly bounded Hartogs domains we now turn to sets of the form

$$\Omega = \{(z,t) \in \omega \times \mathbf{C};\ |t| < R(z)\} = \{(z,t) \in \omega \times \mathbf{C};\ |t|^2 < h(z)\}, \qquad (9.58)$$

where ω is a given open set in $\mathbf{C}$ and h is a C^2 function in the closure of ω satisfying $h > 0$ and the Behnke–Peschl differential condition (9.54). Its boundary is smooth enough over points in ω, but is only Lipschitz at points over $\partial \omega$. It turns out that when ω is a disk, then the Behnke–Peschl differential condition implies lineal convexity: we shall study this question in Subsection 9.4.9. On the other hand, if ω is a set such that $\overline{\omega}$ is not a disk, then the Behnke–Peschl differential condition does not imply lineal convexity. This is the topic of the present subsection.

The property of being a disk is invariant under Möbius mappings, and disks are the only sets that remain convex under all Möbius mappings. This is a kind of explanation for the phenomenon we encounter here, and it is therefore natural to study how the Behnke–Peschl differential condition (9.52) behaves under Möbius mappings. This is explained in the next lemma.

Lemma 9.4.34 *Let Ω be a Hartogs domain in $\mathbf{C}^2$ defined by $|t| < R(z)$, let a, b, c, d be four complex numbers with $ad - bc \neq 0$, and let Ω_1 be the Hartogs domain defined by $|t| < R_1(z) = |c + dz| R((a + bz)/(c + dz))$. Then Ω and Ω_1 are lineally convex simultaneously. The two functions h and $h_1(z) = |c + dz|^2 h((a + bz)/(c + dz))$ satisfy the Behnke–Peschl differential condition (9.54) simultaneously.*

Proof Take constants α, β and c of which not both of α and β are zero, and consider the mapping

$$(\mathbf{C} \smallsetminus \{0\}) \times \mathbf{C} \times \mathbf{C} \ni (z_0, z_1, t) \mapsto (z_1/z_0, t/z_0) \in \mathbf{C}^2.$$

Under it the pull-back of the hyperplane of equation $c + \alpha z + \beta t = 0$ is the hyperplane of equation $c z_0 + \alpha z_1 + \beta t = 0$. It follows that the pull-back of a lineally convex set in $\mathbf{C}^2$ is a complex homogeneous lineally convex set in $\mathbf{C}^3$. Now any linear mapping of the form

$$\mathbf{C}^3 \ni (z_0, z_1, t) \mapsto (c z_0 + d z_1, a z_0 + b z_1, t) \in \mathbf{C}^3$$

with $ad - bc \neq 0$ preserves lineal convexity, and mappings

$$\mathbf{C}^3 \ni (z_0, z_1, t) \mapsto \left(1, \frac{a z_0 + b z_1}{c z_0 + d z_1}, \frac{t}{c z_0 + d z_1}\right) \in \mathbf{C}^3$$

preserve lineally convex sets which are complex homogeneous. If we transport this back to $\mathbf{C}^2$ we get a mapping of the form

$$(z,t) \mapsto \left(\frac{a + bz}{c + dz}, \frac{t}{c + dz}\right).$$

This proves that Ω and Ω_1 as defined in the statement of the lemma are lineally convex at the same time. The statement about the differential condition for h and h_1 can be verified directly, perhaps most easily if we check it for the special mappings $z \mapsto c + dz$ and $z \mapsto 1/z$, which together generate all Möbius mappings.

Lemma 9.4.35 *Let K be a compact subset of $\mathbf{C}$ with connected complement. Assume that K is not a disk. Then there exists a closed disk D_1 containing K such that $K \cap \partial D_1$ has at least two components.*

Proof Let D_0 be the closed disk of minimal radius that contains K. By hypothesis $K \neq D_0$ and $\mathbf{C} \smallsetminus K$ is connected, so there exists a point $a_0 \in \partial D_0 \smallsetminus K$. Let H be an open half plane that contains K but is such that $a_0 \notin \overline{H}$. Now consider the closed disk D_1 of minimal radius among those that contain K and have ∂H as a tangent. We claim that there are four points $a, b, c, d \in \partial D_1$ which are in that order along the circumference and with $a, c \notin K$, $b, d \in K$. This will show that b and d belong to different components of $K \cap \partial D_1$. To find these points we argue as follows. Let a be the point of ∂D_1 at which ∂H is tangent; thus $a \in \partial D_1$ and $a \notin K$. Next, $D_1 \not\subset D_0$, so there is a point $c \in \partial D_1 \smallsetminus D_0$. Thus $c \notin K$. Finally we claim that there are two points $b, d \in \partial D_1 \cap K$ on either side of the segment $[a, c]$. This is so because if one of the arcs from a to c were disjoint from K, then it can easily be seen that D_1 would not be minimal among the disks that contain K and are tangent to ∂H. This completes the proof.

Theorem 9.4.36 *Let ω be a bounded connected open subset of $\mathbf{C}$ such that the complement $S^2 \smallsetminus \overline{\omega}$ of its closure with respect to the Riemann sphere $S^2 = \mathbf{C} \cup \{\infty\}$ has at least one component which is not a disk. Then there exists a Hartogs domain defined by a smooth function and with base ω such that it is not lineally convex, although $\omega = \omega_0 \cup \omega_1$ and the Hartogs domains over ω_0 and ω_1 are both lineally convex. In particular the function defining Ω satisfies the Behnke–Peschl differential condition (9.54).*

Proof Let K be the complement of a component of $S^2 \smallsetminus \overline{\omega}$ which is not a disk; thus K contains $\overline{\omega}$. Moreover the complement of K is connected and $\partial K \subset \partial \omega$. We may assume that K is compact: if not we use a Möbius mapping to reduce ourselves to that case. Let $a, b, c, d \in \partial D_1$ be the four points whose existence is guaranteed by Lemma 9.4.35; recall that $b, d \in K$ and $a, c \notin K$. Now take a new closed disk D_2 which does not contain a, b, or d, but contains c in its interior, and is so close to D_1 that b and d belong to different components of $K \smallsetminus D_2$. This is possible because a does not belong to K. Now we map D_2 onto the closed right half plane, taking a to 0 and some point outside K and near c to infinity. We are thus reduced to a situation where K is still compact in $\mathbf{C}$, whereas ∂D_2 is the imaginary axis, with $a = 0$ and $\text{Im}\, b$ and $\text{Im}\, d$ of different signs, say for definiteness $\text{Im}\, b < 0$ and $\text{Im}\, d > 0$. Moreover we can take D_2 so close to D_1 that the points in K which are not in D_2 are never real. Then we can define a function R as follows. First take a smooth concave function ψ of a real variable such that $\psi(s) = 1$ when $s \geqslant 0$ and $\psi(s) < 1$ for $s < 0$, but still so that $\psi(\operatorname{Re} z) > 0$ for all points $z \in \overline{\omega}$. Then define

$$R(z) = \begin{cases} \psi(\operatorname{Re} z) & \text{when } z \in \omega, \ \operatorname{Re} z < 0, \ \operatorname{Im} z < 0; \\ 1 & \text{at other points in } \omega. \end{cases}$$

This function is continuous, even identically one, in a neighborhood of the intersection of ω and the real axis.

The tangent plane at a point $(z_0, t_0) \in \partial\Omega$ with $z_0 \in \omega$ has the equation (9.46). In particular, we may take $t_0 = R(z_0)$ and get

$$t = R(z_0) + 2R_z(z_0)(z - z_0).$$

In the present case R is locally a function of $\operatorname{Re} z$, say $R(z) = k(x)$, so that $R_z = k_x/2$ is real. Thus the tangent plane is

$$t = R(z_0) + k_x(x_0)(z - z_0) = R(z_0) + k_x(x_0)(x - x_0) + \mathrm{i}k_x(x_0)(y - y_0),$$

and, writing $z = z_0 + z_1$, we obtain

$$|t|^2 = R(z_0)^2 + 2k_x(x_0)R(z_0)x_1 + k_x(x_0)^2 x_1^2 + k_x(x_0)^2 y_1^2.$$

When $x_1 < 0$ and $k_x(x_0)$ is positive and small,

$$|t|^2 \approx R(z_0)^2 + 2k_x(x_0)R(z_0)x_1 < R(z_0)^2.$$

Since ω is connected and has the point b on its boundary, we can choose z_0 such that $y_0 < 0$ and $x_0 < 0$ with $k_x(x_0)$ arbitrarily small, so small that indeed $|t| < R(z_0)$. Then we choose $z = z_0 + z_1 \in \omega$ with $\operatorname{Im} z > 0$. Thus $R(z) = 1$, so the tangent plane at $(z_0, R(z_0))$ cuts Ω in a point above z. This proves that Ω is not lineally convex. However, if we look at the parts of ω where $\operatorname{Im} z > -\varepsilon$ and $\operatorname{Im} z < \varepsilon$ respectively, then R is the restriction of a globally concave function in each of them and therefore defines a lineally convex set.

Theorem 9.4.37 *Let ω be a bounded open set in $\mathbf{C}$ such that $S^2 \smallsetminus \overline{\omega}$ is not connected. Then there is a function $h \in C^\infty(\overline{\omega})$, $h > 0$, which satisfies the Behnke–Peschl differential condition (9.54) but is such that the Hartogs domain it defines over ω is not lineally convex.*

Proof If one of the components of $S^2 \smallsetminus \overline{\omega}$ is not a disk, we already know the result by Theorem 9.4.36. The case when all components of $S^2 \smallsetminus \overline{\omega}$ are disks remains to be considered. This means that $\overline{\omega}$ is a disk from which countably many disks (at least one) have been removed. Any one of these holes can be moved by a Möbius transformation so that it becomes concentric with the outer circumference of $\overline{\omega}$; in other words $\overline{\omega}$ is an annulus $r_0 \leqslant |z| \leqslant r_1$ from which possibly a number of disks have been removed. It is clearly enough to consider the case of the annulus, for the possible presence of other holes will not destroy our conclusion.

So assume $\overline{\omega}$ is the annulus $r_0 \leqslant |z| \leqslant r_1$ and define $R_0(z) = 1 - ax^2 - by^2$, where $0 < a < b$ and b is so small that $R_0 > 0$ in $\overline{\omega}$. Next define φ to be a concave C^∞ function of one real variable such that $\varphi(s) = s$ for all $s \leqslant 1 - br_0^2 + \varepsilon$ and $\varphi(s) = c$ when $s \geqslant 1 - ar_0^2 - \varepsilon$ for some positive ε and a suitable constant c; by necessity we must have $c < 1 - ar_0^2$. Define $R_1(z) = \varphi(R_0(z))$. We observe that $R_0 = R_1$ in a neighborhood of the intersection of the imaginary axis and $\overline{\omega}$. Both R_0 and R_1 are concave in $\mathbf{C}$, so the corresponding Hartogs domains over $|z| < r_1$ are convex and therefore lineally convex. It follows that the Hartogs domains over ω are lineally convex. Now define R to agree with R_0 in the right half plane and with R_1 in the left half plane. Note that $R(z) = R_1(z) = c$ at points $z \in \omega$ close to $-r_0$, so that the tangent plane at a boundary point over such a point has the equation $t = t_0$ with $|t_0| = c < 1 - ar_0^2$. But over a point z in ω close to r_0 we have $R(z) = R_0(z) > c$, so the tangent plane $t = t_0$ cuts Ω. This proves that Ω cannot be lineally convex.

9.4.9 Hartogs domains over a disk

The Behnke–Peschl differential condition over a disk remains to be studied. We shall see that it is then equivalent to lineal convexity.

We shall write $D(c, r)$ for the open disk in the complex plane with center c and radius r, and just D for the open unit disk $D(0, 1)$.

Proposition 9.4.38 *Let $h \in C^2(D)$, $h > 0$, be a real-valued function which satisfies the Behnke–Peschl differential condition*

$$\frac{|h_z|^2}{h} \geqslant h_{z\overline{z}} + |h_{zz}|, \qquad |z| < 1. \tag{9.59}$$

Let $\varphi \in C^2(\mathbf{R})$ be real-valued, decreasing and satisfy $\varphi \leqslant 1$ everywhere and $\varphi'' < 0$ wherever $\varphi < 1$. Assume that there are constants a and A such that

$$\mathrm{Re}\left[\frac{2zh_z(z)}{h(z)}\right] \leqslant a < 1 \tag{9.60}$$

and

$$\left|\frac{2zh_z(z)}{h(z)}\right| \leqslant A < +\infty \tag{9.61}$$

wherever $0 < \varphi(z\overline{z}) < 1$. Then $g(z) = \varphi(z\overline{z})h(z)$ satisfies the differential condition wherever $\varphi(z\overline{z}) > 0$ and $|z| < 1$, provided φ'/φ'' is small enough, more precisely if either $A \leqslant 1$ or else

$$\frac{\varphi'(s)}{s\varphi''(s)} \leqslant \frac{2(1-a)}{A^2 - 1} \text{ when s is such that } 0 < \varphi(s) < 1.$$

Proof With $g(z) = \varphi(z\bar{z})h(z)$ we have

$$g_z = \varphi' \cdot \bar{z}h + \varphi h_z,$$

$$g_{zz} = \varphi'' \cdot \bar{z}^2 h + 2\varphi'\bar{z}h_z + \varphi h_{zz},$$

$$g_{z\bar{z}} = \varphi'' \cdot |z|^2 h + \varphi'h + 2\varphi'\operatorname{Re} zh_z + \varphi h_{z\bar{z}}.$$

Thus what we have to prove is, writing r for $|z|$,

$$\frac{|\varphi'\bar{z}h + \varphi h_z|^2}{\varphi h} \geqslant r^2\varphi''h + \varphi'h + 2\varphi'\operatorname{Re} zh_z + \varphi h_{z\bar{z}} + |\varphi''\bar{z}^2 h + 2\varphi'\bar{z}h_z + \varphi h_{zz}|.$$

We expand the left-hand side and find that the term $2\varphi'\operatorname{Re} zh_z$ appears on both sides. We shall therefore prove

$$\frac{r^2\varphi'^2 h}{\varphi} + \frac{\varphi|h_z|^2}{h} \geqslant r^2\varphi''h + \varphi'h + \varphi h_{z\bar{z}} + |\varphi''\bar{z}^2 h + 2\varphi'\bar{z}h_z + \varphi h_{zz}|.$$

This formula follows from $|h_z|^2/h \geqslant h_{z\bar{z}} + |h_{zz}|$, which holds by hypothesis, and

$$\frac{r^2\varphi'^2 h}{\varphi} \geqslant r^2\varphi''h + \varphi'h + |\varphi''\bar{z}^2 h + 2\varphi'\bar{z}h_z|, \tag{9.62}$$

which we shall prove now. We divide both sides of this inequality by the positive quantity $-r^2\varphi''h$ (if φ'' is zero there is nothing to prove), and find the equivalent inequality

$$-\frac{\varphi'^2}{\varphi\varphi''} \geqslant -1 - \frac{\varphi'}{r^2\varphi''} + \left|-\frac{\bar{z}^2}{r^2} - 2\frac{\varphi'\bar{z}h_z}{r^2\varphi''h}\right| = -1 - \frac{\varphi'}{r^2\varphi''} + \left|1 + \frac{\varphi'}{r^2\varphi''}\frac{2zh_z}{h}\right|.$$

Since $-(\varphi')^2/\varphi\varphi''$ is positive, it suffices to prove that

$$1 + t \geqslant |1 + tw| \quad \text{when} \quad t = \frac{\varphi'(r^2)}{r^2\varphi''(r^2)} \quad \text{and} \quad w = \frac{2zh_z(z)}{h(z)}.$$

This inequality, in turn, follows from

$$(1 + t)^2 \geqslant |1 + tw|^2 = 1 + 2t\operatorname{Re} w + t^2|w|^2,$$

which holds as soon as $2 + t \geqslant 2\operatorname{Re} w + t|w|^2$. By hypothesis $\operatorname{Re} w \leqslant a < 1$ and $|w| \leqslant A$, so (9.62) follows as soon as either $A \leqslant 1$ or else $A > 1$ and $t \leqslant 2(1 - a)/(A^2 - 1)$. This proves the proposition.

Example 9.4.39 *As an example of the function φ in Proposition 9.4.38 we let s_0 be an arbitrary number such that $0 < s_0 < 1$ and take a smooth function φ satisfying $\varphi(s) = 1$ for $s \leqslant s_0$ and whose derivative is $\varphi'(s) = -C\exp(-1/(s - s_0))$ for $s > s_0$. Then we determine C to make $\varphi(1) = 0$; this means that we choose C to satisfy*

$$C\int_{s_0}^1 e^{-1/(s - s_0)}\mathrm{d}s = 1.$$

We note that $\varphi'(s)/s\varphi''(s) = (s-s_0)^2/s$, which varies between 0 and $(1-s_0)^2$. Thus if $1-s_0$ is small enough, we can conclude that the new function $\varphi(z\overline{z})h(z)$ satisfies the Behnke–Peschl differential condition (9.59) over the open unit disk and it agrees with h when $|z| \leqslant \sqrt{s_0}$. $\qquad\square$

We need to study condition (9.60) more closely. In fact it has a simple geometric meaning.

Definition 9.4.40 *Let a complete Hartogs domain*

$$\Omega = \{(z,t) \in \omega \times \mathbf{C};\ |t|^2 < h(z)\}$$

*be defined over a bounded domain ω in $\mathbf{C}$ by a function $h \in C^1(\omega)$, $h > 0$. Denote by $(b(z),0)$ the point at which the tangent at a point $(z,t) \in \partial\Omega$ with $z \in \omega$ intersects the plane $t = 0$ (put $b(z) = \infty$ if there is no such point in $\mathbf{C}$). We shall say that Ω satisfies the **tangent condition** if*

$$\inf_{z\in\omega} d(b(z),\omega) > 0, \qquad (9.63)$$

where d denotes the distance from a point to a set. $\qquad\square$

If Ω is defined by a function $h \geqslant c > 0$ and is lineally convex, then it must satisfy the tangent condition, but not only that—we can deduce important quantitative information from its lineal convexity:

Lemma 9.4.41 *Let $R \in C^1(\omega)$ be such that the set Ω defined by (9.58) is lineally convex. Then*

$$\inf_{z\in\omega} d(b(z),\omega) \geqslant \frac{\inf_\omega R}{2\sup_\omega |R_z|} \geqslant \frac{\inf_\omega h}{\sup_\omega |h_z|}. \qquad (9.64)$$

If $R \geqslant c > 0$ in ω, then Ω satisfies the tangent condition (9.63).

Proof The tangent plane at a point $(z_0,t_0) \in \partial\Omega$ with $z_0 \in \omega$ is given by equation (9.46), and $b(z)$ is given by equation (9.47). The equation for the tangent can also be written as $t = \alpha(z - b(z_0))$. If Ω is lineally convex, then this tangent cannot intersect Ω, so we must have $|t| \geqslant R(z)$ whenever $z, z_0 \in \omega$. Thus

$$|t| = |\alpha(z - b(z_0))| \geqslant R(z) \quad \text{for all } z, z_0 \in \omega;$$

inserting the value of $|\alpha| = 2|R_z(z_0)| = |h_z(z_0)|/\sqrt{h(z_0)}$ we obtain

$$|z - b(z_0)| \geqslant \frac{R(z)}{2|R_z(z_0)|} = \frac{\sqrt{h(z)h(z_0)}}{|h_z(z_0)|}.$$

We now let z, z_0 vary in ω to get the desired conclusion.

The idea is to prove that the tangent condition is not only necessary as in Lemma 9.4.41, but also sufficient if ω is a disk, which we shall do in Proposition 9.4.42. We then proceed to prove that Ω does satisfy the tangent condition under our hypotheses if ω is a disk.

Proposition 9.4.42 *Assume that $h \in C^2\left(\overline{D}\right)$, $h > 0$, satisfies the Behnke–Peschl differential condition (9.59) and that Ω satisfies the tangent condition. Let φ be the function constructed in Example 9.4.39. Then $\varphi(z\overline{z})h(z)$ satisfies the differential condition if s_0 is sufficiently close to 1. Therefore, by Theorem 9.4.33, the open set $\{(z,t) \in D \times \mathbf{C};\ |t|^2 < \varphi(z\overline{z})h(z)\}$, which has a C^2 boundary, is lineally convex; as a consequence also its limit as s_0 tends to 1, viz. Ω itself, is lineally convex.*

Proof Using formula (9.47) for $b(z)$, the relation between the inequality (9.60) used in the proof of Proposition 9.4.38 and the tangent condition is easy to establish. We observe that $|b(z)| = |z - h(z)/h_z(z)| > |z|$ if and only if $\operatorname{Re} 2zh_z(z)/h(z) < 1$. Thus if Ω satisfies the tangent condition, then h satisfies (9.60) for some $a < 1$ and all z in some sufficiently narrow annulus $\sqrt{s_0} \leqslant |z| \leqslant 1$.[4]

Define

$$A = \sup_{|z| \leqslant 1} \left| \frac{2zh_z(z)}{h(z)} \right| \quad \text{and} \quad a(s_0) = \sup_{\sqrt{s_0} \leqslant |z| \leqslant 1} \operatorname{Re}\left[\frac{2zh_z(z)}{h(z)} \right].$$

If $A \leqslant 1$ we are done; otherwise we can choose $s_0 < 1$ so close to 1 that $(1 - s_0)^2 \leqslant 2(1 - a(s_0))/(A^2 - 1)$. Proposition 9.4.38 can be applied and shows that $\varphi(z\overline{z})h(z)$ satisfies the differential condition.

We shall now prove that it can never happen that $\operatorname{Re} 2zh_z(z)/h(z) \geqslant 1$ for any z with $|z| \leqslant 1$.

Proposition 9.4.43 *If $h \in C^2\left(\overline{D}\right)$, $h > 0$, satisfies the Behnke–Peschl differential condition (9.59), then Ω satisfies the tangent condition (9.63).*

Proof Let us define

$$b_0(r) = \inf_{|z| \leqslant r} |b(z)|, \qquad 0 < r \leqslant 1.$$

This is a decreasing function and it is continuous where it is finite. The tangent condition for $\Omega_r = \{(z,t) \in D(0,r) \times \mathbf{C};\ |t|^2 < h(z)\}$ means precisely that $b_0(r) > r$. It is clear that the condition is satisfied for a very small r. Indeed, $b(0) = -h(0)/h_z(0)$ is either ∞ or a non-zero complex number; in view of the continuity, $|b(z)| > r$ if $|z| \leqslant r$ and r is small enough.

If the tangent condition is satisfied for a particular Ω_r, then by Proposition 9.4.42 the set Ω_r is lineally convex, so Lemma 9.4.41 can be applied and shows that $b_0(r) \geqslant r + \varepsilon$, where

$$\varepsilon = \frac{\inf_{|z| \leqslant 1} R}{2 \sup_{|z| \leqslant 1} |R_z|} > 0.$$

---

[4]Here we could remark that it would be enough to require that $b(z) \notin \overline{\omega}$ only for all $z \in \partial\omega$, supposing that $h \in C^2(\overline{\omega})$. The stronger condition used in Definition 9.4.40 is however easier to handle in the proof of Proposition 9.4.43.

We know that $b_0(r) > r$ for small values of r, and we have just seen that if $b_0(r) > r$, then also $b_0(r) \geqslant r + \varepsilon$, for a positive ε that does not depend on r. Therefore that function cannot assume any value in the interval $]r, r + \varepsilon[$: it must satisfy $b_0(r) > r$ all the way up to and including $r = 1$. This means that Ω satisfies the tangent condition.

Theorem 9.4.44 *Let $h \in C^2(D)$, $h > 0$, satisfy the Behnke–Peschl differential condition (9.59). Then the open set $\Omega = \{(z, t) \in D \times \mathbf{C};\ |t|^2 < h(z)\}$ is lineally convex.*

Proof If $h \in C^2(\overline{D})$ with $h > 0$ in $\overline{D}$ we see from Proposition 9.4.43 that Ω satisfies the tangent condition, so that Proposition 9.4.42 can be applied. In the general case with $h \in C^2(D)$, $h > 0$, we apply this result to a smaller disk rD, $r < 1$, to conclude that the domain over rD is lineally convex. Then we let $r \to 1$.

9.5 Weak Lineal Convexity

We start this section with a general presentation of weak lineal convexity. We then discuss local variants of this property.

A locally weakly lineally convex open set with boundary of class C^1 is also (globally) weakly lineally convex provided that it is bounded. But, as shown by Yuriĭ Zelinskiĭ, this is not true for unbounded domains. The purpose here is to construct explicit examples, Hartogs domains, showing this. Their boundary can have regularity $C^{1,1}$ or C^∞.

Obstructions to constructing smoothly bounded domains with certain homogeneity properties will be discussed.

9.5.1 Introduction

After the main definitions about variants of lineal convexity, we shall approach the comparison global vs. local. In my paper (1998) I proved that a differential condition that I called the Behnke–Peschl differential condition implies that a bounded and connected open subset of $\mathbf{C}^n$ with boundary of class C^2 is weakly lineally convex. The proof relied on a result by Yužakov and Krivokolesko (1971a, 1971b), proved also in (Hörmander 1994: Proposition 4.6.4).

Yuriĭ Zelinskij (2002a, 2002b) published an example of an unbounded set that is locally lineally convex but not lineally convex. His example is not very explicit. We shall construct here an explicit example—actually a Hartogs domain, which has the advantage of being easily visualized in three real dimensions. We construct domains with boundary of class $C^{1,1}$ and a certain homogeneity property (Example 9.5.13), and show that this cannot be done with a boundary of class C^2 (Proposition 9.5.18). However, the boundary can be of class C^∞ if the homogeneity requirement is dropped (Example 9.5.14).

9.5.2 Lineal convexity

The property of being lineally convex was defined in Definition 9.4.1 on page 279. To wit:

Definition 9.5.1 *A subset of $\mathbf{C}^n$ is said to be **lineally convex** if its complement is a union of complex affine hyperplanes.* □

To every set A there exists a smallest lineally convex subset $\mu(A)$ that contains A. Clearly the mapping $\mu\colon \mathscr{P}(\mathbf{C}^n) \to \mathscr{P}(\mathbf{C}^n)$, where $\mathscr{P}(\mathbf{C}^n)$ denotes the family of all subsets of $\mathbf{C}^n$ (the power set), is increasing and idempotent, in other words an ethmomorphism (morphological filter). It is also larger than the identity, so that μ is a cleistomorphism (closure operator) in the ordered set $\mathscr{P}(\mathbf{C}^n)$.

This kind of complex convexity was introduced by Heinrich Behnke (1898–1997) and Ernst Ferdinand Peschl (1906–1986). I learnt about it from André Martineau (1930–1972) when I was in Nice during the academic year October 1967 through September 1968. See Martineau's papers (1966, 1967, 1968), also in (*Œuvres de André Martineau* 1977).

Are there lineally convex sets which are not convex? This is obvious in one complex variable, and from there we can easily construct, by taking Cartesian products, lineally convex sets in any dimension that are not convex. But these sets do not have smooth boundaries. Hörmander (1994:293, Remark 3) constructs open connected sets in $\mathbf{C}^n$ with boundary of class C^2 as perturbations of a convex set. These sets are lineally convex and close to a convex set in the C^2 topology, and therefore starshaped with respect to some point if the perturbation is small. Also the symmetrized bidisk

$$\{(z_1 + z_2, z_1 z_2) \in \mathbf{C}^2; |z_1|, |z_2| < 1\},$$

studied by Agler & Young (2004) and Pflug & Zwonek (2012), is not convex—not even biholomorphic to a convex domain (Nikolov et al. (2008)—but it is starshaped with respect to the origin (Agler & Young 2004: Theorem 2.3). So we may ask:

Question 9.5.2 *Does there exist a lineally convex set in $\mathbf{C}^n$, $n \geqslant 2$, with smooth boundary that is not starshaped with respect to any point?* □

We shall return to this question in Subsection 9.5.10.

9.5.3 Weak lineal convexity

Definition 9.5.3 *An open subset Ω of $\mathbf{C}^n$ is said to be **weakly lineally convex** if there passes, through every point on the boundary of Ω, a complex affine hyperplane which does not cut Ω.* □

It is clear that every lineally convex open set is weakly lineally convex. The converse does not hold. This is not difficult to see if we allow sets that are not connected:

Example 9.5.4 *Given a number c with $0 < c < 1$, define an open set Ω_c in $\mathbf{C}^2$ as the union of the set*

$$\{z = (x_1 + iy_1, x_2 + iy_2) \in \mathbf{C}^2; \; c < |x_1| < 1, \; |y_1| < 1, \; |x_2| < c, \; |y_2| < c\}$$

with the two sets obtained by permuting x_1, x_2 and y_2. Thus Ω_c consists of six boxes. It is easy to see that it is weakly lineally convex, but there are many points in its complement such that every complex line passing through that point hits Ω_c.

Any complex line intersects the real hyperplane defined by $y_1 = 0$ in the empty set or in a real line or in a real two-dimensional plane, and the three-dimensional set $\{z; \; y_1 = 0\} \cap \Omega_c$ is easy to visualize. □

It is less easy to construct a connected set with these properties, but this has been done by Yužakov & Krivokolesko (1971b:325, Example 2). See also an example due to Hörmander in the book by Andersson, Passare & Sigurdsson (2004:20–21, Example 2.1.7).

However, the boundary of the constructed set is not of class C^1, and this is essential. Indeed, Yužakov & Krivokolesko (1971b:323, Theorem 1) proved that a connected bounded open set with "smooth" boundary is locally weakly lineally convex in the sense of Definition 9.5.8 below if and only if it is lineally convex. It is then even **C**-convex (1971b:324, Assertion). See also Corollary 4.6.9 in (Hörmander 1994), which states that a connected bounded open set with boundary of class C^1 is locally weakly lineally convex if and only if it is **C**-convex (and every **C**-convex open set is lineally convex).

There cannot be any cleistomorphism connected with the notion of weak lineal convexity for the simple reason that the property is defined only for open sets. We might therefore want to define weak lineal convexity for arbitrary sets. We may ask:

Question 9.5.5 *Is there a reasonable definition of weak lineal convexity for all sets which keeps the definition for open sets and is such that there is a cleistomorphism associating to any $A \subset \mathbf{C}^n$ the smallest set that contains A and is weakly lineally convex?*   □

Question 9.5.6 *The operation $L \mapsto L \cap \Omega$ associating to a complex line L its intersection with an open set Ω has continuity properties that seem to be highly relevant for weak lineal convexity. Here the family of complex lines can arguably have only one topology, but for the family of sets $L \cap \Omega$ there is a choice of several topologies, especially if Ω is unbounded.*

Can an interesting theory be built starting from this remark?

9.5.4 Local weak lineal convexity

Definition 9.5.7 *We shall say that an open set $\Omega \subset \mathbf{C}^n$ is **locally weakly lineally convex** if for every point p there exists a neighborhood V of p such that $\Omega \cap V$ is weakly lineally convex.*   □

Obviously, a weakly lineally convex open set has this property, but the converse does not hold, which is obvious for sets that are not connected: Take the union of two open balls whose closures are disjoint. Also for connected sets the converse does not hold as we showed in Example 9.4.8. In that example it is essential that the boundary is not smooth.

Zelinskij (1993:118, Example 13.1) constructs an open set which is locally weakly lineally convex but not weakly lineally convex. The set is not equal to the interior of its closure.

Definition 9.5.8 *Let us say that an open set Ω is **locally weakly lineally convex in the sense of Yužakov and Krivokolesko** (1971b:323) if for every boundary point p there exists a complex hyperplane Y passing through p and a neighborhood V of p such that Y does not meet $V \cap \Omega$.* □

Zelinskij (1993:118, Definition 13.1) uses this definition and calls the property локальная линейная выпуклость (*lokal'naja linejnaja vypuklost'*). As we shall see, this property is strictly weaker than the local weak lineal convexity defined above in Definition 9.5.7.

Hörmander (1994: Proposition 4.6.4) and Andersson, Passare & Sigurdsson (2004: Proposition 2.5.8) use this property only for open sets with boundary of class C^1. Then the hyperplane Y is unique.

For all open sets, local weak lineal convexity obviously implies local weak lineal convexity in the sense of Yužakov and Krivokolesko. In the other direction, Hörmander's Proposition 4.6.4 shows that for bounded open sets with boundary of class C^1, local weak lineal convexity in the sense of Yužakov and Krivokolesko implies local weak lineal convexity (even weak lineal convexity if the set is connected).

Nikolov (2012: Proposition 3.7.1) and Nikolov et al. (2013: Proposition 3.3) have a local result in the same direction: If Ω has a boundary of class C^k, $2 \leqslant k \leqslant \infty$, and $\Omega \cap B_<(p, r)$, where p is a given point, is locally weakly lineally convex in the sense of Yužakov and Krivokolesko at all points near p, then there exists a **C**-convex open set ω (hence lineally convex) with boundary of class C^k such that $\omega \cap B_<(p, r') = \Omega \cap B_<(p, r')$ for some positive r'.

However, in general, the two properties are not equivalent:

Example 9.5.9 *The bounded connected open subset Ω^r of $\mathbf{C}^2$, taking $r = 2$, which was defined in Example 9.4.8 on page 281, has Lipschitz boundary and is locally weakly lineally convex in the sense of Yužakov and Krivokolesko but not locally weakly lineally convex. While Ω^r is locally weakly lineally convex for $2 < r < \sqrt{5}$, the set Ω^2 is not locally weakly lineally convex: The point $(0, 2)$ does not have a neighborhood with the desired property. But it does satisfy the property of Yužakov and Krivokolesko.* □

9.5.5 Approximation by smooth sets

Let $A_j \subset \mathbf{C}^{n_j}$ be two lineally convex sets in $\mathbf{C}^{n_j}$, $j = 1, 2$. Then it is easy to see that their Cartesian product $A_1 \times A_2 \subset \mathbf{C}^{n_1 + n_2}$ is lineally convex. In particular, if $n_1 = n_2 = 1$, then every Cartesian product in $\mathbf{C}^2$ is lineally convex. However, these sets cannot always be approximated by lineally convex sets with smooth boundaries.

If Ω_j, $j = 1, 2$, are convex open sets, then $\Omega = \Omega_1 \times \Omega_2$ is convex and can be approximated from within by convex open set $\Omega_{[\varepsilon]}$ with C^∞ boundaries, $\Omega_{[\varepsilon]} \nearrow \Omega$ as $\varepsilon \searrow 0$.

But if we let Ω_1 be an annulus and Ω_2 a disk, e.g.,

$$\Omega = \Omega_1 \times \Omega_2 = \{z \in \mathbf{C}^2;\ 1 < |z_1| < 3,\ |z_2| < 1\},$$

then it cannot be approximated by smooth weakly convex sets from the inside as we shall see in the next proposition and its corollary.

Proposition 9.5.10 *Let ω be a nonempty bounded open subset of $\mathbf{R}^2$ with boundary of class C^1. Suppose that $\inf_{x\in\omega}|x_1| > 0$. Define a Reinhardt domain Ω as*

$$\Omega = \{z \in \mathbf{C}^2;\ (|z_1|, |z_2|) \in \omega\}.$$

Then Ω is not locally weakly lineally convex.

Proof. Take a point $q = (q_1, q_2) \in \Omega$ with $q_1 < 0$, $q_2 \geqslant 0$. Denote by Ω^+ the set of all $x \in \Omega$ such that $x_1 > 0$ and $x_2 > 0$, and by L_α the complex line of equation $z_2 - q_2 = \alpha(z_1 - q_1)$, $\alpha \geqslant 0$. For $\alpha = 0$, the line cuts Ω^+ in $(-q_1, q_2)$; for large α it does not cut Ω^+. Now choose the smallest α such that L_α does not cut Ω^+. Then L_α contains at least one point $p \in \partial\Omega^+$, and L_α is the tangent plane of $\partial\Omega$ at p. Since this line meets Ω in q, Ω is not weakly lineally convex. But we can say more: It is not even locally weakly lineally convex. To see this, first note that $\alpha = (p_2 - q_2)/(p_1 - q_1) > 0$. Then there are points $z \in \Omega$ belonging to the tangent at p arbitrarily close to p. Indeed, since α is positive, a point z satisfying

$$|z_1| > p_1 \text{ and } |z_2| < p_2$$

belongs to Ω if it is close enough to p. In terms of z_1 this means that

$$|z_1| > p_1 \text{ and } |q_2 + \alpha(z_1 - q_1)| < p_2;$$

in other words that $z_1 \notin D_{\leqslant}(0, p_1)$ and that $z_1 \in D_{<}(c_1, r_1)$, the open disk with center at $c_1 = q_1 - q_2/\alpha$ and radius $r_1 = p_2/\alpha = p_1 - c_1$. Since $r_1 = p_1 - c_1$, there are points $z_1 \in D_{<}(c_1, r_1) \setminus D_{\leqslant}(0, p_1)$ which are arbitrarily close to p_1. □

Corollary 9.5.11 *A Reinhardt domain*

$$\Omega = \{z \in \mathbf{C}^2;\ r_1 < |z_1| < R_1,\ |z_2| < R_2\}$$

with $r_1 > 0$ is lineally convex but cannot be approximated by lineally convex domains with boundary of class C^1.

Proof. If a domain $\Omega_{[\varepsilon]}$ approximates Ω from the inside in the sense that

$$\Omega_{[\varepsilon]} \subset \Omega \subset \Omega_{[\varepsilon]} + B_{<}(0, \varepsilon),$$

then there is also a Reinhardt domain with this property: We may construct such a set by averaging over all rotations.

We can now apply the proposition to $\Omega_{[\varepsilon]}$. □

9.5.6 The Behnke–Peschl and Levi conditions

We refer to Subsection 9.4.5 for the definitions of the real and complex Hessian, the Levi form as well as the Levi condition and the strong Levi condition. Moreover, we defined there the Behnke–Peschl differential condition and the strict Behnke–Peschl differential condition.

The Behnke–Peschl differential condition says that the restriction of the real Hessian to the complex tangent space at any boundary point shall be positive semidefinite; for the strong case, positive definite.

Because of the different homogeneity of $H^{\mathbf{C}}$ and L, the inequality $\operatorname{Re} H_\rho^{\mathbf{C}} + L_\rho \geqslant 0$ is equivalent to $L \geqslant |H^{\mathbf{C}}|$. The inequality $L \geqslant |H^{\mathbf{C}}| \geqslant 0$ shows that the Behnke–Peschl differential condition implies the Levi condition.

In my paper (1998) I proved that a bounded connected open set with boundary of class C^2 is weakly lineally convex if it satisfies the Behnke–Peschl differential condition.

That this condition is necessary for weak lineal convexity was known since Behnke and Peschl (1935); the sufficiency was unknown at the time.

9.5.7 Yužakov and Krivokolesko: Passage from local to global

Let us quote the part of Proposition 4.6.4 in (Hörmander 1994) which is important for us:

Proposition 9.5.12 *Let $\Omega \subset \mathbf{C}^n$ be a bounded connected open set with boundary of class C^1 and assume that Ω is locally weakly lineally convex in the sense of Yužakov and Krivokolesko. Then Ω is weakly lineally convex.* $\qquad\square$

The result was proved by Yužakov & Krivokolesko (1971a, 1971b) under the condition that the boundary is "smooth."

9.5.8 A new example

We shall construct explicit Hartogs domains here with the properties mentioned in Zelinskij's example. We start with the simplest.

Example 9.5.13 *Define a function $\varphi^\diamond \colon \mathbf{C} \to \mathbf{R}$ by*

$$\varphi^\diamond(z_1) = \begin{cases} -x_1^2 - y_1^2, & x_1 \leqslant 0 \ or \ y_1 \leqslant 0; \\ -x_1^2 + y_1^2, & 0 \leqslant y_1 \leqslant x_1; \\ x_1^2 - y_1^2, & 0 \leqslant x_1 \leqslant y_1. \end{cases}$$

Then $\Omega_{\varphi^\diamond} = \{z \in \mathbf{C}^2;\ 1 + \varphi^\diamond(z_1) + |z_2|^2 < 0\}$ has boundary of class $C^{1,1}$ and is locally weakly lineally convex but not weakly lineally convex. $\qquad\square$

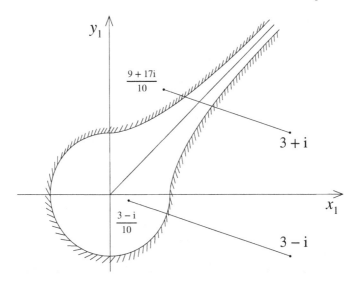

FIGURE 9.2
The set $\Omega_{\varphi^\diamond} \cap \{z \in \mathbf{C}^2;\ z_2 = 0\}$.

The properties of the set in this example will be discussed now and the properties will be seen to hold from Proposition 9.5.15.

The tangent plane at a boundary point $p = (p_1, p_2)$ with $\operatorname{Re} p_1 > 0$, $\operatorname{Im} p_1 > 0$, and $(\operatorname{Re} p_1)^2 > (\operatorname{Im} p_1)^2 + 1$, has the equation $-p_1(z_1 - p_1) + \bar{p}_2(z_2 - p_2) = 0$ and it passes through the point $q = (p_1 - |p_2|^2/p_1, 0)$. Choosing $p = (3+\mathrm{i}, \sqrt{7})$ we get $q = (\frac{9}{10} + \mathrm{i}\frac{17}{10}, 0) \in \Omega_{\varphi^\diamond}$, proving that $\Omega_{\varphi^\diamond}$ is not lineally convex.

We note that the tangent plane at a boundary point p with $\operatorname{Re} p_1 \leqslant 0$ or $\operatorname{Im} p_1 \leqslant 0$ is contained in the complement of $\Omega_{\varphi^\diamond}$; in particular, it hits the plane $z_2 = 0$ at the point $q = (p_1/|p_1|^2, 0) \notin \Omega_{\varphi^\diamond}$. We also note that the part of $\Omega_{\varphi^\diamond}$ where $0 < x_1 < y_1$ is convex, so any tangent plane of this part does not intersect it. Similarly, the part where $0 < y_1 < x_1$ is convex. Therefore $\Omega_{\varphi}^\diamond$ is the union of two lineally convex sets, taking the subsets where $x_1 < \max(y_1, 0)$, and $y_1 < \max(x_1, 0)$, respectively.

When $x_1 < 0$ or $y_1 < 0$ we get $\varphi_{z_1}^\diamond(z_1) = -\bar{z}_1$, $\varphi_{z_1 z_1}^\diamond(z_1) = 0$, $\varphi_{z_1 \bar{z}_1}^\diamond(z_1) = -1$; when $0 < y_1 \leqslant x_1$ we have $\varphi_{z_1}^\diamond(z_1) = -z_1$, $\varphi_{z_1 z_1}^\diamond(z_1) = -1$ and $\varphi_{z_1 \bar{z}_1}^\diamond(z_1) = 0$; when $0 < x_1 \leqslant y_1$ we have $\varphi_{z_1}^\diamond(z_1) = z_1$, $\varphi_{z_1 z_1}^\diamond(z_1) = 1$ and $\varphi_{z_1 \bar{z}_1}^\diamond(z_1) = 0$. In all three cases $|\varphi_{z_1 z_1}| - \varphi_{z_1 \bar{z}_1} = 1$. An application of Proposition 9.5.15 below now gives the result, except that it does not give anything at the exceptional points, where the function is not of class C^∞, i.e., those with $y_1 = 0$, $x_1 > 0$ or $x_1 = 0$, $y_1 > 0$. However, we have already seen that at these points, the tangent plane does not cut $\Omega_{\varphi^\diamond}$.

The boundary of $\Omega_{\varphi^\diamond}$ is not of class C^2 at the points where $y_1 = 0$, $x_1 > 0$ or $x_1 = 0$, $y_1 > 0$. The passage from $-x_1^2 - y_1^2$ for $y_1 \leqslant 0$ to $-x_1^2 + y_1^2$ for $0 \leqslant y_1 \leqslant x_1$ cannot be made analytically.

The function φ° is not of class $C^{1,1}$ at the points where $x_1 = y_1$, $x_1 > 0$, but this is of no consequence, since these points do not belong to the closure of the set it defines.

We note that the function φ° in the example is homogeneous of degree two:

$$\varphi^\circ(z_1) = \varphi^\circ(|z_1|e^{it}) = |z_1|^2\psi(t), \qquad z_1 \in \mathbf{C}, \; t \in \mathbf{R}.$$

It is therefore natural to ask if there is a C^∞ homogeneous function φ with the same properties. More precisely, we may ask for functions $\varphi\colon \mathbf{C} \to \mathbf{R}$ which yield a locally weakly lineally convex domain that is not weakly lineally convex in four different cases.

1.1. *Is there a C^∞ function φ with these properties?*

1.2. *Is there a homogeneous C^∞ function φ with these properties?*

2.1. *Is there an analytic function φ with these properties?*

2.2. *Is there a homogeneous analytic function φ with these properties?*

As we shall see, the answer to the first question is in the affirmative (Example 9.5.14). But the answer to Question 1.2 is in the negative (Proposition 9.5.18).

Example 9.5.14 *Now define $\varphi^\star\colon \mathbf{C} \to \mathbf{R}$ by*

$$\varphi^\star(z_1) = \begin{cases} -x_1^2 + \chi(y_1), & x_1 \geqslant y_1; \\ -y_1^2 + \chi(x_1), & x_1 \leqslant y_1, \end{cases}$$

where $\chi \in C^\infty(\mathbf{R})$ is a function of one real variable such that χ' is convex and which satisfies

$$\chi(y_1) = \begin{cases} -y_1^2 + \rho, & y_1 \leqslant -\tfrac{1}{2}; \\ y_1^2 + \sigma, & y_1 \geqslant \tfrac{1}{2}. \end{cases}$$

The convexity of χ' implies that $2|y_1| \leqslant |\chi'(y_1)| \leqslant \max(2|y_1|, 1)$ with equality to the left for $|y_1| \geqslant \tfrac{1}{2}$. This implies that we must have $\tfrac{1}{2} < \chi(\tfrac{1}{2}) - \chi(-\tfrac{1}{2}) < 1$, and we can actually choose χ so that $\chi(\tfrac{1}{2}) - \chi(-\tfrac{1}{2})$ is any given number in that interval.

For definiteness we now choose $\rho = -\tfrac{1}{4}$, $\sigma = 0$, $\chi(\tfrac{1}{2}) - \chi(-\tfrac{1}{2}) = \tfrac{3}{4}$, which implies that $\varphi^\circ - \tfrac{1}{4} \leqslant \varphi^\star \leqslant \varphi^\circ$, that $\Omega_{\varphi^\star}$ contains Ω_{φ°, and that the set of points $z \in \Omega_{\varphi^\star}$ with $\operatorname{Re} z_1 \geqslant \tfrac{1}{2}$ and $\operatorname{Im} z_1 \geqslant \tfrac{1}{2}$ is unchanged compared to Ω_{φ°. We choose χ as a suitable third primitive of

$$\chi'''(y_1) = C\exp(1/(y_1 - c) - 1/(y_1 + c)), \qquad -c < y_1 < c,$$

for a number c, $0 < c \leqslant \tfrac{1}{2}$, and a positive constant C, taking $\chi'''(y_1)$ equal to zero when $|y_1| \geqslant c$. This implies that χ' is even and that $\chi(0) = \tfrac{1}{2}\rho + \tfrac{1}{2}\sigma = -\tfrac{1}{8}$. Then

$$\Omega_{\varphi^\star} = \{z \in \mathbf{C}^2; \; 1 + \varphi^\star(z_1) + |z_2|^2 < 0\}$$

has boundary of class C^∞ and is locally weakly lineally convex but not lineally convex, since, just as for Ω_{φ°, the tangent plane at the boundary point $p = (3 + \mathrm{i}, \sqrt{7})$ passes through $q = (\frac{9}{10} + \mathrm{i}\frac{17}{10}, 0) \in \Omega_{\varphi^\star}$.   □

The properties mentioned in these two examples will follow from the next proposition and its corollary.

Proposition 9.5.15 *Let $\varphi \colon \mathbf{C} \to \mathbf{R}$ be a function of class C^k, $k = 2, 3, \ldots, \infty, \omega$ (C^ω denoting the family of all real analytic functions), and define an open set in $\mathbf{C}^2$ as*

$$\Omega_\varphi = \{z \in \mathbf{C}^2; \ 1 + \varphi(z_1) + |z_2|^2 < 0\}.$$

We assume that

$$\varphi_{z_1} \neq 0 \ \text{wherever} \ \varphi = -1, \tag{9.65}$$

and that

$$(-\varphi - 1)\left(|\varphi_{z_1 z_1}| - \varphi_{z_1 \bar{z}_1}\right) \leqslant |\varphi_{z_1}|^2 \ \text{in the set where} \ -\varphi - 1 \geqslant 0. \tag{9.66}$$

Then Ω_φ has boundary of class C^k and satisfies the Behnke–Peschl differential condition at every boundary point. If the inequality is strict at a certain point, we get the strict Behnke–Peschl differential condition at that point.

Proof In Lemma 9.4.26 on page 288 I described the domain by an inequality of the form $|z_2|^2 < h(z_1)$ and found that the Behnke–Peschl differential condition takes the form $h(h_{z_1 \bar{z}_1} + |h_{z_1 z_1}|) \leqslant |h_{z_1}|^2$, which, with $h(z_1) = -\varphi(z_1) - 1$, yields (9.66).

Corollary 9.5.16 *Let φ have the form $\varphi(z_1) = -x_1^2 + \chi(y_1)$ for $x_1 \geqslant y_1$ and $\varphi(z_1) = -y_1^2 + \chi(x_1)$ for $y_1 \geqslant x_1$. We assume that $\chi \in C^k(\mathbf{R})$, $k \geqslant 2$, with $-2 \leqslant \chi''$ and such that $\chi(y_1) > -1$ when $\chi'(y_1) = 0$. Then the conclusion of Proposition 9.5.15 holds under the assumption*

$$\tfrac{1}{4}\chi'(y_1)^2 + \chi(y_1) + 1 \geqslant 0, \qquad y_1 \in \mathbf{R}. \tag{9.67}$$

Proof The condition (9.65) is satisfied, since the gradient of φ in this case vanishes only when $x_1 = 0$ and $\chi'(y_1) = 0$. Then $1 + \varphi(z_1) + |z_2|^2 = 1 + \chi(y_1) + |z_2|^2 > 0$, so $z = (\mathrm{i}y_1, z_2)$ cannot be a boundary point of Ω.
  Condition (9.66) reduces to

$$(x_1^2 - \chi(y_1) - 1)\left(\left|-\tfrac{1}{2} - \tfrac{1}{4}\chi''(y_1)\right| + \tfrac{1}{2} - \tfrac{1}{4}\chi''(y_1)\right) \leqslant \left|-x_1 - \tfrac{1}{2}\mathrm{i}\chi'(y_1)\right|^2 = x_1^2 + \tfrac{1}{4}\chi'(y_1)^2,$$

provided $x_1^2 - \chi(y_1) - 1 \geqslant 0$. If $-2 \leqslant \chi''$, we have

$$\left|-\tfrac{1}{2} - \tfrac{1}{4}\chi''(y_1)\right| + \tfrac{1}{2} - \tfrac{1}{4}\chi''(y_1) = \tfrac{1}{2} + \tfrac{1}{4}\chi''(y_1) + \tfrac{1}{2} - \tfrac{1}{4}\chi''(y_1) = 1,$$

which gives (9.67). We then see that in this case the inequality holds also if $x_1^2 - \chi(y_1) - 1 < 0$.

In Example 9.5.14, the defining function $1 - x_1^2 + \chi(y_1) + |z_2|^2$ has nonvanishing gradient everywhere since $\chi' > 0$ everywhere. Smoothness follows.

The function $\varphi^\star$ is not of class C^∞ in the set where $x_1 = y_1$, $x_1 > 0$, but again this is unimportant since these points do not belong to the closure of $\Omega_{\varphi^\star}$. An application of Corollary 9.5.16 now gives the result. In fact, with the choice of $\rho = -\frac{1}{4}$, $\sigma = 0$, we need only note that $\chi(y_1) \geqslant -y_1^2 - \frac{1}{4}$ everywhere, and that $\chi'(y_1) \geqslant 2|y_1|$, so that

$$\tfrac{1}{4}\chi'(y_1)^2 + \chi(y_1) + 1 \geqslant \tfrac{3}{4} > 0, \qquad x_1 \geqslant y_1,$$

thus with strict inequality in (9.67) and (9.66); similarly for $x_1 \leqslant y_1$.

Remark 9.5.17 *It might be of interest to understand where the proof of Hörmander's Proposition 4.6.4 quoted above breaks down in the unbounded case. An important step in the proof is to see that, if we have a continuous family $(L_t)_{t\in[0,1]}$ of complex lines, the set T of parameter values t such that $L_t \cap \Omega$ is connected is both open and closed. Thus, if $0 \in T$, then also $1 \in T$. We shall see that closedness is no longer true for the sets in Examples 9.5.13 and 9.5.14.*

Define complex lines

$$L_t = \{z \in \mathbf{C}^2;\ z_2 = t(z_1 - 1 - \mathrm{i})\}, \qquad t \in [0,1],$$

which all pass through $(1+\mathrm{i}, 0) \notin \Omega_{\varphi^\star}$. Then $L_t \cap \Omega_{\varphi^\star}$ is connected for $0 \leqslant t < 1$ while $L_1 \cap \Omega_{\varphi^\star}$ is not.

We shall first see that $L_1 \cap \Omega_{\varphi^\star}$ is disconnected. If $z \in L_1 \cap \Omega_{\varphi^\star}$ and $x_1 \leqslant 0$ or $y_1 \leqslant 0$, then

$$f(z) = 1 + \varphi^\star(z_1) + |z_2|^2 \geqslant 1 - |z_1|^2 + \rho + |z_1 - 1 - \mathrm{i}|^2 = 3 + \rho - 2(x_1 + y_1).$$

Since we have chosen $\rho = -\frac{1}{4}$, the quantity $f(z)$ can be negative only if $x_1 + y_1 > 0$, which implies that z_1 satisfies either $x_1 > |y_1|$ or $y_1 > |x_1|$. Therefore the real hyperplane of equation $x_1 = y_1$ divides $L_1 \cap \Omega_{\varphi^\star}$ into two sets, which are nonempty since $(2, 1 - \mathrm{i})$ and $(2\mathrm{i}, \mathrm{i} - 1)$ both belong to L_1, the first with $y_1 < x_1$, the second with $y_1 > x_1$, and that both belong to $\Omega_{\varphi^\star}$ in view of the fact that $\chi(0) \leqslant 0$.

Next we shall see that $L_t \cap \Omega_{\varphi^\star}$ is connected when $0 \leqslant t < 1$. Given t such that $0 \leqslant t < 1$, we obtain for $z \in L_t \cap \Omega_{\varphi^\star}$ with $x_1, y_1 \leqslant 0$,

$$1 + \varphi^\star(z_1) + |z_2|^2 \leqslant 1 - (1 - t^2)|z_1|^2 - 2t^2(x_1 + y_1) + 2t^2.$$

This yields the estimate

$$1 + \varphi^\star(z_1) + |z_2|^2 \leqslant 3 + 4|z_1| - (1 - t^2)|z_1|^2,$$

which is negative when $|z_1| = R_t$ for a large enough number R_t, which depends on t. Obviously R_t tends to $+\infty$ as $t \to 1$, which explains that $L_1 \cap \Omega_{\varphi^\star}$ is disconnected. Let Γ_t be the arc in L_t with $|z_1| = R_t$ and $x_1 \leqslant 0$ or $y_1 \leqslant 0$, thus contained in $\Omega_{\varphi^\star}$.

An arbitrary point $a \in L_t \cap \Omega_{\varphi^\star}$ can be joined to a point in Γ_t by a straight-line segment contained in $\Omega_{\varphi^\star}$ and therefore also contained in $L_t \cap \Omega_{\varphi^\star}$. If $\operatorname{Re} a_1 \leqslant 0$ or $\operatorname{Im} a_1 \leqslant 0$ this follows from the fact that the set of points in $\Omega_{\varphi^\star}$ with argument of z_1 equal to that of a_1 is convex; otherwise from the fact that the points in $\Omega_{\varphi^\star}$ with $0 \leqslant \operatorname{Im} z_1 < \operatorname{Re} z_1$ is convex, as is the set of points with $0 \leqslant \operatorname{Re} z_1 < \operatorname{Im} z_1$. □

9.5.9   An impossibility result

Proposition 9.5.18 *Let $\Omega_\varphi = \{z \in \mathbf{C}^2;\ 1 + \varphi(z_1) + |z_2|^2 < 0\}$, where φ is positively homogeneous of degree two and of class C^2 where it is negative. Then either φ is constant and Ω_φ is lineally convex; or φ is not constant and Ω_φ is not connected.*

The set Ω_{φ° in Example 9.5.13 has the properties mentioned here except that its boundary is not of class C^2: We have a striking contrast between the regularity classes $C^{1,1}$ and C^2.

Proof. For functions $\varphi \colon \mathbf{C} \to \mathbf{R}$ which are positively homogeneous of degree two, i.e., of the form $\varphi(z_1) = |z_1|^2 \psi(t)$, $z_1 = |z_1|e^{it} \in \mathbf{C}$, $t \in \mathbf{R}$, condition (9.66) on φ takes the form

$$(-r^2\psi - 1)\left[-\psi + \sqrt{(\tfrac{1}{4}\psi'')^2 + (\tfrac{1}{2}\psi')^2} - \tfrac{1}{4}\psi''\right] \leqslant r^2(\psi^2 + \tfrac{1}{4}\psi'^2),$$

to hold in the set where $-r^2\psi - 1 \geqslant 0$; equivalently

$$4\psi + (-r^2\psi - 1)\left[\sqrt{\psi''^2 + 4\psi'^2} - \psi''\right] \leqslant r^2\psi'^2.$$

From this we obtain, if we divide by r^2 and let r tend to $+\infty$,

$$(-\psi)\left[\sqrt{\psi''^2 + 4\psi'^2} - \psi''\right] \leqslant \psi'^2. \tag{9.68}$$

But this condition is also sufficient, which follows on multiplication by r^2 and adding the trivial inequality $4\psi - \left[\sqrt{\psi''^2 + 4\psi'^2} - \psi''\right] \leqslant 0$.

To get rid of the square root in (9.68) we rewrite it as

$$\psi'^2\left(\psi'^2 + 2(-\psi)\psi'' - 4\psi^2\right) \geqslant 0,$$

where the left-hand side is of degree four.

We now introduce a function g by defining $g(t)$ as the positive square root of $-\psi(t)$ if $\psi(t)$ is negative and as 0 at all other points. Thus g is of class C^2 where it is positive, and $\psi = -g^2$ there. The points where $g = 0$, equivalently $\psi \geqslant 0$, are not of interest, since for them $1 + |z_1|^2\psi(t) + |z_2|^2 \geqslant 1 > 0$, implying that z does not belong to the closure of Ω_φ.

We get an inequality of degree eight but which is easy to analyze:

$$g^5 g'^2 (g + g'') \leqslant 0. \tag{9.69}$$

Thus, for each t such that $g(t) > 0$, either $g'(t) = 0$ or $g(t) + g''(t) \leqslant 0$. If g' is zero everywhere, i.e., if g is constant, it is known that Ω_φ is lineally convex, in particular weakly lineally convex. Wherever g is positive and g' is nonzero we get $g + g'' \leqslant 0$. This implies that any local maximum of g is isolated and that there can only be one point where the maximum is attained.[5] Hence, unless g' vanishes everywhere, $g + g'' \leqslant 0$ everywhere. We define $h = g + g'' \leqslant 0$ and obtain for any $a \in \mathbf{R}$

$$g(t) = g(a) \cos(t - a) + g(a) \int_a^t \sin(t - s) h(s) \mathrm{d}s, \qquad t \in \mathbf{R}.$$

The function g attains its maximum at some point which we may call a, and the formula then shows that $g(t) \leqslant g(a) \cos(t-a)$ for all t with $a \leqslant t \leqslant a+\pi/2$. In particular, g must have a zero t_0 in the interval $]a, a + \pi/2]$. By symmetry, g has a zero t_1 also in the interval $[a - \pi/2, a[$, hence at least two zeros in a period. This means that Ω is not connected, since the union of the rays $\arg z_1 = t_0$ and $\arg z_1 = t_1$ divides the z_1-plane. $\qquad\square$

9.5.10 A set which is not starshaped

A subset A of a vector space is said to be ***starshaped with respect to a point*** $a \in A$ if the segment $[a, b]$ is contained in A as soon as b belongs to A.

In answer to Question 9.5.2 we mention a modification of the set Ω_{φ° which is not starhaped.

Example 9.5.19 *Define* $\varphi^\sharp \colon \mathbf{C} \to \mathbf{R}$ *by*

$$\varphi^\sharp(z_1) = \begin{cases} -x_1^2 - y_1^2, & x_1 + y_1 \leqslant 0; \\ -\frac{1}{2}(x_1 - y_1)^2, & x_1 + y_1 \geqslant 0. \end{cases}$$

Then $\Omega_{\varphi^\sharp} = \{z \in \mathbf{C}^2;\ 1 + \varphi^\sharp(z_1) + |z_2|^2 < 0\}$ *has boundary of class* $C^{1,1}$ *and is lineally convex, but it is not starshaped with respect to any point.* $\qquad\square$

This set can conceivably be modified to have a boundary of class C^∞ like in Example 9.5.14. However, it is unbounded.

Question 9.5.20 *Does there exist a bounded set with boundary of class* C^2 *which is lineally convex but not starshaped?*

---

[5] Defining a function $g \colon \mathbf{R} \to \mathbf{R}$ of period 2π by $g(t) = |\cos 2t|$ when $0 \leqslant t \leqslant \pi$ and $g(t) = 1$ when $\pi < t < 2\pi$, we get a function which satisfies inequality (9.69) in $\mathbf{R} \setminus \pi\mathbf{Z}$, but not the conclusions we have drawn from it. This function is not of class C^2 (but of class $C^{1,1}$).

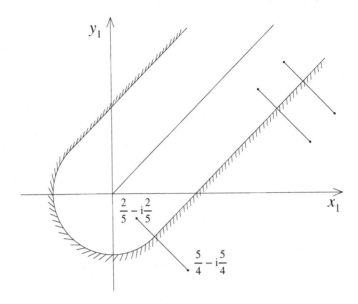

FIGURE 9.3
The set $\Omega_{\varphi^\sharp} \cap \{z \in \mathbf{C}^2;\ z_2 = 0\}$.

Question 9.5.21 *What about a Hartogs domain with these properties?*

The set Ω defined in Corollary 9.5.11 is bounded, lineally convex and not starshaped, but it has only a Lipschitz boundary. It cannot be approximated by a lineally convex set with smooth boundary.

9.6 A Differential Inequality Characterizing Weak Lineal Convexity

Abstract of this section

Behnke and Peschl introduced in 1935 the notion of *Planarkonvexität*, now called weak lineal convexity. They showed that, for domains with smooth boundary, it implies that a differential inequality is satisfied at every boundary point. We shall prove the converse here.

9.6.1 Introduction

In an article published in the *Mathematische Annalen* in 1935, Heinrich Behnke (1898–1979) and Ernst Peschl (1906–1986) introduced a notion of convexity called *Planarkonvexität,* nowadays known as *weak lineal convexity.* They showed that for domains in the space of two complex variables with boundary of class C^2, this property implies that a differential inequality is satisfied at every boundary point. Here we shall prove that, conversely, the differential inequality is sufficient for weak lineal convexity.

Behnke and Peschl (1935:170) proved that for sets with smooth boundary, weak lineal convexity is a local property (see Theorem 9.6.2 below).

Both usual convexity and pseudoconvexity can be characterized infinitesimally. The simplest example of such a result is that a C^2 function of one real variable is convex if and only if its second derivative is nonnegative. More generally, a domain in $\mathbf{R}^n$ with boundary of class C^2 is convex if and only if the Hessian of a defining function is positive semidefinite in the tangent space at every boundary point. Similarly, an open set in $\mathbf{C}^n$ with boundary of class C^2 is pseudoconvex if and only if the Levi form of a defining function is positive semidefinite in the complex tangent space at every boundary point (the Levi condition).

In analogy with these two classical results, we shall prove in the present section that a bounded connected open subset of $\mathbf{C}^n$ with boundary of class C^2 is weakly lineally convex if and only if the real Hessian of a defining function is positive semidefinite in the complex tangent space at every boundary point (the Behnke–Peschl condition).

It is easy to see that semidefiniteness is necessary. It is also known—indeed, this is the *Hauptsatz* of Behnke and Peschl (1935)—that the corresponding strong condition, i.e., that the real Hessian be positive definite, is sufficient. Thus what we have proved is that semidefiniteness is sufficient.

In the case of convexity and pseudoconvexity, the best way to deal with semidefiniteness is to approximate the domain by domains which satisfy the corresponding stronger condition of definiteness. This is not how we approach the problem here, at least not directly. The idea of proof of the main result here is instead to construct Hartogs domains which share a tangent plane with the given domain.

Question 9.6.1 *Can a weakly lineally convex domain with smooth boundary be approximated from the inside by domains satisfying the strong Behnke–Peschl condition? (For Hartogs domains this is known; see Theorem 9.4.33.*

$\square$

9.6.2   The main result

To be able to characterize sets by infinitesimal conditions, we shall describe boundaries and their curvature using defining functions and the Hesse and Levi forms. We refer to the definitions already given in Subsection 9.4.5.

As noted in the introduction, lineal convexity is not a local condition. Simple examples of sets which are locally lineally convex but not weakly lineally convex can be found in Section 9.4. However, weak lineal convexity is a local condition for sets with smooth boundary. The precise result is as follows.

Theorem 9.6.2 *Let Ω be a bounded connected open set in $\mathbf{C}^n$ with boundary of class C^1. Assume that for every boundary point a, the closure of the intersection of Ω with the complex tangent plane at a does not contain a. Then Ω is weakly lineally convex.* $\square$

For sets in $\mathbf{C}^2$ or $\mathbf{P}^2$ with boundary of class C^2, this was proved by Behnke and Peschl (1935:170). For a proof under the hypotheses stated here, see (Hörmander 1994: Proposition 5.6.4.) See also (Andersson, Passare & Sigurdsson 2004: Proposition 2.5.8). We shall need this result in our proof.

We recall two lemmas from Section 9.4: Lemmas 9.4.24 and 9.4.25, both due to Behnke and Peschl (1935: Theorems 7 and 8); local weak lineal convexity is called *Planarkonvexität im kleinen* by them. Cf. also (Zinov'ev 1971), (Hörmander 1994: Corollary 5.6.5).

Combining Lemma 9.4.25 and Theorem 9.6.2 we can deduce that the strict Behnke–Peschl differential condition (9.53) at all boundary points is sufficient for weak lineal convexity. This is the *Hauptsatz* of Behnke and Peschl (1935:170) (for sets in $\mathbf{C}^2$ or $\mathbf{P}^2$). We now state our main result, that in fact also the weaker condition (9.52) is sufficient:

Theorem 9.6.3 *Let Ω be a bounded connected open set in $\mathbf{C}^n$ with boundary of class C^2. Then Ω is weakly lineally convex if and only if Ω satisfies the Behnke–Peschl differential condition condition (9.52) at every boundary point.*

If Ω is locally weakly lineally convex, has a C^1 boundary, and in addition is bounded, then Ω is also $\mathbf{C}$-convex and lineally convex. This follows from (Andersson, Passare & Sigurdsson 2004: Proposition 2.5.8), who consider sets in projective space. I do not know how their result can be applied to unbounded domains in $\mathbf{C}^n$ with smooth boundary; such domains are not necessarily smoothly bounded in $\mathbf{P}^n$.

9.6.3 Results for Hartogs domains

Lineal convexity for Hartogs sets is easier to handle than in the general case. For some results, see Section 9.4, in particular Theorems 9.4.33 and 9.4.44.

Proposition 9.6.4 *Let Ω be an open set in $\mathbf{C}^n$ and define*

$$\Omega_H = \{z \in \mathbf{C}^n; \ (z_1, \ldots, z_{n-1}, \lambda z_n) \in \Omega \text{ for all } \lambda \in \mathbf{C} \text{ with } |\lambda| \leqslant 1\}. \quad (9.70)$$

This is the largest complete Hartogs set contained in Ω. If Ω is lineally convex, then Ω_H is lineally convex; similarly for weak lineal convexity. If $\partial\Omega$ is of class C^2 except perhaps where $z_n = 0$, then so is the boundary of Ω_H at all points z with $z_n \neq 0$ and satisfying the condition

$$2M|z_n| < |\rho_{z_n}(z)|, \quad (9.71)$$

where M is a bound for the second derivatives $\rho_{z_n z_n}$ and $\rho_{z_n \bar{z}_n}$. If, in addition, Ω satisfies the Behnke–Peschl differential condition (9.52) at all boundary points with $z_n \neq 0$, then so does Ω_H at all boundary points with $z_n \neq 0$ satisfying (9.71).

Proof If Ω is lineally convex, then also Ω_H, as an intersection of lineally convex sets, has this property:

$$\Omega_H = \bigcap_{|\lambda| \leqslant 1} \Omega_\lambda, \quad \text{where} \quad \Omega_\lambda = \{z \in \mathbf{C}^n; \ (z_1, \ldots, z_{n-1}, \lambda z_n) \in \Omega\}.$$

Assume now that Ω is only weakly lineally convex, and let a point a on the boundary of Ω_H be given. Then for some λ with $|\lambda| = 1$, a is on the boundary of Ω_λ defined above, and a hyperplane through a which does not intersect Ω_λ does not intersect Ω_H either. (The argument is valid for all a; if $a_n = 0$ we even have $a \in \partial\Omega_\lambda$ for all λ.)

If ρ defines Ω, then

$$\rho_H(z) = \sup_{\theta \in \mathbf{R}} \rho(z_1, \ldots, z_{n-1}, e^{i\theta} z_n) \quad (9.72)$$

defines Ω_H in a neighborhood of its closure. Define

$$\varphi(z_1, \ldots, z_n, \theta) = \rho(z_1, \ldots, z_{n-1}, e^{i\theta} z_n), \quad (z, \theta) \in \mathbf{C}^n \times \mathbf{R}.$$

We can calculate

$$\varphi_\theta = -2\mathrm{Im}\,(\rho_{z_n} e^{i\theta} z_n);$$

$$\varphi_{\theta\theta} = -2\mathrm{Re}\,(\rho_{z_n} e^{i\theta} z_n) - 2\mathrm{Re}\,(\rho_{z_n z_n} e^{2i\theta} z_n^2) + 2\rho_{z_n \bar{z}_n} |z_n|^2.$$

The value of θ which defines the supremum in (9.72) solves the equation $\varphi_\theta'' = 0$, and the implicit function theorem can be applied if $\varphi_{\theta\theta}'' \neq 0$ there. This condition is fulfilled if

$$|\mathrm{Re}\,(\rho_{z_n} e^{i\theta} z_n)| > 2M|z_n|^2, \quad (9.73)$$

where M is a bound for the second derivatives of ρ as defined in the statement of the proposition. However, when $\varphi'_\theta = 0$, the expression $\rho_{z_n} e^{i\theta} z_n$ is real, so that (9.73) simplifies to (9.71). The implicit function theorem then says that the boundary of Ω_H is as smooth as that of Ω where the condition is satisfied.

Now assume that Ω satisfies the Behnke–Peschl condition at a boundary point a of Ω_H with $a_n \neq 0$. Then a is on the boundary of some Ω_λ, $|\lambda| = 1$, as already noted above. Consider the functions

$$\varphi_\lambda(s) = \rho_\lambda(a + st), \qquad \varphi_H(s) = \rho_H(a + st), \qquad s \in \mathbf{R}, \quad t \in T_{\mathbf{C}}(a),$$

where $\rho_\lambda(z) = \rho(z_1, \ldots, z_{n-1}, \lambda z_n)$, the defining function for Ω_λ obtained by rotating ρ in the last coordinate.

The Behnke–Peschl condition holds for Ω_λ, which means that $(\varphi_\lambda)''(0) \geqslant 0$. Now $\varphi_H \geqslant \varphi_\lambda$ and both functions vanish at the origin, which implies $(\varphi_H)''(0) \geqslant (\varphi_\lambda)''(0)$. Thus the condition holds for Ω_H. This completes the proof.

In an application of this proposition in the next subsection we shall let Ω be defined near an arbitrarily given point by an inequality $y_n < f(z', x_n)$ for some real-valued function f of $n-1$ complex variables and one real variable. Then $\rho(z) = y_n - f(z', x_n)$ is a defining function for Ω near the given point. (Here $x_n = \operatorname{Re} z_n$, $y_n = \operatorname{Im} z_n$, and $z' = (z_1, \ldots, z_{n-1})$.) We see that $\rho_{z_n} = -\frac{1}{2}(f_{x_n} + i)$, so that $|\rho_{z_n}| \geqslant \frac{1}{2}$. Moreover

$$\rho_{z_n z_n} = \rho_{z_n \bar{z}_n} = -\tfrac{1}{4} f_{x_n x_n}.$$

This implies that a sufficient condition for (9.71) to hold is

$$C|z_n| < 1, \tag{9.74}$$

where C is a bound for $f_{x_n x_n}$.

Remark 9.6.5 *Condition (9.71) has a simple geometric meaning. With the defining function $\rho(z) = y_n - f(z', x_n)$ it says that the intersection of the boundary of Ω with the subspace $z' = $ constant has smaller curvature than the intersection of the boundary of Ω_H with the same subspace where the two boundaries meet. For simplicity we shall use the stronger condition (9.74) instead.*

9.6.4   Proof of the main result

We shall now prove Theorem 9.6.3. In view of Theorem 9.6.2 it is enough to prove that the complex tangent plane $a + T_{\mathbf{C}}(a)$ does not cut Ω near a. We shall assume that $a + T_{\mathbf{C}}(a)$ cuts Ω in a point b and then show that this leads to a contradiction if b is close to a.

First of all we may assume that $n = 2$ by looking at the two-dimensional affine complex subspace that contains a, b and a third point on the normal to

$\partial\Omega$ through a. We may also assume that the coordinate system is chosen so that $a = 0$ and the real tangent plane $a + T_{\mathbf{R}}(a)$ has the equation $\operatorname{Im} z_2 = 0$. We recall that both weak lineal convexity and the Behnke–Peschl condition (9.52) are invariant under complex affine mappings. The complex tangent plane at a then has the equation $z_2 = 0$, so that $b_2 = 0$. We shall consider a neighborhood W of a such that three conditions are satisfied. Let

$$W = \{z \in \mathbf{C}^2; \ |z_1| < R_1, |z_2| < R_2\}$$

and let V be its intersection with $\mathbf{C} \times \mathbf{R}$:

$$V = \{(z_1, x_2) \in \mathbf{C} \times \mathbf{R}; \ |z_1| < R_1, |x_2| < R_2\}.$$

The three conditions are:

(A) First of all the set Ω shall be defined in W by an inequality $\operatorname{Im} z_2 < f(z_1, \operatorname{Re} z_2)$ for some function f which is of class C^2 in a neighborhood of the closure of V.

(B) Next we shall assume that condition (9.74)) is satisfied for all $z \in W$ with some margin:

$$R_2 \sup_V |f_{x_2 x_2}| < \tfrac{2}{3} < 1.$$

(This is to allow a change of coordinates later.)

(C) Third, R_1 shall be so small that $MR_1 + C(1+M^2)R_1^2 < \tfrac{1}{4}R_2$, where $M = \tfrac{1}{2}CR_2$ and C is defined below.

To satisfy these conditions we have to specify the numbers R_1, R_2 and C. We first choose R_1 and R_2 so that (A) and (B) hold, and then define a constant C as follows. Since f is a function of class C^2 defined in a neighborhood of the closure of V and with vanishing derivatives of order up to one at the origin, there exists a constant C such that

$$|f(z_1, x_2)| \leqslant C(|z_1|^2 + x_2^2),$$
$$|f_{x_2}(z_1, x_2)| \leqslant C(|z_1| + |x_2|), \text{ and}$$
$$|f_{x_2 x_2}(z_1, x_2)| \leqslant C$$

for all $(z_1, x_2) \in V$. We finally shrink R_1 if necessary to make (C) hold.

With the choice of coordinate system we have made, the normal at a is the y_2-axis. Let c be a point on that axis with $\operatorname{Im} c_2 < 0$; it is convenient to take $c = -\tfrac{1}{4}iR_2$. Thus $c = (0, c_2)$ and $|c| = -\operatorname{Im} c_2 = \tfrac{1}{4}R_2$. The circle in the plane $z_1 = 0$ with center at c and radius $|c|$ passes through a and is tangent to the x_2-axis at that point.

We shall prove that $f(b) \leqslant 0$ (hence that $b \notin \Omega$) for all b with $|b_1| < R_1$. Assume the contrary: $f(b) > 0$. Consider the plane $z_1 = b_1$ and the graph of f restricted to that plane. Draw the normal to the graph of $f(b_1, \cdot)$ through the point $z_2 = \mathrm{i}f(b_1, 0)$ in the z_2-plane. This normal intersects the line $y_2 = \mathrm{Im}\,c_2$ at a point, which we call p_2. Define $p_1 = b_1$ and $p = (p_1, p_2) \in \mathbf{C}^2$. The slope of the normal is determined by the slope of the graph at $z_1 = b_1$, $x_2 = 0$, i.e., by $f_{x_2}(b_1, 0)$. This derivative can however be controlled: we know that $f_{x_2}(b_1, 0)$ is not more than $C|b_1|$ in modulus. The distance between p and c is

$$|p - c| = |f_{x_2}(b_1, 0)|(|c| + f(b_1, 0)) \leqslant C|b_1|(\tfrac{1}{4}R_2 + C|b_1|^2) \leqslant \tfrac{1}{2}CR_2|b_1|,$$

where the last estimate is a consequence of (C). Thus $|p - c| \leqslant M|b_1|$ with $M = \tfrac{1}{2}CR_2$.

We have constructed a disk D_0 in the plane $z_1 = 0$ with center at c_2 and with $z_2 = 0$ on its boundary, and now let D_1 be the disk in the plane $z_1 = b_1$ with center at p_2 and $\mathrm{i}f(b_1, 0)$ on its boundary (and therefore containing $z_2 = 0$):

$$D_0 = \{z \in \mathbf{C}^2;\ z_1 = 0, |z_2 - c_2| < |c|\};$$

$$D_1 = \{z \in \mathbf{C}^2;\ z_1 = b_1, |z_2 - p_2| < |\mathrm{i}f(b_1, 0) - p_2|\}.$$

Both disks are moreover contained in $\Omega \cap W$. For D_0 this is obvious from the construction; for D_1 this can be seen as follows. The center of D_1 is p_2 and its radius r_1 is $|\mathrm{i}f(b_1) - p_2|$. The disk is contained in W if $|p_2| + r_1 \leqslant R_2$. This inequality follows from the estimates we already have:

$$|p_2| + r_1 \leqslant 2|p_2| + C|b_1|^2 \leqslant 2|c_2| + 2|p_2 - c_2| + C|b_1|^2 \leqslant \tfrac{1}{2}R_2 + 2MR_1 + CR_1^2 \leqslant R_2,$$

where the last inequality follows from (C). Thus $D_1 \subset W$. That $D_1 \subset \Omega$ now follows from (B); cf. Remark 9.6.5.

If we construct a Hartogs domain by rotating Ω around an axis which passes through c and p, then this Hartogs domain will have a on its boundary and contain b. This is precisely what we shall do.

We introduce new coordinates (w_1, w_2) so that the w_1-axis, i.e., the plane $w_2 = 0$, passes through c and p. The w_2-axis need not be changed. This means that the new coordinates shall be defined as

$$w_1 = z_1, \qquad w_2 = z_2 - c_2 - (p_2 - c_2)z_1/b_1.$$

Indeed $z = c$ gives $w = 0$ and $z = p$ yields $w = b = (b_1, 0)$. We now define Ω_H in the w-coordinates. The tangent plane with equation $z_2 = 0$ has the equation $w_2 = -c_2 - (p_2 - c_2)w_1/b_1$ and is also the tangent plane to $\partial\Omega_H$ at the point $w = (0, -c_2)$. It intersects Ω_H at the point $z = b$, i.e., $w = (b_1, -p_2)$. That this point is an element of Ω_H follows from the construction of D_1.

We shall now apply Theorem 9.4.33 to Ω_H over the disk $|w_1| < R_1$ in the w_1-plane. To be able to do so we have to check that there is a point of Ω_H over every point w_1 with $|w_1| < R_1$, or equivalently that $(w_1, 0) \in \Omega_H$ for all w with $|w_1| < R_1$.

In the new coordinate system, the inequality defining Ω becomes

$$\mathrm{Im}\,w_2 < -\mathrm{Im}\,c_2 - \mathrm{Im}\,(p_2 - c_2)w_1/b_1 + f(w_1, \mathrm{Re}\,w_2 + \mathrm{Re}\,(p_2 - c_2)w_1/b_1).$$

Denote the right-hand side by $g(w_1, \operatorname{Re} w_2)$. In particular

$$g(w_1, 0) = -\operatorname{Im} c_2 - \operatorname{Im}(p_2 - c_2)w_1/b_1 + f(w_1, \operatorname{Re}(p_2 - c_2)w_1/b_1).$$

Recalling the estimate $|p_2 - c_2| \leqslant M|b_1|$ above, we get

$$g(w_1, 0) \geqslant \tfrac{1}{4}R_2 - M|w_1| - C(1 + M^2)|w_1|^2 \geqslant \tfrac{1}{4}R_2 - MR_1 - C(1 + M^2)R_1^2 > 0,$$

the last inequality coming from (C). This ensures that every point $(w_1, 0)$ with $|w_1| < R_1$ lies in Ω and therefore also in Ω_H.

We know that Ω_H satisfies the Behnke–Peschl differential condition at all boundary points if the condition in the w-coordinates corresponding to (9.74) is valid. Note that $|w_2 - z_2 + c_2| \leqslant M|z_1|$ independently of the choice of $b \in W$, from which we deduce

$$|w_2| \leqslant |z_2| + \tfrac{1}{4}R_2 + MR_1 \leqslant \tfrac{3}{2}R_2.$$

The second derivative of g with respect to $\operatorname{Re} w_2$ is the same as the second derivative of f with respect to $x_2 = \operatorname{Re} z_2$, so from (B) we can conclude that the condition (4.5) is satisfied also in the w-coordinates for all points $w \in \partial\Omega_H$ with $|w_1| < R_1$.

It now follows from Theorem 9.4.44 that Ω_H is lineally convex, which contradicts the fact that the tangent plane at the point $w = -c$ intersects Ω_H in $w = (b_1, -p_2)$. This completes the proof.

9.7 Generalized Convexity

Abstract of this section
Inspired by mathematical morphology we study generalized convexity and prove that certain subsets of Hartogs domains are convex in a generalized sense.

9.7.1 Introduction to this section

By the Hahn–Banach theorem, an open convex set in $\mathbf{R}^m$ is an intersection of open half-spaces; its complement a union of closed half-spaces. What if we replace the latter by balls? We shall study here a kind of generalized convexity where a set is called concave if it is a union of closed balls; its complement thus being an intersection of complements of closed balls. This will be done in particular for Hartogs domains which are lineally convex.

Lineal convexity is a kind of complex convexity intermediate between usual convexity and pseudoconvexity. More precisely, if A is a convex set in $\mathbf{C}^n$ which is either open or closed, then A is lineally convex (this is true also in the real category), and if Ω is a lineally convex open set in $\mathbf{C}^n$, then Ω is pseudoconvex.

As mentioned on page 263, there are several different notions of convexity related to lineal convexity.

The main results are presented in Subsections 9.7.8 and 9.7.10. It is shown there that certain subsets of Hartogs domains have convexity properties originating in mathematical morphology. We also study external tangent planes of sets that do not necessarily have a smooth boundary.

9.7.2 Hyperplanes, tangent planes, and multifunctions

Hyperplanes are affine subspaces with real or complex codimension 1, and they will play an important role in the sequel.

To any real hyperplane Y in $\mathbf{C}^n$ and every point $a \in Y$ there is a unique complex hyperplane $Y_{[a]}$ that contains a and is contained in Y. In fact

$$Y_{[a]} = Y \cap (\mathrm{i}(Y - a) + a).$$

We note that $Y_{[a]}$ depends continuously on (Y, a) for the natural topology on hyperplanes and points.

Conversely, every complex hyperplane Z in $\mathbf{C}^n$ is contained in a real hyperplane, but there are now several choices. If a complex hyperplane Z is given and is defined by the equation $\beta \cdot (z - a) = 0$, then for any complex number θ with $|\theta| = 1$ the real hyperplane $Z^{[\theta]}$ defined by $\mathrm{Re}\,\theta(\beta \cdot (z - a)) = 0$ contains Z. The real hyperplane $Z^{[\theta]}$ does not depend on the choice of $a \in Z$ and satisfies $(Z^{[\theta]})_{[b]} = Z$ for every $b \in Z$.

If a real hyperplane Y and a point $a \in Y$ are given, then $(Y_{[a]})^{[\theta]} = Y$ for two values of θ with $|\theta| = 1$. Explicitly, if Y is given by the equation $\operatorname{Re}\beta \cdot (z - a) = 0$, then $Y_{[a]}$ is given by $\beta \cdot (z - a) = 0$ and $(Y_{[a]})^{[\theta]}$ by $\operatorname{Re}\theta(\beta \cdot (z - a)) = 0$; the two choices $\theta = \pm 1$ give us Y back.

For the definition of the real and complex tangent spaces to an open subset Ω of $\mathbf{C}^n$ with boundary of class C^1, as well as the real and complex tangent planes, we refer to Definition 9.4.19.

Clearly $T_{\Omega,\mathbf{C}}(b) = T_{\Omega,\mathbf{R}}(b)_{[0]}$; for the tangent planes,

$$b + T_{\Omega,\mathbf{C}}(b) = (b + T_{\Omega,\mathbf{R}}(b))_{[b]}.$$

Definition 9.7.1 *If A is a subset of $\mathbf{C}^n$, we shall denote by $\Gamma_A(a)$ the set of all complex hyperplanes Z which pass through the origin and are such that $a + Z$ does not intersect A.* □

Definition 9.7.2 *A mapping $F\colon X \to \mathscr{P}(Y)$ will be called a **multifunction** from X into Y and will be written $F\colon X \rightrightarrows Y$. (This means that the value, image, or fiber $F(x)$ of F at a point x is a subset of Y, possibly empty.) The **graph** of a multifunction F, denoted by $\mathbf{graph}(F)$, is the set $\{(x,y) \in X \times Y;\ y \in F(x)\}$.* □

If X and Y are topological spaces, we can equip $X \times Y$ with the Cartesian product topology. In all cases considered here, X is a T_1 space—equivalently, all singleton sets are closed. If so, for $\mathbf{graph}(F)$ to be a closed subset of $X \times Y$, it is necessary but not sufficient that the fiber

$$F(a) = (\{a\} \times Y) \cap \mathbf{graph}(F)$$

be a closed subset of Y for every $a \in X$.

Thus Γ_A is a multifunction $\Gamma_A\colon \mathbf{C}^n \rightrightarrows \mathbf{Gr}_{n-1}(\mathbf{C}^n) = M_{n,n-1}(\mathbf{C})$ with values in the Grassmann manifold of all complex hyperplanes in $\mathbf{C}^n$ passing through the origin. If Ω is open, $\Gamma_\Omega(a)$ is closed for every $a \in \mathbf{C}^n$. See also Proposition 9.7.15.

Lineal convexity of a set A means that $\Gamma_A(a)$ is nonempty for every $a \in \mathbf{C}^n \setminus A$; weak lineal convexity of an open set Ω that $\Gamma_\Omega(b)$ is nonempty for every $b \in \partial\Omega$.

Let us agree to say that a topological space is **connected** if the only sets which are both open and closed are the empty set and the whole space (not necessarily distinct).[6] A subset of a topological space is said to be **connected** if it is connected as a topological subspace.

Zelinskiĭ (1981) has proved that a bounded lineally convex open set Ω is $\mathbf{C}$-convex if and only if $\Gamma_\Omega(b)$ is connected for every boundary point b. See also (Andersson, Passare & Sigurdsson 2004:46, Theorem 2.5.2) for the corresponding result on subsets of projective space.

---

[6]We follow here Bourbaki (1961: I: §11: 1) in that the empty space is defined to be connected. Adrien Douady (personal communication, 2000 June 26) argued for the empty space not to be connected. The difference is important in Definition 5, where $\mathbf{C}$-convexity is defined.

9.7.3   Accessibility

We refer to Subsection 9.3.7 for the set-theoretical operations δ_S, ε_S, κ_S and α_S with respect to a structuring element S. It is convenient to express closedness and openness for some of these operators in terms of accessibility:

Definition 9.7.3 *If A is a subset of $\mathbf{R}^m$ or $\mathbf{C}^n$ and b a point in this space, we shall say that b is S-**accessible from the outside** if b belongs to the closure of $\alpha_S(\complement A)$. In particular we shall speak about accessibility from the outside by balls of radius r if S is equal to $B_{\leqslant}(0,r)$ or $B_{<}(0,r)$.* ☐

Remark 9.7.4 *If b is S-accessible from the outside of a certain class, then there is also a set T of the same class such that $\overline{A} \cap \overline{T} = \{b\}$. Indeed, if S satisfies*

$$\{x;\ f(x) < 0\} \subset S \subset \{x;\ f(x) \leqslant 0\},$$

then T can be taken as the set of all x such that $f(x) + \|x - b\|_2^2 \leqslant 0$. ☐

We shall consider regularity classes $C^{k,\beta}$, where $k \in \mathbf{N}$ and $0 \leqslant \beta \leqslant 1$, meaning that the functions considered are of class C^k and all derivatives of order k are Hölder continuous of order β, with the understanding that $C^{k,0} = C^k$.

Definition 9.7.5 *If $b \in \partial A$ is accessible from the outside by a structuring element S having boundary of class $C^{k,\beta}$ with $k \geqslant 1$, then we shall say that the unique tangent plane to S at b is an **external tangent space** of A at b. The set of all external tangent spaces at a point b, a subset of the Grassmann manifold $\mathbf{Gr}_{m-1}(\mathbf{R}^m) = M_{m,m-1}(\mathbf{R})$ of all real hyperplanes passing through the origin, will be denoted by $\Theta_{A,\mathbf{R}}^{k,\beta}(b)$, and the corresponding multifunction $\partial A \rightrightarrows \mathbf{Gr}_{m-1}(\mathbf{R}^m)$ by $\Theta_{A,\mathbf{R}}^{k,\beta}$.*

*If Ω is an open subset of $\mathbf{C}^n$, we shall denote by $\Theta_{\Omega,\mathbf{C}}^{k,\beta}(b)$ the set of all complex hyperplanes through the origin contained in planes in $\Theta_{\Omega,\mathbf{R}}^{k,\beta}(b)$; we call them **complex external tangent spaces**. It is the set of all complex hyperplanes $Z = Y_{[0]}$, $Y \in \Theta_{\Omega,\mathbf{R}}^{k,\beta}(b)$.*

When the class is clear from the context or is unimportant, we shall omit the superscripts $^{k,\beta}$. ☐

It is easy to see that $\Theta_{\Omega,\mathbf{R}}^{1,1} = \Theta_{\Omega,\mathbf{R}}^2 = \Theta_{\Omega,\mathbf{R}}^{\infty}$.

The relation between $\Gamma_\Omega(b)$ and $\Theta_{\Omega,\mathbf{C}}^2(b)$, $b \in \partial\Omega$, seems to be of interest.

Definition 9.7.6 *Let us say that Ω is **tangentially lineally convex at** $b \in \partial\Omega$ if no complex external tangent plane of class C^2 at b meets Ω, i.e., if $\Theta_{\Omega,\mathbf{C}}^2(b) \subset \Gamma_\Omega(b)$.* ☐

Proposition 9.7.7 *Let $b \in A \subset \mathbf{R}^m$ be accessible from the outside by balls of radius $r > 0$. Then $\Theta_{A,\mathbf{R}}^{k,\beta}(b)$ is connected.*

Proof Take $b = 0$ and assume that $\overline{A} \cap \overline{U_j} = \{0\}$, $j = 0, 1$, where U_j is the set of all points x such that $f_j(x) < 0$, and f_j is a function of a given regularity and with nonvanishing gradient wherever it is zero. This is justified by Remark 9.7.4. We now form $f_s = (1 - s)f_0 + sf_1$, $0 \leqslant s \leqslant 1$, and claim that the set where f_s is negative defines an open set U_s which serves to prove that all gradients

$$(\mathbf{grad}\, f_s)(0) = (1 - s)(\mathbf{grad}\, f_0)(0) + s(\mathbf{grad}\, f_1)(0)$$

can occur, implying that there is a curve connecting the hyperplane defined by f_0 to that defined by f_1. We note that the gradient of f_s is nonzero at the origin except in the case when $(\mathbf{grad}\, f_1)(0)$ is a negative multiple of $(\mathbf{grad}\, f_0)(0)$. In that case, however, the hyperplanes defined by the two gradients are the same, so there is nothing to prove. We modify f_s outside a neighborhood of the origin if necessary to make sure that it satisfies the requirement that its gradient be nonzero everywhere where the function itself vanishes.

If $x \in \overline{A} \smallsetminus \{0\}$, then $x \notin \overline{U_j}$, $j = 0, 1$, so that $f_j(x) > 0$, $j = 0, 1$. This implies that $f_s(x) > 0$, so that $x \notin \overline{U_s}$. Thus we have proved that $\overline{A} \cap \overline{U_s} \subset \{0\}$; obviously $\overline{A} \cap \overline{U_s} \supset \{0\}$. In conclusion, we have proved that the tangent plane of U_s at $b = 0$ belongs to $\Theta^{k,\beta}_{A,\mathbf{R}}(0)$ for all s with $0 \leqslant s \leqslant 1$.

Example 9.7.8 *Let us define a cleistomorphism*

$$\kappa_r \colon \mathscr{P}(\mathbf{C}^n) \to \mathscr{P}(\mathbf{C}^n) \ \text{or} \ \kappa_r \colon \mathscr{P}(\mathbf{R}^m) \to \mathscr{P}(\mathbf{R}^m)$$

as the cleistomorphism with structuring element $U = \complement B_<(0, r)$ for some positive radius r. It follows that $\kappa_r(A)$ is closed for any set A, perhaps most easily seen by observing that its complement, denoted by $\alpha_r(\complement A)$, is the union of all open balls $B_<(x, r)$ that are contained in $\complement A$.

Thus $\kappa_r(A)$ is the smallest invariance set containing A whose boundary points are all accessible by balls of radius r, and we see that the boundary points of a closed set F are accessible by such balls if and only if $\kappa_r(F) = F$.

To treat open sets, we define $\lambda_r(A)$ as the interior of $\kappa_r(A)$. In view of Proposition 9.3.8 the operation $A \mapsto \lambda_r(A) = (\kappa_r(A))^\circ$ is an ethmomorphism. If we restrict it to open sets, it is larger than the identity, i.e., $\lambda_r(\Omega) \supset \Omega$ for all open sets Ω. So accessibility for open sets is defined by the fixed points of λ_r.

The infimum of all the κ_r, $r > 0$, is just the topological closure. $\square$

9.7.4 Concavity and convexity with respect to a structuring element or a family of structuring elements

Just as it is sometimes easier to look at lineally concave sets rather than lineally convex sets, it can be more convenient to define accessibility from the inside than from the outside. We shall do this in terms of concavity and convexity with respect to a structuring element, treating both properties in parallel:

Definition 9.7.9 *Given a subset* S *(called **structuring element**) of an abelian group* G, *we shall say that a subset* A *of* G *is* S-***concave*** *if it is a union of translates* $x + S$ *with* x *in some subset* X *of* G. *We shall say that it is* S-***convex*** *if its complement is* S-*concave.*

We define the S-***kernel*** *of a set* A, *denoted by* $\alpha_S(A)$, *as the union of all translates* $x + S$ *contained in* A. *We define the* S-***hull*** *of a set* B, *denoted by* $\kappa_S(B)$, *as the complement of the* S-*kernel of* $\complement B$. $\square$

Obviously $\complement \alpha_S(A) = \kappa_S(\complement A)$.

The anoiktomorphism α_S

$$\alpha_S(A) = \bigcup_{x \in G} (x + S; \; x + S \subset A), \qquad A \in \mathscr{P}(G),$$

has as fixed points the S-concave sets. We have $\alpha_S(A) \subset A \subset \kappa_S A)$ for all nonempty sets $S \in \mathscr{P}(G)$ and all A.

We can consider the sets $x + S$ as voxels or pixels, and see that no smaller sets are allowed to build up a S-concave set. Or we can think of elements x as atoms and sets $x + S$ as molecules—no free atoms are allowed; they must all be part of a molecule.

What we have done so far is define concavity and convexity with respect to a single set S. Let us also consider families $\mathscr{S}$ of structuring sets:

Definition 9.7.10 *Given a family* $\mathscr{S}$ *of subsets of an abelian group* G, *we shall say that a subset* A *of* G *is* $\mathscr{S}$-***concave*** *if it is a union of translates* $x + S$ *with* $x \in X \subset G$, $S \in \mathscr{S}$. *We shall say that* B *is* $\mathscr{S}$-***convex*** *if its complement is* $\mathscr{S}$-*concave.*

We define

$$\alpha_\mathscr{S}(A) = \bigcup_{\substack{x \in G \\ S \in \mathscr{S}}} (x + S; \; x + S \subset A),$$

called the $\mathscr{S}$-***kernel*** *of* A, *and* $\kappa_\mathscr{S}(B) = \complement \alpha_\mathscr{S}(\complement B)$, *called the* $\mathscr{S}$-***hull*** *of* B. $\square$

Thus $\{S\}$-concavity is the same as S-concavity.

Classical examples are when we take $\mathscr{S}$ as the family $\mathscr{U}$ of all open half-spaces in $\mathbf{R}^m$, defined by an inequality $\xi \cdot x > c$, or the the family $\mathscr{C}$ of all closed half-spaces in $\mathbf{R}^m$, defined by an inequality $\xi \cdot x \geqslant c$, with $\xi \in \mathbf{R}^m \smallsetminus \{0\}$, $c \in \mathbf{R}$. We can also consider the set of all real or complex hyperplanes, or intersections of complex hyperplanes with balls.

Example 9.7.11 *The set* $A = \,]0,1[^2 \cup \{(0,0)\} \subset \mathbf{R}^2$ *(an open square with a vertex added) is convex, but is not an intersection of open half planes, nor of closed half planes; in other words, it is not evenly convex in the sense of Fenchel (1952)—see the discussion about this class of sets in Subsection 9.2.4, page 257. We obtain*

$$A^\circ = \,]0,1[^2 \subsetneqq A \subsetneqq \kappa_\mathscr{U}(A) = [0,1[^2 \subsetneqq \kappa_\mathscr{C}(A) = [0,1]^2 = \overline{A},$$

indicating that more general half planes are needed. □

In view of the above example we now define more general half-spaces, called here **refined half-spaces**, by which we mean convex sets Y such that

$$\{x \in \mathbf{R}^m;\ \xi \cdot x < c\} \subset Y \subset \{x \in \mathbf{R}^m;\ \xi \cdot x \leqslant c\}$$

for some $\xi \in \mathbf{R}^m \setminus \{0\}$ and $c \in \mathbf{R}$. Let us denote by $\mathscr{Y}$ the family of all such sets Y.

Obviously $\kappa_{\mathscr{C}}(A)$ is always a closed set. In view of the Hahn–Banach theorem it is equal to the closed convex hull of A. The mapping $\kappa_{\mathscr{Y}}$ takes an open set to its convex hull (which is open) and a compact set to its convex hull (which is closed).

This is convexity viewed from the outside. We can also work with convexity from the inside: We defined in Definition 9.2.7 on page 256 the convex hull of a set $A \subset \mathbf{R}^m$. It can easily be proved that $\mathbf{cvxh} = \kappa_{\mathscr{Y}}$, showing that the refined half-spaces serve also for convex sets which are not evenly convex—see Section 9.2, page 257, for these sets.

The operation **cvxh** maps any set to its convex hull, which need not be closed even if A is closed. The composition $\mathbf{clos} \circ \mathbf{cvxh}$ takes any set to its closed convex hull. (The composition $\mathbf{cvxh} \circ \mathbf{clos}$ is not idempotent if $m \geqslant 2$.)

Definition 9.7.12 *We shall say that an open subset of* $\mathbf{R}^m$ *or* $\mathbf{C}^n$ *is* r-*concave if it is a union of open balls of radius* r. *A closed subset is called* r-*concave if it is a union of closed balls of radius* r. *A set is called* r-**convex** *if its complement is* r-*concave.* □

This definition agrees for open sets in $\mathbf{C}$ with that of Sergey Favorov and Leonid Golinskii (2015:3). They defined the r-*convex hull* of a set $E \subset \mathbf{C}$, denoted by $\mathbf{conv}_r(E)$, as the set

$$\mathbf{conv}_r(E) = \bigcap \left(\complement D_<(z,r);\ E \subset \complement D_<(z,r) \right), \qquad E \subset \mathbf{C},\ r > 0.$$

Thus $\complement \mathbf{conv}_r(E)$ is a union of open disks. They call a set r-*convex if* $\mathbf{conv}_r(E) = E$. Such a set is always closed. The generalization to $\mathbf{R}^m$ or $\mathbf{C}^n$ is obvious, and we see that $\mathbf{conv}_r(E)$ is exactly the set $\kappa_{B_<(0,r)}(E)$ with the notation from Definition 9.7.9. When r tends to $+\infty$, we get the closed convex hull $\overline{\mathbf{cvxh}(E)}$ as a limiting case.[7]

---

[7]The notion of r-convex closed sets is used by these authors as an hypothesis in results on Blaschke-type conditions for the Riesz measure of a subharmonic function, thus in a context quite different from the one studied here. Since I worked on generalized convexity during the period 1996–2001 see for example Proposition 4.9 in my paper (1996) and then again since 2014, and with quite different problems, our respective studies are independent.

9.7.5   Lineal convexity viewed from mathematical morphology

Lineal concavity is an example of $\mathscr{S}$-concavity, taking $\mathscr{S}$ equal to the family $\mathscr{Z}$ of all complex hyperplanes in $\mathbf{C}^n$ containing the origin. Weak lineal convexity means that $\kappa_{\mathscr{Z}}(\Omega)$ does not meet the boundary of Ω.

There are also local variants of these definitions: we take $\mathscr{S} = \mathscr{Z}_r$ as the family of all intersections $Z \cap B_<(0, r)$, where Z is a complex hyperplane passing through the origin. The corresponding $\mathscr{Z}_r$-convexity, for some positive r, can be called **uniform local lineal convexity**.

Let us take again the family $\mathscr{S}$ of structuring elements in Definition 9.7.10 as the set $\mathscr{Z} \subset \mathscr{P}(\mathscr{P}(\mathbf{C}^n))$ of all complex affine hyperplanes in $\mathbf{C}^n$. We define a dilation $\psi \colon \mathscr{P}(\mathscr{Z}) \to \mathscr{P}(\mathbf{C}^n)$ by

$$\psi(\mathscr{B}) = \bigcup_{Z \in \mathscr{B}} Z, \qquad \mathscr{B} \in \mathscr{P}(\mathscr{Z}). \tag{9.75}$$

Its lower inverse $\psi_{[-1]} \colon \mathscr{P}(\mathbf{C}^n) \to \mathscr{P}(\mathscr{Z})$ is defined by

$$\psi_{[-1]}(A) = \bigcup_{\mathscr{B} \in \mathscr{Z}} (\mathscr{B}; \, \psi(\mathscr{B}) \subset A) = \{Z \in \mathscr{Z}; \, Z \subset A\}, \qquad A \in \mathscr{P}(\mathbf{C}^n). \tag{9.76}$$

We note that $\varepsilon = \psi_{[-1]}$ is an erosion—as the lower inverse of a dilation, but also easily seen directly. There is a relation between Γ_A and ε:

$$\Gamma_A(b) = \{Z \in \varepsilon(\complement A); \, b \in Z\}.$$

The upper inverse $\varepsilon^{[-1]} \colon \mathscr{P}(\mathscr{Z}) \to \mathscr{P}(\mathbf{C}^n)$ of ε is a dilation defined by

$$\varepsilon^{[-1]}(\mathscr{B}) = \bigcap_{A \in \mathscr{P}(\mathbf{C}^n)} (A; \varepsilon(A) \supset \mathscr{B}) = \bigcup_{Z \in \mathscr{B}} Z = \psi(\mathscr{B}), \qquad \mathscr{B} \in \mathscr{P}(\mathscr{Z}). \tag{9.77}$$

By composition we obtain an anoiktomorphism $\alpha_{\mathscr{Z}} \colon \mathscr{P}(\mathbf{C}^n) \to \mathscr{P}(\mathbf{C}^n)$:

$$\alpha_{\mathscr{Z}}(A) = (\varepsilon^{[-1]} \circ \varepsilon)(A) = (\psi \circ \psi_{[-1]})(A) = \bigcup (Z; \, Z \subset A), \qquad A \in \mathscr{P}(\mathbf{C}^n),$$

the union of all complex affine hyperplanes contained in A. We can also form

$$\kappa_{\mathscr{Z}}(\mathscr{B}) = (\varepsilon \circ \varepsilon^{[-1]})(\mathscr{B}) = (\psi_{[-1]} \circ \psi)(\mathscr{B}), \qquad \mathscr{B} \in \mathscr{P}(\mathscr{Z}).$$

We have $\alpha_{\mathscr{Z}}(A) = A$ (equivalently $\alpha_{\mathscr{Z}}(A) \supset A$) if and only if A is lineally concave, which happens if and only if $\complement A$ is lineally convex. If Ω is open, it is lineally convex if and only if $\alpha_{\mathscr{Z}}(\complement \Omega) \supset \complement \Omega$, and weakly lineally convex if and only if $\alpha_{\mathscr{Z}}(\complement \Omega) \supset \partial \Omega$.

9.7.6 Exterior accessibility of Hartogs domains

We shall now study Hartogs domains in $\mathbf{C}^n \times \mathbf{C}$, where we write coordinates as $(z, t) \in \mathbf{C}^n \times \mathbf{C}$.

To define complete Hartogs sets, we may use either the function R, the function $h = R^2$, or the function $f = -\log R$. An open complete Hartogs set is then defined equivalently by $|t| < R(z)$; $|t|^2 < h(z)$; $|t| < e^{-f}$, and we are free to choose whichever is convenient for a specific calculation. We note that if f is plurisubharmonic, then Ω, defined by $\log|t| + f(z) < 0$, is pseudoconvex.

Complex hyperplanes in $\mathbf{C}^n \times \mathbf{C}$ are of three kinds:

1. A hyperplane can be given by an equation $\beta \cdot (z - z^0) = 0$ for some $\beta \in \mathbf{C}^n \smallsetminus \{0\}$ and some point $z^0 \in \mathbf{C}^n$ (we shall call it a **vertical hyperplane**).

2. It can have the equation $t = c$ for some complex constant c (we shall call it a **horizontal hyperplane**).

3. Finally it can have the equation $t = \beta \cdot (z - z^0)$, where β is nonzero. Such a hyperplane intersects the hyperplane $t = 0$ in a hyperplane in $\mathbf{C}^n$ containing z^0.

The projection $\mathbf{C}^n \times \mathbf{C} \ni (z, t) \mapsto (z, |t|) \in \mathbf{C}^n \times \mathbf{R}$ can be used to visualize the set. Equivalently, we can look at the intersection of Ω with the set $\{(z, t); \ z \in \mathbf{C}^n, t \geqslant 0\}$. A hyperplane is then represented in $\mathbf{C}^n \times \mathbf{R}$ by either

1. a vertical plane;

2. a horizontal plane $|t| = |c|$; or

3. a cone $|t| = |\beta \cdot (z - z^0)|$ with vertices at all the points z satisfying $\beta \cdot (z - z^0) = 0$; when $n = 1$ just the unique point z^0.

If $b = (z^0, t^0)$ is a boundary point with $t^0 = 0$, then there is a complex line of equation $z = z^0$ in the complement of Ω, and there may or may not exist a hyperplane in $\Gamma_\Omega(b)$—if the set ω in $\mathbf{C}^n$ where R is positive is lineally convex, there is such a hyperplane. If on the other hand $b = (z^0, t^0)$ is a boundary point satisfying $|t^0| = R(z^0) > 0$, then a hyperplane $Z \in \Gamma_\Omega(b)$ is given by an equation $t/t^0 = \beta \cdot z$; the parallel hyperplane $b + Z$ passing through b has the equation $t/t^0 = 1 + \beta \cdot (z - z^0)$. It may happen that all real hyperplanes containing $b + Z$ cuts Ω, but if this is not the case, the only real hyperplane containing $b + Z$ and not cutting Ω is that of equation $\operatorname{Re}(t/t^0) = 1 + \operatorname{Re}\beta \cdot (z - z^0)$.

Theorem 9.7.13 *Let a function $R\colon \mathbf{C}^n \to [-\infty, +\infty]$ be given and consider the complete Hartogs set Ω defined as in Definition 9.4.2. Assume that Ω is open and weakly lineally convex. Then R is continuous at every point where it is finite and positive, and all boundary points of Ω satisfying (z^0, t^0) with $|t^0| = R(z^0) > 0$ are accessible from the outside of class C^2. In fact, every complex hyperplane which passes through a boundary point (z^0, t^0) with $|t^0| = R(z^0) > 0$ and does not meet Ω is contained in a real external tangent plane.*

In particular $\Gamma_\Omega(b) \subset \Theta_{\Omega,\mathbf{C}}(b)$ *for all points* $b = (z^0, t^0)$ *with* $|t^0| = R(z^0) > 0$
($\Theta_{\Omega,\mathbf{C}}(b)$ *is defined in Definition 9.7.5*).

Proof Any point (z^0, t^0) with $|t^0| = R(z^0) > 0$ belongs to the boundary
of Ω, so there exists by hypothesis a vector $\beta \in \mathbf{C}^n$ such that the complex
hyperplane defined by $t/t^0 = 1 + \beta \cdot (z - z^0)$ lies entirely in the complement
of Ω. We shall prove that there is a real external tangent plane of class C^2
containing it.

That the complex hyperplane does not meet Ω means that

$$\frac{R(z)}{|t^0|} \leqslant |1 + \beta \cdot (z - z^0)|, \qquad z \in \mathbf{C}^n.$$

Now

$$|1 + z| \leqslant \tfrac{1}{2} + \tfrac{1}{2}|1 + z|^2 = 1 + \operatorname{Re} z + \tfrac{1}{2}|z|^2, \qquad z \in \mathbf{C},$$

with equality if and only if $|1 + z| = 1$. It follows that for any $\gamma > \tfrac{1}{2}$,

$$|1 + z| \leqslant 1 + \operatorname{Re} z + \gamma |z|^2, \qquad z \in \mathbf{C},$$

with equality only when $z = 0$. Hence

$$|1 + \beta \cdot (z - z^0)| \leqslant 1 + \operatorname{Re} \beta \cdot (z - z^0) + \gamma|\beta \cdot (z - z^0)|^2$$
$$\leqslant 1 + \operatorname{Re} \beta \cdot (z - z^0) + \gamma \|\beta\|_2^2 \cdot \|z - z^0\|_2^2,$$

with equality between the first and last expression only when $z = z^0$ or $\beta = 0$.
Therefore, if we choose $c > \tfrac{1}{2}\|\beta\|_2^2$,

$$R(z)/|t^0| \leqslant 1 + \operatorname{Re} \beta \cdot (z - z^0) + c\|z - z^0\|_2^2, \qquad z \in \mathbf{C}^n,$$

with equality only when $z = z^0$.

So the set

$$U = \left\{(z, t); \ \operatorname{Re}(t/t^0) > 1 + \operatorname{Re} \beta \cdot (z - z^0) + c\|z - z^0\|_2^2\right\},$$

taking $c > \tfrac{1}{2}\|\beta\|_2^2$, is a set with smooth boundary and the real hyperplane
defined by $\operatorname{Re} t/t^0 = 1 + \operatorname{Re} \beta \cdot (z - z^0)$ is an external tangent plane of class
C^2 of Ω at (z^0, t^0).

From what we just proved it follows in particular that R is upper semicontinuous where positive. On the other hand, Ω is open by hypothesis, which,
as we noted, implies that the restriction $R|_\omega$ is lower semicontinuous.

9.7.7   Unions of increasing squences of domains

If an increasing family $(V_j)_{j \in \mathbf{N}}$ of open sets in $\mathbf{R}^m$ is given with union V and if
$b \in \partial V$, let us denote by $\limsup \Theta_{V_j,\mathbf{R}}(b)$, understood as $\left(\limsup \Theta_{V_j,\mathbf{R}}\right)(b)$,

all limits of real hyperplanes $Y_j \in \Theta_{V_j,\mathbf{R}}(b^{(j)})$ at points $b^{(j)} \in \partial V_j$ such that $b^{(j)} \to b$ as $j \to \infty$. Here $\Theta_V(b)$ is defined in Definition 9.7.5. We shall use a similar notation for the complex hyperplanes: $\limsup \Theta_{\Omega_j,\mathbf{C}}(b)$ when Ω_j increases to Ω, and also $\limsup \Gamma_{\Omega_j}(b)$.

Proposition 9.7.14 *Let* $(V_j)_{j\in\mathbf{N}}$ *be an increasing family of open subsets of* $\mathbf{R}^m$. *Define* $\Theta_{V,\mathbf{R}}$ *as in Definition 9.7.5 using as structuring element a set* S *with boundary of class* $C^{k,\beta}$ *with* $k \geqslant 1$. *Then* $\Theta_{V,\mathbf{R}}(b) \subset \limsup \Theta_{V_j,\mathbf{R}}(b)$ *for all points* $b \in \partial V$. *A similar result holds for the complex external tangent planes* $\Theta_{\Omega,\mathbf{C}}(b)$ *of an open subset* Ω *of* $\mathbf{C}^n$. *Here the inclusion can be strict. The limit superior is always nonempty.*

Proof Take $b = 0$ and let U be an open set with boundary of the class in question such that $\overline{V} \cap \overline{U} = \{0\}$, defined as the set of all points x where $\varphi(x)$ is negative, φ being of the right class and with nonvanishing gradient where it is zero. Let φ_s, $s > 0$, be the function

$$\varphi_s(x) = \varphi(x) - s + \|x\|_2^2, \qquad x \in \mathbf{R}^m,$$

and let U_s be the set where φ_s is negative. We note that when $x \in V$, then $x \notin U$, so that $\varphi(x) \geqslant 0$. If $x \in V \cap U_s$, then $\varphi(x) \geqslant 0$ while $\varphi_s(x) < 0$. So $\|x\|_2^2 < s - \varphi(x) \leqslant s$. Since φ is of class C^1, its gradient at any point in $V \cap U_s$ is close to its gradient at the origin. For every large enough j there is a smallest s_j such that U_{s_j} and V_j have a common boundary point $b^{(j)}$. Necessarily, then, $\|b^{(j)}\|_2^2 \leqslant s$. For large j, s_j is small, so small that the external tangent plane of U_{s_j} at $b^{(j)}$ is as close as we like to the tangent plane of U at the origin. This shows that any hyperplane in $\Theta_{V,\mathbf{R}}(0)$ can be approximated by hyperplanes in $\Theta_{V_j,\mathbf{R}}(b^{(j)})$.

Proposition 9.7.15 *Let* $(\Omega_j)_{j\in\mathbf{N}}$ *be an increasing family of lineally convex open subsets of* $\mathbf{C}^n$ *and denote their union by* Ω. *Then* $\limsup \Gamma_{\Omega_j}(b) = \Gamma_\Omega(b)$. *In particular the graph of* Γ_Ω *is closed.*

Proof If $Z \notin \Gamma_\Omega(b)$, then $b + Z$ intersects Ω. Take a compact ball K in Ω that contains a point of $b + Z$ in its interior. Then for all sufficently large j, Ω_j contains K. All hyperplanes which are close enough to $b + Z$ intersect K and hence also Ω_j for these j. Therefore, if Z_j tends to Z and $b^{(j)} \in \partial \Omega_j$ tends to b, then $b^{(j)} + Z_j$ intersects Ω_j for large j. This means that hyperplanes $b^{(j)} + Z_j$ with $Z_j \in \Gamma_{\Omega_j}(b^{(j)})$ cannot approach $b + Z$. So we have $\limsup \Gamma_{\Omega_j}(b) \subset \Gamma_\Omega(b)$.
   The opposite inclusion is trivially true.

Lemma 9.7.16 *If* A *is a closed set in* $\mathbf{R}^m$ *and* $b \in \partial A$, *then* $\overline{\Theta^2_{A,\mathbf{R}}}(b)$, *where we use a Euclidean ball as structuring element, is nonempty.*

Proof Given $b \in \partial A$ and a positive number s, take $c \notin A$ with $\|c - b\|_2 < s$. Take then $r > 0$ maximal so that $B_<(c,r)$ does not cut A. Clearly $r \leqslant s$. On the boundary of this ball, there must exist a point $p \in A$. Then $\|p - b\|_2 \leqslant$

$r + s \leqslant 2s$, and p is accessible from the outside of class C^2, which means that $\Theta^2_{A,\mathbf{R}}(p)$ is nonempty. Since s is arbitrarily small, the closure of the the graph of $\Theta^2_{A,\mathbf{R}}$ has a nonempty fiber over b.

Theorem 9.7.17 *Let Ω be an open subset of $\mathbf{C}^n$ such that it is equal to the interior of its closure. If Ω is tangentially lineally convex at all points b in some open subset B of $\partial\Omega$ (see Definition 9.7.6), then $\Theta^1_{\Omega,\mathbf{C}}(b) \subset \overline{\Theta^2_{\Omega,\mathbf{C}}(b)} \subset \Gamma_\Omega(b)$, and $\Gamma_\Omega(b)$ is nonempty for all $b \in B$. In particular, tangential lineal convexity at all points $b \in \partial\Omega$ implies weak lineal convexity.*

Proof We apply Lemma 9.7.16 to $A = \overline{\Omega}$. Then the interior of A is equal to Ω. Moreover, $\mathbf{graph}(\Gamma_\Omega)$ is closed; see Proposition 9.7.15.

If Ω is lineally convex, $\Gamma_\Omega(b)$ is not necessarily connected, not even when Ω is a Hartogs set, as is shown by the example below as well as by Example 9.4.8 in Section 9.4, page 281.

Example 9.7.18 *Let Ω be the Cartesian product of an annulus and a disk,*

$$\Omega = \{(z_1, z_2) \in \mathbf{C}^2; \ 1 < |z_1| < 2, \ |z_2| < 1\},$$

a lineally convex set. We define complex hyperplanes Z_β passing through 0 by the equations $\beta z_1 = (1 - \beta)z_2$, $z \in \mathbf{C}^2$, $\beta \in [0,1]$. Use a ball $B_{\leqslant}(0, r)$ with $0 < r < 1$ as structuring element. Then $\Theta^2_{\Omega,\mathbf{C}}(b)$, where $b = (1,1)$, consists of all the Z_β, $\beta \in [0,1]$, whereas $\Gamma_\Omega(b)$ consists of Z_0 and Z_β for $\frac{1}{2} \leqslant \beta \leqslant 1$. Thus $\Gamma_\Omega(b)$ does not contain $\Theta^2_{\Omega,\mathbf{C}}(b)$. We also note that $\Gamma_\Omega(b)$ is not connected; it has two components, $\{Z_0\}$ and $\{Z_\beta; \ \frac{1}{2} \leqslant \beta \leqslant 1\}$.

If $0 < \beta < \frac{1}{2}$, then there are points $z = (-1 - s, z_2) \in Z_\beta \cap \Omega$ far from $b = (1,1)$ (take $s > 0$, $s = -1 - z_1 < 2/\beta - 4$) as well as points in $Z_\beta \cap \Omega$ arbitrarily close to b.

The set Ω is lineally convex, but if we approximate it from the inside by a set with boundary of class C^1 containing all points in Ω with distance to $\partial\Omega$ at least equal to $\varepsilon > 0$, then we get a set which is $\mathscr{L}_1$-convex but not $\mathscr{L}_r$-convex for $r \geqslant 1 + \varepsilon > 1$. (For $\mathscr{S}$-convexity, se Definition 9.7.10; for $\mathscr{L}_r$-convexity, see the beginning of Subsection 9.7.5.) $\qquad\square$

9.7.8 Convexity properties of superlevel sets

Definition 9.7.19 *Given any function f on a set X and with values in the set $\mathbf{R}_!$ of extended real numbers and an element c of $[-\infty, +\infty]$, we define its (non-strict) **superlevel set** as $\{x \in X; \ f(x) \geqslant c\}$. Analogously we define its (non-strict) **sublevel set** as $\{x \in X; \ f(x) \leqslant c\}$.* $\qquad\square$

Given a complete Hartogs set with radius function R, we shall denote by M_c the superlevel set $\{z \in \mathbf{C}^n; \ R(z) \geqslant c\}$.

Example 9.7.20 *Consider the lineally convex Hartogs set* $\Omega \subset \mathbf{C} \times \mathbf{C}$ *defined by the radius function*

$$R(z) = \min(|z - 2|, |z + 2|), \quad |z| < 1; \qquad R(z) = 0, \quad |z| \leqslant 1.$$

Then $(0, 2)$ *belongs to the boundary of* Ω *and* $\Gamma_\Omega((0, 2))$ *consists of precisely two elements, the hyperplanes defined by* $t = -z$ *and* $t = z$, *respectively; thus it is not connected. (This shows that* Ω *is not* $\mathbf{C}$*-convex in view of Zelinskiĭ's criterion mentioned near the end of Subsection 9.7.2.) However, the union of all the* $\Gamma_\Omega(b)$ *with* $b \in \partial\Omega$ *is connected. We note that* $\Gamma_\Omega((i, \sqrt{5}))$ *is connected and contains* $\Gamma_\Omega((0, 2))$. *See Example 9.4.8 in Section 9.4.*

The boundary points of Ω *are accessible from the outside by balls of a not too large radius, and* $\Theta_{\Omega, \mathbf{C}}((0, 2))$ *consists of all hyperplanes* $t = \lambda z$, *with* $\lambda \in [-1, 1]$. *We also note that the intersection of* Ω *with the complex line* $t = c$ *has two components if* $2 \leqslant |c| < \sqrt{5}$.

Also, for $a = s + i(1 - s/2)$ *with a small positive number* s, *the superlevel set* $M_{R(a)}$ *is* $B_<(0, r)$*-convex for* r *slightly smaller than* $\sqrt{5}$, *whereas for* $s = 0$, $a = i$, *the superlevel set* $M_{R(i)}$, *now equal to* $\{i, -i\}$, *is* $B_\leqslant(0, r)$*-convex for any* r *but not convex. (This is a warning that* r*-convexity is not so meaningful for sets that are not regular open or regular closed.)* □

For simplicity we shall assume below that $n = 1$.

Theorem 9.7.21 *Let* $\Omega \subset \mathbf{C} \times \mathbf{C}$ *be a complete lineally convex Hartogs domain defined as in (9.42) with* $n = 1$. *Assume that a point* $a \in \omega$ *is such that* $R(a) < \sup R$. *Then there exists an* $r > 0$ *such that if* a *belongs to the erosion* $\varepsilon_{D_<(0,r)}(\omega)$, *then* $a \in \alpha_{D_\leqslant(0,r)}(\complement M_{R(a)})$. *In other words, since* a *belongs to the open set* ω, *the distance* r *to* $\complement\omega$ *is positive, and* a *is exterior accessible in* $M_{R(a)}$ *by disks of radius* r.

Proof There is a complex hyperplane (thus a complex line in the present situation) in the complement of Ω which passes through $(a, R(a))$. It cannot be vertical since $a \in \omega$ and it cannot be horizontal since $R(a) < \sup R$, so it must have an equation of the form

$$\frac{t}{R(a)} = 1 + \beta(z - a) = \beta(z - a_\beta),$$

where $\beta \neq 0$ and $a_\beta = a - 1/\beta$ is the point where the line hits the line $t = 0$.

This implies that the cone in $\mathbf{C} \times \mathbf{R}$ defined by $|t|/R(a) \geqslant |\beta(z - a_\beta)|$ does not meet any point $(z, |t|) \in \Omega$, in particular that a belongs to the disk $D_\leqslant(a_\beta, s)$ with center at a_β and radius $s = |a - a_\beta| = 1/|\beta|$. As noted, this disk does not meet ω, so $a \in \alpha_{D_\leqslant(0,s)}(\complement M_{R(a)})$. We note finally that $s = |a - a_\beta| \geqslant d(a, \complement\omega) = r$.

There is no uniformity here: r depends on a. But if Ω is bounded and we restrict attention to points a in a compact subset of ω and with $R(a) \geqslant c > 0$, we can choose a fixed $r > 0$. Thus $M_{R(a)}$ is $D_<(0, r)$-convex.

There may be several lines of the form $t = \beta(z - a_\beta)$ as mentioned in the proof. Then among all the possible values of $\beta \in \mathbf{C}$ we can take the infimum of their absolute values, and any limit of these numbers must also define a line in the complement of Ω, since the complement is closed. This gives the largest possible value to $r = 1/|\beta|$.

In Example 9.7.20 we see that, for a real such that $0 < a < 1$,

$$r = 2 - a > d(a, \complement \omega) = 1 - a,$$

implying that the number r obtained in the proof can be smaller than it is in an actual situation.

Remark 9.7.22 *In the other direction, if a closed r-convex set M in $\mathbf{C}$ is given, then there exists a lineally convex open set in $\mathbf{C}^n \times \mathbf{C}$ with radius function R such that $M_{\sup R} = M$; see Proposition 9.4.12 on page 284.* □

Corollary 9.7.23 *If Ω is lineally convex and bounded, and its boundary is of class C^1 at the set where $R > 0$, then a point $a \in \omega$ belongs to $\alpha_{D_{\leqslant}(0,r)}(\complement M_{R(a)})$ if*

$$r \leqslant \frac{1}{\|(\mathbf{grad}\, R)(a)\|_2}.$$

This is the case for all points z with $R(z) = R(a)$ if

$$r \leqslant \frac{1}{\sup_{z \in \omega} \left(\|(\mathbf{grad}\, R)(z)\|_2; \; R(z) = R(a) \right)}.$$

We see that $r \nearrow +\infty$ when $R(a) \nearrow \sup R$, meaning that the superlevel set becomes more and more convex. We shall make this precise in Theorem 9.7.29.

Proof In this situation there is only one line in the complement of Ω passing through $(a, R(a))$, and the absolute value of the coefficient β is $\|(\mathbf{grad}\, R)(a)\|_2 = 2|R_z|$. The radius r depends on a and may vary, but among all the points z with $R(z) = R(a)$ its lower bound is positive.

We now consider a situation with two levels, $R(a)$ and $R(a) + s \geqslant R(a)$.

Theorem 9.7.24 *Let $\Omega \subset \mathbf{C} \times \mathbf{C}$ be a lineally convex Hartogs domain defined as in Definition 9.4.2 with $n = 1$ and take a point $a \in \omega \subset \mathbf{C}$ with $R(a) < \sup R$. Then there exists a number $r > 0$ such that for all $s \geqslant 0$,*

$$d(a, M_{R(a)+s}) \geqslant \frac{sr}{R(a)},$$

where the inequality means that any point w with $R(w) = R(a) + s$ is outside the disk $D_<(a_\beta, r_1)$ with $r_1 = r + sr/R(a)$.

It follows that w is accessible with disks of radius r_1 in the complement of the superlevel set $M_{R(a)+s}$.

Proof As in the proof of Theorem 9.7.21, we see that the cone defined by $|t|/R(a) \geqslant |\beta(z - a_\beta)|$, where β is the coefficient in the equation of the line in the complement of Ω passing through $(a, R(a))$, viz. $t/R(a) = 1 + \beta(z - a)$, does not contain any point of the form $(z, |t|)$ in Ω. We take $r = 1/|\beta|$. In particular the disk $D_<(a_\beta, r_1)$ with $r_1 = r + sr/R(a)$ for any w with $R(w) = R(a) + s$ does not meet $M_{R(a)+s}$.

Since also $(w, R(w))$ admits a line $t/R(w) = 1 + \gamma(z - w)$ in the complement of Ω, we must have $|\gamma| \leqslant |\beta|$, so the corresponding radius $r_2 = 1/|\gamma|$ is not smaller than r_1.

9.7.9 Admissible multifunctions

Definition 9.7.25 *Let Ω be an open subset of $\mathbf{C}^n$ and $\gamma\colon B \rightrightarrows \mathbf{Gr}_{n-1}(\mathbf{C}^n)$ a multifunction defined on a subset B of the boundary of Ω and with values in the Grassmann manifold of all hyperplanes through the origin. Consider the following three conditions on γ.*

$$\gamma(b) \subset \Gamma_\Omega(b) \text{ for all } b \in B; \tag{9.78}$$

$$\text{the graph of } \gamma \text{ is closed; and} \tag{9.79}$$

$$\gamma(b) \text{ is nonempty and connected for every } b \in B. \tag{9.80}$$

*We shall say that γ is **admissible** if these conditions are satisfied.* □

It follows that $\mathbf{graph}(\gamma)$ is connected if $\partial\Omega$ is connected; see Lemma 9.7.27 below.

An example of an admissible multifunction is $\Gamma_\Omega\colon \partial\Omega \rightrightarrows \mathbf{Gr}_{n-1}(\mathbf{C}^n)$ provided $\Gamma_\Omega(b)$ is connected for every $b \in \partial\Omega$. (In particular, this is the case if the boundary is of class C^1.) The graph is then automatically closed in view of Proposition 9.7.15. It is easy to see that in Examples 9.7.18 and 9.7.20, there is no admissible multifunction γ in any neighborhood of the points $(1, 1)$ and $(0, 2)$, respectively.

If Ω is tangentially lineally convex, a candidate for γ might be the closure of $\Theta_{\Omega,\mathbf{C}}$. Then property (9.78) holds by hypothesis, (9.79) by construction, and (9.80) may hold if the boundary of Ω is sufficiently regular.

Example 9.7.26 *Let Ω be a convex open set in $\mathbf{C}^n$. If Ω is empty or equal to the whole space, then its boundary is empty. If Ω is a slice, then its boundary has two components.*

In all other cases, $\partial\Omega$ is connected, and we know that the set of all real hyperplanes passing through a fixed boundary point b and not intersecting Ω is connected. Then also the set of all complex hyperplanes containing b and contained in such a real hyperplane is connected—the mapping $Y \mapsto Y_{[b]}$ is continuous as we noted in Section 9.7.2. Thus Γ_Ω is an admissible multifunction except in the first-mentioned cases, even if the boundary is not of class C^1. □

If a lineally convex open set Ω has a C^1 boundary, $\Gamma_\Omega(b)$, a singleton set, depends continuously on b. When $\Gamma_\Omega(b)$ is no longer a singleton, the following result will serve instead of the continuity.

Lemma 9.7.27 *Let Ω be an open set in $\mathbf{C}^n$ and $\gamma\colon A \rightrightarrows \mathbf{Gr}_{n-1}(\mathbf{C}^n)$ an admissible multifunction on a subset A of $\partial\Omega$. Then the graph of γ over B,*

$$\mathbf{graph}_B(\gamma) = \{(b, Z);\ b \in B \text{ and } Z \in \gamma(b)\},$$

is connected for every connected subset B of A. In particular the graph of γ is connected if the boundary of Ω is connected and γ is defined on all of it.

Proof Assume that $\mathbf{graph}_B(\gamma) = V_0 \cup V_1$, where the V_j are disjoint and closed relative to $\mathbf{graph}_B(\gamma)$. Define B_j as the set of all points b such that some hyperplane in $\gamma(B)$ belongs to V_j, $j = 0, 1$. Then B_0 and B_1 are disjoint, since by hypothesis every $\gamma(b)$ is connected. Moreover B_0 and B_1 are closed relative to B, since the graph of γ is closed and the manifold $\mathbf{Gr}_{n-1}(\mathbf{C}^n)$ is compact. By hypothesis B is connected, so either B_0 or B_1 must be empty. Hence V_0 or V_1 is empty, proving that the graph of γ over B is connected.

   We note that $\gamma_*(B)$ is connected as a continuous image of the graph (it is the projection of the graph on the target space $\mathbf{Gr}_{n-1}(\mathbf{C}^n)$).

Proposition 9.7.28 *Let Ω be an open subset of $\mathbf{C}^n$ and F an affine subspace of $\mathbf{C}^n$. Denote by Ω_F the set $\Omega \cap F$ considered as an open subset of F. Every complex hyperplane Z in $\mathbf{C}^n$ which does not contain F gives rise to a complex hyperplane $\psi(Z) = Z \cap F$ in F. Let an admissible multifunction $\gamma\colon B \rightrightarrows \mathbf{Gr}_{n-1}(\mathbf{C}^n)$ be given and define a multifunction γ_F on $B \cap \partial\Omega_F$ by $\gamma_F(b) = \{\psi(Z);\ Z \in \gamma(b)\}$. Then γ_F is an admissible multifunction on $B \cap \partial\Omega_F$.*

Proof Since $\psi(Z) \subset Z$, it is clear that $\gamma_F(b) \subset \Gamma_{\Omega_F}(b)$; thus (9.78) in Definition 9.7.25 holds. The graph of γ over any compact subset of $\partial\Omega$ is compact; hence the graph of γ_F over any compact subset of $\partial\Omega_F$ is compact, thus closed: property (9.79) holds. Finally (9.80) follows since ψ is continuous and thus maps connected subsets onto connected subsets.

The proposition can in particular be applied to Γ_Ω if $\Gamma_\Omega(b)$ is connected for all $b \in \partial\Omega$.

9.7.10   Links to ordinary convexity

Theorem 9.7.29 *Let R be a continuous real-valued function defined on $\mathbf{C}^n$ and define Ω as in Definition 9.4.2. Assume that Ω is connected and that its boundary is of class C^1 (at least in a neighborhood of $M_{\sup R}$). Then the set $M_{\sup R}$ where R attains its maximum,*

$$M_{\sup R} = \{z \in \mathbf{C}^n;\ R(z) = \sup R\}, \tag{9.81}$$

is convex.

Proof A set is convex if and only if its intersection with every one-dimensional complex affine subspace is convex. Therefore it is enough to prove the theorem for $n = 1$.

So let $n = 1$ and let a belong to the boundary of $M_{\sup R}$. We shall prove that there is an open half plane with a on its boundary which does not meet $M_{\sup R}$, proving the convexity of that set.

We have $(\mathbf{grad}\, R)(a) = 0$, and near a there are points c with $(\mathbf{grad}\, R)(c)$ nonzero and arbitrarily small. In view of Corollary 9.7.23 this means that there is a disk of arbitrarily large radius with c on its boundary. The disk is of the form $D_{\leqslant}(c_\beta, r)$, where $r = |c - c_\beta|$, $c_\beta = c - 1/\beta$ being the point where the line $t/R(c) = 1 + \beta(z - c)$ hits the plane $t = 0$. The normalized vectors $(c - c_\beta)/|c - c_\beta|$ have an accumulation point, and this proves that the union of all the disks $D_{\leqslant}(c_\beta, r)$ when c varies in an arbitrarily small neighborhood of a contains an open half plane with a on its boundary. We are done.

The assumption that the boundary be of class C^1 can be weakened, as we shall now show.

Theorem 9.7.30 *Let R be a continuous real-valued function defined on $\mathbf{C}^n$ and define Ω by (9.42). Assume that Ω is bounded and connected and that there exists an admissible multifunction γ defined at all points (z^0, t^0) with $|t^0| = R(z^0) > 0$, thus on the boundary over the base of Ω (see Definition 9.7.25). Then the set $M_{\sup R}$ where R attains its maximum is convex.*

For a Hartogs domain Ω we always have $\Gamma_\Omega(b) \subset \Theta_{\Omega,\mathbf{C}}(b)$ when $b = (z^0, t^0)$ with $|t^0| = R(z^0) > 0$ (Theorem 9.7.13); if the domain is tangentially lineally convex, we have $\Gamma_\Omega(b) = \Theta_{\Omega,\mathbf{C}}(b)$. For such domains we therefore have an admissible multifunction $\gamma = \Gamma_\Omega = \Theta_{\Omega,\mathbf{C}}$: (9.78) is obvious; (9.79) follows from Proposition 9.7.15; (9.80) follows from Proposition 9.7.7.

We note that the hypothesis is satisfied in particular if Ω is lineally convex and R is of class C^1. In (Kiselman 1996, Theorem 4.8) the result was proved under this hypothesis, and even under the weaker one that R can be approximated from below by C^1 functions.

In view of Zelinskiĭ's characterization of $\mathbf{C}$-convex sets mentioned near the end of Section 9.7.2, the hypotheses are satisfied for $\mathbf{C}$-convex sets, again taking $\gamma = \Gamma_\Omega$. There are easy examples which show that $M_{\sup R}$ need not be convex if we drop the hypothesis of connectedness; see Example 9.4.8.

Proof of Theorem 9.7.30. Again, the set $M_{\sup R}$ is convex if its intersection with every one-dimensional complex affine subspace is convex. Proposition 9.7.28 shows that if we have an admissible multifunction on a subset of $\partial\Omega$, then there is one also on a corresponding subset of $\partial\Omega_F$, F being any affine subspace of $\mathbf{C}^n \times \mathbf{C}$. Therefore, taking F as the Cartesian product of a complex line in $\mathbf{C}^n$ and the line $z = 0$, we see that it is enough to prove the theorem for $n = 1$.

So let $n = 1$. To prove that $M_{\sup R}$ is convex means to prove that the segment $[s_0, s_1]$ is contained in $M_{\sup R}$ if $s_0, s_1 \in M_{\sup R}$. There is no loss in generality if we assume that $s_0 = -1$ and $s_1 = 1$.

A non-vertical and non-horizontal complex line through (a, t^0) with $t^0 \neq 0$ has the equation

$$\frac{t}{t^0} = 1 + \beta(z - a) = \beta(z - a_\beta), \qquad z \in \mathbf{C},$$

where $a_\beta = a - 1/\beta$ is the point where the line hits the plane $t = 0$. We define

$$q(a, \beta) = \begin{cases} a - 1/\beta \text{ if } \beta \neq 0, \\ \infty \text{ if } \beta = 0. \end{cases}$$

In case R is differentiable at the point a, β is uniquely determined if we require that the line be in $\Gamma_\Omega((a, t^0))$.

We denote as before by ω the set of all points $z \in \mathbf{C}$ such that $R(z) > 0$. In general the external tangent is not unique and we shall denote by $Q(a)$ the set of all points $a - 1/\beta$ that can be obtained from complex lines in $\gamma((a, t^0))$, thus

$$Q(a) = \{q(a, \beta); \ \beta \in \gamma(a, R(a))\} \subset S^2 = \mathbf{C} \cup \{\infty\}, \qquad a \in \omega. \quad (9.82)$$

We define $Q(a) = \{a\}$ when $a \notin \omega$. Thus Q is a multifunction, $Q \colon S^2 \rightrightarrows S^2 \setminus \omega$; its images $Q(a)$ are compact and connected.

The radius can always be estimated by

$$R(z) \leqslant R(a)|\beta| \cdot |z - a_\beta|, \qquad z \in \mathbf{C}, \ a \in \omega, \ \beta \in \gamma(a, R(a)), \ a_\beta = q(a, \beta),$$

with equality for $z = a$, assuming $\beta \neq 0$. In particular, if $w \in M_{\sup R}$, then

$$R(a)|\beta| \cdot |a - q(a, \beta)| = R(a) \leqslant R(w) \leqslant R(a)|\beta| \cdot |w - q(a, \beta)|.$$

If $a_\beta \in Q(a) \setminus \{\infty\}$, then necessarily $\beta \neq 0$, so that

$$|a - a_\beta| \leqslant |w - a_\beta|, \qquad a \in \omega, \ w \in M_{\sup R}, \ a_\beta \in Q(a) \setminus \{\infty\}. \quad (9.83)$$

Assume that -1 and 1 belong to $M_{\sup R}$; we shall then prove that any point $c \in [-1, 1]$ belongs to $M_{\sup R}$. Consider $Q(c + iy)$ for real y. We know from Lemma 9.7.27 that the set $Q_*(c + i\mathbf{R})$ is connected. If ω is bounded and y or $-y$ is very large, then $Q(c + iy) = \{c + iy\}$. In general we can prove that $\operatorname{Im} a > 1$ implies that $\operatorname{Im} b > 0$ for all $b \in Q(a)$, and similarly $\operatorname{Im} a < -1$ implies $\operatorname{Im} b < 0$ for all $b \in Q(a)$. This follows from the following lemma.

Lemma 9.7.31 *If Ω is a complete Hartogs domain in $\mathbf{C}^2$ with radius function R and if $\pm 1 \in M_{\sup R}$, then for all $b \in \mathbf{C}$ with $|\operatorname{Re} a| \leqslant 1$ and all $b \in Q(a) \setminus \{\infty\}$ we have*

$$\operatorname{Im} a \geqslant 1 \ \text{implies} \ \operatorname{Im} b \geqslant \tfrac{1}{2}(\operatorname{Im} a - 1) \quad \text{and}$$

$$\operatorname{Im} a \leqslant -1 \ \text{implies} \ \operatorname{Im} b \leqslant \tfrac{1}{2}(\operatorname{Im} a + 1).$$

Proof We know from (9.83) that $|a - b| \leqslant |\pm 1 - b|$. Expanding $|\pm 1 - b|^2 - |a - b|^2 \geqslant 0$, we get

$$2(\operatorname{Re} b)(\operatorname{Re} a \mp 1) + 1 - (\operatorname{Im} a)(\operatorname{Im} a - 2\operatorname{Im} b) \geqslant (\operatorname{Re} a)^2 \geqslant 0,$$

from which we deduce that $1 \geqslant (\operatorname{Im} a)(\operatorname{Im} a - 2\operatorname{Im} b)$, an inequality which implies those in the lemma.

Proof of Theorem 9.7.30, cont'd. So $Q(c + iy)$ must pass from the upper half plane to the lower half plane when y goes from large positive values to large negative values, c being fixed. But it can never pass the real axis at points with $x \geqslant 1$ or $x \leqslant -1$. Indeed, if b is real and larger than or equal to 1, we get from (9.83), taking $a = c + iy$,

$$|a - b| \leqslant |1 - b| = b - 1,$$

implying $\operatorname{Re} a \geqslant 1$, so that $c \geqslant 1$ contrary to assumption. Likewise, $Q(c + iy)$ cannot pass the real axis at a point with $x \leqslant -1$.

However, $Q(c+iy)$ cannot pass from numbers with arbitrarily large positive imaginary part to numbers with large negative imaginary part in the strip $-1 < \operatorname{Re} z < 1$ either. In fact, ω is connected, so there exists a curve contained in ω connecting -1 to 1, and $Q(c + iy)$ cannot cross that curve.

Hence it is impossible for $Q(c + iy)$ to pass from the upper half plane to the lower half plane if it has only finite values. So it must have an infinite value, which means that $c + iy^0 \in M_{\sup R}$ for at least one y^0.

We thus know that there is a y^0 such that $c + iy^0 \in M_{\sup R}$; without loss of generality we may assume that it is nonnegative. Choose y^0 as small as possible. If $y^0 = 0$ we are done: $c \in M_{\sup R}$. Let us assume that $y^0 > 0$ and try to reach a contradiction.

By (9.83) any point $b \in Q(a) \setminus \{\infty\}$ must lie in each of the three half planes

$$|a - b| \leqslant |1 - b| \qquad |a - b| \leqslant |-1 - b|, \qquad |a - b| \leqslant |c + iy^0 - b|.$$

The intersection of these three half planes is a triangle, and the union of these triangles when $a = c + iy$ with $y \in [\frac{1}{2}y^0, y^0]$ is bounded. Thus the possible finite values for b when a varies as indicated is bounded, and for $a = c + iy$ with $\frac{1}{2}y^0 \leqslant y < y^0$ the point b cannot be infinity. On the other hand, when $a = c+iy^0 \in M_{\sup R}$, then $Q(a)$ must contain ∞. This means that the set of all points $b \in Q(a)$ originating from points $a = c + iy$ with $y \in [\frac{1}{2}y^0, y^0]$ consists of ∞ and a nonempty bounded set; it is not connected, in contradiction to Lemma 9.7.27. This contradiction shows that we must have $c \in M_{\sup R}$ and proves the theorem. □

It is easy to modify Theorem 9.7.30 using Möbius mappings, at least if $n = 1$. In fact, any mapping

$$\mathbf{C} \times \mathbf{C} \ni (z, t) \mapsto \left(\frac{a + bz}{c + dz}, \frac{t}{c + dz} \right) = (z', t') \in \mathbf{C} \times \mathbf{C}$$

preserves lineal convexity, as was shown in (Kiselman 1996: Lemma 8.1). Denote by a_β the point where a line $t/t^0 = 1 + \beta(z - z^0)$ intersects the z-plane. The line can be mapped by a Möbius mapping to a line $t' = $ constant. This mapping takes the point a_β to infinity, and all circles in the z-plane which pass through a_β are mapped onto straight lines. Convex sets are transformed accordingly:

Definition 9.7.32 *Let b be a complex number or ∞. Let us say that a subset A of the Riemann sphere $\mathbf{C} \cup \{\infty\}$ is b-**convex** if*

(10.4.1). $b \notin A$; and

(10.4.2). $\varphi_*(A)$ is convex if φ is a Möbius mapping which maps b to infinity. □

Corollary 9.7.33 (to Theorem 9.7.30) *Let Ω and γ be as in Theorem 9.7.30, assume that $n = 1$ and let π denote the projection defined by $\pi(z, t) = z$. Consider a line $Z \in \gamma(a)$, where $a = (z^0, t^0)$, $|t^0| = R(z^0) > 0$, and let b be the point such that $(b, 0) \in a + Z$. Then the set $\pi_*((a + Z) \cap \overline{\Omega})$ is b-convex.* □

9.8 Duality of Functions Defined in Lineally Convex Sets

Abstract of this section

The term *duality* represents a collection of ideas where two sets of mathematical objects confront each other. A most successful duality is that between the space $\mathscr{D}(\Omega)$ of test functions (smooth functions of compact support) and its dual $\mathscr{D}'(\Omega)$ of distributions.

Similarly, the theory of analytic functionals, developed by André Martineau in his doctoral thesis (1963)—also in (*Œuvre de André Martineau* 1977:47–210)—is based on a duality, now between the Fréchet space of holomorphic functions $\mathscr{O}(\Omega)$ in an open set Ω and its dual $\mathscr{O}'(\Omega)$. In many, but not all respects, it is analogous to distribution theory.

In complex geometry, lineal concavity and lineal convexity can be treated successfully using concepts of duality.

9.8.1 Introduction to this section

Lineal convexity, a kind of complex convexity intermediate between usual convexity and pseudoconvexity, appears naturally in the study of Fantappiè transforms of analytic functionals. A set is called lineally convex if its complement is a union of complex hyperplanes. This property can be most conveniently defined in terms of the notion of dual complement: the dual complement of a set in $\mathbf{C}^n$ is the set of all hyperplanes that do not intersect the set. It is natural to add a hyperplane at infinity and consider $\mathbf{C}^n$ as an open subset of $\mathbf{P}^n$, complex projective space of dimension n. The definition of dual complement is then the same, and somewhat more natural: the set of all hyperplanes is again a projective space. In this setting, the dual complement is often called the *projective complement*. Indeed, Martineau (1966) called it *le complémentaire projectif*; the term *dual complement* used here was introduced by Andersson, Passare and Sigurdsson in a preprint from 1991 of their forthcoming book (2004).

We can now simply define a lineally convex set as a set which is the dual complement of its dual complement (here it becomes obvious that we should identify the hyperplanes in the space of all hyperplanes with the points in the original space). So this duality works well for sets. What about functions?

In convexity theory, a convenient dual object of a set is its support function as defined in Section 9.2. For functions, we have the Fenchel transformation, defined as well in Section 9.2.

Is there a duality for functions that generalizes the duality for sets defined by the dual complement? In this section we shall study such a duality. We call it the logarithmic transformation. It has many properties in common with the Fenchel transformation. However, there are some striking differences. The effective domain, defined by formula (9.2.5), of a Fenchel transform is always convex, but the effective domain of a logarithmic transform need not be linearly convex (Example 9.8.16). This is connected with the fact that the union of an increasing sequence of linearly convex sets is not necessarily linearly convex (Example 9.8.17). However, the interior of the effective domain of a logarithmic transform is always linearly convex (Theorem 9.8.14), and the transform is plurisubharmonic there (Theorem 9.8.18).

Working with functions defined on $\mathbf{P}^n$ is the same as working with functions defined on $\mathbf{C}^{1+n} \smallsetminus \{0\}$ which are constant on complex lines, i.e., homogeneous of degree zero. For instance a plurisubharmonic function on an open subset of $\mathbf{P}^n$ can be pulled back to an open cone in $\mathbf{C}^{1+n} \smallsetminus \{0\}$ and the pullback is plurisubharmonic for the $1 + n$ coordinates there. However, I cannot define a duality for such functions. I have been led to consider instead functions defined on subsets of $\mathbf{C}^{1+n} \smallsetminus \{0\}$ which are homogeneous in another sense: they satisfy $f(tz) = -\log|t| + f(z)$. Such functions are not pullbacks of functions on projective space, but the duality works for them. In a coordinate patch like $z_0 = 1$ we can identify them with functions on a subset of $\mathbf{P}^n$. Given any function F on $\mathbf{C}^n$, we can define a function f on $\mathbf{C}^{1+n} \smallsetminus \{0\}$ by $f(z) = F(z_1/z_0, \dots, z_n/z_0) + c \log|z_0|$ when $z_0 \neq 0$ and $f(z) = +\infty$ when $z_0 = 0$, where c is an arbitrary real constant; this function is homogeneous in the sense that $f(tz) = c \log|t| + f(z)$, so we can choose any type of homogeneity. In other words, locally all kinds of homogeneity are equivalent, and there is no restriction in imposing the homogeneity we have here, viz. $c = -1$.

As mentioned in Section 9.2, there are several other notions related to lineal convexity. The property called *Planarkonvexität* in German (see Behnke & Peschl 1935), or *weak lineal convexity* is weaker than lineal convexity: an open connected set is called weakly linearly convex if through any boundary point there passes a complex hyperplane which does not intersect the set. Aïzenberg (1967) proved that these domains are precisely the components of Ω^{**} (for notation see Subsection 9.8.2 below).

Strong lineal convexity was defined by Martineau (1966: Definition 2.2) as a topological property of the space of holomorphic functions in a domain. Martineau (1966: Theorem 2.2) and Aïzenberg (1966) proved independently that convex sets are strongly linearly convex. The property was given a geometric characterization by Znamenskij (1979). This geometric property is now called **C**-*convexity*. Its relation to lineal convexity has been studied by Zelinskij (1988) and others. For these two properties we refer also to the survey by Andersson, Passare and Sigurdsson (2004) and the monograph by Hörmander (1994).

Another generalization is the notion of *m-lineal convexity* to be studied in the next section.

9.8.2 Notation

Let A be a subset of $\mathbf{C}^{1+n} \smallsetminus \{0\}$, where $n \geqslant 1$. We shall say that A is ***homogeneous*** if $tz \in A$ as soon as $z \in A$ and $t \in \mathbf{C} \smallsetminus \{0\}$. To any homogeneous subset A of $\mathbf{C}^{1+n} \smallsetminus \{0\}$ we define its ***dual complement*** A^* as the set of all hyperplanes passing through the origin which do not intersect A. Since any such hyperplane has an equation $\zeta \cdot z = \zeta_0 z_0 + \cdots + \zeta_n z_n = 0$ for some $\zeta \in \mathbf{C}^{1+n} \smallsetminus \{0\}$, we can define

$$A^* = \{\zeta \in \mathbf{C}^{1+n} \smallsetminus \{0\}; \ \zeta \cdot z \neq 0 \text{ for every } z \in A\}. \tag{9.84}$$

Strictly speaking, we should have two copies of $\mathbf{C}^{1+n} \smallsetminus \{0\}$ (a Greek and a Latin one), and consider A^* as a subset of the dual (i.e., the Greek) space. A homogeneous set is called ***lineally convex*** if $\mathbf{C}^{1+n} \smallsetminus A$ is a union of complex hyperplanes passing through the origin. A dual complement A^* is always lineally convex, and we always have $A^{**} \supset A$. The set A^{**} is called the ***lineally convex hull*** of A. A set A is lineally convex if and only if $A = A^{**}$.

The operation of taking the dual complement is an example of a Galois correspondence, and the operation of taking the lineally convex hull defines a cleistomorphism in the ordered set of all subsets of $\mathbf{C}^{1+n} \smallsetminus \{0\}$. For the general definitions of these concepts, see Section 9.3.

We shall write $z = (z_0, z') = (z_0, z_1, \dots, z_n)$ for points in $\mathbf{C}^{1+n} \smallsetminus \{0\}$, with $z_0 \in \mathbf{C}$ and $z' = (z_1, \dots, z_n) \in \mathbf{C}^n$. Homogeneous sets in $\mathbf{C}^{1+n} \smallsetminus \{0\}$ correspond to subsets of projective n-space $\mathbf{P}^n$, and we can transfer the notions of dual complement and lineal convexity to $\mathbf{P}^n$. In the open set where $z_0 \neq 0$ we can use z' as coordinates in $\mathbf{P}^n$.

We shall denote by

$$Y_\zeta = \{z \in \mathbf{C}^{1+n} \smallsetminus \{0\}; \ \zeta \cdot z = 0\}, \qquad \zeta \in \mathbf{C}^{1+n} \smallsetminus \{0\}, \tag{9.85}$$

the hyperplane defined by ζ. Then the dual complement can be conveniently defined as

$$A^* = \{\zeta; \ Y_\zeta \cap A = \varnothing\}, \tag{9.86}$$

and its set-theoretical complement in $\mathbf{C}^{1+n} \smallsetminus \{0\}$ is

$$\complement A^* = \left(\mathbf{C}^{1+n} \smallsetminus \{0\}\right) \smallsetminus A^* = \{\zeta; \ Y_\zeta \cap A \neq \varnothing\}. \tag{9.87}$$

The complement of the lineally convex hull A^{**} can be written as

$$\complement A^{**} = \bigcup_{\alpha \in A^*} Y_\alpha.$$

We shall use this idea in the following lemma.

Lemma 9.8.1 *For any subset Γ of $\mathbf{C}^{1+n} \smallsetminus \{0\}$ we define*

$$A = \complement \left(\bigcup_{\gamma \in \Gamma} Y_\gamma \right) = \bigcap_{\gamma \in \Gamma} \complement Y_\gamma.$$

Then A is lineally convex. Moreover $A^{**} = A = \Gamma^*$ and $A^* = \Gamma^{**} \supset \Gamma$.

Proof Clearly A as the complement of a union of hyperplanes is linearly convex, so $A^{**} = A$. The statement $a \in A$ is equivalent to $\gamma \cdot a \neq 0$ for all $\gamma \in \Gamma$, which by definition means that $a \in \Gamma^*$; thus $A = \Gamma^*$. As a consequence, $A^* = \Gamma^{**}$.

How does the operation of taking the dual complement intertwine with the topological operations of taking the interior and closure? The answer is the following (we write A° for the interior and $\overline{A}$ for the closure of a set A).

Proposition 9.8.2 *For any homogeneous subset A of $\mathbf{C}^{1+n} \setminus \{0\}$ we have*
  (A) *If A is open, then A^* is closed.*
  (B) *If A is closed, then A^* is open.*
  (C) $\overline{A^*} \subset A^{\circ *}$.
  (D) $A^{*\circ} = \overline{A}^*$.
  (E) *If A is the closure of an open set, then A^* is the interior of a closed set.*

Proof (A) and (B). To see that A^* is closed if A is open we only have to look at (9.86). The same formula shows that A^* is open if A is closed.
  (C). Since A° is open, $A^{\circ *}$ is closed according to (A), so that $A^{\circ *} = \overline{A^{\circ *}} \supset \overline{A^*}$.
  (D). Since $\overline{A}^*$ is open according to (B), and since $\overline{A}^* \subset A^*$, we get $\overline{A}^* \subset A^{*\circ}$. To prove the inclusion $A^{*\circ} \subset \overline{A}^*$ we argue as follows. If $\zeta \in A^{*\circ}$, then $Y_\theta \cap A = \emptyset$ for all θ near ζ. The union of these hyperplanes Y_θ is a neighborhood of Y_ζ, so $\zeta \in \overline{A}^*$.
  (E). If $A = \overline{B}$ with B open, then according to (D), $A^* = \overline{B}^* = B^{*\circ}$, the interior of the closed set B^*.
  This proves the proposition.

If A is the interior of a closed set C, we get $A^* = C^{\circ *} \supset \overline{C^*}$, possibly strictly.

Corollary 9.8.3 *If a subset A of $\mathbf{C}^{1+n} \setminus \{0\}$ is strongly contained in a set B in the sense that $\overline{A} \subset B^\circ$, then B^* is strongly contained in A^*.*

Proof Using (C) and (D) in Proposition 9.8.2 we see that $\overline{A} \subset B^\circ$ implies $\overline{B^*} \subset B^{\circ *} \subset \overline{A}^* = A^{*\circ}$.

Corollary 9.8.4 *If a subset A of $\mathbf{C}^{1+n} \setminus \{0\}$ is lineally convex, then its interior A° is also lineally convex.*

Proof If $A = B^*$, then $A^\circ = B^{*\circ} = \overline{B}^*$ by (D) in Proposition 9.8.2, which shows that A° is lineally convex.

By way of contrast, the closure of a lineally convex set is not necessarily lineally convex if $n \geq 2$. It turns out that the lineal convexity of the closure is connected with the question whether we have equality in (C) in Proposition 9.8.2, as shown by the following result.

Corollary 9.8.5 *Let B be any lineally convex subset of $\mathbf{C}^{1+n} \smallsetminus \{0\}$. Then its closure $\overline{B}$ is lineally convex if and only if its dual complement $A = B^*$ satisfies (C) in Proposition 9.8.2 with equality.*

Proof Using the lineal convexity of B and then (C) and (D) in Proposition 9.8.2, we get
$$\overline{B} = \overline{A^*} \subset A^{\circ *} = B^{* \circ *} = \overline{B}^{**}.$$
Thus equality in (C) is equivalent to $\overline{B}$ being lineally convex.

The inclusion (C) in Proposition 9.8.2 can be strict simply for dimensionality reasons. This will be clear from the following result, where we use the interior relative to a subspace instead of the interior with respect to the whole space.

Proposition 9.8.6 *Let A be a homogeneous set in $\mathbf{C}^{1+n} \smallsetminus \{0\}$ which is contained in a complex subspace F of $\mathbf{C}^{1+n}$. Let $A_F = \mathbf{relint}(A)$ denote the relative interior of A, i.e., the interior taken with respect to F. Then $\overline{A^*} \subset (A_F)^* \cup F^\Diamond$, where $F^\Diamond$ is the set*

$$F^\Diamond = \{\zeta \in \mathbf{C}^{1+n} \smallsetminus \{0\}; \ F \smallsetminus \{0\} \subset Y_\zeta\}.$$

If A is open in F, then $A^* \cup F^\Diamond$ is closed.

Note that when $F = \mathbf{C}^{1+n}$, then $F^\Diamond$ is empty and we are reduced to Proposition 9.8.2.

Proof Take a point $\zeta \notin (A_F)^* \cup F^\Diamond$. Then there is a point $a \in A_F \cap Y_\zeta$ and a non-zero vector $b \in F \smallsetminus Y_\zeta$. If θ is close to ζ, then the hyperplane Y_θ cuts the complex line $\{a + tb; \ t \in \mathbf{C}\}$ in a unique point $a(\theta, t)$ close to a, and since a is in the relative interior of A, $a(\theta, t)$ belongs to A as soon as θ is close enough to ζ. Therefore $\theta \notin A^*$ for all these θ, which means that $\zeta \notin \overline{A^*}$.

Finally, if A is open in F, then $\overline{A^* \cup F^\Diamond} = \overline{A^*} \cup F^\Diamond \subset (A_F)^* \cup F^\Diamond = A^* \cup F^\Diamond$, since $A_F = A$ and $F^\Diamond$ is closed.

Example 9.8.7 *It is now obvious that the inclusion in (C) can be strict. Take a nonempty relatively open set $A \subset F \neq \mathbf{C}^{1+n}$. Then $A^\circ = \emptyset$, and $A^{\circ *} = \mathbf{C}^{1+n} \smallsetminus \{0\}$. But $\overline{A^*} \subset (A_F)^* \cup F^\Diamond = A^* \cup F^\Diamond \neq \mathbf{C}^{1+n} \smallsetminus \{0\}$.* □

Example 9.8.8 *Also the inclusion in Proposition 9.8.6 can be strict. There are sets A such that $A^\circ = \emptyset$, $\overline{A}^\circ = B \neq \emptyset$, and $B^* = A^*$. Thus $A^{\circ *} = \mathbf{C}^{1+n} \smallsetminus \{0\}$ and $\overline{A^*} = \overline{B^*} = B^* \neq \mathbf{C}^{1+n} \smallsetminus \{0\}$. Such a set is the set A of all $z \in \mathbf{C}^{1+2}$ with $|z_1|^2 + |z_2|^2 < |z_0|^2$ and either z_1 is a complex rational or $z_2 = 0$. (Here the only choice for F is the whole space, so that $F^\Diamond$ is empty.)* □

Example 9.8.9 *A simple example of a lineally convex set whose closure is not lineally convex is the following. Define*

$$A = \{z \in \mathbf{C}^{1+2} \smallsetminus \{0\}; \ |z_1| < |z_2|\}; \qquad \overline{A} = \{z \in \mathbf{C}^{1+2} \smallsetminus \{0\}; \ |z_1| \leqslant |z_2|\}$$

Then A is linearly convex. Any hyperplane which avoids A must pass through the point $(z_0, z_1, z_2) = (1, 0, 0)$. But this point belongs to $\overline{A}$. This shows that $\overline{A}$ is not linearly convex.

More generally, let Γ be a linearly convex subset of the Greek copy of $\mathbf{C}^{1+n} \setminus \{0\}$ and define A as in Lemma 9.8.1. We can easily choose Γ without interior points but still such that the union

$$\bigcup_{\gamma \in \Gamma} Y_\gamma = \complement A$$

has interior points. Thus $\Gamma^\circ = \emptyset$, $\complement \overline{A} = (\complement A)^\circ \neq \emptyset$. Then $(\overline{A})^{**} = A^{*\circ*} = \Gamma^{\circ*} = \mathbf{C}^{1+n} \setminus \{0\}$ (see Lemma 9.8.1 and (D) in Proposition 9.8.2), but $\overline{A} = \complement((\complement A)^\circ) \neq \mathbf{C}^{1+n} \setminus \{0\}$. This shows that $\overline{A}$ cannot be linearly convex. □

9.8.3 Duality for functions

A function $f : \mathbf{C}^{1+n} \setminus \{0\} \to \mathbf{R}_!$ with values in the extended real line will be called (-1)-**homogeneous** if

$$f(tz) = -\log|t| + f(z), \qquad z \in \mathbf{C}^{1+n} \setminus \{0\}, \quad t \in \mathbf{C} \setminus \{0\}. \tag{9.88}$$

For such functions we define the **logarithmic transform** $\mathscr{L}f$:

$$(\mathscr{L}f)(\zeta) = \sup_{z \in \mathbf{dom}(f)} \left(-\log|\zeta \cdot z| - f(z) \right), \qquad \zeta \in \mathbf{C}^{1+n} \setminus \{0\}. \tag{9.89}$$

We define $\log 0 = -\infty$. The difference $-\log|\zeta \cdot z| - f(z)$ is well-defined if $f(z) < +\infty$; another way to formulate the definition is to use lower addition $\dotplus$:

$$(\mathscr{L}f)(\zeta) = \sup_z \left((-\log|\zeta \cdot z|) \dotplus (-f(z)) \right), \qquad \zeta \in \mathbf{C}^{1+n} \setminus \{0\}. \tag{9.90}$$

Lower and upper addition are defined in Section 9.2 on page 253.

Proposition 9.8.10 *For any homogeneous function $f : \mathbf{C}^{1+n} \setminus \{0\} \to \mathbf{R}_!$ its logarithmic transform $\mathscr{L}f$ is a homogeneous function with*

$$\mathbf{dom}(\mathscr{L}f) \subset (\mathbf{dom}(f))^*, \tag{9.91}$$

where $\mathbf{dom}(f)$ denotes the effective domain of f.

Proof The homogeneity of $\mathscr{L}f$ is obvious from its definition (9.89). To prove (9.91), we note that $\zeta \notin (\mathbf{dom}(f))^*$ means by definition that the hyperplane Y_ζ and the effective domain $\mathbf{dom}(f)$ have a common point z (cf. (9.87)), so that $(\mathscr{L}f)(\zeta) \geq -\log|\zeta \cdot z| - f(z) = +\infty$, thus $\zeta \notin \mathbf{dom}(\mathscr{L}f)$. The inclusion (9.91) may be strict as will be shown below: see Example 9.8.16 and Remark 9.8.21.

The analogue of Fenchel's inequality holds:

$$-\log|\zeta \cdot z| \leqslant f(z) \dotplus (\mathscr{L}f)(\zeta), \qquad \zeta, z \in \mathbf{C}^{1+n} \smallsetminus \{0\}. \tag{9.92}$$

Moreover the usual rules for a Galois correspondence hold: $f \leqslant g$ implies $\mathscr{L}f \geqslant \mathscr{L}g$, and we always have $\mathscr{L}(\mathscr{L}f) \leqslant f$. As a consequence of these two properties, $\mathscr{L} \circ \mathscr{L} \circ \mathscr{L} = \mathscr{L}$. A function f will be called $\mathscr{L}$-*closed* if $\mathscr{L}(\mathscr{L}f) = f$ (equivalently, if it belongs to the range of $\mathscr{L}$). Some simple examples follow.

Example 9.8.11 *If f assumes the value $-\infty$, then $\mathscr{L}f$ is $+\infty$ identically. The same is true if f never takes the value $+\infty$ $(n \geqslant 1)$. If f is $+\infty$ identically, then $\mathscr{L}f$ is $-\infty$ identically. If $f(z) = -\log|t|$ when $z = ta$ for a fixed $a \in \mathbf{C}^{1+n} \smallsetminus \{0\}$ and $+\infty$ otherwise, then $(\mathscr{L}f)(\zeta) = -\log|\zeta \cdot a|$. If*

$$f(z) = -\log|\alpha \cdot z|$$

for some α, then $(\mathscr{L}f)(\zeta) = -\log|t|$ when $\zeta = t\alpha$ and $+\infty$ otherwise. All these functions are $\mathscr{L}$-closed. ☐

As a consequence of (9.90), we note that $\sup_j \mathscr{L}f_j = \mathscr{L}(\inf_j f_j)$ for any indexed family (f_j) of functions. Indeed this follows from the rule $\sup_j(c + a_j) = c + \sup_j a_j$, which is valid also for any constant $c \in \mathbf{R}_!$. This implies that any supremum of $\mathscr{L}$-closed functions is $\mathscr{L}$-closed; in fact, we have

$$\sup_j f_j = \sup_j \mathscr{L}(\mathscr{L}f_j) = \mathscr{L}(\inf_j \mathscr{L}f_j) \tag{9.93}$$

if the f_j are $\mathscr{L}$-closed.

Homogeneous functions appear rather naturally in complex analysis. Let μ be an analytic functional in an open subset ω of $\mathbf{C}^n$, $\mu \in \mathscr{O}'(\omega)$. Its Fantappiè transform is

$$(\mathscr{F}\mu)(\zeta) = \mu\big(z \mapsto (\zeta_0 + \zeta_1 z_1 + \cdots + \zeta_n z_n)^{-1}\big),$$

which is a holomorphic function of $\zeta \in \Omega^*$, where Ω is the set of all $z \in \mathbf{C}^{1+n} \smallsetminus \{0\}$ such that $z_0 \neq 0$ and $(z_1/z_0, \ldots, z_n/z_0) \in \omega$. This implies that $\log|\mathscr{F}\mu|$ is plurisubharmonic in Ω^*, and it is moreover homogeneous in the sense of (9.88). (We define it as $+\infty$ outside Ω^*.)

Given f defined in $\mathbf{C}^{1+n} \smallsetminus \{0\}$, we can define a function F in $\mathbf{C}^n$ by putting $F(z') = f(1, z_1, \ldots, z_n)$, $z' \in \mathbf{C}^n$. Conversely, if F is defined in $\mathbf{C}^n$, we can define a homogeneous function f in $\mathbf{C}^{1+n} \smallsetminus \{0\}$ by

$$f(z) = \begin{cases} F(z_1/z_0, \ldots, z_n/z_0) - \log|z_0|, & z \in \mathbf{C}^{1+n} \smallsetminus \{0\}, \quad z_0 \neq 0; \\ +\infty & z \in \mathbf{C}^{1+n} \smallsetminus \{0\}, \quad z_0 = 0. \end{cases}$$

The transform (9.89) then takes the form

$$(\mathscr{L}F)(\zeta') = \sup_{F(z')<+\infty} \big(-\log|1 + \zeta' \cdot z'| - F(z')\big), \qquad \zeta' \in \mathbf{C}^n. \tag{9.94}$$

In particular, if F is radial (i.e., a function of $\|z'\|_2 = r$), then the transform becomes

$$(\mathscr{L}F)(\rho) = \sup_{F(r)<+\infty} \left(-\log(1 - \rho r) - F(r) \right), \qquad \rho = \|\zeta'\|_2 \geqslant 0. \qquad (9.95)$$

Example 9.8.12 *Take* $F(r) = 0$ *when* $r \leqslant R$ *and* $F(r) = +\infty$ *otherwise in* (9.95). *Then* $(\mathscr{L}F)(\rho) = -\log(1 - R\rho)$, $\rho < 1/R$, *and* $(\mathscr{L}F)(\rho) = +\infty$, $\rho \geqslant 1/R$. *The second transform is* $\mathscr{L}(\mathscr{L}F) = F$, *so that* F *is* $\mathscr{L}$-*closed.* □

Example 9.8.13 *The radial function*

$$F(r) = -\frac{1}{2}\log(1 - r^2)$$

is selfdual, i.e., $(\mathscr{L}F)(\rho) = -\frac{1}{2}\log(1 - \rho^2)$. *Going back to* $\mathbf{C}^{1+n} \smallsetminus \{0\}$, *we see that the function*

$$f(z) = \begin{cases} -\frac{1}{2}\log(|z_0|^2 - \|z'\|_2^2), & z \in \mathbf{C}^{1+n} \smallsetminus \{0\}, \quad |z_0| > \|z'\|_2; \\ +\infty, & z \in \mathbf{C}^{1+n} \smallsetminus \{0\}, \quad |z_0| \leqslant \|z'\|_2 \end{cases}$$

has this property. This function therefore plays the same role as the convex function $f(x) = \frac{1}{2}\|x\|_2^2$, $x \in \mathbf{R}^n$, *for usual convexity.* □

Now let A be a homogeneous set in $\mathbf{C}^{1+n} \smallsetminus \{0\}$. We define a function d_A, the distance to the complement of A relative to $\mathbf{C}^{1+n} \smallsetminus \{0\}$, as

$$d_A(z) = \inf \left(\|z - w\|_2; \ w \in (\mathbf{C}^{1+n} \smallsetminus \{0\}) \smallsetminus A \right), \qquad z \in \mathbf{C}^{1+n} \smallsetminus \{0\}. \quad (9.96)$$

The function $-\log d_A$ is homogeneous, and it is less than $+\infty$ precisely in the interior of A. Analogously we define a function d_{A^*} by

$$d_{A^*}(\zeta) = \inf \left(|\zeta - \theta|; \ \theta \in (\mathbf{C}^{1+n} \smallsetminus \{0\}) \smallsetminus A^* \right), \qquad \zeta \in \mathbf{C}^{1+n} \smallsetminus \{0\}, \quad (9.97)$$

where A^* is the dual complement of A defined by (9.84). If A is empty, then $d_A = 0$ identically, whereas $d_{A^*} = +\infty$ identically.

Theorem 9.8.14 *Let* $f \colon \mathbf{C}^{1+n} \smallsetminus \{0\} \to \mathbf{R}_!$ *be any homogeneous function. Then*

$$C - \log \|\zeta\|_2 \leqslant (\mathscr{L}f)(\zeta) \leqslant C - \log d_{A^*}(\zeta), \qquad \zeta \in \mathbf{C}^{1+n} \smallsetminus \{0\}, \quad (9.98)$$

where d_{A^*} *is defined by* (9.97) *taking* $A = \mathbf{dom}(f)$, *and* $C = -\inf_{\|z\|_2=1} f(z) \leqslant +\infty$. *We have* $C = -\infty$ *if and only if* f *is* $+\infty$ *identically; in this case* $\mathscr{L}f$ *is* $-\infty$ *identically. We have* $C = +\infty$ *if and only if* f *is unbounded from below on the unit sphere* S; *then* $\mathscr{L}f$ *is* $+\infty$ *identically. If* f *is bounded from below on* S, *then* $C < +\infty$ *and* (9.98) *shows that* $\mathscr{L}f$ *has at most logarithmic growth at the boundary of* $(\mathbf{dom}(f))^*$; *moreover*

$$\overline{\mathbf{dom}(f)}^* = (\mathbf{dom}(f))^{*\circ} = (\mathbf{dom}(\mathscr{L}f))^\circ \subset \mathbf{dom}(\mathscr{L}f) \subset (\mathbf{dom}(f))^*, \tag{9.99}$$

and

$$\overline{\mathbf{dom}(\mathscr{L}f)} \subset \overline{(\mathbf{dom}(f))^*} \subset (\mathbf{dom}(f))^{\circ*}. \tag{9.100}$$

In particular $\mathbf{dom}(\mathscr{L}f) = (\mathbf{dom}(f))^*$ *if* $\mathbf{dom}(f)$ *is closed.*

Lemma 9.8.15 *For any subset* A *of* $\mathbf{C}^{1+n} \smallsetminus \{0\}$ *we have*

$$|\zeta \cdot z| \geqslant d_{A^*}(\zeta) \|z\|_2, \qquad \zeta \in \mathbf{C}^{1+n} \smallsetminus \{0\}, \quad z \in A, \tag{9.101}$$

and

$$|\zeta \cdot z| \geqslant \|\zeta\|_2 d_A(z), \qquad \zeta \in A^*, \quad z \in \mathbf{C}^{1+n} \smallsetminus \{0\}. \tag{9.102}$$

Proof Given $\zeta \in \mathbf{C}^{1+n} \smallsetminus \{0\}$ and $z \in A$ we define $\alpha = \zeta + t\bar{z}$ where $t = -\|z\|_2^{-2}(\zeta \cdot z)$. Then $\alpha \cdot z = 0$, which, if $\alpha \neq 0$, means that $\alpha \in \complement A^*$ since $z \in A$. Therefore $d_{A^*}(\zeta) \leqslant \|\zeta - \alpha\|_2 = |\zeta \cdot z|/\|z\|_2$, which proves the first inequality except when $\zeta = \|z\|_2^{-2}(\zeta \cdot z)\bar{z}$. Since d_{A^*} is continuous, this restriction can be removed. If we now interchange the role of ζ and z, we get $|\zeta \cdot z| \geqslant \|\zeta\|_2 d_{A^{**}}(z)$. But $A^{**} \supset A$, so $d_{A^{**}}(z) \geqslant d_A(z)$. This proves the lemma. (Interchanging z and ζ once more, we see that (9.101) holds even for all $z \in A^{**}$.)

Proof of Theorem 9.8.14. By the Schwarz inequality and (9.101) applied to $A = \mathbf{dom}(f)$ we get

$$-\log \|\zeta\|_2 \leqslant -\log |\zeta \cdot z| \leqslant -\log d_{A^*}(\zeta), \qquad \zeta \in \mathbf{C}^{1+n} \smallsetminus \{0\}, \quad z \in A \cap S.$$

Thus

$$(\mathscr{L}f)(\zeta) = \sup_{z \in A \cap S} \left(-\log|\zeta \cdot z| - f(z) \right) \begin{cases} \leqslant (-\log d_{A^*}(\zeta)) \;\dotplus\; \sup_{A \cap S}(-f); \\ \geqslant -\log \|\zeta\|_2 + \sup_{A \cap S}(-f). \end{cases}$$

The cases where $+$ and $\dotplus$ give different results never occur, so we can replace $\dotplus$ by usual addition. This proves (9.98); note that $\sup_{A \cap S}(-f) = \sup_S(-f) = -\inf_S f$.

We already know that $\mathbf{dom}(\mathscr{L}f) \subset (\mathbf{dom}(f))^*$; see (9.91). If $\zeta \in (\mathbf{dom}(f))^{*\circ}$ and $C < +\infty$, then $d_{A^*}(\zeta) > 0$ and $(\mathscr{L}f)(\zeta) \leqslant C - \log d_{A^*}(\zeta) < +\infty$, so that $\zeta \in \mathbf{dom}(\mathscr{L}f)$. This proves that $(\mathbf{dom}(f))^{*\circ} \subset \mathbf{dom}(\mathscr{L}f) \subset (\mathbf{dom}(f))^*$. Taking the interior of these sets we get (9.99); taking the closure we get (9.100) (cf. Proposition 9.8.2).

Example 9.8.16 *The effective domain of* $\mathscr{L}f$ *may fail to be lineally convex, although it is squeezed in between the two lineally convex sets* $(\mathbf{dom}(f))^{*\circ} = \overline{\mathbf{dom}(f)}^*$ *and* $(\mathbf{dom}\,f)^*$; *see (9.99). Indeed, let* $w^k = (k^{-2}, k^{-1}, 1) \in \mathbf{C}^{1+2}$ *and define* $f(w^k) = \log k$, $k = 1, 2, 3, \dots$, *and* $f(z) = +\infty$ *when* $z \notin \complement w^k$. *Then*

$$(\mathscr{L}f)(\zeta) = \sup_k (-\log |\zeta_0/k + \zeta_1 + k\zeta_2|), \qquad \zeta \in \mathbf{C}^{1+2} \smallsetminus \{0\}.$$

Put $\alpha = (1, 0, 0)$ and $\beta = (1, 1, 0)$. Then

$$(\mathscr{L}f)(\alpha) = \sup_k(-\log|k^{-1}|) = +\infty,$$

so that $\alpha \notin \mathbf{dom}(\mathscr{L}f)$, whereas

$$(\mathscr{L}f)(\beta) = \sup_k(-\log|k^{-1} + 1|) = 0,$$

showing that $\beta \in \mathbf{dom}(\mathscr{L}f)$. The points w^k define hyperplanes

$$Y_{w^k} = \{\zeta;\ \zeta_0 k^{-2} + \zeta_1 k^{-1} + \zeta_2 = 0\},$$

which converge to a hyperplane $Y_w = \{\zeta; \zeta_2 = 0\}$ with $w = \lim w^k = (0, 0, 1)$. By (9.99),

$$\overline{(\mathbf{dom}\,f)}^* = \complement\big(Y_w \cup \big(\bigcup Y_{w^k}\big)\big) \subset \mathbf{dom}(\mathscr{L}f) \subset (\mathbf{dom}(f))^* = \complement\big(\bigcup Y_{w^k}\big).$$

Both α and β belong to Y_w, but as $k \to +\infty$, the hyperplanes Y_{w^k} approach α more rapidly than β (note that $\alpha \cdot w^k = 1/k^2$, while $\beta \cdot w^k = 1/k + 1/k^2$). This explains why $\alpha \notin \mathbf{dom}(\mathscr{L}f)$ while $\beta \in \mathbf{dom}(\mathscr{L}f)$. A hyperplane which avoids $\mathbf{dom}(\mathscr{L}f)$ must be either one of the hyperplanes Y_{w^k}, or (possibly) their limit Y_w. However, the hyperplanes Y_{w^k} do not contain α, and the hyperplane Y_w intersects $\mathbf{dom}(\mathscr{L}f)$ in β. Therefore there is no hyperplane which passes through α and avoids $\mathbf{dom}(\mathscr{L}f)$. This shows that $\mathbf{dom}(\mathscr{L}f)$ is not lineally convex. In particular we must have $(\mathbf{dom}(f))^{*\circ} \neq \mathbf{dom}(\mathscr{L}f) \neq (\mathbf{dom}(f))^*$; cf. (9.99). □

Example 9.8.17 *A fundamental property of convexity is that the union of an increasing sequence of convex sets is convex. (More generally, this is true for the union of a directed family.) This is not so with lineal convexity. Let A_k be the set of all ζ such that $(\mathscr{L}f)(\zeta) \leqslant k$. It is easy to see that this is a lineally convex set; indeed,*

$$A_k = \bigcap_{z \in \mathbf{dom}(f)} \{\zeta \in \mathbf{C}^{1+n} \smallsetminus \{0\};\ -\log|\zeta \cdot z| - f(z) \leqslant k\}.$$

The union of the A_k is $\mathbf{dom}(\mathscr{L}f)$. If we let f be the function constructed in Example 9.8.16 we get an example where the A_k are lineally convex but their union is not. □

Theorem 9.8.18 *Let f be a function on $\mathbf{C}^{1+n} \smallsetminus \{0\}$ which is bounded from below on the unit sphere and let $\mathscr{L}f$ be its transform defined by (9.89). Then $\mathscr{L}f$ is plurisubharmonic in the interior of $\mathbf{dom}(\mathscr{L}f)$, which is a lineally convex set. Moreover $\mathscr{L}f$ is locally Lipschitz continuous in $(\mathbf{dom}(\mathscr{L}f))^\circ$; more precisely*

$$\limsup_{t \to 0+} \frac{(\mathscr{L}f)(\zeta + t\theta) - (\mathscr{L}f)(\zeta)}{t} \leqslant \frac{\|\theta\|_2}{d_{A^*}(\zeta)}, \quad \zeta \in (\mathbf{dom}(\mathscr{L}f))^\circ,\ \theta \in \mathbf{C}^{1+n},$$

where d_{A^*} is the distance to the complement of $\mathbf{dom}(\mathscr{L}f)$.

Proof Consider the function $g(\zeta) = -\log|\zeta \cdot z|$. Its gradient has length $\|z\|_2/|\zeta \cdot z|$. At the point $\alpha = \zeta + t\bar{z}$, where $t = -\|z\|_2^{-2}(\zeta \cdot z)$, g takes the value $+\infty$, so

$$d_{\mathbf{dom}(g)}(\zeta) \leqslant \|\alpha - \zeta\|_2 \leqslant \frac{|\zeta \cdot z|}{\|z\|_2} = \frac{1}{\|\,\mathbf{grad}\,g(\zeta)\|_2}.$$

Now $\mathscr{L}f$ is a supremum of functions of the form g plus a constant for various choices of z. All competing functions must satisfy $\mathbf{dom}(g) \supset \mathbf{dom}(\mathscr{L}f)$, so that $d_{\mathbf{dom}(g)} \geqslant d_{A^*}$. Therefore they have a gradient whose length is at most $1/d_{A^*}(\zeta)$, which implies that $\mathscr{L}f$ is Lipschitz continuous as indicated. That $\mathscr{L}f$ is plurisubharmonic now follows from standard properties of such functions: f is a continuous supremum of plurisubharmonic functions.

Finally (9.99) shows that $(\mathbf{dom}(\mathscr{L}f))^\circ$ is lineally convex: it is equal to the dual complement of the closure of $\mathbf{dom}(f)$.

9.8.4 Examples of functions in duality

In this subsection we shall make a detailed study of the functions

$$f_c(z) = \begin{cases} -(1-c)\log\|z\|_2 - c\log d_A(z), & z \in A; \\ +\infty, & z \in (\mathbf{C}^{1+n} \smallsetminus \{0\}) \smallsetminus A, \end{cases} \tag{9.103}$$

and

$$\varphi_c(\zeta) = \begin{cases} -(1-c)\log\|\zeta\|_2 - c\log d_{A^*}(\zeta), & \zeta \in A^*; \\ +\infty, & \zeta \in (\mathbf{C}^{1+n} \smallsetminus \{0\}) \smallsetminus A^*, \end{cases} \tag{9.104}$$

where $0 \leqslant c \leqslant 1$, A is any homogeneous subset of $\mathbf{C}^{1+n} \smallsetminus \{0\}$, A^* its dual complement, and where d_A and d_{A^*} are defined by (9.96) and (9.97), respectively.

We shall call $f_0 = I_A$ the **logarithmic indicator function** of the set A. Its restriction to the unit sphere is the indicator function in the usual sense. And $\mathscr{L}f_0 = \mathscr{L}I_A$ is analogous to the support function of A, thus preserving the situation from convex analysis where the support function is the Fenchel transform of the indicator function. We shall determine this function explicitly: it is $\varphi_1 = -\log d_{A^*}$.

More generally, it turns out that the function φ_{1-c} is essentially dual to f_c. It might seem strange to consider functions like f_0 which are not plurisubharmonic. We must have $\mathscr{L}(\mathscr{L}f_0) < f_0$ in the interior of A. From this point of view it is more natural to consider

$$g_c(z) = \begin{cases} -(1-c)\log|z_0| - c\log d_A(z), & z \in A; \\ +\infty, & z \in (\mathbf{C}^{1+n} \smallsetminus \{0\}) \smallsetminus A, \end{cases} \tag{9.105}$$

and

$$\psi_c(\zeta) = \begin{cases} -(1-c)\log|\zeta_0| - c\log d_{A^*}(\zeta), & \zeta \in A^*; \\ +\infty, & \zeta \in (\mathbf{C}^{1+n} \smallsetminus \{0\}) \smallsetminus A^*. \end{cases}$$
$$(9.106)$$

If A is contained in a coordinate patch $z_0 \neq 0$ and if moreover $\|z\|_2/|z_0|$ is bounded when $z \in A$, then f_c and g_c are finite in the same set and differ there by a bounded function. If moreover $(1, 0, \ldots, 0)$ is an interior point of A, then $\zeta_0 \neq 0$ when $\zeta \in A^*$ and $\|\zeta\|_2/|\zeta_0|$ is bounded there, so φ_c and ψ_c are finite in the same set and their difference is bounded there. Therefore our results on f_c and φ_c can easily be translated into inequalities for g_c and ψ_c.

The first result is a simple inequality.

Proposition 9.8.19 *With f_c and φ_c defined by (9.103) and (9.104) we have $\mathscr{L}f_c \leqslant \varphi_{1-c}$ for $0 \leqslant c \leqslant 1$.*

Proof If $\zeta \notin A^*$, then $\varphi_{1-c}(\zeta) = +\infty$, so the inequality certainly holds. If on the other hand $\zeta \in A^*$, we can estimate $(\mathscr{L}f_c)(\zeta)$ using Lemma 9.8.15:

$$(\mathscr{L}f_c)(\zeta)$$
$$= \sup_{z \in A(c)} \left[-\log|\zeta \cdot z| + (1-c)\log\|z\|_2 + c\log d_A(z) \right]$$
$$= \sup_{z \in A(c)} \left[-(1-c)\log|\zeta \cdot z| + (1-c)\log\|z\|_2 - c\log|\zeta \cdot z| + c\log d_A(z) \right]$$
$$\leqslant \sup_{z \in A(c)} \left[-(1-c)\log(d_{A^*}(\zeta)\|z\|_2) + (1-c)\log\|z\|_2 \right.$$
$$\left. - c\log(\|\zeta\|_2 d_A(z)) + c\log d_A(z) \right]$$

$$\leqslant -(1-c)\log d_{A^*}(\zeta) - c\log\|\zeta\|_2 = \varphi_{1-c}(\zeta).$$

The supremum is over the set $A(c)$ of all z such that $f_c(z) < +\infty$, that is

$$A(c) = \begin{cases} A & \text{for } c = 0; \\ A^\circ & \text{for } 0 < c \leqslant 1. \end{cases} \qquad (9.107)$$

($A(c)$ can be empty; in that case $\mathscr{L}f_c$ is $-\infty$ identically.)

We now study inequalities in the other direction. The cases $c = 0$ and $c = 1$ are easy and will be considered first.

Proposition 9.8.20 *For any homogeneous subset A of $\mathbf{C}^{1+n} \smallsetminus \{0\}$ we have $\mathscr{L}I_A = \mathscr{L}f_0 = \varphi_1 = -\log d_{A^*}$. (The analogue of the support function of A.)*

Remark 9.8.21 *Note that here $\mathrm{dom}(\mathscr{L}f_0) = A^{*\circ} = \overline{A}^* = (\mathrm{dom}(f)_0)^{*\circ}$ is open and lineally convex, whereas $(\mathrm{dom}(f)_0)^* = A^*$; again we see that the inclusion $\mathrm{dom}(\mathscr{L}f) \subset (\mathrm{dom}(f))^*$ may be strict (cf. (D) in Proposition 9.8.2).* $\qquad \Box$

Lemma 9.8.22 *Assume that A is homogeneous and not empty. For every $\zeta \in A^*$ there is a point $z \in \partial A$, $z \neq 0$, such that $|\zeta \cdot z| \leqslant d_{A^*}(\zeta)\|z\|_2$.*

Proof For every $\zeta \in A^*$ there is a point $\alpha \in \partial A^*$, $\alpha \neq 0$, such that $\|\alpha - \zeta\|_2 = d_{A^*}(\zeta)$. Thus $\alpha \notin A^{*\circ} = \overline{A}^*$ (cf. (D) in Proposition 9.8.2). Now $\alpha \notin \overline{A}^*$ means that $Y_\alpha \cap \overline{A} \neq \varnothing$ (see (9.85) and (9.87)). On the other hand $\alpha \in \partial A^* \subset \overline{A}^* \subset A^{\circ*}$ (cf. (C) in Proposition 9.8.2), so that $Y_\alpha \cap A^\circ = \varnothing$. Therefore Y_α meets the boundary of A, and we can choose $z \in S \cap \partial A$ such that $\alpha \cdot z = 0$. Then $|\zeta \cdot z| = |(\zeta - \alpha) \cdot z| \leqslant \|\zeta - \alpha\|_2 \cdot \|z\|_2 = d_{A^*}(\zeta)\|z\|_2$.

Proof of Proposition 9.8.20. If $A = \varnothing$, we have $\mathscr{L}I_A = -\log d_{A^*} = -\infty$. Otherwise the lemma provides us, given any $\zeta \in A^*$, with a point $z \in \partial A \cap S$ such that

$$(\mathscr{L}f_0)(\zeta) = \sup_{w \in A}\left(-\log|\zeta \cdot w| + \log\|w\|_2\right) \geqslant -\log|\zeta \cdot z| + \log\|z\|_2$$
$$\geqslant -\log d_{A^*}(\zeta) = \varphi_1(\zeta).$$

For $\zeta \in (\mathbf{C}^{1+n} \setminus \{0\}) \setminus A^*$ both $\mathscr{L}f_0$ and φ_1 take the value $+\infty$. Thus $\mathscr{L}f_0 \geqslant \varphi_1$ everywhere. The inequality $\mathscr{L}f_0 \leqslant \varphi_1$ was proved already in Proposition 9.8.19.

Proposition 9.8.23 *Assume A that is open and not empty. Then there is a constant M, which depends on the geometry of A, such that*

$$\varphi_0 = I_{A^*} \geqslant \mathscr{L}f_1 = \mathscr{L}(-\log d_A) \geqslant I_{A^*} - M.$$

In fact M can be taken as $\inf_S f_1 = \inf_S(-\log d_A)$, where as before S is the unit sphere.

Here $\mathbf{dom}(\mathscr{L}f_1) = A^* = (\mathbf{dom}(f)_1)^*$ is closed and lineally convex; cf. (9.99).

Lemma 9.8.24 *Assume A has a nonempty interior and take any point $z \in A^\circ$. Then there is a constant C such that $|\zeta \cdot z| \leqslant C\|\zeta\|_2 d_A(z)$ for all ζ.*

Proof Given $z \in A^\circ$ define $C = \|z\|_2/d_A(z)$. We have

$$|\zeta \cdot z| \leqslant \|\zeta\|_2\|z\|_2 = C\|\zeta\|_2 d_A(z).$$

The best choice is a point $z \in S$ such that $d_A(z) = \sup_S d_A$, so that $C = 1/\sup_S d_A$.

Proof of Proposition 9.8.23 Using the lemma above we get for any $\zeta \in A^*$,

$$(\mathscr{L}f_1)(\zeta) = \sup_{w \in A}(-\log|\zeta \cdot w| + \log d_A(w)) \geqslant -\log|\zeta \cdot z| + \log d_A(z)$$
$$\geqslant -\log C - \log\|\zeta\|_2 = \varphi_0(\zeta) - M.$$

When $\zeta \notin A^*$, there is a point $z \in A$ such that $\zeta \cdot z = 0$, and since A is open, $f_1(z) < +\infty$, so that $(\mathscr{L}f_1)(\zeta) = +\infty$. Thus we have $\mathscr{L}f_1 \geqslant \varphi_0 - M$ everywhere. The inequality $I_{A^*} \geqslant \mathscr{L}f_1$ was already proved in Proposition 9.8.19.

Theorem 9.8.25 *Let A be an open homogeneous set. Then A is lineally convex if and only if $-\log d_A$ is $\mathscr{L}$-closed.*

Proof If $A = B^*$, then $\mathscr{L}I_B = -\log d_{B^*} = -\log d_A$ by Proposition 9.8.20, so $-\log d_A$ is $\mathscr{L}$-closed. Conversely, Proposition 9.8.23 shows that $\mathscr{L}(-\log d_A) \geqslant I_{A^*} - M$, which implies $\mathscr{L}(-\log d_A) \leqslant -\log d_{A^{**}} + M$. Therefore, if z belongs to the open set A^{**} (cf. Proposition 9.8.2), then $\mathscr{L}(-\log d_A(z))$ is finite. If $-\log d_A$ is $\mathscr{L}$-closed, this is equivalent to $-\log d_A(z)$ being finite, which implies $z \in A$. Thus $A^{**} \subset A$; this inclusion means that A is lineally convex.

Theorem 9.8.26 *A closed lineally convex set A can be recovered from $\mathscr{L}I_A$. Indeed, if A is a set with these properties different from $\mathbf{C}^{1+n} \smallsetminus \{0\}$, then $I_A \geqslant \mathscr{L}I_A \geqslant I_A - M$, so that A is the set where $\mathscr{L}I_A$ is finite. If A is equal to $\mathbf{C}^{1+n} \smallsetminus \{0\}$, then $\mathscr{L}I_A$ is $-\infty$ identically. If A is a closed and lineally convex set such that $\|z'\|_2 \leqslant R|z_0|$ for all $z \in A$, then $\mathscr{L}I_A \geqslant I_A - \log \sqrt{1 + R^2}$.*

This theorem is thus analogous to the result in convexity theory which states that a closed convex set can be recovered from its support function. By way of contrast, an open set A can be recovered from $\mathscr{L}I_A$ only under special conditions, since $\mathscr{L}I_A = \mathscr{L}I_{\overline{A}}$. If A is open and equal to the interior of its closure, and if its closure is lineally convex, then A is the interior of the set where $\mathscr{L}I_A$ is finite. But an open lineally convex set can always be recovered from $\mathscr{L}(-\log d_A)$; see Proposition 9.8.23.

Proof If A is closed, lineally convex and not equal to all of $\mathbf{C}^{1+n} \smallsetminus \{0\}$, then A^* is open and nonempty, so we can apply Proposition 9.8.23 to A^* and obtain

$$I_{A^{**}} \geqslant \mathscr{L}(-\log d_{A^*}) \geqslant I_{A^{**}} - M.$$

From Proposition 9.8.20 we have $\mathscr{L}I_A = -\log d_{A^*}$. Combining this information we deduce that

$$I_{A^{**}} \geqslant \mathscr{L}(-\log d_{A^*}) = \mathscr{L}I_A \geqslant I_{A^{**}} - M.$$

Since A is lineally convex, $A = A^{**}$, and we see that $\mathscr{L}I_A$ and I_A are finite in the same set (and differ there at most by a bounded function).

The last statement follows if we keep track of the constant in Lemma 9.8.24. Alternatively we can compare I_A with the function $F(z) = -\log|z_0|$ when $\|z'\|_2 \leqslant R|z_0|$, defining $F(z) = +\infty$ otherwise. This gives $F(z) \geqslant I_A(z) \geqslant F + \log|z_0| - \log\|z\|_2$ for $z \in A$, so that $I_A \geqslant F - \log\sqrt{1 + R^2}$ everywhere, implying that $\mathscr{L}I_A \geqslant \mathscr{L}F - \log\sqrt{1 + R^2}$. Since $\mathscr{L}F = F$ (cf. Example 9.8.12), we can conclude that $\mathscr{L}I_A \geqslant F - \log\sqrt{1 + R^2} \geqslant I_A - \log\sqrt{1 + R^2}$ in A.

Finally we shall deduce an estimate from below for $\mathscr{L}f_c$ when $0 < c < 1$. We shall need a definition.

Definition 9.8.27 *We shall say that A satisfies the* **homogeneous interior cone condition** *if there exist positive numbers γ and R such that for every $b \in \partial A$ and every $r \leqslant R$, the inequality*

$$\sup_{z} \left(d_A(z); \; \|z - b\|_2 \leqslant r\|b\|_2\right) \geqslant \gamma r\|a\|_2$$

holds for every $b \in \partial A$ and every $r \leqslant R$. □

Proposition 9.8.28 *Assume that A is open, nonempty, and satisfies the homogeneous interior cone condition just defined. Then there is a constant M such that $\varphi_{1-c} \geqslant \mathscr{L} f_c \geqslant \varphi_{1-c} - M$ for every $c \in [0,1]$.*

Here $\mathbf{dom}(\mathscr{L} f_c) = (\mathbf{dom}(f_c))^{*\circ}$ for $0 \leqslant c < 1$, whereas it is closed for $c = 1$ as already noted.

In particular a set with Lipschitz boundary satisfies the homogeneous interior cone condition. To prove this proposition we shall need a lemma which combines Lemmas 9.8.22 and 9.8.24. The requirements concerning the point z are somewhat contradictory, since $z \in \partial A$ in the first and $z \in A^\circ$ in the second. Nevertheless, we can find a compromise:

Lemma 9.8.29 *With A as in Proposition 9.8.28, there exists a constant C such that for every $\zeta \in A^*$ there is a point $z = z_\zeta \in A$ such that*

$$|\zeta \cdot z| \leqslant C\|\zeta\|_2 d_A(z) \text{ and } |\zeta \cdot z| \leqslant C d_{A^*}(\zeta)\|z\|_2.$$

Proof First pick any point $w \in A$. It will serve as the point z_ζ for all ζ such that $d_{A^*}(\zeta) \geqslant R\|\zeta\|_2$:

$$|\zeta \cdot w| \leqslant \|\zeta\|_2\|w\|_2 = \frac{\|w\|_2}{d_A(w)}\|\zeta\|_2 d_A(w) \leqslant C\|\zeta\|_2 d_A(w)$$

and

$$|\zeta \cdot w| \leqslant \|\zeta\|_2\|w\|_2 = \frac{\|\zeta\|_2}{d_{A^*}(\zeta)}d_{A^*}(\zeta)\|w\|_2 \leqslant \frac{1}{R}d_{A^*}(\zeta)\|w\|_2 \leqslant C d_{A^*}(\zeta)\|w\|_2$$

for a constant $C \geqslant \max(R^{-1}, \|w\|_2/d_A(w))$.

The case $d_{A^*}(\zeta) \leqslant R\|\zeta\|_2$ remains to be considered. To a given $\zeta \in A^*$ we choose $\alpha \in \partial A^*$, $\alpha \neq 0$, such that $\|\alpha - \zeta\|_2 = d_{A^*}(\zeta) = r\|\zeta\|_2$, $r \leqslant R$. Since Y_α meets ∂A (cf. the proof of Lemma 9.8.22), we can choose $a \in \partial A$, $a \neq 0$, such that $\alpha \cdot a = 0$. The homogeneous interior cone condition now implies the existence of a point $z = z_\zeta \in A$ such that $d_A(z) \geqslant \gamma r\|a\|_2$ and $\|z - a\|_2 \leqslant r\|a\|_2 \leqslant R\|a\|_2$. Then $|\zeta \cdot a| = |(\zeta - \alpha) \cdot a| \leqslant \|\zeta - \alpha\|_2\|a\|_2 = d_{A^*}(\zeta)\|a\|_2$ and $\|z - a\|_2 \leqslant r\|a\|_2 = \|a\|_2 d_{A^*}(\zeta)/\|\zeta\|_2$, so that

$$|\zeta \cdot z| = |\zeta \cdot a + \zeta(z - a)| \leqslant |\zeta \cdot a| + \|\zeta\|_2|z - a| \leqslant 2d_{A^*}(\zeta)\|a\|_2 \leqslant \frac{2}{1-R}d_{A^*}(\zeta)\|z\|_2.$$

Here the last inequality follows from $\|z - a\|_2 \leqslant R\|a\|_2$; it is no restriction to assume that $R < 1$. On the other hand $d_A(z) \geqslant \gamma r\|a\|_2 = \gamma\|a\|_2 d_{A^*}(\zeta)/\|\zeta\|_2$ so that

$$|\zeta \cdot z| \leqslant 2d_{A^*}(\zeta)\|a\|_2 \leqslant \frac{2}{\gamma}d_A(z)\|\zeta\|_2.$$

With the constant

$$C = \max\left[\frac{1}{R}, \frac{\|w\|_2}{d_A(w)}, \frac{2}{1-R}, \frac{2}{\gamma}\right]$$

this proves the lemma.

Proof of Proposition 9.8.28. Since A is open, $\mathbf{dom}(f_c) = A$ for all c; cf. (9.107). Using the lemma we get for any $\zeta \in A^*$, taking $z \in A$ from Lemma 9.8.29,

$$
\begin{aligned}
(\mathscr{L}f_c)(\zeta) &= \sup_{w \in A}\big(-\log|\zeta \cdot w| + (1-c)\log\|w\|_2 + c\log d_A(w)\big) \\
&\geqslant -\log|\zeta \cdot z| + (1-c)\log\|z\|_2 + c\log d_A(z) \\
&= (1-c)(\log\|z\|_2 - \log|\zeta \cdot z|) + c(\log d_A(z) - \log|\zeta \cdot z|) \\
&\geqslant (1-c)(\log\|z\|_2 - \log(Cd_{A^*}(\zeta)\|z\|_2) \\
&\qquad\qquad + c(\log d_A(z) - \log(C\|\zeta\|_2 d_A(z))) \\
&= -(1-c)\log d_{A^*}(\zeta) - c\log\|\zeta\|_2 - \log C = \varphi_{1-c}(\zeta) - M.
\end{aligned}
$$

If $\zeta \notin A^*$, then $(\mathscr{L}f_c)(\zeta) = \varphi_{1-c}(\zeta) = +\infty$.

9.8.5   The support function of a convex set

Let A_0 be a subset of $\mathbf{C}^n$. Its support function is

$$H_{A_0}(\zeta') = \sup_{z' \in A_0} \mathrm{Re}\,(\zeta' \cdot z'), \qquad \zeta' = (\zeta_1, \ldots, \zeta_n) \in \mathbf{C}^n. \tag{9.108}$$

If A_0 is closed and convex, then H_{A_0} determines A_0; in fact, A_0 is the set of all $z' = (z_1, \ldots, z_n)$ such that $\mathrm{Re}\,(\zeta' \cdot z') \leqslant H_{A_0}(\zeta')$ for all ζ'. Now let A be the homogeneous set of all $z \in \mathbf{C}^{1+n} \smallsetminus \{0\}$ such that $z_0 \neq 0$ and $z'/z_0 \in A_0$. With the set A we associate the function $\mathscr{L}I_A$ defined by

$$(\mathscr{L}I_A)(\zeta) = -\log d_{A^*}(\zeta) = \sup_{z \in A}(-\log|\zeta \cdot z| + \log\|z\|_2), \qquad \zeta \in \mathbf{C}^{1+n} \smallsetminus \{0\}, \tag{9.109}$$

the projective analogue of the support function. What is the relation between these two support functions? To answer this question we first modify I_A a little and define

$$h_A(\zeta) = \sup_{z \in A}(-\log|\zeta \cdot z| + \log|z_0|), \qquad \zeta \in \mathbf{C}^{1+n} \smallsetminus \{0\}. \tag{9.110}$$

If A_0 is bounded, then h_A and $\mathscr{L}I_A$ are finite in the same set and differ there by a bounded function.

We shall express h_A in terms of H_{A_0}. We first formulate an auxiliary result, which we shall need for convex sets in the complex plane only, but which is also valid in $\mathbf{R}^n$. We shall therefore use the real support function

$$H_A(\xi) = \sup_{x \in A} \xi \cdot x, \qquad \xi \in \mathbf{R}^n. \tag{9.111}$$

Lemma 9.8.30 *Let A be a convex set in $\mathbf{R}^n$. Then*

$$- \inf_{x \in A} \|x\|_2 \leqslant \inf_{\|\xi\|_2 = 1} H_A(\xi) \leqslant \inf_{x \notin A} \|x\|_2 \tag{9.112}$$

with equality on the left if $0 \notin A^\circ$, and on the right if $0 \in \overline{A}$.

Proof For any set A we have, writing S for the unit sphere,

$$\inf_S H_A = \inf_{\xi \in S} \sup_{x \in A} \xi \cdot x \geqslant \sup_{x \in A} \inf_{\xi \in S} \xi \cdot x = \sup_{x \in A}(-\|x\|_2) = - \inf_{x \in A} \|x\|_2.$$

If A is convex and $x \notin A$, then there is a $\xi \in S$ such that $\xi \cdot x \geqslant H_A(\xi)$; thus $\|x\|_2 \geqslant \xi \cdot x \geqslant H_A(\xi)$, so that $\|x\|_2 \geqslant \inf_S H_A$. This shows that $\inf_{x \notin A} \|x\|_2 \geqslant \inf_S H_A$ and proves (9.112) for all convex sets.

Now assume that $0 \notin A^\circ$. Then A must be contained in a half-space $\{x; \; \xi \cdot x \geqslant c\}$ for some $\xi \in S$ and $c \geqslant 0$, which shows that $H_A(-\xi) \leqslant -c \leqslant 0$. If A is empty, (9.112) has the form $-\infty \leqslant -\infty \leqslant 0$, so the result is true. If A is not empty, then we can choose $c = \inf_{x \in A} \|x\|_2$, so that $\inf_S H_A \leqslant -c = -\inf_{x \in A} \|x\|_2$; we have proved equality on the left in (9.112).

On the other hand, if $0 \in \overline{A}$ and $H_A(\xi) < c$ for some $\xi \in S$ and some c, then necessarily $c > 0$ and the vector $c\xi$ cannot belong to A, so that $\inf_{x \notin A} \|x\|_2 \leqslant \|c\xi\|_2 = c$. Thus $\inf_{x \notin A} \|x\|_2 \leqslant \inf_S H_A$; we have proved equality on the right in (9.112).

Lemma 9.8.31 *For any convex set A in $\mathbf{R}^n$ we have*

$$\inf_{x \in A} \|x\|_2 = \sup_{\|\xi\|_2 = 1} H_A(\xi)^-; \tag{9.113}$$

$$\inf_{x \notin A} \|x\|_2 = \inf_{\|\xi\|_2 = 1} H_A(\xi)^+, \tag{9.114}$$

where $t^+ = \max(t, 0)$, $t^- = \max(-t, 0)$ for $t \in \mathbf{R}$.

Proof If $0 \notin A$, then (9.113) is just the first part of (9.112) with equality, and (9.114) reduces to $0 = 0$. If $0 \in A$, then (9.113) reduces to $0 = 0$ while (9.114) is the second part of (9.112) with equality.

Proposition 9.8.32 *Let A_0 be any convex set in $\mathbf{C}^n$, and A the homogeneous set of all $z \in \mathbf{C}^{1+n} \setminus \{0\}$ such that $z_0 \neq 0$ and $z'/z_0 \in A_0$. Define H_{A_0} and h_A by (9.108) and (9.110), respectively. Then*

$$h_A(\zeta) = \inf_{|t|=1} \log \left(H_{A_0}(t\zeta') + \operatorname{Re}(t\zeta_0) \right)^-, \qquad \zeta \in \mathbf{C}^{1+n} \setminus \{0\}. \tag{9.115}$$

Proof If A is empty, (9.115) certainly holds, because both sides are equal to $-\infty$. Fix $\zeta \in \mathbf{C}^{1+n} \setminus \{0\}$ and denote by L the linear mapping $z' \mapsto \zeta' \cdot z'$. If A is not empty, then

$$e^{-h_A(\zeta)} = \inf_{z' \in A_0} |\zeta_0 + \zeta' \cdot z'| = \inf_{s \in L(A_0)} |\zeta_0 + s| = |\zeta_0 + a(\zeta_0)|, \qquad (9.116)$$

where $a(\zeta_0)$ denotes the point in the closure of $L(A_0)$ which is closest to $-\zeta_0$. Here the first equality holds because, in view of the homogeneity, it is enough to let z vary with $z_0 = 1$ in the definition of h_A.

We now note that

$$H_{A_0}(t\zeta') = \sup_{z' \in A_0} \operatorname{Re}(t\zeta' \cdot z') = \sup_{z' \in A_0} \operatorname{Re} tL(z') = \sup_{s \in L(A_0)} \operatorname{Re} ts = H_{L(A_0)}(t),$$

which shows that the support function of the set $M = L(A_0) + \zeta_0$ is

$$H_M(t) = H_{A_0}(t\zeta') + \operatorname{Re}(t\zeta_0), \qquad t \in \mathbf{C}.$$

We can apply (9.113) to the convex set M. This yields

$$e^{-h_A(\zeta)} = |\zeta_0 + a(\zeta_0)| = \sup_{|t|=1} H_M(t)^- = \sup_{|t|=1} \left(H_{A_0}(t\zeta') + \operatorname{Re}(t\zeta_0) \right)^-,$$

where the first equality is that of (9.116), and thus proves (9.115).

Conversely, we can express H_{A_0} in terms of h_A.

Proposition 9.8.33 *Let A_0 be a bounded but not necessarily convex set in $\mathbf{C}^n$, and A the set of all $z \in \mathbf{C}^{1+n} \setminus \{0\}$ such that $z_0 \neq 0$ and $z'/z_0 \in A_0$. Define H_{A_0} and h_A by (9.108) and (9.110), respectively. Then*

$$H_{A_0}(\zeta') = \lim_{\substack{\zeta_0 \in \mathbf{R} \\ \zeta_0 \to -\infty}} \left(-\zeta_0 - e^{-h_A(\zeta)} \right), \qquad \zeta' \in \mathbf{C}^n. \qquad (9.117)$$

Proof We can still use (9.116) even though A_0 now is perhaps not convex, if we let $a(\zeta_0)$ denote one of the closest points to $-\zeta_0$ in the closure of $L(A_0)$. Let ζ_0 be real and tend to $-\infty$. Then

$$e^{-h_A(\zeta)} + \zeta_0 = |\zeta_0 + a(\zeta_0)| + \zeta_0 \to -\operatorname{Re} a(-\infty),$$

where $a(-\infty)$ is an accumulation point of $a(\zeta_0)$ as $\zeta_0 \in \mathbf{R}$ and $\zeta_0 \to -\infty$. It is a point in the closure of $L(A_0)$ which satisfies

$$H_{A_0}(\zeta') = \sup_{z' \in A_0} \operatorname{Re}(\zeta' \cdot z') = \sup_{s \in L(A_0)} \operatorname{Re} s = \operatorname{Re} a(-\infty).$$

This implies (9.117).

9.8.6 The dual functions expressed as a dual complement

In convexity theory, the Fenchel transform generalizes the support function: the support function (9.111) is just the Fenchel transform of the indicator function. Conversely, we can express the Fenchel transform $\tilde{f}$ of a function f in terms of the support function if we add one dimension: by definition we have $\tilde{f}(\xi) = \sup_x(\xi \cdot x - f(x))$, and we see that $\tilde{f}(\xi) = H_{\mathbf{epi}\,f}(\xi, -1)$, where $H_{\mathbf{epi}\,f}$ is the support function of the finite epigraph of f, i.e.,

$$H_{\mathbf{epi}\,f}(\xi, \eta) = \sup_{(x,y)\in\mathbf{R}^n\times\mathbf{R}} (\xi \cdot x + \eta y; \, f(x) \leqslant y), \qquad (\xi, \eta) \in \mathbf{R}^n \times \mathbf{R}.$$

We already know that the dual complement A^* of a closed set A can be expressed in terms of the dual function (indeed, A^* is the set where $\mathscr{L}I_A$ is finite; see Proposition 9.8.20). Conversely, we shall see here that we can express the dual function in terms of a dual complement if we go up one step in dimension (cf. (9.130) below). The functions will then give rise to Hartogs sets, which we proceed to discuss.

Hartogs domains and complete Hartogs domains were defined in Section 9.4 above. We shall generalize this in two ways: first we shall need to study sets that are not necessarily open; second, it is natural to add a hyperplane at infinity and look at subsets of projective space. Thus we consider sets $A \subset (\mathbf{C}^{1+n} \smallsetminus \{0\}) \times \mathbf{C}$ that are homogeneous in the sense of Subsection 9.8.3, i.e., such that $(sz, st) \in A$ if $(z, t) \in A$ and $s \in \mathbf{C} \smallsetminus \{0\}$. We shall say that A is a **complete Hartogs set** if (z, t') belongs to A as soon as $(z, t) \in A$ and $|t'| \leqslant |t|$. Such a set is therefore defined by an inequality $|t| < R(z)$ or $|t| \leqslant R(z)$ for some function R with $0 \leqslant R \leqslant +\infty$. We shall however use $f = -\log R$ to indicate the radius of the disks.

Definition 9.8.34 *Let* $f \colon \mathbf{C}^{1+n} \smallsetminus \{0\} \to \mathbf{R}_!$ *be a homogeneous function and* X *a homogeneous subset of* $\mathbf{C}^{1+n} \smallsetminus \{0\}$. *We associate to* f *and* X *a homogeneous complete Hartogs set* $E(X; f)$ *in* $\mathbf{C}^{1+n+1}$: *it is the set of all* $(z, t) \in (\mathbf{C}^{1+n} \smallsetminus \{0\}) \times \mathbf{C}$ *such that* $|t| \leqslant e^{-f(z)}$ *when* $z \in X$, *and* $|t| < e^{-f(z)}$ *when* $z \notin X$. □

The fiber of $E(X; f)$ over z is thus the whole t-plane if $f(z) = -\infty$; it is a closed disk of finite positive radius if $z \in (\mathbf{dom}(f)) \cap X$ and $f(z) > -\infty$; it is an open disk of finite positive radius if $z \in (\mathbf{dom}(f)) \smallsetminus X$ and $f(z) > -\infty$; it is the origin if $z \in X \smallsetminus \mathbf{dom}(f)$; finally, the fiber is empty if $z \notin X \cup \mathbf{dom}(f)$. If $X_1 \subset X_2$ and $f_1 \geqslant f_2$, we have an obvious inclusion $E(X_1; f_1) \subset E(X_2; f_2)$.

Every complete Hartogs set A is of the form $E(X; f)$ for some X and some f: we can take X as the set of all z such that the fiber is not open and define $f(z)$ as the infimum of all real numbers c such that (z, e^{-c}) belongs to A. A complete Hartogs set defines the set $X \cap \{z; \, f(z) > -\infty\}$ uniquely: if $f(z) > -\infty$, then $z \in X$ if and only if the fiber over z is closed and nonempty. On the other hand the choice of $X \cap \{z; \, f(z) = -\infty\}$ is immaterial in the definition of $E(X; f)$.[8]

---

[8]To get uniqueness, one could for example require that X always contain $\{z; \, f(z) = -\infty\}$

Theorem 9.8.35 *Consider the dual complement of* $E(X; f)$,

$$E(X; f)^* = \{(\zeta, \tau) \in \mathbf{C}^{1+n+1} \smallsetminus \{0\};\ \zeta \cdot z + \tau t \neq 0 \text{ for all } (z, t) \in E(X; f)\}. \tag{9.118}$$

If both X and $\mathbf{dom}(f)$ are empty, then also $E(X; f)$ is empty and $E(X; f)^*$ is equal to the whole set $\mathbf{C}^{1+n+1} \smallsetminus \{0\}$. If on the other hand $X \cup \mathbf{dom} f \neq \emptyset$, then $E(X; f)$ is nonempty and its dual complement $E(X; f)^*$ is a subset of $(\mathbf{C}^{1+n} \smallsetminus \{0\}) \times \mathbf{C}$ and a complete Hartogs set, thus

$$E(X; f)^* = E(\Xi; \varphi)$$

for some set Ξ and some function φ. Here the function φ is uniquely determined:

$$\varphi(\zeta) = \begin{cases} (\mathscr{L}f)(\zeta) & \text{when } \zeta \in (X \cup \mathbf{dom}(f))^*, \text{ and} \\ +\infty & \text{when } \zeta \in (\mathbf{C}^{1+n} \smallsetminus \{0\}) \smallsetminus (X \cup \mathbf{dom} f)^*; \end{cases} \tag{9.119}$$

thus $\varphi = \mathscr{L}f$ as soon as $\mathbf{dom}(\mathscr{L}f) \subset (X \cup \mathbf{dom}(f))^*$, in particular if $X \subset \mathbf{dom}(f)$.

We define the set Ξ as follows. We let $\zeta \in \Xi$ if and only if $\zeta \in (X \cup \mathbf{dom}(f))^*$ and either f takes the value $-\infty$ or

$$\inf_{z \in \mathbf{dom}(f) \cap X} |\zeta \cdot z| e^{f(z)} \geqslant \inf_{w \in \mathbf{dom}(f) \smallsetminus X} |\zeta \cdot w| e^{f(w)} \tag{9.120}$$

or else

$$\text{for all } z^0 \in X \cap \mathbf{dom}(f) \text{ we have } |\zeta \cdot z^0| e^{f(z^0)} > \inf_{z \in \mathbf{dom}(f) \cap X} |\zeta \cdot z| e^{f(z)}. \tag{9.121}$$

If f is not $+\infty$ identically, then Ξ is uniquely determined, so that this is the only set which satisfies $E(X; f)^* = E(\Xi; \varphi)$. Moreover, we always have

$$E(X; f)^* \cap \left((\mathbf{C}^{1+n} \smallsetminus \{0\}) \times \{0\} \right) = (X \cup \mathbf{dom}(f))^* \times \{0\}, \tag{9.122}$$

which proves that $\Xi \cup \mathbf{dom}(\varphi) = (X \cup \mathbf{dom}(f))^*$. The particular cases when X is empty or equal to $\mathbf{dom}(f)$ are of interest. If $\mathbf{dom} f \neq \emptyset$ we have

$$E(\emptyset; f)^* = E((\mathbf{dom}(f))^*; \mathscr{L}f) \supset E(\mathbf{dom}(\mathscr{L}f); \mathscr{L}f). \tag{9.123}$$

If $\mathbf{dom}(f)$ is closed and nonempty and f is lower semicontinuous and never takes the value $-\infty$, then

$$E(\mathbf{dom}(f); f)^* = E(\emptyset; \mathscr{L}f). \tag{9.124}$$

---

or that these two sets be disjoint, or else take the Riemann sphere as the fiber over points in X such that $f(z) = -\infty$, but we shall refrain from doing so.

Remark 9.8.36 *If $X \cup \mathbf{dom}(f) \neq \emptyset$, then $E(X;f)^*$ contains*

$$\left(E(\emptyset; \mathscr{L}f) \cap \left((X \setminus \mathbf{dom}(f))^* \times (\mathbf{C} \setminus \{0\})\right)\right) \cup \left((X \cup \mathbf{dom}(f))^* \times \{0\}\right)$$

and is contained in

$$E\left((\mathbf{dom}(\mathscr{L}f)) \cup (X \cup \mathbf{dom}(f))^*; \mathscr{L}f\right).$$

If X is a subset of $\mathbf{dom}(f)$, then $\varphi = \mathscr{L}f$ and these inclusions simplify to:

$$E(\emptyset; \mathscr{L}f) \cup ((\mathbf{dom}(f))^* \times \{0\}) \subset E(X;f)^* \subset E((\mathbf{dom}\, f)^*; \mathscr{L}f). \quad (9.125)$$

We also note the following two special cases. If $f = +\infty$ identically, then

$$E(X;+\infty) = X \times \{0\} \text{ and } E(X;+\infty)^* = X^*.$$

In this case the definition of Ξ in the theorem yields $\Xi = X^*$.
We have

$$E(X;f)^* = (X \cup \mathbf{dom}(f))^* \times \{0\} = E((X \cup \mathbf{dom}(f))^*; \mathscr{L}f), \quad (9.126)$$

if f assumes the value $-\infty$. □

Proof of Theorem 9.8.35. If $X \cup \mathbf{dom}(f)$ is empty, then $E(X;f)$ is empty, and its dual complement is the whole space except the origin. If $X \cup \mathbf{dom}(f)$ is not empty, then the hyperplane $(\mathbf{C}^{1+n} \setminus \{0\}) \times \{0\}$ cuts $E(X;f)$, so that no point $(0,\tau)$ belongs to $E(X;f)^*$, which therefore is contained in $(\mathbf{C}^{1+n} \setminus \{0\}) \times \mathbf{C}$.

We need to find the conditions for (ζ, τ) to belong to $E(X;f)^*$. This happens precisely when $\zeta \cdot z + \tau t$ is non-zero for all $(z,t) \in E(X;f)$. The case $\tau = 0$ is easy: we find that $(\zeta, 0) \in E(X;f)^*$ if and only if $\zeta \cdot z \neq 0$ for all $z \in X \cup \mathbf{dom}(f)$, thus if and only if $\zeta \in (X \cup \mathbf{dom}(f))^*$. This proves (9.122). Now let $\tau \neq 0$. Then we see that $(\zeta, \tau) \in E(X;f)^*$ precisely when the following three conditions hold:

$$|\tau| < |\zeta \cdot z| e^{f(z)} \text{ for all } z \in (\mathbf{dom}(f)) \cap X; \quad (9.127)$$

$$|\tau| \leqslant |\zeta \cdot w| e^{f(w)} \text{ for all } w \in (\mathbf{dom}\, f) \setminus X; \quad (9.128)$$

$$|\zeta \cdot z| \neq 0 \text{ for all } z \in X \setminus \mathbf{dom}(f). \quad (9.129)$$

Fix $\zeta \in (X \cup \mathbf{dom}(f))^*$. We see that the three formulas (9.127)–(9.129) imply that

$$|\tau| \leqslant \exp\left(-(\mathscr{L}f)(\zeta)\right),$$

and that they are implied by $|\tau| < \exp\left(-(\mathscr{L}f)(\zeta)\right)$. This shows that φ is as described in (9.119), and it only remains to be seen when the inequality $|\tau| \leqslant \exp\left(-(\mathscr{L}f)(\zeta)\right)$ is strict. The condition on τ means that it shall belong to all open disks of radius $|\zeta \cdot z| e^{f(z)}$ for $z \in (\mathbf{dom}(f)) \cap X$, and all closed disks of radius $|\zeta \cdot w| e^{f(w)}$ for $w \in \mathbf{dom}(f) \setminus X$. Now an intersection of a family of closed disks is always closed, and an intersection of nonempty concentric open disks with finite radii is closed exactly when it contains, along with any disk, also a disk of strictly smaller radius. This is what is expressed by conditions (9.120) and (9.121). Finally (9.123) and (9.124) follow from an analysis of (9.120)–(9.121) in the special cases $X = \varnothing$ and $X = \mathbf{dom}(f)$.

Corollary 9.8.37 *The logarithmic transform $\mathscr{L}f$ of any function f can be obtained from the dual complement of $E(\varnothing; f)$: it is minus the logarithm of a certain distance, viz. the distance from $(\zeta, 0)$ to the complement of $E(\varnothing; f)^*$ in the direction $(0, \ldots, 0, 1)$:*

$$(\mathscr{L}f)(\zeta) = -\log\left(\inf_{\tau}(|\tau|; \ (\zeta, \tau) \in \mathbf{C}^{1+n+1} \smallsetminus E(\varnothing; f)^*)\right). \qquad (9.130)$$

Proof This follows from (9.123). The result also explains why we cannot expect these functions to be pullbacks of functions on projective space.

9.8.7 Lineally convex Hartogs sets

Intuitively, it seems that $E(\varnothing; f)$ and $E(\mathbf{dom}(f); f)$ ought to be lineally convex simultaneously. This is not quite true. We shall note three results in the positive direction, Propositions 9.8.38–9.8.40 below, and one result in the negative direction, Example 9.8.41. Then we shall establish conditions under which it is true that f is $\mathscr{L}$-closed if and only if $E(\mathbf{dom}(f); f)$ is lineally convex (Corollary 9.8.43), as well as conditions which guarantee that f is $\mathscr{L}$-closed if and only if $E(\varnothing; f)$ is lineally convex (Theorem 9.8.49).

Proposition 9.8.38 *If $E(X; f)$ is lineally convex, then also $X \cup \mathbf{dom}(f)$ and $E(X \cup \mathbf{dom}(f); f)$ are lineally convex. In particular, if $E(\varnothing; f)$ is lineally convex, then so are $\mathbf{dom}(f)$ and $E(\mathbf{dom}(f); f)$.*

Proof Suppose that $E(X; f)$ is lineally convex. That $X \cup \mathbf{dom}(f)$ is lineally convex then follows from the easily proved result that the intersection of a lineally convex set and a complex subspace is lineally convex as a subset of the latter. If $E(X; f)$ is lineally convex, then also $E(X; f + a)$ is lineally convex for any real number a. Any intersection of lineally convex sets has the same property, so we only need to note that $E(X \cup \mathbf{dom}(f); f)$ is equal the intersection of all $E(X; f - a)$, $a > 0$.

Proposition 9.8.39 *If f is upper semicontinuous and there exists a set X such that $E(X; f)$ is lineally convex, then $E(\varnothing; f)$ is lineally convex.*

Proof We know from Corollary 9.8.4 that $E(X; f)^\circ$ is lineally convex if $E(X; f)$ is lineally convex. Now $E(X; f)^\circ = E(\varnothing; f)$ if f is upper semicontinuous, hence the result.

However, the semicontinuity of f is not important—it is the fact that the effective domain is open which is relevant. This is shown by the following result.

Proposition 9.8.40 *If $X \cup \mathbf{dom}(f)$ is open and $E(X; f)$ is lineally convex, then $E(X \setminus \mathbf{dom}(f); f)$ is lineally convex. In particular, $E(\emptyset; f)$ is lineally convex if $\mathbf{dom}(f)$ is open and $E(X; f)$ is lineally convex for some subset X of $\mathbf{dom}\, f$.*

Proof Assume that $E(X; f)$ is lineally convex. Then also $E(X; f + a)$ is lineally convex for any real a, and we shall prove that the union of all $E(X; f + 1/k)$, $k = 1, 2, \ldots$, which equals $E(X \setminus \mathbf{dom}(f); f)$, is lineally convex.

The hyperplanes in $\mathbf{C}^{1+n+1}$ will be denoted by $Y_{(\zeta, \tau)}$ in analogy with (9.85), thus

$$Y_{(\zeta, \tau)} = \{(z, t) \in (\mathbf{C}^{1+n} \times \mathbf{C}) \setminus \{0\};\ \zeta \cdot z + \tau t = 0\}. \tag{9.131}$$

Let $(z, t) \notin E(X \setminus \mathbf{dom}(f); f)$ be given with $z \in X \cup \mathbf{dom}(f)$. For any k there is a hyperplane $Y_{(\zeta^k, \tau^k)}$ that contains the point (z, t) and which does not meet the set $E(X; f + 1/k)$. We may assume that $\|\zeta^k\|_2^2 + |\tau^k|^2 = 1$. Take an accumulation point (ζ, τ) of the sequence (ζ^k, τ^k). Since $X \cup \mathbf{dom}(f)$ is open, we can be sure that $\tau \neq 0$. The hyperplane $Y_{(\zeta, \tau)}$ passes through (z, t) and does not meet $E(X \setminus \mathbf{dom}(f); f)$ since $\tau \neq 0$. If on the other hand $z \notin X \cup \mathbf{dom}(f)$, there is a hyperplane $Y_{(\zeta, 0)}$ which passes through (z, t) and does not cut $E(X; f)$.

The openness in Proposition 9.8.40 cannot be dispensed with as we shall see now.

Example 9.8.41 *There is a function f such that $E(\mathbf{dom}(f); f)$ is lineally convex while $E(\emptyset; f)$ is not. Define*

$$R(z) = \inf_{k \in \mathbf{N} \setminus \{0\}} |(k + 1)z_1 - z_0 - z_0/k|, \qquad z = (z_0, z_1) \in \mathbf{C}^{1+1} \setminus \{0\},$$

and let $f = -\log R$. Then $\mathbf{dom}(f)$ consists of the complement of the hyperplanes Y_ζ, $\zeta = (1, -k)$, and

$$E(\mathbf{dom}(f); f) = \bigcap_{k=1}^{\infty} \{(z, t);\ z \notin Y_{(1, -k)}\ \text{and}\ |t| \leqslant |(k + 1)z_1 - z_0 - z_0/k|\}$$

is lineally convex. (Note, however, that $E(X; f)$ is not lineally convex if X contains $\mathbf{dom}(f)$ strictly; cf. Example 9.8.9.) The function f is $\mathcal{L}$-closed; cf. (9.93) and Theorem 9.8.42 below. To prove that $E(\emptyset; f)$ is not lineally convex, let us note that $(1, 0, 1) \notin E(\emptyset; f)$, for $R(1, 0) = 1$. Suppose there exists a hyperplane $Y_{(\zeta, \tau)}$ which passes through the point $(1, 0, 1)$ but does not cut $E(\emptyset; f)$. Then $\zeta_0 + \tau = 0$. We must also have $\tau \neq 0$ since $(1, 0, 0) \in E(\emptyset; f)$ as well as $\zeta_1 \neq 0$ since $(1, 2, 1) \in E(\emptyset; f)$. Moreover

$$\frac{|\zeta \cdot z|}{|\tau|} = \frac{|\zeta \cdot z|}{|\zeta_0|} \geqslant \inf_k |(k+1)z_1 - z_0 - z_0/k| \qquad \text{for all } z.$$

Taking $z = (\zeta_1, -\zeta_0)$ we see that there is a number m such that $\zeta_1 = -m\zeta_0$, and we can conclude that, taking $z_0 = 1$,

$$\frac{|\zeta \cdot z|}{|\zeta_0|} = m|z_1 - 1/m| \geqslant \inf_k |(k+1)z_1 - 1 - 1/k|.$$

However, for z_1 close to $1/m$ we must have

$$\inf_k |(k+1)z_1 - 1 - 1/k| = |(m+1)z_1 - 1 - 1/m|,$$

so that

$$m|z_1 - 1/m| \geqslant (m+1)|z_1 - 1/m|$$

for all z_1 close to $1/m$. This is impossible, which shows that there is no such hyperplane. $\qquad\square$

Theorem 9.8.42 *If* $f = \mathscr{L}\mathscr{L}f$ *in* $(X \cup \mathbf{dom}(f))^{**}$, *then* $E((X \cup \mathbf{dom}(f))^{**}; f)$ *is lineally convex. In particular, if we assume* $\mathbf{dom}\, f$ *to be lineally convex and* $X \subset \mathbf{dom}(f)$, *then* $\mathscr{L}\mathscr{L}f = f$ *in* $\mathbf{dom}(f)$ *implies that* $E(\mathbf{dom}(f); f)$ *is lineally convex.*

Conversely, if $E(X; f)$ *is lineally convex, then* $f = \mathscr{L}(\mathscr{L}f)$ *in* $X \cup \mathbf{dom}(f)$, *which is a lineally convex set. If* f *is bounded from below on the unit sphere, then* $f = \mathscr{L}(\mathscr{L}f) = +\infty$ *outside* $\overline{\mathbf{dom}(f)}^{**}$. *Thus in this case* $\mathscr{L}(\mathscr{L}f) = f$ *everywhere if* $X \supset \overline{\mathbf{dom}(f)}^{**} \setminus \mathbf{dom}(f)$.

Proof Suppose that $f = \mathscr{L}\mathscr{L}f$ in $(X \cup \mathbf{dom}(f))^{**}$. Take

$$(z^0, t^0) \notin E((X \cup \mathbf{dom}(f))^{**}; f).$$

We shall then prove that there is a hyperplane $Y_{(\zeta, \tau)}$ (see (9.131)) which contains (z^0, t^0) and does not cut $E((X \cup \mathbf{dom}\, f)^{**}; f)$. Consider first the case

$$z^0 \in (X \cup \mathbf{dom}(f))^{**}.$$

We know that $|t^0| > e^{-f(z^0)}$. By the definition of $\mathscr{L}f$ and since $(\mathscr{L}\mathscr{L}f)(z^0) = f(z^0) > -\log|t^0|$, we can choose ζ such that

$$-\log|\zeta \cdot z^0| - (\mathscr{L}f)(\zeta) > -\log|t^0|.$$

Then we take $\tau = -\zeta \cdot z^0/t^0$, so that $(z^0, t^0) \in Y_{(\zeta, \tau)}$ and $-(\mathscr{L}f)(\zeta) > \log|\tau|$. Moreover, for any $(z, t) \in Y_{(\zeta, \tau)}$ we have

$$f(z) \geqslant (\mathscr{L}\mathscr{L}f)(z) \geqslant -\log|\zeta \cdot z| - (\mathscr{L}f)(\zeta) = -\log|\tau t| - (\mathscr{L}f)(\zeta) > -\log|t|,$$

which shows that $(z,t) \notin E((X \cup \mathbf{dom}(f))^{**}; f)$. The case $z^0 \notin (X \cup \mathbf{dom}(f))^{**}$ remains to be considered. In this case there is a hyperplane Y_ζ that contains z^0 and does not meet $(X \cup \mathbf{dom}(f))^{**}$, so the hyperplane $Y_{(\zeta,0)}$ does not cut $E((X \cup \mathbf{dom}(f))^{**}; f)$.

Now assume that $E(X; f)$ is linearly convex. We already know that $X \cup \mathbf{dom}(f)$ is linearly convex (cf. Proposition 9.8.38). If f assumes the value $-\infty$, then $E(X; f) = (X \cup \mathbf{dom}(f)) \times \mathbf{C}$, and $f = \mathscr{L}\mathscr{L}f = -\infty$ in $X \cup \mathbf{dom}(f)$. If $f > -\infty$, let z^0 be any point in $X \cup \mathbf{dom}(f)$ and take t^0 such that $|t^0| > e^{-f(z^0)}$, thus $(z^0, t^0) \notin E(X; f)$. By hypothesis there is a hyperplane $Y_{(\zeta,\tau)}$ which passes through (z^0, t^0) and does not meet $E(X; f)$. Since $(z^0, 0) \in E(X, f)$, we must have $\tau \neq 0$, so we obtain a minorant of f of the form $-\log|\zeta \cdot z| + \log|\tau| \leqslant f(z)$, where the left-hand side takes the value $-\log|t^0| < f(z^0)$ at the point z^0 and moreover can be chosen larger than any number less than $f(z^0)$. Thus $(\mathscr{L}\mathscr{L}f)(z^0) \geqslant f(z^0)$ and we conclude that $\mathscr{L}\mathscr{L}f = f$ in all of $X \cup \mathbf{dom}(f)$.

Finally, assume that $f \geqslant -C$ on the unit sphere without any further assumption. Thus, putting $A = \mathbf{dom}(f)$, we have $f \geqslant g = I_A - C$, so that $\mathscr{L}f \leqslant \mathscr{L}g = C - \log d_{A^*}$ by Proposition 9.8.20. We now note that $d_{A^*} = d_{A^{*\circ}}$ and take the transformation once again, this time using Proposition 9.8.23. We get $\mathscr{L}\mathscr{L}f \geqslant \mathscr{L}\mathscr{L}g \geqslant I_{A^{*\circ*}} - M - C$. In particular $\mathscr{L}f(z) = +\infty$ if $z \notin A^{*\circ*} = \overline{A}^{**} = \overline{\mathbf{dom}(f)}^{**}$ (cf. (D) in Proposition 9.8.2). This finishes the proof.

Corollary 9.8.43 *Assume that f is bounded from below on the unit sphere and that $\mathbf{dom}(f)$ is closed and linearly convex. Then f is $\mathscr{L}$-closed if and only if $E(\mathbf{dom}(f); f)$ is linearly convex.* □

We now proceed to study the case when $\mathbf{dom}(f)$ is open and f tends to $+\infty$ at the boundary. Propositions 9.8.44 and 9.8.47 below are applicable when f tends rather fast to $+\infty$, and Theorem 9.8.49 in a more general situation.

Proposition 9.8.44 *Let f be a homogeneous function on $\mathbf{C}^{1+n} \smallsetminus \{0\}$ which tends to $+\infty$ at the boundary of $\mathbf{dom}(f) = A$ in the strong sense that $f \geqslant -C - \log d_A$ for some constant C, where d_A is defined by (9.96). Then f is $\mathscr{L}$-closed if and only if $E(\emptyset; f)$ is linearly convex.*

Proof If f is $\mathscr{L}$-closed, then its effective domain $A = A^\circ$ must be linearly convex by Theorem 9.8.14, for $A = (\mathbf{dom}(\mathscr{L}\mathscr{L}f))^\circ = \overline{\mathbf{dom}(\mathscr{L}f)}^*$ in view of (9.99) applied to $\mathscr{L}f$ (this function is bounded from below on the unit sphere unless f is $+\infty$ identically, a trivial case). Theorem 9.8.42 now shows that $E(A; f)$ is linearly convex and Proposition 9.8.40 implies that $E(\emptyset; f)$ is linearly convex.

Conversely, assume that $E(\emptyset; f)$ is linearly convex. In view of Theorem 9.8.42 it only remains to be proved that $\mathscr{L}\mathscr{L}f = f = +\infty$ outside $A = \mathbf{dom}(f)$. Now if $f \geqslant -C - \log d_A$, then $\mathscr{L}f \leqslant C + \mathscr{L}(-\log d_A) \leqslant C + I_{A^*}$.

by Proposition 9.8.23. We take the transformation again and obtain $\mathscr{L}\mathscr{L}f \geqslant -C+\mathscr{L}I_{A^*} = -C - \log d_{A^{**}}$, using Proposition 9.8.20. But $\mathbf{dom}(f)$ is linearly convex, so $A^{**} = A$. Hence $\mathscr{L}\mathscr{L}f = +\infty$ in the complement of A.

Functions with bounded logarithmic transforms exhibit the behavior studied in Proposition 9.8.44:

Proposition 9.8.45 *Let f be a homogeneous function on $\mathbf{C}^{1+n} \smallsetminus \{0\}$ such that $\mathbf{dom}(f) = A$ equals the interior of its closure and such that $\overline{\mathbf{dom}(f)}$ is linearly convex. Assume that f is bounded from below on the unit sphere and that $\mathscr{L}f$ is bounded from above in $S \cap \mathbf{dom}(\mathscr{L}f)$. Then $f \geqslant -C - \log d_A$, where C is a constant.*

Proof We have $\mathscr{L}f \leqslant C + I_B$, where $B = \mathbf{dom}(\mathscr{L}f)$. Therefore $f \geqslant \mathscr{L}(\mathscr{L}f) \geqslant -C + \mathscr{L}I_B = -C - \log d_{B^*} = -C - \log d_{B^{*\circ}}$. The next lemma shows that $B^{*\circ} = A$.

Lemma 9.8.46 *Let f be a homogeneous function on $\mathbf{C}^{1+n} \smallsetminus \{0\}$ such that*

$$\overline{\mathbf{dom}(f)}^{\;\circ} = \mathbf{dom}(f).$$

Assume that f is bounded from below on the unit sphere and that $\overline{\mathbf{dom}(f)}$ is linearly convex. Then $(\mathbf{dom}(\mathscr{L}f))^{*\circ} = \mathbf{dom}(f)$.

Proof From (9.99) we deduce, recalling that f is bounded from below on S, that

$$\overline{\mathbf{dom}(f)}^{\;**} \supset (\mathbf{dom}(\mathscr{L}f))^* \supset (\mathbf{dom}(f))^{**}.$$

Now, since $\overline{\mathbf{dom}\,f}$ is linearly convex, so is its interior $\overline{\mathbf{dom}(f)}^{\;\circ} = \mathbf{dom}(f)$ (Corollary 9.8.4). Therefore $\overline{\mathbf{dom}(f)} \supset (\mathbf{dom}(\mathscr{L}f))^* \supset \mathbf{dom}(f)$. Taking the interior of these sets we get $(\mathbf{dom}(\mathscr{L}f))^{*\circ} = \mathbf{dom}(f)$.

Under a regularity assumption we can let f tend to infinity at a slower pace:

Proposition 9.8.47 *Let f be a homogeneous function on $\mathbf{C}^{1+n} \smallsetminus \{0\}$ which tends to $+\infty$ at the boundary of $\mathbf{dom}(f) = A$ in the sense that $f \geqslant -C - c\log d_A$ on the unit sphere for some constants C and c with $0 < c \leqslant 1$. Assume that A^* satisfies the homogeneous interior cone condition (Definition 9.8.27). Then f is $\mathscr{L}$-closed if and only if $E(\varnothing; f)$ is linearly convex.*

Remark 9.8.48 *It can be easily proved that if A is linearly convex and its dual complement A^* satisfies the homogeneous interior cone condition, then so does its set-theoretic complement $\complement A$.*   □

Proof In view of Theorem 9.8.42 and the proof of Proposition 9.8.44, it only remains to be proved that $\mathscr{L}\mathscr{L}f = f = +\infty$ outside $\mathbf{dom}(f)$ if $E(\varnothing; f)$ is linearly convex. Now if $f \geqslant -C - c\log d_A$ on the unit sphere S, then we obtain $f \geqslant -C + f_c$ everywhere, introducing the function f_c of (9.103). We take the logarithmic transformation once to obtain $\mathscr{L}f \leqslant C + \mathscr{L}f_c \leqslant C + \varphi_{1-c}$ (Proposition 9.8.19), and then again to get $\mathscr{L}\mathscr{L}f \geqslant -C + \mathscr{L}\varphi_{1-c} \geqslant -C - M + f_c$, this time applying Proposition 9.8.28 to the function φ_{1-c} and using the homogeneous interior cone condition (Definition 9.8.27) on A^*. This shows that $\mathscr{L}\mathscr{L}f$ equals $+\infty$ in the complement of A.

We finally come to the general case of a function which tends to infinity at the boundary.

Theorem 9.8.49 *Let f be a homogeneous function on $\mathbf{C}^{1+n} \setminus \{0\}$. Assume that f is bounded from below on the unit sphere and tends to $+\infty$ at the boundary of $A = \mathbf{dom}(f)$ in the sense that $A_s = \{z \in S;\ f(z) < s\}$ is strongly contained in $\mathbf{dom}(f)$ for all numbers s, i.e., the closure of A_s is contained in the interior of A. (This implies that A is open.) Assume moreover that A^* satisfies the homogeneous interior cone condition. Then $E(\varnothing; f)$ is lineally convex if and only if f is $\mathscr{L}$-closed.*

Proof For the proof we shall need the functions $f_{A,r}$, where A is a homogeneous set and r is a positive number, defined as $f_{A,r} = -\log r + I_A$, thus $f_{A,r} = -\log r - \log\|z\|_2$ when $z \in A$ and $f_{A,r} = +\infty$ otherwise. We note that $\mathscr{L}f_{A,r} = \log r - \log d_{A^*}$ (Proposition 9.8.20, so that (ζ, τ) belongs to $E(A^*; \mathscr{L}f_{A,r})$ if and only if $\zeta \in A^*$ and $r|\tau| \leqslant d_{A^*}(\zeta)$.

What remains to be done, considering Theorem 9.8.42 and the proof of Proposition 9.8.44, is the following, assuming $E(\varnothing; f)$ to be lineally convex. Given any $z^0 \notin \mathbf{dom}(f) = A$ it is required to find a hyperplane $Y_{(\zeta,\tau)}$ with $\tau \neq 0$ which does not cut $E(\varnothing; f)$ and passes through (z^0, t^0) with $|t^0|$ arbitrarily small; the problem is to avoid the vertical hyperplanes, those with $\tau = 0$. Since f is bounded from below on S, there is a number R such that $E(\varnothing; f)$ is contained in $E(\varnothing; f_{A,R})$. On the other hand, given any $\varepsilon > 0$, there is a homogeneous set K which is strongly contained in A and such that $f \geqslant -\log \varepsilon$ on $S \setminus K$; this means that $E(\varnothing; f)$ is contained in $E(\varnothing; f_{A,\varepsilon}) \cup E(\varnothing; f_{K,R})$. We shall find a hyperplane which does not cut the latter set for a suitable choice of ε. This amounts to finding (ζ, τ) in $E(\varnothing; f_{A,\varepsilon})^* \cap E(\varnothing; f_{K,R})^*$, equivalently in $E(A^*; \mathscr{L}f_{A,\varepsilon}) \cap E(K^*; \mathscr{L}f_{K,R})$; cf. (9.123). The hyperplane shall also contain a point (z^0, t^0) with $|t^0| = \delta$ positive but arbitrarily small.

We shall thus find ζ and τ such that

$$0 \neq |\tau| \leqslant \frac{1}{\varepsilon} d_{A^*}(\zeta) \text{ and } |\tau| \leqslant \frac{1}{R} d_{K^*}(\zeta).$$

We take $0 \neq \tau = -(\zeta \cdot z^0)/t^0$ to ensure that (z^0, t^0) belongs to $Y_{(\zeta,\tau)}$, and then the problem is reduced to finding ζ such that

$$0 \neq |\zeta \cdot z^0| \leqslant \frac{\delta}{\varepsilon} d_{A^*}(\zeta) \text{ and } |\zeta \cdot z^0| \leqslant \frac{\delta}{R} d_{K^*}(\zeta). \tag{9.132}$$

Since by hypothesis $z^0 \notin A = A^{**}$, there is a point $\zeta^0 \in A^*$ such that $\zeta^0 \cdot z^0 = 0$. If ζ^0 is in the interior of A^*, then finding ζ is easy: we have $d_{K^*}(\zeta^0) \geqslant$

$d_{A^*}(\zeta^0) > 0 = |\zeta^0 \cdot z^0|$ and can take ζ close to ζ^0. If on the other hand $\zeta^0 \in \partial A^*$, we argue as follows. By the homogeneous interior cone condition,

$$\sup_{\|\zeta - \zeta^0\|_2 \leqslant s\|\zeta^0\|_2} d_{A^*}(\zeta) \geqslant \gamma s \|\zeta^0\|_2$$

for some positive constant γ. On the other hand

$$\sup_{\|\zeta - \zeta^0\|_2 \leqslant s\|\zeta^0\|_2} |\zeta \cdot z^0| = s\|\zeta^0\|_2 \|z^0\|_2.$$

Given any positive δ, it is thus enough to choose ε such that $\varepsilon |z^0| \leqslant \delta\gamma$ to satisfy the first inequality in (9.132) for some ζ close to ζ^0, more precisely satisfying $|\zeta - \zeta^0| \leqslant s|\zeta^0|$ for any given sufficiently small positive s. The second is then satisfied strictly when $\zeta = \zeta^0$, because A^* is strongly contained in K^* by Corollary 9.8.3, so that $d_{K^*}(\zeta^0) > d_{A^*}(\zeta^0) = 0$, and it must therefore also be satisfied for all ζ satisfying $|\zeta - \zeta^0| \leqslant s|\zeta^0|$ for all sufficiently small positive s. This completes the proof.

9.8.8 A necessary differential condition for $\mathscr{L}$-closed functions

It is well known that convex functions as well as plurisubharmonic functions of class C^2 can be characterized by differential conditions. Is the same true for $\mathscr{L}$-closed functions? We shall first establish a necessary differential condition.

Proposition 9.8.50 *Suppose that f is an $\mathscr{L}$-closed function of class C^2 in some open set Ω of $\mathbf{C}^{1+n} \setminus \{0\}$. Then*

$$\left| \sum (f_{z_j z_k} - 2f_{z_j} f_{z_k}) b_j b_k \right| \leqslant \sum f_{z_j \bar{z}_k} b_j \bar{b}_k, \qquad \text{in } \Omega \text{ for all } b \in \mathbf{C}^{1+n}. \tag{9.133}$$

In particular, if $n = 1$ and we define $F(z) = f(1, z)$, $z \in \mathbf{C}$, then

$$|F_{zz} - 2F_z^2| \leqslant F_{z\bar{z}}. \tag{9.134}$$

Proof Define $g(z) = -\log |\beta \cdot z|$. For every point a where $f(a)$ is finite there is a vector β such that $\mathbf{grad}\, g(a) = \mathbf{grad}\, f(a)$. Indeed, let us first note that by homogeneity $\sum a_j f_{z_j}(a) = -1/2$ for all a. If we choose $\beta_j = f_{z_j}(a)$, then $\beta \cdot a = -1/2$ and

$$\frac{\partial g}{\partial z_j}(z) = -\frac{1}{2} \frac{\beta_j}{\beta \cdot z}$$

takes the value

$$\frac{\partial g}{\partial z_j}(a) = -\frac{1}{2} \frac{\beta_j}{\beta \cdot a} = \beta_j$$

at $z = a$. Then by $\mathscr{L}$-closedness $f(z) \geqslant f(a) + g(z) - g(a)$ for all z, for the definition of the $\mathscr{L}$-transformation uses precisely the functions g plus a constant. Take a curve $t \mapsto \gamma(t)$ such that $\gamma(0) = a$ and compare the two functions $\varphi = f \circ \gamma$ and $\psi = f(a) + g \circ \gamma - g(a)$. We have $\varphi(0) = \psi(0)$ and $\varphi'(0) = \psi'(0)$ and must therefore have $\varphi''(0) \geqslant \psi''(0)$. We calculate φ'':

$$\tfrac{1}{2}\varphi''(t) = \operatorname{Re} \sum f_{z_j z_k}(\gamma(t))\gamma'_j(t)\gamma'_k(t)$$

$$+ \sum f_{z_j \bar{z}_k}(\gamma(t))\gamma'_j(t)\overline{\gamma'_k(t)} + \operatorname{Re} \sum f_{z_j}(\gamma(t))\gamma''_j(t).$$

The corresponding formula for ψ simplifies to

$$\tfrac{1}{2}\psi''(t) = \operatorname{Re} \sum g_{z_j z_k}(\gamma(t))\gamma'_j(t)\gamma'_k(t) + \operatorname{Re} \sum g_{z_j}(\gamma(t))\gamma''_j(t),$$

since g_{z_j} is holomorphic, i.e., $g_{z_j \bar{z}_k} = 0$. Also

$$g_{z_j z_k}(z) = \frac{1}{2}\frac{\beta_j \beta_k}{(\beta \cdot z)^2} = 2g_{z_j}(z)g_{z_k}(z).$$

Moreover $g_{z_j} = f_{z_j}$ at $z = a$, so that

$$\tfrac{1}{2}\psi''(0) = 2\operatorname{Re} \sum f_{z_j}(a)f_{z_k}(a)\gamma'_j(0)\gamma'_k(0) + \operatorname{Re} \sum f_{z_j}(a)\gamma''_j(0).$$

The inequality $\varphi''(0) \geqslant \psi''(0)$ then means that

$$\operatorname{Re} \sum f_{z_j z_k}(a)\gamma'_j(0)\gamma'_k(0) \ + \sum f_{z_j \bar{z}_k}(a)\gamma'_j(0)\overline{\gamma'_k(0)} + \operatorname{Re} \sum f_{z_j}(a)\gamma''_j(0)$$

$$\geqslant 2\operatorname{Re} \sum f_{z_j}(a)f_{z_k}(a)\gamma'_j(0)\gamma'_k(0) + \operatorname{Re} \sum f_{z_j}(a)\gamma''_j(0).$$

If we now let the direction $\gamma'(0)$ vary, this means that (9.133) holds.

We shall now prove that the differential condition (9.134) is not sufficient for $\mathscr{L}$-closedness in a simply connected domain which is not a disk.

Lemma 9.8.51 *Let f be an $\mathscr{L}$-closed function which is of class C^1 in a neighborhood of a point $a \in \mathbf{C}^{1+1} \setminus \{0\}$. We consider its restriction to $z_0 = 1$, and write z for the coordinate there, thus $F(z) = f(1, z)$. Let $G(z) = \log|1 + \beta z|$, $z \in \mathbf{C}$. We now choose β such that $\partial F(a)/\partial z = \partial G(a)/\partial z = -\beta/(1 + \beta a)$. With this value of β we have $F(z) \geqslant F(a) + G(z) - G(a)$ for all z. In particular $F(z) \geqslant F(a)$ at all points on the circle $|1 + \beta z| = |1 + \beta a|$.*
The proof is easy.

Proposition 9.8.52 *Let ω be a connected open subset of the Riemann sphere $S^2 = \mathbf{C} \cup \{\infty\}$ such that $S^2 \setminus \overline{\omega}$ has at least one component which is not a disk. Then there exists a function F which is $+\infty$ in $S^2 \setminus \omega$, C^∞ in ω, and satisfies the differential condition (9.134) in ω such that $\mathscr{L}\mathscr{L}f \neq f$ at some point in ω for the corresponding function f defined by $f(z_0, z_1) = F(z_1/z_0) - \log|z_0|$.*

Proof Let K be the complement of a component of $S^2 \setminus \overline{\omega}$ which is not a disk; thus K contains $\overline{\omega}$. Moreover the complement of K is connected and $\partial K \subset \partial \omega$. Let a, b, c, d be the four points whose existence is guaranteed by

Lemma 9.4.35; recall that $b, d \in K$ and $a, c \notin K$. Since they are on a circle, we can move them by a Möbius transformation: we can move a to 0 and c to ∞, and b, d to some points on the real axis, $d < 0 < b$. Also $b, d \in \partial\omega$ since $b, d \in \partial K$. Now define

$$
F(z) = \begin{cases}
\exp(-1/(\operatorname{Im} z + \varepsilon)) & \text{when } z \in \omega, \ \operatorname{Re} z > 0, \ \operatorname{Im} z > -\varepsilon, \\
0 & \text{at other points in } \omega, \\
+\infty & \text{otherwise.}
\end{cases}
$$

Here we choose $\varepsilon > 0$ so small that the disk $\|z\|_2 \leqslant \varepsilon$ does not intersect K; this means that the function is continuous, even identically zero in a neighborhood of the intersection of ω and the imaginary axis. Moreover we choose ε so small that F satisfies the differential condition $(\partial F/\partial y)^2 \leqslant \partial^2 F/\partial y^2$, and such that the point $b_1 = b - i\varepsilon$ belongs to ω. This means that F satisfies (9.134). Now at a point b_2 near b_1, F takes arbitrarily small positive values. If there is a function G tangent to F at such a point, it forces F to be positive at some point with negative real part. Indeed, the circle $|1 + \beta z| = |1 + \beta b_2|$ passes through points arbitrarily close to the line $\operatorname{Im} z = \operatorname{Im} b_2$ (see Lemma 9.8.51). This is a contradiction, since we defined F to be identically zero at all points in ω in the left half plane, and there are such points by construction.

9.9 Lineal Convexity in Infinite Dimension

Abstract of this section

The purpose of this section is to study some problems on lineal convexity in spaces of infinite dimension, in particular to prove that a pseudoconvex or lineally convex set which is open in a subspace can be extended to an open set in the whole space which is pseudoconvex or lineally convex, respectively.

Given a pseudoconvex open set ω in $\mathbf{C}^k$ regarded as a subspace $\mathbf{C}^k \times \{0\}$ of $\mathbf{C}^k \times \mathbf{C}^{n-k} = \mathbf{C}^n$, we can fatten it into pseudoconvex set which is open in the whole space by taking a Cartesian product $\omega \times \omega'$ for a suitable set ω' in $\mathbf{C}^{n-k}$. How can this be done in an infinite-dimensional space? It is also of interest to construct a set which tapers off at the boundary of ω.

In an infinite-dimensional Hausdorff topological vector space the subspaces of finite dimension (which are always closed) and those subspaces of finite codimension which are closed are of particular interest. Here the codimension of a subspace F of E is the dimension of the quotient space E/F.

9.9.1 Generalizing lineal convexity

We start by generalizing the concept of lineally convex set to higher codimenions:

Definition 9.9.1 *Let E be a topological vector space and A a subset of E. We shall say that A is m-**lineally convex** if its complement is a union of closed affine subspaces of codimension m.* □

Thus 1-lineal convexity is just lineal convexity as defined in Definition 9.4.1 on page 279. An m-lineally convex set is also k-lineally convex for $k \geqslant m$. An m-lineally convex open set in $\mathbf{C}^n$ is $(n-m)$-pseudoconvex in the sense of Rothstein (1955:130).

9.9.2 Inverse images of m-lineally convex sets

The intersection of an m-lineally convex set with a subspace is also m-lineally convex in that subspace—more generally, we have the following result.

Proposition 9.9.2 *Let $f \colon E \to F$ be a continuous linear mapping of a topological vector space E into another one, F. If Y is an m-lineally convex subset of F, then its inverse image $A = f^*(Y)$ under f is also m-lineally convex.*

Proof Take $x \in E \smallsetminus X$. Then $y = f(x)$ belongs to $F \smallsetminus Y$ and by hypothesis we can find a linear subspace $K \subset F$ of codimension m such that $y + K$ does not meet Y. The inverse image $H = f^*(K)$ is a vector subspace of E and E/H is of finite dimension. For the codimensions we obtain

$$\text{codim}_E(H) = \dim E/H = \dim f_*(E)/K \leqslant \dim F/K = \text{codim}_F K = m.$$

So $E \smallsetminus X$ is a union of closed affine subspaces of codimension at most m, thus X is m-lineally convex.

9.9.3   Constructing thicker sets

Theorem 9.9.3 *Let E be a normed complex vector space, V an open cone in E which is m-lineally convex, and F a vector subspace of E of dimension m. Assume that $V \cap F = F \smallsetminus \{0\}$. If ω is a pseudoconvex open set in F, then*

$$\Omega = \bigcap_{x \in F \smallsetminus \omega} (x + V)$$

is open and pseudoconvex in E, and $\Omega \cap F = \omega$.

Proof We have

$$E \smallsetminus \Omega = (F \smallsetminus \omega) + (E \smallsetminus V), \tag{9.135}$$

so that $E \smallsetminus \Omega$ is the vector sum of two closed sets in E. Let B be the unit ball in E. Since F is of finite dimension and V contains $F \smallsetminus \{0\}$, there is a positive number s such that

$$(F \smallsetminus B) + sB \subset V;$$

in view of the homogeneity of F and V we have for all $r > 0$

$$\left[F \smallsetminus (rs^{-1}B) \right] + rB \subset V. \tag{9.136}$$

Let now $x \in E \smallsetminus \Omega$ with $\|x\| \leqslant r/s$. There is, according to (9.135) a representation $x = y + z$ of x with $y \in F \smallsetminus \omega$ and $z \in E \smallsetminus V$. It follows that $\|y\| \leqslant r/s$, for if we have $\|y\| > r/s$, it would follow from (9.136) that $y \in F \smallsetminus (rs^{-1}B)$. This proves that

$$rB \smallsetminus \Omega \subset \left[(rs^{-1}) \cap (F \smallsetminus \omega) \right] + (E \smallsetminus V).$$

Since $(rs^{-1}B) \cap (F \smallsetminus \omega)$ is compact and the vector sum of a compact set and a closed set is closed, we have finally proved that $(rB) \smallsetminus \Omega$ is closed; consequently, since r is arbitrary, that Ω is open.

In order to prove that Ω is pseudoconvex, we shall find, given any point $y \in E \smallsetminus \Omega$, a continuous linear projection $\pi \colon E \to F$ such that $\pi^*(\omega)$ contains Ω but not y.

So let y belong to $E \smallsetminus \Omega$. We then have $y \notin x + V$ for some $x \in F \smallsetminus \omega$. By hypothesis there exists a closed vector subspace H of E of codimension m and such that $y - x + H$ does not meet V. Lemma 9.9.4 below shows that

$y - x + H = H$ and since F is of codimension m and $F \cap H = \{0\}$, we have $E = F + H$, a direct sum. The projection $\pi \colon E \to F$ with kernel H and image F is continuous. Since $y - x \in H$, we have $\pi(y) = \pi(x) = x \in F$; that is, $y \notin \pi^*(\omega)$. On the other hand, $\pi_*(V) = F \smallsetminus \{0\}$, which after a translation proves that $\pi_*(x + V) = F \smallsetminus \{x\}$, hence

$$\pi_*(\Omega) \subset \bigcap_x (F \smallsetminus \{x\}; \ x \in F \smallsetminus \omega) = \omega.$$

Summing up, $y \notin \pi^*(\omega)$ and $\pi^*(\omega) \supset \pi^*(\pi_*(\Omega) = \Omega$, the conclusion we wanted.

During the proof above we needed the following lemma.

Lemma 9.9.4 *Let V be an m-lineally convex open cone in a topological vector space E. Assume that V contains $F \smallsetminus \{0\}$, where F is a vector subspace of E of dimension m. Then $A \smallsetminus V$ is a union of closed vector subspaces of dimension m; in fact, every closed affine vector subspace of E which is contained in $E \smallsetminus V$ and of codimension m contains the origin and is a linear subspace.*

Proof Let H be a subspce of E of codimension m and such that $(a+H) \cap V = \emptyset$. We shall thus prove that $a + H = H$. If we had $F + H \neq E$, there would exist $x \in F \cap H$, $x \neq 0$. Then $a + rx \in a + H \subset E \smallsetminus V$. But $a + rx$ belongs to V for all sufficiently large r since $x \in F \smallsetminus \{0\} \subset V$. This contradiction shows that $F + H = E$; it follows that a can be represented as $a = y + z$, where $y \in F$, $z \in H$ and $y = a - z \in a + H$. Hence

$$y \in (y + H) \cap F \subset (a + H) \cap (V \cup \{0\}) \subset \{0\};$$

in other words that $0 \in a + H$, which is a vector subspace of E. This proves the lemma and so completes the proof of Theorem 9.9.3.

Let us note that we also have the next result, which is closely related to the theorem just proved.

Theorem 9.9.5 *Let E, F, V and Ω be as in Theorem 9.9.3. Assume that that ω is m-lineally convex in F. Then Ω is also m-lineally convex.*

Proof Let $y \in E \smallsetminus \Omega$. As in the proof of Theorem 9.9.3, there is an $x \in F \smallsetminus \omega$ and a projection $\pi \colon E \to F$ such that $\pi(y) = \pi(x) = x$. According to Proposition 9.9.2, $\pi^*(\omega)$ is m-lineally convex and it follwos that Ω, the intersection of all the sets $\pi^*(\omega)$ for the various choices of π, is also m-lineally convex.

9.9.4 Constructing convex cones

Finally we shall indicate how it is possible to construct cones V that can serve in Theorems 9.9.3 and 9.9.5.

Proposition 9.9.6 *Let E be a topological vector space and let A_j, $j = 1, \ldots, p$, be m_j-lineally convex subsets of E. Then the intersection $A = A_1 \cup \cdots \cup A_p$ is m-lineally convex, where $m = m_1 + \cdots + m_p$.*

Proof Take any point $a \in E \smallsetminus A$. Then $a \in E \smallsetminus A_j$ for every j and there exists a closed subspace F of E of codimension m_j such that $a + F_j$ does not meet A_j. The intersection $F = F_1 \cap \cdots \cap F_p$ is closed and of codimension at most equal to m and does not cut A. We are ready.

Let us denote by $\mathbf{T}$ the unit circle $\mathbf{T} = \{t \in \mathbf{C}; \; |t| = 1\}$, and by $\mathbf{T}A = \{tx; \; x \in A\}$ the circled set generated by a set A in a vector space.

Lemma 9.9.7 *If V is an open convex cone in a complex vector space E, then $\mathbf{T}V$ is 1-lineally convex.*

Proof If $\mathbf{T}V$ is equal to E there is nothing to prove, so let us suppose that $\mathbf{T}V \neq E$. If $a \in E \smallsetminus \mathbf{T}V$, let us denote by $F = \mathbf{C}a$ the vector subspace generated by a. Then F does not meet V (note that the origin does not belong to V), so the Hahn–Banach theorem gives us a closed hyperplane H that contains F and does not meet V. Now $\mathbf{T}H = H$, so H does not meet $\mathbf{T}V$ either, and, since $a \in H$, this proves that $\mathbf{T}V$ is 1-lineally convex.

Proposition 9.9.8 *Let $V_1, \ldots, V_m$ be open convex cones in a topological vector space E. Then*

$$V = \mathbf{T}V_1 \cup \cdots \cup \mathbf{T}V_m$$

is m-lineally convex.

Proof We combine Proposition 9.9.2 and Lemma 9.9.7.

Proposition 9.9.9 *Let E be a normed complex vector space, F a vector subspace of E and $\xi_1, \ldots, \xi_m$ nonzero linear forms on E such that their restrictions to F have norm at most equal to 1. Then*

$$V = \left\{x \in E; \; \|x - y\| < \sup_{j=1,\ldots,m} |\xi_j(y)| \text{ for some } y \in F\right\} \qquad (9.137)$$

is an open m-lineally convex cone in E.

Proof Let us define

$$u_j(x) = \inf_{y \in F} \left(\|x - y\| - \operatorname{Re} \xi_j(y)\right), \qquad x \in E, \quad j = 1, \ldots, m.$$

This is a convex function and it never takes the value $-\infty$, since

$$\|x - y\| - \operatorname{Re} \xi_j(y) \geqslant \|x - y\| - \|y\| \geqslant -\|x\| > -\infty$$

for every $x \in E$ and every $y \in F$. Hence we have $-\|x\| \leqslant u_j \leqslant \|x\|$. The cone

$$V_j = \{x \in E; \; u_j(x) < 0\}$$

is convex and open in view of the continuity of u_j. From Proposition 9.9.8 we see that $V = \mathbf{T}V_1 \cup \cdots \cup \mathbf{T}V_m$ is m-lineally convex.

Let us also note that the cone V defined by (9.137) contains the cone

$$V_0 = F \setminus \bigcap_{j=1}^{m} \ker \xi_j.$$

This cone is open in F but not in E. If we make it a bit thicker to obtain an open cone in E, it can serve in Theorems 9.9.3 and 9.9.5 provided F is a subspace of dimension m and

$$F \cap \ker \xi_1 \cap \cdots \cap \ker \xi_m = \{0\};$$

this latter condition means that the restrictions of the ξ_j to F form a base in the dual of F.

9.10   References

Agler, J.; Young, N. J. 2004. The hyperbolic geometry of the symmetrized bidisc. *J. Geom. Anal.* **14**, no. 3, 375–403.

Aĭzenberg, Lev A. 1966. Общий вид линейного непрерывного функционала в пространствах функций, голоморфных в выпуклых областях $\mathbf{C}^n$ [Obščij vid linejnogo nepreryvnogo funkcionala v prostranstvah funkcij, golomorfnyh v vypuklyh oblastjah $\mathbf{C}^n$]. *Dokl. Akad. Nauk SSSR* **166**, 1015–1018. (English translation in *Soviet Math. Dokl.* **7**, (1966), 198–202.)

Aĭzenberg, Lev A. 1967. О разложении голоморфных функций многих комплексных переменных на простейшие дроби [O razloženii golomorfnyh funkcij mnogih kompleksnyh peremennyh na prostejšie drobi]. *Sib. Mat. Ž.* **8**, 1124–1142. (English translation in *Siberian Math. J.* **8**, (1967), 859–872.)

Andersson, Mats; Passare, Mikael; Sigurdsson, Ragnar. 1991. *Analytic functionals and complex complex convexity.* Preprint, May 1991, 28 pp. Reykjavik: University of Iceland, Science Institute.

Andersson, Mats; Passare, Mikael; Sigurdsson, Ragnar. 2004. *Complex Convexity and Analytic Functionals.* Progress in Mathematics, 225. Basel et al.: Birkhäuser.

Behnke, H[einrich]; Peschl, E[rnst]. 1935. Zur Theorie der Funktionen mehrerer komplexer Veränderlichen. Konvexität in bezug auf analytische Ebenen im kleinen und großen. *Math. Ann.* **111**, 158–177.

Bourbaki, N[icolas]. 1954. *Éléments de mathématique. Théorie des ensembles.* Chapitres I et II. Paris: Hermann & Cie.

Bourbaki, N[icolas]. 1961. *Topologie générale.* Éléments de mathématique, première partie, livre III, chapitres 1 & 2. Third edition. Paris: Hermann.

Bourbaki, N[icolas]. 1963. *Théorie des ensembles.* Chapter 3. Second edition. Paris: Hermann.

Diederich, Klas; Fischer, Bert. 2006. Hölder estimates on lineally convex domains of finite type. *Michigan Math. J.* **54**, no. 2, 341–352.

Diederich, Klas; Fornæss, John Erik. 2003. Lineally convex domains of finite type: holomorphic support functions. *Manuscripta Math.* **112**, no. 4, 403–431.

Favorov, Sergey; Golinskii, Leonid. 2015. Blaschke-type conditions on unbounded domains, generalized convexity, and applications to perturbation theory. *Rev. Mat. Iberoam.* **31**, No. 1, 1–33.

Fenchel, Werner. 1949. On conjugate convex functions. *Canad. J. Math.* **1**, 73–77.

Fenchel, W. 1952. A remark on convex sets and polarity. **In:** *Communications du Séminaire de l'Université de Lund. Tome supplémentaire dédié à Marcel Riesz*, pp. 82–89. Lund: C. W. K. Gleerup.

Fenchel, W. 1953. *Convex cones, sets, and functions.* Notes of lectures at Princeton University, Spring Term 1951. 155 pp.

Fenchel, W. 1983. Convexity through the ages. **In:** Gruber, Peter M.; Wills, Jörg M., Eds., 1983. *Convexity and Its Applications*, pp. 120–130. Basel et al.: Birkhäuser Verlag.

Heijmans, Henk J. A. M. 1994. *Morphological Image Operators*, xv + 509 pp. Boston: Academic Press.

Heijmans, Henk J. A. M. 1995. Mathematical morphology: A modern approach in image processing based on algebra and geometry. *SIAM Review* **37**, no. 1, 1–36.

Heijmans, H. J. A. M.; Ronse, C. 1990. The algebraic basis of mathematical morphology. I. Dilations and erosions. *Comput. Graphics Image Vision Process.* **50**, 245–295.

Helton, J. William; Marshall, Donald E. 1990. Frequency domain design and analytic selections. *Indiana Univ. Math. J.* **39**, no. 1, 157–184.

Hiriart-Urruty, Jean-Baptiste; Lemaréchal, Claude. 1993. *Convex Analysis and Minimization Algorithms I. Fundamentals*, vxii + 417 pp. Berlin et al.: Springer-Verlag.

Hiriart-Urruty, Jean-Baptiste; Lemaréchal, Claude. 2001. *Fundamentals of Convex Analysis*, X + 259 pp. Berlin et al.: Springer.

Hörmander, Lars. 1990. *An Introduction to Complex Analysis in Several Variables*. Third Edition (Revised). Amsterdam et al.: North-Holland.

Hörmander, Lars. 1994. *Notions of convexity*. viii + 414 pp. Boston: Birkhäuser.

Jacquet, David. 2008. *On complex convexity*. PhD Thesis, defended at Stockholm University on 2008-04-14. Stockholm: Stockholm University. 70 pp.

Kerzman, Norberto; Rosay, Jean-Pierre. 1981. Fonctions plurisousharmoniques d'exhaustion bornées et domaines taut. *Math. Ann.* **257**, no. 2, 171–184.

Kiselman, Christer O. 1978. Sur la convexité linéelle. *An. Acad. Brasil. Ciênc.* **50** (4), 455–458.

Kiselman, Christer O. 1996. Lineally convex Hartogs domains. *Acta Math. Vietnamica* **21**, 69–94.

Kiselman, Christer O. 1997. Duality of functions defined in lineally convex sets. *Universitatis Iagellonicae Acta Mathematica* **35** (1997), 7–36.

Kiselman, Christer O. 1998. A differential inequality characterizing weak lineal convexity. *Math. Ann.* **311**, 1–10.

Kiselman, Christer O. 2007. Division of mappings between complete lattices. *Mathematical Morphology and its Applications to Signal and Image Processing. Proceedings of the 8th International Symposium on Mathematical Morphology, Rio de Janeiro, RJ, Brazil, October 10–13, 2007*, pp. 27–38. G. J. F. Banon; J. Barrera; U. de Mendonça Braga-Neto, Eds. São José dos Campos, SP: MCT/INPE.

Kiselman, Christer O. 2010. Inverses and quotients of mappings between ordered sets. *Image and Vision Computing* **28**, 1429–1442.

Kiselman, Christer O. 2016. Weak lineal convexity. **In:** Białas-Cież, Leokadia; Kosek, Marta, Eds. 2016. *Constructive Approximation of Functions*. Banach Center Publications, Polish Academy of Sciences, volume **107**, pp. 159–174. Proceedings from a conference in Będlewo 2014.

Kiselman, Christer Oscar. 2019. Generalized convexity: The case of linearly convex Hartogs domains. *Ann. Polon. Math.* **123**, 319–344.

Kiselman, Christer Oscar. ms 2021. *Elements of Digital Geometry, Mathematical Morphology, and Discrete Optimization*, xxiv + 461 pp. Accepted for publication by World Scientific, Singapore.

Krantz, Steven G. 2001. *Function Theory of Several Complex Variables*, xvi + 564 pp. Providence, RI: American Mathematical Society; AMS Chelsea Publishing.

Martineau, André. 1966. Sur la topologie des espaces de fonctions holomorphes. *Math. Ann.* **163**, 62–88. (Also in *Œuvres de André Martineau* 1977:215–246.)

Martineau, André. 1968. Sur la notion d'ensemble fortement linéellement convexe. *An. Acad. Brasil. Ciênc.* **40** (4), 427–435. (Also in *Œvres de André Martineau* 1977:323–334.)

Moreau, Jean-Jacques. 1963. Remarques sur les fonctions à valeurs dans $[-\infty, +\infty]$ définies sur un demi-groupe. *C. R. Acad. Sci. Paris* **257**, 3107–3109.

Moreau, J.-J. 1970. Inf-convolution, sous-additivité, convexité des fonctions numériques. *J. Math. Pures et Appl.* **49**, 109–154.

Najman, Laurent; Talbot, Hugues, Eds. 2008. *Morphologie mathématique 1. Approches déterministes*, 260 pp. Paris: Lavoisier.

Najman, Laurent; Talbot, Hugues, Eds. 2010. *Mathematical Morphology. From Theory to Applications*, 507 pp. London; Hoboken, NJ: ISTE Ltd; John Wiley & Sons.

Nikolov, Nikolai [Marinov]. 2012. Invariant functions and metrics in complex analysis. *Dissertationes Math.* **486**, 1–100. (Based on the author's D.Sc. Dissertation, written in Bulgarian and defended 2010-10.)

Nikolov, Nikolai [Marinov]; Pflug, Peter; Thomas, Pascal J. 2013. On different extremal bases for **C**-convex domains. *Proc. Amer. Math. Soc.* **141**, no. 9, 3223–3230.

Nikolov, Nikolai [Marinov]; Pflug, Peter; Zwonek, Włodzimierz. 2008. An example of a bounded **C**-convex domain which is not biholomorphic to a convex domain. *Math. Scand.* **102**, no. 1, 149–155.

Œuvres de André Martineau. Paris: Centre National de la Recherche Scientifique. 879 pp. ISBN 2-222-01846-3.

Pflug, Peter; Zwonek, Włodzimierz. 2012. Exhausting domains of the symmetrized bidisc. *Ark. mat.* **50**, 397–402.

Rockafellar, R. Tyrrell. 1970, 1997. *Convex Analysis.* Princeton, NJ: Princeton University Press.

Ronse, Christian. 1990. Why mathematical morphology needs complete lattices. *Signal Processing* **21**, 129–154.

Ronse, Christian; Serra, Jean. 2008. Fondements algébriques de la morphologie. **In:** (Najman & Talbot, Eds., 2008:49–96).

Ronse, Christian; Serra, Jean. 2010. Algebraic foundations of morphology. **In:** (Najman & Talbot, Eds., 2010:35–80).

Rothstein, Wolfgang. 1955. Zur Theorie der analytischen Mannigfaltigkeiten im Raume der n komplexen Veränderlichen. *Math. Ann.* **129**, 96–138.

Salinas, Norberto. 1976. Extensions of C^*-algebras and essentially n-normal operators. *Bull. Amer. Math. Soc.* **82**, no. 1, 143–146.

Salinas, Norberto. 1979. Hypoconvexity and essentially n-normal operators. *Trans. Amer. Math. Soc.* **256**, 325–351.

Schmitt, Michel; Mattioli, Juliette. 1994. *Morphologie mathématique.* Paris: Masson. xiv + 213 pp. ISBN 2-225-84385-6.

Serra, Jean. Ed., 1988. *Image Analysis and Mathematical Morphology. Volume 2: Theoretical Advances.* London et al.: Academic Press.

Slatyer, Harry J. 2016. The Levi problem in $\mathbb{C}^n$: a survey. *Surv. Math. Appl.* **11**, 33–75.

Webster's Ninth Collegiate Dictionary. 1983. Springfield, MA: Merriam-Webster Inc.

Whittlesey, Marshall A. 2000. Polynomial hulls and H^∞ control for a hypoconvex constraint. *Math. Ann.* **317**, no. 4, 677–701.

Южаков, А. П.; Кривоколеско, В. П. [Yužakov, A. P.; Krivokolesko, V. P.] 1971. Некоторые свойства линейно выпуклых областей с гладкими границами в $\mathbb{C}^n$ [Nekotorye svojstva linejno vypuklyh oblastej s gladkimi granicami v $\mathbb{C}^n$]. *Sib. Mat. Zhurn.* **12**, No 2, 452–458. (English translation in *Siberian Math. J.* **12**, no. 2, 323–327.)

Zelinskiĭ, Ju. B. 1981. О геометрических критериях сильной выпуклости [O geometriqeskih kriterijah sil'noj vypuklosti]. *Dokl. Akad. Nauk SSSR* **261**, 11–13. English translation: Geometric criteria for strong linear convexity, *Soviet Math. Dokl.* **24** (1981), no. 3, 449–451 (1982).

Zelinskii, Yu. B. 1984. On the geometric criteria of strong linear convexity. **In:** *Комплексный анализ и приложения '81, Complex Analysis and Applications'81*, pp. 533–536.

Зелинский, Ю. Б. [Zelinskij, Ju. B.] 1988. О линейно выпуклых областях с гладкими границами [O linejno vypuklyh oblastyah s gladkimi granicami]. *Укр. мат. Ж.* [*Ukr. mat. Ž.*] **40**, 53–58. (English translation in *Ukrainian Math. J.*, 40, (1988), 44–48.)

Зелинский, Ю. Б. [Zelinskij, Ju. B.] 1993. *Многозначные отображения в анализе* [*Mnogoznačnye otobraženija v analize*]. Kiev: Наукова Думка [Naukova Dumka], 264 pp.

Зелинский, Ю. Б. [Zelinskij, Ju. B.] 2002. О локально линейно выпуклых областях [O lokal′no linejno vypuklyh oblastjah]. *Укр. мат. журн.* [*Ukr. mat. zhur.*] **54**, no. 2, 280–284. (English translation: On locally linearly convex domains, in *Ukrainian Math. J.* **54**, no. 2, 345–349.)

Zinov′ev, B. S. 1971. Аналитические условия и некоторые вопросы аппроксимации линейно-выпуклых областей с гладкими границами в пространстве $\mathbf{C}^n$ [Analitičeskie uslovija i nekotorye voprosy approksimacii linejno-vypuklyh oblastej s gladkami granicami v prostranstve $\mathbf{C}^n$]. *Izv. vuzov. Matematika* **11**, No. 6, 61–69.

Знаменский, С. В. [Znamenskij, S. V.] 1979. Геометрический критерий сильной линейной выпуклости [Geometričeskij kriterij sil′noj linejnoj vypuklosti]. *Функциональный анализ и его приложения* [*Funkcional′nyj analiz i evo priloženija*]. **13**(3), 83–84. (English translation in *Functional Anal. Appl.*, 13, 224–225.)

Знаменский, С. В. [Znamenskij, S. V.] 1990. Томография в пространствах аналитических функционалов. [Tomografija v prostranstvah analitičeskih funkcionalov]. *Доклады Академии Наук СССР* [*Doklady Akademii Nauk SSSR*]. **312**(5), 1037–1040.

Знаменский, С. В. [Znamenskij, S. V.] 2001. Семь задач о ℂ-выпуклости [Sem′ zadač o ℂ-vypuklosti]. **In:** Чирка, Е. (Ed.), *Комплексный анализ в современной математике: к 80-летию со дня рождения Б. В. Шабата* [Čirka, E. (Ed.), *Kompleksnyj analiz v sovremennoj matematike: k 80-letiju so. dnja rozhdeniya B. V. Šabata*], pp. 123–131. Moscow: FAZIS.

Знаменский, С. В.; Знаменская, Л. Н. [Znamenskij, S. V.; Znamenskaja, L. N.] 1996. Спиральная связность сечений и проекций ℂ-выпуклых множеств. [Spiral′naja svyaznost′ sečenij i proekcij ℂ-vypuklyh množestv.] *Математические заметки* [*Matematičeskie zametki*] **59**(3), 359–369.

Author's address: Uppsala University, Department of Information Technology, P. O. Box 337, SE-751 05 Uppsala, Sweden.
email addresses: kiselman@it.uu.se, christer@kiselman.eu
URL: http://www.cb.uu.se/~kiselman
ORCiD, *Open Researcher and Contributor ID:* 0000-0002-0262-8913

10

Reproducing Kernels in Complex Analysis

Steven G. Krantz

CONTENTS

10.1   Introduction

In the complex analysis of one variable, the Cauchy integral formula plays a central role. Indeed, virtually all of the fundamental results in the subject are derived from that simple formula.

In the realm of harmonic functions, the Poisson kernel and integral formula play a central role. Fundamental results, such as the Fatou theorem about nontangential boundary behavior of harmonic and holomorphic functions, depend critically on the nature of the Poisson kernel, and particularly on the shape of its singularity.

In the complex analysis of several variables—where the quality of the analysis depends on the geometry of the domain in question—the nature of kernels plays an even more critical role. On the one hand, it is generally very difficult

DOI: 10.1201/9781315160658-10

to calculate the relevant kernels. On the other hand, knowledge of the kernels is a valuable tool.

It is a pleasure to thank Joseph A. Cima for a careful reading of this paper and for many constructive suggestions.

10.2   The Cauchy Formula in One Complex Variable

Throughout this paper we use the term *domain* to mean a connected open set in $\mathbb{C}$ or $\mathbb{C}^n$.

Viewed properly, the Cauchy integral formula in one complex variable is *not* a result about holomorphic functions. Rather, it is a result about *arbitrary smooth functions*. And it is a corollary of Stokes's theorem. We now explicate this point of view.

The language of differential forms is needed in order to formulate Stokes's theorem. We say just a few words about the matter here. We shall only need differential 1-forms. You know from calculus that, in the plane, these are generated by dx and dy. In complex analysis, we let

$$dz = dx + idy \qquad \text{and} \qquad d\bar{z} = dx - idy \,.$$

Correspondingly, we define

$$\frac{\partial}{\partial z} = \frac{1}{2}\left(\frac{\partial}{\partial x} - i\frac{\partial}{\partial y}\right) \qquad \text{and} \qquad \frac{\partial}{\partial \bar{z}} = \frac{1}{2}\left(\frac{\partial}{\partial x} + i\frac{\partial}{\partial y}\right) \,.$$

It is worth noting that

$$\frac{\partial}{\partial z}z = 1 \ , \quad \frac{\partial}{\partial z}\bar{z} = 0$$

and

$$\frac{\partial}{\partial \bar{z}}\bar{z} = 1 \ , \quad \frac{\partial}{\partial \bar{z}}z = 0 \,.$$

There are corresponding intuitively appealing pairings between $\partial/\partial z$ and dz and between $\partial/\partial\bar{z}$ and $d\bar{z}$.

If $\omega = u\,dz + v\,d\bar{z}$ is a 1-form, then

$$d\omega = \frac{\partial u}{\partial \bar{z}}d\bar{z}\wedge dz + \frac{\partial v}{\partial z}dz\wedge d\bar{z} = \left(\frac{\partial v}{\partial z} - \frac{\partial u}{\partial \bar{z}}\right)dz\wedge d\bar{z}. \qquad (10.1)$$

Of course, 1-forms can be integrated over 1-manifolds, such as the circle, and 2-forms can be integrated over 2-manifolds, such as the sphere or torus or a planar domain.

Finally, we recall the exterior differential operator d. In the complex setting we can decompose $d = \partial + \bar{\partial}$, where

$$\partial\omega = \frac{\partial v}{\partial z}dz\wedge d\bar{z}$$

and

$$\overline{\partial}\omega = \frac{\partial u}{\partial \overline{z}} d\overline{z} \wedge dz.$$

Refer back to equation (10.1).

Theorem 10.2.2 (Stokes) *Let* $\Omega \subseteq \mathbb{C}$ *be a bounded open set with* C^1 *boundary. Let* ω *be a differential form of degree 1 with coefficients in* $C^1(\overline{\Omega})$. *Then*

$$\int_{\partial\Omega} \omega = \int_{\Omega} d\omega = \int_{\Omega} \partial\omega + \overline{\partial}\omega.$$

Standard references for Stokes's theorem are [RUD] and [DER]. The Stokes's theorem that we have recorded here is the standard one, simply expressed in complex notation.

Now we have

Proposition 10.2.3 *Let* $\Omega \subseteq \mathbb{C}$ *be a simply connected domain with* C^1 *boundary. Let* f *be a* C^1 *function on* $\overline{\Omega}$. *Then it holds, for* $z \in \Omega$, *that*

$$f(z) = \frac{1}{2\pi i} \int_{\partial\Omega} \frac{f(\zeta)}{\zeta - z} d\zeta - \frac{1}{2\pi i} \iint_{\Omega} \frac{\partial f/\partial\overline{\zeta}}{\zeta - z} d\overline{\zeta} \wedge d\zeta. \qquad (10.2)$$

Proof: Fix a point $z \in \Omega$ and let $\epsilon > 0$ be so small that $\overline{D}(z, \epsilon) \subseteq \Omega$. We apply Stokes's theorem to the 1-form $f(\zeta)/(\zeta - z) \, d\zeta$ on the domain $D_\epsilon \equiv \Omega \setminus \overline{D}(z, \epsilon)$.
Thus we obtain

$$\frac{1}{2\pi i} \int_{\partial\Omega_\epsilon} \frac{f(\zeta)}{\zeta - z} d\zeta = \frac{1}{2\pi i} \iint_{\Omega_\epsilon} \frac{\partial f/\partial\overline{\zeta}}{\zeta - z} d\overline{\zeta} \wedge d\zeta. \qquad (10.3)$$

Now, as $\epsilon \to 0^+$, the right-hand side of equation (10.3) tends to

$$\frac{1}{2\pi i} \iint_{\Omega} \frac{\partial f/\partial\overline{\zeta}}{\zeta - z} d\overline{\zeta} \wedge d\zeta,$$

just because $1/(\zeta - z)$ is integrable. Let us examine the left-hand side of (10.3).
We write

$$\frac{1}{2\pi i} \int_{\partial\Omega_\epsilon} \frac{f(\zeta)}{\zeta - z} d\zeta$$

$$= \frac{1}{2\pi i} \int_{\partial\Omega} \frac{f(\zeta)}{\zeta - z} d\zeta - \frac{1}{2\pi i} \int_{\partial D(0,\epsilon)} \frac{f(\zeta)}{\zeta - z} d\zeta.$$

It is worth noting that $\partial\Omega$ and $\partial D(z, \epsilon)$ have opposite orientations because of their different roles in the boundary of Ω_ϵ. Now the first expression on the right-hand side of this last display is the first term of the right-hand side of

(10.2). It remains to analyze the second term on the right-hand side of this last display.

In fact, we have

$$
\frac{1}{2\pi i}\int_{\partial D(0,\epsilon)}\frac{f(\zeta)}{\zeta - z}\,d\zeta = \frac{1}{2\pi i}\int_0^{2\pi}\frac{f(z + \epsilon e^{it})}{\epsilon e^{it}}\,i\epsilon e^{it}\,dt
$$

$$
= \frac{1}{2\pi}\int_0^{2\pi}f(z + \epsilon e^{it})\,dtv
$$

$$
\to\quad f(z)
$$

as $\epsilon \to 0^+$.

Putting together all the pieces, we find that we have derived equation (10.2).   □

It is important to note now that, if Ω is as in the statement of the proposition and $f \in C^1(\overline{\Omega})$ and holomorphic on the interior, then equation (10.2) simplifies to

$$
f(z) = \frac{1}{2\pi i}\int_{\partial\Omega}\frac{f(\zeta)}{\zeta - z}\,d\zeta\,.
$$

This, of course, is the Cauchy integral formula that one can find in any complex analysis text.

10.3 Harmonic Functions and the Poisson Kernel

If we take it on faith that the Dirichlet problem on the disc always has a solution (see, for instance, [GRK]), then we may discover a formula for that solution in the following fashion.

Let f be an L^2 function on ∂D, the boundary of the unit disc. Then we may expand f, in the L^2 topology, as a Fourier series:

$$
f(e^{i\theta}) = \sum_{j=-\infty}^{\infty}\widehat{f}(j)e^{ij\theta}
$$

$$
= \sum_{j=-\infty}^{\infty}\frac{1}{2\pi}\int_0^{2\pi}f(t)e^{-itj}\,dt\cdot e^{ij\theta}
$$

$$
= \int_0^{2\pi}f(t)\left(\frac{1}{2\pi}\sum_{j=-\infty}^{\infty}e^{ij(\theta-t)}\right)dt\,.
$$

Now each character e^{ijt} has a harmonic extension to the disc given by

$\varphi_j(re^{it}) \equiv r^{|j|}e^{ijt}$. So it is natural to consider that the harmonic extension of f to the disc D is given by

$$f(re^{i\theta}) = \int_0^{2\pi} f(t) \left(\frac{1}{2\pi} \sum_{j=-\infty}^{\infty} r^{|j|}e^{ij(\theta-t)} \right) dt$$

It behooves us then to calculate the sum

$$P_r(e^{i(\theta-t)}) = \frac{1}{2\pi} \sum_{j=-\infty}^{\infty} r^{|j|}e^{ij(\theta-t)}.$$

Now we have

$$P_r(e^{i(\theta-t)}) = \frac{1}{2\pi} \sum_{-\infty}^{-1} r^{-j}e^{ij(\theta-t)} + \frac{1}{2\pi} \sum_{1}^{\infty} r^j e^{ij(\theta-t)} + \frac{1}{2\pi} \equiv I + II + III.$$

We see that

$$I = \frac{1}{2\pi} \sum_{1}^{\infty} r^j e^{-ij(\theta-t)}$$

$$= \frac{1}{2\pi} \sum_{0}^{\infty} r^{j+1} e^{-i(j+1)(\theta-t)}$$

$$= \frac{1}{2\pi} re^{-i(\theta-t)} \sum_{j=0}^{\infty} \left(re^{-i(\theta-t)} \right)^j$$

$$= \frac{1}{2\pi} re^{-i(\theta-t)} \cdot \frac{1}{1 - re^{-i(\theta-t)}}$$

$$= \frac{1}{2\pi} \frac{r}{e^{i(\theta-t)} - r}.$$

A similar calculation shows that

$$II = \frac{1}{2\pi} \frac{r}{e^{-i(\theta-t)} - r}.$$

As a result,

$$P_r(e^{i(\theta-t)}) = \frac{1}{2\pi} \cdot \frac{r}{e^{i(\theta-t)} - r} + \frac{1}{2\pi} \cdot \frac{r}{e^{-i(\theta-t)} - r} + \frac{1}{2\pi} = \frac{1}{2\pi} \cdot \frac{1 - r^2}{1 - 2r\cos(\theta - t) + r^2}.$$

This is the Poisson kernel.

The basic result about this kernel is as follows.

Theorem 10.3.1 *Let f be a continuous function on the unit circle $\mathbb{T} = \partial D$. Then the function*

$$F(re^{i\theta}) = P_r * f(e^{i\theta}).$$

is continuous on $\overline{D}$ and harmonic on D. Note that $F(e^{i\theta}) = f(e^{i\theta})$.

10.4   Introduction to the Bergman Theory

Stefan Bergman introduced his reproducing kernel in 1921. It was anticipated somewhat by the thesis of S. Bochner, which was later published in [BOC]. Since Hilbert space was a very new idea at the time, Bergman's approach made quite an impact.

In this section we work only in $\mathbb{C}$, the complex plane. Let $\Omega \subseteq \mathbb{C}$ be a domain. Define the *Bergman space*

$$A^2(\Omega) = \left\{ f \text{ holomorphic on } \Omega : \int_\Omega |f(z)|^2 \, dA(z)^{1/2} \equiv \|f\|_{A^2(\Omega)} < \infty \right\}.$$

Observe that $A^2(\Omega)$ is a closed subspace of $L^2(\Omega)$. The orthogonal subspace B is also of interest. Of course, $L^2(\Omega) = A^2(\Omega) \oplus B$.

Lemma 10.4.1 *Let $K \subseteq \Omega$ be compact. There is a constant $C_K > 0$, depending on K, such that*

$$\sup_{z \in K} |f(z)| \leq C_K \|f\|_{A^2(\Omega)} \ , \quad \text{all } f \in A^2(\Omega).$$

Proof: In what follows, $D(z, r)$ is the open disc with center z and radius r. Also dA denotes 2-dimensional area measure.

Since K is compact, there is an $r(K) = r > 0$ so that, for any $z \in K, D(z, r) \subseteq \Omega$. Therefore, for each $z \in K$ and $f \in A^2(\Omega)$, the mean value theorem tells us that

$$
\begin{aligned}
|f(z)| &= \frac{1}{A(D(z, r))} \left| \int_{D(z,r)} f(t) \, dA(t) \right| \\
&\leq (A(D(z, r)))^{-1/2} \|f\|_{L^2(D(z,r))} \\
&\leq C \cdot r^{-1} \|f\|_{A^2(\Omega)} \\
&\equiv C_K \|f\|_{A^2(\Omega)}.
\end{aligned}
$$
$\square$

Lemma 10.4.2 *The space $A^2(\Omega)$ is a Hilbert space with the inner product $\langle f, g \rangle \equiv \int_\Omega f(z)\overline{g(z)} \, dA(z)$.*

Proof: Everything is clear except for completeness. Let $\{f_j\} \subseteq A^2$ be a sequence that is Cauchy in norm. Since L^2 is complete there is an L^2 limit function f. We need to see that f is holomorphic. But Lemma 4.1 yields that norm convergence implies normal (uniform on compact sets) convergence. And holomorphic functions are closed under normal limits. Therefore, f is holomorphic and $A^2(\Omega)$ is complete.
$\square$

Lemma 10.4.3 *For each fixed $z \in \Omega$, the functional*

$$\Phi_z : f \mapsto f(z), \quad f \in A^2(\Omega)$$

is a continuous linear functional on $A^2(\Omega)$.

Proof: This is immediate from Lemma 4.1 if we take K to be the singleton $\{z\}$. □

We may now apply the Riesz representation theorem to see that there is an element $k_z \in A^2(\Omega)$ such that the linear functional Φ_z is represented by inner product with k_z: if $f \in A^2(\Omega)$, then for all $z \in \Omega$ we have

$$f(z) = \Phi_z(f) = \langle f, k_z \rangle.$$

Definition 10.4.4 The Bergman kernel is the function $K(z, \zeta) = \overline{k_z(\zeta)}$, $z, \zeta \in \Omega$. It has the reproducing property

$$f(z) = \int K(z, \zeta) f(\zeta) \, dA(\zeta), \quad \forall f \in A^2(\Omega).$$

Proposition 10.4.5 *The Bergman kernel $K(z, \zeta)$ is conjugate symmetric: $K(z, \zeta) = \overline{K(\zeta, z)}$.*

Proof: By its very definition, $\overline{K(\zeta, \cdot)} \in A^2(\Omega)$ for each fixed ζ. Therefore, the reproducing property of the Bergman kernel gives

$$\int_\Omega K(z, t) \overline{K(\zeta, t)} \, dA(t) = \overline{K(\zeta, z)}.$$

On the other hand,

$$\int_\Omega K(z, t) \overline{K(\zeta, t)} \, dA(t) = \overline{\int K(\zeta, t) \overline{K(z, t)} \, dA(t)}$$

$$= \overline{\overline{K(z, \zeta)}}$$

$$= K(z, \zeta). \qquad \square$$

Proposition 10.4.6 *The Bergman kernel is uniquely determined by the properties that it is an element of $A^2(\Omega)$ in z, is conjugate symmetric and reproduces $A^2(\Omega)$.*

Proof: Let $K'(z, \zeta)$ be another such kernel. Then

$$K(z, \zeta) = \overline{K(\zeta, z)} = \int K'(z, t) \overline{K(\zeta, t)} \, dA(t)$$

$$= \overline{\int K(\zeta, t) \overline{K'(z, t)} \, dA(t)}$$

$$= \overline{\overline{K'(z, \zeta)}} = K'(z, \zeta). \qquad \square$$

Since $L^2(\Omega)$ is a separable Hilbert space, then so is its subspace $A^2(\Omega)$. Thus there is a complete orthonormal basis $\{\phi_j\}_{j=1}^\infty$ for $A^2(\Omega)$.

Proposition 10.4.7 *Let K be a compact subset of Ω. Then the series*

$$\sum_{j=1}^\infty \phi_j(z)\overline{\phi_j(\zeta)}$$

sums uniformly on $K \times K$ to the Bergman kernel $K(z,\zeta)$.

Proof: Let $\{\phi_j\}$ be a complete orthonormal basis for the Bergman space. By the Riesz-Fischer and Riesz representation theorems, we obtain

$$\sup_{z \in K} \left(\sum_{j=1}^\infty |\phi_j(z)|^2 \right)^{1/2} = \sup_{z \in K} \left\| \{\phi_j(z)\}_{j=1}^\infty \right\|_{\ell^2}$$

$$= \sup_{\substack{\|\{a_j\}\|_{\ell^2}=1 \\ z \in K}} \left| \sum_{j=1}^\infty a_j \phi_j(z) \right|$$

$$= \sup_{\substack{\|f\|_{A^2}=1 \\ z \in K}} |f(z)|$$

$$\leq C_K. \tag{10.4}$$

In the last lines we have used Lemma 4.1 and the Hahn-Banach theorem. Therefore,

$$\sum_{j=1}^\infty \left| \phi_j(z)\overline{\phi_j(\zeta)} \right| \leq \left(\sum_{j=1}^\infty |\phi_j(z)|^2 \right)^{1/2} \left(\sum_{j=1}^\infty |\phi_j(\zeta)|^2 \right)^{1/2}$$

and the convergence is uniform over $z, \zeta \in K$. For fixed $z \in \Omega$, (10.4) shows that $\{\phi_j(z)\}_{j=1}^\infty \in \ell^2$. Hence we have that $\sum \phi_j(z)\overline{\phi_j(\zeta)} \in \overline{A^2(\Omega)}$ as a function of ζ. Let the sum of the series be denoted by $K'(z,\zeta)$. Notice that K' is conjugate symmetric by its very definition. Also, for $f \in A^2(\Omega)$, we have

$$\int K'(\cdot, \zeta)f(\zeta)\, dA(\zeta) = \sum \hat{f}(j)\phi_j(\cdot) = f(\cdot),$$

where convergence is in the Hilbert space topology. [Here $\hat{f}(j)$ is the jth Fourier coefficient of f with respect to the basis $\{\phi_j\}$.] But Hilbert space convergence dominates pointwise convergence (Lemma 4.1) so

$$f(z) = \int K'(z,\zeta)f(\zeta)\, dA(\zeta), \quad \text{all } f \in A^2(\Omega).$$

Therefore, K' is the Bergman kernel.   □

REMARK 10.4.8 It is worth noting explicitly that the proof of Proposition 4.7 shows that

$$\sum \phi_j(z)\overline{\phi_j(\zeta)}$$

equals the Bergman kernel $K(z,\zeta)$ *no matter what the choice* of complete orthonormal basis $\{\phi_j\}$ for $A^2(\Omega)$. $\qquad\square$

Proposition 10.4.9 *If Ω is a bounded domain in $\mathbb{C}$ then the mapping*

$$P : f \mapsto \int_\Omega K(\cdot,\zeta)f(\zeta)\,dA(\zeta)$$

is the Hilbert space orthogonal projection of $L^2(\Omega, dA)$ onto $A^2(\Omega)$.

Proof: Notice that, by Fubini's theorem, P is idempotent and self-adjoint and that $A^2(\Omega)$ is precisely the set of elements of L^2 that are fixed by P. $\square$

Given a holomorphic mapping $\Phi : \Omega_1 \to \Omega_2$, we may ask to compare the complex derivative Φ' with the real Jacobian determinant $J_\mathbb{R}(\Phi)$. The fact of the matter, and this is easy to check just using the Cauchy–Riemann equations, is that

$$\det J_\mathbb{R}(\Phi) = |\Phi'|^2.$$

A holomorphic mapping $f : \Omega_1 \to \Omega_2$ of domains $\Omega_1 \subseteq \mathbb{C}, \Omega_2 \subseteq \mathbb{C}$ is said to be *biholomorphic* (or *conformal*) if it is one-to-one, onto, and $f'(z) \neq 0$ for every $z \in \Omega_1$.

In what follows we denote the Bergman kernel for a given domain Ω by K_Ω.

Proposition 10.4.10 *Let Ω_1, Ω_2 be domains in $\mathbb{C}$. Let $f : \Omega_1 \to \Omega_2$ be conformal. Then*

$$f'(z)K_{\Omega_2}(f(z), f(\zeta))\overline{f'(\zeta)} = K_{\Omega_1}(z,\zeta).$$

Proof: Let $\phi \in A^2(\Omega_1)$. Then, by change of variable,

$$\int_{\Omega_1} f'(z)K_{\Omega_2}(f(z), f(\zeta))\overline{f}'(\zeta)\phi(\zeta)\,dA(\zeta)$$

$$= \int_{\Omega_2} f'(z)K_{\Omega_2}(f(z), \widetilde{\zeta})\overline{f}'(f^{-1}(\widetilde{\zeta}))\phi(f^{-1}(\widetilde{\zeta}))$$

$$\times \det J_\mathbb{R} f^{-1}(\widetilde{\zeta})\,dA(\widetilde{\zeta}).$$

By the remark following Proposition 4.9, this simplifies to

$$f'(z)\int_{\Omega_2} K_{\Omega_2}(f(z), \widetilde{\zeta})\left\{\left(f'(f^{-1}(\widetilde{\zeta}))\right)^{-1}\phi\left(f^{-1}(\widetilde{\zeta})\right)\right\}dA(\widetilde{\zeta}).$$

By change of variables, the expression in braces { } is an element of $A^2(\Omega_2)$. So the reproducing property of K_{Ω_2} applies, and the last line equals

$$= f'(z) \left(f'(z)\right)^{-1} \phi\left(f^{-1}(f(z))\right) = \phi(z).$$

By the uniqueness of the Bergman kernel, the proposition follows.   □

Proposition 10.4.11 *For $z \in \Omega$ it holds that $K_\Omega(z,z) > 0$.*

Proof: Now

$$K_\Omega(z,z) = \sum_{j=1}^\infty |\phi_j(z)|^2 \geq 0.$$

If in fact $K(z,z) = 0$ for some z then $\phi_j(z) = 0$ for all j hence $f(z) = 0$ for every $f \in A^2(\Omega)$ by Proposition 4.7. This is absurd.   □

Definition 10.4.12 *For any $\Omega \subseteq \mathbb{C}$, we define a Hermitian metric on Ω by*

$$g(z) = \frac{\partial^2}{\partial z \partial \bar{z}} \log K(z,z), \quad z \in \Omega.$$

This means that the square of the length of a tangent vector ξ at a point $z \in \Omega$ is given by

$$|\xi|^2_{B,z} = g(z) \xi \bar{\xi}.$$

The metric that we have defined is called the *Bergman metric*.

In a Hermitian metric g, the length of a C^1 curve $\gamma : [0,1] \to \Omega$ is given by

$$\ell(\gamma) = \int_0^1 \left(g(\gamma(t)) \, \gamma'(t) \, \bar{\gamma}'(t)\right)^{1/2} dt.$$

If P, Q are points of Ω then their distance $d_\Omega(P,Q)$ in the metric is defined to be the infimum of the lengths of all piecewise C^1 curves connecting the two points.

It is not *a priori* obvious that the Bergman metric for a bounded domain Ω is given by a positive definite matrix at each point.

Proposition 10.4.13 *Let $\Omega_1, \Omega_2 \subseteq \mathbb{C}$ be domains and let $f : \Omega_1 \to \Omega_2$ be a biholomorphic mapping. Then f induces an isometry of Bergman metrics:*

$$|\xi|_{B,z} = |f'(z)\xi|_{B,f(z)}$$

for all $z \in \Omega_1, \xi \in \mathbb{C}$. Equivalently, f induces an isometry of Bergman distances in the sense that

$$d_{\Omega_2}(f(P), f(Q)) = d_{\Omega_1}(P,Q)$$

for all $P, Q \in \Omega_1$.

Proof: This is a formal exercise, but we include it for completeness: From the definitions, it suffices to check that

$$g^{\Omega_2}(f(z))\,(f'(z)w)\left(\overline{f'(z)w}\right) = g^{\Omega_1}(z)w\overline{w} \tag{10.5}$$

for all $z \in \Omega, w \in \mathbb{C}$. But, by Proposition 4.10,

$$\begin{aligned}
g^{\Omega_1}(z) &= \frac{\partial^2}{\partial z \partial \overline{z}}\log K_{\Omega_1}(z,z)\\
&= \frac{\partial^2}{\partial z \partial \overline{z}}\log\left\{|f'(z)|^2 K_{\Omega_2}(f(z),f(z))\right\}\\
&= \frac{\partial^2}{\partial z \partial \overline{z}}\log K_{\Omega_2}(f(z),f(z)) \tag{10.6}
\end{aligned}$$

since $\log|f'(z)|^2$ is locally

$$\log(f') + \log(\overline{f'}) + C$$

hence is annihilated by the mixed second derivative. But line (10.6) is nothing other than

$$g^{\Omega_2}(f(z))\frac{\partial f(z)}{\partial z}\frac{\partial \overline{f(z)}}{\partial \overline{z}}$$

and (10.5) follows. $\qquad\square$

Proposition 10.4.14 *Let* $\Omega \subset\subset \mathbb{C}$ *be a domain. Let* $z \in \Omega$. *Then*

$$K(z,z) = \sup_{f \in A^2(\Omega)}\frac{|f(z)|^2}{\|f\|_{A^2}^2} = \sup_{\|f\|_{A^2(\Omega)}=1}|f(z)|^2.$$

Proof: Now

$$\begin{aligned}
K(z,z) &= \sum|\phi_j(z)|^2\\
&= \left(\sup_{\|\{a_j\}\|_{\ell^2}=1}\left|\sum\phi_j(z)a_j\right|\right)^2\\
&= \sup_{\|f\|_{A^2}=1}|f(z)|^2,
\end{aligned}$$

by the Riesz-Fischer theorem,

$$= \sup_{f \in A^2}\frac{|f(z)|^2}{\|f\|_{A^2}^2}. \qquad\square$$

10.4.1   Calculating the Bergman kernel

We present three methods for calculating the Bergman kernel on the unit disc $D \subseteq \mathbb{C}$. For the first, we use Proposition 4.7.

Construction of the Bergman Kernel with an Orthonormal Basis

Now we first notice that $\{\zeta^j\}_{j=0}^{\infty}$ forms an orthogonal basis for $A^2(D)$. This follows just from circular symmetry. In order to use 4.7, we need an *orthonormal* basis. So we calculate that

$$\int_D |\zeta^j|^2 \, dA(\zeta) = \int_0^{2\pi} \int_0^1 r^{2j} \cdot r \, dr d\theta = \frac{2\pi}{2j+2} \, .$$

Hence a complete orthonormal system for $A^2(D)$ is

$$\left\{ \frac{\sqrt{2j+2} \cdot \zeta^j}{\sqrt{2\pi}} \right\}_{j=0}^{\infty} \, .$$

We then may calculate that

$$K_D(z, \zeta) = \sum_{j=0}^{\infty} \frac{z^j \overline{\zeta}^j (2j+2)}{2\pi} \, .$$

Set $\alpha = z\overline{\zeta}$. Then we have

$$
\begin{aligned}
K_D(z, \zeta) &= \frac{1}{2\pi} \sum_{j=0}^{\infty} (2j+2)\alpha^j \\
&= \frac{1}{\pi} \sum_{j=0}^{\infty} (j+1)\alpha^j \\
&= \left(\frac{\alpha}{\pi} \sum_{j=0}^{\infty} \alpha^j \right)' \\
&= \left(\frac{\alpha}{\pi} \cdot \frac{1}{1-\alpha} \right)' \\
&= \frac{1}{\pi} \cdot \frac{1}{(1-\alpha)^2} \\
&= \frac{1}{\pi} \cdot \frac{1}{(1-z\overline{\zeta})^2} \, .
\end{aligned}
$$

This is our first calculation of the Bergman kernel for D.

Construction of the Bergman Kernel by Way of Differential Equations

It is actually possible to obtain the Bergman kernel of a domain in the plane from the Green's function for that domain (see [KRA3, Section 1.3.3]). Let us now summarize the key ideas. Unlike the first Bergman kernel construction, the present one will work for *any* domain with C^2 boundary. Thanks to work of Garabedian [GAR], one can say rather precisely what the Green's function of any planar domain is (see also [JAK]).

First, the fundamental solution for the Laplacian in the plane is the function

$$\Gamma(\zeta, z) = \frac{1}{2\pi} \log |\zeta - z|.$$

This means that $\triangle_\zeta \Gamma(\zeta, z) = \delta_z$ in the sense of distributions. [Observe that δ_z denotes the Dirac "delta mass" at z and $\triangle_\zeta$ is the Laplacian in the ζ variable.] In more prosaic terms, the condition is that

$$\int \Gamma(\zeta, z) \cdot \triangle \varphi(\zeta) \, d\xi d\eta = \varphi(z)$$

for any C^∞ function φ with compact support. We write, as usual, $\zeta = \xi + i\eta$. [This topic is treated in detail in [KRA1, Ch. 0, 1].]

Given a domain $\Omega \subseteq \mathbb{C}$, the *Green's function* is posited to be a function $G(\zeta, z)$ on $\Omega \times \Omega$ that satisfies

$$G(\zeta, z) = \Gamma(\zeta, z) - F_z(\zeta),$$

where $F_z(\zeta) = F(\zeta, z)$ is a particular harmonic function in the ζ variable (to be specified momentarily). Moreover, it is mandated that $G(\,\cdot\,, z)$ vanish on the boundary of Ω. One constructs the function $F(\,\cdot\,, z)$, for each fixed z, by solving a Dirichlet problem with boundary data $\Gamma(\,\cdot\,, z)$. Again, the reference [KRA1, p. 40] has all the particulars. It is worth noting, and this point is not completely obvious but is discussed in [KRA1, Ch. 1], that the Green's function is a symmetric function of its arguments.

The next proposition establishes a striking connection between the Bergman kernel and the classical Green's function.

Proposition 10.4.15 *Let $\Omega \subseteq \mathbb{C}$ be a bounded domain with C^2 boundary. Let $G(\zeta, z)$ be the Green's function for Ω and let $K(z, \zeta)$ be the Bergman kernel for Ω. Then*

$$K(z, \zeta) = 4 \cdot \overline{\frac{\partial^2}{\partial \zeta \partial \overline{z}} G(\zeta, z)}. \tag{10.7}$$

Proof: Our proof will use a version of Stokes's theorem written in the notation of complex variables. This matter was discussed earlier in the paper. It says that, if $u \in C^1(\overline{\Omega})$, then

$$\oint_{\partial U} u(\zeta) \, d\zeta = 2i \cdot \iint_U \frac{\partial u}{\partial \overline{\zeta}} \, d\xi \, d\eta, \tag{10.8}$$

where again $\zeta = \xi + i\eta$. The reader is invited to convert this formula to an expression in ξ and η and to confirm that the result coincides with the standard real-variable version of Stokes's theorem that can be found in any calculus book (see, e.g., [THO], [BLK]).

Now we already know that

$$G(\zeta, z) = \frac{1}{4\pi} \log(\zeta - z) + \frac{1}{4\pi} \log \overline{(\zeta - z)} + F(\zeta, z). \qquad (10.9)$$

Here we think of the logarithm as a multivalued holomorphic function; after we take a derivative, the ambiguity (which comes from an additive multiple of $2\pi i$) goes away.

Differentiating with respect to z (and using subscripts to denote derivatives), we find that

$$G_z(\zeta, z) = \frac{1}{4\pi} \frac{-1}{\zeta - z} + F_z(\zeta, z).$$

We may rearrange this formula to read

$$\frac{1}{\zeta - z} = -4\pi \cdot G_z(\zeta, z) + 4\pi F_z(\zeta, z).$$

We know that G, as a function of ζ, vanishes on $\partial\Omega$. Hence so does G_z. Let $f \in C^2(\overline{\Omega})$ be holomorphic on Ω. It follows that the Cauchy formula

$$f(z) = \frac{1}{2\pi i} \oint_{\partial\Omega} \frac{f(\zeta)}{\zeta - z} \, d\zeta$$

can be rewritten as

$$f(z) = -2i \oint_{\partial\Omega} f(\zeta) F_z(\zeta, z) \, d\zeta.$$

Now we apply Stokes's theorem (in the complex form) to rewrite this last as

$$f(z) = 4 \cdot \iint_\Omega (f(\zeta) F_z)_{\overline{\zeta}}(\zeta, z) \, d\xi \, d\eta,$$

where $\zeta = \xi + i\eta$. Since f is holomorphic and F is real-valued, we may conveniently write this last formula as

$$f(z) = 4 \cdot \iint_\Omega f(\zeta) \overline{F_{\overline{\zeta}z}}(\zeta, z) \, d\xi \, d\eta.$$

Now formula (10.9) tells us that $F_{\zeta\overline{z}} = G_{\zeta\overline{z}}$. Therefore, we have

$$f(z) = \iint_\Omega f(\zeta) 4 \overline{G_{\zeta\overline{z}}}(\zeta, z) \, d\xi \, d\eta. \qquad (10.10)$$

With a suitable limiting argument, we may extend this formula from functions f that are holomorphic and in $C^2(\overline{\Omega})$ to functions in $A^2(\Omega)$.

It is straightforward now to verify that $4\overline{G_{\zeta\bar z}}$ satisfies the first three characterizing properties of the Bergman kernel, just by examining our construction. The crucial reproducing property is, of course, formula (10.10). Then it follows that

$$K(z,\zeta) = 4 \cdot \overline{\frac{\partial^2}{\partial\zeta\partial\bar z}G(\zeta,z)}.$$

That is the desired result. $\qquad\square$

It is worth noting that the proposition we have just established gives a practical method for confirming the existence of the Bergman kernel—by relating it to the Green's function, whose existence is elementary. See [HAP1] for a version of these techniques in the several complex variables context.

Now let us calculate. Of course, the Green's function of the unit disc D is

$$G(\zeta,z) = \frac{1}{2\pi}\log|\zeta-z| - \frac{1}{2\pi}\log|1-\zeta\bar z|,$$

as a glance at any classical complex analysis text will tell us (see, for example, [AHL] or [HIL]). Verify the defining properties of Green's function for yourself.

With formula (10.7) in mind, we can make life a bit easier by writing

$$
\begin{aligned}
G(\zeta,z) \;=\;& \frac{1}{4\pi}\log(\zeta-z) + \frac{1}{4\pi}\log(\overline{\zeta-z}) \\
& -\frac{1}{4\pi}\log(1-\zeta\bar z) - \frac{1}{4\pi}\log\left(\overline{1-\zeta\bar z}\right).
\end{aligned}
$$

Here we think of the expression on the right as the concatenation of four multivalued functions, in view of the ambiguity of the logarithm function. This ambiguity is irrelevant for us because the derivative of Green's function is still well defined (i.e., the derivative annihilates additive constants).

Now we readily calculate that

$$\frac{\partial G}{\partial\bar z} = \frac{1}{4\pi}\cdot\frac{-1}{\zeta-z} + \frac{1}{4\pi}\cdot\frac{\zeta}{1-\zeta\bar z}$$

and

$$\frac{\partial^2 G}{\partial\zeta\partial\bar z} = \frac{1}{4\pi}\cdot\frac{1}{(1-\zeta\bar z)^2}.$$

In conclusion, we may apply Proposition 4.15 to see that

$$K(z,\zeta) = \frac{1}{\pi}\cdot\frac{1}{(1-z\cdot\bar\zeta)^2}.$$

This result is consistent with that obtained in the other two calculations (see above and below).

Construction of the Bergman Kernel by Way of Conformal Invariance

Let $D \subseteq \mathbb{C}$ be the unit disc. First we notice that, if either $f \in A^2(D)$ or $\overline{f} \in A^2(D)$, then

$$f(0) = \frac{1}{\pi} \iint_D f(\zeta)\, dA(\zeta). \tag{10.11}$$

This is the standard, two-dimensional area form of the mean-value property for holomorphic or harmonic functions.

Of course, the constant function $u(z) \equiv 1$ is in $A^2(D)$, so it is reproduced by integration against the Bergman kernel. Hence, for any $w \in D$,

$$1 = u(w) = \iint_D K(w,\zeta)u(\zeta)\, dA(\zeta) = \iint_D K(w,\zeta)\, dA(\zeta),$$

or

$$\frac{1}{\pi} = \frac{1}{\pi} \iint_D K(w,\zeta)\, dA(\zeta).$$

By (10.11), we may conclude that

$$\frac{1}{\pi} = K(w,0)$$

for any $w \in D$.

Now, for $a \in D$ fixed, consider the Möbius transformation

$$h(z) = \frac{z-a}{1-\overline{a}z}.$$

We know that

$$h'(z) = \frac{1-|a|^2}{(1-\overline{a}z)^2}.$$

We may thus apply Proposition 4.10 with $\phi = h$ to find that

$$\begin{aligned}
K(w,a) &= h'(w) \cdot K(h(w),h(a)) \cdot \overline{h'(a)} \\
&= \frac{1-|a|^2}{(1-\overline{a}w)^2} \cdot K(h(w),0) \cdot \frac{1}{1-|a|^2} \\
&= \frac{1}{(1-\overline{a}w)^2} \cdot \frac{1}{\pi} \\
&= \frac{1}{\pi} \cdot \frac{1}{(1-w\overline{a})^2}.
\end{aligned}$$

This is our formula for the Bergman kernel.

10.4.2 The Poincaré-Bergman metric on the disc

If $D \subseteq \mathbb{C}$ is the unit disc, $z \in D$, then Definition 4.12 shows that

$$|w|_{B,z} = \left\{ \frac{2|w|^2}{(1 - |z|^2)^2} \right\}^{1/2} = \frac{\sqrt{2}|w|}{1 - |z|^2},$$

where the subscript B indicates that we are working in the Bergman metric. We now use this formula to derive an explicit expression for the Poincaré distance from $0 \in D$ to $r + i0 \in D, 0 < r < 1$. Call this distance $d(0, r)$. Then

$$d(0, r) = \inf \left\{ \int_0^1 |\gamma'(t)|_{B,\gamma(t)} dt : \right.$$

$$\left. \gamma \text{ is a curve in } D, \gamma(0) = 0, \gamma(1) = r + i0 \right\}.$$

Elementary comparisons show that, among curves of the form $\psi(t) = t + iw(t), 0 \le t \le 1$, the curve $\gamma(t) = tr + i0$ is the shortest in the Poincaré metric. Further elementary arguments show that a general curve of the form $\psi(t) = v(t) + iw(t)$ is always longer than some corresponding curve of the form $t + i\widetilde{w}(t)$. We leave the details of these assertions to the reader. Thus

$$d(0, r) = \int_0^1 \frac{\sqrt{2}r}{(1 - (rt)^2)} dt$$

$$= \sqrt{2} \int_0^r \frac{1}{1 - t^2} dt$$

$$= \frac{1}{\sqrt{2}} \log \left(\frac{1 + r}{1 - r} \right).$$

Since rotations are conformal maps of the disc, we may next conclude that

$$d(0, re^{i\theta}) = \frac{1}{\sqrt{2}} \log \left(\frac{1 + r}{1 - r} \right).$$

Finally, if w_1, w_2 are arbitrary, then the Möbius transformation

$$\phi : z \mapsto \frac{z - w_1}{1 - \overline{w}_1 z}$$

satisfies $\phi(w_1) = 0, \phi(w_2) = (w_2 - w_1)/(1 - \overline{w}_1 w_2)$. Then Proposition 4.15 yields that

$$d(w_1, w_2) = d \left(0, \frac{w_2 - w_1}{1 - \overline{w}_1 w_2} \right)$$

$$= \frac{1}{\sqrt{2}} \left(\frac{1 + \left| \frac{w_2 - w_1}{1 - \overline{w}_1 w_2} \right|}{1 - \left| \frac{w_2 - w_1}{1 - \overline{w}_1 w_2} \right|} \right).$$

We note in passing that the expression $\rho(w_1, w_2) \equiv |(w_1 - w_2)/(1 - \overline{w}_1 w_2)|$ is called the *pseudohyperbolic distance*. It is also conformallly invariant, but it does *not* arise from integrating an infinitesimal metric (i.e. lengths of tangent vectors at a point). A fuller discussion of both the Poincaré metric and the pseudohyperbolic metric on the disc may be found in [GARN].

10.5   Preliminary Facts about Boundary Smoothness

The Riemann mapping theorem is a powerful result. It allows us to transfer the function theory of one proper, simply connected domain in the plane to another. But for many applications we need to know something about the boundary regularity of a conformal mapping. Does the mapping extend univalently and bicontinuously to the boundary? Does it extend smoothly to the boundary?

Curiously, the second of these questions was answered first—by Paul Painlevé in his thesis [PAI]. About 35 years later, Constantin Carath'eodory [CAR] answered the first question (and it is the first question that is typically treated in textbooks, not the second—although [GRK] is one of the exceptions).

For many purposes, continuity to the boundary is insufficient. In the recent work [APF], studies were made of the Bergman kernel on a variety of domains with different boundary geometries. The behavior of the kernel on a simply connected domain Ω was studied by considering a conformal mapping $\varphi : \Omega \to D$ and the corresponding transformation formula

$$K_\Omega(z, \zeta) = K_D(\varphi(z), \varphi(\zeta))\varphi'(z)\overline{\varphi'(\zeta)}$$

(see Proposition 4.10). In order to pass back and forth, it was essential to know that the expression $|\varphi'(z)|$ is bounded above and below:

$$c_1 \le |\varphi'(z)| \le c_2. \tag{10.12}$$

Note also that a set of inequalities like (5.1) tells us that the amount of stretching that occurs under the map φ is bounded above and below; this fact was also crucial in the results of [APF].

A result that implies the kind of estimates we have been discussing is the following

Theorem 10.5.2 *Let $\Omega \subseteq \mathbb{C}$ be a bounded, simply connected domain in $\mathbb{C}$ with sufficiently smooth boundary. Then any conformal mapping $\varphi : \Omega \to D$ will extend univalently and continuously differentiably to $\overline{\Omega}$, and the inverse mapping φ^{-1} will extend univalently and continuously differentiably to $\overline{D}$.*

With this theorem in hand, we know that $|\varphi'|$ extends continuously to the compact set $\overline{\Omega}$. Thus it is bounded. In other words,

$$|\varphi'(z)| \leq C \quad \text{for all } z \in \Omega.$$

Likewise, $|(\varphi^{-1})'|$ extends continuously to the compact set $\overline{D}$. So it is bounded. But then

$$\left| \frac{1}{\varphi'(\varphi^{-1}(z))} \right| = \left| (\varphi^{-1})'(z) \right| \leq C,$$

and hence[1]

$$\left| \varphi'(\varphi^{-1}(z)) \right| \geq \frac{1}{C}.$$

In the statement of Theorem 5.2, we have intentionally been imprecise about what "sufficiently smooth" means, since that is one of the main points of the rest of this section. For the moment, we shall content ourselves with proving the next result (see [BEK]). First, some definitions.

We shall require a brief discussion of the concept of boundary smoothness. We will offer two approaches to the question.

Boundary Smoothness by Way of Calculus: We will only consider domains with finitely many boundary components. Each boundary component will be a simple closed curve—see Figure 10.1.

If S^1 is the unit circle in the plane, parametrized by $t \mapsto e^{it}$, then we may think of each boundary curve γ_j as given by

$$\gamma_j : S^1 \to \mathbb{C}.$$

Let $k \in \{0, 1, 2, \dots\}$. We say that $\partial\Omega$ is C^k if γ_j' is never zero, each j, and the first k partial derivatives of the function γ_j exist, each j, and each derivative is continuous.

Boundary Smoothness by Way of a Defining Function: In geometric analysis it is frequently useful to think of a domain Ω in space as the sublevel set of a function $\rho = \rho_\Omega$. For example, the unit disc is given by

$$D = \{z \in \mathbb{C} : \rho(z) \equiv |z|^2 - 1 < 0\}. \tag{10.13}$$

It is a nice exercise with the implicit function theorem to see that any domain whose boundary consists of C^1 curves can be written as the sublevel set of some function ρ, as in equation (10.13). We usually demand that $\nabla\rho \neq 0$ on $\partial\Omega$ in order to prevent degeneracies and, more particularly, so that $\nabla\rho(P)$

---

[1]Implicit in this discussion is an important point that ought to be explicitly enunciated. Namely, if Ω_1 and Ω_2 are domains with C^2 boundary and $\varphi : \Omega_1 \to \Omega_2$ is a conformal map, then there are constants $c, C > 0$ such that $c \cdot \text{dist}(\varphi(z), \partial\Omega_2) \leq \text{dist}(z, \partial\Omega_1) \leq C \cdot \text{dist}(\varphi(z), \partial\Omega_2)$. In other words, φ preserves—qualitatively speaking—the distance to the boundary. This assertion follows from the Hopf lemma, a result that we treat below.

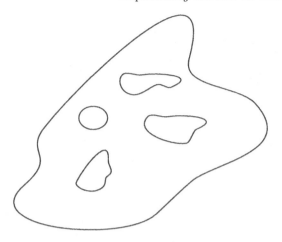

FIGURE 10.1
A domain with finitely many boundary components.

gives a well-defined outward normal vector at each boundary point P. We say
that $\Omega = \{z \in \mathbb{C} : \rho(z) < 0\}$ has C^k boundary, $k \in \{0, 1, 2, \dots\}$, if there is a
defining function ρ that is C^k.

We say that a bounded domain $\Omega \subseteq \mathbb{C}$ has C^∞ (or *smooth*) boundary if
there is a function $\rho : \mathbb{C} \to \mathbb{R}$ such that ρ is C^∞,

$$\Omega = \{z \in \mathbb{C} : \rho(z) < 0\},$$

and $\nabla\rho$ is nowhere zero on $\partial\Omega$.

It is worth noting that possession of a defining function says nothing about
the topology of the domain. The domain could be simply connected or not.

Example 10.5.4 Let

$$\Omega = \left\{ z \in \mathbb{C} : \rho(z) = \left(\frac{z + \bar{z}}{2}\right)^2 + \left(\frac{z - \bar{z}}{8i}\right)^2 - 1 < 0 \right\}.$$

Then Ω is the region bounded by an ellipse. Certainly, Ω has C^∞ boundary.

Theorem 10.5.5 *Let $k \in \{0, 1, 2, \dots\}$. Then there is an integer $N = N(k)$
such that, if $\Omega \subseteq \mathbb{C}$ is a bounded domain with C^N boundary, then any conformal mapping $\varphi : \Omega \to D$ will extend to be univalent and C^k from $\overline{\Omega}$ to $\overline{D}$.
Likewise, the inverse mapping φ^{-1} will extend to be univalent and C^k from $\overline{D}$
to $\overline{\Omega}$. It may be shown—although it requires considerable extra effort—that
N may be taken to be $k + 1$, or even $k + \epsilon$.*

This theorem is properly attributed to P. Painlevé, who proved it in his thesis [PAI]. Over the years, it has been sharpened through work of Kellogg, Warschawski, Pommerenke, and others. We shall indicate some of these results in what follows.

Let $\Omega \subseteq \mathbb{C}$ be a domain. We say that Ω satisfies *Bell's Condition R* if, for each $k \in \{0, 1, 2, \dots\}$, there is an $\ell \in \{0, 1, 2, \dots\}$ such that, whenever $f : \Omega \to \mathbb{C}$ has bounded derivative up to order ℓ, then $P_\Omega f$ (the Bergman projection of f) has bounded derivatives up to order k. Condition R has proved to be of historical importance in the study of holomorphic mappings. In this section we prove that $D = D(0, 1)$ satisfies a version of Condition R (see Theorem 5.13). This will be the key to the proof below of Theorem 5.5.

If Ω is a domain and $\Phi : \Omega \to \mathbb{C}$, then we say that Φ is k times boundedly continuously differentiable, and we write $\Phi \in C_b^k(\Omega)$, if all partial derivatives $(\partial/\partial x)^s (\partial/\partial y)^t \Phi$ of order $s + t \leq k$ on Ω (the interior of Ω, not the boundary) exist, are continuous, and are bounded. We let $C_b^\infty(\Omega) = \cap_k C_b^k(\Omega)$. If $\Phi \in C_b^k(\Omega)$, then we set

$$\|\Phi\|_{C_b^k(\Omega)} = \sum_{s+t \leq k} \sup_\Omega \left\| \left(\frac{\partial}{\partial x}\right)^s \left(\frac{\partial}{\partial y}\right)^t \Phi(x + iy) \right\|.$$

We also need an idea of continuity of derivatives at the boundary: A function $\Phi : \overline{\Omega} \to \mathbb{C}$ is called C^k if all partial derivatives of Φ up to and including order k extend continuously to $\overline{\Omega}$. As before, $C^\infty(\overline{\Omega})$ is defined to be $\cap_k C^k(\overline{\Omega})$.

Definition 10.5.6 If $f, g \in C^k(\overline{\Omega})$, then we say that f and g *agree up to order k on $\partial\Omega$* if

$$\left(\frac{\partial}{\partial x}\right)^s \left(\frac{\partial}{\partial y}\right)^t (f(z) - g(z))\bigg|_{\partial\Omega} = 0$$

for all $s + t \leq k$.

Lemma 10.5.7 *For every $k = 0, 1, 2, \dots$ there is a function $\lambda_k \in C^k(\mathbb{R})$ such that*

1. $\lambda_k(x) = 0$ *if* $x \leq \frac{1}{3}$,
2. $\lambda_k(x) = 1$ *if* $x \geq \frac{2}{3}$,
3. $0 \leq \lambda_k(x) \leq 1$ *for all* $x \in \mathbb{R}$.

Proof: Let

$$\psi(x) = \begin{cases} x^{k+1} & \text{if} \quad x \geq 0, \\ 0 & \text{if} \quad x < 0. \end{cases}$$

Obviously $\psi \in C^k(\mathbb{R})$. Define

$$\phi(x) = \psi\left(x - \frac{1}{3}\right) \cdot \psi\left(-x + \frac{2}{3}\right).$$

Then $\phi \in C^k$ and $\phi(x) \neq 0$ only if $1/3 < x < 2/3$. Observe that $\phi(x) \geq 0$ for all x. Let

$$u(x) = \int_{-\infty}^{x} \phi(t)\, dt.$$

Then $u \in C^k(\mathbb{R}), u(x) = 0$ for $x \leq 1/3$, and $u \equiv c$ a positive constant for $x \geq 2/3$. Define

$$\lambda_k(x) = \frac{1}{c} u(x).$$

This function has all the required properties. $\qquad\square$

It is actually possible to find a C^∞ function with properties **1**, **2**, and **3** of Lemma 5.7.

Lemma 10.5.8 *If $\Omega \subseteq \mathbb{C}$ is a domain with C^k boundary, $\Omega = \{z \in \mathbb{C} : \rho(z) < 0\}$, and if U is a neighborhood of $\partial\Omega$ on which $|\nabla\rho| \geq c > 0$, then there is a C^k function α_k on $\overline{\Omega}$ such that*

1. $\alpha_k = 0$ *on* $\Omega \setminus U$,
2. $\alpha_k = 1$ *in a neighborhood of* $\partial\Omega$.

Proof: Choose a number $\epsilon > 0$ such that, if $-\epsilon < \rho(z) < 0$, then $z \in U \cap \Omega$. Let λ_k be as in the last lemma. Define

$$\alpha_k(z) = \lambda_k\left(1 + \frac{\rho(z)}{\epsilon}\right).$$

Then α_k has all the desired properties. $\qquad\square$

Lemma 10.5.9 *Let $\Omega = \{z \in \mathbb{C} : \rho(z) < 0\}$ be a bounded domain with C^∞ boundary. Given $g \in C^\infty(\overline{\Omega})$, choose $f(z) = \rho(z)g(z)$. If $h \in A^2(\Omega)$ (the Bergman space—see Section 4), then*

$$\int_\Omega \overline{h(\zeta)} \frac{\partial}{\partial\zeta} f(\zeta)\, d\xi\, d\eta = 0.$$

Proof: We want to integrate by parts, but h is not defined on $\partial\Omega$ so we need a limiting argument.

Let $\epsilon > 0$ be small and let $\Omega_\epsilon = \{z : \rho(z) < -\epsilon\}$. Then define $f_\epsilon(z) = (\rho(z) + \epsilon)g(z)$. Notice that $f_\epsilon \in C^\infty(\overline{\Omega_\epsilon})$ and $f_\epsilon\big|_{\partial\Omega_\epsilon} = 0$. Also $h \in C^\infty(\overline{\Omega_\epsilon})$. So, by Green's theorem (or just the one-variable fundamental theorem of calculus, applied one variable at a time),

$$\int_{\Omega_\epsilon} \overline{h(\zeta)} \frac{\partial}{\partial\zeta} f_\epsilon(\zeta)\, d\xi\, d\eta = -\int_{\Omega_\epsilon} \left(\frac{\partial}{\partial\zeta}\overline{h(\zeta)}\right) \cdot f_\epsilon(\zeta)\, d\xi\, d\eta. \qquad (10.14)$$

There is no boundary term since $h \cdot f_\epsilon|_{\partial \Omega_\epsilon} \equiv 0$. But $h \in A^2(\Omega)$ so $(\partial / \partial \zeta)\overline{h} \equiv 0$ and the last expression is 0.

Finally, let

$$\chi_\epsilon(\zeta) = \begin{cases} 1 & \text{if } \zeta \in \Omega_\epsilon, \\ 0 & \text{if } \zeta \notin \Omega_\epsilon. \end{cases}$$

Then, using (10.14),

$$
\begin{aligned}
\left| \int_\Omega \overline{h(\zeta)} \frac{\partial}{\partial \zeta} f(\zeta) \, d\xi \, d\eta \right| &= \left| \int_\Omega \overline{h(\zeta)} \frac{\partial}{\partial \zeta} f(\zeta) \, d\xi \, d\eta - \int_\Omega \overline{h(\zeta)} \chi_\epsilon(\zeta) \frac{\partial}{\partial \zeta} f_\epsilon(\zeta) \, d\xi \, d\eta \right| \\
&\leq \int_\Omega \left| \overline{h(\zeta)} \right| \left| \frac{\partial f}{\partial \zeta}(\zeta) \right| |1 - \chi_\epsilon(\zeta)| \, d\xi \, d\eta \\
&\quad + \int_\Omega \left| \overline{h(\zeta)} \right| |\chi_\epsilon(\zeta)| \cdot \left| \frac{\partial f}{\partial \zeta}(\zeta) - \frac{\partial f_\epsilon}{\partial \zeta}(\zeta) \right| \, d\xi \, d\eta.
\end{aligned}
$$

By the Cauchy–Schwarz inequality, this is

$$
\begin{aligned}
&\leq \left(\int_\Omega |\overline{h(\zeta)}|^2 \left| \frac{\partial f}{\partial \zeta}(\zeta) \right|^2 d\xi \, d\eta \right)^{1/2} \cdot \left(\int_\Omega |1 - \chi_\epsilon(\zeta)|^2 \, d\xi \, d\eta \right)^{1/2} \\
&\quad + \left(\int_\Omega |\overline{h(\zeta)}|^2 \, d\xi \, d\eta \right)^{1/2} \cdot \left(\int_\Omega \left| \frac{\partial f}{\partial \zeta}(\zeta) - \frac{\partial f_\epsilon}{\partial \zeta}(\zeta) \right|^2 \, d\xi \, d\eta \right)^{1/2}.
\end{aligned}
$$

Now $\int_\Omega |1 - \chi_\epsilon(\zeta)|^2 \, d\xi \, d\eta$ clearly tends to zero as $\epsilon \to 0^+$. Also we see that $\partial f_\epsilon / \partial \zeta \to \partial f / \partial \zeta$ uniformly. So, letting $\epsilon \to 0$, we obtain

$$\int_\Omega \overline{h(\zeta)} \frac{\partial f}{\partial \zeta}(\zeta) \, d\xi \, d\eta = 0.$$

That is the desired conclusion. □

If $\Phi : \Omega_1 \to \Omega_2$ is conformal, then we might expect the Bergman kernel for Ω_1 to be related to that for Ω_2. This is indeed the case, as our Proposition 4.10 showed. We review it now:

Theorem 10.5.10 *If $\Phi : \Omega_1 \to \Omega_2$ is conformal then*

$$K_{\Omega_1}(z, w) = \Phi'(z) K_{\Omega_2}(\Phi(z), \Phi(w)) \overline{\Phi'(w)}.$$

We conclude this section with Bell's projection formula, which is based on Theorem 5.10.

If $\Omega \subseteq \mathbb{C}$ is a domain and K its Bergman kernel, then for any square integrable f on Ω we define

$$P_\Omega f(z) = \int_\Omega f(\zeta) K_\Omega(z, \zeta) \, d\xi \, d\eta.$$

This is the *Bergman projection.*

Recall that, if Ω is a bounded domain in $\mathbb{C}$, then the mapping

$$P : f \mapsto \int_{\Omega} K(\cdot, \varsigma) f(\varsigma) \, dA(\varsigma)$$

is the Hilbert space orthogonal projection of $L^2(\Omega, dA)$ onto $A^2(\Omega)$.

Next is Bell's formula:

Proposition 10.5.11 *Let Ω_1, Ω_2 be domains and let*

$$\Phi : \Omega_1 \to \Omega_2$$

be a conformal map. Let $f \in L^2(\Omega_1)$. Then

$$P_{\Omega_2}((\Phi^{-1})' \cdot (f \circ \Phi^{-1})) = (\Phi^{-1})' \cdot ((P_{\Omega_1} f) \circ \Phi^{-1}).$$

Proof: We sketch the proof. For any $f \in L^2(\Omega_1)$ and $g \in L^2(\Omega_2)$,

$$
\begin{aligned}
\langle P_{\Omega_2}((\Phi^{-1})' \cdot (f \circ \Phi^{-1})), g \rangle_2
&= \langle (\Phi^{-1})' \cdot (f \circ \Phi^{-1}), P_{\Omega_2} g \rangle_2 \\
&= \langle f, ((P_{\Omega_2} g) \circ \Phi) \cdot \Phi' \rangle_1 \\
&= \langle P_{\Omega_1} f, ((P_{\Omega_2} g) \circ \Phi) \cdot \Phi' \rangle_1 \\
&= \langle ((P_{\Omega_1} f) \circ \Phi^{-1}) \cdot (\Phi^{-1})', P_{\Omega_2} g \rangle_2 \\
&= \langle ((P_{\Omega_1} f) \circ \Phi^{-1}) \cdot (\Phi^{-1})', g \rangle_2 .
\end{aligned}
$$

Since $g \in L^2(\Omega_2)$ was arbitrary, we may conclude that Bell's identity is proved.
$\square$

The next lemma, due to Bell, is central to the theory.

Proposition 10.5.12 *If $\Omega = \{z \in \mathbb{C} : \rho(z) < 0\}$ is a bounded domain with C^∞ boundary, if $u \in C^{k+1}(\overline{\Omega})$, and if $k \geq 1$, then there is a $g \in C^k(\overline{\Omega})$ that agrees with u up to order k on $\partial\Omega$ and such that $P_\Omega g = 0$.*

Proof: Let α_k be as in Lemma 5.8. We define g by induction. For the C^1 case, let

$$v_1(z) = \frac{\partial}{\partial z} w_1(z),$$

where

$$w_1(z) = \frac{\alpha_1(z) \cdot u(z) \cdot \rho(z)}{(\partial \rho / \partial z)(z)}.$$

Then

$$
\begin{aligned}
v_1(z) &= \alpha_1(z) \cdot u(z) + \rho(z) \cdot \frac{\partial}{\partial z}\left(\frac{\alpha_1(z) \cdot u(z)}{(\partial \rho / \partial z)(z)} \right) \\
&\equiv \alpha_1(z) \cdot u(z) + \rho(z) \cdot \eta_1(z),
\end{aligned}
$$

where η_1 is defined by this identity and is continuous on $\overline{\Omega}$. Then

$$v_1 - u = \rho(z) \cdot \eta_1(z) \quad \text{near} \quad \partial\Omega.$$

(So v_1 and u agree to order *zero* on $\partial\Omega$.) In particular, $v_1 - u\big|_{\partial\Omega} = 0$. Also

$$P_\Omega v_1(z) = \int_\Omega K(z,\zeta) \frac{\partial}{\partial \zeta} w_1(\zeta) \, d\xi \, d\eta = 0$$

by Lemma 5.9.

Suppose inductively that we have constructed $w_{\ell-1}$ and $v_{\ell-1} = \frac{\partial}{\partial z} w_{\ell-1}$ such that $v_{\ell-1}$ agrees to order $(\ell-1) - 1$ with u on $\partial\Omega$ and $P_\Omega v_{\ell-1} = 0$. We shall now construct a function w_ℓ of the form

$$w_\ell = w_{\ell-1} + \theta_\ell \cdot \rho^\ell \tag{10.15}$$

such that $v_\ell = \frac{\partial}{\partial z} w_\ell$ agrees with u up to order $\ell - 1$ on $\partial\Omega$ and $P_\Omega v_\ell = 0$.

Let α_ℓ be as in Lemma 5.8 and define a differential operator $\mathcal{D}$ on $\overline{\Omega}$ by

$$\mathcal{D}(\phi) = \frac{\alpha_\ell}{|\partial\rho/\partial z|^2} \operatorname{Re}\left(\frac{\partial\rho}{\partial z} \frac{\partial\phi}{\partial \overline{z}}\right).$$

Notice that

$$\mathcal{D}\rho(z) = 1 \tag{10.16}$$

when $z \in \partial\Omega$. We define

$$\theta_\ell = \frac{\alpha_\ell \mathcal{D}^{\ell-1}(u - v_{\ell-1})}{\ell! \partial\rho/\partial z}. \tag{10.17}$$

Then, with w_ℓ defined as in (10.15) and $v_\ell = \frac{\partial}{\partial z} w_\ell$, we have

$$\begin{aligned}
\mathcal{D}^{\ell-1}(u - v_\ell) &= \mathcal{D}^{\ell-1}u - \mathcal{D}^{\ell-1}\frac{\partial}{\partial z}(w_{\ell-1} + \theta_\ell \cdot \rho^\ell) \\
&= \mathcal{D}^{\ell-1}(u - v_{\ell-1}) - \theta_\ell \frac{\partial\rho}{\partial z} \cdot (\mathcal{D}\rho)^{\ell-1} \cdot \ell! \\
&\quad + (\text{terms that involve a factor of } \rho).
\end{aligned}$$

If $z \in \partial\Omega$, then (using (10.16), (10.17))

$$\mathcal{D}^{\ell-1}(u - v_\ell - 1) - \mathcal{D}^{\ell-1}(u - v_{\ell-1}) + 0 = 0. \tag{10.18}$$

Since any directional derivative at $P \in \partial\Omega$ is a linear combination of

$$\mathcal{D} = a(P)\frac{\partial}{\partial x} + b(P)\frac{\partial}{\partial y} \quad \text{and} \quad \tau = -b(P)\frac{\partial}{\partial x} + a(P)\frac{\partial}{\partial y},$$

we may re-express our task as follows: We need to see that

$$(\tau)^s \mathcal{D}^t (u - v_\ell)\Big|_{\partial\Omega} = 0$$

for all $s+t = \ell-1$ (notice that the case $s+t < \ell-1$ follows from the inductive hypothesis and the explicit form of w_ℓ in (10.15)). The case $s = 0, t = \ell-1$ was treated in (10.18). If $s \geq 1$, then we write

$$\tau^s \mathcal{D}^t(u - v_\ell) = \tau(\tau^{s-1}\mathcal{D}^t(u - v_\ell)). \qquad (10.19)$$

Since $(s-1)+t = \ell-2$, the expression in parentheses is 0 on $\partial\Omega$ by the inductive hypothesis. But τ is a directional derivative *tangent* to $\partial\Omega$ (because $\mathcal{D}$ is normal); hence (10.19) is 0.

The induction is now complete, and v_ℓ has been constructed. We set $v_\ell = g$. Then

$$P_\Omega g(z) = \int_\Omega K_\Omega(z,\zeta)g(\zeta)d\xi\,d\eta = \int_\Omega K_\Omega(z,\zeta)\frac{\partial}{\partial\zeta}w_k(\zeta)\,d\xi\,d\eta = 0$$

by Lemma 5.9.   □

We shall use Bell's lemma (Proposition 5.10) twice. Our first use right now is on the disc. First, note the following two simple facts:

(a)   If K_D is the Bergman kernel for the disc, then

$$\left|\left(\frac{\partial}{\partial z}\right)^k K(z,w)\right| = \left|\frac{(k+1)!\overline{w}^k}{\pi(1 - z\cdot\overline{w})^{k+2}}\right| \leq \frac{(k+1)!}{(1 - |w|)^{k+2}}.$$

(b)   If $u \in C^k(\overline{D})$ and if u *vanishes to order* k at ∂D (i.e., u agrees with the zero function to order k at ∂D), then there is a $C > 0$ such that $|u(z)| \leq C \cdot (1 - |z|)^k$.

Theorem 10.5.13 ([Condition R for the disc]) *If* $k \geq 1$ *and* $u \in C^{k+2}(\overline{D})$, *then* $\|P_D u\|_{C_b^{k-1}(D)} < \infty$.

Proof: Use Proposition 5.12 to find a function $v \in C^{k+1}(\overline{D})$ that agrees with u to order $k+1$ on ∂D and such that $Pv = 0$. Then $P_D(u-v) = P_D u$, and $u - v$ vanishes to order $k+1$ on ∂D. In particular, by observation (b) above, $|u(\zeta) - v(\zeta)| \leq C \cdot (1 - |\zeta|)^{k+1}$. Then, for $j \leq k-1$, we have

$$\left|\left(\frac{\partial}{\partial z}\right)^j P_D u(z)\right| = \left|\left(\frac{\partial}{\partial z}\right)^j P_D(u-v)(z)\right|$$

$$= \left|\int_D \left(\frac{\partial}{\partial z}\right)^j K_D(z,\zeta)(u-v)(\zeta)\,d\xi\,d\eta\right|$$

$$\leq \int_D (j+1)!(1 - |\zeta|)^{-j-1}C \cdot (1 - |\zeta|)^{k+1}\,d\xi\,d\eta,$$

where we have used observation (a) above. This last integral is clearly bounded, independent of z.   □

REMARK 10.5.14 Item **(b)** above actually holds on any bounded domain Ω with C^k boundary: If $u \in C^k(\overline{\Omega})$ vanishes to order k on $\partial\Omega$, then there is a $C > 0$ such that

$$|u(z)| \leq C \cdot \delta_\Omega(z)^k.$$

Here, for $z \in \Omega$,

$$\delta_\Omega(z) = \inf_{w \notin \Omega} |z - w|.$$

Of course, δ_Ω could also be replaced here by any C^k-smooth defining function. This is immediate from the definitions and the Taylor expansion in the normal direction (in terms of powers of the defining function ρ).

10.6   Smoothness to the Boundary of Conformal Mappings

Let $\Omega = \{z \in \mathbb{C} : \rho(z) < 0\}$ be a bounded and simply connected domain with C^∞ boundary. Let $F : D \to \Omega$ be a conformal mapping. We wish to show that the one-to-one continuation of F to $\overline{D}$ (provided by Theorem 5.2) is actually in $C^\infty(\overline{D})$. For this we need a few lemmas. The first is a classical result from the theory of partial differential equations due to Hopf.

Lemma 10.6.1 (Hopf's lemma) *Let $U \subseteq \mathbb{C}$ be smoothly bounded. Let u be a harmonic function on U, continuous on the closure $\overline{U}$. Suppose that u assumes a local maximum value at $P \in \partial U$. Let ν be the unit outward normal vector to ∂U at P. Then the one-sided lower derivative $\partial u/\partial \nu$, defined to be*

$$\frac{\partial u}{\partial \nu} = \liminf_{t \to 0^+} \frac{u(P) - u(P - t\nu)}{t},$$

is positive.

Proof: It is convenient to make the following normalizations: Assume that u assumes the value 0 at P and is negative nearby and inside U; finally take the negative of our function so that u has a local *minimum* at P.

Now, since U has smooth boundary, there is an internally tangent disc at P. After scaling, we may as well suppose that it is the unit disc and that $P = 1 + i0$. Thus we may restrict our positive function u, with the minimum value 0 at $P = 1$, to the closed unit disc. Note in particular that $u(0) > 0$. Set $C = u(0) > 0$.

The Harnack inequality shows that $u(r) \geq [(1 - r)/(1 + r)]u(0)$, hence

$$\frac{u(1) - u(r)}{1 - r} = \frac{-u(r)}{1 - r} \leq -\frac{u(0)}{1 + r} \equiv \frac{-C}{1 + r} \leq -\frac{C}{2}.$$

The desired inequality for the normal derivative of u now follows.   □

REMARK 10.6.2 It is worth noting that the definition of the derivative and the fact that P is a local maximum guarantee—just from first principles—that the indicated one-sided lower normal derivative will be nonnegative. The Hopf lemma asserts that this derivative is actually positive.

Lemma 10.6.3 *If Ω is a bounded, simply connected domain with C^∞ boundary and if $F : D \to \Omega$ is a biholomorphic mapping, then there is a constant $C > 0$ such that*

$$\delta_\Omega(F(z)) \leq C(1 - |z|), \quad \text{all} \ z \in \Omega.$$

Here $\delta_\Omega(z) = \inf_{w \notin \Omega} |z - w|$.

Proof: The issue has to do only with points z near the boundary of D, and thus with points where $F(z)$ is near the boundary of Ω. Consider the function $w \mapsto \log |F^{-1}(w)|$. This function is defined for all w sufficiently near $\partial\Omega$ and indeed on $\Omega \setminus \{F(0)\}$. And it is harmonic there. Moreover, it is continuous on $(\Omega \setminus \{F(0)\}) \cup \partial\Omega$, with value 0 on $\partial\Omega$. In particular, it attains a (global) maximum at every point of $\partial\Omega$, since $|F^{-1}(w)| < 1$ if $w \in \Omega \setminus \{F(0)\}$. So the Hopf lemma applies. The logarithm function has nonzero derivative at all points. The conclusion of the lemma follows from combining this fact with the "normal derivative" conclusion of the Hopf lemma. That is, we show that

$$\left. \frac{\partial}{\partial\nu}|F^{-1}| \right|_P \geq c > 0$$

at each $P \in \partial D$. [Note that here, as in the Hopf lemma, no differentiability at boundary points is assumed: The derivative estimates are on the "lower derivative" only, which, as a lim inf, always exists and has value in the extended reals.]   □

Lemma 10.6.4 *With F, Ω as above and $k \in \{0, 1, 2, \dots\}$ it holds that*

$$\left| \left(\frac{\partial}{\partial z} \right)^k F(z) \right| \leq C_k (1 - |z|)^{-k}.$$

Proof: Since F takes values in Ω, it follows that F is bounded. Now apply the Cauchy estimates on $D(z, (1 - |z|))$.   □

Lemma 10.6.5 *If $\psi \in C^{2k+2}(\overline{\Omega})$ vanishes to order $2k + 1$ on $\partial\Omega$, then $F' \cdot (\psi \circ F) \in C_b^k(D)$. That is, $F' \cdot (\psi \circ F)$ has bounded derivatives up to and including order k.*

Proof: For $j \leq k$ we have

$$\left| \left(\frac{\partial}{\partial z} \right)^j (F' \cdot (\psi \circ F)) \right| = \left| \sum_{\ell=0}^j \binom{j}{\ell} \left(\frac{\partial}{\partial z} \right)^\ell (F') \cdot \left(\frac{\partial}{\partial z} \right)^{j-\ell} (\psi \circ F) \right|.$$

But

$$\left(\frac{\partial}{\partial z} \right)^{j-\ell} (\psi \circ F) \tag{10.20}$$

is a linear combination, with complex coefficients, of terms of the form

$$\left[\left(\left(\frac{\partial}{\partial z} \right)^m \psi \right) (F(z)) \right] \cdot \left(\frac{\partial}{\partial z} \right)^{n_1} F(z) \cdots \left(\frac{\partial}{\partial z} \right)^{n_k} F(z)$$

where $m \leq j - \ell$ and $n_1 + \cdots + n_k \leq j - \ell$. So (10.20) is dominated by

$$C \cdot \delta_\Omega (F(z))^{2k+1-(j-\ell)} (1 - |z|)^{-n_1} \cdots (1 - |z|)^{-n_k}$$
$$\leq C \cdot \delta_\Omega (F(z))^{k+1} \cdot (1 - |z|)^{-k-1}.$$

By Lemma 6.3, this is

$$\leq C \cdot (1 - |z|)^{k+1} \cdot (1 - |z|)^{-k-1} \leq C.$$

$\square$

REMARK 10.6.6 There is a remarkable formula of Faà di Bruno that formalizes the expansion for the higher derivatives of a composition. While the identity was first discovered in the eighteenth century, it is still being studied today (in the higher-dimensional version).

Lemma 10.6.7 *Let $G : D \to \mathbb{C}$ be holomorphic and have the property that*

$$\left| \left(\frac{\partial}{\partial z} \right)^j G(z) \right| \leq C_j < \infty$$

for $j = 0, \ldots, k+1$. Then each $(\partial/\partial z)^j G$ extends continuously to $\overline{D}$ for $j = 1, \ldots, k$.

Proof: It is enough to treat the case $k = 0$. The general case follows inductively.

If $P \in \partial D$, then we define

$$G(P) = \int_0^1 G'(tP) \cdot P dt + G(0).$$

It is clear that this defines a continuous extension of G to ∂D. $\square$

Theorem 10.6.8 ([Painlevé]) *If $\Omega = \{z \in \mathbb{C} : \rho(z) < 0\}$ is a bounded, simply connected domain with C^∞ boundary and $F : D(0,1) \to \Omega$ is a conformal mapping, then $F \in C^\infty(\overline{D})$ and $F^{-1} \in C^\infty(\overline{\Omega})$.*

Proof: Let $k \in \{1, 2, \dots\}$. By Proposition 5.12 applied to the function $u = 1$ on Ω, there is a function $v \in C^{2k+8}(\overline{\Omega})$ such that v agrees with u up to order $2k + 8$ on $\partial\Omega$ and $P_\Omega v = 0$. Then $\phi \equiv 1 - v$ satisfies $P_\Omega(\phi) = P_\Omega 1 - P_\Omega v = 1$ and ϕ vanishes to order $2k + 8$ on $\partial\Omega$. By Lemma 6.5, $F' \cdot (\phi \circ F) \in C_b^{k+3}(D)$. By Lemma 6.7, $F' \cdot (\phi \circ F) \in C^{k+2}(\overline{D})$. By Theorem 5.13, $P_D(F' \cdot (\phi \circ F)) \in C^{k-1}(D)$ and has $k - 1$ bounded derivatives. But the transformation law (Proposition 5.11) tells us that

$$P_D(F' \cdot (\phi \circ F)) = F' \cdot ((P_\Omega\phi) \circ F) = F' \cdot 1 = F'.$$

Thus F' has bounded derivatives up to order $k - 1$. By Lemma 6.7, all derivatives of F up to order $k - 2$ extend continuously to $\overline{D}$. Since k was arbitrary, we may conclude that $F \in C^\infty(\overline{D})$.

To show that $F^{-1} \in C^\infty(\overline{\Omega})$, it is enough to show that the Jacobian determinant of F as a real mapping on $\overline{D}$ does not vanish at any boundary point of D (we already know it is everywhere nonzero on D). Since F is holomorphic, F continues to satisfy the Cauchy–Riemann equations on $\overline{D}$. So it is enough to check that, at each point of $\overline{D} \setminus D$, some first derivative of F is nonzero. This assertion follows from a Hopf lemma argument analogous to the proof of Lemma 6.1. Details are left as an exercise. $\qquad\square$

Classically, Theorem 6.8 was proved by studying Green's potentials. The result dates back to P. Painlevé's thesis. All the ideas in the proof we have presented here are due to S. Bell and E. Ligocka [BELL]. An account of Bell's approach, in a more general context, can be found in [BEK].

10.7   The Bochner–Martinelli Formula in Several Complex Variables

If ω is a differential form on $\Omega \subseteq \mathbb{R}^N$ then, in coordinates, ω can be written as a finite sum of terms of the form $\omega_\alpha dx^\alpha$, where α is a multi-index and ω_α is a smooth function. Differential forms on $\mathbb{C}^n$ may be written in this fashion also, since $\mathbb{C}^n$ is canonically identified with $\mathbb{R}^{2n}$. However, it is much more convenient to use complex notation. Thus if $\Omega \subseteq \mathbb{C}^n$ and ω is a differential form on Ω, then ω is a sum of terms of the form $\omega_{\alpha\beta} dz^\alpha \wedge d\overline{z}^\beta$, where α, β are multi-indices with $|\alpha| \leq n, |\beta| \leq n$. If $0 \leq p, q \leq n$ and

$$\omega = \sum_{|\alpha|=p,|\beta|=q} \omega_{\alpha\beta} \wedge dz^\alpha d\overline{z}^\beta,$$

then ω is said to be a differential form of *type* (or *bidegree*) (p, q).

In classical advanced calculus, only a differential form of total degree m may be integrated on a space or surface or manifold of (real) dimension m. Likewise, in our new notation, only forms of type (p, q) with $p + q = m$ may be integrated on a space or surface or manifold of (real) dimension m.

If $\omega = \sum_{\alpha,\beta} \omega_{\alpha\beta} dz^\alpha \wedge d\bar{z}^\beta$, then we define

$$\partial\omega = \sum_{j=1}^{n} \sum_{\alpha,\beta} \frac{\partial \omega_{\alpha\beta}}{\partial z_j} dz_j \wedge dz^\alpha \wedge d\bar{z}^\beta,$$

$$\bar{\partial}\omega = \sum_{j=1}^{n} \sum_{\alpha,\beta} \frac{\partial \omega_{\alpha\beta}}{\partial \bar{z}_j} d\bar{z}_j \wedge dz^\alpha \wedge d\bar{z}^\beta.$$

Letting d denote the usual exterior differential operator on forms (see [RUD], [FED]), we see by a straightforward calculation that $d = \partial + \bar{\partial}$.

Notice that when f is a C^1 *function* (or $(0,0)$ form) then

$$\bar{\partial}f = \sum_j \frac{\partial f}{\partial \bar{z}_j} d\bar{z}_j.$$

Since the differentials $d\bar{z}_j$ are linearly independent, we conclude that $\bar{\partial}f \equiv 0$ on Ω if and only if $\partial f / \partial \bar{z}_j \equiv 0$ for $j = 1, \ldots, n$. That is, $\bar{\partial}f \equiv 0$ on Ω if and only if f is holomorphic in each variable separately.

The language of differential forms is needed in order to formulate Stokes's theorem:

Theorem 10.7.1 (Stokes) *Let $\Omega \subseteq \mathbb{C}^n$ be a bounded open set with C^1 boundary. Let ω be a differential form of bidegree (p, q) with coefficients in $C^1(\overline{\Omega})$. Then*

$$\int_{\partial\Omega} \omega = \int_{\Omega} d\omega = \int_{\Omega} \partial\omega + \bar{\partial}\omega.$$

Standard references for Stokes's theorem are W. Rudin [1] and G. de Rham [1]. The Stokes's theorem that we have recorded here is the standard one simply expressed in complex notation.

The *full Cauchy integral formula* in $\mathbb{C}$ is a formula not just about holomorphic functions but about all continuously differentiable functions. We now derive this more general result for all $\mathbb{C}^n, n \geq 1$, and learn what consequences it has for the function theory of both one and several variables.

Definition 10.7.2 On $\mathbb{C}^n$ we let

$$\omega(z) \equiv dz_1 \wedge dz_2 \wedge \cdots \wedge dz_n$$

$$\eta(z) \equiv \sum_{j=1}^{n} (-1)^{j+1} z_j dz_1 \wedge \cdots \wedge dz_{j-1} \wedge dz_{j+1} \wedge \cdots \wedge dz_n.$$

The form η is sometimes called the *Leray form*. We will often write $\omega(\bar{z})$ to mean $d\bar{z}_1 \wedge \cdots \wedge d\bar{z}_n$ and likewise $\eta(\bar{z})$ to mean $\sum_{j=1}^{n}(-1)^{j+1}\bar{z}_j d\bar{z}_1 \wedge \cdots \wedge d\bar{z}_{j-1} \wedge d\bar{z}_{j+1} \wedge \cdots \wedge d\bar{z}_n$.

The genesis of the Leray form is explained by the following lemma.

Lemma 10.7.3 For any $z_0 \in \mathbb{C}^n$, any $\epsilon > 0$, we have

$$\int_{\partial B(z^0,\epsilon)} \eta(\bar{z}) \wedge \omega(z) = n \int_{B(z^0,\epsilon)} \omega(\bar{z}) \wedge \omega(z).$$

Proof: Notice that $d\eta(\bar{z}) = \bar{\partial}\eta(\bar{z}) = n\omega(\bar{z})$. Therefore, by Stokes's theorem,

$$\int_{\partial B(z^0,\epsilon)} \eta(\bar{z}) \wedge \omega(z) = \int_{B(z^0,\epsilon)} d[\eta(\bar{z}) \wedge \omega(z)].$$

Of course, the expression in [ ] is saturated in dz's so, in the decomposition $d = \partial + \bar{\partial}$, only the term $\bar{\partial}$ will not die. Thus the last line equals

$$\int_{B(z^0,\epsilon)} [\bar{\partial}(\eta(\bar{z}))] \wedge \omega(z) = n \int_{B(z^0,\epsilon)} \omega(\bar{z}) \wedge \omega(z). \qquad \square$$

Remark:

Notice that, by change of variables,

$$
\begin{aligned}
\int_{B(z^0,\epsilon)} \omega(\bar{z}) \wedge \omega(z) &= \int_{B(0,\epsilon)} \omega(\bar{z}) \wedge \omega(z) \\
&= \epsilon^{2n} \int_{B(0,1)} \omega(\bar{z}) \wedge \omega(z).
\end{aligned}
$$

A straightforward calculation shows that

$$
\begin{aligned}
&\int_{B(0,1)} \omega(\bar{z}) \wedge \omega(z) \\
&= (-1)^{q(n)} \cdot (2i)^n \cdot (\text{volume of the unit ball in } \mathbb{C}^n \approx \mathbb{R}^{2n}),
\end{aligned}
$$

where $q(n) = [n(n-1)]/2$. We denote the value of this integral by $W(n)$. $\square$

Theorem 10.7.4 (Bochner–Martinelli) Let $\Omega \subseteq \mathbb{C}^n$ be a bounded domain with C^1 boundary. Let $f \in C^1(\overline{\Omega})$. Then, for any $z \in \Omega$, we have

$$
\begin{aligned}
f(z) &= \frac{1}{nW(n)} \int_{\partial\Omega} \frac{f(\zeta)\eta(\bar{\zeta}-\bar{z}) \wedge \omega(\zeta)}{|\zeta-z|^{2n}} \\
&\quad - \frac{1}{nW(n)} \int_{\Omega} \frac{\bar{\partial}f(\zeta)}{|\zeta-z|^{2n}} \wedge \eta(\bar{\zeta}-\bar{z}) \wedge \omega(\zeta).
\end{aligned}
$$

Proof: Fix $z \in \Omega$. We apply Stokes's theorem to the form

$$L_z(\zeta) \equiv \frac{f(\zeta)\eta(\overline{\zeta} - \overline{z}) \wedge \omega(\zeta)}{|\zeta - z|^{2n}}$$

on the domain $\Omega_{z,\epsilon} \equiv \Omega \backslash \overline{B}(z, \epsilon)$, where $\epsilon > 0$ is chosen so small that $\overline{B}(z, \epsilon) \subseteq \Omega$. Note that Stokes's theorem does not apply to forms that have a singularity; thus, we may not apply the theorem to L_z on any domain that contains the point z in either its interior or its boundary. This observation helps to dictate the form of the domain $\Omega_{z,\epsilon}$. As the proof develops, we shall see that it also helps to determine the outcome of our calculation.

Notice that

$$\partial(\Omega_{z,\epsilon}) = \partial\Omega \cup \partial B(z, \epsilon)$$

but that the two pieces are equipped with opposite orientations.

Thus, by Stokes,

$$\int_{\partial\Omega} L_z(\zeta) - \int_{\partial B(z,\epsilon)} L_z(\zeta) = \int_{\partial\Omega_{z,\epsilon}} L_z(\zeta)$$

$$= \int_{\Omega_{z,\epsilon}} d_\zeta(L_z(\zeta)). \tag{10.21}$$

Notice that we consider z to be fixed and ζ to be the variable. Now

$$d_\zeta L_z(\zeta) = \overline{\partial}_\zeta L_z(\zeta)$$

$$= \frac{\overline{\partial} f(\zeta) \wedge \eta(\overline{\zeta} - \overline{z}) \wedge \omega(\zeta)}{|\zeta - z|^{2n}}$$

$$+ f(\zeta) \cdot \left[\sum_{j=1}^{n} \frac{\partial}{\partial\overline{\zeta}_j} \left(\frac{\overline{\zeta}_j - \overline{z}_j}{|\zeta - z|^{2n}} \right) \right] \omega(\overline{\zeta}) \wedge \omega(\zeta). \tag{10.22}$$

Observing that

$$\frac{\partial}{\partial\overline{\zeta}_j} \left(\frac{\overline{\zeta}_j - \overline{z}_j}{|\zeta - z|^{2n}} \right) = \frac{1}{|\zeta - z|^{2n}} - n \frac{|\overline{\zeta}_j - \overline{z}_j|^2}{|\zeta - z|^{2n+2}},$$

we find that the second term on the far right of (10.22) dies and we have

$$d_\zeta L_z(\zeta) = \frac{\overline{\partial} f(\zeta) \wedge \eta(\overline{\zeta} - \overline{z}) \wedge \omega(\zeta)}{|\zeta - z|^{2n}}.$$

Substituting this identity into (10.21) yields

$$\int_{\partial\Omega} L_z(\zeta) - \int_{\partial B(z,\epsilon)} L_z(\zeta) = \int_{\Omega_{z,\epsilon}} \frac{\overline{\partial} f(\zeta) \wedge \eta(\overline{\zeta} - \overline{z}) \wedge \omega(\zeta)}{|\zeta - z|^{2n}}. \tag{10.23}$$

Next we remark that

$$\int_{\partial B(z,\epsilon)} L_z(\zeta) = f(z) \int_{\partial B(z,\epsilon)} \frac{\eta(\bar\zeta - \bar z) \wedge \omega(\zeta)}{|\zeta - z|^{2n}}$$
$$+ \int_{\partial B(z,\epsilon)} \frac{(f(\zeta) - f(z))\,\eta(\bar\zeta - \bar z) \wedge \omega(\zeta)}{|\zeta - z|^{2n}}$$
$$\equiv T_1 + T_2. \tag{10.24}$$

Since $|f(\zeta) - f(z)| \le C|\zeta - z|$ (and since each term of $\eta(\bar\zeta - \bar z)$ has a factor of some $\bar\zeta_j - \bar z_j$) it follows that the integrand of T_2 is of size $O(|\zeta - z|)^{-2n+2} \approx \epsilon^{-2n+2}$. Since the surface over which the integration is performed has area $\approx \epsilon^{2n-1}$, it follows that $T_2 \to 0$ as $\epsilon \to 0^+$.

By Lemma 7.3, we also have

$$T_1 = \epsilon^{-2n} f(z) \int_{\partial B(z,\epsilon)} \eta(\bar\zeta - \bar z) \wedge \omega(\zeta)$$
$$= n\epsilon^{-2n} f(z) \int_{B(0,\epsilon)} \omega(\bar\zeta) \wedge \omega(\zeta)$$
$$= nW(n)f(z) \tag{10.25}$$

Finally, (10.23)-(10.25) yield that

$$\left(\int_{\partial\Omega} L_z(\zeta)\right) - nW(n)f(z) + o(1) = \int_{\Omega_{z,\epsilon}} \bar\partial f(\zeta) \wedge \left[\frac{\eta(\bar\zeta - \bar z)}{|\zeta - z|^{2n}}\right] \wedge \omega(\zeta).$$

Since

$$\left|\frac{\eta(\bar\zeta - \bar z)}{|\zeta - z|^{2n}}\right| = O(|\zeta - z|^{-2n+1}),$$

the last integral is absolutely convergent as $\epsilon \to 0^+$ (remember that $\bar\partial f$ is bounded). Thus we finally have

$$f(z) = \frac{1}{nW(n)} \int_{\partial\Omega} L_z(\zeta) - \frac{1}{nW(n)} \int_\Omega \bar\partial f(\zeta) \wedge \frac{\eta(\bar\zeta - \bar z)}{|\zeta - z|^{2n}} \wedge \omega(\zeta).$$

This is the Bochner–Martinelli formula. $\square$

Corollary 10.7.5 *If $\Omega \subseteq \mathbb{C}$ is a bounded domain with C^1 boundary and if $f \in C^1(\bar\Omega)$ then, for any $z \in \Omega$,*

$$f(z) = \frac{1}{2\pi i} \int_{\partial\Omega} \frac{f(\zeta)}{\zeta - z}\,d\zeta - \frac{1}{2\pi i} \int_\Omega \frac{(\partial f(\zeta)/\partial\bar\zeta)}{\zeta - z}\,d\bar\zeta \wedge d\zeta.$$

Proof: It is necessary only to note that, when $n = 1$,

$$\omega(\zeta) = d\zeta \quad, \quad \eta(\bar{\zeta} - \bar{z}) = \bar{\zeta} - \bar{z} \quad, \quad \text{and} \quad nW(n) = 2\pi i. \qquad \square$$

Corollary 10.7.6 *With hypotheses as in Corollary 7.5, and the additional assumption that $\bar{\partial}f = 0$ on Ω, we have*

$$f(z) = \frac{1}{2\pi i} \int_{\partial\Omega} \frac{f(\zeta)}{\zeta - z} d\zeta.$$

10.8   The Bergman Kernel in Several Complex Variables

The basic theory of the Bergman kernel in several complex variables is exactly like that in one complex variable. The proofs are all soft analysis and proceed step by step in just the same way.

All these ideas have been gathered under the rubric of "Hilbert space with reproducing kernel," an idea first formulated by Aronszajn in [ARO]. All that is really needed to make the theory run is a version of our Lemma 4.1. For the Bergman space, the proof of this lemma in one variable works just as well in several variables. For the Szegő space, an application of the Bochner–Martinelli formula does the job.

10.9   The Szegő and Poisson-Szegö Kernels

The basic theory of the Szegö kernel is similar to that for the Bergman kernel— they are both special cases of a general theory of "Hilbert spaces with reproducing kernel" (see [ARO]). Thus we only outline the basic steps here, leaving details to the reader.

Let $\Omega \subseteq \mathbb{C}^n$ be a bounded domain with C^2 boundary. Let $A(\Omega)$ be those functions continuous on $\bar{\Omega}$ that are holomorphic on Ω. Let $H^2(\partial\Omega)$ be the space consisting of the closure in the $L^2(\partial\Omega, d\sigma)$ topology of the restrictions to $\partial\Omega$ of elements of $A(\Omega)$ (here $d\sigma$ is $(2n-1)$-dimensional area measure on the boundary). Then $H^2(\partial\Omega)$ is a proper Hilbert subspace of $L^2(\partial\Omega)$. Each element $f \in H^2(\partial\Omega)$ has a natural holomorphic extension to Ω given by its Poisson integral Pf. It is the case (see [KRA1, Ch. 8]) that, for σ-almost every $\zeta \in \partial\Omega$, it holds that

$$\lim_{\epsilon \to 0^+} f(\zeta - \epsilon\nu_\zeta) = f(\zeta).$$

Here, as usual, ν_ζ is the unit outward normal vector to $\partial\Omega$ at the point ζ.

For each fixed $z \in \Omega$ the functional

$$\psi_z : H^2(\partial\Omega) \ni f \mapsto Pf(z)$$

is continuous (why?). Let $k_z(\zeta)$ be the Hilbert space representative for the functional ψ_z. Define the Szegö kernel $S(z, \zeta)$ by the formula

$$S(z, \zeta) = \overline{k_z(\zeta)} \quad, z \in \Omega, \quad \zeta \in \partial\Omega.$$

If $f \in H^2(\partial\Omega)$ then

$$Pf(z) = \int_{\partial\Omega} S(z, \zeta) f(\zeta) d\sigma(\zeta)$$

for all $z \in \Omega$. We shall not explicitly formulate and verify the various uniqueness and extremal properties for the Szegö kernel. The reader is invited to consider these topics.

Let $\{\phi_j\}_{j=1}^{\infty}$ be an orthonormal basis for $H^2(\partial\Omega)$. Define

$$S'(z, \zeta) = \sum_{j=1}^{\infty} \phi_j(z) \overline{\phi_j(\zeta)} \quad, \quad z, \zeta \in \Omega.$$

For convenience we tacitly identify here each function with its Poisson extension to the interior of the domain. Then, for $K \subseteq \Omega$ compact, the series defining S' converges uniformly on $K \times K$. By a Riesz-Fischer argument, $S'(\cdot, \zeta)$ is the Poisson integral of an element of $H^2(\partial\Omega)$, and $S'(z, \cdot)$ is the conjugate of the Poisson integral of an element of $H^2(\partial\Omega)$. So S' extends to $(\overline{\Omega} \times \Omega) \cup (\Omega \times \overline{\Omega})$, where it is understood that all functions on the boundary are defined only almost everywhere. The kernel S' is conjugate symmetric. Also, by Riesz-Fischer theory, S' reproduces $H^2(\partial\Omega)$. Since the Szegö kernel is unique, it follows that $S = S'$.

The Szegö kernel may be thought of as representing a map

$$\mathbf{S} : f \mapsto \int_{\partial\Omega} f(\zeta) S(\cdot, \zeta) d\sigma(\zeta)$$

from $L^2(\partial\Omega)$ to $H^2(\partial\Omega)$. Since $\mathbf{S}$ is self-adjoint and idempotent, it is the Hilbert space projection of $L^2(\partial\Omega)$ to $H^2(\partial\Omega)$.

The Poisson-Szegö kernel is obtained by a formal procedure from the Szegö kernel: this procedure manufactures a *positive* reproducing kernel from one that is not necessarily positive. Note in passing that, just as we argued for the Bergman kernel in the last section, $S(z, z)$ is never 0 when $z \in \Omega$.

Proposition 10.9.1 *Define*

$$\mathcal{P}(z, \zeta) = \frac{|S(z, \zeta)|^2}{S(z, z)}, \quad z \in \Omega, \ \zeta \in \partial\Omega.$$

Then for any $f \in A(\Omega)$ and $z \in \Omega$ it holds that

$$f(z) = \int_{\partial\Omega} f(\zeta) \mathcal{P}(z, \zeta) d\sigma(\zeta).$$

Proof: Fix $z \in \Omega$ and $f \in A(\Omega)$ and define

$$u(\zeta) = f(\zeta)\frac{\overline{S(z,\zeta)}}{S(z,z)}, \quad \zeta \in \partial\Omega.$$

Then $u \in H^2(\partial\Omega)$ hence

$$
\begin{aligned}
f(z) &= u(z) = \int_{\partial\Omega} S(z,\zeta)u(\zeta)d\sigma(\zeta) \\
&= \int_{\partial\Omega} \mathcal{P}(z,\zeta)f(\zeta)d\sigma(\zeta).
\end{aligned}
$$

This is the desired formula. □

REMARK 10.9.2 In passing to the Poisson-Szegö kernel, we gain the advantage of positivity of the kernel (for more on this circle of ideas, see [KRA1, Ch. 1, 8] and also [KAT, Ch. 1]). However, we lose something in that $\mathcal{P}(z,\zeta)$ is no longer holomorphic in the z variable nor conjugate holomorphic in the ζ variable. The literature on this kernel is rather sparse and there are many unresolved questions. □

As an exercise, use the paradigm of Proposition 9.1 to construct a positive kernel from the Cauchy kernel on the disc (be sure to first change notation in the usual Cauchy formula so that it is written in terms of arc length measure on the boundary). What familiar kernel results?

Like the Bergman kernel, the Szegö and Poisson-Szegö kernels can almost never be explicitly computed. They can be calculated asymptotically in a number of important instances, however (see [FEF], [BOS]). We will give explicit formulas for these kernels on the ball.

Lemma 10.9.3 *The functions $\{z^\alpha\}$, where α ranges over multi-indices, are pairwise orthogonal and span $H^2(\partial B)$.*

Proof: The orthogonality follows from symmetry considerations. For the completeness, notice that it suffices to see that the span of $\{z^\alpha\}$ is dense in $A(B)$ in the uniform topology on the boundary. By the Stone-Weierstrass theorem, the closed algebra generated by $\{z^\alpha\}$ and $\{\overline{z}^\alpha\}$ is all of $C(\partial B)$. But the monomials $\overline{z}^\alpha, \alpha \neq 0$, are orthogonal to $A(B)$ (use the power series expansion about the origin to see this). The claimed density follows. □

Lemma 10.9.4 *Let $\mathbf{1} = (1,0,\ldots,0)$. Then*

$$S(z,\mathbf{1}) = \frac{(n-1)!}{2\pi^n}\frac{1}{(1-z_1)^n}.$$

Proof: We have that

$$S(z, \mathbf{1}) = \sum_\alpha \frac{z^\alpha \cdot \mathbf{1}^\alpha}{\|z_1^\alpha\|_{L^2(\partial B)}^2} = \sum_{k=0}^\infty \frac{z_1^k}{\eta(k)} = \frac{1}{2\pi^n} \sum_{k=0}^\infty \frac{z_1^k(k+n-1)!}{k!}$$

$$= \frac{(n-1)!}{2\pi^n} \sum_{k=0}^\infty \binom{k+n-1}{n-1} z_1^k$$

$$= \frac{(n-1)!}{2\pi^n} \frac{1}{(1-z_1)^n}. \qquad \square$$

Lemma 10.9.5 *Let ρ be a unitary rotation on $\mathbf{C}^n$. For any $z \in \overline{B}, \zeta \in \partial B$, we have that $S(z, \zeta) = S(\rho z, \rho \zeta)$.*

Proof: This is a standard change of variables argument, and we omit it. $\square$

Theorem 10.9.6 *The Szegő kernel for the ball is*

$$S(z, \zeta) = \frac{(n-1)!}{2\pi^n} \frac{1}{(1 - z \cdot \overline{\zeta})^n}.$$

Proof: Let $z \in B$ be arbitrary. Let ρ be the unique unitary rotation such that ρz is a multiple of $\mathbf{1}$. Then, by 10.4,

$$S(z, \zeta) = S(\rho^{-1}\mathbf{1}, \zeta) = S(\mathbf{1}, \rho\zeta) = \overline{S(\rho\zeta, \mathbf{1})} = \frac{(n-1)!}{2\pi^n} \frac{1}{\left(1 - \overline{(\rho\zeta) \cdot \mathbf{1}}\right)^n}$$

$$= \frac{(n-1)!}{2\pi^n} \frac{1}{\left(1 - \overline{\zeta} \cdot (\rho^{-1}\mathbf{1})\right)^n} = \frac{(n-1)!}{2\pi^n} \frac{1}{(1 - z \cdot \overline{\zeta})^n}.$$

$$\square$$

Corollary 10.9.7 *The Poisson-Szegő kernel for the ball is*

$$\mathcal{P}(z, \zeta) = \frac{(n-1)!}{2\pi^n} \frac{(1 - |z|^2)^n}{|1 - z \cdot \overline{\zeta}|^{2n}}.$$

10.10 A Variety of Different Kernels

A kernel is *constructible* if it can be written down explicitly. A kernel is *nonconstructible* (or *canonical*) if it is produced abstractly by some functional-analytic argument.

In the theory of one complex variable, all canonical kernels (the Cauchy kernel, the Poisson kernel) are constructible, and all constructible kernels are canonical.

Not so in several complex variables. Certainly in $\mathbb{C}^n$ we may produce the Bergman and Szegő kernels abstractly. But, in general (except on domains with a great deal of symmetry like the ball or the polydisc or the bounded symmetric domains of Cartan), these kernels are quite difficult to construct. Fefferman [FEF] produced an asymptotic expansion for the Bergman kernel on a strongly pseudoconvex domain (and Boutet de Monvel/Sjöstrand [BOS] did something similar for the Szegö kernel), but these arguments are quite deep and difficult.

It was considered to be quite exciting around 1970 when, independently, G. M. Henkin [HEN1], Grauert and Lieb [GRL], and Ramirez [RAM] were able to actually *construct* reproducing kernels on strongly pseudoconvex domains. These constructions were highly nontrivial, but they produced a formula quite analogous to the Cauchy kernel of one complex variable. In particular, these new reproducing kernels $K(z, \zeta)$ had the important property that *they were holomorphic in the free z variable*. This meant that the new kernels $K(z, \zeta)$ would not only *reproduce* holomorphic functions; they would also *create* holomorphic functions. That is to say, if $d\mu$ were any finite measure, then

$$ f(z) \equiv \int K(z, \zeta) \, d\mu(\zeta) $$

will be a holomorphic function in z.

The new kernels proved to be particularly useful. For instance, they could be used to produce explicit solutions to the equation $\overline{\partial}u = f$ (see [HEN2], for instance, or [KRA1, Ch. 5]). They could also be used to attack various problems in the theory of function algebras.

Here we sketch the Henkin construction for a domain Ω which is strongly convex (all boundary curvatures positive). For the full details of the matter, consult [KRA1, Ch. 5].

We begin with an idea called the Cauchy–Fantappié formula or formalism. This is a device for creating reproducing formulas in complex analysis. Good references for this notion are [RAN] and [KOP].

Recall the notation $\omega(z)$ and $\eta(z)$ from our discussion of the Bochner–Martinelli formula in Definition 7.2, ff. Further recall that a straightforward calculation shows that

$$ \int_{B(0,1)} \omega(\overline{z}) \wedge \omega(z) $$
$$ = (-1)^{q(n)} \cdot (2i)^n \cdot (\text{volume of the unit ball in } \mathbf{C}^n \approx \mathbf{R}^{2n}), $$

where $q(n) = [n(n-1)]/2$. We denote the value of this integral by $W(n)$.

Theorem 10.10.1 *Let $\Omega \subset\subset \mathbf{C}^n$ be a domain with C^1 boundary. Let $w(z, \zeta) = (w_1(z, \zeta), \ldots, w_n(z, \zeta))$ be a C^1, vector-valued function on $\overline{\Omega} \times \overline{\Omega} \setminus \{\text{diagonal}\}$ that satisfies*

$$\sum_{j=1}^{n} w_j(z, \zeta)(\zeta_j - z_j) \equiv 1.$$

Then we have, for any $f \in C^1(\overline{\Omega}) \cap \{\text{holomorphic functions on } \Omega\}$ and any $z \in \Omega$, the formula

$$f(z) = \frac{1}{nW(n)} \int_{\partial\Omega} f(\zeta)\eta(w) \wedge \omega(\zeta).$$

Of course, it should be understood here that ζ is the variable of integration and z is a free variable.

Sketch of Proof: For convenience we restrict attention to complex dimension 2. We may assume that $z = 0 \in \Omega$.

(a) If $\alpha^1 = (a_1^1, a_2^1), \alpha^2 = (a_1^2, a_2^2)$ are 2–tuples of C^1 functions on $\overline{\Omega}$ that satisfy $\sum_{j=1}^{2} a_j^k(\zeta) \cdot (\zeta_j - z_j) = 1$, let

$$B(\alpha^1, \alpha^2) = \sum_{\sigma \in S_2} \epsilon(\sigma) a_{\sigma(1)}^1 \wedge \overline{\partial}(a_{\sigma(2)}^2),$$

where S_2 is the symmetric group on 2 letters (which, of course, has just two elements) and $\epsilon(\sigma)$ is the signature of the permutation σ. Prove that B is independent of α^1.

(b) It follows that $\overline{\partial}B = 0$ on $\overline{\Omega} \setminus \{0\}$ (indeed $\overline{\partial}B$ is an expression like B with the expression $a_{\sigma(1)}^1$ replaced by $\overline{\partial}a_{\sigma(1)}^1$).

(c) Use (b), especially the paranthetical remark, to prove inductively that if $\beta^1 = (b_1^1, b_2^1), \beta^2 = (b_1^2, b_2^2)$, then there is a form γ on $\Omega \setminus \{0\}$ such that

$$\left[B(\alpha^1, \alpha^2) - B(\beta^1, \beta^2)\right] \wedge \omega(\zeta) = \overline{\partial}\gamma = d\gamma.$$

(d) Prove that if $\alpha^1 = \alpha^2 = (w_1, w_2)$ then $B(\alpha^1, \alpha^2)$ simplifies as

$$B(\alpha^1, \alpha^2) \wedge \omega(\zeta) = (2 - 1)!\eta(w) \wedge \omega(\zeta).$$

(e) Let $\mathcal{S}$ be a small sphere of radius $\epsilon > 0$ centered at 0 such that $\mathcal{S} \subseteq \Omega$. Use part (c) to see that

$$\int_{\partial\Omega} f(\zeta)\eta(w) \wedge \omega(\zeta) = \int_{\mathcal{S}} f(\zeta)\eta(w) \wedge \omega(\zeta).$$

(f) Now use **(c)** and **(d)** to see that

$$\int_S f(\zeta)\eta(w) \wedge \omega(\zeta) = \int_S f(\zeta)\eta(v) \wedge \omega(\zeta),$$

where

$$v(z,\zeta) = \frac{\bar{\zeta}_j - \bar{z}_j}{|\zeta - z|^2}.$$

[**Warning:** Be careful if you decide to apply Stokes's theorem.] From the theory of the Bochner–Martinelli kernel, we know that the last line is $2 \cdot W(2) \cdot f(0)$. □

Now let

$$\Omega = \{z \in \mathbf{C}^2 : \rho(z) < 0\}.$$

We assume that Ω is strongly convex, i.e., that the matrix of (real) second partial derivatives of ρ induces a positive definite quadratic form.

Define the complex tangent plane to $\partial\Omega$ at a point $P \in \partial\Omega$ to be the zero set of the function

$$\mathcal{L}_P(z) = \rho(P) + \frac{\partial\rho}{\partial z_1}(P)(z_1 - P_1) + \frac{\partial\rho}{\partial z_2}(P)(z_2 - P_2).$$

Now the Taylor expansion of ρ about a point $P \in \partial\Omega$, expressed in complex notation, is

$$\rho(z) = \rho(P) + \frac{\partial\rho}{\partial z_1}(P)(z_1 - P_1) + \frac{\partial\rho}{\partial z_2}(P)(z_2 - P_2) + \frac{\partial\rho}{\partial\bar{z}_1}(P)(\bar{z}_1 - \bar{P}_1)$$

$$+ \frac{\partial\rho}{\partial\bar{z}_2}(P)(\bar{z}_2 - \bar{P}_2) + (\text{quadratic terms}) + \mathcal{E}(z).$$

Here $\mathcal{E}(z)$ is an error term of order 3 and higher.

This last may be rewritten (because $P \in \partial\Omega$ so $\rho(P) = 0$) as

$$\rho(z) = 2\mathrm{Re}\,\mathcal{L}_P(z) + (\text{quadratic terms}) + \mathcal{E}(z).$$

And, because Ω is hypothesized to be strongly convex, we find that

$$\rho(z) = 2\mathrm{Re}\,\mathcal{L}_P(z) + (\text{positive term}).$$

We conclude that the variety $\{z : \mathcal{L}_P(z) = 0\}$ only intersects $\overline{\Omega}$ at the point P. For this reason, we may use the notation of the Cauchy–Fantappié formula to define

$$w_1(z, P) = \frac{(\partial\rho/\partial z_1)(P)}{\mathcal{L}_P(z)}$$

and

$$w_2(z, P) = \frac{(\partial\rho/\partial z_2)(P)}{\mathcal{L}_P(z)}.$$

It follows then that

$$\sum_{j=1}^{2} w_j \cdot (z_j - P_j) = 1\,.$$

So Cauchy–Fantappié theory applies, using this (w_1, w_2), and we get a reproducing kernel.

In the case that Ω is the unit ball B, the reader may check that the resulting kernel is the Szegő kernel.

10.11   Constructible Kernels vs. Non-Constructible Kernels

We close our discussion of reproducing kernels with a consideration of how the Szegő kernel is related to the more explicit kernel that arises from the Cauchy–Fantappié formalism. This set of ideas is due to Kerzman and Stein [KES].

C. Fefferman [FEF] made an important contribution in 1974 when he produced an asymptotic expansion for the Bergman kernel of a strongly pseudoconvex domain. Basically he was able to write

$$K(z, \zeta) = P(z, \zeta) + \mathcal{E}(z, \zeta)\,,$$

where P (the principal term) is, in suitable local coordinates, the Bergman kernel of the ball and $\mathcal{E}$ (the error term) is a term of strictly lower order (in some measurable sense). This powerful formula gives one a means for calculating mapping properties of the Bergman integral. Fefferman himself used the formula to calculate the boundary asymptotics of Bergman metric geodesics (for the purpose of proving the smooth boundary extension of biholomorphic mappings). Fefferman states in his paper—although the details have never been worked out—that there is a similar asymptotic expansion for the Szegő kernel of a strongly pseudoconvex domain.

At about the same time, Boutet de Monvel and Sjöstrand [BOS] used the technique of Fourier integral operators [HOR] to directly derive an asymptotic expansion for the Szegő kernel of a strongly pseudoconvex domain. This expansion is quite similar to Fefferman's: there is a principal term, which in suitable local coordinates is the Szegő kernel of the ball, and there is an error term which is of lower order. It is not known whether the techniques of [BOS] can be used to derive an asymptotic expansion for the Bergman kernel.

The main purpose of the present section is to consider another method, due to Kerzman and Stein, for deriving asymptotic expansions for the canonical kernels that is more elementary and uses less machinery. Fefferman's rather complicated argument uses Kohn's solution of the $\overline{\partial}$-Neumann problem as well as the theory of nonisotropic singular integrals. Boutet de Monvel and Sjöstrand's argument uses the theory of Fourier integral operators. The method of Kerzman and Stein [KES] that we treat here uses only basic complex function theory and a little functional analysis.

At this time there are virtually no results about asymptotic expansions for the canonical kernels on weakly pseudoconvex domains. Some interesting partial results appear in [HAN].

The ideas that we present now have thus far only been developed on strongly pseudoconvex domains. It is an important open problem to determine how to carry out a similar program on finite type domains or more general domains.

In previous sections, we have defined the Szegő projection $\mathbf{S} : L^2(\partial\Omega) \to H^2(\Omega)$. We also have a mapping $\mathbf{H} : L^2(\partial\Omega) \to H^2(\Omega)$ that is determined by the constructive Henkin kernel discussed above. We note that $\mathbf{H}$ defines a bounded operator from $L^2(\partial\Omega)$ to $H^2(\Omega)$ (the Hardy space—see [KRA, Chapter 8]) for the following reason.

Let Ω be strongly pseudoconvex. This means that the Levi form is positive definite at each boundary point (see [KRA1, Ch. 3] for details). It is known that $\partial\Omega$, when equipped with balls coming from the complex structure and the usual boundary area measure (see [STE], [KRA1, Ch. 8]), is a space of homogeneous type in the sense of Coifman and Weiss [COW]. Further, it is straightforward to verify that the Henkin operator $\mathbf{H}$ satisfies the hypotheses of the David-Journé $T1$ theorem for spaces of homogeneous type (see [CHR] for a nice exposition of these ideas). Thus we may conclude that the Henkin operator maps $L^2(\partial\Omega)$ to $L^2(\partial\Omega)$. Since the Henkin kernel also obviously maps $L^2(\partial\Omega)$ to holomorphic functions, we may conclude that the Henkin integral maps $L^2(\partial\Omega)$ to $H^2(\partial\Omega)$.

Now of course $\mathbf{S}$, being a projection, is self-adjoint. So $\mathbf{S} = \mathbf{S}^*$. It is not at all true that $\mathbf{H} = \mathbf{H}^*$, but one may calculate (see below for the details) that $\mathbf{A} \equiv \mathbf{H}^* - \mathbf{H}$ is small in a measurable sense.

We also have

$$\mathbf{HS} = \mathbf{S} , \qquad \mathbf{SH}^* = \mathbf{S} ,$$

$$\mathbf{SH} = \mathbf{H} , \qquad \mathbf{H}^*\mathbf{S} = \mathbf{H}^* .$$

Let us discuss these four identities for a moment.

For the first, notice that $\mathbf{S}$ is the projection onto H^2, and $\mathbf{H}$ preserves holomorphic functions. So certainly $\mathbf{HS} = \mathbf{S}$. For the second, we calculate that

$$\langle \mathbf{SH}^* x, y \rangle = \langle \mathbf{H}^* x, \mathbf{S}y \rangle = \langle x, \mathbf{HS}y \rangle = \langle x, \mathbf{S}y \rangle$$

(because $\mathbf{H}$ preserves holomorphic functions) and thus $= \langle \mathbf{S}x, y \rangle$. Hence $\mathbf{SH}^* = \mathbf{S}$. For the third, notice that $\mathbf{H}$ maps to the holomorphic functions and $\mathbf{S}$ preserves holomorphic functions. And, for the fourth, we calculate that

$$\langle \mathbf{H}^*\mathbf{S}x, y \rangle = \langle \mathbf{S}x, \mathbf{H}y \rangle = \langle x, \mathbf{SH}y \rangle = \langle x, \mathbf{H}y \rangle = \langle \mathbf{H}^* x, y \rangle .$$

In conclusion, $\mathbf{H}^*\mathbf{S} = \mathbf{H}^*$.

Now we see that

$$\mathbf{SA} = \mathbf{S}(\mathbf{H}^* - \mathbf{H}) = \mathbf{SH}^* - \mathbf{SH} = \mathbf{S} - \mathbf{H}.$$

As a result,

$$\mathbf{S} = \mathbf{H} + \mathbf{SA}$$

so

$$\mathbf{S}(\mathbf{I} - \mathbf{A}) = \mathbf{H}.$$

In conclusion,

$$\mathbf{S} = \mathbf{H}(\mathbf{I} - \mathbf{A})^{-1}.$$

If, indeed, we can show that $\mathbf{A}$ is norm small in a suitable sense, then $(\mathbf{I} - \mathbf{A})^{-1}$ is well defined by a Neumann series. Thus we may write

$$\mathbf{S} = \mathbf{H} + \mathbf{HA} + \mathbf{HA}^2 + \cdots + \mathbf{HA}^{\mathbf{j}} + \mathbf{HA}^{\mathbf{j}+1} + \cdots.$$

Hence we have expressed the Szegő projection $\mathbf{S}$ as an asymptotic expansion in terms of the Henkin projection $\mathbf{H}$. By applying this asymptotic expansion to the Dirac delta mass, this last formula can be translated into saying that the Szegő *kernel S* can be written as an asymptotic expansion in terms of the Henkin *kernel*.

Some new ideas connected with the discussion here appear in [KRA4].

It should be noted that Ewa Ligocka [LIG] has shown that these same ideas may be applied to expand the Bergman kernel in an asymptotic expansion in terms of the Henkin kernel. We shall not treat the details of her argument here.

10.12   Concluding Remarks

Reproducing kernels are part of the bedrock of modern analysis. A detailed treatment of reproducing kernels in the real variable setting appears in [BIN1], [BIN2]. Of course, there are many books about integral kernels in the complex analysis setting. Certainly, [RAN] is one of the most comprehensive. The book [KRA1] also has considerable treatment of the Bergman, Szegő, and Poisson-Szegő kernels.

We expect that the ideas presented here will be central to our subject for a good many years to come.

References

[**AHL**]   L. Ahlfors, *Complex Analysis*, 3rd ed., McGraw-Hill, New York, 1979.

[APF] L. Apfel, Localization properties and boundary behavior of the Bergman kernel, thesis, Washington University, 2003.

[ARO] N. Aronszajn, Theory of reproducing kernels, *Trans. Am. Math. Soc.* 68(1950), 337–404.

[BEK] S. R. Bell and S. G. Krantz, Smoothness to the boundary of conformal maps, *Rocky Mt. Jour. Math.* 17(1987), 23–40.

[BELL] S. R. Bell and E. Ligocka, A simplification and extension of Fefferman's theorem on biholomorphic mappings, *Invent. Math.* 57(1980), 283–289.

[BIN1] O. Besov, V. Ilin, and S. Nikol'skii, *Integral representations of functions and imbedding theorems*, Vol. I, Scripta Series in Mathematics, V. H. Winston & Sons, Washington, D.C.; Halsted Press [John Wiley & Sons], New York-Toronto, Ont.-London, 1978.

[BIN2] O. Besov, V. Ilin, and S. Nikol'skii, *Integral representations of functions and imbedding theorems*, Vol. II, Scripta Series in Mathematics, V. H. Winston & Sons, Washington, D.C.; Halsted Press [John Wiley & Sons], New York-Toronto, Ont.-London, 1979.

[BOC] S. Bochner, Orthogonal systems of analytic functions, *Math. Z.* 14(1922), 180–207.

[BOS] L. Boutet de Monvel and J. Sjöstrand, Sur la singularité des noyaux de Bergman et Szegö, *Soc. Mat. de France Asterisque* 34-35(1976), 123–164.

[CAR] C. Carathéodory, Untersuchungen über die konformen Abbildungen von festen und veränderlichen Gebieten, *Math. Ann.* 72(1912), 107–144.

[CHR] M. Christ, *Lectures on Singular Integral Operators*, CBMS Regional Conference Series in Mathematics, 77, Published for the Conference Board of the Mathematical Sciences, Washington, DC; by the American Mathematical Society, Providence, RI, 1990.

[COW] R. R. Coifman and G. Weiss, *Analyse harmonique non-commutative sur certains espaces homogénes*, (French) Étude de certaines intégrales singuliéres, Lecture Notes in Mathematics, Vol. 242, Springer-Verlag, Berlin-New York, 1971.

[DER] G. de Rham, *Varieties Differentiables*, 3rd ed., Hermann, Paris, 1973.

[FED] H. Federer, *Geometric Measure Theory*, Springer-Verlag, New York, 1969.

[FEF] C. Fefferman, The Bergman kernel and biholomorphic mappings of pseudoconvex domains, *Invent. Math.* 26(1974), 1–65.

[**GAM**]   T. Gamelin, *Uniform Algebras*, Prentic-Hall, Englewood Cliffs, NJ, 1969.

[**GAR**]   P. Garabedian, *Partial Differential Equations*, Wiley, New York, 1964.

[**GARN**] J. Garnett, *Bounded Analytic Functions*, Academic Press, New York, 1981.

[**GRL**]   H. Grauert and I. Lieb, Das Ramirezsche Integral und die Gleichung $\overline{\partial}u = \alpha$ im Bereich der beschränkten Formen, *Rice University Studies* 56(1970), 29–50.

[**GRK**]   R. E. Greene and S. G. Krantz, *Function Theory of One Complex Variable*, 2nd ed., American Mathematical Society, Providence, RI, 2002.

[**HAN**]   N. Hanges, Explicit formulas for the Szegö kernel for some domains in $\mathbf{C}^2$. *J. Funct. Anal.* 88 (1990), 153–165.

[**HAP1**] R. Harvey and J. Fundamental solutions in complex analysis I and II, *Duke Math. J.* 46(1979), 253–300 and 46(1979), 301–340.

[**HEN1**] G. M. Henkin, Integral representations of functions holomorphic in strictly pseudoconvex domains and some applications, *Mat. Sb.* 78(120)(1969), 611–632; *Math. U.S.S.R. Sb.* 7(1969), 597–616.

[**HEN2**] G. M. Henkin, Integral representations of functions holomorphic in strictly pseudoconvex domains and applications to the $\overline{\partial}$ problem, *Mat. Sb.* 82(124), 300–308 (1970); *Math. U.S.S.R. Sb.* 11(1970), 273–281.

[**HIL**]   E. Hille, E. Hille, *Analytic Function Theory*, 2nd ed., Ginn and Co., Boston, 1973.

[**JAK**]   S. Jakobsson, Weighted Bergman kernels and biharmonic Green functions, Ph.D. thesis, Lunds Universitet, 2000, 134 pages.

[**KAT**]   Y. Katznelson, *An Introduction to Harmonic Analysis*, Wiley, New York, 1968.

[**KES**]   N. Kerzman and E. M. Stein, The Szegő kernel in terms of Cauchy–Fantappiè kernels, *Duke Math. J.* 45(1978), 197–224.

[**KOP**]   W. Koppelman, The Cauchy integral formula for functions of several complex variables, *Bull. A.M.S.* 73(1967), 373–377.

[**KRA1**] S. G. Krantz, *Function Theory of Several Complex Variables*, 2nd ed., American Mathematical Society, Providence, RI, 2001.

[**KRA2**] S. G. Krantz, Canonical Kernels Versus Constructible Kernels, *Rocky Mt. Jour. Math.*, to appear.

[**KRA3**] S. G. Krantz, *Cornerstones of Geometric Function Theory: Explorations in Complex Analysis*, Birkhäuser Publishing, Boston, 2006.

[KRA4] S. G. Krantz, Canonical kernels versus constructible kernels, *Rocky Mountain Journal*, to appear.

[LIG] E. Ligocka, The Hölder continuity of the Bergman projection and proper holomorphic mappings, *Studia Math.* 80(1984), 89–107.

[PAI] P. Painlevé, Sur les lignes singulières des functions analytiques, *Thèse*, Gauthier-Villars, Paris, 1887.

[RAM] E. Ramirez de Arellano, Divisions problem in der komplexen analysis mit einer Anwendung auf Rand integral darstellung, *Math. Ann.* 184(1970), 172–187.

[RAN] R. M. Range, *Holomorphic Functions and Integral Representations in Several Complex Variables*, Springer-Verlag, Berlin, 1986.

[RUD] W. Rudin, *Principles of Mathematical Analysis*, 3rd ed., McGraw-Hill, New York, 1976.

[STE] E. M. Stein, *Boundary Behavior of Holomorphic Functions*, Princeton University Press, Princeton, NJ, 1972.

11

The Green's Function Method for the Riemann Mapping Theorem

Bingyuan Liu

CONTENTS

11.1 Preliminary

The Riemann mapping theorem (Theorem 11.2.1) is the most important and beautiful theorem in the field of one complex variable. It basically asserts that any two reasonable simply connected domains in $\mathbb{C}$ are biholomorphic. This is a non-trivial result. We all know, if two domains are simply connected, then they are homeomorphic, i.e., a bijective continuous mapping from the first domain to the second domain exists. Indeed, for two reasonably simply connected domains, we can even find a bijective smooth mapping between the two domains. This result is classical. However, from smooth mapping to holomorphic function is not trivial at all. After all, a holomorphic function is a harmonic mapping (i.e., both components are harmonic). The generic smooth mapping is far away from being harmonic. This is the point of the Riemann mapping theorem which states, a biholomorphism (i.e., a bijective holomorphic mapping of which the inverse is also holomorphic) can still be found after a careful treatment. This note requires the reader to have a little background on complex analysis and partial differential equations.

For the following, we give several useful and well-known theorems in complex variables. We also include a brief proof for some of them. For the details, readers are referred to standard textbooks of complex variables (see, e.g., Taylor [6]).

The first theorem is called the argument principle.

DOI: 10.1201/9781315160658-11

Theorem 11.1.1 (argument principle). *Let $C \subset \mathbb{C}$ be a closed contour, i.e., a simple closed Jordan curve. Let D be the region enclosed by C. If $f(z)$ is a holomorphic function on $\overline{D}$, and f has no zeros on C, then*

$$\frac{1}{2\pi \sqrt{-1}} \oint_C \frac{f'(z)}{f(z)} \, dz = N(0)$$

where $N(0)$ denotes the number of zeros inside the contour C, with multiplicities counted.

Here, we give a very brief proof. By change of variables $z \mapsto w := f(z)$, we have that

$$\frac{1}{2\pi \sqrt{-1}} \oint_c \frac{f'(z)}{f(z)} \, dz = \frac{1}{2\pi \sqrt{-1}} \oint_{f(C)} \frac{1}{w} \, dw = N(0),$$

where the last equation is due to a direct computation. This completes the proof of the argument principle.

We are not satisfied by knowing how many zeros are inside a contour C. Roughly speaking, for a fixed q in the image of f, we want to compute the number of z so that $f(z) = q$. For this, we observe $f(z) - q$ and obtain the following generalized argument principle.

Corollary 11.1.1 (generalized argument principle). *Let $C \subset \mathbb{C}$ be a closed contour, i.e., a simple closed Jordan curve. Let D be the region enclosed by C. Let $f(z)$ be a holomorphic function on $\overline{D}$. Fix $q \in f(D)$ and assume there is no $z \in C$ so that $f(z) = q$. Then*

$$\frac{1}{2\pi \sqrt{-1}} \oint_C \frac{f'(z)}{f(z) - q} \, dz = N(q)$$

where $N(q)$ denotes the number of z for which $f(z) = q$ inside the contour C, with multiplicities counted.

We observe that

$$\frac{1}{2\pi \sqrt{-1}} \oint_C \frac{f'(z)}{f(z) - q} \, dz \in \mathbb{Z}^+$$

for all holomorphic functions f. Observe that $\frac{f'(z)}{f(z)-q}$ is continuous on C if and only if $f(z) \neq q$ on C. Thus,

$$\frac{1}{2\pi \sqrt{-1}} \oint_C \frac{f'(z)}{f(z) - q} \, dz$$

is continuous. But a continuous function with image of integers is a constant function. This completes the proof. Moreover, the generalized argument principle says $N(q)$ remains constant on each component of $f(D) \backslash f(C)$ as long as no points in $f^{-1}(q)$ cross the boundary.

The generalized argument principle is quite useful for determining the bijectivity of a mapping. We will see it soon.

Theorem 11.1.2 (Strong maximal principle and Hopf lemma). *Let U be a bounded domain with smooth (C^∞) boundary and u be a non-constant harmonic function on $\overline{U}$. Then there exists a point $p \in \partial\Omega$ so that $u(p) = \max_{z\in\overline{U}} u(z)$ and $u(p) > u(z)$ for arbitrary $z \in U$. Moreover, at p, we have that $\frac{\partial u}{\partial n}(p) > 0$, where $\frac{\partial u}{\partial n}(p) > 0$ is the outward directional derivative at p.*

In the proof, we also need the following Schwarz reflection principle.

Theorem 11.1.3 (Schwarz reflection principle). *Assume that $f : U \to \mathbb{C}$, where U is an open subset of $\mathbb{C}$ and ∂U is real analytic, is a holomorphic function which extends continuously to ∂U. If f maps ∂U into another real analytic curve $M \subset \mathbb{C}$, then f extends holomorphically to an open neighborhood of $\overline{U}$.*

11.2 The Riemann Mapping Theorem and Its Proof

Theorem 11.2.1 (Riemann mapping theorem). *Let U be a non-empty simply connected open proper subset with real analytic boundaries ∂U of $\mathbb{C}$, then there exists a biholomorphic mapping f (i.e. a bijective holomorphic mapping whose inverse is also holomorphic) from U onto the open unit disk*

$$\mathbb{D} := \{z \in \mathbb{C} : |z| < 1\}.$$

For simplicity, we will call such a mapping a Riemann mapping.

In this note, we only consider bounded domains with real analytic boundaries.

To construct a Riemann mapping, we assume that

$$f(z) = (z - z_0)e^{u(z)+\sqrt{-1}\,v(z)},$$

for some $z_0 \in U$. Here u and v are real functions that will be defined later. Since the exponential function has no zeros, $f(z)$ will have its only zero at z_0. By the generalized argument principle, f must be bijective, because $N(q) = N(0)$ for an arbitrary $q \in f(U)$. We want to construct f with some constraints as follows.

1. The function f has to be holomorphic.

2. We hope the image of f to be the unit disc $\mathbb{D}$.

3. We need to verify that the f is a biholomorphism.

Among others, we look at condition 2. Since we hope the image of f is the unit disc $\mathbb{D}$, we hope f maps ∂U to $\partial \mathbb{D}$. For this aim, we hope $|f(z)| = 1$, whenever

$z \in \partial U$. It is equivalent to say that $|z - z_0|e^{u(z)} = 1$. We, hence, construct a function f as follows. We want to find a harmonic function u defined in $\overline{U}$ such that

$$u(z) = -\log|z - z_0|,$$

for $z \in \partial U$.

By solving the Dirichlet problem, one can find a harmonic function such that

$$u(z) = -\log|z - z_0|,$$

for $z \in \partial U$. Indeed, Green's kernel extends to a simply connected domain containing $\overline{U}$ by the Schwarz reflection principle, and so u is harmonic on a larger domain containing $\overline{U}$. After finding u, we need to find the conjugate harmonic function v. The v is defined as follows:

$$v = \int_{z_0}^{z} u_x \, dy - u_y \, dx. \tag{11.1}$$

Since U is simply connected and $u_{xx} - -u_{yy}$ by the harmonicity of u, one can see the definition is independent of the path from z_0 to z. The u and v clearly satisfy the following Cauchy–Riemann equation:

$$u_x = v_y \qquad \text{and} \qquad u_y = -v_x.$$

We will use Theorem 11.1.2 to verify $f|_{\partial U}$ is a diffeomorphism from ∂U onto $\partial \mathbb{D}$ and then we can verify by the argument principle, the condition 3, namely the biholomorphism from U onto $\mathbb{D}$.

On ∂U,

$$f(z) = (z - z_0)e^{-\log|z-z_0|+\sqrt{-1}\,v(z)} = \frac{z - z_0}{|z - z_0|}e^{\sqrt{-1}\,v(z)} = e^{\sqrt{-1}(\mathrm{Arg}(z-z_0)+v(z))}.$$

Thus, for arbitrary $z \in \partial U$, $|f(z)| = 1$, i.e., $f(z) \in \partial \mathbb{D}$.

We want to show that $e^{\sqrt{-1}\,\nu} := e^{\sqrt{-1}(\mathrm{Arg}(z-z_0)+v(z))}$ is a covering map from ∂U onto $\partial \mathbb{D}$. Observe that the conjugate harmonic function of ν is $-\mu :=$ $-\log|z - z_0| - u(z)$. Since, by the Cauchy–Riemann equation, the tangential directional derivative of ν is the same as the outward directional derivative of μ, we first verify $\frac{\partial \mu}{\partial n} \neq 0$ on ∂U. By the strong maximal principle and the Hopf lemma, one can see that $\frac{\partial \mu}{\partial n} > 0$ at all maximal points. Indeed, by $u|_{\partial U} = -\log|z-z_0||_{\partial U}$, we see that $\mu|_{\partial U} = 0$. And when $z \to z_0$, $\mu \to -\infty$. So it turns that all points on ∂U are maximal points, and hence, we can use the Hopf lemma on all ∂U. It implies $\frac{\partial \mu}{\partial n}(z) > 0$ for all $z \in \partial U$. Thus, the tangential directional derivative of ν is nonzero and $e^{\sqrt{-1}\,\nu}$ is a local diffeomorphism. So $e^{\sqrt{-1}\,\nu}$ is a covering map. Since $e^{\sqrt{-1}\,\nu}$ is a covering map, so does $f|_{\partial U}$.

At the last, we use the generalized argument principle to verify that f is a bijective holomorphic function from U onto $\mathbb{D}$. Indeed, $f|_{\partial U}$ maps ∂U to $\partial \mathbb{D}$ and thus, if $|q| < 1$, there is no $z \in \partial U$ so that $f(z) = q$. By inspecting

$$f(z) = (z - z_0)e^{u(z)+\sqrt{-1}\,v(z)},$$

one finds z_0 is the only zero of f, and it is simple, namely $N(0) = 1$. Consequently, $N(q) = 1$ for all $q \in \mathbb{D}$ by the discussion after Corollary 11.1.1. So f is a biholomorphism by Corollary 11.1.1.

11.3   An Example

We now bring the Riemann mapping theorem to some precise examples to see what these Riemann mappings look like. To get a precise computation, we need a bounded domain of smooth boundary with an explicit Green's function. So we mainly work on the unit disk because otherwise, the Green function is hard to obtain explicitly.

Example 11.3.1. Let U be the unit disc $\mathbb{D}$ and $z_0 \in \mathbb{D}$.

One can check that $u(z) = -\log|1 - \bar{z}_0 z|$ solves the following Dirichlet problem:

$$\begin{cases} \Delta u & = 0 \qquad z \in \mathbb{D} \\ u(z) & = -\log|z - z_0| \qquad z \in \partial \mathbb{D}. \end{cases}$$

Indeed, on $\partial \mathbb{D}$, we have that

$$\begin{aligned} &|1 - \bar{z}_0 z|^2 \\ =& 1 + |z|^2|z_0|^2 - 2\operatorname{Re} z_0 \bar{z} \\ =& |z|^2 + |z_0|^2 - 2\operatorname{Re} z_0 \bar{z} \\ =& |z - z_0|^2. \end{aligned}$$

It is also easy to check u is harmonic because the only singular point of $\log|1 - \bar{z}_0 z|$ is outside $\mathbb{D}$ and $\frac{\partial}{\partial z}\log|1 - \bar{z}_0 z|^2 = \frac{-\bar{z}_0(1 - z_0 \bar{z})}{|1 - z_0 \bar{z}|^2} = -\frac{\bar{z}_0}{1 - \bar{z}_0 z}$ which is holomorphic.

We are going to use u to look for v. Letting $z_0 = x_0 + \sqrt{-1}\,y_0$ and $z = x + \sqrt{-1}\,y$, we let $g(z) = 1 - \bar{z}_0 z$ and thus, $u(z) = -\log|g(z)|$. One can definitely use the knowledge of Calculus III (finding potential) to find the conjugate v of u. Here, we give another approach.

Observe that $g(z)$ maps the closed unit disk into the right half plane $\mathbb{K} = \{w \in \mathbb{C} : \operatorname{Re} w > 0\}$. The principal branch $\operatorname{Log}(w) = \log|w| + \sqrt{-1}\arg w$ of the logarithm with $-\frac{\pi}{2} < \arg w < \frac{\pi}{2}$ is holomorphic on $\mathbb{K}$. Therefore, the composite function $\operatorname{Log} g(z) = \log|g(z)| + \sqrt{-1}\arg g(z)$ is holomorphic on $\mathbb{D}$. Hence, $\arg g(z)$ is the harmonic conjugate for $\log|g(z)|$. Thus,

$$v = \arg(1 - z_0\bar{z}).$$

This gives us

$$f(z) = (z - z_0)e^{-\log|1 - z_0\bar{z}| + \sqrt{-1}\arg(1 - z_0\bar{z})} = (z - z_0)e^{\log\frac{1}{1 - \bar{z}_0 z}} = \frac{z - z_0}{1 - \bar{z}_0 z}.$$

One can see this is actually a Möbius transformation that maps z_0 to 0.

For the cases of unbounded domains and rough boundary, readers may consult the book Taylor [5] (Page 398-402). Other references include Conway [1] (Page 277), Fuchs [2] (Page 31-32), Needham [3] (Page 550-554) and Nevanlinna–Paatero [4] (Page 325-327).

Acknowledgments. The author thanks the anonymous referee for constructive suggestions.

References

[1] John B. Conway. *Functions of One Complex Variable I*. Second. Vol. 11. Graduate Texts in Mathematics. Springer New York, Apr. 7, 1995, pp. xiii+317. 336 pp. isbn: 0387903283.

[2] W. H. J. Fuchs. *Topics in the theory of functions of one complex variable*. Manuscript prepared with the collaboration of Alan Schumitsky. Van Nostrand Mathematical Studies, No. 12. D. Van Nostrand Co., Inc., Princeton, N.J.-Toronto, Ont.-London, 1967, pp. vi+193.

[3] Tristan Needham. *Visual complex analysis*. The Clarendon Press, Oxford University Press, New York, 1997, pp. xxiv+592. isbn: 0-19-853447-7.

[4] Rolf Nevanlinna and Veikko Paatero. *Introduction to complex analysis*. second. Translated from the 1965 German original by T. Kövari and G. S. Goodman. AMS Chelsea Publishing, Providence, RI, 2007, pp. x+350. isbn: 978-0-8218-4399-4.

[5] Michael E. Taylor. *Partial differential equations I. Basic theory*. Second. Vol. 115. Applied Mathematical Sciences. Springer, New York, 2011, pp. xxii+654. isbn: 978-1-4419-7054-1. doi: 10.1007/978-1-4419-7055-8.

[6] Michael E. Taylor. *Partial differential equations. I*. Vol. 115. Applied Mathematical Sciences. Basic theory. Springer-Verlag, New York, 1996, pp. xxiv+563. isbn: 0-387-94653-5. doi: 10. 1007/978-1-4684-9320-7.

12

Polynomial Trace Identities in $SL(2, \mathbf{C})$, Quaternion Algebras, and Two-Generator Kleinian Groups

---

T. H. Marshall and Gaven Martin

CONTENTS

DOI: 10.1201/9781315160658-12

12.1 Introduction

In his work on automorphic functions [6] Fricke shows that if $\Gamma \subset SL(2, \mathbb{C})$ is a subgroup, then the trace of any word in the generators is a polynomial with integral coefficients in the finitely many variables consisting of the traces of the generators of Γ together with finitely many of their products. The case Γ has two generators, $\Gamma = \langle A, B \rangle$ has been particulary well studied. There the trace of any word $w(A, B) \in \Gamma$ is a polynomial in the three complex variables x, y and z where

$$x = \text{tr}(A), \quad y = \text{tr}(B), \quad z = \text{tr}(AB). \tag{12.1}$$

As a consequence every conjugacy class of an element $w(A, B)$ in Γ uniquely determines a polynomial $P(x, y, z)$ with integral coefficients: define P by

$$P(x, y, z) = \text{tr}(w(A, B)).$$

Horowitz [17] showed that the polynomial P may not determine the conjugacy class of $w(A, B)$ uniquely, although, for a given polynomial, there are only finitely many conjugacy classes represented by cyclically reduced words $w(A, B)$ giving P. These results are largely based around the conjugacy invariance of trace and Fricke's simple identity

$$\text{tr}(AB) + \text{tr}(AB^{-1}) = \text{tr}(A) \cdot \text{tr}(B).$$

Traina [29, Corollary 1.] develops a family of trace identities to establish the following theorem.

Theorem 12.1.1 (Uniqueness) *Cyclically reduced words $w_1 = w_1(a, b)$ and $w_2 = w_2(a, b)$ can have the same trace polynomial only if the absolute values of the exponents of the generators of a in w_2 arise from those in w_1 by a permutation, and the same must be true for the exponents of b*

Our initial interest lies in understanding these polynomial trace identities further and their connection with discrete groups of Möbius transformations. The Möbius group acts as linear fractional transformations of the Riemann sphere;

$$PSL(2, \mathbb{C}) \ni \pm \begin{bmatrix} a & b \\ c & d \end{bmatrix} \leftrightarrow \frac{az + b}{cz + d} \in \text{Möb}(\hat{\mathbb{C}})$$

and through the Poincaré extension we identify Möb($\hat{\mathbb{C}}$) with $Isom^+(\mathbb{H}^3)$ the group of orientation preserving isometries of hyperbolic 3-space. A thorough discussion of these things can be found in Beardon's book [2]. The numbers x, y, z defined above at (12.1) are not well defined in $PSL(2, \mathbb{C})$, and so we first identify new parameters.

Given two matrices $A, B \in SL(2, \mathbb{C})$ we define the parameters

$$\gamma(A, B) = \text{tr}[A, B] - 2, \quad \beta(A) = \text{tr}^2(A) - 4, \quad \beta(B) = \text{tr}^2(B) - 4. \tag{12.2}$$

Here $[A, B] = ABA^{-1}B^{-1}$ is the multiplicative commutator. These parameters depend only on the conjugacy class of $\langle A, B \rangle$ and are well defined in the projective group $PSL(2, \mathbb{C})$. They determine the group $\langle A, B \mid \cdots \rangle$ uniquely up to conjugacy if $\gamma(A, B) \neq 0$, [12].

Note that $\gamma(A, B)$ is unchanged by Nielson moves (automorphisms of the free group of rank 2) on the generating pair $\{A, B\}$, so for instance

$$\gamma(A, B) = \gamma(B, A) = \gamma(A, A^m B^{-1} A^n)$$

and so forth. The parameters are set up so that if $\langle A, B \rangle = \mathbf{I}$, the trivial group, then $(\gamma, \beta, \tilde{\beta}) = 0 \in \mathbb{C}^3$.

In this article we are primarily interested in a special family of words, called *good words*, and the family of trace polynomials they generate. These words are defined in Section 12.5 below. This remarkable family $\mathcal{W}$ has the following properties reminiscent of the Chebyshev polynomials:

1. **[Semigroup structure]** $\mathcal{W}$ forms a semigroup under the operation

$$w_1(a, b) * w_2(a, b) = w_1(a, w_2(a, b)) \tag{12.3}$$

2. **[Polynomials and composition]** For each $w(a, b) \in \mathcal{W}$ there is an associated monic polynomial with integer coefficients in *two* complex variables $P_w(\gamma, \beta)$. These polynomials have the property that if $w_1(a, b), w_2(a, b) \in \mathcal{W}$, then

$$P_{w_1 * w_2}(\gamma, \beta) = P_{w_1}(P_{w_2}(\gamma, \beta), \beta) \tag{12.4}$$

That is the semigroup operation above induces polynomial composition.

3. **[Commutators and bounded roots]** Given a representation of

$$\Gamma = \langle a, b \mid b^2 = 1 \rangle$$

into $SL(2, \mathbb{C})$, $a \mapsto A$, $b \mapsto B$, set $\gamma = \gamma(A, B)$ and $\beta = \beta(A)$. Then for $w = w(a, b) \in \Gamma$, w is a good word and we have

(a) Commutator independence from the third complex variable $\beta(B)$,

$$\gamma(A, w(A, B)) = P_w(\gamma, \beta)$$

(b) Suppose A is not an irrational rotation, equivalently

$$\beta \notin \{-4\sin^2(r\pi) : r \in \mathbb{R} \setminus \mathbb{Q}\},$$

and denote the zero set of the polynomials by

$$\mathcal{Z}_\beta = \{z \in \mathbb{C} : P_w(z, \beta) = 0 \text{ for some } w \in \Gamma\}$$

Then $\overline{\mathcal{Z}}$ is compact and

$$\mathbb{C} \setminus \overline{\mathcal{Z}_\beta} = \mathcal{R}_\beta$$

where $\mathcal{R}_\beta$ is nonempty, unbounded and conformally equivalent to the punctured disk.

(c) The group $\langle A, B \rangle$ is discrete and free on generators

$$\langle A, B \rangle \cong \langle A \rangle * \langle B \rangle$$

if and only if $\gamma(A, B) \in \overline{\mathcal{R}_\beta}$.

Here we will prove the density of the roots of good word polynomials in the exterior of the moduli space of discrete and faithful representations of Γ, that is 3 (b). The hard part of 3 (c) concerns the structure of the boundary, and the only proof we have relies on some very deep results concerning the geometry of discrete groups, such as the density and the ending lamination theorems, see [26]. This is because $\mathcal{R}_\beta$ can be identified with the (moduli) space of discrete and faithful geometrically finite representations of $\mathbb{Z} * \mathbb{Z}_2$, with the generator of $\mathbb{Z} = \langle A \rangle$ and $\beta(A) = \beta$. The "pants" decomposition of a geometrically finite Riemann surface with fundamental group $\mathbb{Z} * \mathbb{Z}_2$ or $\mathbb{Z}_p * \mathbb{Z}_2$ shows that $\mathcal{R}_\beta$ is topologically a punctured disk. These are obtained from the disk with two cone points of order two glued along its boundary to a disk with two holes (or punctures). In the case $\beta = 0$, $\overline{\mathcal{R}}$ is known in the literature as the Riley slice and the boundary $\partial\mathcal{R}$ is a topological circle [1]. This is expected to persist for all other $\beta \in \mathbb{C}$ as well. The geometrically infinite faithful representations lie in the continuum (topological circle ?) $\partial\mathcal{R}$.

The complement $\mathbb{C} \setminus \overline{\mathcal{R}_\beta}$ consists of nondiscrete groups apart from a countable discrete set in $\mathcal{R}_\beta$ of points which are the roots of polynomials corresponding to relators in groups that are discrete but not splitting. There are some conjectures about the structure of the polynomials and the words they come from - basically that they are associated with Dehn surgeries on two bridge knots and links and associated Hecke groups (obtained by adding an unknotting tunnel). The cusp points on the boundary arise from pinching a geodesic (arising from a Farey word) of the Riemann surface with fundamental group $\mathbb{Z} * \mathbb{Z}_2$ as in [19], giving a ray in the unbounded region ending on $\partial\mathcal{R}$, while from the bounded region $\mathbb{C} \setminus \partial\mathcal{R}$ cusps are associated with Dehn surgery limits (via Thurston's Dehn Surgery theorem) from the inside.

Indeed it is the strong connection between these representation spaces of discrete groups, low dimensional hyperbolic geometry and topology and the good word polynomials that motivates our consideration of them. The geometry of commutators plays an important role in understanding the geometry and topology of discrete groups and their associated quotients, hyperbolic 3-manifolds and 3-orbifolds. For instance if $A, B \in \Gamma$ where Γ is a discrete subgroup of $SL(2, \mathbb{C})$, we put $\beta = \beta(A)$ and then suppress it writing $P_w(\gamma)$ for $P_w(\gamma, \beta)$ to find that $\{P_w(\gamma) : w \in \mathcal{W}\}$ is a collection of traces of commutators in Γ. Further, if $w \in \mathcal{W}$, then the semigroup operation gives $w * w * \cdots * w \in \mathcal{W}$ and

$$P_w(\gamma), P_{w*w}(\gamma) = P_w(P_w(\gamma)), \ldots, P_{w*w*\cdots*w}(\gamma) = P_w^{on}(\gamma)$$

gives a sequence of commutator traces from the holomorphic dynamical system given by iteration of the polynomial P_w. As perhaps the simplest nontrivial

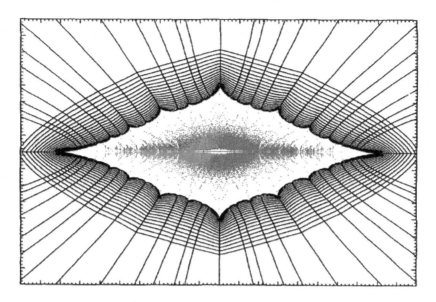

FIGURE 12.1
The Riley slice (shaded region) is symmetric about the origin and its complement meets the real line in the interval $[-4, 4]$ and the imaginary axis in $[-2, 2]$. From David Wright. In [21] it is shown that the exterior - the closure of the complement of the Riley slice - is the Julia set of the good word semigroup for groups generated by a parabolic and an elliptic of order two. Also illustrated are the first few thousand roots of the good word polynomials.

example, with $w(a, b) = bab^{-1}$, we have $P_w(\gamma, \beta) = \gamma(\gamma - \beta)$. If $\beta = 0$, then we see $\gamma, \gamma^2, \ldots, \gamma^n \ldots$ is a sequence of commutator traces. If $0 < |\gamma| < 1$, then the sequence $\{\gamma^n\}_{n \geq 0}$ accumulates on 0. It is not a particularly difficult exercise to show this can't happen in a discrete group, and we therefore obtain the classical Shimitzu-Leutbecher inequality.

Theorem 12.1.2 (Shimitzu-Leutbecher inequality) *If* $\langle A, B \rangle \subset SL(2, \mathbb{C})$ *is discrete and A is parabolic* $(\beta = 0)$, *then* $|\gamma| = |\mathrm{tr}[A, B] - 2| \geq 1$.

Jørgensen's inequality [18] follows in the same way if $|\beta| < 1$ for then 0 is an attracting fixed point for the iterates of P_w and the disk $\mathbb{D}(0, 1 - |\beta|)$ lies in the Fatou set so $\gamma \notin \mathbb{D}(0, 1 - |\beta|)$ and so $|\gamma| + |\beta| \geq 1$. We will give other examples later.

In order to fully exploit these polynomials in low dimensional topology and geometry, it is crucial to understand more about them and develop a systematic approach to uncovering the inequalities and regions of moduli space where their roots lie. For instance, to understand and extend the important

$\frac{1}{2}\log 3$ theorem of Gabai, Meyerhoff and Thurston [8], used to prove the topological rigidity of hyperbolic three manifolds [7], an ad hoc approach required rigorous estimates on the computation of 100+ matrix multiplications—these words were called *killer words* as they removed small regions of moduli space using discreteness criteria, such as Jørgensen's inequality or other criteria, such as contradicting a choice of shortest geodesic. An approach based on good words is far simpler since estimates are required for the roots of a polynomial equation with integer coefficients of lesser degree. Such searches have been used to resolve a number of problems such as:

1. The unique minimal volume 3-orbifold (co-volume lattice of hyperbolic isometries) identified as the arithmetic Coxeter reflection group 3-5-3, extended by the order two symmetry induced from the diagram [14, 22].

2. Structure of the singular set. Tables 6-10 of [9, 14] give sharp bounds for the distance between components of the singular set of a hyperbolic 3-orbifold and the distance between tetraheral, octahedral and icosahedral points in a Kleinian group.

3. Automorphism groups of 3-manifolds and 3-dimensional Hurwitz groups. Sharp bounds for the order of the automorphism group of a hyperbolic 3-manifold group in terms of the volume and analogous to the $84(g-1)$ Theorem of Hurwitz [5]

4. Margulis constant. The Margulis constant is achieved in a two- or three- generator group, the case of two generator groups is completely resolved, [13] and the only remaining case concerns Kleinian groups generated by three elements of order two.

5. Geodesic length spectrum of 3-folds. Inequalities are used to find bounds on the length of intersecting closed geodesics, or non-simple geodesics, which are within a factor of 2 of being sharp. These together with estimates on the Margulis constant, yield good bounds for the thick and thin decompositions of hyperbolic 3-manifolds.

6. Explicit examples of small volume hyperbolic 3-manifolds and 3-orbifolds with various geometric properties including a sequence of orbifolds with torsion of order q interpolating between the smallest volume cusped orbifold ($q = 6$) and the smallest volume limit orbifold $q \to \infty$, hyperbolic 3-manifolds with automorphism groups with large orders in relation to volume and in arithmetic progression, and the smallest volume hyperbolic manifolds with totally geodesic surfaces in [4].

In this paper we uncover a group $\mathcal{V}$ of elements of unit norm in a quaternion algebra $\mathcal{Q}$ with associated indeterminates, which maps under an "evaluation homomorphism" $\rho : \mathcal{V} \to PSL(2, \mathbb{C})$ to a group which includes these good words on two generators. Further $\mathcal{V}$ naturally extends to a larger group which gives a corresponding extension of the isometry group $\rho(\mathcal{V})$. Roughly, polynomials R, S, T and W in the indeterminates u and v form a "quaternion" $(R, S, T, W) \in \mathcal{Q}$, which has norm 1 when

$$R^2 - (u^2 - 1)S^2 - (v^2 - 1)T^2 + (u^2 - 1)(v^2 - 1)W^2 = 1 \qquad (12.5)$$

A special case of interest occurs when u or v is ± 1, in which case this equation reduces to the polynomial Pell equation,

$$P^2(x) - (x^2 - 1)Q^2(x) = 1, \qquad (12.6)$$

An obvious similarity between the two equations is that the solution sets have a natural group structure; this is what we will exploit to begin to understand the structure of good words. However, there are significant differences. For instance while the solutions $P(x)$, $Q(x)$ of the polynomial Pell equation must have integer coefficients, there are members of $\mathcal{V}$ whose polynomials have coefficients which need not even be rational (see Section 12.8.2 for an example). We also note that (12.5) has some solutions with strictly complex coefficients: a simple example is $R(u, v) = uv$, $S(u, v) = T(u, v) = 1$, $W(u, v) = i$. However, when we confine ourselves to solutions with rational coefficients, some remarkable properties emerge; in particular, it turns out all that such solutions actually have half-integer coefficients Section 8.

In order to study these word polynomials more fully, as well as justify the sorts of results we are seeking, we need to develop a few ideas from hyperbolic geometry and, in particular, from the geometry of discrete groups of hyperbolic isometries of hyperbolic 3-space.

12.2   Background in Hyperbolic Geometry

Let $\mathrm{Isom}^+(\mathbb{H}^3)$ be the group of orientation preserving isometries of $\mathbb{H}^3$, hyperbolic 3-space,

$$\mathbb{H}^3 = \{x = (x_1, x_2, x_3) \in \mathbb{R}^3, x_3 > 0\}, \text{ with metric } ds = \frac{|dx|}{x_3}$$

of constant negative curvature equal to -1.

We briefly review some well-known facts about the group $\mathrm{Isom}^+(\mathbb{H}^3)$; see e.g. [2, 12] or [23] for more details.

Each $f \in \mathrm{Isom}^+(\mathbb{H}^3)$ is the Poincaré extension of a Möbius transformation of the boundary $\partial\mathbb{H}^3$ which we identify as $\hat{\mathbb{C}} = \mathbb{C} \cup \{\infty\}$, the Riemann sphere. Hence there is a natural isomorphism between $\mathrm{Isom}^+(\mathbb{H}^3)$ and $PSL(2, \mathbb{C})$.

Using the definition at (12.2) we can thus define the trace and β and γ parameters for isometries $f, g \in Isom^+(\mathbb{H}^3)$, simply by setting $\mathrm{tr}(f) = \mathrm{tr}(A)$, $\beta(f) = \beta(A)$ and $\gamma(f,g) = \gamma(A,B)$, where $A, B \in PSL(2,\mathbb{C})$ represent f and g respectively.

Each non-identity $f \in \mathrm{Isom}^+(\mathbb{H}^3)$ has either one or two fixed-points on the boundary $\hat{\mathbb{C}}$. If there is just one, then f is called *parabolic*; if there are two, then we define the *axis* of f, $\mathrm{ax}(f)$ to be the hyperbolic geodesic line joining them. Now f leaves $\mathrm{ax}(f)$ invariant, and its action on this geodesic is a translation along by a distance $\tau = \tau(f) \geq 0$, the *translation length* of f, together with a rotation through an angle $\eta = \eta(f)$, the *holonomy* of f around $\mathrm{ax}(f)$. If $\tau(f) > 0$, then $\eta \in (-\pi, \pi]$, is taken anticlockwise around $\mathrm{ax}(f)$, as determined by the direction of the translation of $\mathrm{ax}(f)$ performed by f, and the right-hand rule; in this case f is called *loxodromic*. If $\tau(f) = 0$, that is if f fixes $\mathrm{ax}(f)$ pointwise, then f is called *elliptic*, in which case the distinction between clockwise and anticlockwise disappears, and we may assume that $\eta \geq 0$, that is $\eta \in (0, \pi]$.

When f is elliptic or loxodromic the parameters $\tau(f)$ and $\eta(f)$ together determine f up to conjugacy.

When f is parabolic or the identity, we set $\tau(f) = \eta(f) = 0$.

The following lemma classifies the isometries in $\mathrm{Isom}^+(\mathbb{H}^3)$ up to conjugacy, and identifies, for each isometry, the conjugations which leave it unchanged.

Lemma 12.2.1 *A non-identity isometry* $f \in \mathrm{Isom}^+(\mathbb{H}^3)$ *is conjugate to* $z+1$ *if* f *is parabolic, and otherwise to a unique isometry of the form* $f(z) = re^{i\theta}z$, *where* $r = e^{\tau(f)} \geq 1$, $-\pi < \theta \leq \pi$ *if* $r > 1$, *and* $0 \leq \theta \leq \pi$ *if* $r = 1$.

If $gfg^{-1} = f$, *then either* g *is the identity,* g *and* f *have exactly the same fixed points on* $\hat{\mathbb{C}}$, *or,* f *is an elliptic of order 2, and* g *is an elliptic of order 2 which interchanges the endpoints of* $\mathrm{ax}(f)$.

As previously remarked, both parameters $\beta(f)$ and $\gamma(f,g)$ are invariant under conjugacy. Conversely, if $\beta(f) \neq 0$, then $\beta(f)$ determines f up to congugacy, and if $\gamma(f,g) \neq 0$, then $\beta(f)$, $\beta(g)$ and $\gamma(f,g)$ together determine the group $\langle f, g \rangle$ up to conjugacy [12]. We prove this result in Theorem 12.4.1 below by identifying a canonical representation.

Both the parameters $\gamma(f,g)$ and $\beta(f)$ encode geometric information. For instance:

$$\beta(f) = 4\sinh^2\left(\frac{\tau + i\eta}{2}\right), \tag{12.7}$$

and, when f is elliptic or loxodromic,

$$\gamma(f,g) = \frac{1}{4}\beta(f)\beta(g)\sinh^2(\Delta), \tag{12.8}$$

where $\Delta = \Delta(\mathrm{ax}(f), \mathrm{ax}(gfg^{-1}))$ represents the *complex distance* between

$\mathrm{ax}(f)$ and $\mathrm{ax}(gfg^{-1})$ (the imaginary part of this distance, which represents the angle between the two axes, is defined modulo π, so the right-hand side of (12.8) is well defined). It is an elementary fact (see e.g. [2] or Theorem 12.4.1 below) that $\gamma(f,g) = 0$ if and only if f and g share a fixed point on the boundary $\hat{\mathbb{C}}$ of $\mathbb{H}^3$; indeed for non-parabolic f and g, this follows immediately from (12.8), Δ being 0 when the axes of f and gfg^{-1} either meet at a point of $\hat{\mathbb{C}}$ or coincide. In applications we often want to distinguish between these two cases. We develop an algebraic test in Section 12.6.

12.3   Matrix Identities

We collect some matrix identities for later use. Let

$$M = \begin{bmatrix} k & m \\ 0 & k^{-1} \end{bmatrix}, \; P = \begin{bmatrix} 1 & 1 \\ 0 & 1 \end{bmatrix},$$

$$Q = \begin{bmatrix} 0 & i\sqrt{k} \\ i/\sqrt{k} & 0 \end{bmatrix}, \; N = \begin{bmatrix} a & b \\ c & d \end{bmatrix}, \tag{12.9}$$

where $ad - bc = 1$. Then

$$QNQ^{-1} = \begin{bmatrix} d & kc \\ b/k & a \end{bmatrix}, \tag{12.10}$$

$$MNM^{-1} = \begin{bmatrix} a + k^{-1}mc & -m^2c + mk(d-a) + k^2b \\ k^{-2}c & d - k^{-1}mc \end{bmatrix}, \tag{12.11}$$

and when $m = 0$

$$[M,N] = \begin{bmatrix} ad - k^2bc & ab(k^2 - 1) \\ cd(k^{-2} - 1) & ad - k^{-2}bc, \end{bmatrix} \tag{12.12}$$

$$[P,N] = \begin{bmatrix} 1 + c^2 + ac & 1 - a^2 - ac \\ c^2 & 1 - ac \end{bmatrix}. \tag{12.13}$$

In particular we have the useful trace identities, when $m = 0$

$$\mathrm{tr}[M,N] = 2 - (k - k^{-1})^2 bc \tag{12.14}$$

$$\mathrm{tr}[P,N] = 2 + c^2 \tag{12.15}$$

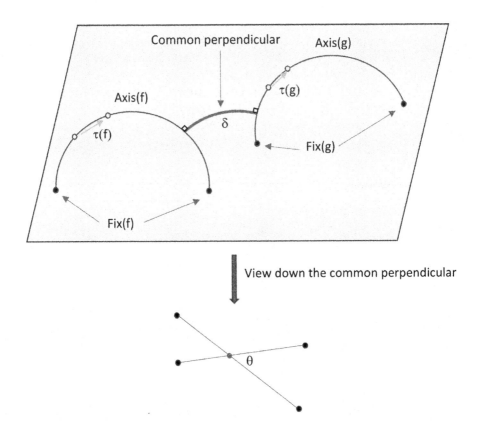

FIGURE 12.2

Illustrated are two loxodromic elements f and g. The axes are the hyperbolic lines connecting the fixed points in $\hat{\mathbb{C}}$. Each of f and g has a translation length—the distance a point is moved along the axes—denoted $\tau(f)$ and $\tau(g)$. The common perpendicular has hyperbolic length δ. When viewed along this common perpendicular, the two axes form an angle θ. All of these quantities are related by the formula.

$$\frac{4\gamma(f,g)}{\beta(f)\beta(g)} = \sinh^2(\delta + i\theta) \tag{3}$$

12.4   Two-Generator Groups

We now classify up to conjugacy all two-generator subgroup of $\mathrm{Isom}^+(\mathbb{H}^3)$, by finding a canonical representative for each conjugacy class. Throughout we always use the principle values of square roots.

Theorem 12.4.1 *Every group generated by two non-identity isometries in* $\text{Isom}^+(\mathbb{H}^3)$ *is conjugate to a group of the form* $\langle f, g \rangle$, *where* f *and* g *have matrix representatives* A *and* B *respectively in* $PSL(2, \mathbb{C})$ *such that either:*

Case 1.

$$A = \begin{bmatrix} \frac{\sqrt{\beta(f)(\beta(f)+4)}+\beta(f)}{2\sqrt{\beta(f)}} & 0 \\ 0 & \frac{\sqrt{\beta(f)(\beta(f)+4)}-\beta(f)}{2\sqrt{\beta(f)}} \end{bmatrix}, \quad B = \begin{bmatrix} a & b \\ c & d \end{bmatrix} \quad (12.16)$$

$$\begin{aligned} a &= \frac{1}{2}\left(\sqrt{\beta(g)+4} + \sqrt{\frac{4\gamma(f,g)+\beta(f)\beta(g)}{\beta(f)}}\right) \\ d &= \frac{1}{2}\left(\sqrt{\beta(g)+4} - \sqrt{\frac{4\gamma(f,g)+\beta(f)\beta(g)}{\beta(f)}}\right), \end{aligned} \quad (12.17)$$

and

$$c = -b = \sqrt{\frac{\gamma(f,g)}{\beta(f)}} \quad (12.18)$$

when $\gamma(f,g) \neq 0$, *and either*

$$b = 0, c = 1 \quad or \quad b = 1, c = 0 \quad or \quad b = c = 0 \quad (12.19)$$

when $\gamma(f,g) = 0$; *or*

Case 2.

$$A = \begin{bmatrix} 1 & 1 \\ 0 & 1 \end{bmatrix}, \quad B = \begin{bmatrix} 0 & -1/\sqrt{\gamma(f,g)} \\ \sqrt{\gamma(f,g)} & \sqrt{\beta(g)+4} \end{bmatrix}, \quad (12.20)$$

or

Case 3.

$$\begin{aligned} A &= \begin{bmatrix} 1 & 1 \\ 0 & 1 \end{bmatrix}, \\ B &= \begin{bmatrix} \frac{1}{2}[\sqrt{\beta(g)+4}+\sqrt{\beta(g)}] & \ell \\ 0 & \frac{1}{2}[\sqrt{\beta(g)+4}-\sqrt{\beta(g)}] \end{bmatrix} \quad (12.21) \end{aligned}$$

where $\ell = 0$ *when* $\beta(g) \neq 0$, *and can take any complex value when* $\beta(g) = 0$.

The three cases are respectively the cases $\beta(f) \neq 0$ (f non-parabolic), $\beta(f) = 0$ and $\gamma(f,g) \neq 0$ and $\beta(f) = \gamma(f,g) = 0$ (f parabolic).

Proof. We set $\beta = \beta(f)$, $\gamma = \gamma(f, g)$. Suppose first that f is loxodromic or elliptic ($\beta \neq 0$). By Lemma 12.2.1 we can conjugate f so that its matrix representative A is diagonal.

By (12.14) we have that $\gamma = -bc\beta$, whence

$$bc = \frac{-\gamma}{\beta}, \quad ad = 1 - \frac{\gamma}{\beta} \tag{12.22}$$

We have $(a + d)^2 - 4 = \beta(g)$, and since B is determined only up to sign, we may thus assume that $a + d = \sqrt{\beta(g) + 4}$. Together with (12.22), this gives that a and d are either as given by (12.17) or are obtained from these by interchanging the values of a and d. Using (12.10), we may then conjugate A and B if necessary, to interchange a and d, so that (12.17) holds, and A is still diagonal.

Let r and s be the diagonal entries of A. We have $(r + s)^2 - 4 = \beta$, and since A is determined only up to sign, we may assume that $r + s = \sqrt{\beta(\beta + 4)}/\sqrt{\beta}$. Together with the condition $rs = 1$, this gives that either A or A^{-1} takes the form given by (12.16). Since $\langle A^{-1}, B \rangle$ and $\langle A, B \rangle$ are the same group, we may assume that A satisfies (12.16).

Finally, we apply a conjugacy of the type (12.11) (with $m = 0$) to A and B to adjust the values of b and c, leaving A unchanged. If $b, c \neq 0$ (i.e. when $\gamma \neq 0$), we can use such a conjugacy to give b are c any values subject to (12.22); in particular, we can make (12.18) hold. If exactly one of the values of b and c is nonzero, then we conjugate to make it 1. The only other possibility is $b = c = 0$, so the options given in (12.18) and (12.19) are exhaustive.

Now we suppose that f is parabolic. Using Lemma 12.2.1 we conjugate so that $f(z) = z + 1$, so that its matrix representative $A = \begin{bmatrix} 1 & 1 \\ 0 & 1 \end{bmatrix}$. By (12.15) $c^2 = \gamma$, and we may assume, since B is determined only up to sign, that $c = \sqrt{\gamma}$.

Now we have two subcases, determined by whether or not $\gamma = 0$. If $\gamma \neq 0$, then $c \neq 0$, and we apply the conjugation (12.11) with $k = 1$ and $m = -a/c$ to A and B, leaving A the same, and changing a to 0, whence

$$B = \begin{bmatrix} 0 & -1/\sqrt{\gamma} \\ \sqrt{\gamma} & \pm\sqrt{\beta(g) + 4} \end{bmatrix}. \tag{12.23}$$

Let $B_{(+)}$ and $B_{(-)}$ be the matrices obtained by taking the $+$ and $-$ signs respectively in (12.23). We show that $\langle A, B_{(+)} \rangle$ and $\langle A, B_{(-)} \rangle$ are conjugate groups in $PSL(2, \mathbb{C})$. A conjugation of the form (12.11) with $k = i$ and $m = 0$ takes A to A^{-1} and $B_{(-)}$ to $-B_{(+)}$. Thus $\langle A, B_{(-)} \rangle$ is conjugate to $\langle A^{-1}, -B_{(+)} \rangle = \langle A, B_{(+)} \rangle$ in $PSL(2, \mathbb{C})$ as required. Thus, without loss of generality, we take the $+$ sign in (12.23).

If $\gamma = 0$, then $c = 0$ so $ad = 1$. As in previous cases, we may assume that $a + d = \sqrt{\beta(g) + 4}$, so that B must take the form (12.21), up to an interchange of the diagonal entries. If $\beta(g) = 0$, then these entries are the same, and we are done. Otherwise $a \neq d$, and we can apply a further conjugation of the form (12.11), with $k = 1$, so as to get both $\ell = 0$ and to leave A unchanged. Now B is diagonal, and interchanging a and d replaces B by B^{-1}. Since this operation leaves the group $\langle A, B \rangle$ unchanged, we may assume that B is given by (12.21). $\qquad\square$

Remarks. We can characterize geometrically the four ways of assigning values to b and c given by (12.18) and (12.19). In (12.18) $\gamma \neq 0$, so $b, c \neq 0$ and f and g have no common fixed points in $\hat{\mathbb{C}}$. In this case, as remarked in the proof, b and c can be made to take any values whose product is $-\gamma/\beta$, and the exact choice is rather arbitrary. However, the normalization that we have chosen is quite natural from a geometric viewpoint; it makes the fixed points of g mutually reciprocal, and (consequently), when g is non-parabolic, the common perpendicular of $\mathrm{ax}(f)$ and $\mathrm{ax}(g)$ is the geodesic with endpoints ± 1. When g is parabolic, the fixed point is $z = 1$. See [22] for more details.

The first two cases of (12.19), when $\{b, c\} = \{0, 1\}$ occur when f and g have a single common fixed point in $\hat{\mathbb{C}}$. If f is loxodromic, then this point is repulsive when $b = 0$, $c = 1$, and attractive when $b = 1$, $c = 0$. If f is elliptic, then f rotates $\mathbb{H}^3$ anticlockwise (resp. clockwise) around $\mathrm{ax}(f)$ oriented away from the shared fixed point when $b = 0$, $c = 1$ (resp. $b = 1$, $c = 0$). (When the elliptic f is order two, these two cases are conjugate.) Finally $b = c = 0$ when f and g are both elliptic or loxodromic and have the same axis.

Note that, when $\gamma = 0$, although $\beta(f)$, $\beta(g)$ and γ do not determine the conjugacy class of $\langle f, g \rangle$, when $\beta(f) \neq 0$ (and symmetrically when $\beta(g) \neq 0$) then there are only three possibilities. Only when $\beta(f) = \beta(g) = \gamma = 0$ (Case 3 of the theorem with $\beta(g) = 0$) do the same parameters give an infinite family of non-conjugate groups.

12.5   Good Words

A *good word* on the letters a and b is a word of the form

$$w(a, b) = b^{s_1} a^{r_1} b^{s_2} a^{r_2} \dots b^{s_{m-1}} a^{r_{m-1}} b^{s_m}$$

where $s_1 = \pm 1$, $s_j = (-1)^{j+1} s_1$, and the r_j take integer values.

Thus the powers of b in a good word alternate in sign. By setting $r_1 = 0$ (resp. $r_{m-1} = 0$), we obtain a good word which begins (resp. ends) with a power of a.

A good word is *even* if $r_1 + r_2 + \dots r_{m-1}$ is even, *odd* otherwise, *balanced* if m is even, *unbalanced* otherwise, and *regular* if $s_1 = 1$, *irregular* otherwise. If $r_j = 0$ for any $1 < j < m - 1$, then $w(a, b)$ collapses into a shorter good word (which has the same balance, parity and regularity as the original word), so we may assume that these interior powers are non-zero.

The following easy observation is quite useful. We leave the proof to the reader.

Theorem 12.5.1 *The regular balanced words in* $\langle a, b \rangle$, *say* Γ_{reg}, *comprise a subgroup of the free group on* a *and* b, *of which the regular balanced even words form an index-two subgroup.*

Lemma 12.5.1 *The group of regular balanced even words, say* Γ_{reg}^{even}, *on* a *and* b *is generated by* a^2, ba^2b^{-1} *and* $[b, a] = bab^{-1}a^{-1}$.

Proof. Let $w = ba^{r_1}b^{-1}a^{r_2} \dots ba^{r_{2m-1}}b^{-1}$ be regular, balanced and even. We use induction on m. We have

$$ba^i b^{-1} a^j = (ba^2 b^{-1})^{i/2} (a^2)^{j/2} \tag{12.24}$$

when i and j are both even, and

$$ba^i b^{-1} a^j = (ba^2 b^{-1})^{(i-1)/2} [b, a] (a^2)^{(j+1)/2} \tag{12.25}$$

when i and j are both odd. This deals with the case $m = 1$ and the induction step when r_1 and r_2 have the same parity.

If r_1 and r_2 have the opposite parity, then $m > 1$, and we use the same identities together with

$$ba^{r_1} b^{-1} a^{r_2} ba^{r_3} b^{-1} a^{r_4} = (ba^{r_1} b^{-1} a^{r_2-1})[b, a]^{-1} (ba^{r_3+1} b^{-1} a^{r_4})$$

This completes the proof.   □

The next corollary is also immediate.

Corollary 12.5.1 *Suppose that* a *has order three,* $a^3 = 1$. *Then the group of regular balanced even words on* a *and* b *is a two generator group generated by* a *and* bab^{-1}.

We recall here the well known identity

$$[b, a] \quad = \quad (ba)^2 (a^{-1}b^{-1}a)^2 (a^{-1})^2$$

This tells us that the regular balanced words in $\Gamma = \langle a, b \rangle$ lie in the group $\Gamma^{(2)}$ generated by squares of elements.

Corollary 12.5.2 *The group of regular balanced even words on* a *and* b *lies in the group generated by the four squares*

$$\Gamma_{reg}^{even} < \langle a^2, (bab^{-1})^2, (ba)^2, (aba^{-1})^2 \rangle$$

The remark following the next result, Theorem 12.5.2, shows that the index between these two groups is infinite. In fact, for any representation in $SL(2,\mathbb{C})$, the trace fields are $\mathbb{Q}(\mathrm{tr}\Gamma_{reg}^{even}) = \mathbb{Q}(\beta,\gamma)$ and $\mathbb{Q}(\beta,\gamma,\beta(ba))$, using our earlier notation.

We can now state our first main theorem.

Theorem 12.5.2 *Let* $w = w(a,b) = ba^{r_1}b^{-1}a^{r_2}\ldots ba^{r_{2m-1}}b^{-1}$ *be a regular balanced even word, then there are polynomials* r_w, s_w, t_w, w_w, *such that*

$$2r_w, 2s_w, 2t_w, 2w_w, r_w - s_w, t_w - w_w \in \mathbb{Z}[x,z], \tag{12.26}$$

$r_w(0,0) = 1$ *and*

$$g_w(x,z) := \frac{s_w(x,z) - zw_w(x,z)}{x}$$

is also a polynomial, and if $f, g \in \mathrm{Isom}^+(\mathbb{H}^3)$ *are not the identity, have parameters* $\beta = \beta(f)$, $\beta' = \beta(g)$ *and* $\gamma = \gamma(f,g)$, *and if* A *and* B *are the matrices from Theorem 12.4.1 which represent (up to conjugacy)* f *and* g *respectively, then: for* f *non-parabolic* $(\beta \neq 0)$

$$w(A,B) = \begin{bmatrix} r_w(\beta,\gamma) + \frac{s_w(\beta,\gamma)Q}{\beta} & ab[\beta t_w(\beta,\gamma) + w_w(\beta,\gamma)Q] \\ cd[\beta t_w(\beta,\gamma) - w_w(\beta,\gamma)Q] & r_w(\beta,\gamma) - \frac{s_w(\beta,\gamma)Q}{\beta} \end{bmatrix} \tag{12.27}$$

where $Q = \sqrt{\beta(\beta+4)}$ *and* a, b, c, d *are as in (12.17) and (12.18); and for* f *parabolic* $(\beta = 0)$, *with* $\gamma \neq 0$

$$w(A,B) = \begin{bmatrix} r_w(0,\gamma) + \gamma t_w(0,\gamma) & 4g_w(0,\gamma) + 2w_w(0,\gamma) \\ 2\gamma w_w(0,\gamma) & r_w(0,\gamma) - \gamma t_w(0,\gamma) \end{bmatrix}, \tag{12.28}$$

and for f *parabolic, with* $\gamma = 0$

$$w(A,B) = \begin{bmatrix} 1 & 4g_w(0,0) - \left(\beta' + \sqrt{\beta'}\sqrt{\beta'+4}\right)w_w(0,0) \\ 0 & 1 \end{bmatrix} \tag{12.29}$$

In particular, the trace $\mathrm{tr}(w(f,g)) = 2r_w(\beta,\gamma) \in \mathbb{Z}[x,z]$.

Remark. A key feature here is that the polynomials r_w, s_w, t_w and w_w, and in case (12.28) the whole matrix, are independent of $\beta' = \beta(g)$. In particular this is true of traces of the matrix representations above.

We prove the above result in Section 12.7. We can also use it to find $w(A,B)$ when w is unbalanced; we do this next for non-parabolic f.

Corollary 12.5.3 *If* $w = w(a,b) = ba^{r_1}b^{-1}a^{r_2}\ldots b^{-1}a^{r_{2m-2}}b$ *is a regular unbalanced even word, then there are polynomials* r_w, s_w, t_w, w_w *with half-integer coefficients such that for* f *non-parabolic*

$$w(A,B) = \begin{bmatrix} a(r_w(\beta,\gamma) + s_w(\beta,\gamma)Q) & b(t_w(\beta,\gamma) + w_w(\beta,\gamma)Q) \\ c(t_w(\beta,\gamma) - w_w(\beta,\gamma)Q) & d(r_w(\beta,\gamma) - s_w(\beta,\gamma)Q) \end{bmatrix}, \tag{12.30}$$

where a, b, c, d *are as in (12.17) and (12.19).*

Proof. First note that $\tilde{w} = w * b^{-1}$ is balanced, thus

$$
w(A,B) = \tilde{w}(A,B)B = \begin{bmatrix} r_{\tilde{w}} + \frac{s_{\tilde{w}}Q}{\beta} & ab(\beta t_{\tilde{w}} + w_{\tilde{w}}Q) \\ cd(\beta t_{\tilde{w}} - w_{\tilde{w}}Q) & r_{\tilde{w}} - \frac{s_{\tilde{w}}Q}{\beta} \end{bmatrix} \begin{bmatrix} a & b \\ c & d \end{bmatrix} =
$$

$$
\begin{bmatrix} a[r_{\tilde{w}} + \frac{s_{\tilde{w}}Q}{\beta} - \frac{\gamma}{\beta}(\beta t_{\tilde{w}} + w_{\tilde{w}}Q)] & b[r_{\tilde{w}} + \frac{s_{\tilde{w}}Q}{\beta} + \left(1 - \frac{\gamma}{\beta}\right)(\beta t_{\tilde{w}} + w_{\tilde{w}}Q)] \\ c[r_{\tilde{w}} - \frac{s_{\tilde{w}}Q}{\beta} + \left(1 - \frac{\gamma}{\beta}\right)(\beta t_{\tilde{w}} - w_{\tilde{w}}Q)] & d[r_{\tilde{w}} - \frac{s_{\tilde{w}}Q}{\beta} - \frac{\gamma}{\beta}(\beta t_{\tilde{w}} - w_{\tilde{w}}Q)] \end{bmatrix}
$$

$$
= \begin{bmatrix} a\left((r_{\tilde{w}} - \gamma t_{\tilde{w}}) + g_{\tilde{w}}Q\right) & b[r_{\tilde{w}} + (\beta - \gamma)t_{\tilde{w}} + (g_{\tilde{w}} + w_{\tilde{w}})Q] \\ c[r_{\tilde{w}} + (\beta - \gamma)t_{\tilde{w}} - (g_{\tilde{w}} + w_{\tilde{w}})Q] & d\left((r_{\tilde{w}} - \gamma t_{\tilde{w}}) - g_{\tilde{w}}Q\right) \end{bmatrix}
$$

using (12.22). We have now found the polynomials for w explicitly in terms of those for $\tilde{w}$:

$$
\begin{aligned}
r_w &= r_{\tilde{w}} - \gamma t_{\tilde{w}} \\
s_w &= g_{\tilde{w}} \\
t_w &= r_{\tilde{w}} + (\beta - \gamma)t_{\tilde{w}} \\
w_w &= g_{\tilde{w}} + w_{\tilde{w}}
\end{aligned}
$$

$\square$

We now extend the definition of the polynomials r_w, s_w, t_w and w_w to arbitrary good words.

Definition. Let w be a good word. If w is regular and even, then then the polynomials r_w, s_w, t_w and w_w are as defined in Theorem 12.5.2 for balanced w, and Corollary 12.5.3 for unbalanced w. If w is regular and odd, then $v := w.a$ is regular and even and we define $r_w = r_v$, $s_w = s_v$, $t_w = t_v$ and $w_w = w_v$. Finally, if $w = w(a,b)$ is irregular, then $w' = w(a, b^{-1})$ is regular, and we define $r_w = r_{w'}$ etc. Note that $(w.a)' = w'.a$, so that if w is irregular *and* odd, we have $r_w = r_{(w.a)'} = r_{w'.a}$ and so forth.

If $w = w(a,b)$ is an irregular even word, then w' is regular and even. Since $\beta(g^{-1}) = \beta(g)$ and $\gamma(f, g^{-1}) = \gamma(f, g)$, Theorem 12.5.2 and Corollary 12.5.3 show that $w(f, g) = w'(f, g^{-1})$ is conjugate to $w'(f, g)$; in particular these have the same trace $2r_w(\beta, \gamma) = 2r_{w'}(\beta, \gamma)$.

Since for any good word $w = w(a, b)$, the commutator $[a, w]$ is balanced and even, it follows easily that $\gamma(f, w(f, g)) = p_w(\beta, \gamma)$, for some polynomial $p_w \in \mathbb{Z}[x, z]$ as first observed in [12]. The next result expresses these polynomials in terms of t_w and w_w.

Theorem 12.5.3 *Let* $w = w(a, b)$ *be a good word,* $\beta = \beta(f)$, $\gamma = \gamma(f, g)$ *and* $h = w(f, g)$, *then there is a polynomial* $p_w \in \mathbb{Z}[x, z]$, *such that*

$$
\gamma(f, h) = p_w(\beta, \gamma) \tag{12.31}
$$

If w is balanced, then

$$p_w(x, z) = z(x - z)(x t_w^2(x, z) - (x + 4) w_w^2(x, z)), \qquad (12.32)$$

If w is unbalanced, then

$$p_w(x, z) = z(t_w^2(x, z) - x(x + 4) w_w^2(x, z)). \qquad (12.33)$$

Proof for non-parabolic f. We may assume that f and g have matrix representatives A and B, respectively as given by (12.16)-(12.19). Suppose first that w is regular and even, and set $W = w(A, B)$. The required identities follow from (12.14) and (12.22), together with (12.27) when W is balanced and (12.30) when W is unbalanced. For balanced w, $p_w \in \mathbb{Z}[x, z]$ follows from (12.26). For unbalanced w, using the results and notation of the proof of Corollary 12.5.3, we have

$$
\begin{aligned}
t_w - x w_w &= r_{\tilde{w}} + (x + z) t_{\tilde{w}} - x(g_{\tilde{w}} + w_{\tilde{w}}) \\
&= r_{\tilde{w}} + (x + z) t_{\tilde{w}} - (s_{\tilde{w}} - z w_{\tilde{w}}) - x w_{\tilde{w}} \\
&= (r_{\tilde{w}} - s_{\tilde{w}}) + (x + z)(t_{\tilde{w}} - w_{\tilde{w}}) \in \mathbb{Z}[x, z],
\end{aligned}
$$

again by (12.26). Consequently, also $t_w + x w_w \in \mathbb{Z}[x, z]$, and the same then follows for p_w, as given by (12.33).

If w is odd, the result follows from the previous case, together with the identity $[A, W] = [A, WA]$. Finally suppose that w is irregular, then the commutator $[a, w]$ is irregular and even, and $[a, w]' = [a, w']$ is regular. Since, as previously noted, $\text{tr}([a, w]) = \text{tr}([a, w]') = \text{tr}([a, w'])$, we are reduced to the case where w is regular. $\qquad \square$

It is not difficult to prove the above directly when f is parabolic. However, we will instead use a limiting argument in Section 12.7 (Corollary 12.7.1).

12.5.1 Examples of word polynomials

Before going too much further it is worthwhile giving a few examples of polynomials. These appear in Table 12.1 below.

TABLE 12.1
Some examples of word polynomials.

Polynomial	word
$\gamma(\gamma - \beta)$	bab
$(\beta + 4)(\gamma - \beta)\gamma$	ba^2b
$(\beta - \gamma + 1)^2 \gamma$	$babab$
$\gamma(1 - 2\beta + \gamma^2 - (\beta - 2)\gamma)$	$baba^{-1}b$
$\gamma(1 + \beta(\beta + 1)(\beta + 4) - (\beta + 4)(2\beta + 1)\gamma + (\beta + 4)\gamma^2)$	$baba^2b$
$(\beta^2 - (\gamma - 4)\beta - 4\gamma + 1)^2 \gamma$	ba^2ba^2b
$\gamma(\gamma - \beta)(\beta - \gamma + 2)^2$	$bababab$
$\gamma(\beta^2 + \gamma^3 - 2\beta\gamma^2 + (\beta - 1)\beta\gamma)$	$bababa^{-1}b$
$\gamma(\beta + 4)(\beta^2 + \gamma^3 - 2\beta\gamma^2 + (\beta - 1)\beta\gamma)$	$baba^2ba^{-1}b$
$\gamma^3(\gamma - \beta)(\beta + 4)(\beta(\gamma^2 - 3\gamma - 4) - \beta^2(\gamma + 1) + 4\gamma^2 + 4\gamma + 1)$	$ba^{-2}bababa^{-2}bab$

The last polynomial here is quite long, but it has the remarkable property that it has $\gamma = 0$ as a super-attracting fixed point. We will need this fact later.

12.6   Applications

Inequalities such as Jørgensen's inequality hold quite generally for discrete groups of isometries of negatively curved metrics in an appropriate form as it is essentially based on Zassenhaus' lemma (see eg. [20]). However, it is the explicit and sharp nature of these and other inequalities in terms of natural parameters which yield good information in three dimensional hyperbolic geometry.

If $\langle f, g \rangle$ is discrete and non-elementary, which in this setting the latter means that $\langle f, g \rangle$ is not virtually abelian, then we may recover Jørgensen's inequality, [18] as follows. Recall we sketched a proof for this earlier at Theorem 12.1.2.

Theorem 12.6.1 *Let* $\langle f, g \rangle$ *be a discrete nonelementary subgroup of* $SL(2, \mathbb{C})$. *Then*

$$|\beta(f)| + |\gamma(f, g)| \geq 1. \tag{12.34}$$

If w is a good word and if $h = w(f, g)$, then we have $\gamma(f, h) = p_w(\gamma, \beta)$, where p_w is the polynomial of Theorem 12.5.3 and so we deduce from Jørgensen's inequality applied to the group $\langle f, h \rangle$, that

$$|\beta(f)| + |\gamma(f, h)| = |\beta| + |p_w(\beta, \gamma)| \geq 1$$

unless $\gamma(f, h) = 0$, in which case $\langle f, h \rangle$ is elementary.

We would like to understand how this exception happens more generally.

If $|\beta(f)| + |\gamma(f, h)| < 1$ and $\gamma(f, h) \neq 0$, then $\langle f, g \rangle$ is non-elementary since f is either loxodromic or elliptic of order at least 7. Then Jørgensen's inequality implies that $\langle f, g \rangle$ cannot be discrete. On the other hand, this group may be elementary if $\gamma(f, h) = 0$. However, in this case, we know that the fixed point sets of f and of h on $\hat{\mathbb{C}}$ intersect, and they may coincide; for non-parabolic f and h, coincidence means that they have the same axis. We can use polynomials to determine when this happens.

Proposition 12.6.1 *Let* $\langle f, g \rangle$ *be discrete and non-elementary, and* $h = w(f, g)$ *for a good word* w. *Set* $\beta = \beta(f) \neq -4$, *and* $\gamma = \gamma(f, g) \neq \beta, 0$. *Then the fixed point sets of* f *and* h *on* $\hat{\mathbb{C}}$ *coincide if and only if* $t_w(\beta, \gamma) = w_w(\beta, \gamma) = 0$

Proof. We suppose first that w is regular and even. We may suppose that f and g are normalized so that their matrix representatives are as given in Theorem 12.4.1 (specifically, by (12.16)-(12.19) for $\beta \neq 0$, and by (12.21) for $\beta = 0$), and consequently that the matrix representative for h is given by Theorem 12.5.2.

Suppose first that $\beta \neq 0$, so that f is loxodromic or elliptic with fixed points 0 and ∞ on the boundary. Since $abcd = \gamma(\gamma - \beta)/\beta^2 \neq 0$, (12.27) and (12.30) show that h shares these fixed points if and only if

$$\beta t_w(\beta, \gamma) \pm w_w(\beta, \gamma)Q = 0,$$

when w is balanced, and

$$t_w(\beta, \gamma) \pm w_w(\beta, \gamma)Q = 0,$$

when it is not. Since also $\beta \neq 0$ and $Q \neq 0$, this is equivalent to $t_w(\beta, \gamma) = w_w(\beta, \gamma) = 0$ in both cases.

Now suppose first that $\beta = 0$, so that $f(z) = z + 1$ with fixed point ∞. If ∞ is also the only fixed point of h, then h must also be parabolic, and by (12.28), we must have $r_w(0, \gamma) = 1$ and $w_w(0, \gamma) = 0$, in order to get the right trace and the right fixed point respectively. The determinant condition then gives $t_w(0, \gamma) = 0$. The converse is clear.

If w is regular and odd, then the fixed point sets of f and h on $\hat{\mathbb{C}}$ coincide if and only if the same is true of the fixed point sets of f and hf. Since $w.a$ is even this occurs exactly when $t_{w.a} = w_{w.a} = 0$, by the previous case, and since $t_w = t_{w.a}$ and $w_w = w_{w.a}$, we are done. If w is irregular then $h = w'(f, g^{-1})$, so that, since $\beta(g^{-1}) = \beta(g)$, $\gamma(f, g^{-1}) = \gamma(f, g)$ and w' is regular, the previous cases give that the fixed point sets of f and h on $\hat{\mathbb{C}}$ coincide if and only if $t_{w'}(\beta, \gamma) = w_{w'}(\beta, \gamma) = 0$, and we are done, since by definition, $t_w = t_{w'}$ and $w_w = w_{w'}$. $\square$

The above, together with (12.32) and (12.33), gives the following corollary which is useful. It allows us to obtain a contradiction to discreteness unless we can identify a multiple root, which is easy to do computationally, using the vanishing of a discriminant.

Corollary 12.6.1 *Let f, g and h be as above and let $\beta \neq -4$, $\beta \neq \gamma$ and $\gamma \neq 0$. If f and h have the same fixed points in $\hat{\mathbb{C}}$, then $z = \gamma$ is a multiple root of $p_w(\beta, z)$.*

Remark. The converse of this result is false. For example, when f is parabolic, f and h cannot have the same fixed points on the boundary unless h is also parabolic, but inspection of (12.32) and (12.33) shows that $p_w(0, z)$ has no simple roots apart from $z = 0$.

12.7   Quaternion Algebras

In order to prove Theorem 12.5.2, we switch from matrices into the slightly
more abstract setting of quaternion algebras. The objects we will be dealing
with here are essentially the same as matrices of the form (12.27), but with
numbers replaced by indeterminates (we will let x and z correspond to β and
γ respectively), and with square roots defined abstractly. We first recall some
basic facts; see [24] or [30] for more details.

For each field $\mathbb{F}$ of characteristic $\neq 2$ and non-zero $a, b \in \mathbb{F}$, the quaternion
algebra

$$\mathcal{A} = \left(\frac{a, b}{\mathbb{F}} \right), \tag{12.35}$$

is defined to be the associative algebra over the field $\mathbb{F}$ with multiplicative
identity 1 and basis $\{1, \mathbf{i}, \mathbf{j}, \mathbf{k}\}$, with multiplication determined by $\mathbf{i}^2 = a$,
$\mathbf{j}^2 = b$, $\mathbf{ij} = -\mathbf{ji} = \mathbf{k}$, whence also $\mathbf{k}^2 = -ab$, $\mathbf{jk} = -b\mathbf{i}$ and $\mathbf{ki} = -a\mathbf{j}$. The
generic member of $\mathcal{A}$ is thus $x + y\mathbf{i} + z\mathbf{j} + w\mathbf{k}$, where $x, y, z, w \in \mathbb{F}$; we may
abbreviate this to (x, y, z, w).

$\mathcal{A}$ is also isomorphic to the algebra of matrices of the form

$$\begin{bmatrix} x + y\xi_1 & (z + w\xi_1)\xi_2 \\ (z - w\xi_1)\xi_2 & x - y\xi_1 \end{bmatrix} \tag{12.36}$$

over the extension field $\mathbb{F}(\xi_1, \xi_2)$, where $\xi_1^2 = a$ and $\xi_2^2 = b$.

If $n = (x, y, z, w) \in \mathcal{A}$, then the *conjugate* of n is given by

$$\overline{n} = (x, -y, -z, -w),$$

and the *norm* of n by

$$N(n) = n\overline{n} = x^2 - ay^2 - bz^2 + abw^2 = x^2 - ay^2 - b(z^2 - aw^2) \in \mathbb{F}.$$

Note that the norm becomes the determinant under the mapping (12.36).

For reference, multiplication in the quaternion algebra (12.35) is given
explicitly by

$$\begin{aligned} (x, y&, z, w)(x', y', z', w') \\ &= \ (xx' + ayy' + bzz' - abww', \ xy' + yx' + b(wz' - zw'), \\ &\qquad xz' + zx' + a(yw' - wy'), \ xw' + wx' + yz' - zy'). \end{aligned} \tag{12.37}$$

If the field $\mathbb{F}$ is the field of fractions of an integrally closed integral domain
R, we define an *R-lattice* in $\mathcal{A}$ to be a finitely generated R-module L in $\mathcal{A}$;
L is an *ideal* if $\mathbb{F}L = \mathcal{A}$. An element $\alpha \in \mathcal{A}$ is an integer (over R) if $R[\alpha]$ is

an R-lattice. An *order* in $\mathcal{A}$ is an ideal which is also ring with 1. By contrast with the commutative case, the set of all integers in $\mathcal{A}$ is not generally a ring.

We will be particularly concerned with elements of norm 1. Note that these are units, and in any order which is closed under conjugation, they form a multiplicative group.

We consider the following quaternion algebra over the field of rational functions in two indeterminates.

$$\mathcal{Q}_0 := \left(\frac{(x+4)/x, z(z-x)}{\mathbb{R}(x,z)} \right)$$

Theorem 12.7.1 *The set $\mathcal{V}_0$ of elements of norm 1 of $\mathcal{Q}_0$ of the form*

$$[r(x,z) + s(x,z)\mathbf{i} + t(x,z)\mathbf{j} + w(x,z)\mathbf{k}],$$

where

$$2r, 2s, 2t, 2w \in \mathbb{Z}[x,z], \tag{12.38}$$

$$r(0,0) = 1 \tag{12.39}$$

and

$$s(x,z) \equiv zw(x,z) \bmod x, \tag{12.40}$$

is a group. This group is not trivial as, for instance,

$$\left\{ \frac{1}{2}\left(x+2+x\mathbf{i}\right), \frac{1}{2}\left(x+2+(x-2z)\mathbf{i}-2\mathbf{k}\right), \frac{1}{2}\left(z+2-z\mathbf{i}-\mathbf{j}-\mathbf{k}\right) \right\} \subset \mathcal{V}_0$$

Proof. Let $\mathbf{u} = (r,s,t,w) = (1/2)(r_1,s_1,t_1,w_1) \in \mathcal{V}_0$. The fact that this has norm 1 gives $\mathbf{u}^{-1} = \bar{\mathbf{u}} = (r,-s,-t,-w)$, so clearly $\mathcal{V}_0$ is closed under inversion. We need only show that it is also closed under multiplication. We have

$$r_1^2 - \left(\frac{x+4}{x}\right)s_1^2 - z(z-x)t_1^2 + \left(\frac{x+4}{x}\right)z(z-x)w_1^2 = 4 \tag{12.41}$$

Reducing modulo 2 gives

$$\begin{aligned}(r_1 - s_1)^2 &\equiv r_1^2 - s_1^2 \\ &\equiv z(z-x)(w_1^2 - t_1^2) \\ &\equiv z(z-x)(w_1 - t_1)^2 \quad \bmod 2,\end{aligned}$$

whence $r_1 - s_1 \equiv w_1 - t_1 \equiv 0 \bmod 2$, and so

$$r_1(x,z) \equiv s_1(x,z) \bmod 2, \tag{12.42}$$

$$t_1(x,z) \equiv w_1(x,z) \bmod 2. \tag{12.43}$$

Let $\mathbf{u_2} = (1/2)(r_2, s_2, t_2, w_2), \mathbf{u_3} = (1/2)(r_3, s_3, t_3, w_3) \in U$, then using (12.37), $\mathbf{u_2}\mathbf{u_3} = (1/2)(r, s, t, w)$, where

$$
\begin{aligned}
2r &= r_2 r_3 + \left(\frac{x+4}{x}\right) s_2 s_3 + z(z-x)t_2 t_3 - z(z-x)\left(\frac{x+4}{x}\right) w_2 w_3 \\
&= r_2 r_3 + z(z-x)t_2 t_3 + (x+4)z w_2 w_3 \\
&\quad + (x+4)\left(\frac{s_2(s_3 - z w_3)}{x} + \frac{z w_3(s_2 - z w_2)}{x}\right) \\
2s &= r_2 s_3 + s_2 r_3 + z(z-x)(w_2 t_3 - t_2 w_3) \\
2t &= r_2 t_3 + t_2 r_3 + \left(\frac{x+4}{x}\right)(s_2 w_3 - w_2 s_3) \\
2w &= r_2 w_3 + w_2 r_3 + s_2 t_3 - t_2 s_3
\end{aligned}
\tag{12.44}
$$

The congruence (12.40) applied to $\mathbf{u_2}$ and $\mathbf{u_3}$ shows that each of these is a polynomial. Moreover, since (12.40) also gives $s_2(0,0) = 0$, setting $x = z = 0$ in (12.44) then gives (12.39). Next

$$
2(s - zw) = (r_2 + zt_2)(s_3 - zw_3) + (r_3 - zt_3)(s_2 - zw_2) + xz(t_2 w_3 - w_2 t_3),
$$

so (12.40) holds for $\mathbf{u_2}\mathbf{u_3}$. It remains to show that the polynomials r, s, t and w have integer coefficients. We have

$$
\begin{aligned}
2r &= r_2 r_3 + \left(\frac{x+4}{x}\right) s_2 s_3 + z(z-x)t_2 t_3 - z(z-x)\left(\frac{x+4}{x}\right) w_2 w_3 \\
&\equiv r_2 r_3 + s_2 s_3 + z(z-x)(t_2 t_3 - w_2 w_3) \mod 2 \\
&\equiv 0 \mod 2,
\end{aligned}
\tag{12.45}
$$

using (12.42) and (12.43), whence $r \in \mathbb{Z}[x, z]$. Similar, and easier, arguments give the same conclusion for s, t and w. $\qquad\square$

Remark. It is not difficult to show that $r(0,0) = \pm 1$ and $s(x,z) \equiv \pm zw(x,z) \mod x$ follow from (12.41), so that (12.39) and (12.40) are just normalizing choices of sign.

For each fixed $\beta, \beta', \gamma \in \mathbb{C}$, with $\beta \neq 0$, if we let $Q = \sqrt{\beta(\beta+4)}$ and D_1 and D_2 be any fixed numbers such that $D_1 D_2 = \gamma(\gamma - \beta)/\beta^2$, then the evaluation map

$$
\begin{aligned}
&\phi_{\beta,\beta',\gamma}([r(x,z) + s(x,z)\mathbf{i} + t(x,z)\mathbf{j} + w(x,z)\mathbf{k}]) \\
&= \begin{bmatrix} r(\beta,\gamma) + \frac{s(\beta,\gamma)Q}{\beta} & D_1(\beta t(\beta,\gamma) + w(\beta,\gamma)Q) \\ D_2(\beta t(\beta,\gamma) - w(\beta,\gamma)Q) & r(\beta,\gamma) - \frac{s(\beta,\gamma)Q}{\beta} \end{bmatrix}
\end{aligned}
\tag{12.46}
$$

is an algebra homomorphism from $\mathcal{Q}_0$ to $M_2(\mathbb{C})$, the algebra of 2×2 matrices over $\mathbb{C}$.

For $\beta = 0$, $\gamma \neq 0$, we set

$$\phi_{0,\beta',\gamma}([r(x,z) + s(x,z)\mathbf{i} + t(x,z)\mathbf{j} + w(x,z)\mathbf{k}])$$
$$= \begin{bmatrix} r(0,\gamma) + \gamma t(0,\gamma) & 4g(0,\gamma) + 2w(0,\gamma) \\ 2\gamma w(0,\gamma) & r(0,\gamma) - \gamma t(0,\gamma) \end{bmatrix}, \qquad (12.47)$$

where g (a polynomial by (12.40)) is given by

$$g(x,z) := \frac{s(x,z) - zw(x,z)}{x}, \qquad (12.48)$$

and

$$\phi_{0,\beta',0}([r(x,z) + s(x,z)\mathbf{i} + t(x,z)\mathbf{j} + w(x,z)\mathbf{k}])$$
$$= \begin{bmatrix} r(0,0) & 4g(0,0) - \left(\beta' + \sqrt{\beta'}\sqrt{\beta'+4}\right)w(0,0) \\ 0 & r(0,0) \end{bmatrix} \qquad (12.49)$$

The next theorem shows that the maps $\phi_{0,\beta',\gamma}$ arise as limits of maps $\phi_{\beta,\beta',\gamma}$ (after conjugating $\phi_{\beta,\beta',\gamma}$ in such a way as to make its fixed points approach a common limit as $\beta \to 0$). It follows that the maps $\phi_{\beta,\beta',\gamma}$ are all algebra homomorphisms (this is also not difficult to show directly). In particular, each $\phi_{\beta,\beta',\gamma}$ restricted to $\mathcal{V}_0$ is a group homomorphism to $SL(2,\mathbb{C})$, and thence by projection to $PSL(2,\mathbb{C})$.

Theorem 12.7.2 *Suppose that $\beta \neq 0$, $\gamma \neq 0$, $k^2 = 1/\sqrt{\beta}$, and that*

$$Q = \sqrt{\beta(\beta+4)}, \quad Q' = \sqrt{\beta}\sqrt{\beta+4},$$

$$m = -k^{-1}\left[\frac{1}{\sqrt{\beta}} + \frac{\sqrt{\beta'+4}}{2\sqrt{\gamma}}\right], \quad m_1 = \frac{\sqrt{\beta'+4} + \sqrt{\beta'}}{2\sqrt{\gamma}},$$

$$D_1 = -\frac{1}{2}\sqrt{\frac{\gamma}{\beta}}\left(\sqrt{\beta'+4} + \sqrt{\frac{4\gamma + \beta\beta'}{\beta}}\right),$$

$$D_2 = \frac{1}{2}\sqrt{\frac{\gamma}{\beta}}\left(\sqrt{\beta'+4} - \sqrt{\frac{4\gamma + \beta\beta'}{\beta}}\right)$$

(i.e. $D_1 = ab$ and $D_2 = cd$, where a, b, c, d are given by (12.17) and (12.18)),

$$D_1' = -\left(\frac{\sqrt{\gamma}}{2\sqrt{\beta}}\right)\left(\sqrt{\beta'+4} + \frac{2\sqrt{\gamma}\sqrt{1+\beta\beta'/(4\gamma)}}{\sqrt{\beta}}\right),$$

$$D_2' = \left(\frac{\sqrt{\gamma}}{2\sqrt{\beta}}\right)\left(\sqrt{\beta'+4} - \frac{2\sqrt{\gamma}\sqrt{1+\beta\beta'/(4\gamma)}}{\sqrt{\beta}}\right),$$

$$C = \begin{cases} \begin{bmatrix} \sqrt{D_1'/D_1} & 0 \\ 0 & \sqrt{D_2'/D_2} \end{bmatrix}, & \text{if } Q' = Q \\[2em] \begin{bmatrix} 0 & i\sqrt{D_1'/D_2} \\ i\sqrt{D_2'/D_1} & 0 \end{bmatrix}, & \text{if } Q' = -Q \end{cases}$$

$$M = \begin{bmatrix} k & m \\ 0 & k^{-1} \end{bmatrix} C, \quad M_1 = \begin{bmatrix} 1 & m_1 \\ 0 & 1 \end{bmatrix}.$$

Then for $\gamma \neq 0$, $\mathbf{x} \in \mathcal{Q}$,

$$\lim_{\beta \to 0} M\phi_{\beta,\beta',\gamma}(\mathbf{x})M^{-1} = \phi_{0,\beta',\gamma}(\mathbf{x}) \tag{12.50}$$

and

$$\lim_{\gamma \to 0} M_1\phi_{0,\beta',\gamma}(\mathbf{x})M_1^{-1} = \phi_{0,\beta',0}(\mathbf{x}) \tag{12.51}$$

Proof. Since $D_1'D_2' = D_1D_2$, the diagonal entries of C are the same, in the case $Q' = Q$, and the off-diagonal entries of C divided by i are mutually reciprocal, otherwise. Thus we can apply (12.11) and (12.10) respectively to obtain

$$C\phi_{\beta,\beta'\gamma}(\mathbf{x})C^{-1} = \begin{bmatrix} r(\beta,\gamma) + \frac{s(\beta,\gamma)Q'}{\beta} & D_1'(\beta t(\beta,\gamma) + w(\beta,\gamma)Q') \\ D_2'(\beta t(\beta,\gamma) - w(\beta,\gamma)Q') & r(\beta,\gamma) - \frac{s(\beta,\gamma)Q'}{\beta} \end{bmatrix}$$

Thus

$$M\phi_{\beta,\beta'\gamma}(\mathbf{x})M^{-1} = \begin{bmatrix} a_{11} & a_{12} \\ a_{21} & a_{22} \end{bmatrix}, \tag{12.52}$$

where, using (12.11), and writing $r(\beta,\gamma) = r$ etc.

$$a_{11} = r + \frac{sQ'}{\beta} - k^{-2}\left[\frac{1}{\sqrt{\beta}} + \frac{\sqrt{\beta'+4}}{2\sqrt{\gamma}}\right]\left(\frac{\sqrt{\gamma}}{2\sqrt{\beta}}\right) \cdot$$
$$\cdot \left(\sqrt{\beta'+4} - \frac{2\sqrt{\gamma}\sqrt{1+\beta\beta'/(4\gamma)}}{\sqrt{\beta}}\right)(\beta t - wQ')$$
$$= r + \frac{sQ'}{\beta} - \frac{\sqrt{\gamma}}{2}\left[1 + \frac{\sqrt{\beta}\sqrt{\beta'+4}}{2\sqrt{\gamma}}\right] \cdot$$
$$\cdot \left(\sqrt{\beta'+4} - 2\frac{\sqrt{\gamma}}{\sqrt{\beta}} + O(\sqrt{\beta})\right)\left(\sqrt{\beta}t - 2w + O(\beta)\right)$$

Thus

$$\lim_{\beta \to 0} a_{11} = \lim_{\beta \to 0} \left[r + \frac{sQ'}{\beta} - \frac{\sqrt{\gamma}}{2} \left(1 + \frac{\sqrt{\beta}\sqrt{\beta'+4}}{2\sqrt{\gamma}} \right) \left(\sqrt{\beta'+4} - 2\frac{\sqrt{\gamma}}{\sqrt{\beta}} \right) (\sqrt{\beta}t - 2w) \right]$$

$$= \lim_{\beta \to 0} \left[r + \frac{2s}{\sqrt{\beta}} + \sqrt{\gamma} \left(1 + \frac{\sqrt{\beta}\sqrt{\beta'+4}}{2\sqrt{\gamma}} \right) \left(t\sqrt{\gamma} + w\sqrt{\beta'+4} - 2w\frac{\sqrt{\gamma}}{\sqrt{\beta}} \right) \right]$$

$$= \lim_{\beta \to 0} \left[r + \frac{2s}{\sqrt{\beta}} + \sqrt{\gamma} \left(t\sqrt{\gamma} + w\sqrt{\beta'+4} - w\sqrt{\beta'+4} \right) - \frac{2w\gamma}{\sqrt{\beta}} \right]$$

$$= \lim_{\beta \to 0} \left[r + \frac{2(s - \gamma w)}{\sqrt{\beta}} + t\gamma \right] = r(0,\gamma) + \gamma t(0,\gamma),$$

using (12.40) at the last step. Since conjugation preserves traces, we then have $\lim_{\beta \to 0} a_{22} = r(0,\gamma) - \gamma t(0,\gamma)$.

$$a_{21} = k^{-2} \left(\frac{\sqrt{\gamma}}{2\sqrt{\beta}} \right) \left(\sqrt{\beta'+4} - \frac{2\sqrt{\gamma}\sqrt{1 + \beta\beta'/(4\gamma)}}{\sqrt{\beta}} \right) (\beta t - wQ')$$

$$= \left(\frac{\sqrt{\gamma}}{2} \right) \left(\sqrt{\beta'+4} - \frac{2\sqrt{\gamma}\sqrt{1 + \beta\beta'/(4\gamma)}}{\sqrt{\beta}} \right) (\beta t - wQ')$$

so

$$\lim_{\beta \to 0} a_{21} = \left(\frac{\sqrt{\gamma}}{2} \right) \lim_{\beta \to 0} \left(\frac{-2\sqrt{\gamma}}{\sqrt{\beta}} \right) \left(-2w\sqrt{\beta} \right) = 2\gamma w(0,\gamma) = 2s(0,\gamma), \quad (12.53)$$

again using (12.40).

Finally we show that $\lim_{\beta \to 0} a_{12} = 2w(0,\gamma) + 4g(0,\gamma)$. Since $\text{Det}(\phi_{\beta,\beta',\gamma}) = 1$, and this determinant is preserved under conjugation and limits, it suffices to show that

$$(r(0,\gamma) + \gamma t(0,\gamma))(r(0,\gamma) - \gamma t(0,\gamma)) - 2s(0,\gamma)(2w(0,\gamma) + 4g(0,\gamma)) = 1.$$

This is readily verified by letting $\beta \to 0$ in (12.41), keeping in mind the definition of g, (12.48). This completes the proof of (12.50). A similar, but much easier, calculation gives (12.51). Forming the conjugate

$$\begin{bmatrix} 1 & m_1 \\ 0 & 1 \end{bmatrix} \begin{bmatrix} r(0,\gamma) + \gamma t(0,\gamma) & 4g(0,\gamma) + 2w(0,\gamma) \\ 2\gamma w(0,\gamma) & r(0,\gamma) - \gamma t(0,\gamma) \end{bmatrix} \begin{bmatrix} 1 & -m_1 \\ 0 & 1 \end{bmatrix},$$

using (12.11), and letting $\gamma \to 0$, gives the matrix at (12.49). $\qquad \square$

12.7.1 Proof of Theorem 12.5.2

Let

$$\mathbf{w_1} = \frac{1}{2}(x + 2, x, 0, 0),$$

$$\mathbf{w_2} = \frac{1}{2}(x + 2, x - 2z, 0, -2)$$

$$\mathbf{w_3} = \frac{1}{2}(z + 2, -z, -1, -1). \quad (12.54)$$

Polynomial Trace Identities and Kleinian Groups

Each $\mathbf{w_i} \in \mathcal{V}_0$. Let A, B be as in Theorem 12.4.1, $Q = \sqrt{\beta(\beta + 4)}$. In the first case, $\beta \neq 0$, we calculate

$$A^2 = \frac{1}{2} \begin{bmatrix} \beta + 2 + Q & 0 \\ 0 & \beta + 2 - Q \end{bmatrix} \tag{12.55}$$

$$BA^2B^{-1} = \frac{1}{2} \begin{bmatrix} \beta + 2 + (\beta - 2\gamma)Q/\beta & -2abQ \\ 2cdQ & \beta + 2 - (\beta - 2\gamma)Q/\beta \end{bmatrix} \tag{12.56}$$

$$[B, A] = \frac{1}{2} \begin{bmatrix} \gamma + 2 - \gamma Q/\beta & -ab(\beta + Q) \\ -cd(\beta - Q) & \gamma + 2 + \gamma Q/\beta, \end{bmatrix}. \tag{12.57}$$

In the second case, $\beta = 0$, $\gamma \neq 0$, we have

$$A^2 = \begin{bmatrix} 1 & 2 \\ 0 & 1 \end{bmatrix}, \tag{12.58}$$

$$BA^2B^{-1} = \begin{bmatrix} 1 & 0 \\ -2\gamma & 1 \end{bmatrix}, \tag{12.59}$$

$$[B, A] = \begin{bmatrix} 1 & -1 \\ -\gamma & \gamma + 1 \end{bmatrix} \tag{12.60}$$

Finally, if $\beta = \gamma = 0$, then A^2 is still given by (12.58), and

$$BA^2B^{-1} = \begin{bmatrix} 1 & 2 + \beta(g) + \sqrt{\beta(g)}\sqrt{\beta(g) + 4} \\ 0 & 1 \end{bmatrix} \tag{12.61}$$

and

$$[B, A] = \begin{bmatrix} 1 & \left(\beta(g) + \sqrt{\beta(g)}\sqrt{\beta(g) + 4}\right)/2 \\ 0 & 1 \end{bmatrix} \tag{12.62}$$

In this last case the matrices do depend on $\beta(g)$, but are independent of the parameter ℓ.

Thus, in all cases, a straightforward calculation using the evaluation map at (12.46) gives

$$A^2 = \phi_{\beta,\beta',\gamma}(\mathbf{w}_1), \quad BA^2B^{-1} = \phi_{\beta,\beta',\gamma}(\mathbf{w}_2), \quad [B, A] = \phi_{\beta,\beta',\gamma}(\mathbf{w}_3),$$

where $\beta = \beta(f)$, $\beta' = \beta(g)$ and $\gamma = \gamma(f, g)$, and where (in the case $\beta \neq 0$), we set $D_1 = ab$ and $D_2 = cd$ with a, b, c, d given by (12.17) and (12.19). Since, by Lemma 12.5.2, these words generate all regular balanced even words in A and B, it follows immediately that every such word is $\phi_{\beta,\beta',\gamma}(\mathbf{w})$ for some $\mathbf{w} \in \mathcal{V}_0$. Theorem 12.5.2 then follows, using Theorem 12.7.1, (12.42) and (12.43). $\quad\square$

Corollary 12.7.1 *Let $A(\beta)$ and $B(\beta, \beta', \gamma)$ be the matrix representatives of f and g respectively given by Theorem 12.4.1, where now we have made the dependency on parameters $\beta = \beta(f)$, $\beta' = \beta(g)$ and $\gamma = \gamma(f, g)$ explicit. Let $w(a, b)$ be a regular balanced even word. Let $\beta \neq 0$, and k, M and M_1 as in Theorem 12.7.2, then for $\gamma \neq 0$,*

$$\lim_{\beta \to 0} M(w(A(\beta), B(\beta, \beta', \gamma)))M^{-1} = w(A(0), B(0, \beta', \gamma)) \quad (12.63)$$

and

$$\lim_{\gamma \to 0} M_1(w(A(0), B(0, \beta', \gamma)))M_1^{-1} = w(A(0), B(0, \beta', 0)) \quad (12.64)$$

Since polynomials and the trace function are continuous, and trace is preserved under conjugation, Theorem 12.5.3 for $\beta = 0$ follows from the case $\beta \neq 0$, by letting $\beta \to 0$.

12.7.2 A change of variable

We now introduce two new parameters which can be used to describe 2-generator groups (up to conjugacy), and which, when $\beta \neq 0$, can be used interchangeably with β and γ and will simplify many formulas in what follows. For $f, g \in Isom^+(\mathbb{H}^3)$ we define

$$\lambda = \lambda(f) = (\text{tr}^2(f) - 2)/2 = \cosh(\tau + i\eta), \quad (12.65)$$

so $\lambda = (\beta(f) + 2)/2$. When f is elliptic or loxodromic,

$$\mu = \mu(f, g) = \frac{\text{tr}^2(f) - 2\text{tr}[f, g]}{\text{tr}^2(f) - 4}. \quad (12.66)$$

In terms of our earlier parameters

$$\mu = 1 - \frac{2\gamma(f, g)}{\beta(f)}$$

Rewriting (12.7) and (12.8) in terms of $\lambda(f)$ and $\mu(f, g)$ gives

$$\lambda(f) = \cosh(\tau(f) + i\eta(f)),$$

and

$$\mu(f, g) = 1 - (\lambda(g) - 1)\sinh^2(\Delta),$$

where we recall Δ—the complex distance between axes—is defined above at (12.8). An important special case is captured by the next lemma.

Lemma 12.7.1 *If g is order 2, $\lambda(g) = -1$, and we obtain the particularly simple form:*

$$\mu(f, g) = \cosh(2\Delta).$$

Unwinding these parameters gives

$$\beta(f) = 2(\lambda(f) - 1), \quad \text{and} \quad \gamma(f, g) = -(\lambda(f) - 1)(\mu(f, g) - 1) \qquad (12.67)$$

If $\lambda = \lambda(f) \neq 1$ then it determines f up to conguacy, and if further, $\mu = \mu(f, g) \neq 1$, then λ and μ together determine the group $\langle f, g \rangle$. When $\lambda \neq 1$, we can rewrite the matrices A and B at (12.16) in terms of the new parameters as

$$A = \begin{bmatrix} \frac{\sqrt{\lambda^2 - 1} + (\lambda - 1)}{\sqrt{2(\lambda - 1)}} & 0 \\ 0 & \frac{\sqrt{\lambda^2 - 1} - (\lambda - 1)}{\sqrt{2(\lambda - 1)}} \end{bmatrix}, \quad B = \begin{bmatrix} a & b \\ c & d \end{bmatrix}, \qquad (12.68)$$

where now, writing $\lambda(g) = \lambda'$,

$$a = \frac{1}{\sqrt{2}} \left(\sqrt{\lambda' + 1} + \sqrt{\lambda' - \mu} \right), \quad d = \frac{1}{\sqrt{2}} \left(\sqrt{\lambda' + 1} - \sqrt{\lambda' - \mu} \right), \qquad (12.69)$$

and b and c are given by (12.19) when $\mu = 1$, and otherwise

$$c = -b = \sqrt{\frac{1 - \mu}{2}} \qquad (12.70)$$

for $\mu \neq 1$.

Given a regular balanced even word w, we can now rewrite the matrix $w(A, B)$ at (12.27) as

$$\begin{bmatrix} R_w(\lambda, \mu) + S_w(\lambda, \mu)\sqrt{\lambda^2 - 1} & 2ab(T_w(\lambda, \mu) + W_w(\lambda, \mu)\sqrt{\lambda^2 - 1}) \\ 2cd(T_w(\lambda, \mu) - W_w(\lambda, \mu)\sqrt{\lambda^2 - 1} & R_w(\lambda, \mu) - S_w(\lambda, \mu)\sqrt{\lambda^2 - 1} \end{bmatrix}$$
$$(12.71)$$

where a, b, c, d are given by (12.69) and (12.70), and, setting $x = 2(u - 1)$, $z = -(u - 1)(v - 1)$, the polynomials R_w, S_w, T_w, W_w are given by

$$R_w(u, v) = r_w(x, z), \quad W_w(u, v) = w_w(x, z) \qquad (12.72)$$

$$S_w(u, v) = 2s_w(x, z)/x, \quad T_w(u, v) = xt_w(x, z)/2. \qquad (12.73)$$

The congruence (12.40) ensures that S_w is a polynomial. For arbitrary balanced words we take the above as *definitions* of R_w etc. Recall that, in this case, there is a regular even balanced word v such that $r_w = r_v$, $s_w = s_v$ etc., so that it remains true that S_w is a polynomial in this case.

Each of the polynomials $2R_w$, $2S_w$, $2T_w$, and $2W_w$ has integer coefficients, and corresponding to these new parameters, we define the quaternion algebra $\mathcal{Q}$ by

$$\mathcal{Q} = \left(\frac{u^2 - 1, v^2 - 1}{\mathbb{R}(u, v)} \right), \qquad (12.74)$$

Here, the indeterminates u and v correspond to λ and μ respectively. It is straightforward to show that the map $\rho : \mathcal{Q}_0 \to \mathcal{Q}$ given by

$$\rho(r, s, t, w) = (r, \frac{s}{u-1}, (u-1)t, w) \tag{12.75}$$

is an isomorphism. On the right-hand side, x and z are converted into terms of u and v by the formulae

$$x = 2(u-1), \quad z = -(u-1)(v-1), \quad \left(u = \frac{x+2}{2}, \quad v = 1 - \frac{2z}{x} \right), \tag{12.76}$$

these conversions being just the same as those relating β and γ to λ and μ. The proof uses the observations that

$$u^2 - 1 = \frac{x^2}{2} \left(\frac{x+4}{x} \right) \quad \text{and} \quad v^2 - 1 = \frac{2}{x^2} z(z-x).$$

It is then easy to see that the inverse map is given by

$$\rho^{-1}(R, S, T, W) = (R, \frac{x}{2} S, \frac{2}{x} T, W) \tag{12.77}$$

Where we now use the second pair of equations in (12.76) to convert the right-hand side back into terms of x and z.

We can now characterize the image under ρ of the group $\mathcal{V}_0$ defined in Theorem 12.7.1. This result is a direct consequence of (12.75), (12.76), (12.77) and the definition of $\mathcal{V}_0$.

Theorem 12.7.3 $\mathcal{V} := \rho(\mathcal{V}_0)$ *comprises the elements*

$$(R, S, T, W) = (R(u, v), S(u, v), T(u, v), W(u, v))$$

of $\mathcal{Q}$ for which $R(1, 1) = 1$, and each of $2R$, $2(u-1)S$, $2(u-1)^{-1}T$, $2W$ and $S + (v-1)W$ is a polynomial of the form

$$\sum a_{n,m} (u-1)^m (v-1)^n$$

such that, for each term, $m \geq n$ and $a_{n,m}$ is an integer multiple of 2^{m-n} (in particular, each $a_{n,m}$ is an integer).

Remark. The condition on $S(u, v) + (v-1)W(u, v)$ is equivalent to (12.40): we have $g(x, z) := (s(x, z) - zw(x, z))/x = (S(u, v) + (v-1)W(u, v))/2$. Note also that this condition insures that $S(u, v)$ is a polynomial.

If $\mathbf{x} = (r(x, z), s(x, z), t(x, z), w(x, z)) \in \mathcal{V}_0$, and if

$$\rho(\mathbf{x}) = (R(u, v), S(u, v), T(u, v), W(u, v)) \in \mathcal{V},$$

then by the definition of ρ, the matrix at (12.71) is $\phi_{\beta,\beta',\gamma}(\mathbf{x})$ ($\beta \neq 0$). Accordingly we define, for each $\lambda \neq 1, \lambda', \mu \in \mathbb{C}$, the algebra homomorphisms $\psi_{\lambda,\lambda',\mu} : \mathcal{Q} \to M_2(\mathbb{C})$ by

$$\psi_{\lambda,\lambda',\mu}(R, S, T, W) \tag{12.78}$$

$$= \left[\begin{array}{cc} R(\lambda, \mu) + S(\lambda, \mu)\sqrt{\lambda^2 - 1} & 2ab(T(\lambda, \mu) + W(\lambda, \mu)\sqrt{\lambda^2 - 1}) \\ 2cd(T(\lambda, \mu) + W(\lambda, \mu)\sqrt{\lambda^2 - 1}) & R(\lambda, \mu) - S(\lambda, \mu)\sqrt{\lambda^2 - 1} \end{array} \right],$$

where a, b, c, d are given by (12.69) and (12.70). We thus have

Proposition 12.7.1 *For $\beta \neq 0, \beta', \gamma \in \mathbb{C}, x \in \mathcal{V}_0$,*

$$\psi_{\lambda,\lambda',\mu}(\rho(\mathbf{x})) = \phi_{\beta,\beta',\gamma}(\mathbf{x}),$$

where $\lambda = 1 + \beta/2$, $\lambda' = 1 + \beta'/2$ and $\mu = 1 - 2\gamma/\beta$.

In some respects λ and μ are better parameters to use than β and γ: they have a simpler geometrical interpretation, the matrix representations and quaternion algebras are simpler and neater, and there is an obvious symmetry between λ and μ, corresponding to the symmetry between two loxodromics with perpendicular axes (Subsection 12.8.2 below). They also have a major drawback: μ is undefined when f is parabolic, so to deal with this case we still need β and γ.

12.8 Elements of Unit Norm in $\mathcal{Q}$

We have found a group $\mathcal{V}$ of elements of norm 1 in $\mathcal{Q}$, which maps under each evaluation homomorphism $\psi_{\lambda,\lambda',\mu}$ to a group which includes the regular balanced even words in two generators f and g, where f is elliptic or loxodromic. In this section will show how $\mathcal{V}$ naturally extends to a larger group which we will denote $\mathcal{U}$, and this gives a corresponding extension of the isometry group $\psi_{\lambda,\lambda',\mu}(\mathcal{V})$. We look further at this group in Section 12.8.2. To begin with we consider properties of elements of norm 1 in general.

The requirement that $(R, S, T, W) \in \mathcal{Q}$ has norm 1 is given explicitly by

$$R^2 - (u^2 - 1)S^2 - (v^2 - 1)T^2 + (u^2 - 1)(v^2 - 1)W^2 = 1 \tag{12.79}$$

We will confine our attention to the solutions of (12.79) for which R, S, T, W are all polynomials, with the additional normalizing condition that $R(1, 1) = 1$. These solutions clearly form a group, which we denote by $\mathcal{U}_1$.

We define the *degree* of $\mathbf{u} = (R, S, T, W) \in \mathcal{U}_1$ by

$$\deg(\mathbf{u}) = \max\{\deg(R), \deg(S) + 1, \deg(T) + 1, \deg(W) + 2\}.$$

TABLE 12.2
$(R, S, T, W) \in \mathcal{U}_1$ of degree at most 2.

R	S	T	W
1	0	0	0
u	1	0	0
v	0	1	0
u	v	0	1
v	0	u	1
uv	1	u	0
uv	v	1	0
uv	v	u	1
$2u^2 - 1$	$2u$	0	0
$2v^2 - 1$	0	$2v$	0
$(1+u+v-uv)/2$	$(v-1)/2$	$(u-1)/2$	$1/2$
$(1+u-v+uv)/2$	$(v+1)/2$	$(u-1)/2$	$1/2$
$(1-u+v+uv)/2$	$(v-1)/2$	$(u+1)/2$	$1/2$
$(-1+u+v+uv)/2$	$(v+1)/2$	$(u+1)/2$	$1/2$

It is easy to check that $\deg(\mathbf{uv}) \leq \deg(\mathbf{u}) + \deg(\mathbf{v})$. For any fixed degree it is possible in principle to evaluate all members of $\mathcal{U}_1$ of any fixed degree d, by equating coefficients in (12.79), and we have done this for $d \leq 4$. Table 12.2 lists all members of $\mathcal{U}_1$ of degree at most 2 (up to sign changes of the components S, T and W). In this case, the polynomials, like those of $\mathcal{V}$, all have integer or half-integer coefficients. Our main result in this section is that all members of $\mathcal{Q}$ with components in $\mathbb{Q}[u, v]$ and norm in $\mathbb{Z}[u, v]$ have this property, and that they form an order (with reference to the underlying ring $\mathbb{Z}[u, v]$). We define

$$\mathcal{O} = \{\mathbf{u} = (R, S, T, W) \in \mathcal{Q} \mid R, S, T, W \in \mathbb{Q}[u, v], N(\mathbf{u}) \in \mathbb{Z}[u, v]\}$$

Theorem 12.8.1

1. $\mathcal{O}$ is the set of quaternions of the form $(R, S, T, W) + \frac{1}{2}P(u, v)((u+1)(v+1), v+1, u+1, 1)$, where $R, S, T, W, P \in \mathbb{Z}[u, v]$.

2. $\mathcal{O}$ is the unique maximal order of $\mathcal{Q}$, which contains $\mathbf{i}$ and $\mathbf{j}$.

It follows (since $\mathcal{O}$ is clearly closed under conjugation) that the elements of $\mathcal{O}$ of norm 1 form a group, which we denote $\mathcal{U}$.

The appearance of half-integer coefficients in $\mathcal{O}$ is reminiscent of the *Hurwitz order* $\mathcal{H}$ in $\mathbb{H}$, the quaternions of Hamilton, defined as $\mathcal{H} = \{\frac{1}{2}(n_1 + n_2\mathbf{i} + n_3\mathbf{j} + n_4\mathbf{k}) \in \mathbb{H} \mid n_1, n_2, n_3, n_4 \in \mathbb{Z}, n_1 \equiv n_2 \equiv n_3 \equiv n_4 \bmod 2\}$. Theorem 12.8.1 has no analog for $\mathcal{H}$; there are plenty of quaternions with integer norm and components which are rational but not half-integers, for example $(3/5, 4/5, 0, 0)$. However, both the characterizations of $\mathcal{O}$ given in the Theorem have their counterparts for $\mathcal{H}$. It is easy to see that $\mathcal{H}$ comprises the quaternions of the form $\mathbf{u} + \frac{1}{2}(1,1,1,1)$, where $\mathbf{u} \in \mathbb{Z}^4$, and corresponding to Theorem 12.8.1 (2), we have the following classical result (see e.g. [30]).

Theorem 12.8.2 $\mathcal{H}$ *is the unique maximal order of* $\mathbb{H}$, *which contains* $\mathbf{i}$ *and* $\mathbf{j}$.

We will take an axiomatic approach which covers both of these theorems.

We may also characterize $\mathcal{H}$ as the set of quaternions with integer norm and half-integer components; this amounts to the simple observation that, for integers a, b, c, d

$$a^2 + b^2 + c^2 + d^2 \equiv 0 \bmod 4 \Rightarrow a \equiv b \equiv c \equiv d \bmod 2. \tag{12.80}$$

The integers also satisfy the similar property

$$a^2 + b^2 + c^2 + d^2 \equiv 0 \bmod 8 \Rightarrow a \equiv b \equiv c \equiv d \equiv 0 \bmod 2. \tag{12.81}$$

As a simple application of this we observe that a quaternion $\mathbf{u}$ with rational components and integer norm has no component with denominator divisible by 4. For if this occurred we would have, clearing denominators, a quaternion with integer components not all even, and norm divisible by 16, contrary to (12.81).

We say that a commutative ring which satisfies (12.80) or (12.81) has the *four squares* property and the *strong four squares* property, respectively. To justify this terminology, we show that (12.81) $\Rightarrow$ (12.80). Suppose that (12.81) holds, and that $a^2 + b^2 + c^2 + d^2 \equiv 0 \bmod 4$, then $(a-b)^2 + (a+b)^2 + (c-d)^2 + (c+d)^2 = 2(a^2 + b^2 + c^2 + d^2) \equiv 0 \bmod 8$, and so applying (12.81), we get $a \equiv b \bmod 2$ and $c \equiv d \bmod 2$. The same argument with b and c interchanged gives $a \equiv c \bmod 2$, and so $a \equiv b \equiv c \equiv d \bmod 2$, proving (12.80). The converse fails in general (consider, for example, $R = \mathbb{Z}_4$), but holds when R is an integral domain. In this case, if R has characteristic 2, then $1^2 + 1^2 + 0^2 + 0^2 = 0$, so that (12.80) fails, and the implication is vacuous. Otherwise suppose that (12.80) holds, and $a^2 + b^2 + c^2 + d^2 \equiv 0 \bmod 8$, then by (12.80) $a \equiv b \equiv c \equiv d \bmod 2$, and we have $\left(\frac{a-b}{2}\right)^2 + \left(\frac{a+b}{2}\right)^2 + \left(\frac{c-d}{2}\right)^2 + \left(\frac{c+d}{2}\right)^2 = \frac{1}{2}(a^2 + b^2 + c^2 + d^2) \equiv 0 \bmod 4$. Applying (12.80) again gives $\left(\frac{a-b}{2}\right) \equiv \left(\frac{a+b}{2}\right) \equiv \left(\frac{c-d}{2}\right) \equiv \left(\frac{c+d}{2}\right) \equiv 0 \bmod 2$, whence it easily follows that $a \equiv b \equiv c \equiv d \equiv 0 \bmod 2$.

The main step in our proofs is to show that, if R has the (strong) four squares property, then the polynomial rings $R[x_1, x_2, \ldots, x_n]$ also have this property, together with some generalizations thereof.

Lemma 12.8.1 *Suppose R is a commutative ring and suppose φ_i $(1 \leq i \leq 4)$ are fixed polynomials in $R[x_1, x_2, \ldots, x_n]$ such that*

> *1. the constant term in each φ_i is 1,*
>
> *2. for each non-constant monomial $r x_1^{p_1} x_2^{p_2} \ldots x_n^{p_n}$ in each φ_i, at least one of the powers p_i is odd,*

3. $\varphi_1 \equiv \varphi_2 \equiv \varphi_3 \equiv \varphi_4 \equiv 0 \bmod 2$,

Then for all $k \geq 1$, if

$$\text{For all } a \in R, \ a^2 \equiv 0 \bmod 2 \Rightarrow a \equiv 0 \bmod 2, \text{ when } k = 1 \quad (12.82)$$

$$R \text{ has the four squares property, when } k = 2 \quad (12.83)$$

$$R \text{ has the strong four squares property, when } k \geq 3 \quad (12.84)$$

and, for $p_1, p_2, p_3, p_4 \in R[x_1, x_2, \ldots, x_n]$

$$\varphi_1 p_1^2 + \varphi_2 p_2^2 + \varphi_3 p_3^2 + \varphi_4 p_4^2 \equiv 0 \bmod 2^k,$$

then

- $p_1 \equiv p_2 \equiv p_3 \equiv p_4 \bmod 2^{k/2}$ *when k is even*
- $p_1 \equiv p_2 \equiv p_3 \equiv p_4 \equiv 0 \bmod 2^{(k-1)/2}$ *and*
  $p_1 + p_2 + p_3 + p_4 \equiv 0 \bmod 2^{(k+1)/2}$ *when k is odd*

Proof. We first note that $(12.84) \Rightarrow (12.83) \Rightarrow (12.82)$. We have already seen the first of these implications; for the second note that if (12.82) fails then there is $r \in R$ with $r^2 \equiv 0 \bmod 2$, $r \not\equiv 0 \bmod 2$, in which case (12.83) fails with $a = b = r$, $c = d = 0$. Thus the hypotheses of the lemma for any k imply those for all smaller k.

For convenience, we suppose that $n = 2$ (the proof for $n > 2$ is an obvious generalization of this). Throughout this proof we order $\mathbb{R}^2$ lexicographically, that is $(a, b) < (c, d)$ when either $a < c$ or $a = c$ and $b < d$.

Suppose that polynomials $\varphi_i(x, y) = \sum c_i(m, n) x^m y^n \in R[x, y]$ are as in the statement of the lemma, and that $p_i(x, y) = \sum a_i(m, n) x^m y^n \in R[x, y]$.

We use induction on k. Let $k \in \mathbb{N}$, and suppose that the theorem holds for smaller values. The hypotheses are

$$\varphi_1(x, y) p_1^2(x) + \varphi_2(x, y) p_2^2(x) + \varphi_3(x, y) p_3^2(x) + \varphi_4(x, y) p_4^2(x) \equiv 0 \bmod 2^k.$$
$$(12.85)$$

together with the conditions (12.82), (12.83) and (12.84) on R according as $k = 1$, $k = 2$ or $k \geq 3$. First suppose that $k = k_1 > 3$, then by the case $k = 3$ each $p_i \equiv 0 \bmod 2$, and by the induction hypothesis we apply the case $k = k_1 - 2$ to the $p_i/2 \in R[x, y]$ to get the required result. We suppose then that $k \leq 3$.

For $k = 1, 2, 3$ respectively, the required result can be stated in terms of coefficients as, for all $n, m \in \mathbb{Z}$,

- $a_1(n, m) + a_2(n, m) + a_3(n, m) + a_4(n, m) \equiv 0 \bmod 2$ (12.86)
- $a_1(n, m) \equiv a_2(n, m) \equiv a_3(n, m) \equiv a_4(n, m) \bmod 2$ (12.87)
- $a_1(n, m) \equiv a_2(n, m) \equiv a_3(n, m) \equiv a_4(n, m) \equiv 0 \bmod 2$ and
  $a_1(n, m) + a_2(n, m) + a_3(n, m) + a_4(n, m) \equiv 0 \bmod 4$ (12.88)

We set $p_i^2(x,y) = \sum s_i(m,n)x^m y^n$ $(1 \leq i \leq 4)$, and define vectors

$$\mathbf{c}(m,n) = (c_1(m,n), c_2(m,n), c_3(m,n), c_4(m,n))$$
$$\mathbf{a}(m,n) = (a_1(m,n), a_2(m,n), a_3(m,n), a_4(m,n))$$
$$\mathbf{s}(m,n) = (s_1(m,n), s_2(m,n), s_3(m,n), s_4(m,n)).$$

We have, for p and q even,

$$s_i(p,q) = 2 \sum_{(s,t) < (p/2, q/2)} a_i(s,t)a_i(p-s, q-t) + a_i^2(p/2, q/2) \quad (12.89)$$

and, for p or q odd

$$s_i(p,q) = 2 \sum_{(s,t) < (p/2, q/2)} a_i(s,t)a_i(p-s, q-t), \quad (12.90)$$

whereupon summing gives us

$$\sum_{i=1}^{4} s_i(p,q) = 2 \sum_{(s,t)<(p/2,q/2)} \mathbf{a}(s,t) \cdot \mathbf{a}(p-s, q-t) + \sum_{i=1}^{4} a_i^2(p/2, q/2), \quad (12.91)$$

with the last sum only present when p and q are even.

Equating the coefficient of $x^{2n}y^{2m}$ in the left side of (12.85) to 0 mod 2^k gives, using the second hypotheses on the φ_i,

$$\sum_{\substack{p + p' = 2n, q + q' = 2m \\ p \text{ or } q \text{ is odd}}} \mathbf{s}(p,q) \cdot \mathbf{c}(p',q') + \sum_{i=1}^{4} s_i(2n, 2m) \equiv 0 \bmod 2^k. \quad (12.92)$$

By (12.90), each term in the first sum is even, whence using (12.91) with $p = 2n$, $q = 2m$, $\sum_{i=1}^{4} a_i^2(n,m) \equiv 0 \bmod 2$. Since $\sum a_i^2(n,m) \equiv \left(\sum a_i(n,m) \right)^2 \bmod 2$, (12.82) gives the congruence (12.86). For $k = 1$, this completes the proof.

We now prove (12.87) for $k = 2$ and (12.88) for $k = 3$ by induction on (n,m). Suppose the result holds for all $(s,t) < (n,m)$. We first show, for $p \leq 2n$, $q \leq 2m$ with p or q odd, that

$$s_i(p,q) \equiv 0 \bmod 2^{k-1}, \quad (i = 1, 2, 3, 4) \quad (12.93)$$

$$\sum_{i=1}^{4} s_i(2n, 2m) \equiv \sum_{i=1}^{4} a_i^2(n,m) \bmod 2^k, \quad \text{and} \quad (12.94)$$

$$\sum_{i=1}^{4} s_i(p,q) \equiv 0 \bmod 2^k, \quad (12.95)$$

The first of these follows from (12.90) and the induction hypothesis, and the other two from (12.91) since, for $(s,t) < (p/2, q/2) \leq (n,m)$, and $k = 2$

$$\mathbf{a}(s,t) \cdot \mathbf{a}(p-s, q-t) \equiv a_1(s,t) \sum_{i=1}^{4} a_i(p-s, q-t)$$

$$\equiv 0 \bmod 2,$$

using the induction hypothesis at the first congruence, and (12.86) at the second.

For $k = 3$, by the induction hypothesis, each $a_i(s,t)$ is even, for $(s,t) < (p/2, q/2) \leq (n,m)$, and

$$\frac{1}{2}\mathbf{a}(s,t) \cdot \mathbf{a}(p-s, q-t) \equiv a_1(p-s, q-t) \sum_{i=1}^{4} \frac{a_i(s,t)}{2}$$

$$\equiv 0 \bmod 2 \quad \text{by (12.88)}.$$

Here we use the lemma for $k = 2$, which gives (12.87), at the first step and the induction hypothesis at the second.

Recalling that $c_1(p', q') \equiv c_2(p', q') \equiv c_3(p', q') \equiv c_4(p', q') \bmod 2$, by the third hypothesis on the φ_i, and using (12.93), the summand in the first sum of (12.92) is

$$\mathbf{s}(p,q) \cdot \mathbf{c}(p', q') = \sum_{i=1}^{4} s_i(p,q)c_i(p', q')$$

$$= 2^{k-1} \sum_{i=1}^{4} \left(\frac{s_i(p,q)}{2^{k-1}} \right) c_i(p', q')$$

$$\equiv 2^{k-1} c_1(p', q') \sum_{i=1}^{4} \frac{s_i(p,q)}{2^{k-1}} \bmod 2^k$$

$$= c_1(p', q') \sum_{i=1}^{4} s_i(p,q)$$

$$\equiv 0 \bmod 2^k \quad \text{(by (12.95))}$$

This, together with (12.92) and (12.94) gives $a_1^2(n,m) + a_2^2(n,m) + a_3^2(n,m) + a_4^2(n,m) \equiv 0 \bmod 2^k$, whence by hypothesis we get (12.87) for $k = 2$ and the first half of (12.88) for $k = 3$,

When $k = 3$ we have $((a_1(n,m) + a_2(n,m) + a_3(n,m) + a_4(n,m))/2)^2 \equiv (a_1(n,m)/2)^2 + (a_1(n,m)/2)^2 + (a_1(n,m)/2)^2 + (a_1(n,m)/2)^2 \equiv 0 \bmod 2$, whence using (12.82) we complete the proof of (12.88). $\qquad \square$

We first note a simple special case ($\varphi_1 = \varphi_2 = \varphi_3 = \varphi_4 = 1$, $k = 2, 3$).

Corollary 12.8.1 *If the ring R satisfies the (strong) four squares property (12.80), then so does the polynomial ring $R[x_1, x_2, \ldots, x_n]$. In particular this is true of $\mathbb{Z}[x_1, x_2, \ldots, x_n]$.*

Corollary 12.8.2 *If $a, b \in \mathbb{Z}[u_1, u_2 \ldots u_k]$ can be written $a \equiv \alpha\alpha'$ mod 4, $b \equiv \beta\beta'$ mod 4, with $\alpha, \alpha', \beta, \beta' \in \mathbb{Z}[u_1, u_2 \ldots u_k]$ satisfying $\alpha \equiv \alpha'$ mod 2, $\beta \equiv \beta'$ mod 2, and $\varphi_1 = \alpha\beta$, $\varphi_2 = -\alpha'\beta$, $\varphi_3 = -\alpha\beta'$, $\varphi_4 = \alpha'\beta'$ satisfy the hypotheses of Lemma 12.8.1, then for $R, S, T, W \in \mathbb{Z}[u_1, u_2 \ldots u_k]$,*

$$R^2 - aS^2 - bT^2 + abW^2 \equiv 0 \text{ mod } 4 \Rightarrow R \equiv aS \equiv bT \equiv abW \text{ mod } 2 \quad (12.96)$$

and if $a, b \not\equiv 0$ mod 2,

$$R^2 - aS^2 - bT^2 + abW^2 \equiv 0 \text{ mod } 8 \Rightarrow R \equiv S \equiv T \equiv W \equiv 0 \text{ mod } 2 \quad (12.97)$$

Proof. Multiplying the left-hand side of (12.96) through by $\alpha\beta$, and setting $r = R$, $s = \alpha S$, $t = \beta T$ and $w = \alpha\beta W$, gives the equivalent form

$$\alpha\beta r^2 - \alpha'\beta s^2 - \alpha\beta' t^2 + \alpha'\beta' w^2 \equiv 0 \text{ mod } 4. \quad (12.98)$$

Lemma 12.8.1 then gives (12.96) with $k = 2$, and (12.97) with $k = 3$. $\square$

Theorem 12.8.3 *Let R be an integral domain of characteristic $\neq 2$, with field of fractions K. Let $a, b, \alpha, \beta \in R$ be such that $\alpha^2 \equiv a$ mod 2 and $\beta^2 \equiv b$ mod 2, and let $\mathcal{Q}$ be the quaternion algebra $\frac{a,b}{K}$. Let O comprise the quaternions of the form $\mathbf{r} + \frac{r}{2}\mathbf{c}$, where $\mathbf{c} = (\alpha\beta, \beta, \alpha, 1)$, and r and the components of $\mathbf{r}$ are in R, then*

>   *1. O is an order in $\mathcal{Q}$, and $N(\mathbf{u}) \in R$ for each $\mathbf{u} \in O$. If moreover R has the property that $2|x^2 \Rightarrow 2|x$ for all $x \in R$, then O is independent of the choice of α and β.*

>   *2. If further*

>   *(a) R is integrally closed*

>   *(b) 2 is prime in R*

>   *(c) a and b are not divisible by 2;*

>   *(d) If a and b divide x, then ab divides x.*

>   *(e) If $b|y^2 - ax^2$, then $b|x, y$, and if $a|y^2 - bx^2$, then $a|x, y$.*

>   *(f) If $\mathbf{u} = (x, y, z, w) \in R^4$, and $N(\mathbf{u}) = x^2 - ay^2 - bz^2 + abw^2 \equiv 0$ mod 4, then $x \equiv \alpha y \equiv \beta z \equiv \alpha\beta w$ mod 2,*

>   *then every quaternion in $\frac{1}{2}R^4$ with norm in R is in O, and every order which contains $\mathbf{i}$ and $\mathbf{j}$ lies in O. In particular, O is maximal.*

Proof. Clearly O is an ideal. For $\mathbf{u} = (x, y, z, w) \in R^4$, a straightforward calculation gives $\mathbf{uc} \equiv \mathbf{cu} \equiv (x + \alpha y + \beta z + \alpha\beta w)\mathbf{c}$ mod 2 and $\mathbf{c}^2 = -(b - \beta^2)(a - \alpha^2)1 + 2\alpha\beta\mathbf{c} \equiv 2\alpha\beta\mathbf{c}$ mod 4, from which it follows that O is also a ring. Since $N(\mathbf{c}) = (\alpha^2 - a)(\beta^2 - b) \equiv 0$ mod 4, it readily follows that $N(\mathbf{u}) \in R$ for $\mathbf{u} \in O$.

If $2|x^2 \Rightarrow 2|x$ for all $x \in R$, and $\alpha_1, \beta_1 \in R$ satisfy $\alpha_1^2 \equiv a \bmod 2$ and $\beta_1^2 \equiv b \bmod 2$, then

$$(\alpha_1 - \alpha)^2 \equiv (\alpha_1 - \alpha)(\alpha_1 + \alpha) = \alpha_1^2 - \alpha^2 \equiv 0 \bmod 2,$$

so by hypothesis $\alpha_1 \equiv \alpha \bmod 2$, and similarly $\beta_1 \equiv \beta \bmod 2$. It follows that $(\alpha_1\beta_1, \beta_1, \alpha_1, 1) = (\alpha\beta, \beta, \alpha, 1) \bmod 2$, so that the definition of O is independent of the choice of α and β.

Now suppose that (2a)–(2f) hold. If $\mathbf{u} = (x, y, z, w) \in \frac{1}{2}R^4$ and $N(\mathbf{u}) \in R$, then (2f) applied to $2\mathbf{u}$, gives $2x \equiv \alpha 2y \equiv \beta 2z \equiv \alpha\beta 2w \bmod 2$. By (2b) and (2c) we may cancel modulo 2 to obtain $2x \equiv 2\alpha\beta w$, $2y \equiv 2\beta w$ and $2z \equiv 2\alpha w$ (all mod 2). That is $2\mathbf{u} \equiv 2w\mathbf{c} \bmod 2$, so $\mathbf{u} \in O$. (so far using only (2b), (2c) and (2f))

Now let O' be an order which contains $\mathbf{i}$ and $\mathbf{j}$, and suppose $\mathbf{v} = (x, y, z, w) \in O'$, then because R is integrally closed, $\mathrm{tr}(\mathbf{v})$, $\mathrm{tr}(\mathbf{iv})$, $\mathrm{tr}(\mathbf{jv})$, $\mathrm{tr}(\mathbf{kv})$ and $N(\mathbf{v})$ are all in R ([30], Corollary 3.6). These give in turn $2x \in R$, $2ay \in R$, $2bz \in R$, $2abw \in R$ and $x^2 - ay^2 - bz^2 + abw^2 \in R$. Setting $X = 2x$, $Y = 2ay$, $Z = 2bz$, $W = 2abw$, multiplying the last equation by $4ab$ gives

$$abX^2 - bY^2 - aZ^2 + W^2 \in 4abR \tag{12.99}$$

whence

$$W^2 - bY^2 \in aR \qquad W^2 - aZ^2 \in bR \tag{12.100}$$

By (2e) it follows that $a|W, Y$ and $b|W, Z$, which together with (2d) also gives $ab|W$. It follows that $x, y, z, w \in \frac{1}{2}R$, and since $N(\mathbf{v}) \in R$, the first statement then gives $\mathbf{v} \in O$. $\qquad\square$

Theorem 12.8.3 with $R = \mathbb{Z}$ and $a = b = -1$, $\alpha = \beta = 1$ gives Theorem 12.8.2. In this case (2f) is the statement that $\mathbb{Z}$ has the four squares property.

Proof of Theorem 12.8.1. First, we show that any element of $\mathcal{O}$ has half-integer coefficients. Let $\mathbf{u} = (R, S, T, W) \in \mathcal{O}$, let d be the lowest common denominator of all the coefficients (reduced as far as possible) of the components of $\mathbf{u}$, then $d\mathbf{u} \in \mathbb{Z}[u, v]^4$ and $N(d\mathbf{u}) \in d^2\mathbb{Z}[u, v]$. If d is divisible by an odd prime p, then reducing the coefficients in $d\mathbf{u}$ mod p we obtain a nonzero quaternion in $\mathcal{Q}_p := \left(\frac{u^2-1, v^2-1}{\mathbb{Z}_p[u,v]} \right)$, which has zero norm, but this is impossible as we will show that $\mathcal{Q}_p$ is a division algebra. By [24], Theorem 2.3.1, it suffices to show that the equation

$$(u^2 - 1)p^2(u, v) + (v^2 - 1)q^2(u, v) = 1$$

has no solution with $p, q \in \mathbb{Z}_p(u, v)$. Setting $v = 1$ this equation becomes $(u^2 - 1)p^2(u, 1) = 1$, which clearly has no solution, as $(u^2 - 1)$ is not a square. So we conclude that d is a power of 2. If d were a multiple of 4, then $d\mathbf{u}$ would have integer coefficients, not all even, and norm divisible by 16, but the second part of Corollary 12.8.2, with $a = u^2 - 1$, $b = v^2 - 1$, $\alpha = u + 1$, $\beta = v + 1$, $\alpha' = u - 1$, $\beta' = v - 1$, shows that this is impossible.

To complete the proof, we apply Theorem 12.8.3 with $R = \mathbb{Z}[u, v]$, $a = u^2 - 1$, $b = v^2 - 1$, $\alpha = u + 1$, $\beta = v + 1$. In this case, we can easily verify (2a)-(2d). To prove (2e), let $u^2 - 1 | p^2(u, v) + (1 - v^2)q^2(u, v)$, where $p, q \in \mathbb{Z}[u, v]$. For all $v \in (-1, 1)$, both summands on the right-hand side are nonnegative. Hence, when $u = \pm 1$, both vanish. It follows that $u^2 - 1$ divides p and q. Together with the corresponding statement obtained by interchanging u and v, this gives (2e). Finally, the first part of Corollary 12.8.2 gives (2f). $\square$

Lemma 12.8.2 *There is a member of $u \in \mathcal{U}_1$ with irrational coefficients.*

An example is the quartic

$$\mathbf{u} = [(1 - u^2)(a - av^2 + v^2) + u^2 v, (v - 1)((b - au)(v + 1) + uv),$$
$$(1 - a)(1 - v)(1 - u^2) + u, a + bv - u(a - 1)(v - 1)],$$

which has norm 1 whenever $2a - 3a^2 - b^2 = 1 - 2a + a^2 - ab = 0$. A routine calculation shows that these have real solutions $a = b = 1/2$, and where a and b are the (unique) real roots of $2x^3 - 2x^2 + 2x - 1$ and $2x^3 + 6x^2 + 4x - 1$ respectively. These roots are not rational.

12.8.1   Generation

Here we consider the question as to whether or not $\mathcal{U}$ finitely generated. We thank Alan Reid for providing us with a simpler proof than our earlier argument based on arithmetic Kleinian groups.

Theorem 12.8.4 *The group $\mathcal{U}$ is not finitely generated.*

Proof. We may identify $\mathcal{U}$ is the obvious way with the group of elements of norm 1 in the quaternion algebra

$$\mathcal{Q}_{\mathbb{Q}} = \left(\frac{u^2 - 1, v^2 - 1}{\mathbb{Q}(u, v)} \right), \tag{12.101}$$

Suppose that O is a (maximal) order in $\mathcal{Q}_{\mathbb{Q}}$ and simply specialize u, v as follows. Put $u = 0$ and, for p a prime $p \geq 3$,

$$v = \frac{1}{2}(p + \sqrt{p^2 - 4}),$$

which has conjugate $\bar{v} = \frac{1}{2}\left(p - \sqrt{p^2 - 4}\right) \in (-1, 1)$. Then $\mathcal{Q}_{\mathbb{Q}}$ has homomorphic image

$$\left(\frac{-1, \frac{1}{2}(p^2 - 2 + p\sqrt{p^2 - 4})}{\mathbb{Q}(\sqrt{p^2 - 4})} \right), \tag{12.102}$$

Apart from the identity, the other real embedding is $\sigma(v) = \bar{v}$ and so $\sigma(v^2 - 1) = \bar{v}^2 - 1 < 0$. Hence the group of elements of norm 1 in the order O so specialized is some arithmetic Fuchsian group coming from a division algebra over $\mathcal{Q}_\mathbb{Q}(v)$, see Theorem 12.9.1 below.

Now the rank of this group must go to infinity with p as there are only finitely many arithmetic Fuchsian groups whose quotients are surfaces of a given topological type, [24]. In particular, this implies the group of elements of norm 1 of O cannot be finitely generated. $\qquad\square$

Calculation shows that the 5-element set

$$\left\{(u,1,0,0),\ (v,0,1,0),\ (u,v,0,-1),\tfrac{1}{2}(1+u+v-uv,v-1,1-u,-1),\right.$$

$$\left.\tfrac{1}{2}(1+u+v-uv,v-1,1-u,1)\right\}, \tag{12.103}$$

each of which is of degree 1 or 2, generates every member of $\mathcal{U}$ of degree at most 4. However, we also have for example (proof omitted) that $\mathbf{u}$ below does not lie in the subgroup generated by these elements.

$$\mathbf{u} = \frac{1}{2}(-1+u^2-2u^3-v^2+3u^2v^2+2u^3v^2, 1-u+2u^2-v^2+5uv^2-2u^2v^2,$$

$$1-u^2+v-2uv+u^2v+4u^3v, 1+u-v+3uv-4u^2v)$$

12.8.2 Quaternions as isometries

We now look at what happens to the members of $\mathcal{U}$ under the evaluation map $\psi_{\lambda,\lambda',\mu}$. We will assume for the moment that $\lambda,\mu \notin [-1,1]$ and $\lambda' = -1$, and abbreviate $\psi_{\lambda,-1,\mu}$ to ψ. We set $\Gamma = \Gamma(\lambda,\mu) = \psi(\mathcal{U})$. Now (12.69) and (12.70) become

$$a = -d = \frac{1}{\sqrt{2}}\sqrt{-1-\mu}, \quad c = -b = \frac{1}{\sqrt{2}}\sqrt{1-\mu} \tag{12.104}$$

and in addition

$$ab = cd = \frac{1}{2}\sqrt{1-\mu}\sqrt{-1-\mu} = \pm\frac{1}{2}\sqrt{\mu^2-1},$$

and so $\psi((R,S,T,W))$ is

$$\begin{bmatrix} R_w + S_w\sqrt{\lambda^2-1} & \pm(T_w + W_w\sqrt{\lambda^2-1})\sqrt{\mu^2-1} \\ \pm(T_w - W_w\sqrt{\lambda^2-1})\sqrt{\mu^2-1} & R_w - S_w\sqrt{\lambda^2-1} \end{bmatrix}.$$

$$\tag{12.105}$$

First we revisit the three quaternions $\mathbf{w}_i \in \mathcal{V}_0$ $(i = 1,2,3)$ defined at

(12.54), which have images $\tilde{\mathbf{w}}_i := \rho(\mathbf{w_i}) \in \mathcal{V}$, namely

$$
\begin{aligned}
\tilde{\mathbf{w}}_1 &= (u, 1, 0, 0), \\
\tilde{\mathbf{w}}_2 &= (u, v, 0, -1), \\
\tilde{\mathbf{w}}_3 &= \frac{1}{2}(1 + u + v - uv, v - 1, 1 - u, -1),
\end{aligned}
\tag{12.106}
$$

As we have already seen (or directly from (12.105)), these map respectively to the isometries f^2, gf^2g^{-1} and $[g, f]$, where $\lambda(f) = \lambda$, $\mu(f, g) = \mu$, f is loxodromic (since $\lambda \notin [-1, 1]$), $\mathrm{ax}(f) = (0, \infty)$, and g is an order 2 elliptic whose axis is disjoint from $\mathrm{ax}(f)$ (since $\mu \notin [-1, 1]$), and has mutually reciprocal endpoints. As noted in the remarks after the proof of Theorem 12.4.1, this means that the common perpendicular of $\mathrm{ax}(f)$ and $\mathrm{ax}(gf^2g^{-1})$ has endpoints ± 1.

We have now got back the subgroup of Γ comprising the balanced even words in f and g (since g is order 2, the distinction between regular and irregular words now vanishes). We can now extend this subgroup. Let $\varphi_f(z) = -z$, $\varphi_h(z) = 1/z$ $\varphi(z) = -1/z$; these three isometries are each of order 2, have mutually orthogonal axes and generate a Klein 4-group, K. We define $h = g\varphi_f$. Recall (12.68) that g has matrix representative $B = \begin{bmatrix} a & b \\ c & d \end{bmatrix}$. Thus, using (12.104), h and h^2 have respective matrix representatives

$$
M_h = \frac{1}{\sqrt{2}} \begin{bmatrix} \sqrt{\mu+1} & \pm\sqrt{\mu-1} \\ \pm\sqrt{\mu-1} & \sqrt{\mu+1} \end{bmatrix}, \quad M_{h^2} = \begin{bmatrix} \mu & \pm\sqrt{\mu^2-1} \\ \pm\sqrt{\mu^2-1} & \mu \end{bmatrix}
$$

The axis of h has endpoints ± 1, and $\lambda(h) = \mu$. Also $h^2 \in \Gamma$; specifically $h^2 = \psi((v, 0, \pm 1, 0))$.

At this point a certain symmetry between f and h is becoming apparent. Both are loxodromic, both have squares in Γ and their axes are mutually perpendicular. To develop this symmetry further we express h, like f, as a product of two order 2 elliptics. Set $\tilde{g} = f\varphi_h$; explicitly, $\tilde{g}(z) = A/z$, where $f(z) = Az$, so that $\tilde{g}$ has order 2. We now have

$$
\begin{aligned}
g\varphi_f &= h, & \tilde{g}\varphi_h &= f, \\
\lambda(h) &= \mu(f, g), & \lambda(f) &= \mu(h, \tilde{g}), \\
\mathrm{ax}(\varphi_f) &= \mathrm{ax}(f), & \mathrm{ax}(\varphi_h) &= \mathrm{ax}(h).
\end{aligned}
$$

We can summarise all this by saying that the pair $(h, \tilde{g})$ is obtained from (f, g) (up to conjugacy) by interchanging the parameters λ and μ.

Theorem 12.8.5 *The subgroup P of $\langle f, h \rangle$ comprising the isometries of the form $f^{n_1} h^{m_1} f^{n_2} h^{m_2} \dots f^{n_k} h^{m_k}$, where $n_1 + n_2 + \dots n_k$ and $m_1 + m_2 + \dots m_k$ are both even, is a subgroup of Γ.*

Sketch of Proof. We first show that $P = \langle f^2, h^2, fh^2f^{-1}, hf^2h^{-1} \rangle$. This can be done using induction along the same lines as the proof of Lemma 12.5.2. We have already seen that $f^2, h^2 \in \Gamma$. The proof is completed by showing that fh^2f^{-1}, hf^2h^{-1} have respective matrix representatives

$$M_{fh^2f^{-1}} = \begin{bmatrix} \mu & \pm(\lambda + \sqrt{\lambda^2 - 1})\sqrt{\mu^2 - 1} \\ \pm(\lambda - \sqrt{\lambda^2 - 1})\sqrt{\mu^2 - 1} & \mu \end{bmatrix}$$
$$= \psi((v, 0, \pm u, \pm 1))$$

$$M_{hf^2h^{-1}} = \begin{bmatrix} \lambda + \mu\sqrt{\lambda^2 - 1} & \mp\sqrt{\lambda^2 - 1}\sqrt{\mu^2 - 1} \\ \pm\sqrt{\lambda^2 - 1}\sqrt{\mu^2 - 1} & \lambda - \mu\sqrt{\lambda^2 - 1} \end{bmatrix}$$
$$= \psi((u, v, 0, \mp 1)).$$

$\square$

Clearly P is a finite-index subgroup of $\langle f, h \rangle$, and it follows in particular that if Γ is discrete, then so is $\langle f, h \rangle$. Further (see (12.105)) a sufficient condition for this is that λ and μ both lie in a discrete subring of $\mathbb{C}$ (i.e. a subring of the ring of integers of some imaginary quadratic field).

Corollary 12.8.3 *If R is a discrete subring of $\mathbb{C}$, f and h are non-parabolic non-identity isometries in* $\mathrm{Isom}^+(\mathbb{H}^3)$ *with perpendicular axes, and* $\lambda(f), \lambda(h) \in R$, *then* $\langle f, h \rangle$ *is discrete.*

In particular we have discreteness when $\lambda(f), \lambda(h)$ are integers. Another discrete example is $\lambda(f) = \lambda(h) = \frac{-1+\sqrt{3}}{2}$, which minimizes $\max\{\tau_f, \tau_h\}$ among all two generator non-elementary groups having loxodromic generators with perpendicular axes, [23]. We discuss discreteness criteria further in Section 12.9.

Additionally, we can add in all the order 2 elliptics and preserve discreteness. These elliptics fall into three Klein 4-groups: $K := \{\varphi_f, \varphi_h, \varphi\}$, $K_f := \{\varphi_f, \tilde{g}, \tilde{g}\varphi_f\}$ and $K_h := \{\varphi_h, g, g\varphi_h\}$.

Theorem 12.8.6 *The group P_1 generated by K K_f and K_h is an extension of* $\langle f, h \rangle$ *of index at most 2.*

Proof. Every $\alpha \in P_1$ can be represented by a word in $\{\varphi_f, \varphi_h, \varphi, g, \tilde{g}, f, h\}$ which we suppose to have the fewest possible elliptic letters, and with the first elliptic letter occurring as close to the right as possible. If $a \in K \cup K_f$ then $afa = f^{\pm 1}$, so that $af = f^{\pm 1}a$, and similarly $af^{-1} = f^{\mp 1}a$. If $a = g$ then $af^{\pm 1} = gf^{\pm 1} = g(\varphi_f f^{\pm 1}\varphi_f^{-1})g^{-1}g = hf^{\pm 1}h^{-1}g$, because f and φ_f commute. By our assumptions about the word, it follows that no elliptic letter can immediately precede an $f^{\pm 1}$, and symmetrically it cannot immediately precede an $h^{\pm 1}$ either. It follows that all of the elliptic letters are at the right of the word. But the product of any two elliptics is either another elliptic or a product of (at most two) of the loxodromics $f^{\pm 1}$ and $h^{\pm 1}$, so the word contains at most one elliptic letter. Thus P_1 is an extension of $\langle f, h \rangle$ of index at most 2. $\square$

12.9 Arithmeticity

In this section we first recall some further terminology concerning quaternion algebras with an aim to extending the discreteness conditions described above. This section is adapted from Section 4 of [9].

Let k be a number field. A *place* ν of k is an equivalence class of valuations on k. Such a place is real (complex) if it is associated to a real embedding (conjugate pair of complex embeddings) of k. We denote by k_ν the completion of k at the place ν. If Q is a quaternion algebra over k, we say that Q is ramified at ν if $Q \otimes_k k_\nu$ is a division algebra of quaternions. Otherwise ν is unramified. If ν is a real place, then Q is ramified if and only if $Q \otimes_k k_\nu \equiv \mathbb{H}$. It is straightforward to check whether a quaternion algebra $Q = \left(\frac{a,b}{k}\right)$ is ramified at a real place ν; if ν corresponds to the real embedding σ, then Q is ramified at ν if and only if $\sigma(a)$ and $\sigma(b)$ are both negative.

We can now define an arithmetic Kleinian group. Let k be a number field with one complex place and Q a quaternion algebra over k ramified at all real places. Next let ρ be an embedding of Q into $SL(2, \mathbb{C})$, let O be an order of Q and O_1 the elements of norm 1 in O. Then $\rho(O_1)$ is a discrete subgroup of $SL(2, \mathbb{C})$ and its projection to $PSL(2, \mathbb{C})$ is an Kleinian group. Kleinian groups so constructed, together with those which are commensurable to them, are *arithmetic*. We note in passing that arithmetic Fuchsian groups arise in a similar manner. However, in that case, the number field is totally real and the algebra ramified at all real places except the identity.

For a subgroup Γ of $SL(2, \mathbb{C})$ the invariant trace field is defined as

$$k\Gamma = \mathbb{Q}(\{\operatorname{tr}^2(g) : g \in \Gamma\}) \tag{12.107}$$

Then we set

$$Q\Gamma = \left\{ \sum a_i\, g_i : a_i \in \mathbb{Q}(\operatorname{tr}(\Gamma)), g_i \in \Gamma \right\}$$

Then $Q\Gamma$ is a quaternion algebra over $\mathbb{Q}(\operatorname{tr}(\Gamma))$. Additionally, if $\operatorname{tr}(\Gamma)$ consists of algebraic integers we see that

$$O\Gamma = \left\{ \sum a_i\, g_i : a_i \in R_{\mathbb{Q}(\operatorname{tr}(\gamma))}, g_i \in \Gamma \right\}$$

is an order in $Q\Gamma$. Here $R_{\mathbb{Q}(\operatorname{tr}(\gamma))}$ is the ring of integers in $\mathbb{Q}(\operatorname{tr}(\Gamma))$. Then Γ is arithmetic if and only if the following conditions are satisfied:

1. $k\Gamma$ is an algebraic number field;

2. $\operatorname{tr}(\Gamma)$ consists of algebraic integers;

3. for every $\mathbb{Q}$-isomorphism $\sigma : k\Gamma \to \mathbb{C}$, other than the identity or complex conjugation, $\sigma(\operatorname{tr}(\Gamma^{(2)}))$ is bounded in $\mathbb{C}$.

In practice, it is hard to apply this characterization directly, the problem being to establish the boundedness of the traces at real embeddings. However, in [9] we obtained the following more useful method for proving groups discrete.

Theorem 12.9.1 *Let Γ be a finitely generated non-elementary subgroup of $SL(2, \mathbb{C})$ such that*

1. *$k\Gamma$ has exactly one complex place or is totally real;*

2. *$\mathrm{tr}(\Gamma)$ consists of algebraic integers;*

3. *$Q\Gamma^{(2)}$ is ramified at all non-identity real places of $k\Gamma$,*

then Γ is a subgroup of an arithmetic Kleinian or Fuchsian group.

Corollary 12.9.1 *A group of elements $\mathcal{G}$ of norm 1 in an order $\mathcal{O}$ of the quaternion algebra $\mathcal{Q}_{\mathbb{Q}}$,*

$$\mathcal{Q}_{\mathbb{Q}} = \left(\frac{u^2 - 1, v^2 - 1}{\mathbb{Q}(u, v)} \right)$$

is a discrete subgroup of an arithmetic Kleinian group if

- *u is a complex algebraic integer with irreducible polynomial of degree n, which has $n - 2$ real conjugates, $r_1, r_2, \ldots, r_{n-2}$, all of which lie in the interval $(-1, 1)$.*

- *v is an algebraic integer in $\mathbb{Q}(u)$.*

- *for each non-identity real embedding $\sigma_i : \mathbb{Q}(u) \to \mathbb{Q}$, $\sigma|\mathbb{Q} = \mathrm{id}$, defined by $\sigma_i(u) = r_i$, the image $\sigma(v) \in (-1, 1)$.*

Proof. The first condition gives $\mathbb{Q}(u)$ a number field of degree n over $\mathbb{Q}$ with one complex place, and $v \in \mathbb{Q}(u)$ then gives $\mathbb{Q}(u, v) = \mathbb{Q}(u)$. If σ_i is a real embedding, then $\sigma_i(u^2 - 1) = \sigma_i(u)^2 - 1 = r_i^2 - 1 < 0$, and $\sigma_i(v^2 - 1) < 0$ by hypothesis, so the quaternion algebra is ramified at all the real places. Next, the trace is $2R(u, v) \in \mathbb{Z}(u, v)$, which must be an algebraic integer since both u and v are. $\qquad\square$

12.10 Discreteness: Necessary Conditions

Let $\Gamma := \psi_{\lambda, \lambda', \mu}(\mathcal{U})$, a subgroup of $SL(2, \mathbb{C})$ which, we recall, comprises the matrices of the form

$$W = \begin{bmatrix} R(\lambda, \mu) + S(\lambda, \mu)\sqrt{\lambda^2 - 1} & 2ab(T(\lambda, \mu) + W(\lambda, \mu)\sqrt{\lambda^2 - 1}) \\ 2cd(T(\lambda, \mu) - W(\lambda, \mu)\sqrt{\lambda^2 - 1}) & R(\lambda, \mu) - S(\lambda, \mu)\sqrt{\lambda^2 - 1} \end{bmatrix}$$

where the polynomials R, S, T and W are polynomials with half-integer coefficients satisfying (12.79), with

$$a = \frac{1}{\sqrt{2}}\left(\sqrt{\lambda'+1}+\sqrt{\lambda'-\mu}\right), \quad d = \frac{1}{\sqrt{2}}\left(\sqrt{\lambda'+1}-\sqrt{\lambda'-\mu}\right),$$

and b and c are given by (12.19) when $\mu = 1$, and $c = -b = \sqrt{\frac{1-\mu}{2}}$ otherwise.

We know that Γ extends the group of regular even balanced words in A and B given by

$$A = \begin{pmatrix} \frac{\lambda+\sqrt{\lambda^2-1}-1}{\sqrt{2}\sqrt{\lambda-1}} & 0 \\ 0 & \frac{-\lambda+\sqrt{\lambda^2-1}+1}{\sqrt{2}\sqrt{\lambda-1}} \end{pmatrix}, \quad B = \begin{pmatrix} a & b \\ c & d \end{pmatrix}, \qquad (12.108)$$

As we have observed (Theorem 12.5.1) this group is an index two subgroup of the group of all regular balanced words in A and B. Similarly we can show that Γ an index two subgroup of the group $\tilde{\Gamma}$ generated by Γ and A. To see this, it is enough to show that, for $G \in \Gamma$, $AGA^{-1} \in \Gamma$, a straightforward calculation using (12.11). It follows in particular that Γ is discrete if and only if $\tilde{\Gamma}$ is.

Another routine calculation, using (12.14) and the facts that $\beta(A) = 2(\lambda - 1)$ and

$$4abcd = bc(1+bc) = 4\left(\frac{\mu-1}{2}\right)\left(1+\frac{\mu-1}{2}\right) = \mu^2 - 1,$$

gives

$$\gamma(A, W) = \operatorname{tr}[A, W] - 2 = -8(\lambda-1)abcd\left(T^2(\lambda, \mu) - W^2(\lambda, \mu)(\lambda^2 - 1)\right)$$
$$= -2(\lambda-1)(\mu^2-1)\left(T^2(\lambda, \mu) - W^2(\lambda, \mu)(\lambda^2 - 1)\right)$$

Since $R^2 - (\lambda^2 - 1)S^2 - (\mu^2 - 1)(T^2 - (\lambda^2 - 1)W^2) = 1$ we then have

$$\gamma(A, W) - \beta(A) = -2(\lambda-1)\left[R^2(\lambda, \mu) - (\lambda^2 - 1)S^2(\lambda, \mu)\right]$$

We write $\beta = \beta(A)$ and $\tilde{\gamma} = \gamma(A, W)$ to obtain the following three identities:

$$|\beta| + |\tilde{\gamma}| = 2|\lambda - 1|\left(1 + |\mu^2 - 1||T^2(\lambda, \mu) - W^2(\lambda, \mu)(\lambda^2 - 1)|\right)$$
$$|\beta| + |\beta - \tilde{\gamma}| = 2|\lambda - 1|\left(1 + |R^2(\lambda, \mu) - (\lambda^2 - 1)S^2(\lambda, \mu)|\right)$$

and $|\tilde{\gamma}||\beta - \tilde{\gamma}|$

$$= 4|\lambda - 1|^2|\mu^2 - 1||R^2(\lambda, \mu) - (\lambda^2 - 1)S^2(\lambda, \mu)||T^2(\lambda, \mu) - W^2(\lambda, \mu)(\lambda^2 - 1)|$$

The three equations above enable us to use the following test for the discreteness of $\tilde{\Gamma}$ (and so of Γ).

Theorem 12.10.1 *Let $c_0 = 2 - 2\cos(\pi/7) \approx 0.198062$ and $\lambda^2, \mu^2 \neq 1$. Then with the notation above, Γ is discrete if and only if for every $W \in \Gamma$ the following three inequalities hold.*

$$|\beta| + |\tilde{\gamma}| \geq 1, \qquad \text{if } \tilde{\gamma} \neq 0, \text{ and} \qquad (12.109)$$

$$|\beta| + |\beta - \tilde{\gamma}| \geq 1, \qquad \text{if } \tilde{\gamma} \neq \beta, \text{ and} \qquad (12.110)$$

$$|\tilde{\gamma}||\tilde{\gamma} - \beta| \geq c_0, \qquad \text{if } \tilde{\gamma} \neq 0, \beta. \qquad (12.111)$$

Proof. [$\Rightarrow$] First suppose $\{W_i\}_{i=1}^{\infty} \subset \Gamma$ is an infinite sequence, that $W_i \to id$ as $i \to \infty$, and that (with the obvious notation) $\gamma_i \neq 0, \beta$. Then of course ultimately the last inequality is violated since $\gamma_i = \gamma(A, W_i) = \text{tr}[A, W_i] - 2 \to 0$. To remove the assumption that $\gamma_i \neq 0, \beta$ we consider two cases.

Case 1. $\gamma_i = 0$ for infinitely many i. Then W_i has a fixed point in common with A in $\hat{\mathbb{C}}$. Now $X = BAB^{-1} \in \Gamma$ (only if B is order 2). If X shares a fixed point with A, or maps one fixed point to another, then A and XAX^{-1} have a common fixed point in $\hat{\mathbb{C}}$ and hence

$$0 = \gamma(A, X) = \gamma(A, BAB^{-1}) = \gamma(A, B)(\gamma(A, B) - \beta)$$

so $\gamma(A, B) = 0$ or $\gamma(A, B) = \beta$. However, $\mu = 1 - 2\gamma/\beta \in \{\pm 1\}$ in either case, and this is excluded by hypothesis. We now deduce that $X^{-1}W_iX$ does not share a fixed point with A for infinitely many i, and $XW_iX^{-1} \to id$.

Case 2. $\gamma_i = \beta$ for infinitely many i. Then $V_i = W_iAW_i^{-1} \to A$, $0 = \gamma(A, V_i) = \gamma(A, A^{-1}V_i)$ and so we reduce to the first case by replacing W_i by $[A^{-1}, W_i] \to id$.

[$\Leftarrow$] Next, suppose the group Γ is discrete, but one of these inequalities is violated for some $W \in \Gamma$. The first two inequalities are Jørgensen's inequality and a well known variant of it [10]. These are necessary conditions for the discreteness of the group $\langle A, B \rangle$ provided this group is not virtually abelian. The last condition is a result of Cao [3] improving other versions of inequalities Jørgensen found [18, 11] for discrete groups generated by two elements of the same trace. We state this in the following lemma.

Lemma 12.10.1 *If $\langle f, g \rangle$ is Kleinian and $\beta(f) = \beta(g)$, then $|\gamma(f, g)| \geq c_0$ where*

$$c_0 = 2 - 2\cos\left(\frac{\pi}{7}\right)$$

This bound is sharp and achieved in the $(2, 3, 7)$-triangle group.

Thus the violation of one of these inequalities shows that $\langle A, W \rangle$ is virtually abelian. If the group is abelian, then $\gamma(A, W) = 0$. If $WAW^{-1} = A^{-1}$, the dihedral case, then $WAW^{-1}A^{-1} = A^{-2}$, and hence

$$\gamma(A, W) = \text{tr}(WAW^{-1}A^{-1}) - 2 = \text{tr}A^2 - 2 = \text{tr}^2(A) - 4 = \beta(A)$$

By hypothesis A is not parabolic ($\lambda \neq 1$). If A is loxodromic, then $\langle A, W \rangle$,

being discrete, is Kleinian unless W fixes or interchanges the fixed points of
A. Otherwise there would be three, and hence uncountably many limit points,
[2]. These reduce to the cyclic or dihedral cases. If A is elliptic, then $|\beta| < 1$
is required to violate either of the first two inequalities. That is A has order
7 or more. The classification of the elementary discrete groups [2] shows this
to reduce to the abelian or dihedral cases as well. What remains is the case
$\langle A, W \rangle$ is a discrete group with the last inequality violated. Then this group
is elementary and as $\gamma(A, WAW^{-1}) = \gamma_w(\gamma_w - \beta)$ a little argument using the
classification of the elementary discrete groups reduces to the previous cases.
□

12.11   Examples

We calculate some of the polynomials for balanced, even, good words in
f and g, and investigate when these have the same axis as f. For $W = fgf^5g^{-1}fgf^2g^{-1}f^{-3}$,

$$2r(u,v) = -1 + 3u - 2u^2 - 10u^3 + 4u^4 + 8u^5 - v - 3uv + 8u^2v + 4u^3v$$
$$- 8u^4v + 2v^2 - 6uv^2 - 6u^2v^2 + 14u^3v^2 + 4u^4v^2 - 8u^5v^2$$
$$2s(u,v) = -1 + 2u + 10u^2 - 4u^3 - 8u^4 - v - 4uv + 4u^2v + 8u^3v + 2uv^2$$
$$- 6u^2v^2 - 4u^3v^2 + 8u^4v^2$$
$$2t(u,v) = (u-1)(-1+2u+4u^2)(-1-2u+4u^2+4u^3-4uv+4u^3v)$$
$$2w(u,v) = (1+6u-4u^2-20u^3+8u^4+16u^5-2v+6uv+8u^2v-20u^3v$$
$$- 8u^4v + 16u^5v)$$

The only solution of $t(u,v) = w(u,v) = 0$ is $u = -1/2$, $v = -1/3$. For these
values we also have $s(u,v) = 0$ and $r(u,v) = -1$, i.e. W is a relator of $\langle f, g \rangle$
for these values. However, we return back to (12.65) and (12.66) to see

$$\beta = 2u - 2 = -3, \qquad \gamma = \beta(1-v)/2 = -2$$

so f has order three and f and g .

Corollary 12.11.1 *Let Γ be a Kleinian group and $f, g \in \Gamma$. Then*

$$fgf^5g^{-1}fgf^2g^{-1}f^{-3} = 1$$

*if and only if f has order 3 and $\langle f, g \rangle$ is a Euclidean triangle group or an
abelian group.*

Of course if f and g commute then $fgf^5g^{-1}fgf^2g^{-1}f^{-3} = 1$, then $f^6 = 1$.

For $W = fgf^5g^{-1}f^{-2}$

$$2r(u, v) = -1 - 3u + 2u^2 + 4u^3 - v + 3uv + 2u^2v - 4u^3v$$
$$2s(u, v) = 1 + 2u - 4u^2 - v + 2uv + 4u^2v$$
$$2t(u, v) = (u - 1)(1 + 2u)(-1 + 2u + 4u^2)$$
$$2w(u, v) = (-1 + 2u)(-1 + 2u + 4u^2)$$

This time t and w have a common factor $-1 + 2u + 4u^2$, so that they vanish simultaneously when $u = 1/4(-1 \pm \sqrt{5})$, and for all values of v. However, $s(1/4(-1 \pm \sqrt{5}), v) = 1/2(-1 + \sqrt{5}) \neq 0$, so W can never be a relator of $\langle f, g \rangle$.

For $W = fgfg^{-1}f^2gfg^{-1}fgf^{-1}g^{-1}f^2gf^{-1}g^{-1}$

$$r(u, v) = -u - u^2 + u^3 + u^4 + u^5 - v^2 + uv^2 - u^2v^2 + u^3v^2 + 2u^4v^2$$
$$\qquad - 2u^5v^2 + v^4 - 3uv^4 + 2u^2v^4 + 2u^3v^4 - 3u^4v^4 + u^5v^4$$
$$s(u, v) = -1 + u^2 + 2u^3 + u^4 + v^2 - 2uv^2 + 3u^2v^2 - 2u^4v^2 - v^4$$
$$\qquad + 2uv^4 - 2u^3v^4 + u^4v^4$$
$$t(u, v) = (u - 1)(u + 1)(u - v + uv)(-1 - u - u^2 - v + uv + v^2$$
$$\qquad - 2uv^2 + u^2v^2)$$
$$w(u, v) = (u - 1)(-u - u^2 - u^3 - v - uv - u^2v - u^3v + v^2 - uv^2 - u^2v^2$$
$$\qquad + u^3v^2 + v^3 - uv^3 - u^2v^3 + u^3v^3)$$

Here W is a relator in the group that minimizes the maximum of the two translation lengths $\max\{\tau_f, \tau_h\}$, when f and h are two loxodromics with perpendicular axes [23]. Now t and w have a common factor $(u - 1)$, so that they vanish simultaneously when $u = 1$. However, $t = w = 0$ also holds when $v = 0$ (perpendicular axes), when $u = 0$ and when $u = -1 \pm i\sqrt{3}$. In the last two cases $r = 1$ and (consequently) $s = 0$.

These examples raise some general questions:

1. Which words can be relators? (ie for which words do $s = t = w = 0$, $r = \pm 1$ have a solution, apart from the trivial solutions $u = 1$ ($f = Identity$) and $v = 1$ (f and g have the same axis)?)

2. For which words do t and w have a non-constant common factor in which neither of the variables u and v is absent? (giving an infinite family of solutions for $t = w = 0$)

3. For which words do t and w have no such common factor, so that $t = w = 0$ has only finitely many roots, and one of these roots also makes $s = 0$ and (hence) $r = \pm 1$ (ignoring the trivial cases $u = 1$, $v = 1$). Is any such group discrete?

12.11.1 Explicit formulae

We now give (without going into details of the computation) explicit values for the polynomials $R = R_w$, $S = S_w$, $T = T_w$ and W_w of (12.71) associated with the even word $w = f^{n_1}gf^{n_2}g^{-1}f^{n_3}gf^{n_4}g^{-1}f^{n_5}$, which are expressed in terms of Chebyshev polynomials indexed by various combinations of the powers n_i. For such a word, and for $S \subseteq \{1,2,3,4,5\}$, we let $T_S(u) = T_{(\epsilon_1 n_1 + \epsilon_2 n_2 + \epsilon_3 n_3 + \epsilon_4 n_4 + \epsilon_5 n_5)/2}(u)$, $U_S(u) = U_{(\epsilon_1 n_1 + \epsilon_2 n_2 + \epsilon_3 n_3 + \epsilon_4 n_4 + \epsilon_5 n_5)/2 - 1}(u)$, where $\epsilon_i = -1$ if $i \in S$, $\epsilon_i = 1$ if $i \notin S$, e.g. $T_{\{2,3\}}(u) = T_{(n_1 - n_2 - n_3 + n_4 + n_5)/2}(u)$. (We set $U_{-1}(x) = 0$, and, for $n < 0$, $T_n(x) = T_{|n|}(x)$, $U_{n-1}(x) = -U_{|n|-1}(x)$).

We have calculated:

$$
\begin{aligned}
R(u,v) \;=\; & \frac{1}{4}\left(\sum_{S \subseteq \{2,3,4\}} (-1)^{|S|} T_S(u)\right) v^2 \\
& + \frac{1}{2}\left(T_\emptyset(u) - T_{\{2,4\}}(u)\right) v \\
& + \frac{1}{4}\left(T_\emptyset(u) + T_{\{2\}}(u) + T_{\{3\}}(u) + T_{\{4\}}(u) + T_{\{2,4\}}(u) + T_{\{2,3,4\}}(u)\right)
\end{aligned}
$$

$S(u,v)$ is the same, but with U_S substituted for T_S throughout in (12.112) above,

$$
\begin{aligned}
T(u,v) \;=\; & \frac{1}{4}\left(\sum_{S \subseteq \{2,3,4\}} (-1)^{|S|} T_{S \cup \{5\}}(u)\right) v \\
& + \frac{1}{4}\big(T_{\{3,5\}}(u) - T_{\{1,3\}}(u) + T_{\{2,5\}}(u) - T_{\{1,4\}}(u) \\
& \quad - T_{\{1,2,3\}}(u) + T_{\{3,4,5\}}(u) - T_{\{1\}}(u) + T_{\{5\}}(u)\big).
\end{aligned}
$$

$$
\begin{aligned}
W(u,v) \;=\; & \frac{1}{4}\left(\sum_{S \subseteq \{2,3,4\}} (-1)^{|S|} U_{S \cup \{5\}}(u)\right) v \\
& + \frac{1}{4}\big(U_{\{1,3\}}(u) + U_{\{3,5\}}(u) + U_{\{1,4\}}(u) + U_{\{2,5\}}(u) \\
& \quad + U_{\{1,2,3\}}(u) + U_{\{3,4,5\}}(u) + U_{\{1\}}(u) + U_{\{5\}}(u)\big).
\end{aligned}
$$

Remark: These formulae exhibit the general fact that if the sequence $(n_1, n_2, n_3, n_4, n_5)$ is reversed, the sign of t is changed, and r, s, w are unchanged.

We thus have, for the shorter word $A^{n_1}BA^{n_2}BA^{n_3}$ $(n_4 = n_5 = 0)$

$$
\begin{aligned}
R(u,v) \;=\; & \frac{1}{2}\big[\left(T_{(n_1+n_2+n_3)/2}(u) - T_{(n_1-n_2+n_3)/2}(u)\right) v \\
& + \left(T_{(n_1+n_2+n_3)/2}(u) + T_{(n_1-n_2+n_3)/2}(u)\right)\big],
\end{aligned}
$$

$S(u, v)$ the same, but with U_{n-1} substituted for T_n throughout, and

$$T(u, v) = \frac{1}{2}\left[T_{(n_1+n_2-n_3)/2}(u) - T_{(n_1-n_2-n_3)/2}(u)\right], \quad (12.112)$$

$W(u, v)$ the same, but with U_{n-1} substituted for T_n throughout.

12.12   Roots of Trace Polynomials

The purpose of this section is to establish a theorem which shows that the zero sets of the "good word" trace polynomials discussed in Section 12.5 are dense in the complement of the space of discrete and faithful representations of $\mathbb{Z}_p * \mathbb{Z}_2$ for $3 \leq p \leq \infty$. Indeed we show that the complement of the representations which are discrete and free on marked generators is the Julia set of the semigroup of good word polynomials, where we define the Julia set of any family $\mathcal{P}$ of analytic functions mapping an open set $U \subseteq \mathbb{C}$ into itself, as the set of $z \in U$ such that $\mathcal{P}$ is not a normal family in any neighbourhood of z.

Theorem 12.12.1 *Let f, g be Möbius transformations with $\beta = \beta(f) \neq -4$, $\beta(g) = -4$, $\gamma = \gamma(f, g)$ and suppose that $\langle f, g \rangle$ is not discrete and free on the two generators f and g. Then for any open set U, $\gamma \in U \in \mathbb{C}$ there is a good word $w = w(f, g)$ for which the polynomial $q_w(z) = p_w(z, \beta)$, given by Theorem 12.5.3, has a root in U.*

Proof. Since g is of order two, every member of $\langle f, g \rangle$ can be represented as a good word in f and g. There are two cases.

12.12.1 $\langle f, g \rangle$ is discrete but not free on generators

In this case there is a nontrivial good word $w \in \langle f, g \rangle$ representing the identity, whence $0 = \gamma(w, f) = p_w(\gamma, \beta)$ so that γ itself is the root of a good word polynomial.

12.12.2 $\langle f, g \rangle$ is not discrete

Let U be a neighbourhood of γ and define the good word polynomial zero set as

$$\mathcal{Z} = \{z \in \mathbb{C} : \text{ there is a good word } w \text{ so that } p_w(z, \beta) = 0 \}$$

In Section 12.5.1 we gave a few examples of good words. From that table we quickly deduce that among many other points

$$\{0, \beta, 1 + \beta, 2 + \beta\} \subset \mathcal{Z}.$$

Thus $\mathcal{Z}$ contains at least three finite points. To simplify notation we suppress the β variable in our word polynomials. Next, suppose that for some good word v we have $p_v(U) \cap \mathcal{Z} \neq \emptyset$. Then there is a word $w \in \langle f, g \rangle$ and $z \in U$ such that $p_w(p_v(z)) = 0$. However, we know that the set of good words is closed under composition, and $p_w(p_v(z)) = p_{w*v}(z)$. Thus $z \in \mathcal{Z}$. We are left to consider the subcase.

12.12.2.1   For all good words v, $p_v(U) \cap \mathcal{Z} = \emptyset$

Let $\mathcal{F} = \{p_v : v$ is a good word$\}$. We have seen that on U $\mathcal{F}$ omits $\mathcal{Z}$ which contains at least three points. Thus Montel's criterion shows that the functions of $\mathcal{F}$ restricted to U is a normal family. In other words U does not meet the Julia set of $\mathcal{F}$. Since $\langle f, g \rangle$ is not discrete, there is a sequence of good words $\{w_i\}_{i=1}^{\infty}$ in $\langle f, g \rangle$ with $w_i \to$ *identity* as $i \to \infty$ (this convergence is in the topology of $PSL(2, \mathbb{C})$, that is in each entry of representative matrices).

$$p_{w_i}(\gamma) = \text{tr}[f, w_i] - 2 \to \text{tr}[f, identity] - 2 = 0$$

It follows that each neighbourhood of 0 meets $p_{w_i}(U)$ for some i. We will thus be done if we can show that Julia set of $\mathcal{F}$ contains some neighbourhood of 0.

12.12.2.2   Density of roots near 0

We analyse this case in a fairly general framework using some of the theory of the dynamics of polynomial semigroups. Much more can be found about this subject, see for instance [15, 16, 25, 27, 28] and the references therein. The point here is that, for each nonzero $\beta = \beta(f)$, we can find a good word polynomial which has 0 as a repelling fixed point, and another which has zero as a superattracting fixed point under iteration. In such a setting, the Julia set of the semigroup generated by these two polynomials contains a neighbourhood of 0 and the preimages of 0 are dense in it.

Lemma 12.12.1 *Let p and q be entire functions, with a common fixed point c, which is superattractive for p and repulsive for q. Let $\mathcal{P}$ be the semigroup $\langle p, q \rangle$, then the Julia set of $\mathcal{P}$ contains a neighbourhood of c.*

Proof. This is a standard "push me, pull you" argument which we sketch. We may assume that $c = 0$. We have $p(z) = az^m + O(z^{m+1})$, $q(z) = \mu(z + bz^2 + O(z^3))$, where $a \neq 0$, $m \geq 2$ and $|\mu| > 1$. We construct a sequence of functions $\{f_n\}$ inductively by $f_0(z) = z$ and $f_{n+1}(z) = g(f(z))$, where g is either p or q. We choose $r > 0$ to be sufficiently small that we can ignore higher degree terms in p and q. Let $z_0 = z$ be chosen with $|z| < r$, suppose that $f_0, f_1, \ldots f_k$ have already been defined, and set $z_i = f_i(z)$. If $|z_k| \geq r$, then let $f_{n+1}(z) = p(f(z))$; otherwise let $f_{n+1}(z) = q(f(z))$. As soon as $|z_k| \geq r$, the next number z_{k+1} is much smaller; then the z_i gradually increase in size (because $|\mu| > 1$), until eventually it exceeds r, and the process begins again. The sequence $\{z_i\}$ is bounded above and below $|a|r^m \leq z_i \leq |\mu|r$.

If $f_{n+1}(z) = p(f_n(z))$, then the logarithmic derivative

$$\frac{f'_{n+1}(z)}{f_{n+1}(z)} = p'(z_n)\frac{z_n}{p(z_n)}\frac{f'_n(z)}{f_n(z)} \simeq m\frac{f'_n(z)}{f_n(z)}$$

If $f_{n+1}(z) = q(f_n(z))$, then

$$\frac{f'_{n+1}(z)}{f_{n+1}(z)} = q'(z_n)\frac{z_n}{q(z_n)}\frac{f'_n(z)}{f_n(z)} \simeq \frac{z + 2bz^2}{z + bz^2}\frac{f'_n(z)}{f_n(z)} \simeq (1 + bz)\frac{f'_n(z)}{f_n(z)}$$

If $|z_{n-1}| \geq r$, then $|z_n| \geq |a|r^m$, and it takes t applications of q to get the size of z_i over r again, where t is at most about $\log_{|\mu|}(1/(|a|r^{m-1}))$, in the course of which we multiply the absolute value of the logarithmic derivative by at least

$$(1 - |bz_n|)(1 - |bz_{n+1}|)\ldots(1 - |bz_{n+t}|) \geq (1 - |b|r)^t \geq (1 - |b|r)^{C\log(1/r)}$$

which can be made as close to 1 as we like by taking r sufficiently small. Each time we apply p, we multiply the logarithmic derivative by approximately m. It follows that $|\frac{f'_n(z)}{f_n(z)}| \to \infty$ as $n \to \infty$. Since the $|f_n(z)|$ is bounded below, it also follows that $|f'_n(z)| \to \infty$ as well. Thus no subsequence of $\{f_n(z)\}$ can converge to an analytic function. Since $|f_n(z)|$ is bounded above, $\{f_n(z)\}$ cannot converge to ∞ either. Thus $\langle p, q \rangle$ is not a normal family on any neighbourhood of z. $\qquad\square$

To complete our proof we recall from Section 12.5.1 the trace polynomials $\gamma(\gamma - \beta)$ from the word bab, $(\beta - \gamma + 1)^2\gamma$ from the word $babab$, $\gamma(1 - 2\beta + \gamma^2 - (\beta - 2)\gamma)$ from the word $baba^{-1}b$ and $\gamma^3(\gamma - \beta)(\beta + 4)(\beta(\gamma^2 - 3\gamma - 4) - \beta^2(\gamma + 1) + 4\gamma^2 + 4\gamma + 1)$ from the word $ba^{-2}bababa^{-2}bab$.

The last polynomial here is superattractive at $z = 0$ and the rest have multipliers at 0 of $-\beta$, $(1 + \beta)^2$ and $1 - 2\beta$ respectively, so that for each $\beta \neq 0$ at least one of them has $z = 0$ as a repulsive fixed point. Thus, by the lemma, the Julia set of the trace polynomials contains a neighbourhood of 0, and so a zero-free region U has an image under a trace polynomial into a region which intersects the Julia set, contradicting the fact that these polynomials generate a normal family on U. The case $\beta = 0$ is separately dealt with in [21].

References

[1] H. Akiyoshi, M. Sakuma, M. Wada and Y. Yamashita, *Punctured torus groups and two bridge knot groups (I)*, Lecture Notes in Mathematics 1909, Springer-Verlag Berlin Heidelberg, 2007.

[2] A. Beardon, *The geometry of discrete groups*, Springer–Verlag, 1983.

[3] C. Cao, *Some trace inequalities for discrete groups of Möbius transformations*, Proc. Amer. Math. Soc., **123**, (1995), 3807–3815.

[4] M. D. E. Conder and G. J. Martin, *Cusps, triangle groups and hyperbolic 3-folds*, Journal of the Australian Mathematical Society, **55**, (1993), 149–182.

[5] M. D. E. Conder, G. J. Martin and A. Torstensson, *Maximal symmetry groups of hyperbolic 3-manifolds*, New Zealand J. Math., **35**, (2006), 37–62.

[6] R. Fricke and F. Klein, *Vorlesungen über die Theorie der automorphen Functionen*, Chapter 2, Teubner, Leipzig, 1897.

[7] D. Gabai, *On the Geometric and Topological Rigidity of Hyperbolic 3-Manifolds*, J. American Math. Soc., **10**, (1997), 37–74.

[8] D. Gabai, R. Meyerhoff and N. Thurston, *Homotopy hyperbolic 3-manifolds are hyperbolic*, Ann. of Math., **157**, (2003), 335–431.

[9] F. W. Gehring, C.Maclachlan, G. J. Martin and A. W. Reid *Arithmeticity, Discreteness and Volume*, Trans. Amer. Math. Soc., **349**, (1997), 3611–3643.

[10] F. W. Gehring and G. J. Martin, *Iteration theory and inequalities for Kleinian groups*, Bull. Amer. Math. Soc., **21**, (1989), 57–63.

[11] F. W. Gehring and G. J. Martin, *Some universal constraints for discrete Möbius groups*, Paul Halmos; Celebrating 50 Years of Mathematics, Springer-Verlag, New York, pp. 205–220.

[12] F. W. Gehring and G. J. Martin, *Commutators, collars and the geometry of Möbius groups* , J. d'Analyse Math., **63**, (1994), 175–219.

[13] F. W. Gehring and G. J. Martin, *(p,q,r)-Kleinian groups and the Margulis constant*, Complex analysis and dynamical systems II, Contemp. Math., 382, (2005), 149–169.

[14] F. W. Gehring and G. J. Martin, *Minimal covolume lattices* **I**: *spherical points of a Kleinian group*, Annals of Math., **170**, (2009), 123–161.

[15] A. Hinkkanen and G. J. Martin, *The Dynamics of Semigroups of Rational Functions I*, Proc. London Math. Soc., **73**, (1996), 358–384.

[16] A. Hinkkanen and G. J. Martin, *Julia sets of rational semigroups*, Math. Z. **222**, (1996), 161–169.

[17] R. Horowitz, *Characters of free groups represented in the two-dimensional linear group*, Comm. Pure Appl. Math., **25**, (1972), 635–649.

[18] T. Jørgensen, *On discrete groups of Mobius transformations*, Amer. J. Math., **98**, (1976), 739–749.

[19] L. Keen and C. Series, *The Riley Slice of Schottky Space*, Proc. London Math. Soc., **69**, (1994), 72–90.

[20] G.J. Martin, *On discrete isometry groups of negative curvature*, Pacific journal of mathematics, **160**, (1993), 109–127.

[21] G.J. Martin, *Nondiscrete parabolic characters of the free group F_2: supergroup density and Nielsen classes in the complement of the Riley slice*, Math. ArXiv 2001.10077.

[22] T. H. Marshall and G. J. Martin, *Volumes of hyperbolic 3-manifolds: Notes on a paper of D. Gabai, G. Meyerhoff and P. Milley*, J. Conf. Geom. and Dynamics, **7**, 34–48, 2003.

[23] T. H. Marshall and G. J. Martin, *Minimal co-volume hyperbolic lattices, II: Simple torsion in a Kleinian group*, Ann. Math., **176**, (2012), 261–301.

[24] C. Maclachlan and A. Reid, *The arithmetic of hyperbolic 3-manifolds*, Springer–Verlag, **219**, 2003.

[25] R. Stankewitz and H. Sumi, *Dynamical properties and structure of Julia sets of postcritically bounded polynomial semigroups*, Trans. Amer. Math. Soc., **363**, (2011), 293–5319.

[26] H. Namazi and J. Souto, *Non-realizability and ending laminations: Proof of the density conjecture,* Acta Math., **209**, (2012), 323–395.

[27] H. Sumi, *On dynamics of hyperbolic rational semigroups*, J. Math. Kyoto Univ. 37 (1997) 717–733.

[28] H. Sumi, *Rational semigroups, random complex dynamics, and singular functions on the complex plane*, SUGAKU **61**, (2009), 133–161.

[29] C. R. Traina, *Trace polynomial for two generator subgroups of $SL(2,\mathbb{C})$*, Proc. Amer. Math.Soc., **79**, (1980), 369–372.

[30] John Voight, *Quaternion Algebras*, unpublished lecture notes, available at https://math.dartmouth.edu/ jvoight/quat-book.pdf

13

Boundary Value Problems on Klein Surfaces

Vicentiu Radulescu and Monica Rosiu

CONTENTS

13.1 Introduction

In the preface to the first edition of Courant-Hilbert's "Methoden der mathematischen Physik" (see [12]), R. Courant noted the danger that mathematical research would lose the initial link between the problems and methods of analysis and the physical and geometric intuition, the tendencies being to refine the methods and to extreme generalize the existing concepts.

Over the years, these trends led to an increasing distinction between pure and applied mathematicians, who severely criticized each other. This constructive criticism gave rise to the theory of real numbers and to many topological concepts including non-orientable surfaces. It is obvious that some areas that

use mathematical methods but their object is derived from physical and geometric intuition are disadvantaged in such a discussion.

The present paper is a piece of the bridge between the theoretical approach
of the pure mathematician and the practical interest of the engineer, physicist
and applied mathematician. The main purpose is to bring together various
geometrical and physical concepts relating to surfaces that have motivated
the development of the theory of Klein surfaces.

Riemann surfaces, in the form of domains spread out over the complex
plane were introduced in Riemann's dissertation whose methods were developed much further in the first edition of Riemann's paper on Abelian functions
"Theorie der Abel'schen Functionen" (see [27]), in 1857. Riemann's works
provided the basic tools to classify all compact orientable surfaces and, more
generally, to study the topology of manifolds. They are equally important for
the development of algebraic geometry and the geometric treatment of complex analysis. As for the importance that was attached to this topic, it suffices
to say that Albert Einstein's "general theory of relativity" is wholly based on
Riemann's ideas.

In his "Extremale quasikonforme Abbildungen und quadratische Differentiale" (see [35]) Teichmüller considered the cases of oriented bordered Riemann surfaces and non-orientable Riemann surfaces. He defined the double
of a oriented bordered Riemann surfaces or of a non-orientable Riemann surfaces. These are closed Riemann surfaces with genus depending on the original
surfaces. He introduced the notions of meromorphic functions, n-differentials
and divisors on bordered non-orientable Riemann surfaces.

Teichmüller defined, through two examples, the notion of conformal invariant for non-orientable regions. It is important to note that according to
Teichmüller's definition, a conformal mapping (even in the orientable case)
preserves only the angles, but not necessarily the orientation. Thus, such a
mapping is also defined in the non-orientable case. Teichmüller exhibited complete systems of conformal invariants for some special surfaces. For example,
a simply connected domain with two distinguished points in the interior has
one conformal invariant, namely the Green function.

Teichmüller considered special cases of the fact that one deals with non-
orientable surfaces by passing to the orientation double cover. To treat the
most general surfaces, he considered a symmetrization process on the corresponding doubles. This idea was used to solve the main problem of extremality,
known as the Teichmüller theorem. By using the two-sheeted covering (an annulus) of the Möbius strip, he gets the Teichmüller distance. He showed that
problems on the projective plane with two distinguished points can be reduced
to similar problems on the sphere with four distinguished points. In two examples of non-orientable surfaces, Möbius strip and the projective plane, the
problem of finding the conformal invariants is lifted to the oriented double
cover. Teichmuller considered problems on the Klein bottle and lifted them
to problems concerning the case of the torus. The torus is the two-sheeted
orientation covering of the Klein bottle.

The genesis and development of the idea of symmetry are related to Lie's and Klein's research that were inspired by their deep interest in the theory of groups and in various aspects of the notion of symmetry. According to Klein's Erlangen program, a geometry is determined by a "domain of action" (the plane, space, etc.) and a "group of automorphisms" (or a symmetry group) acting on the domain. When we change the symmetry group we change the geometric scheme under consideration, namely we obtain a new "geometry."

Thus, the main difference between, say, Euclidean and hyperbolic geometry is not the possibility of constructing one or more lines passing through a point and not intersecting a given line, but the difference in the structure of the respective groups of symmetries of Euclidean and hyperbolic geometry. Therefore, the object of the geometry is the study of those properties of a domain that are preserved by the transformations in a symmetry group. The description of all possible geometries is an open problem.

In the same way, classifying non-classical topological compact surfaces is the same thing as classifying all orientation reversing involutions of a classical compact surface.

This brings us to the interesting question of the possible global forms of various (say, two-dimensional) geometric systems (Euclidean, hyperbolic, elliptic) first stated (in connection with Euclidean geometry) by the outstanding geometer W. K. Clifford. Today this question is known as the Clifford-Klein problem and the possible global forms of geometries are called Clifford-Klein forms.

It is known that there are only two spatial forms of two-dimensional elliptic geometry (the sphere and the elliptic plane), but there are as many as five forms of two-dimensional Euclidean geometry (the ordinary Euclidean plane, the infinite Möbius strip, the infinite cylinder, the torus and the so-called Klein bottle). Finally, there are infinitely many forms of two-dimensional hyperbolic geometry.

In this context, non-orientable surfaces are a possible geometric system capable of "modelling" the real shape of the universe surrounding us. For more details, we refer to the book by Weeks [37], which fills the gap between the simplest examples, such as the Mobius strip and the Klein bottle, and the sophisticated mathematics found in upper-level college courses.

Spencer and Schiffer in their advanced monograph "Functionals of Finite Riemann Surfaces," extended the investigation of finite Riemann surfaces from the point of view of functional analysis, that is, the study of the various Abelian differentials of the surface in their dependence on the surface itself.

The methodology that Schiffer and Spencer employed is characterized by Ahlfors' next comment: "such a surface has a double, obtained by reflection across the boundary, and one of the main features of the book is the systematic use of this symmetrization process."

The notion of *Klein surface* goes back to Felix Klein due to his closing remarks in [15], even though one does not find a definition of a Klein surface there. Klein surfaces generalize Riemann surfaces, and they are dianalytic manifolds of complex dimension 1. Roughly speaking, a Klein surface is a surface on which the notion of angle between two tangent vectors at a given point is well-defined, and so is the angle between two intersecting curves on the surface.

Basic function theory on Klein surfaces and the relation between compact Klein surfaces and real algebraic function fields were developed in the monograph "Foundations of the Theory of Klein surfaces" (see [3]) by N. Alling and N. Greenleaf. They showed that every Klein surface can be represented as the quotient of a Riemann surface by a conjugate analytic involution. Thus, it is natural to extend on Klein surfaces the most fundamental problems in engineering, physics and other sciences. Alling and Greenleaf were the ones who introduced the name "Klein surface."

Our approach is an alternative theory to the standard theory given by Alling and Greenleaf and aims at the natural imbedded of calculus on Klein surfaces in the well-known Cartan's model of calculus on manifolds. We have developed this theory because of the unusual behavior from the analytical point of view of the Alling and Greenleaf's results. For instance, functions are not usual functions but equivalence classes of families of meromorphic functions relative to dianalytic atlases. Such a family defines an usual function if and only if all its members are the same real constant. The meromorphic differentials are also equivalence classes of families of functions satisfying some compatibility conditions that lead to the impossibility of defining a consistent integral on Klein surfaces (see [3, Theorem 1.10.4]).

We follow Schiffer and Spencer's method to study the objects on Klein surfaces by means of the complex double, whose existence and uniqueness are demonstrated in [3].

We are enabled to bring together systematically and concisely the concepts of the Green and Neumann functions, the harmonic kernel function and the harmonic measure and to build from them an elegant generalization for the basic ideas of boundary value problems on Klein surfaces.

The main objectives of study in this paper are the Dirichlet problem and the Neumann problem for harmonic functions on Klein surfaces. The technique is based on the fact that according to a classical result due to Klein, the boundary value problems on a Klein surface can be reduced to similar problems on its complex double. This process has many advantages, starting from the fact that complex double is a symmetric Riemann surface, that is, a Riemann surface endowed with a fixed point free antianalytic involution. Consequently, we obtain harmonic functions on a Klein surface by adding together a pair of harmonic functions on the symmetric Riemann surface, whose singularities lye at symmetric points. In our study, we use methods that have wide applicability in function theory and partial differential equations.

The symmetric conditions on the boundary determine symmetric solutions on the complex double, which lead to solutions for the similar problems on the Klein surface. Specifically, in the case of Klein surfaces, the formula for the solution of the Dirichlet problem is expressed in terms of an analogue of the Green function, which has the symmetry in argument and parameter. In these terms, we extend the use of the Green function to the study of the harmonic measure on a Klein surface. That is why we distinguish the method to solve the Dirichlet problem for harmonic functions on a Klein surface, once the harmonic measure on a symmetric Riemann surface is known. This procedure generates an explicit formula for the solution of the Dirichlet problem on a Klein surface, which is similar to the Poisson integral. At the same time, we rewrite the Radon–Nikodym derivative of harmonic measure against the symmetric arc length.

The corresponding solution of the Neumann problem for harmonic functions on a Klein surface is expressed in terms of an analogue of a Neumann function.

The harmonic kernel function is related to the classical domain functions, such as the Green function and the Neumann function on a Klein surface. In such a way it is possible to solve both boundary value problems of potential theory on a Klein surface, once the harmonic kernel function on a symmetric Riemann surface is known.

We refer to Krantz [18, 16, 17, 20] for an excellent exposition of various topics at the interplay between complex analysis and partial differential equations.

The study of objects on Klein surfaces is an important part of surface topology due to the applications of these surfaces in several fields of science, such as quantum physics, chemistry and biology. Indeed, for a physicist, Möbius's band and Klein's bottle are essential elements in the so-called annulment of divergences. In chemistry, the recent synthesis and the "half-cutting" of a molecular Möbius strip (see [36]) was considered as a spectacular event, described as "the most topologically stimulating molecular structure synthesized to date" it catalyzed the birth of extrinsic graph theory, dealing with topological chirality, a field now burgeoning in mathematics.

The study of liquid crystals is another field where Klein surfaces have surprisingly materialized themselves. In the so-called nematic liquids, the molecules form ribbons that may or may not be orientable (see [9]). A systematic topological analysis highlighted the double topological character of distortions in liquid crystals differentiated for "energetic reasons" (see [10]). If we consider the potential function of some form of internal energy of the ribbons, then the normal derivative on the border characterizes the flow of energy across the border. It may be necessary to determine this potential knowing the respective flow or the values of the potential on the border. These are boundary value problems that will be solved in this paper.

The natural tendency of some macromolecules to store energy through distortions, a fact well-known to chemists, might be the cause itself for the formation of non-orientable strings, thus making obvious the practical need of dealing with boundary value problems related to them.

A unified principle for science that works with dualism is presented in terms of torsion fields and the non-orientable surfaces, namely the Klein Bottle, the Möbius strip and the projective plane, in (see [26]). This principle is applied to the complex numbers and cosmology, to non-linear systems integrating the issue of hyperbolic divergences with the change of orientability, to the biomechanics of vision and the mammal heart, to the morphogenesis of crustal shapes on Earth in connection to the wavefronts of gravitation, elasticity and electromagnetism, to pattern recognition of artificial images and visual recognition, to neurology and the topographic maps of the sensorium, to perception, in particular of music.

As it is noticed in (see [14]), these are the types of problems that contribute to a unifying treatment of orientable and non-orientable surfaces, not only from the topological point of view (see [11] and [34]) but also from an analytical point of view (see [33] and [3]).

13.2   Klein Surfaces and Symmetric Riemann Surfaces

Klein surfaces are the most general two-manifolds that support harmonic functions. In order to be able to extend results about boundary value problems for harmonic functions on Riemann surfaces to Klein surfaces, we have to review some results on the topology of surfaces, Klein surfaces, and the uniformization of Riemann surfaces. The history of Klein surfaces is going back to Klein (see [15]) who considered the group of conformal maps of the Klein bottle and other non-orientable surfaces. In their monograph, Schiffer and Spencer (see [33]) did the first modern study of the surfaces endowed with dianalytic structures. Much of the material of this Section will be presented without proofs and will be completed with references to proofs. The main reference to topology of surfaces is the monograph of Ahlfors and Sario (see [2]).

A connected topological Hausdorff space $\mathcal{X}$ is a *surface with boundary* if every point $\widetilde{P} \in \mathcal{X}$ has an open neighborhood $\widetilde{U}$, which is homeomorphic to a relatively open subset of the closed upper half-plane. A homeomorphism $h : \widetilde{U} \to h(\widetilde{U})$ is called a *local parameter* at the point $\widetilde{P} \in \widetilde{U}$. The boundary $\partial \mathcal{X}$ of $\mathcal{X}$ consists of those points $\widetilde{P} \in \mathcal{X}$, such that $h(\widetilde{P}) \in \mathbb{R}$, for all the local parameters h at the point $\widetilde{P}$. The pair $(\widetilde{U}, h)$ is called a *chart*. Let $h_i : \widetilde{U}_i \to h_i(\widetilde{U}_i)$ and $h_j : \widetilde{U}_j \to h_j(\widetilde{U}_j)$ be two local parameters, such that $\widetilde{U}_i \cap \widetilde{U}_j \neq \emptyset$, $i, j \in I$. The mapping $h_i \circ h_j^{-1} : h_j(\widetilde{U}_i \cap \widetilde{U}_j) \to h_i(\widetilde{U}_i \cap \widetilde{U}_j)$ is called a *transition function*.

Let A and B be non-empty open sets in the closed upper half-plane. A continuous map of A into B is analytic on A (resp., antianalytic on A) if it extends to an analytic (resp., antianalytic) function on some neighborhood of A in $\mathbb{C}$ into $\mathbb{C}$. If f or the complex conjugate of f is analytic on each connected component of the set A, then f is called *dianalytic* on A.

An *atlas* of the surface $\mathcal{X}$ is a family $\mathcal{A} = \{(\widetilde{U}_i, h_i) \,|\, i \in I\}$ of charts, where $(\widetilde{U}_i)_{i \in I}$ is an open cover of $\mathcal{X}$. The atlas $\mathcal{A}$ is *dianalytic* if all of its transition functions are dianalytic. Two dianalytic atlases $\mathcal{A}$ and $\mathcal{B}$ are called *equivalent* if $\mathcal{A} \cup \mathcal{B}$ is a dianalytic atlas as well. An equivalence class A of dianalytic atlases of $\mathcal{X}$ is called a *dianalytic structure* on X.

A *Klein surface* is a surface $\mathcal{X}$ with boundary endowed with a dianalytic structure A and will be denoted by X. Observe that a classical Riemann surface is an orientable Klein surface with empty boundary.

The main tool in our study is the complex double of a Klein surface. Details about the history of this concept may be found in [33] and for some of its applications see [3], [11] and [34].

Let X be a Klein surface endowed with the maximal atlas $A = \{(\widetilde{U}_i, h_i) \,|\, i \in I\,\}$. We recall the construction of the complex double of X, which we shall use in order to apply results about Riemann surfaces to Klein surfaces. We consider the disjoint union $S = \bigcup_{i \in I} \widetilde{U}_i$. Let $(\widetilde{U}_i, h_i)$ and $(\widetilde{U}_j, h_j)$ be two charts, such that $\widetilde{U}_i \cap \widetilde{U}_j \neq \emptyset$. For a point $\widetilde{P} \in \widetilde{U}_i \cap \widetilde{U}_j$, the set S has two points, which both correspond to the point $\widetilde{P}$, namely the point $\widetilde{P} \in \widetilde{U}_i$ and the same point $\widetilde{P} \in \widetilde{U}_j$. We denote the latter one with $\widetilde{P}^*$. Next, we identify the points $\widetilde{P}$ and $\widetilde{P}^*$, if the corresponding transition function is analytic. If $\partial X \neq \emptyset$, then there are two points lying over each boundary point of X. Identifying these two points, we obtain a surface O_2, which is called the *complex double* of the surface X. The two points of O_2, which lie over the same point of X are called *symmetric points* of O_2. For more details, see Alling and Greenleaf [3].

Similar to the orientable case it is obtained that the cover group of the double cover $\pi : O_2 \to X$ is generated by an orientation reversing involution. For details, see Seppala and Sorvali [34].

The next theorem relates a Klein surface to its complex double. We refer to [3] for the proof and more details.

Theorem 13.1. *Given a Klein surface X, there exist a double cover $\pi : O_2 \to X$ of the Klein surface X by a Riemann surface O_2 and an antianalytic involution $k : O_2 \to O_2$, with $\pi \circ k = \pi$, such that X is dianalytically equivalent with $O_2/\langle k \rangle$, where $\langle k \rangle$ is the group generated by k. Conversely, given a pair (O_2, k) consisting of a Riemann surface X and an antianalytic involution k, the orbit space $O_2/\langle k \rangle$ admits a unique structure of Klein surface, such that $f : O_2 \to O_2/\langle k \rangle$ is a morphism of Klein surfaces, provided that one regards O_2 as a Klein surface.*

The mapping π is a local homeomorphism at all points $P \in O_2$, for which $\pi(P) \notin \partial X$. At points lying over the boundary of X, the mapping π is a folding map similar to the mapping $x + iy \to x + i\,|y|$ at the real axes. For more details about the folding map and the morphisms of Klein surfaces, see [3] and [4].

By Poincaré's uniformization theorem, each compact Riemann surface of algebraic genus $g \geqslant 2$ can be represented as an orbit space H/Γ^+ of the upper half complex plane H. Next, H is endowed with the conformal structure induced by the group M of the Möbius transformations and the acting group

Γ^+ is a Fuchsian group, that is, a discrete subgroup of M. The group Γ^+ can be chosen with no elements of finite order. For details, we refer to Poincaré [23].

In his unpublished thesis, Preston proved the real counterpart of Poincaré's uniformization theorem: for a Klein surface X of algebraic genus $g \geqslant 2$, there exists a non-euclidean crystallographic (NEC in short) group Γ, that is, a discrete subgroup of the extended modular group, such that X and H/Γ are isomorphic as Klein surfaces. This NEC group can be assumed having no orientation preserving mapping of finite order. For details, see [24] and [28].

By Klein's definition, a *symmetric Riemann surface*, (O_2, k), is a Riemann surface O_2, together with an orientation reversing involution $k : O_2 \to O_2$. The involution k is called a *symmetry* of O_2. For more details about symmetries of a topologic surface, see [34].

A set D of O_2 is called *symmetric* if $k(D) = D$. Thus, given Ω a subset of X, then $\pi^{-1}(\Omega) = D$ is a symmetric subset of O_2.

A function f defined on a symmetric set is called a *symmetric function* if it is k-invariant, that is, $f = f \circ k$.

Next, we identify X with the orbit space $O_2/\langle k \rangle$ obtained by identifying P with $k(P)$, for all $P \in O_2$. If $\widetilde{U}$ is a parametric disk on X, then $\pi^{-1}(\widetilde{U}) = U \cup k(U)$ is a pair of symmetric disks of O_2, hence it is natural to consider restrictions on $U \cup k(U)$ for the local study of the objects on O_2. Since k is an involution without fixed points, one can suppose that $U \cap k(U) = \emptyset$.

We identify the points of O_2, respectively X, with their images on $\mathbb{C}$ from the corresponding local parameters, with respect to the relation between the dianalytic atlas on X and the analytic atlases on O_2. Let z be the local parameter on U. Then $k(z)$ is the local parameter on $k(U)$ and $\widetilde{z} = \widetilde{k(z)} = \pi(z) = \pi(k(z)) = \{z, k(z)\}$ is the local parameter on $\widetilde{U}$.

Let $\mathcal{F}(X)$ be the vector space of the complex functions on the Klein surface X and $\mathcal{F}_s(O_2)$ the vector space of the symmetric functions on O_2. By Theorem 13.1, we conclude that there exists an isomorphism $\pi^* : F(X) \to F_s(O_2)$, between the vector spaces $F(X)$ and $F_s(O_2)$. Indeed, let $F : X \to \overline{\mathbb{C}}$ be a complex function on X, that can take the value ∞ only on finite sets. Its lifting f to O_2 is given by

$$f(z) = f(k(z)) = F(\widetilde{z}), \ z \in O_2, \ \widetilde{z} = \pi(z). \tag{13.1}$$

Then, it is easy to see that the function π^*, defined by $\pi^*(F) = f$ is an isomorphism.

Also, to any function $g : O_2 \to \overline{\mathbb{C}}$, we can associate a function $f = g + g \circ k$, which is a symmetric function on O_2. Thus, (13.1) defines a function F on X.

Let $\widetilde{\gamma}$ be a piecewise smooth Jordan curve on a parametric disk $\widetilde{U}$. The curve $\widetilde{\gamma}$ has exactly two lifts from $\pi^{-1}(\widetilde{U})$. If $\widetilde{\gamma}(0) = \widetilde{z}_0 = \{z_0, k(z_0)\}$ and if γ is the lift of $\widetilde{\gamma}$ on O_2 from z_0, then $k \circ \gamma$ is the lift of $\widetilde{\gamma}$ on O_2 from $k(z_0)$. We refer to [1] for details about covering surfaces. By definition of γ, we obtain $\pi \circ \gamma = \pi \circ k \circ \gamma$, hence for any continuous real-valued function F defined on

$\tilde{\gamma}$, the function $f = F \circ \pi$ is a continuous real-valued symmetric function on $\gamma \cup k(\gamma)$.

The Euclidean lengths of the two curves γ and its symmetric $k \circ \gamma$, that is their lengths with respect to the metric $ds = |dz|$, may be different. We modify this metric and get a new metric $d\sigma$ on O_2, such that the lengths of γ and $k \circ \gamma$, with respect to the metric $d\sigma$, will be the same. We define a symmetric metric on O_2 by

$$d\sigma = \frac{1}{2} (ds + ds \circ k).$$

Then the $d\sigma$-lengths of γ and $k \circ \gamma$ are equal. By definition, the length of $\tilde{\gamma}$ is the common $d\sigma$-length of γ and $k \circ \gamma$. Then

$$d\Sigma(\tilde{z}) = d\sigma(z) = d\sigma(k(z)), \tilde{z} = \pi(z) \in X$$

is a metric on X. The metric $d\Sigma$ is invariant with respect to the group of conformal or anticonformal transition functions of X.

By definition,

$$\int_{\tilde{\gamma}} F d\Sigma = \int_{\gamma} f d\sigma = \int_{k \circ \gamma} f d\sigma.$$

For more details about measure and integration on Klein surfaces, see [5].

Any Riemann surface O_2 of class C^1 is endowed with a *Riemannian metric* determined by the line element

$$ds = \nu |dz + \mu d\bar{z}| ,$$

where ν is a positive function. If μ is identically zero, then the metric

$$ds(z) = \nu(z) |dz|$$

and the local parameter z are called *isothermal*.

It is known that the isothermal metric ds defines a natural analytic structure on O_2. Similar to the orientable case, the isothermal metric $d\sigma$ defines a dianalytic structure on the Klein surface X. See [2] for details.

Next, we give an example of a symmetric isothermal metric (see Schiffer and Spencer [33]).

Example 13.1. *The simplest example of a Klein surface is provided by the Möbius strip. Consider $R > 1$ and the annulus*

$$A_R = \left\{ z \in \mathbb{C} \,\middle|\, \frac{1}{R} < |z| < R \right\}$$

of the z-plane. The Möbius strip, denoted by M, is obtained from A_R by identifying the points z and $-1/\bar{z}$. Let $k : A_R \to A_R$ defined by $k(z) = -1/\bar{z}$. Then (A_R, k) is a symmetric Riemann surface and the quotient space $A_R / \langle k \rangle$

is a Möbius strip. The Möbius strip is obtained by cutting the ring along the real axis in the z-plane and joining the two halves together along corresponding boundaries. Thus, the annulus A_R with points z and $-1/\bar{z}$ identified is a canonical form for the Möbius strip. The Euclidean metric

$$ds = |dz|$$

is not symmetric. We define a symmetric isothermal metric on A_R by

$$d\sigma = \frac{1}{2}\left(|dz| + \left|d\left(-\frac{1}{\bar{z}}\right)\right|\right)$$

$$= \frac{1}{2}\left(1 + \frac{1}{|z|^2}\right)|dz|.$$

By definition, the metric on the Möbius strip is

$$d\Sigma(\tilde{z}) = d\sigma(z) = d\sigma(k(z))$$

thus,

$$d\Sigma(\tilde{z}) = \frac{1}{2}\left(1 + \frac{1}{|z|^2}\right)|dz|.$$

The area element da on A_R is

$$da(z) = \frac{1}{4}\left(1 + \frac{1}{|z|^2}\right)^2 dm(x,y),$$

where m is the Lebesgue measure in the complex plane. Then the area element $d\mathcal{A}$ on the Möbius strip is

$$d\mathcal{A}(\tilde{z}) = da(z) = da(k(z)),$$

hence

$$d\mathcal{A}(\tilde{z}) = \frac{1}{4}\left(1 + \frac{1}{|z|^2}\right)^2 dm(x,y).$$

Let $\gamma : [a,b] \to A_R$ be a piecewise continuously differentiable curve and let $f : \gamma([a,b]) \to \mathbb{C}$ be a continuous function. The integral of f on the curve γ, denoted by $\int_\gamma f d\sigma$, is defined by

$$\int_\gamma f d\sigma = \frac{1}{2}\int_a^b f(\gamma(t))\left(1 + \frac{1}{|\gamma(t)|^2}\right)|\gamma'(t)|\,dt$$

and

$$\int\int_M F d\mathcal{A} = \frac{1}{8}\int\int_{A_R} f(z)\left(1 + \frac{1}{|z|^2}\right)^2 dm(x,y).$$

Let γ be a σ-rectifiable Jordan arc γ, parametrized in terms of the arc σ-length. Therefore, $\gamma : z = z(s) = x(s) + iy(s)$, $s \in [0, l]$, where l is the σ-length of γ. Then the unit inward normal vector to γ at $z(s)$ is $n_\sigma = \left(-\dfrac{dy}{d\sigma}, \dfrac{dx}{d\sigma} \right)$ and we denote by $\dfrac{\partial}{\partial n_\sigma}$ the inward normal derivative, with respect to the symmetric metric $d\sigma$. In this way, our approach is consistent with Nevanlinna [22], Bergman [8] and Schiffer and Spencer [33]. For more details about the normal derivative and Green's identities in terms of $d\sigma$, see [7].

13.3   The Dirichlet Problem for Harmonic Functions

This section is devoted to the study of harmonic functions with Dirichlet boundary condition on a Klein surface. The similar analysis in the complex plane has been developed in Krantz [18, Section 1.2].

The notion of harmonic function, as being a solution of the Laplace equation, makes sense on a Klein surface. Moreover, a Klein surface is the most general two-manifold in which this notion of harmonic function makes sense. For details, see [3]. We notice that the notion of analytic function is meaningless on a Klein surface.

The Dirichlet problem on an arbitrary Riemann surface can be solved because the property that a function that is harmonic remains invariant under bi-holomorphic mappings. For the existence of a harmonic function that vanishes on the boundary and has a finite number of isolated singularities with given singular parts in a relatively compact region, which is contained in a chart of a Riemann surface, we refer to Ahlfors and Sario [2].

Any Klein surface X can be regularly imbedded in a border free surface using a duplication process (see [2]). Therefore, for the boundary problems involving a part of ∂X we can consider it as a part of the boundary of a region on a border free surface.

Let O_2 be a region in the complex plane, bounded by a finite number of analytic Jordan curves. Then $\overline{O_2} = O_2 \cup \partial O_2$ can be conceived as a bordered Riemann surface (see [1], [33]). Because the Klein surfaces X and $O_2/\langle k \rangle$ are dianalytically equivalent, a boundary value problem on a region Ω of the Klein surface X, can be replaced by a similar problem on a symmetric region D of its double O_2, as follows.

Consider the Dirichlet problem on X for harmonic functions

$$\begin{cases} \Delta U = 0 \text{ on } \Omega \\ U = F \text{ on } \partial\Omega. \end{cases} \tag{13.2}$$

where Ω is a region of X bounded by a finite number of σ- rectifiable Jordan curves and F is a continuous real-valued function on $\partial\Omega$.

We define $D = \pi^{-1}(\Omega)$ and $f = F \circ \pi$ on ∂D. Then D is a symmetric region of O_2, bounded by a finite number of σ-rectifiable Jordan curves on O_2, some of which may contain part of ∂O_2. Since $\pi \circ k = \pi$, we obtain $f = f \circ k$ on ∂D, hence f is a symmetric, continuous real-valued function on ∂D. The Dirichlet problem (13.2) on X is equivalent with the following Dirichlet problem for harmonic functions on O_2

$$\begin{cases} \Delta u = 0 \text{ on } D \\ u = f \text{ on } \partial D. \end{cases} \tag{13.3}$$

For details about the Dirichlet problem on bordered Riemann surfaces, see Ahlfors and Sario [2].

The Dirichlet problem turned out to be fundamental in many areas of mathematics and physics. For example, if D is a thin, heat-conducting metal plate and f is a continuous temperature distribution on ∂D, then the solution u of problem (13.3) represents the resulting steady-state heat distribution on D (see [12], [19]).

Using the maximum principle for harmonic functions, it follows that the Dirichlet problem (13.3) with continuous boundary values has a unique solution for any region D with only regular points. For some basic monotonicity, analytic and variational methods of the theory of partial differential equations of elliptic type, we refer to [32].

The symmetric conditions on the boundary imply symmetric solutions for the problem (13.3). For more details, see Schiffer and Spencer [33].

Proposition 13.2. *A solution u of the problem* (13.3) *is a symmetric function on D.*

Proof. Let u be a solution of the problem (13.3). We define $\tilde{u} : \overline{D} \to \mathbb{R}$ by $\tilde{u} = \dfrac{1}{2}(u + u \circ k)$. Then $\Delta \tilde{u} = 0$ on D. By hypothesis, $f = f \circ k$ on ∂D, hence

$$\tilde{u} = \frac{1}{2}(f + f \circ k) = f \quad \text{on } \partial D.$$

Thus, $\tilde{u}$ is also a solution of the problem (13.3). The uniqueness of the solution yields $\tilde{u} = u$ on D, therefore, $u = u \circ k$ on D.   □

13.3.1   The symmetric Green function

Let D be a symmetric region bounded by a finite number of σ-rectifiable Jordan curves on the symmetric Riemann surface O_2.

Fix a point $\zeta \in D$. The function $v(z, \zeta) = -\ln|z - \zeta|$ is harmonic at all points $z \neq \zeta$. Let w be the solution of the Dirichlet problem on D, with the boundary condition $w(z) = v(z, \zeta)$ on ∂D. The unique function $G_D(z, \zeta) = v(z, \zeta) - w(z)$ defined on $\overline{D} \setminus \{\zeta\}$ is called the *Green function* of the region D, with singularity at ζ (see [2]).

We assume that u and v are continuously twice differentiable in D and once on the boundary ∂D. We will use the following Green formula:

$$\int_{\partial D} \left(u \frac{\partial v}{\partial n_\sigma} - v \frac{\partial u}{\partial n_\sigma} \right) d\sigma = - \int \int_D (u \Delta v - v \, \Delta u) \, dx dy,$$

where $d\sigma$ is the arc σ-length element on ∂D and the derivatives on the left are taken with respect to the inward normal on ∂D. For more details, see [22].

The next theorem is similar to the Cauchy integral formula for harmonic functions in terms of the metric $d\sigma$.

Proposition 13.3. *(Green representation formula) Let D be a symmetric region bounded by a finite number of σ-rectifiable Jordan curves and let u be a harmonic function in D and continuously differentiable on its boundary ∂D. Then, for all ζ in D,*

$$u(\zeta) = \frac{1}{2\pi} \int_{\partial D} \left(u(z) \frac{\partial v(z, \zeta)}{\partial n_\sigma} - v(z, \zeta) \frac{\partial u(z)}{\partial n_\sigma} \right) d\sigma, \tag{13.4}$$

where the derivatives are taken with respect to the inward normal on ∂D.

Proof. Fix a point $\zeta \in D$ and a positive number ε that is less than the Euclidean distance of ζ to ∂D. Define $D_\varepsilon = D \setminus \overline{D}(\zeta, \varepsilon)$. Let C_ε be the negatively oriented circle of radius ε, centered at ζ. We apply the Green formula for D_ε, with the harmonic functions u and v. It follows that

$$\int_{\partial D} \left(u \frac{\partial v}{\partial n_\sigma} - v \frac{\partial u}{\partial n_\sigma} \right) d\sigma = - \int_{C_\varepsilon} \left(u \frac{\partial v}{\partial n_\sigma} - v \frac{\partial u}{\partial n_\sigma} \right) d\sigma. \tag{13.5}$$

The curve $-C_\varepsilon$ is parameterized by $z = z(\theta) = \zeta + \varepsilon e^{i\theta}$, $0 \leqslant \theta \leqslant 2\pi$. We deduce that

$$- \int_{C_\varepsilon} v \frac{\partial u}{\partial n_\sigma} d\sigma = \int_{-C_\varepsilon} v \frac{\partial u}{\partial n_\sigma} d\sigma = -\varepsilon \int_0^{2\pi} v(z(\theta), \zeta) \frac{\partial u(z(\theta))}{\partial \rho} d\theta.$$

As the function u has continuous partial derivatives in D, there is a constant C such that $\left| \frac{\partial u}{\partial \rho} \right| \leqslant C$ on C_ε. Then, on C_ε, we obtain

$$\left| \int_{-C_\varepsilon} v \frac{\partial u}{\partial n_\sigma} d\sigma \right| \leqslant 2\pi C \varepsilon \left| \ln \varepsilon \right|.$$

We observe that the right-hand side of the last inequality tends to zero as ε tends to zero. Therefore,

$$\lim_{\varepsilon \to 0} \int_{-C_\varepsilon} v \frac{\partial u}{\partial n_\sigma} d\sigma = 0.$$

Using the mean value property, we have

$$\int\limits_{C_\varepsilon} u \frac{\partial v}{\partial n_\sigma} d\sigma = - \int\limits_0^{2\pi} u(z(\theta)) d\theta = -2\pi u(\zeta).$$

Then relation (13.5) becomes

$$\int\limits_{\partial D} \left(u \frac{\partial v}{\partial n_\sigma} - v \frac{\partial u}{\partial n_\sigma} \right) d\sigma = 2\pi u(\zeta).$$

The proof is now complete.   □

Following Nevannlina (see [22]), we obtain that the values of u inside D are determined from its values and the values of the normal derivative of the Green function on the boundary ∂D.

Theorem 13.4. *Let D be a symmetric region, whose boundary ∂D consists of a finite number of σ-rectifiable Jordan curves. If u is harmonic on D and continuously differentiable on ∂D, then for all ζ in D,*

$$u(\zeta) = \frac{1}{2\pi} \int\limits_{\partial D} u(z) \frac{\partial G_D(z,\zeta)}{\partial n_\sigma} d\sigma. \tag{13.6}$$

Proof. Applying Green's formula for D with the harmonic functions u and w, we obtain

$$\int\limits_{\partial D} \left(v \frac{\partial u}{\partial n_\sigma} - u \frac{\partial w}{\partial n_\sigma} \right) d\sigma = 0. \tag{13.7}$$

Dividing (13.7) by 2π, and adding this identity to the Green representation formula, we obtain (13.6).   □

The function

$$P_\zeta(z) = \frac{1}{2\pi} \frac{\partial G_D(z,\zeta)}{\partial n_\sigma}$$

is called the *Poisson kernel* of the Laplace operator and the Dirichlet problem on the region D.

We define $G_D^{(k)}(z,\widetilde{\zeta})$ as

$$G_D^{(k)}(z,\widetilde{\zeta}) = \frac{1}{2} \left[G_D(z,\zeta) + G_D(z,k(\zeta)) \right]$$

on $\overline{D} \backslash \{\zeta, k(\zeta)\}$.

Let w_s be the solution of the Dirichlet problem on D, with the boundary condition $w_s(z) = \frac{1}{2} \left[v(z,\zeta) + v(z,k(\zeta)) \right]$ on ∂D. Then

$$G_D^{(k)}(z,\widetilde{\zeta}) = \frac{1}{2} \left[v(z,\zeta) + v(z,k(\zeta)) \right] - w_s(z).$$

Therefore, $G_D^{(k)}(z, \widetilde{\zeta})$ is a harmonic function of z in $D \backslash \{\zeta, k(\zeta)\}$, with singularities $-\frac{1}{2} \ln |z - \zeta|$ and $-\frac{1}{2} \ln |z - k(\zeta)|$ at ζ and $k(\zeta)$, respectively. Also, $G_D^{(k)}(z, \widetilde{\zeta}) = 0$ for all z on ∂D.

We can derive the following result (see [7]):

Proposition 13.5. *For every symmetric region D, the function $G_D^{(k)}(\cdot, \widetilde{\zeta})$ is symmetric on $\overline{D}$, that is, for all $z \in \overline{D}$,*

$$G_D^{(k)}(z, \widetilde{\zeta}) = G_D^{(k)}(k(z), \widetilde{\zeta}).$$

Consequently, the function $G_D^{(k)}(z, \widetilde{\zeta})$ is called the *symmetric Green function* of the region D, with singularities at ζ and $k(\zeta)$.

An explicit form for the symmetric Green function of the annulus is obtained in [7]. For additional information on this topic we refer to [33].

13.3.2 The symmetric harmonic measure

Let D be a symmetric region bounded by a finite number of σ-rectifiable Jordan curves on O_2 and $\mathcal{B}(\partial D)$ the σ-algebra of Borel sets of ∂D. The σ-algebra of symmetric Borel sets of ∂D is denoted by $\mathcal{B}_s(\partial D)$ and $\mathcal{B}_s(\partial D) = \{U \cup k(U) \, | \, U \in \mathcal{B}(\partial D)\}$.

The *harmonic measure* for D is a function $\omega_D : D \times \mathcal{B}_s(\partial D) \to [0, 1]$ such that:

1. for each $\zeta \in D$, the map $B \mapsto \omega_D(\zeta, B)$ is a Borel probability measure on ∂D;

2. if $f : \partial D \to \mathbb{R}$ is a continuous function, then the solution of the Dirichlet problem, for D and the boundary function f, is the generalized Poisson integral of f on D given by

$$P_D f(\zeta) = \int_{\partial D} f(z) d\omega_D(\zeta, z), \zeta \in D. \qquad (13.8)$$

For details, see [25].

Remark 1. *The uniqueness of ω_D is a consequence of the Riesz representation theorem.*

An extensive study of the harmonic measure is developed in [13].

A method of determining the harmonic measure is given by the following characterization (see [25]):

Proposition 13.6. *The function $\omega_D(\cdot, B)$, is the solution of the generalized Dirichlet problem with boundary function $f = 1_B$.*

The harmonic measure for D is related to another conformal invariant, the Green function for the symmetric region D.

Using Theorem 13.4 and the fact that Borel measures are determined by their actions on continuous functions, we obtain a representation of the harmonic measure in terms of the inward normal derivative of the Green function with respect to $d\sigma$.

Proposition 13.7. *Let D be a symmetric region, whose boundary ∂D consists of a finite number of σ-rectifiable Jordan curves. If $\zeta \in D$, then for any $z \in \partial D$,*

$$d\omega_D(\zeta, z) = \frac{\partial G_D(z, \zeta)}{\partial n_\sigma} \cdot \frac{d\sigma(z)}{2\pi}.$$

Thus, the harmonic measure for $\zeta \in D$ is absolutely continuous to arc σ-length on ∂D and on ∂D, the density being

$$\frac{d\omega_D}{d\sigma} = \frac{1}{2\pi} \frac{\partial G_D(z, \zeta)}{\partial n_\sigma} = P_\zeta(z).$$

Let $\omega_D^{(k)} : D \times \mathcal{B}_s(\partial D) \to [0, 1]$ be the function defined by

$$\omega_D^{(k)}(\widetilde{\zeta}, B) = \frac{1}{2} \left[\omega_D(\zeta, B) + \omega_D(k(\zeta), B) \right],$$

where $\widetilde{\zeta} = \{\zeta, k(\zeta)\}$, $\zeta \in D$, $B \in \mathcal{B}_s(\partial D)$; see [29].

Remark 2. *The symmetry of the region D implies that the function $\omega_D^{(k)}(\widetilde{\zeta}, B)$ is symmetric with respect to B on $\mathcal{B}_s(\partial D)$, that is, for any $B \in \mathcal{B}_s(\partial D)$,*

$$\omega_D^{(k)}(\widetilde{\zeta}, B) = \omega_D^{(k)}(\widetilde{\zeta}, k(B)).$$

The function $\omega_D^{(k)}(\widetilde{\zeta}, B)$ is called the *symmetric harmonic measure* for D. The function

$$P_{\widetilde{\zeta}}^{(k)}(z) = \frac{1}{2\pi} \frac{\partial G_D^{(k)}(z, \widetilde{\zeta})}{\partial n_\sigma}, \quad z \in D$$

is called the *symmetric Poisson kernel* for the region D.

13.3.3   The Dirichlet problem on the complex double

The following Poisson integral formula both reproduces and creates harmonic functions on the complex double. Roughly speaking, the next theorem yields the formula for the solution of the Dirichlet problem (13.3) on a symmetric region D, in terms of the symmetric Green function.

Theorem 13.8. *Let D be a symmetric region bounded by a finite number of σ-rectifiable Jordan curves and let f be a symmetric, continuous function on*

∂D. There is a unique symmetric function u on $\overline{D}$, which is harmonic in D, continuous on $\overline{D}$, such that $u = f$ on ∂D. Moreover, for all ζ in D,

$$u(\zeta) = \frac{1}{2\pi} \int_{\partial D} f(z) \frac{\partial G_D^{(k)}(z,\zeta)}{\partial n_\sigma} d\sigma, \quad \zeta \in D. \tag{13.9}$$

Proof. By Theorem 13.4, for all $\zeta \in D$,

$$u(\zeta) = \frac{1}{2\pi} \int_{\partial D} u(z) \frac{\partial G_D(z,\zeta)}{\partial n_\sigma} d\sigma.$$

Replacing ζ with $k(\zeta)$ we obtain

$$u(k(\zeta)) = \frac{1}{2\pi} \int_{\partial D} u(z) \frac{\partial G_D(z,k(\zeta))}{\partial n_\sigma} d\sigma.$$

Adding the last two equations and dividing by 2, we obtain

$$\frac{u(\zeta) + u(k(\zeta))}{2} = \frac{1}{4\pi} \int_{\partial D} u(z) \left[\frac{\partial G_D(z,\zeta)}{\partial n_\sigma} + \frac{\partial G_D(z,k(\zeta))}{\partial n_\sigma} \right] d\sigma,$$

for all $\zeta \in D$. By Proposition 13.2, u is a symmetric function on D, then the left-hand side of the last equality is $u(\zeta)$. We conclude that for all ζ in D,

$$u(\zeta) = \frac{1}{4\pi} \int_{\partial D} u(z) \left[\frac{\partial G_D(z,\zeta)}{\partial n_\sigma} + \frac{\partial G_D(z,k(\zeta))}{\partial n_\sigma} \right] d\sigma.$$

The uniqueness of the solution of the Dirichlet problem for harmonic functions implies relation (13.9). □

Theorem 13.8 is the equivalent of the Poisson formula for the solution of the Dirichlet problem on the disc in the complex plane, see Krantz [20, Section 7.3]. In such a way, Theorem 13.8 creates a function that agrees with f on the boundary of the domain D and is harmonic inside.

The formula for the solution to the Dirichlet problem on the annulus is obtained in [6].

In a similar way, we obtain the following representation of the solution of the problem (13.3) on a symmetric region D, in terms of the symmetric harmonic measure.

Theorem 13.9. *Let D be a symmetric region bounded by a finite number of σ-rectifiable Jordan curves and let f be a symmetric, continuous function on ∂D. There exists a unique symmetric function u on $\overline{D}$, which is harmonic on D, continuous on $\overline{D}$, such that $u = f$ on ∂D. For all ζ in D, we have*

$$u(\zeta) = \int_{\partial D} f(z) d\omega_D^{(k)}(\tilde{\zeta}, z). \tag{13.10}$$

Proof. Let ζ be a point in D. By (13.8), for all $\zeta \in D$,

$$u(\zeta) = \int_{\partial D} f(z) d\omega(\zeta, z) d\sigma.$$

Replacing ζ with $k(\zeta)$ we get

$$u(k(\zeta)) = \int_{\partial D} u(z) d\omega(k(\zeta), z) d\sigma.$$

Adding the last two equations and dividing by 2, we obtain

$$\frac{u(\zeta) + u(k(\zeta))}{2} = \frac{1}{2} \int_{\partial D} f(z) \left[d\omega(\zeta, z) + d\omega(k(\zeta), z) \right],$$

for all ζ in D. By Proposition 13.2, u is a symmetric function on D, then the left-hand side of the last equality is $u(\zeta)$ and we conclude that for all ζ in D,

$$u(\zeta) = \int_{\partial D} f(z) d\omega_D^{(k)}(\widetilde{\zeta}, z).$$

The proof is now complete. □

By Proposition 13.7, we obtain the Radon-Nikodym derivative of symmetric harmonic measure for D against σ-arc length.

Proposition 13.10. *Let D be a symmetric region whose boundary ∂D consists of a finite number of σ-rectifiable Jordan curves. If $\zeta \in D$, then for any $z \in \partial D$,*

$$d\omega_D^{(k)}(\widetilde{\zeta}, z) = \frac{\partial G_D^{(k)}(z, \widetilde{\zeta})}{\partial n_\sigma} \cdot \frac{d\sigma(z)}{2\pi}.$$

This result shows that the symmetric harmonic measure for D is absolutely continuous to arc σ-length on ∂D and on ∂D, the density being

$$\frac{d\omega_D^{(k)}}{d\sigma} = \frac{1}{2\pi} \frac{\partial G_D^{(k)}(z, \widetilde{\zeta})}{\partial n_\sigma} = P_{\widetilde{\zeta}}^{(k)}(z).$$

13.3.4   The Dirichlet problem on the Klein surface

Let X be a Klein surface and let Ω be a region bounded by a finite number of σ-rectifiable Jordan curves. The Klein surface X is the factor manifold of the symmetric Riemann surface O_2 with respect to the group $\langle k \rangle$. Then, Ω is obtained from the symmetric region D by identifying the corresponding symmetric points.

The Green function of Ω with singularity at $\widetilde{\zeta}$ is defined by

$$G_\Omega(\widetilde{z}, \widetilde{\zeta}) = G_D^{(k)}(z, \widetilde{\zeta}) = G_D^{(k)}(k(z), \widetilde{\zeta}),$$

where $\widetilde{z} = \pi(z)$.

By definition, the function $G_\Omega(\widetilde{z}, \widetilde{\zeta})$ is continuous on $\overline{\Omega}$, harmonic on $\Omega \backslash \left\{\widetilde{\zeta}\right\}$ and has the singularity at $\widetilde{\zeta} = \pi(\zeta)$.

Remark 3. *By Proposition 13.5, it follows that $G_\Omega(\widetilde{z}, \widetilde{\zeta})$ is well-defined on Ω.*

An explicit form for the Green function of the Möbius strip is obtained in [7].

The harmonic measure for Ω, $\omega_\Omega : \Omega \times \mathcal{B}(\partial\Omega) \to [0, 1]$, is defined by

$$\omega_\Omega(\widetilde{\zeta}, \widetilde{B}) = \omega_D^{(k)}(\widetilde{\zeta}, B) = \omega_D^{(k)}(\widetilde{\zeta}, k(B)).$$

for all $\widetilde{\zeta} \in \Omega$ and $\widetilde{B} = \pi(B) \in \mathcal{B}(\partial\Omega)$.

The function

$$P_{\widetilde{\zeta}}(\widetilde{z}) = P_{\widetilde{\zeta}}^{(k)}(z) = P_{\widetilde{\zeta}}^{(k)}(k(z)), \ z \in D$$

is called the *Poisson kernel* for the region Ω.

Remark 4. *By Remark 2, it follows that the function ω_Ω is well-defined. By Proposition 13.5, it follows that the function $P_{\widetilde{\zeta}}$ is well-defined, too.*

The symmetric solutions on O_2 determine the solutions of the similar problems on the Klein surface X.

Consequently, we obtain the solution of the Dirichlet problem on the region Ω, with respect to the Green function of Ω.

Theorem 13.11. *Let F be a continuous real-valued function on the border $\partial\Omega$. The solution of the Dirichlet problem (13.2) with the boundary function F is the function U defined on $\overline{\Omega}$, by the relation $u = U \circ \pi$, where π is the canonical projection of O_2 on X and u is the solution (13.9) of the Dirichlet problem (13.3) on the symmetric region D, with the boundary function f, given by $f = F \circ \pi$.*

Proof. By definition, $\Delta U(\widetilde{\zeta}) = \Delta u(\zeta) = 0$, for all $\widetilde{\zeta} \in \Omega$, where $\widetilde{\zeta} = \pi(\zeta)$. Thus, U is a harmonic function on Ω. The symmetry of the function f on ∂D implies

$$U(\widetilde{\zeta}) = u(\zeta) = f(\zeta) = f(k(\zeta)) = F(\widetilde{\zeta}),$$

for all $\widetilde{\zeta} \in \partial\Omega$. Due to the uniqueness of the solution, the function U defined on $\overline{\Omega}$ by

$$U(\widetilde{\zeta}) = u(\zeta) = u(k(\zeta)),$$

for all $\widetilde{\zeta}$ in $\overline{\Omega}$, where $\widetilde{\zeta} = \pi(\zeta)$, is the solution of the Dirichlet problem (13.2) on Ω. $\square$

In a similar way, we obtain the solution of the problem (13.2) on the region Ω, with respect to the harmonic measure for the region Ω.

Theorem 13.12. *Let F be a continuous real-valued function on the border $\partial\Omega$. The solution of the problem (13.2) with the boundary function F is the function U defined on $\overline{\Omega}$, by the relation $u = U \circ \pi$, where π is the canonical projection of O_2 on X and u is the solution (13.10) of the problem (13.3) on the symmetric region D, with the boundary function f, given by $f = F \circ \pi$.*

By Proposition 13.10, we obtain the Radon-Nikodym derivative of harmonic measure for Ω against Σ-arc length.

Proposition 13.13. *Let Ω be a region bounded by a finite number of σ-rectifiable Jordan curves. If $\widetilde{\zeta} \in \Omega$, then for all $\widetilde{z} \in \partial\Omega$,*

$$d\omega_\Omega(\widetilde{\zeta}, \widetilde{z}) = d\omega_D^{(k)}(\widetilde{\zeta}, z) = d\omega_D^{(k)}(\widetilde{\zeta}, k(z)).$$

This result implies that the harmonic measure for Ω is absolutely continuous to arc Σ-length on $\partial\Omega$ and on $\partial\Omega$, the density being

$$\frac{d\omega_\Omega}{d\Sigma} = P_{\widetilde{\zeta}}(\widetilde{z}).$$

13.4   The Neumann Problem for Harmonic Functions

This section is devoted to the study of harmonic functions with Neumann boundary condition on a Klein surface. The similar analysis in the complex place has been developed by Schiffer and Spencer [33].

Consider the Neumann problem for harmonic functions

$$\begin{cases} \Delta U = 0 \text{ on } \Omega \\ \dfrac{\partial U}{\partial n_\Sigma} = G \text{ on } \partial\Omega, \end{cases} \tag{13.11}$$

where Ω is a region of X bounded by a finite number of σ-rectifiable Jordan curves and G is a continuous real-valued function on $\partial\Omega$.

We define $D = \pi^{-1}(\Omega)$ and $g = G \circ \pi$ on ∂D. Since $\pi \circ k = \pi$, we obtain that D is a symmetric region bounded by a finite number of σ-rectifiable Jordan curves on O_2, some of that may contain part of ∂O_2 and g is a symmetric, continuous real-valued function on the boundary ∂D.

The Neumann problem on X is equivalent with the following Neumann problem on O_2

$$\begin{cases} \Delta u = 0 \text{ on } D \\ \dfrac{\partial u}{\partial n_\sigma} = g \text{ on } \partial D. \end{cases} \tag{13.12}$$

Since k is an antianalytic involution, the symmetry of D and the symmetry of g on ∂D, imply that the prescribed values of the normal derivative satisfy the compatibility condition

$$\int_{\partial D} g d\sigma = 0.$$

Therefore, the Neumann problem on O_2 for the region D and the boundary function g has solutions. For details, see [21].

Proposition 13.14. *If the problem (13.12) admits a solution, then it is unique up to an additive constant.*

Proof. Let u_1 and u_2 be solutions of the problem (13.12). If $u = u_1 - u_2$, then u is harmonic on D and $\dfrac{\partial u}{\partial n_\sigma} = 0$ on ∂D. Applying Green's first identity, we get

$$\int\int_D \left(u_x^2 + u_y^2 \right) dx dy = 0.$$

Therefore, u is constant on D. $\qquad\qquad\square$

Proposition 13.15. *The solution of the problem (13.12) is a symmetric function on D.*

Proof. Let u be a solution of the problem (13.12). We define $\tilde{u} : \overline{D} \to \mathbb{R}$ by $\tilde{u} = \dfrac{1}{2}(u + u \circ k)$. By hypothesis $g = g \circ k$ on ∂D, then $\dfrac{\partial \tilde{u}}{\partial n_\sigma} = \dfrac{\partial u}{\partial n_\sigma} = g$ on ∂D and $\Delta \tilde{u} = 0$ on D. Thus, $\tilde{u}$ is also a solution of the problem (13.12). By Proposition 13.14, there is a constant c such that $\tilde{u} = u + c$ on D. Thus, $u \circ k = u + 2c$ on D and using the symmetry of the region D, we obtain $u = u \circ k + 2c$ on D. Hence $c = 0$, that is, $u \circ k = u$ on D. $\qquad\square$

13.4.1 The symmetric Neumann function

Let ζ be a point inside D. A *Neumann function* $N_D(z, \zeta)$ for the region D, with singularity at ζ, in terms of the metric $d\sigma$, is the function

$$N_D(z, \zeta) = v(z, \zeta) - h(z, \zeta), \quad z \in D, \ z \neq \zeta,$$

where $h(z, \zeta)$ is a solution of the following Neumann problem in terms of the metric $d\sigma$:

$$\begin{cases} \Delta h(z, \zeta) = 0, & z \in D \\ \dfrac{\partial h}{\partial n_\sigma}(z, \zeta) = \dfrac{\partial v}{\partial n_\sigma}(z, \zeta) - \dfrac{2\pi}{l}, & z \in \partial D, \end{cases}$$

where $l = \displaystyle\int_{\partial D} d\sigma$ is the σ-length of ∂D.

Remark 5. *The boundary value of the inward normal derivative of the Neumann function is a constant equal to* $\dfrac{2\pi}{l}$.

Theorem 13.16. *Let D be a symmetric region bounded by a finite number of σ-rectifiable Jordan curves. If u is harmonic in D and continuously differentiable on ∂D then, up to an additive constant,*

$$u(\zeta) = -\frac{1}{2\pi} \int\limits_{\partial D} \frac{\partial u(z)}{\partial n_\sigma} N_D(z,\zeta) d\sigma, \ \zeta \in D.$$

Proof. Fix a point $\zeta \in D$ and a positive number ε that is less than the Euclidean distance of ζ to ∂D. Define $D_\varepsilon = D \setminus \overline{D}(\zeta, \varepsilon)$. Let C_ε be the negatively oriented circle of radius ε, centered at ζ. Applying Green formula for D_ε with the harmonic functions h and u, we obtain

$$\int\limits_{\partial D} \left(h \frac{\partial u}{\partial n_\sigma} - u \frac{\partial h}{\partial n_\sigma} \right) d\sigma = 0. \tag{13.13}$$

Dividing (13.13) by 2π and adding this identity to the Green representation formula, it follows that

$$u(\zeta) = -\frac{1}{2\pi} \int\limits_{\partial D} N(z,\zeta) \frac{\partial u}{\partial n_\sigma} d\sigma + \frac{1}{l} \int\limits_{\partial D} u d\sigma.$$

Thus, u is determined up to the additive constant $\dfrac{1}{l} \int\limits_{\partial D} u(z) d\sigma.$ $\square$

Let $N_D^{(k)}(z,\widetilde{\zeta})$ be the function defined by

$$N_D^{(k)}(z,\widetilde{\zeta}) = \frac{1}{2} \left[N_D(z,\zeta) + N_D(z,k(\zeta)) \right], \ z \in D \setminus \{\zeta, k(\zeta)\},$$

where $N_D(z, k(\zeta))$ is a Neumann function for the region D, with singularity at $k(\zeta)$ and $\widetilde{\zeta} = \{\zeta, k(\zeta)\}$. Therefore,

$$N_D^{(k)}(z,\widetilde{\zeta}) = \frac{1}{2} \left[v(z,\zeta) + v(z,k(\zeta)) \right] - h_s(z,\widetilde{\zeta}), \ z \neq \zeta, \ z \neq k(\zeta),$$

where h_s is a harmonic function on D that satisfies

$$\frac{\partial h_s}{\partial n_\sigma}(z,\widetilde{\zeta}) = \frac{1}{2} \left[\frac{\partial v}{\partial n_\sigma}(z,\zeta) + \frac{\partial v}{\partial n_\sigma}(z,k(\zeta)) \right] - \frac{2\pi}{l}.$$

Therefore, $N_D^{(k)}(z,\widetilde{\zeta})$ is a harmonic function of z in $D \setminus \{\zeta, k(\zeta)\}$, with singularities at ζ and $k(\zeta)$ and $\dfrac{\partial N_D^{(k)}}{\partial n_\sigma}(z,\widetilde{\zeta}) = \dfrac{2\pi}{l}$, for all $z \in \partial D$.

An explicit form for the function $N_D^{(k)}(z,\widetilde{\zeta})$ of the annulus and of the Möbius strip are obtained in [31].

Proposition 13.17. *If D is a symmetric region, then the function $N_D^{(k)}(z,\widetilde{\zeta})$ is symmetric with respect to z on D, that is, for any $z \in D$,*

$$N_D^{(k)}(z,\widetilde{\zeta}) = N_D^{(k)}(k(z),\widetilde{\zeta}).$$

Proof. Let $h^*(\cdot,\zeta)$ be a harmonic function in D, such that

$$\frac{\partial h^*}{\partial n_\sigma}(z,\zeta) = \frac{1}{2}\left(\frac{\partial v}{\partial n_\sigma}(z,\zeta) + \frac{\partial v}{\partial n_\sigma}(k(z),\zeta)\right) - \frac{2\pi}{l}, \quad z \in \partial D.$$

Therefore,

$$\frac{\partial h^*}{\partial n_\sigma}(z,\zeta) = \frac{\partial h^*}{\partial n_\sigma}(k(z),\zeta), \quad \text{for all } z \in \partial D.$$

By Proposition 13.15, $h^*(\cdot,\zeta)$ is a symmetric function. Hence the function

$$M_D^{(k)}(z,\widetilde{\zeta}) = \frac{1}{2}\left[v(z,\zeta) + v(k(z),\zeta)\right] - h^*(z;\zeta)$$

is a symmetric function, harmonic in $D \setminus \{\widetilde{\zeta}\}$ and $\dfrac{\partial M_D^{(k)}}{\partial n}(z,\widetilde{\zeta}) = \dfrac{2\pi}{l}$. So, $N_D^{(k)}(z,\widetilde{\zeta})$ and $M_D^{(k)}(z,\widetilde{\zeta})$ are solutions of the same Neumann problem. Thus, by Proposition 13.14, there is a constant c such that $N_D^{(k)}(z,\widetilde{\zeta}) = M_D^{(k)}(z,\widetilde{\zeta}) + c$. Since $M_D^{(k)}(z,\widetilde{\zeta})$ is a symmetric function, we obtain that $N_D^{(k)}(z,\widetilde{\zeta})$ is also a symmetric function. $\qquad\square$

Let ζ_0 be a point of D. A Neumann function $N_D(z,\zeta)$ is not a conformal invariant, but the difference $N_D(z,\zeta) - N_D(z,\zeta_0)$ is a Neumann function and has a vanishing normal derivative on ∂D, hence it is a conformal invariant. We redefine the difference

$$N_D(z,\zeta,\zeta_0) = N_D(z,\zeta) - N_D(z,\zeta_0)$$

to be a *Neumann function* for the region D on the Riemann surface O_2, see [33].

The function $N_D^{(k)}(z,\widetilde{\zeta},\widetilde{\zeta_0})$ defined by

$$N_D^{(k)}(z,\widetilde{\zeta},\widetilde{\zeta_0}) = \frac{1}{2}\left[N_D(z,\zeta,\zeta_0) + N_D(z,k(\zeta),k(\zeta_0))\right],$$

for all $z \in \overline{D}\setminus\{\widetilde{\zeta},\widetilde{\zeta_0}\}$, is called a *symmetric Neumann function* for the region D.

13.4.2 The symmetric harmonic kernel function

Let D be a symmetric region in the complex plane, bounded by a finite number of σ-rectifiable Jordan curves. In this section, we introduce closed systems

$(\varphi_i)_i$ of harmonic functions in D, which are orthonormal with respect to the Dirichlet integral

$$D\{\varphi_i, \varphi_j\} = \int \int_D \left(\frac{\partial \varphi_i}{\partial x} \frac{\partial \varphi_j}{\partial y} + \frac{\partial \varphi_i}{\partial y} \frac{\partial \varphi_j}{\partial x} \right) dx dy.$$

We recall some notions and results about orthogonal harmonic functions. For more details, see [8].

Let $\Lambda^2(D)$ be the set of harmonic functions φ in D with a finite Dirichlet integral

$$D\{\varphi\} = D\{\varphi, \varphi\} < \infty$$

such that

$$D\{N_D(z, \zeta), \varphi(\zeta)\} = 2\pi\varphi(\zeta),$$

where $N_D(z, \zeta)$ is the Neumann function of D with its singularity at the fixed point ζ, $\zeta \in D$.

Remark 6. *The second condition is imposed to normalize $D\{\varphi, \varphi\}$ to be zero if and only if φ vanishes identically.*

Proposition 13.18. *There exists a closed system $(\varphi_i)_i$ for the class $\Lambda^2(D)$, which is orthonormal with respect to the Dirichlet integral, that is,*

$$D\{\varphi_i, \varphi_j\} = \delta_{ij}, \ \delta_{ii} = 1, \ \delta_{ij} = 0, \ i \neq j.$$

Let ζ be a point inside D. The *harmonic kernel function* $K_D(z, \zeta)$ of the closed orthonormal system $(\varphi_i)_i$ for the region D, with respect to the point ζ, is the function defined by

$$K_D(z, \zeta) = \sum_{i=1}^{\infty} \varphi_i(z) \varphi_i(\zeta), \ z \in \overline{D}.$$

The harmonic kernel function is uniquely characterized by the following properties:

$$K_D(z, \zeta) = K_D(\zeta, z)$$

and

$$D\{K_D(z, \zeta), \varphi(\zeta)\} = \varphi(\zeta), \ \varphi \in \Lambda^2(D).$$

An extensive study of the harmonic kernel function is due to [8].

The representation of the harmonic kernel function in terms of a closed orthonormal system gives the opportunity to solve numerically the Dirichlet problem for arbitrarily multiply connected regions. This is an important tool in physics, in particular in fluid mechanics, elasticity and electricity.

It is known that the harmonic kernel function $K_D(z, \zeta)$, the Green function $G_D(z, \zeta)$ and the Neumann function $N_D(z, \zeta)$ satisfy the relation

$$K_D(z, \zeta) = \frac{1}{2\pi} \left[N_D(z, \zeta) - G_D(z, \zeta) \right], \ z \in \overline{D}. \tag{13.14}$$

We first derive a formula that solves the Dirichlet problem (13.3). We prove that if u is harmonic inside a region D and continuous on ∂D, then we can determine the values of u inside of D by integrating on ∂D the product of u times the inward normal derivative of the harmonic kernel function for the region D, which is a fixed function that depends only on D.

Theorem 13.19. *Let D be a symmetric region bounded by a finite number of σ-rectifiable Jordan curves. If u is harmonic in D and continuous on $\overline{D}$, then, up to an additive constant,*

$$u(\zeta) = -\int_{\partial D} u(z)\frac{\partial K_D(z,\zeta)}{\partial n_\sigma}d\sigma, \ \zeta \in D. \tag{13.15}$$

Proof. From (13.6), the solution of the Dirichlet problem (13.3) is

$$u(\zeta) = \frac{1}{2\pi}\int_{\partial D} u(z)\frac{\partial G_D(z,\zeta)}{\partial n_\sigma}d\sigma, \ \zeta \in D. \tag{13.16}$$

Using (13.14), we obtain

$$\frac{\partial K_D(z,\zeta)}{\partial n_\sigma} = \frac{1}{2\pi}\frac{\partial N_D(z,\zeta)}{\partial n_\sigma} - \frac{1}{2\pi}\frac{\partial g_D(z,\zeta)}{\partial n_\sigma}$$

$$= \frac{1}{l} - \frac{1}{2\pi}\frac{\partial G_D(z,\zeta)}{\partial n_\sigma},$$

for any $z \in \partial D$, where l is the length of ∂D (see [21]). Combining this relation with (13.16), we find

$$u(\zeta) = -\int_{\partial D} u(z)\frac{\partial K_D(z,\zeta)}{\partial n_\sigma}d\sigma + \frac{1}{l}\int_{\partial D} u(z)d\sigma.$$

Thus, u is determined up to the additive constant $\dfrac{1}{l}\displaystyle\int_{\partial D} u(z)d\sigma.$ $\quad\square$

Next, we derive a formula that solves the Neumann problem (13.12).

Theorem 13.20. *Let D be a symmetric region bounded by a finite number of σ-rectifiable Jordan curves. If u is harmonic in D and continuously differentiable on ∂D then, up to an additive constant,*

$$u(\zeta) = -\int_{\partial D}\frac{\partial u(z)}{\partial n_\sigma}K_D(z,\zeta)d\sigma, \ \zeta \in D. \tag{13.17}$$

Proof. By Theorem 13.16, using Green formula, it follows that, up to an additive constant, a solution of the Neumann problem is given by

$$u(\zeta) = -\frac{1}{2\pi}\int_{\partial D}\frac{\partial u(z)}{\partial n_\sigma}N_D(z,\zeta)d\sigma, \ \zeta \in D. \tag{13.18}$$

The constant is chosen such that $u(z)$ is in $\Lambda^2(D)$.

By (13.14), for $\zeta \in \partial D$, we have

$$K_D(z,\zeta) = \frac{1}{2\pi} N_D(z,\zeta).$$

Substituting this in (13.18), we obtain (13.17).   □

Let $K_D^{(k)}(z,\widetilde{\zeta})$ be the function defined by

$$K_D^{(k)}(z,\widetilde{\zeta}) = \frac{1}{2} \left[K_D(z,\zeta) + K_D(z,k(\zeta)) \right], \ z \in \overline{D},$$

where $K_D(z,k(\zeta))$ is the harmonic kernel function of the closed orthonormal system $(\varphi_i)_i$, for the region D, with respect to the point $k(\zeta)$. The function $K_D^{(k)}(z,\widetilde{\zeta})$ is in $\Lambda^2(D)$ (see [8], [30]).

Proposition 13.21. *If D is a symmetric region, then the function $K_D^{(k)}(z,\widetilde{\zeta})$ is symmetric with respect to z on $\overline{D}$, that is, for every $z \in \overline{D}$,*

$$K_D^{(k)}(z,\widetilde{\zeta}) = K_D^{(k)}(k(z),\widetilde{\zeta}).$$

Proof. We use (13.14) and the symmetric properties of the symmetric Green function and symmetric Neumann function.   □

Let ζ_0 be a point of D. Let $\Lambda_0^2(D)$ be the class of harmonic functions φ that satisfy the conditions:

$$D\{\varphi,\varphi\} < \infty$$

and

$$\varphi(\zeta_0) = 0.$$

The harmonic kernel function $K_D(z,\zeta,\zeta_0)$ of the class $\Lambda_0^2(D)$ is related to the harmonic kernel function $K_D(z,\zeta)$ of the class $\Lambda_0^2(D)$ by the following identity:

$$K_D(z,\zeta,\zeta_0) = K_D(z,\zeta) - K_D(\zeta,\zeta_0).$$

The harmonic kernel function $K_D(z,\zeta)$ for the region D, with respect to the point ζ is not a conformal invariant but the harmonic kernel function $K_D(z,\zeta,\zeta_0)$ is invariant under conformal mapping (see [8]), therefore, $K_D(z,\zeta,\zeta_0)$ is well-defined on the Riemann surface O_2.

The function $K_D^{(k)}(z,\widetilde{\zeta},\widetilde{\zeta_0})$ defined by

$$K_D^{(k)}(z,\widetilde{\zeta},\widetilde{\zeta_0}) = \frac{1}{2} \left[K_D(z,\zeta,\zeta_0) + K_D(z,k(\zeta),k(\zeta_0)) \right],$$

for all $z \in \overline{D} \backslash \{\widetilde{\zeta}, \widetilde{\zeta_0}\}$, is called the *symmetric harmonic kernel function* for the region D.

13.4.3 Integral representations on the double cover

We first express the solution of the Neumann problem (13.12) for harmonic functions in terms of $d\sigma$ as a line integral involving the boundary function and a symmetric Neumann function.

Theorem 13.22. *Let D be a symmetric region bounded by a finite number of σ-rectifiable Jordan curves and let g be a symmetric, continuous function on ∂D. If u is harmonic in D and g is its inward normal derivative on ∂D, then up to an additive constant*

$$u(\zeta) = -\frac{1}{2\pi} \int_{\partial D} g(z) N_D^{(k)}(z, \widetilde{\zeta}) d\sigma, \ \zeta \in D. \tag{13.19}$$

Proof. By Theorem 13.16, up to the additive constant $\dfrac{1}{l} \displaystyle\int_{\partial D} u(z) d\sigma$, we have for all $\zeta \in D$,

$$u(\zeta) = -\frac{1}{2\pi} \int_{\partial D} g(z) N_D(z, \zeta) d\sigma .$$

Replacing ζ with $k(\zeta)$ we get

$$u(k(\zeta)) = -\frac{1}{2\pi} \int_{\partial D} g(z) N_D(z, k(\zeta)) d\sigma.$$

Adding the last two equations and dividing by 2, we obtain, up to the additive constant $\dfrac{1}{l} \displaystyle\int_{\partial D} u(z) d\sigma$,

$$\frac{u(\zeta) + u(k(\zeta))}{2} = -\frac{1}{2\pi} \int_{\partial D} g(z) \frac{N_D(z, \zeta) + N_D(z, k(\zeta))}{2} d\sigma.$$

By Proposition 13.15, u is a symmetric function on D, then the left-hand side of the last equality is $u(\zeta)$. Therefore,

$$u(\zeta) = -\frac{1}{4\pi} \int_{\partial D} g(z) \left[N_D(z, \zeta) + N_D(z, k(\zeta)) \right] d\sigma,$$

up to the additive constant $\dfrac{1}{l} \displaystyle\int_{\partial D} u(z) d\sigma$. □

Similarly, we obtain a formula for the symmetric solution of the Neumann problem (13.12) on a symmetric region D, in terms of the symmetric harmonic kernel function.

Theorem 13.23. *Let D be a symmetric region bounded by a finite number of σ-rectifiable Jordan curves. Let g be a symmetric, continuous function on ∂D. If u is harmonic in D and g is its inward normal derivative on ∂D, then up to an additive constant,*

$$u(\zeta) = -\int_{\partial D} g(z) K_D^{(k)}(z, \tilde{\zeta}) d\sigma, \ \zeta \in D. \tag{13.20}$$

Proof. It is similar with the proof of the Theorem 13.22. Here we use Theorem 13.20. $\qquad\square$

The next theorem yields a formula for the symmetric solution of the Dirichlet problem (13.3) on a symmetric region D, in terms of the symmetric harmonic kernel function.

Theorem 13.24. *Let D be a symmetric region bounded by a finite number of σ-rectifiable Jordan curves. Let f be a symmetric, continuous function on ∂D. There is a unique symmetric function u on $\overline{D}$, which is harmonic on D, continuous on $\overline{D}$, such that $u = f$ on ∂D. For all ζ in D,*

$$u(\zeta) = -\int_{\partial D} f(z) \frac{\partial K_D^{(k)}(z, \tilde{\zeta})}{\partial n_\sigma} d\sigma. \tag{13.21}$$

Proof. By Theorem 13.19, for all $\zeta \in D$,

$$u(\zeta) = -\int_{\partial D} u(z) \frac{\partial K_D(z, \zeta)}{\partial n_\sigma} d\sigma.$$

Replacing ζ with $k(\zeta)$ we get, for all $\zeta \in D$,

$$u(k(\zeta)) = -\int_{\partial D} u(z) \frac{\partial K_D(z, k(\zeta))}{\partial n_\sigma} d\sigma.$$

Adding the last two equations and dividing by 2, it follows that

$$\frac{u(\zeta) + u(k(\zeta))}{2} = -\frac{1}{2} \int_{\partial D} u(z) \left[\frac{\partial K_D(z, \zeta)}{\partial n_\sigma} + \frac{\partial K_D(z, k(\zeta))}{\partial n_\sigma} \right] d\sigma,$$

for all $\zeta \in D$.

By Proposition 13.2, u is a symmetric function on D, then the left-hand side of the last equality is $u(\zeta)$ and we conclude that for all ζ in D,

$$u(\zeta) = -\frac{1}{2} \int_{\partial D} u(z) \left[\frac{\partial K_D(z, \zeta)}{\partial n_\sigma} + \frac{\partial K_D(z, k(\zeta))}{\partial n_\sigma} \right] d\sigma.$$

The uniqueness of the solution of the Dirichlet problem for harmonic functions implies (13.21). $\qquad\square$

13.4.4   Integral representations on the Klein surface

Let $\widetilde{\zeta}$ be a point inside Ω. A *Neumann function* $N_\Omega(\widetilde{z}, \widetilde{\zeta})$ for the region Ω, with singularity at $\widetilde{\zeta}$ is defined by

$$N_\Omega(\widetilde{z}, \widetilde{\zeta}) = N_D^{(k)}(z, \widetilde{\zeta}) = N_D^{(k)}(k(z), \widetilde{\zeta}), \tag{13.22}$$

where $\widetilde{z} = \pi(z)$.

Remark 7. *By Proposition 13.17, it follows that* $N_\Omega(\widetilde{z}, \widetilde{\zeta})$ *is well-defined on* Ω.

Therefore, $N_\Omega(\widetilde{z}, \widetilde{\zeta})$ is a harmonic function on $\Omega \backslash \{\widetilde{\zeta}\}$, which has a constant normal derivative $\dfrac{\partial N_\Omega}{\partial n_\Sigma}$ on the boundary $\partial\Omega$ and has a logarithmic pole at the point $\widetilde{\zeta} = \pi(\zeta)$.

Next, we derive the solution of the Neumann problem (13.11) on the region Ω.

Theorem 13.25. *Let G be a continuous real-valued function on $\partial\Omega$. Then, up to an additive constant, the solution of problem (13.11) is the function U defined by the relation $u = U \circ \pi$, where π is the canonical projection of O_2 on X and u is the solution (13.19) of the problem (13.12) on the symmetric region D, with the inward normal derivative g given by $g = G \circ \pi$ on ∂D.*

Proof. The symmetry of the function u on D, yields

$$\Delta U(\widetilde{\zeta}) = \Delta u(\zeta) = \Delta u(k(\zeta)) = 0 \quad \text{for all } \widetilde{\zeta} \in \Omega,$$

where $\widetilde{\zeta} = \pi(\zeta)$.

Using the symmetry of the function g on ∂D, we obtain

$$\frac{\partial U}{\partial n_\Sigma}(\widetilde{\zeta}) = \frac{\partial U}{\partial n_\sigma}(\zeta) = g(\zeta) = g(k(\zeta)) = G(\widetilde{\zeta}),$$

for all $\widetilde{\zeta} \in \partial\Omega$. Then, up to an additive constant, the function U defined on Ω by

$$U(\widetilde{\zeta}) = u(\zeta) = u(k(\zeta)),$$

for all $\widetilde{\zeta}$ in Ω, is the solution of problem (13.11). $\qquad\square$

Let $\widetilde{\zeta}$ be a point inside Ω. The harmonic kernel function $K_\Omega(\widetilde{z}, \widetilde{\zeta})$ of the closed orthonormal system $(\varphi_i)_i$, for the region Ω, with respect to the point $\widetilde{\zeta} = \{\zeta, k(\zeta)\}$ is defined by

$$K_\Omega(\widetilde{z}, \widetilde{\zeta}) = K_D^{(k)}(z, \widetilde{\zeta}) = K_D^{(k)}(k(z), \widetilde{\zeta}), \widetilde{z} = \pi(z) \in \Omega.$$

Remark 8. *By Proposition 13.21, it follows that $K_\Omega(\widetilde{z}, \widetilde{\zeta})$, is well-defined on* Ω.

The symmetric solutions on O_2 determine the solutions of the similar problems on the Klein surface X. Thus, we obtain the solution of the Dirichlet problem (13.2) on the region Ω, with respect to the harmonic kernel function, for the region Ω.

Theorem 13.26. *Let F be a continuous real-valued function on the border $\partial\Omega$. The solution of the problem* (13.2) *with the boundary function F is the function U defined on $\overline{\Omega}$, by the relation $u = U \circ \pi$, where π is the canonical projection of O_2 on X and u is the solution* (13.21) *of the problem* (13.3) *on the symmetric region D, with the boundary function f given by $f = F \circ \pi$.*

Proof. By definition,

$$\Delta U(\widetilde{\zeta}) = \Delta u(\zeta) = 0 \quad \text{for all } \widetilde{\zeta} \in \Omega,$$

where $\widetilde{\zeta} = \pi(\zeta)$, thus U is a harmonic function. The symmetry of the function f on ∂D, implies

$$U(\widetilde{\zeta}) = u(\zeta) = f(\zeta) = f(k(\zeta)) = F(\widetilde{\zeta}) \quad \text{for all } \widetilde{\zeta} \in \partial\Omega.$$

Due to the uniqueness of the solution, the function U defined on $\overline{\Omega}$ by

$$U(\widetilde{\zeta}) = u(\zeta) = u(k(\zeta)),$$

for all $\widetilde{\zeta}$ in $\overline{\Omega}$, where $\widetilde{\zeta} = \pi(\zeta)$, is the solution of problem (13.2) on Ω. $\qquad\square$

The next theorem gives the solution of the Neumann problem (13.11) on the region Ω, with respect to the harmonic kernel function, for the region Ω.

Theorem 13.27. *Let G be a continuous real-valued function on the border $\partial\Omega$. Then, up to an additive constant, the solution of the problem* (13.11) *with the normal derivative G on $\partial\Omega$ is the function U defined on $\overline{\Omega}$, by the relation $u = U \circ \pi$, where π is the canonical projection of O_2 on X and u is the solution* (13.20) *of the problem* (13.12) *on the symmetric region D, with the normal derivative function g given by $g = G \circ \pi$ on ∂D.*

Proof. By definition,

$$\Delta U(\widetilde{\zeta}) = \Delta u(\zeta) = 0 \quad \text{for all } \widetilde{\zeta} \in \Omega,$$

where $\widetilde{\zeta} = \pi(\zeta)$, thus U is a harmonic function. The symmetry of the function g on ∂D, implies

$$\frac{\partial U(\widetilde{\zeta})}{\partial n_\Sigma} = \frac{\partial u(\zeta)}{\partial n_\sigma} = g(\zeta) = g(k(\zeta)) = G(\widetilde{\zeta}),$$

for all $\widetilde{\zeta} \in \partial\Omega$. Thus, up to an additive constant, the function U defined on $\overline{\Omega}$ by

$$U(\widetilde{\zeta}) = u(\zeta) = u(k(\zeta)),$$

is the solution of the problem (13.11) on Ω. $\qquad\square$

Concluding remarks

The methods developed in this paper remain valid in the case of all differential operators associated to *conformal invariant metrics*. Such an example corresponds to the *invariant Laplacian* (or sometimes the *Laplace-Beltrami operator* for the Poincaré-Bergman metric), see Krantz [18, Section 6.5]. We also refer to the *pseudo-hyperbolic metric*, which is conformally invariant, but it does not arise from integrating an infinitesimal metric (that is, lengths of tangent vectors at a point). A comprehensive analysis of the pseudo-hyperbolic metric on the disc may be found in Krantz [16].

To the best of our knowledge, there are not further results involving either *linear* or *non-linear* elliptic equations on Klein surfaces. This study can include qualitative and quantitative properties of solutions but also related singular or degenerate phenomena. We consider that the mathematical analysis of these classes of PDEs on *Klein surfaces* is a very rich and attractive research field at the interplay between complex analysis and non-linear analysis.

Acknowledgments

The research of Vicenţiu D. Rădulescu was supported by a grant of the Ministry of Research, Innovation and Digitization, CNCS/CCCDI–UEFISCDI, project number PCE 137/2021, within PNCDI III.

References

[1] Lars V. Ahlfors. *Conformal invariants: topics in geometric function theory*. McGraw-Hill Series in Higher Mathematics. McGraw-Hill Book Co., New York-Düsseldorf-Johannesburg, 1973.

[2] Lars V. Ahlfors and Leo Sario. *Riemann surfaces*. Princeton Mathematical Series, No. 26. Princeton University Press, Princeton, N.J., 1960.

[3] Norman L. Alling and Newcomb Greenleaf. *Foundations of the theory of Klein surfaces*. Lecture Notes in Mathematics, Vol. 219. Springer-Verlag, Berlin-New York, 1971.

[4] Cabiria Andreian Cazacu. On the morphisms of Klein surfaces. *Rev. Roumaine Math. Pures Appl.*, 31(6):461–470, 1986.

[5] I. Bârză. Integration on nonorientable Riemann surfaces. In *Almost complex structures (Sofia, 1992)*, pages 63–97. World Sci. Publ., River Edge, NJ, 1994.

[6] I. Bârză and D. Ghişa. Boundary value problems on nonorientable surfaces. volume 43, pages 67–79. 1998. Collection of papers in memory of Martin Jurchescu.

[7] I. Bârză and D. Ghişa. Explicit formulas for Green's functions on the annulus and on the Möbius strip. *Acta Appl. Math.*, 54(3):289–302, 1998.

[8] Stefan Bergman. *The Kernel Function and Conformal Mapping.* Mathematical Surveys, No. 5. American Mathematical Society, New York, N. Y., 1950.

[9] Y. Bouligand. Geometry and topology of cell membranes. In *Geometry in condensed matter physics*, volume 9 of *Dir. Condensed Matter Phys.*, pages 191–231. World Sci. Publ., River Edge, NJ, 1990.

[10] Y. Bouligand, B. Derrida, V. Poénaru, Y. Pomeau, and G. Toulouse. Distortions with double topological character: the case of cholesterics. *J. Physique*, 39(8):863–867, 1978.

[11] Emilio Bujalance, José J. Etayo, José M. Gamboa, and Grzegorz Gromadzki. *Automorphism groups of compact bordered Klein surfaces*, volume 1439 of *Lecture Notes in Mathematics*. Springer-Verlag, Berlin, 1990. A combinatorial approach.

[12] R. Courant and D. Hilbert. *Methods of mathematical physics. Vol. I.* Interscience Publishers, Inc., New York, N.Y., 1953.

[13] John B. Garnett and Donald E. Marshall. *Harmonic measure*, volume 2 of *New Mathematical Monographs*. Cambridge University Press, Cambridge, 2005.

[14] M. Jurchescu, K. Spallek, and F. Succi. DC-spaces (double complex spaces). In *Topics in complex analysis, differential geometry and mathematical physics (St. Konstantin, 1996)*, pages 81–93. World Sci. Publ., River Edge, NJ, 1997.

[15] Felix Klein. *Vorlesungen über die Theorie der elliptischen Modulfunktionen. Band I: Grundlegung der Theorie.* Bibliotheca Mathematica Teubneriana, Band 10. Johnson Reprint Corp., New York; B. G. Teubner Verlagsgesellschaft, Stuttgart, 1966. Ausgearbeitet und vervollständigt von Robert Fricke, Nachdruck der ersten Auflage.

[16] Steven G. Krantz. *Complex analysis: the geometric viewpoint*, volume 23 of *Carus Mathematical Monographs*. Mathematical Association of America, Washington, DC, 1990.

[17] Steven G. Krantz. *Complex analysis: the geometric viewpoint*, volume 23 of *Carus Mathematical Monographs*. Mathematical Association of America, Washington, DC, 1990.

[18] Steven G. Krantz. *Partial differential equations and complex analysis.* Studies in Advanced Mathematics. CRC Press, Boca Raton, FL, 1992. Lecture notes prepared by Estela A. Gavosto and Marco M. Peloso.

[19] Steven G. Krantz. *A panorama of harmonic analysis,* volume 27 of *Carus Mathematical Monographs.* Mathematical Association of America, Washington, DC, 1999.

[20] Steven G. Krantz. *A guide to complex variables,* volume 32 of *The Dolciani Mathematical Expositions.* Mathematical Association of America, Washington, DC, 2008. MAA Guides, 1.

[21] Rolf Nevanlinna. *Uniformisierung.* Die Grundlehren der mathematischen Wissenschaften in Einzeldarstellungen mit besonderer Berücksichtigung der Anwendungsgebiete, Band LXIV. Springer-Verlag, Berlin-Göttingen-Heidelberg, 1953.

[22] Rolf Nevanlinna. *Analytic functions.* Die Grundlehren der mathematischen Wissenschaften, Band 162. Springer-Verlag, New York-Berlin, 1970. Translated from the second German edition by Phillip Emig.

[23] H. Poincaré. Sur l'uniformisation des fonctions analytiques. *Acta Math.,* 31(1):1–63, 1908.

[24] Richard Rogers Preston, Jr. *Fundamental Domains and Projective Structures on Compact Klein Surfaces.* ProQuest LLC, Ann Arbor, MI, 1975. Thesis (Ph.D.)–The University of Texas at Austin.

[25] Thomas Ransford. *Potential theory in the complex plane,* volume 28 of *London Mathematical Society Student Texts.* Cambridge University Press, Cambridge, 1995.

[26] Diego L. Rapoport. Surmounting the Cartesian cut through philosophy, physics, logic, cybernetics, and geometry: self-reference, torsion, the Klein bottle, the time operator, multivalued logics and quantum mechanics. *Found. Phys.,* 41(1):33–76, 2011.

[27] B. Riemann. Theorie der Abel'schen Functionen. *J. Reine Angew. Math.,* 54:115–155, 1857.

[28] Monica Roşiu. Associating divisors with quadratic differentials on Klein surfaces. *Complex Var. Elliptic Equ.,* 60(2):181–190, 2015.

[29] Monica Roşiu. Harmonic measures and Poisson kernels on Klein surfaces. *Electron. J. Differential Equations,* pages Paper No. 269, 7, 2017.

[30] Monica Roşiu. Kernel function and integral representations on Klein surfaces. *Electron. J. Differential Equations,* pages Paper No. 132, 8, 2017.

[31] Monica Roşiu. On a class of Neumann problems on Klein surfaces. *Proc. Rom. Acad. Ser. A Math. Phys. Tech. Sci. Inf. Sci.*, 18(2):116–123, 2017.

[32] Vicenţiu D. Rădulescu. *Qualitative analysis of nonlinear elliptic partial differential equations: monotonicity, analytic, and variational methods*, volume 6 of *Contemporary Mathematics and Its Applications*. Hindawi Publishing Corporation, New York, 2008.

[33] Menahem Schiffer and Donald C. Spencer. *Functionals of finite Riemann surfaces*. Princeton University Press, Princeton, N. J., 1954.

[34] Mika Seppälä and Tuomas Sorvali. *Geometry of Riemann surfaces and Teichmüller spaces*, volume 169 of *North-Holland Mathematics Studies*. North-Holland Publishing Co., Amsterdam, 1992.

[35] Oswald Teichmüller. Extremale quasikonforme Abbildungen und quadratische Differentiale. *Abh. Preuss. Akad. Wiss. Math.-Nat. Kl.*, 1939(22):197, 1940.

[36] David M. Walba. Topological stereochemistry: knot theory of molecular graphs. In *Graph theory and topology in chemistry (Athens, Ga., 1987)*, volume 51 of *Stud. Phys. Theoret. Chem.*, pages 23–42. Elsevier, Amsterdam, 1987.

[37] Jeffrey R. Weeks. *The shape of space*, volume 96 of *Monographs and Textbooks in Pure and Applied Mathematics*. Marcel Dekker, Inc., New York, 1985. How to visualize surfaces and three-dimensional manifolds.

Index

Italicized pages refer to figures.

A
Ahlfors maps, 1, 17, 27
Algebra homomorphisms, 462
Algebraic dual, 260
Analytic-ave functions, 13
Analytic-circ function, 10–11
Analytic continuation, theme of,
229–232
    linear PDE
        G. Herglotz' Memoir of 1914,
233–234
        Herglotz question, 234–237
        ODE *vs.* PDE, 232–233
        problems of uniqueness, 237–238
        series of orthogonal polynomials,
238–240
        series of zonal harmonics,
238–240
Analytic-diff function, 11–12
Analytic discs, 192
Analytic functions, 9–14, 47
Analytic polynomials, 2, 12
Angle in $\mathbb{H}$, 154–156
Animations, 133, 140–141
Anoiktomorphism, 270, 277, 324
Antiextensive mapping, 269
Approximate biholomorphisms
    exhaustion, 79–82
    upper semicontinuity of
        automorphism groups,
82–84
Approximating bounded lineally
    convex Hartogs domains,
289–291
Approximation, 303–304

A (right column)
ARBITRARY polynomial, 233
Arbitrary smooth functions, 380
Arithmetic Fuchsian groups, 471, 474
Arithmeticity, 474
Arithmetic Kleinian group, 474
Automorphic functions, 433
Automorphisms, 50, 68, 438
    groups, 67
        and characterization of
            domains, 68–70
        semicontinuity of, 82–84
Averaging property on Ω, 2
Azukawa metric, 216
Azukawa pseudometric, 215

B
Banach space, 53
B-convex, 338
Behnke–Peschl differential condition,
265, 287, 289–308, 314–317
Bell's Condition R, 399
Bell's lemma, 404
Bell's projection formula, 401–402
Bergman integral, 420
Bergman kernel, 22, 65–66, 218–219,
382, 385–386, 390, 393
    conformal Invariance, 394
    differential equations, 390–393
    orthonormal basis, 390
Bergman metric, 218–219, 388
Bergman projection, 66
Bergman pseudodistance, 219
Bergman space, 66, 129
Bergman theory, 219, 384–389
Bézout coefficients, 159–160

521